Environmental Science and Engineering

Environmental Engineering

Series Editors

Ulrich Förstner, Technical University of Hamburg-Harburg, Hamburg, Germany
Wim H. Rulkens, Department of Environmental Technology,
Wageningen, The Netherlands
Wim Salomons, Institute for Environmental Studies, University of Amsterdam,
Haren, The Netherlands

More information about this subseries at http://www.springer.com/series/3172

Zhaojun Wang · Yingxin Zhu · Fang Wang ·
Peng Wang · Chao Shen · Jing Liu
Editors

Proceedings of the 11th International Symposium on Heating, Ventilation and Air Conditioning (ISHVAC 2019)

Volume II: Heating, Ventilation, Air Conditioning and Refrigeration System

Editors
Zhaojun Wang
Department of Building Thermal
Engineering, School of Architecture
Harbin Institute of Technology
Harbin, Heilongjiang, China

Yingxin Zhu
Department of Building Science
School of Architecture
Tsinghua University
Beijing, China

Fang Wang
Department of Building Thermal
Engineering, School of Architecture
Harbin Institute of Technology
Harbin, Heilongjiang, China

Peng Wang
Department of Building Thermal
Engineering, School of Architecture
Harbin Institute of Technology
Harbin, Heilongjiang, China

Chao Shen
Department of Building Thermal
Engineering, School of Architecture
Harbin Institute of Technology
Harbin, Heilongjiang, China

Jing Liu
Department of Building Thermal
Engineering, School of Architecture
Harbin Institute of Technology
Harbin, Heilongjiang, China

ISSN 1863-5520   ISSN 1863-5539 (electronic)
Environmental Science and Engineering
ISSN 1431-2492
Environmental Engineering
ISBN 978-981-13-9523-9   ISBN 978-981-13-9524-6 (eBook)
https://doi.org/10.1007/978-981-13-9524-6

© Springer Nature Singapore Pte Ltd. 2020
This work is subject to copyright. All rights are reserved by the Publisher, whether the whole or part of the material is concerned, specifically the rights of translation, reprinting, reuse of illustrations, recitation, broadcasting, reproduction on microfilms or in any other physical way, and transmission or information storage and retrieval, electronic adaptation, computer software, or by similar or dissimilar methodology now known or hereafter developed.
The use of general descriptive names, registered names, trademarks, service marks, etc. in this publication does not imply, even in the absence of a specific statement, that such names are exempt from the relevant protective laws and regulations and therefore free for general use.
The publisher, the authors and the editors are safe to assume that the advice and information in this book are believed to be true and accurate at the date of publication. Neither the publisher nor the authors or the editors give a warranty, expressed or implied, with respect to the material contained herein or for any errors or omissions that may have been made. The publisher remains neutral with regard to jurisdictional claims in published maps and institutional affiliations.

This Springer imprint is published by the registered company Springer Nature Singapore Pte Ltd.
The registered company address is: 152 Beach Road, #21-01/04 Gateway East, Singapore 189721, Singapore

Committees

Conference Organizing Committee

President

Prof. Zhaojun Wang, Harbin Institute of Technology, China

Technical Program Chair

Prof. Jing Liu, Harbin Institute of Technology, China

Advisory Chair

Prof. Yingxin Zhu, Tsinghua University, China

Secretary-General

Peng Wang, Harbin Institute of Technology, China

Members

Fang Wang, Harbin Institute of Technology, China
Chao Shen, Harbin Institute of Technology, China
Jiankai Dong, Harbin Institute of Technology, China
Yanling Wang, Harbin Institute of Technology, China
Zhigang Zhou, Harbin Institute of Technology, China
Haiyan Wang, Harbin Institute of Technology, China
Xiumei Bai, Harbin Institute of Technology, China

Student Members

Yuchen Ji, Harbin Institute of Technology, China
Qingwen Xue, Harbin Institute of Technology, China
Xiaowen Su, Harbin Institute of Technology, China

Subei Bu, Harbin Institute of Technology, China
Xin Chen, Harbin Institute of Technology, China
Jing Du, Harbin Institute of Technology, China

International Advisory Committee

Edward Arens, University of California, Berkeley, USA
Chungyoon Chun, Yonsei University, South Korea
Richard Corsi, University of Texas at Austin, USA
Richard de Dear, The University of Sydney, Australia
Lin Duanmu, Dalian University of Technology, China
Xiumu Fang, Harbin Institute of Technology, China
Yi Jiang, Tsinghua University, China
Kwang-Woo Kim, Seoul National University, South Korea
Angui Li, Xi'an University of Architecture and Technology, China
Baizhan Li, Chongqing University, China
Yuguo Li, The University of Hong Kong, China
Junjie Liu, Tianjin University, China
Bjarne W. Olesen, Technical University of Denmark, Denmark
Andrew Persily, National Institute of Standards and Technology, USA
Shin-ichi Tanabe, Waseda University, Japan
Sven Werner, Halmstad University, Sweden
Hiroshi Yoshino, Tohoku University, Japan
Guoqiang Zhang, Hunan University, China
Xu Zhang, Tongji University, China
Jianing Zhao, Harbin Institute of Technology, China

International Scientific Committee

Alireza Afshari, University of Aalborg, Denmark
Hazim Awbi, University of Reading, UK
Bin Cao, Tsinghua University, China
Guangyu Cao, Norwegian University of Science and Technology, Norway
Shijie Cao, Guangzhou University, China
Ping Cui, Shandong Jianzhu University, China
Lei Fang, Technical University of Denmark, Denmark
Guohui Feng, Shenyang Jianzhu University, China
Jun Gao, Tongji University, China
Naiping Gao, Tongji University, China
Weijun Gao, University of Kitakyushu, Japan
Tomonobu Goto, Tohoku University, Japan

Runa T. Hellwig, University of Aalborg, Denmark
Jan Hensen, Eindhoven University of Technology, the Netherlands
Sabine Hoffmann, Technische Universität Kaiserslautern, Germany
Tianzhen Hong, Lawrence Berkeley National Laboratory, USA
Yiqiang Jiang, Harbin Institute of Technology, China
Xianting Li, Tsinghua University, China
Xiangli Li, Dalian University of Technology, China
Zhengrong Li, Tongji University, China
Zhiwei Lian, Shanghai Jiao Tong University, China
Borong Lin, Tsinghua University, China
Zhang Lin, City University of Hong Kong, China
Xiaohua Liu, Tsinghua University, China
Zhiwen Luo, University of Reading, UK
Jianlei Niu, The Hong Kong Polytechnic University, China
Atila Novoselac, University of Texas at Austin, USA
Yiqun Pan, Tongji University, China
Menghao Qin, Technical University of Denmark, Denmark
Mattheos Santamouris, University of New South Wales, Australia
Chandra Sekhar, National University of Singapore, Singapore
Stefano Schiavon, University of California, Berkeley, USA
Marcel Schweiker, Karlsruhe Institute of Technology, Germany
Li Song, University of Oklahoma, USA
Fu-Jen Wang, National Chin-Yi University of Technology, China
Gang Wang, University of Miami, USA
Leon Wang, Concordia University, Canada
Shengwei Wang, The Hong Kong Polytechnic University, China
Yi Wang, Xi'an University of Architecture and Technology, China
Wei Xu, China Academy of Building Research, China
Da Yan, Tsinghua University, China
U Yanagi, Kogakuin University, Japan
Hongxing Yang, The Hong Kong Polytechnic University, China
Xudong Yang, Tsinghua University, China
Runming Yao, University of Reading, UK
Yang Yao, Harbin Institute of Technology, China
Yonggao Yin, Southeast University, China
Yanping Yuan, Southwest Jiaotong University, China
Zhiqiang (John) Zhai, University of Colorado at Boulder, USA
Hui Zhang, University of California, Berkeley, USA
Jili Zhang, Dalian University of Technology, China
Linhua Zhang, Shandong Jianzhu University, China
Quan Zhang, Hunan University, China
Tengfei (Tim) Zhang, Dalian University of Technology, China

Yinping Zhang, Tsinghua University, China
Yufeng Zhang, South China University of Technology, China
Xudong Zhao, University of Hull, UK
Xiang Zhou, Tongji University, China

Organization

Organized by

Host and Organizer

Harbin Institute of Technology

Co-organizer

Tsinghua University

Supported by

American Society of Heating, Refrigerating and Air-Conditioning Engineers

Journal of Heating, Ventilation and Air Conditioning

Chinese Association of Refrigeration (CAR)

The Chinese Committee of Heating, Ventilation and Air Conditioning (CCHVAC)

List of Plenary Lectures

Speakers	Affiliation	Topic
Yi Jiang	Tsinghua University, China	The Approach for Building Energy Efficiency in Emerging Economies Countries
Bill Bahnfleth	Pennsylvania State University, USA	Promise and Practicalities of Smart Systems
Bjarne W. Olesen	Technical University of Denmark, Denmark	State of the Art of HVAC Technology in Europe
Yingxin Zhu	Tsinghua University, China	How to Create Comfortable and Healthy Indoor Environment?—A New Challenge for Sustainable Building
Katsunori Nagano	Hokkaido University, Japan	Contribution of Heat Pumping Technologies for the Green Environment and Our Life
Rajan Rawal	CEPT University, India	Thermal Comfort for All: Setting up a Stage
Vladimir G. Gagarin	Moscow State University, Russia	Structure and Development of the System of Regulatory Documents in Russia on the Thermal Performance and Energy Saving of Buildings
Zhaojun Wang	Harbin Institute of Technology, China	Thermal Comfort and Thermal Adaptation in the Severe Cold Area, China

Preface

The International Symposium on Heating, Ventilation and Air Conditioning (ISHVAC) series was initiated by Tsinghua University of China in 1991. It has been a premier International Conference on Heating, Ventilating, and Air Conditioning (HVAC) in China and has played a significant role for presenting newly found research results and exchanging ideas among colleagues from academics, research organizations, and industry in the field. The 11th ISHVAC (ISHVAC 2019) was held from July 12 to 15, 2019, in Harbin, China, hosted by Harbin Institute of Technology (HIT), and Prof. Zhaojun Wang from HIT was President of ISHVAC 2019.

In developed countries, buildings use 40% of the primary energy, while in developing countries like China, buildings use 1/3 of the primary energy. Thus, energy used in buildings is very significant and a majority of the energy used in buildings is by HVAC systems that should create thermally comfortable and healthy indoor environment for their occupants. However, the indoor environment created is far from ideal because of poor design, operation, and occupant intervention. Therefore, ISHVAC 2019 focused on not only HVAC but also indoor environmental quality, health, thermal comfort, human activities, energy efficiency, renewable energy, etc. The conference provided a great opportunity to exchange new ideas and the state-of-the-art sciences and technologies, to identify solutions to problems in built environment, and to build partnerships among engineers, architects, environmental scientists, facility managers, HVAC system manufacturers, and policy makers.

The themes of this conference include indoor and outdoor environment; heating, ventilation, air conditioning, and refrigeration; and building and energy. More than 600 full papers were received, 543 full papers were accepted, and 407 papers were included in the proceedings. There were eight keynote lectures, 15 workshops, 56 parallel sessions, and one poster session. More than 500 experts and scholars from 20 countries or regions attended the conference, of which more than 80 were foreign guests. The participation of many internationally renowned experts and scholars has raised the level of ISHVAC 2019. The conference had the following features:

- The conference invited prominent Indian and Russian scholars to present in plenary sessions, and their presentations were of significance to the study of global climate and energy consumption.
- Many outstanding scholars were invited to organize 15 workshops, focusing on hot issues including thermal environment and thermal comfort, indoor environmental quality, pollutant transmission and control, human behaviors and building energy consumption simulation, smart heating, evaporative cooling, and heat pump.
- During the conference, the Asian HVAC Alliance was established. We hope that the ISHVAC series would become a grand event in the field of HVAC in Asia in the future, like CLIMA conferences in Europe and ASHRAE conferences in North America.

We would like to appreciate the International Advisory Committee and Scientific Committee for their hard work and efforts in reviewing papers and providing support needed. We also would like to thank the co-organizer Tsinghua University and co-sponsors such as ASHRAE, Journal of HV&AC, the Chinese Association of Refrigeration, and the Chinese Committee of Heating, Ventilation and Air Conditioning (CCHVAC).

Harbin, China	Zhaojun Wang
Beijing, China	Yingxin Zhu
Harbin, China	Fang Wang
Harbin, China	Peng Wang
Harbin, China	Chao Shen
Harbin, China	Jing Liu
	Editors of the Proceedings of ISHVAC 2019

Contents

An Experimental Study on the Energy Use during a TES-Based Reverse Cycle Defrosting Method for Cascade Air Source Heat Pumps 1
Minglu Qu, Tongyao Zhang, Rao Zhang, Jianbo Chen, Zhao Li and Tianrui Li

Heating Method and Heat Recovery Potential Prediction of Underground Railway Station Using Waste Heat from Equipment Room 11
Yongjin Chai, Shunian Zhao, Shaoxiong Zhang, Tingting Sun, Lu Jin and Yanfeng Liu

Parametric Analysis and Exergy Analysis for a New Cogeneration System Based on Ejector Heat Pump 23
Jiyou Lin, Chenghu Zhang and Yufei Tan

Modeling Method of Heat Pump System Based on Recurrent Neural Network 33
Yin Zheng, Guiqiang Wang, Guohui Feng and Zhiqiang Kang

Experimental Study on the Influence of Uniformity Liquid Distribution on the Flow Pattern Conversion of Horizontal Tube 41
Zhennan Qu, Zhixian Ma and Jili Zhang

Analysis of the Influence of Atomization Characteristics on Heat Transfer Characteristics of Spray Cooling 49
Jun Bao, Yu Wang, Xinjie Xu, Xuetao Zhou and Jinxiang Liu

Research of Personalized Heating by Radiant Floor Panels in Hot Summer–Cold Winter Area 61
Guoqing Yu, Zhuzheng Diao and Zhaoji Gu

Passive Thermal Protections of Smoke Exhaust Fans for a High-Temperature Heat Source 71
Yixiang Huang, Chengqiang Zhi, Qianru Zhang, Wei Ye and Xu Zhang

Investigation on Polymer Electrolyte Membrane-Based Electrochemical Dehumidification with Photoelectro-Catalyst Anode for Air-Conditioning 79
Mingming Guo and Ronghui Qi

Characteristics of a Natural Heat Transfer Air-Conditioning Terminal Device for Nearly Zero Energy Buildings 87
Haiwen Shu, Xu Bie, Shan Jiang, Zhiqiang Yang, Yang Zhang, Gao Shu, Hongbin Wang and Guangyu Cao

Study on Independent Air Conditioning System Based on the Temperature and Humidity of Rotary Dehumidifier and Heat Pipe 97
Jiangbo Li, Haiwei Ji and Liu Chen

The Influence Factors and Performance Optimization Analysis of the Total Heat Exchanger 105
Jiafang Song, Shuang Liang, Xiangquan Meng and Shuhui Liu

On-Site Operation Performance Investigation of Nocturnal Cooling Radiator for Heat Radiation 115
Yi Man, Shuo Li, Xinyu Zhang, Guoxin Jiang and Tiantian Du

Simulative Investigation of Ground-Coupled Heat Pump System with Spiral Coil Energy Piles 125
Yi Man, Xinyu Zhang, Shuo Li, Tiantian Du and Guoxin Jiang

A Developed CRMC Design Method and Numerical Modeling for the Ejector Component in the Steam Jet Heat Pumps 135
Chenghu Zhang, Yaping Li and Jianli Zhang

Working Fluid Selection and Thermodynamic Performance of the Steam Jet Large-Temperature-Drop Heat Exchange System 145
Chenghu Zhang, Yaping Li and Jianli Zhang

Optimization Method of Reducing Return Water Temperature of Primary Heating Circuit 155
Haiyan Wang, Yanling Wang, Fang Wang, Xin Xu and Shuai Gao

Simulation Study on Nominal Heat Extraction of Deep Borehole Heat Exchanger 165
Tiantian Du, Yi Man, Guoxin Jiang, Xinyu Zhang and Shuo Li

Numerical Study of Data Center Composite Cooling System 173
Xuetao Zhou, Xiaolei Yuan, Jinxiang Liu, Risto Kosonen and Xinjie Xu

Influence of Air Supply Outlet on Displacement Ventilation System for Relic Preservation Area in Archaeology Museum 183
Juan Li, Xilian Luo, Bin Chang and Zhaolin Gu

A Study on Polymer Electrolyte Membrane (PEM)-Based Electrolytic Air Dehumidification for Sub-Zero Environment 193
Tao Li and Ronghui Qi

Experimental Investigation on an Evaporative Cooling System for the Environmental Control in Archaeology Museum 201
Bin Chang, Xilian Luo, Yanqian Shen, Juan Li and Zhaolin Gu

Numerical Simulation of a Double-Layer Kang Based on Chimney Effect 211
Shilin Lei, Bin Chang, Yanqian Shen, Xiaoyu Zhu, Juan Li and Xilian Luo

Analysis on Control Mechanism and Response Characteristics of Residential Thermostats in District Heating Systems 219
Baoping Xu, Xi Wang and Yuekang Liu

The Study on the Influence of the Containing Ice Ratio on the Flow and Heat Transfer Characteristics of Ice Slurry in Coil Tubes 229
Changfa Ji, Chenyang Ji, Huan Zhang, Liu Chen and Xiyuan Yu

Multivariable Linear Regression Model for Online Predictive Control of a Typical Chiller Plant System 241
Jiaming Wang, Tianyi Zhao and Meng Xu

Applied Analysis of the Direct Cooling by Cooling Towers in Lanzhou Area 251
Xiaowei Wang, Wenhen Zhou, Lixin Zhao, Xin Bao, Lu Zheng and Xiaofei Han

Experimental Investigation on Humidity-Sensitive Properties of Polyimide Film for Humidity Sensors 261
Jianyun Wu, Wenhe Zhou, Xiaowei Wang and Shicheng Li

Research of the Influence of Valve Position on Flow Measurement of Butterfly Valve with Differential Pressure Sensor 269
Yuanpeng Mu, Zhixian Ma and Mingsheng Liu

Analysis on Performance of Solar Novel Heat Pipes Radiant Heating System 277
Yaping Zhang, Yongxin Guo, Pei Wang and Yao Chen

Numerical Simulation and Optimization of Air–Air Total Heat Exchanger with Plate-Fin 287
Jiafang Song, Shuhui Liu and Xiangquan Meng

Evaluation of Factors Toward Flow Distribution in the Dividing Manifold Systems with Parallel Pipe Arrays Using the Orthogonal Experiment Design 297
Wanqing Zhang, Angui Li and Feifei Cao

Performance Studies of R134a-Dimethylformamide Absorption Refrigeration System for Utilizing Bus Exhaust Gas Based on Aspen Plus Software 307
Xiao Zhang, Liang Cai, Qiang Zhou and Liping Chen

Study on Influence of Outdoor Airflow on Performance of Evaporative Cooling Composite Air-Conditioning System for Data Center 321
Haotian Wei, Zongwei Han, Chenguang Bai, Qi Fu and Xinwei Meng

Numerical Simulation of Heat Transfer and Pressure Drop Characteristics of Elliptical Tube Perforated Fins Heat Exchanger 329
Jiaen Luo, Zhaosong Fang, Lan Tang and Zhimin Zheng

Research on Cooling Effect of Data Center Cabinet Based on on-Demand Cooling Concept 339
Qi Fu, Zongwei Han, Xiaopeng Bi, Xinwei Meng and Haotian Wei

Analysis for Vibration Characteristics of Water Pump Piping System 347
Tiantian Liu and Zhiyong Liu

Analysis on Transient Thermal Behaviors of the Novel Vapor Chamber 357
Yao Chen, Yaping Zhang, Pei Wang and Yongxin Guo

The Role of Cylinder Obstacles before Air Conditioning Filter in the Quenching of Flame Propagation during Their Gas Deflagration Production Process 365
Lijia Fan, Chenghu Zhang, Jihong Wei and Yufei Tan

An Association Rule-Based Online Data Analysis Method for Improving Building Energy Efficiency 375
Chaobo Zhang, Yang Zhao and Xuejun Zhang

Numerical Study on Two-Phase Flow of Transcritical CO_2 in Ejector 385
Xu Feng, Zhenying Zhang, Jianjun Yang and Dingzhu Tian

3D Numerical Simulation on Flow Field Characteristic Inside the Large-Scale Adjustable Blade Axial-Flow Fan 395
Lin Wang, Kun Wang, Nini Wang, Suoying He, Yuetao Shi and Ming Gao

A Study of Wet-Bulb Temperature and Approach Temperature Based Control Strategy of Water-Side Economizer Free-Cooling System for Data Center 405
Jiajie Li, Zhengwei Li and Hai Wang

Effect of Adding Ambient Air Before Evaporating on Multi-stage Heat Pump Drying System 415
Xu Jin, Peng Xu, Baorui Wang, Zhongyan Liu and Wenpeng Hong

Study on Moisture Transfer Characteristics of Corn during Drying Process at Low Air Temperature 425
Xu Jin, Chen Wang, Qingyue Bi, Zhongyan Liu and Wenpeng Hong

The Study on a Dual Evaporating Temperatures-Based Chilled Water System Applied in Small-to-Medium Residential Buildings 435
Zhao Li, Jianbo Chen, Chunhui Liu, Lei Zhang, Shangqing Yang and Xiaoyu Liu

Study on the Inter-Stage Release Characteristics of Two-Stage Compression Heat Pump System 445
Xu Jin, Zhe Wu, Zhongyan Liu and Wenpeng Hong

Theoretical and Experimental Analysis of a New Evaporative Condenser 453
Yaxiu Gu, Yang Zou, Song Pan, Junwei Wang and Guangdong Liu

Studying the Performance of an Indirect Evaporative Pre-cooling System in Humid Tropical Climates 463
Xin Cui, Le Sun, Weichao Yan, Sicong Zhang, Liwen Jin and Xiangzhao Meng

Preheating Strategy of Intermittent Heating for Public Buildings in Cold Areas of China 471
Chunhua Sun, Jiali Chen, Yuan Liang and Shanshan Cao

On-Site Performance Investigation of the Existing Ground Source Heat Pump Systems in Residential Buildings in Cold Area 479
Lixia He, Han Du, Peng Gao, Zhuangzhuang Zheng and Ping Cui

Analysis of Heat and Mass Exchange Performance of Enthalpy Recovery Wheel 489
Hong Fan and Liu Chen

A New Two-Stage Compression Refrigeration System with Primary Throttling Intermediate Complete Cooling for Defrosting 501
Chaohui Xuan, Yongan Yang and Ruishen Li

Prediction of Pressure Drop in Adsorption Filter Using Friction Factor Correlations for Packed Bed 511
Ruiyan Zhang, Zhenhai Li, Lingjie Zeng and Fei Wang

Evaluations and Optimizations on Practical Performance of the Heat Pump Integrated with Heat-Source Tower in a Residential Area in Changsha, China 521
Fenglin Zhang, Nianping Li, Haijiao Cui, Jikang Jia, Meng Wang and Meiyao Lu

Research on the Effect of Solid Particle Diameter on the Performance of Solid-Liquid Centrifugal Pump 533
Kuanbing CaoZhu and Changfa Ji

Experimental Study on Flow Maldistribution and Performance of Carbon Dioxide Microchannel Evaporator 541
Jing Lv, Guo Li, Tang fuyi Xu and Chenxi Hu

Capture of High-Viscosity Particles: Utilizing Swirling Flow in the Multi-Layer Square Chamber 551
Leqi Tong and Jun Gao

Experimental Study on Vertical Temperature Profiles under Two Forms of Airflow Organization in Large Space during the Heating Season 561
Chenlu Shi, Xin Wang, Gang Li, Hongkuo Li, Minglei Shao, Xin Jiang and Bingyan Song

Experimental Study of Electroosmotic Effect in Composite Desiccant 571
Shanshan Cai, Xu Luo, Xing Zhou, Wanyin Huang, Xu Li and Jiajun Ji

District Heating System Load Prediction Using Machine Learning Method 581
Meng Jia, Chunhua Sun, Shanshan Cao and Chengying Qi

A Case Study on Existing Building HVAC System Optimization of a Five-Star Hotel in Shanghai 589
WeiFeng Zhu, Zhuling Zheng, Mengyuan Liu and Guangwei Deng

Research on Heat and Moisture Transfer Characteristics of Soil in Unsaturated and Saturated Condition with Soil Stratification under Vertical Borehole Ground Heat Exchanger Operation 599
Yao Wang and Songqing Wang

Performance Analysis of a Hybrid Solar Energy, Heat Pump, and Desiccant Wheel Air-Conditioning System in Low Energy Consumption Building 611
Shaochen Tian and Xing Su

Feasibility of Hybrid Ground Source Heat Pump Systems Utilizing Capillary Radiation Roof Terminal in the Yangtze River Basin of China 621
Lu Xing, Chen Ren, Hanbin Luo, Yin Guan, Dongkai Li, Lei Yan, Yuhang Miao and Pingfang Hu

Study on Energy Evolution Characteristics of Metro Environmental Control Equipment in Different Periods.......................... 631
Jie Song, Yi Zheng, Lihui Wang, Shan Zhang, Renyi Gao, Chang Liu and Xuecheng Zou

Energy-Saving and Economic Analysis of Anaerobic Reactor Heating System Based on Biogas and Sewage Source Heat Pump............ 641
Shouwen Sheng and Fang Wang

Air-Conditioner Usage Patterns in Teaching Buildings of Universities by Data Mining Approach 649
Xinyue Li, Shuqin Chen, Jiahe Li and Hongliang Li

Design Optimization of Radiation Cooling Terminal for Ultra-low-Energy Consumption Office Buildings 661
Zhengrong Li, Xiangyun Chen and Dongkai Zhang

Study on Operating Performance of Ground-Coupled Heat Pump with Seasonal Soil Cool Storage System.......................... 671
Chao Lyu, Jiachen Zhong, Ping Jiang, Zhiyi Wang, Feng Yu, Yueqin Liu and Maoyu Zheng

Controlling Technique and Policy of Adjacent Rooms Pressure Difference in High-Level Biosafety Laboratory 679
Peng Gao, Guoqing Cao, Ziguang Chen and Yuming Lu

Investigated on Energy-Saving Measures of HVAC System in High-Level Biosafety Laboratory 689
Peng Tan, Guoqing Cao and Ziguang Chen

The Optimization Design of Sewage Heat Exchanger in Direct Sewage Source Heat Pump System 699
Zhaoyi Zhuang, Jun Xu, Jian Song and Wenzeng Shen

The Optimization of Sensitivity Coefficients for the Virtual in Situ Sensor Calibration in a LiBr–H_2O Absorption Refrigeration System 709
Peng Wang, Kaihong Han, Liangdong Ma, Sungmin Yoon and Yuebin Yu

Research on the Energy-Saving Coefficient and Environmental Effect of the Surface Water Source Heat Pump System 719
Ying Xu, Yuebin Wu, Liang Chen and Qiang Sun

Application of New Evaporative Cooling Air-Conditioning System in a Data Center in Xinjiang 727
Xiang Huang, Zhicheng Guo, Zhenwu Tian, Jingwen Xuan and Jincheng Yan

Trial-and-Error Method for Variable Outdoor Air Volume Setpoint of VAV System Based on Outdoor Air Damper Static Pressure Difference Control 737
Pengmin Hua, Tianyi Zhao, Wuhe Dai and Jili Zhang

An Extension Theory-Based Fault Diagnosis Method for an Air Source Heat Pump 747
Yudong Xia, Qiang Ding, Shu Jiangzhou, Yin Liu and Xuejun Zhang

Effect of Lewis Factor on Performance of Closed Heat Source Tower under Spraying Conditions 759
Fenglin Zhang, Nianping Li, Haijiao Cui, Shengbing Li, Meng Wang and Meiyao Lu

An Experimental Study on the Thermosiphon Loop with a Microchannel Heat Sink Operating with the Phase Change Emulsion 769
Xiaoxu Cai, Shugang Wang, Jihong Wang, Tengfei Zhang and Xiaozhou Wu

Experimental Study on Effect of Water Flow Rate on Heating Performance of a Series Bathing Wastewater Source Heat Pump Hot Water Unit 781
Liangdong Ma, Tixiu Ren, Tianyi Zhao and Jili Zhang

Field Test Analysis of a Novel Continuous Running Dual-Channel Condensation Gasoline Vapor Recovery System 791
Mengmeng Wu and Lin Cao

Matching Characteristics of Two Heat Exchangers for the Direct Sewage Source Heat Pump System 801
Zhaoyi Zhuang, Jian Song, Jun Xu and Wenzeng Shen

Single-Phase Heat Transfer and Pressure Drop of Developing Flow at a Constant Heating Flux Inside Horizontal Helical Finned Tubes 809
Zhixian Ma, Nan Zhao, Anping Zhou and Jili Zhang

Experimental Study on Influence of Outdoor Ambient Temperature on Heating Performance of Two-Stage Scroll Compression Air Source Heat Pump System 817
Yiling Wu and Lin Cao

The Field Survey on Local Heat Island Effect of Precision Air-Conditioning 827
Mo Chen, Zhixian Ma and Mingsheng Liu

Study on Operation Strategy of Cross-Season Solar Thermal Storage Heating System in Alpine Region 835
Haoran Li, Hanyu Yang, Enshen Long, Xin Liu and Yin Zhang

Experimental Research on Performance of VRF-Based Household Radiant Air-Conditioning System 845
Danyang Wang, Jianbo Chen, Chenyue Yan and Meng Zhao

Precise Control for Heating Supply to Households Based on Heating Load Prediction 855
Ruiting Wang, Fulin Wang, Zhaohan Nan, Minjie Xiao and Aijun Ding

Experimental Study on the Influence of Fouling Growth on the Flow and Heat Transfer of Sewage in the Heat Exchange Tube 865
Shunzhi Chen, Liangdong Ma, Zhiyuan Zhang and Jili Zhang

Study on Exhaust Uniformity of a Multi-terminal System 873
Yirui Wang and Jun Gao

Topology Description of HVAC Systems for the Automatic Integration of a Control System Based on a Collective Intelligence System 883
Zhen Yu, Huai Li and Wei Liu

Theoretical Analysis of a Novel Two-Stage Compression System Using Refrigerant Mixtures 893
Zuo Cheng, Baolong Wang, Wenxing Shi and Xianting Li

Optimal Control of Water Valve in AHU Based on Actual Characteristics of Water Valve 903
Xia Wu, Yan Gao and Bin Wang

Study on Heat Recovery Air Conditioning System with Adsorption Dehumidification by Solar Powered 913
Yi Liu and Liu Chen

Study on the Optimal Cooling Power for the Internally Cooled Ultrasonic Atomization Dehumidifier with Liquid Desiccant 923
Ruiyang Tao, Zili Yang, Yanming Kang and Zhiwei Lian

Experimental Study and Energy-Saving Analysis on Cooling Effect with Large Temperature Difference and High Temperature of Chilled Water System in Data Center 933
Zhibo Kang, Zhenhua Shao, Lin Su, Kaijun Dong and Hongxian Liu

Sensitivity and Uncertainty Analysis for Chiller Sequencing Control of the Variable Primary Flow System 943
Zhenbing Cai and Yundan Liao

Theoretical Analysis of Smoke Exhaust System with Ringed Arrangements in the Field of HVAC 953
Minmin Zhang, Yixue Wu and Meiling He

An Experimental Study on the Energy Use during a TES-Based Reverse Cycle Defrosting Method for Cascade Air Source Heat Pumps

Minglu Qu, Tongyao Zhang, Rao Zhang, Jianbo Chen, Zhao Li and Tianrui Li

Abstract Adopting cascade air source heat pumps (CASHPs) is a possible way to widespread application of air source heat pumps in cold area. When CASHPs are operated in winter, frosting/defrosting may become problematic. Thermal energy storage-based reverse cycle defrosting (TES-based RCD) method for CASHPs has been developed to provide heat to defrosting and continuously heat indoor space. In this paper, the energy use during the TES-based RCD for CASHPs was experimentally studied. It was found that most of the heat supplied was dissipating to the ambient air, accounting for 44.3%, and the defrosting efficiency was 30.2%. Meanwhile, most of the heat came from the TES-heat exchanger (TES-HE), accounting for 40%. This study can be utilized to guide the design optimization for the TES-HE and promote energy-saving for TES-based RCD method.

Keywords Defrosting · Energy use · Thermal energy storage · Cascade air source heat pump

1 Introduction

To advance the operating performance of ASHP in cold regions, employing cascade air source heat pumps (CASHPs) is a feasible way. In CASHPs, the two refrigeration cycles, i.e., the high temperature (HT) cycle and low temperature (LT) cycle, adopting different refrigerant, are connected by a cascade heat exchanger. Extensive experimental and theoretical studies on CASHPs have been carried out recently [1–4].

When adopting CASHP unit in cold regions, frosting and defrosting may also be problematic. Nevertheless, seldom research was conducted to develop suitable

M. Qu (✉) · T. Zhang · R. Zhang · J. Chen · Z. Li · T. Li
School of Environment and Architecture, University of Shanghai for Science and Technology, No. 516, Jungong Road, 200093 Shanghai, China
e-mail: quminglu@126.com

© Springer Nature Singapore Pte Ltd. 2020
Z. Wang et al. (eds.), *Proceedings of the 11th International Symposium on Heating, Ventilation and Air Conditioning (ISHVAC 2019)*, Environmental Science and Engineering, https://doi.org/10.1007/978-981-13-9524-6_1

defrost method for CASHPs. To handle this problem, Qu et al. [5, 6] proposed thermal energy storage-based reverse cycle defrosting method (TES-based RCD) for CASHPs to provide heat to both defrosting for LT cycle and indoor heating for HT cycle.

On the other hand, in order to shorten defrost duration and improve defrosting efficiency, it is therefore necessary to identify the energy consumption and energy supply in the defrosting process. Previous research has been conducted to investigate the energy use during reverse cycle defrosting for ASHPs [7, 8]. However, energy use during TES-based RCD for CASHPs is seldom seen. Therefore, an experimental study is conducted to investigate the energy use during a TES-based RCD method for CASHPs in this paper. This study can be utilized to guide the design optimization for the TES-HE and promote energy-saving for TES-based RCD method.

2 Experimental Prototype and Procedures

2.1 Experimental Prototype

In this research, the prototype CASHP unit was modified from a conventional CASHP unit with heating capacity of 10 kW and shown in Fig. 1, and the introduction of this system including the PCM-HE is described in Refs. [5, 6]. Commercial composite organic phase-change material (PCM) RT 10 whose melting point 6.17 and melting latent heat 134.9 kJ/kg was adopted. Its thermophysical

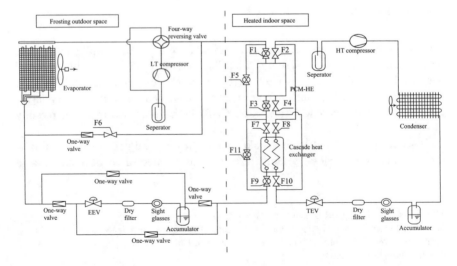

Fig. 1 Schematics of the prototype

properties can be found in Ref. [6]. In this research, 16.7 kg of RT10 was used; thus, about 2255 kJ of latent heat could be stored.

2.2 Experimental Condition and Procedures

During the frosting/TES heating process, the air dry-bulb temperature and RH inside the heated indoor space were maintained at 22 ± 0.1 °C and $50 \pm 3\%$ relative humidity, and those inside the frosting outdoor space at -9 ± 0.3 °C and at $85 \pm 3\%$ relative humidity, respectively. The operating procedures and the details of instrumentation can be found in Ref. [6].

2.3 Data Reductions

During TES-based RCD process, the heat was applied in five parts, i.e., melt the frost on the evaporator surface, vaporize the retained water on the evaporator surface, heat the evaporator, dissipate to the ambient air, and heat the melted frost. The determination of each part was introduced in the following.

The heat used to melt the frost layer, Q_m, can be obtained by:

$$Q_m = m_{fr} L_{sf} \tag{1}$$

where m_{fr} is the total mass of frost layer formed on the evaporator, whose calculation method can be found in Ref. [7], kg, L_{sf} the latent heat of frost melting, kJ/kg. The heat used to vaporize the retained water, Q_v, was:

$$m_v = m_{fr} - m_{me} \tag{2}$$

$$Q_v = m_v L_v \tag{3}$$

where m_v and m_{me} are the mass of vaporized melted frost and the mass of collected melted frost at the end of a defrost duration, respectively, kg. L_v is the latent heat of evaporation of water.

The heat used to heat the evaporator metal, Q_{Me}, could be evaluated by:

$$Q_{Me} = C_{pMe} \times (m_{Cu} + m_{Al}) \times \Delta T_{Me} \tag{4}$$

in which ΔT_{Me} is the temperature difference of the evaporator metal during the TES-based RCD process, °C. C_{pMe} is the averaged specific heat of copper and aluminum.

The heat dissipated to the ambient air, Q_a, could be obtained by:

$$Q_a = C_{pa} \times \rho_a \times V_a \times (t_{out} - t_{in}) \tag{5}$$

where C_{pAl} and ρ_a are the specific heat and density of air, respectively, kJ/(kg K), kg/m^3, V_a the volumetric flow rate of the air across the evaporator, m^3/s, and t_{out} and t_{in} the air temperature at the inlet and outlet of the evaporator, respectively, °C.

The total heat supplies to defrost can be calculated from the LT refrigerant side, Q, as shown:

$$Q = \int_0^{t_d} m_r \left(h_{Eva,o} - h_{Eva,i} \right) dt \tag{6}$$

where m_r is refrigerant mass flow rate, kg/s, $h_{Eva,i}$ and $h_{Eva,o}$ the enthalpies of LT refrigerant at both inlet and exit of the evaporator, respectively, kJ/kg, obtained from the measured compressor discharge pressure and the measured inlet and outlet temperatures, t_d the defrost duration, s.

The heat used to heat the melted frost, Q_r, was obtained from the difference of the total heat supplied and the other four parts of heat:

$$Q_r = Q - Q_m - Q_v - Q_{Me} - Q_a \tag{7}$$

To evaluating defrosting operation, defrosting efficiency, η_d, is an important parameter [7]. It can be evaluated by:

$$\tag{8}$$

On the other hand, during TES-based RCD, the heat supplied to defrost was actually came from the energy stored in the PCM-HE, the LT compressor power input, and the energy stored in the metal of the PCM-HE and the tubes.

Heat supplied to LT cycle from the PCM-HE, Q_{PCM}, was calculated by:

$$Q_{PCM} = \int_0^{t_d} m_r \left(h_{PCM,o} - h_{PCM,i} \right) dt \tag{9}$$

where $h_{PCM,i}$ and $h_{PCM,o}$ the enthalpies of LT refrigerant at both inlet and exit of the PCM-HE, respectively, kJ/kg, obtained from the measured pressure and the measured inlet and outlet temperatures.

Heat from the LT compressor, W_{com}, was obtained from the measured electric input of the LT compressor.

$$W_{com} = P\eta \tag{10}$$

where compressor power consumption, kW, acquired by power meter on the enthalpy difference chamber control cabinet and η the electrical efficiency of the compressor, 0.6.

The heat from the coil and tube metal, Q_{store}, could be then evaluated by:

$$Q_{store} = Q - Q_L - W_{com} \qquad (11)$$

3 Experimental Results

Main experimental results were shown in Table 1. The calculated mass of accumulated frost was 2.515 kg. The collected melted frost in the measuring cylinder was 2.352 kg. The defrosting duration was 510 s, and the total heat supplies to defrost obtained by Eq. (6) was 4099.5 kJ.

3.1 Heat Consumptions

Figure 2 presents the average copper tube surface temperature of the evaporator during TES-based RCD process. As seen, the tube surface temperature began to increase gradually from −5.4 to 0 °C at about 60 s, with the hot refrigerant gas flew into the evaporator (condenser). Afterward, the temperature remained for a while, which can be considered as frost-melting period, until the tube surface temperature increased sharply at 180 s, suggesting most of the frost layer was melted. Thereafter, the supplied heat was utilized to heat and vaporize the melted frost and remained water. At the end of the TES-based RCD process, the tube surface temperature of the evaporator reached 32.5 °C.

Table 1 Main experimental results and the energy consumption

Item	Value
Mass of accumulated frost (kg)	2.515
Mass of collected melted frost in the measuring cylinder (kg)	2.352
Mass of retained water (kg)	0.163
Defrosting duration (s)	510
Heat used to melt the frost layer (kJ)	840.0
Heat used to vaporize retained water (kJ)	398.2
Heat used to heat the melted frost (kJ)	288.3
Heat used to heat evaporator metal (kJ)	757.1
Heat dissipated to ambient air (kJ)	1815.9
Total heat supply to defrost (kJ)	4099.5
Defrosting efficiency (%)	30.2

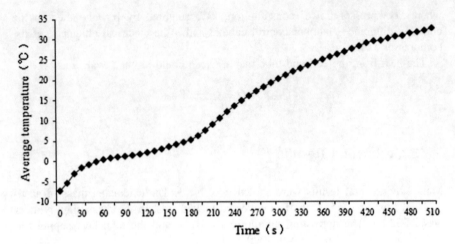

Fig. 2 Average copper tube surface temperature of the evaporator during TES-based RCD process

Figure 3 illustrates measured air temperatures at the inlet and outlet of the evaporator during TES-based RCD process. The air temperature at the outlet of the evaporator was −10.3 °C, with the heat dissipated to the ambient air by natural convection from the evaporator coil; it raised gradually to 2.5 °C at the end of the TES-based RCD process. Meanwhile, the air temperatures at the inlet of the

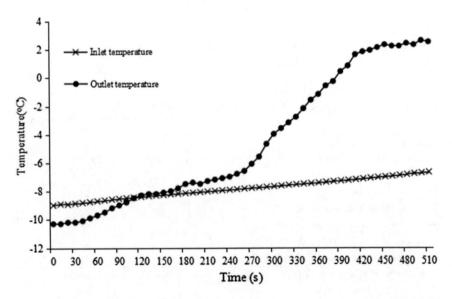

Fig. 3 Measured air temperatures at the inlet and outlet of the evaporator during TES-based RCD process

evaporator began to grow ion from the evaporator coil; it raised gradually to 2.5 °C in at the end of the TES-based RCD process. Meanwhile, the air temperatures at the inlet of the evaporator began to grow from −8.9 to −6.7 °C, with the defrosting proceeded.

Figure 4 shows the measured air velocity of the evaporator during TES-based RCD process. The air velocity was zero due to the shutdown of the evaporator fan and the block of the frost layer on the coil surface at the first 180 s. Thereafter, with the most frost layer melted, the ambient air temperature was therefore heated by the evaporator coil. The ambient air began to rise due to thermal pressure, and the air velocity increased remarkably along with the rise of the coil surface temperature. It reached 0.71 m/s at the end of the TES-based RCD process.

Given the experimental results, the heat consumption during TES-based RCD process was obtained from Eqs. (1–8) and listed in Table 2 and Fig. 5. The largest proportion of the heat supplied went on dissipate to ambient air, 44.3%, which was

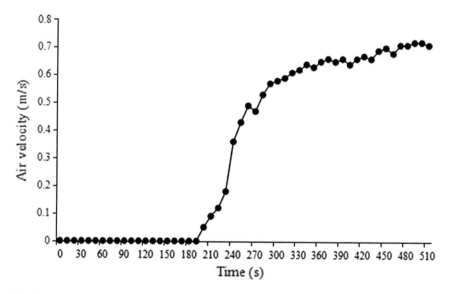

Fig. 4 Measured air velocity variations of the evaporator during TES-based

Table 2 Energy supplies

Total heat supplies to defrost (kJ)	4099.5
Heat from the PCM-HE (kJ)	1637.7
Proportions of heat from the PCM-HE (%)	40.0
Heat from the LT compressor (kJ)	969.5
Proportions of heat from the LT compressor (%)	23.7
Heat from the coil and tubes metal (kJ)	1492.3
Proportions of heat from the coil and tubes metal (%)	36

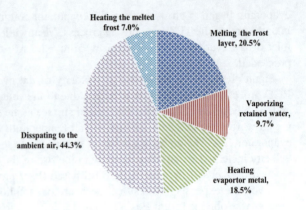

Fig. 5 Energy consumption during TES-based RCD process

caused by uneven defrosting on the multi-circuit evaporator coil [7]. When lower circuits were still undergone frost-melting process, higher circuits were already free of frost or even retained water, and heat would loss to the ambient air. As seen, the heat used to heat the melted frost had the lowest proportion, 7.0%. According to Eq. (8), the defrosting efficiency was evaluated at 30.2%.

3.2 Heat Supplies

Heat supplies during the TES-based RCD process are mainly from the PCM-HE, the LT compressor, and the heat that stored in the coil and tubes metal, and each part was calculated from Eqs. (9–11) and listed in Table 2.

It can be seen that during TES-based RCD process, the heat used to defrost was mainly came from the PCM-HE, accounted for 40%, and from the LT compressor accounted for 23.7% of the total heat supplies to defrost.

Figure 6 presents the heat supplies variations with time during TES-based RCD process. The growth of the heat supplied from the PCM-HE reached 7.65 kW at 110 s. Thereafter, it began to decrease gradually to 5.28 kW at 240 s as most of the PCM solidified, and heat transfer from the PCM to the refrigerant being dominated by convection instead of conduction. And then, the descent rate increased until 440 s and remained at a very low level until the end of the defrosting process.

4 Conclusions

An experimental study on the energy use during TES-based RCD process for CASHPs was conducted and reported. Under the frosting outdoor condition of −9 °C and 85%, and the frost mass of 2.515 kg, it was concluded that most of the heat supplied was dissipating to the ambient air, accounting for 44.3%, during

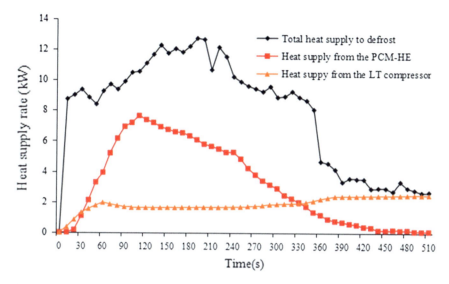

Fig. 6 Heat supplies variations with time during TES-based RCD process

TES-based RCD process and the defrosting efficiency was 30.2%. On the other hand, most of the heat came from the TES-HE, accounting for 40%. This study can be utilized to guide the design optimization for the TES-HE and promote energy-saving for TES-based RCD method.

Acknowledgements Funding for this research was supported by The National Natural Science Foundation of China (Project No.: 51406119).

References

1. Shen, J.B., Guo, T., Tian, Y.F., Xing, Z.W.: Design and experimental study of an air source heat pump for drying with dual modes of single stage and cascade cycle. Appl. Therm. Eng. **129**, 280–289 (2018)
2. Wang, G., Chen, Z.S., Li, C., Jiang, B.: Preliminary theoretical analyses of thermal performance and available energy consumption of two-stage cascade cycle heat pump water heater. Int. J. Refrig. **82**, 381–388 (2017)
3. Jung, H.W., Kang, H., Chung, H., Ahn, J.H., Kim, Y.C.: Performance optimization of a cascade multi-functional heat pump in various operation modes. Int. J. Refrig. **42**, 57–68 (2014)
4. Qu, M.L., Fan, Y.N., Chen, J.B., Li, T.R., Li, Z., Li, H.: Experimental study of a control strategy for a cascade air source heat pump water heater. Appl. Therm. Eng. **110**, 835–843 (2017)
5. Qu, M.L., Qin, R.F., Tang, Y.B., Fan, Y.N., Li, T.R.: Improving defrosting performance of cascade air source heat pump using thermal energy storage based reverse cycle defrosting method. Appl. Therm. Eng. **121**, 728–736 (2017)

6. Qu, M.L., Tang, Y.B., Zhang, T.Y., Li, Z., Chen, J.B.: Experimental investigation on the multi-mode heat discharge process of a PCM heat exchanger during TES based reverse cycle defrosting using in cascade air source heat pumps. Appl. Therm. Eng. **151**, 154–162 (2019)
7. Qu, M.L., Deng, S.M., Jiang, Y.Q.: A study of the reverse cycle defrosting performance on a multi-circuit outdoor coil unit in an air source heat pump—part I experiments. Appl. Therm. Eng. **110**, 835–843 (2012)
8. Song, M.J., Xu, X.G., Mao, N., Deng, S.M., Xu, Y.J.: Energy transfer procession in an air source heat pump unit during defrosting. Appl. Energy **204**, 679–689 (2017)

Heating Method and Heat Recovery Potential Prediction of Underground Railway Station Using Waste Heat from Equipment Room

Yongjin Chai, Shunian Zhao, Shaoxiong Zhang, Tingting Sun, Lu Jin and Yanfeng Liu

Abstract A great deal of waste heat is generated from the equipment room of subway station while heating demand exists in station platform, station hall, and office area. The conventional method is to use ventilation to remove the waste heat of the equipment room and use the heating system to meet the heat load of the corresponding heating area, which will cause a large amount of energy consumption and waste heat. The idea of using water-loop heat pump and multi-connected technology to recover waste heat from equipment room for heating of station is presented in this work. By analyzing the performance of the water-loop multi-connected heat pump system used in subway station, the preliminary design concept of the system is formed, the relevant mathematical model is established, and the optimal water temperature of the water loop is discussed. The simulation analyzes the operating characteristics of the system, and the variation law of auxiliary heating amount, heat recovery energy efficiency ratio, and overall energy efficiency ratio of the system are obtained. Based on this, this paper gives a prediction of the waste heat recovery potential of the subway stations in Harbin, Moscow, Montreal, and Helsinki.

Keywords Waste heat recovery · Water-loop heat pump · Multi-connected heat pump · Underground railway station

1 Introduction

The public areas of the underground railway stations are generally not heated. Only ticket halls with long-term stays are heated and hot-air curtains are installed at the entrances and exits. Most of the energy used for heating is electricity. Coexisting with this problem is that the equipment rooms of the station are ventilated or air conditioned throughout the year to remove waste heat. If the waste heat dissipated

Y. Chai · S. Zhao · S. Zhang · T. Sun (✉) · L. Jin · Y. Liu
Xi'an University of Architecture and Technology, 710055 Xi'an, China
e-mail: suntt@xauat.edu.cn

steadily from the equipment rooms is used to meet the station heating, the problems of removing the waste heat and fulfilling heating could both be solved. The indoor air temperatures of the equipment rooms are much higher than that of outdoor in winter that is very beneficial for the operation of the air-source heat pump. In addition, when designing the air conditioning system in metro, besides setting up an all-air air conditioning system in an important equipment room, a multi-connected air conditioning system is also needed as a standby system to guarantee uninterrupted air conditioning in summer. The multi-connected air conditioning system is completely idle in winter, but still it is indispensable. If the multi-connected air conditioning system is involved in the heat recovery system, a portion of equipment investment can be reduced. However, a multi-connected air conditioning system is not competent to recovery waste heat from the whole subway station because of the scale limitation. For the above reasons, it is proposed to use the water-loop multi-connected heat pump technology to collect the waste heat of the equipment room and distribute it to the needed areas of the station.

The previous research on metro waste heat recovery could be divided into two groups: direct utilization and heat pump heating. The direct utilization method is to deliver the exhaust air with waste heat directly to the energy-demanding area [1, 2]. Heat pump heating method could be divided into three cases: The heat exchanger is located in air duct or air shaft [1, 3–5], in metro tunnel [4–8], and in surrounding soil. The waste heat recovery from equipment rooms has been overlooked by previous studies that makes the considerable waste heat could not be fully reused. The water-loop heat pump technology and the water-loop multi-connected heat pump technology are very mature. However, there is still no attempt to apply this technology to waste heat recovery from equipment room in metro station [9].

There are some particularities when using water-ring multi-connected heat pump in metro station. (1) There are no regularities of distribution between waste heat-generating area and heat-supplying area in the underground railway station. (2) There are significant differences between the design temperatures of different areas. (3) The optimal water temperature of the water loop must be recalculated [10]. (4) Energy saving potential is unknown.

2 Mathematical Model

2.1 Calculating Model of Waste Heat and Heating Load in Metro Station

The calculation of waste heat of equipment room in metro station only considers long-term stable heat dissipation's equipment. The values are provided by technology-related specialty. The unsteady heat productions are treated as the surplus of heating. The heating load of metro station is mainly composed of heat load of envelope structure and fresh air heat load. There are two 60 kW hot-air curtains at each entrance and exit of the station. The heating quantity is fixed. The calculation formula of heating load of metro station is as follows:

$$Q' = \sum_{i=1}^{n} K_i A_i (t_{in} - t_E) + \frac{n \rho_a V_f c_p (t_{in} - t_o)}{3600} \quad (1)$$

where Q' is the heating heat load except the hot air curtain, kW; K_i is the heat transfer coefficient of the envelope structure, kW/(m^2 K); A_i is the heat transfer area of the envelope structure, m^2; ρ_a is the air density, kg/m^3, n is the number of people in the room, p; V_f is the fresh air per capital, m^3/(p h); c_p is the specific pressure heat capacity of the air, kJ/(kg K); t_{in}, t_o, t_E are the indoor air, outdoor air, and the soil temperature outside the envelope structure, °C.

2.2 Model of Water-Cooled Multi-connected Refrigeration System

RWEYQ30MY1 multiplexer cooling conditions:

$$\text{COP} = a_1 + b_1 t_w + c_1 t_w^{0.5} \quad (2)$$

$$k_c = a_2 + \frac{b_2}{k^{0.5}} + \frac{c_2 \ln k}{k^2} \quad (3)$$

RWEYQ20MY1 multiplexer cooling conditions:

$$\text{COP} = a_3 + b_3 t_w^3 + c_3 t_w^{0.5} \quad (4)$$

$$k_c = a_4 + b_4 \ln k + \frac{c_4}{k^2} \quad (5)$$

RWEYQ10MY1 multiplexer heating conditions:

$$\varepsilon_h = a_5 + b_5 t_w^{1.5} \quad (6)$$

$$k_\varepsilon = a_6 k^3 + b_6 k^2 + c_6 k + d_6 \quad (7)$$

RWEYQ10MY1 multiplexer heating conditions:

$$\varepsilon_h = a_7 + b_7 t_w^2 + c_7 t_w^{2.5} \quad (8)$$

$$k_\varepsilon = a_8 k^3 + b_8 k^{2.5} + c_8 \quad (9)$$

Table 1 Coefficients of fitted formula

C	Value	C	Value	C	Value
a_1	18.709500	b_4	−0.5630302	c_7	8.033909×10^{-4}
b_1	0.073603	c_4	6.292635×10^{-2}	a_8	1.252345
c_1	−2.813741	a_5	2.643689	b_8	−1.701688
a_2	−0.024848	b_5	2.267794×10^{-2}	c_8	1.449235
b_2	1.024830	a_6	1.710446	a_9	5.436710
c_2	0.089306	b_6	−3.4707096	b_9	-6.668182×10^{-2}
a_3	17.507195	c_6	1.648514	c_9	-2.583249×10^{-2}
b_3	5.355246×10^{-6}	d_6	1.110142	d_9	4.969697×10^{-3}
c_3	−2.200265	a_7	3.1925529	e_9	-2.567340×10^{-4}
a_4	1.063002	b_7	8.587360×10^{-3}	f_9	3.140000×10^{-2}

In formulas (2)–(9), COP is the coefficient of refrigeration performance; t_w is multi-connected inlet water temperature, °C; k_c is the ratio of refrigeration performance coefficient to full-load refrigeration performance coefficient under partial load; k is partial load rate; ε_h is heating performance coefficient; k_ε is the ratio of heating performance coefficient to full-load heating performance coefficient under partial load; a, b, c are fitting coefficients.

The fitting coefficients of the above formulas are shown in Table 1, and the correlation degree of fitting is more than 99%. The energy consumption principle of multi-connected indoor fan is as follows: Energy consumption of 1 kW refrigeration capacity is 30 W, and energy consumption of 1 kW heating capacity is 20 W.

2.3 Model of Hot-Air Curtain

The hot-air curtain's heating performance coefficient is fitted according to MWHX125CR unit (see Eq. (10)). The fitting coefficient is shown in Table 1, and the correlation degree is more than 99%. Hot-air curtain with fan is treated with constant power in heating season, and the value is 1.5 kW.

$$\varepsilon_h = a_9 + b_9 t_a + c_9 t_a^2 + d_9 t_a^3 + e_9 t_a^4 + f_9 t_w \qquad (10)$$

where t_a is air inlet temperature, °C.

3 Optimum Temperature of Loop Water

Taking a typical underground two-story non-transfer Island subway station in Harbin as an example, the waste heat output and heating heat load are calculated and the system is divided. The general information of the station is as follows: 213 m in length, 19.7 m in standard section width, 16.410 m in foundation pit excavation depth, and 3 m in roof burial depth. There are three entrances, one emergency evacuation exit, two environmental control rooms, two emergency lighting power supply rooms, one signal equipment and power supply room, one communication equipment room, one comprehensive monitoring equipment room, one station control room, one police monitoring room, one screen door equipment and monitoring room, one civil communication room, one high-voltage switchgear room, one low-voltage switchgear room, and one storage battery room.

The station waste heat output and heat load of enclosure structure during the whole heating season are stable, and ventilation can be set to the minimum fresh air volume. Thus, the outdoor temperature t_0 is the only factor affecting the optimal water temperature t_{wb}. Taking the maximum EER_r (heat recovery energy efficiency ratio = heat supply/total power consumption) as the control principle, the t_{wb} at different t_0 is calculated, as shown in Fig. 2. It can be seen that t_{wb} changes linearly with t_0. The higher the t_0, the lower the t_{wb}. The reason is that with the increase of t_0, the heat load decreases and the residual heat offset by fresh air decreases, and the recoverable residual heat increases. That is to say, the heat capacity of heating equipment decreases while the refrigeration capacity of refrigeration equipment increases. Loop water temperature t_w decreases, refrigeration performance coefficient of refrigeration equipment increases, and heating performance coefficient of heating equipment decreases. Figure 1 can be used as a theoretical reference for loop water temperature control of multi-connected heat pump system for metro waste heat recovery.

Figure 2 shows the follow-up relationship between EER_r and t_w. Figure 3 shows that the system can maintain a high EER_r when the water temperature of the loop fluctuates near t_{wb} at any t_0 condition.

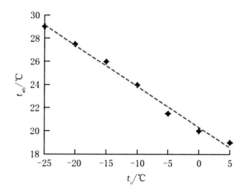

Fig. 1 t_w at different t_0

Fig. 2 Relationship between EER_r and t_w

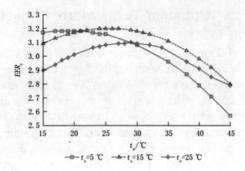

Fig. 3 Relationship between H and t_w

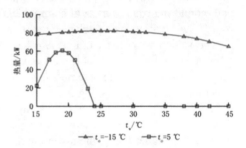

4 Analysis and Prediction of System Performance

4.1 Variation of Auxiliary Heating Load H with t_w

The simulation result shows that the waste heat recovered by the water-loop multi-connected heat pump system can meet the heating load except the hot-air curtain during the whole heating season, but the hot-air curtain load cannot be guaranteed when the t_0 is low. Therefore, auxiliary heat source should be added to the hot-air curtain, which can be supplied by electric heating or municipal heating network. Under the same t_0 condition, the required auxiliary heating amount H varies with the change of t_w, as shown in Fig. 3. For a given t_0, H maximizes at t_{wb}. This is because when t_0 is constant, the recoverable waste heat Q_a and the total heating load Q of the station are fixed values. So if $t_w = t_{wb}$, the power consumption of heat recovery system P is the smallest and the sum of Q_a and P is also the smallest. Only the power consumption of pumps does not fully enter the system in all system power consumption equipments, but the proportion of energy consumption of pumps is very small, so its impact can be neglected. Therefore, there is an approximate relationship between Q, Q_a, P, and H as follows:

$$Q = Q_a + P + H \tag{11}$$

$Q_a + P$ is the smallest; that is, H is the largest. Therefore, for a given t_0, H achieves the maximum at t_{wb}. As shown in Fig. 4, H can be reduced to zero when

Fig. 4 Variation of Q, Q_r, and H_{max} with t_0

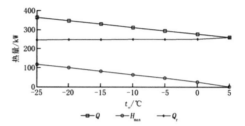

$t_0 = 5\ °C$ and the water temperature of the loop is far away from t_{wb} ($t_w > 24\ °C$). But the calculation shows that when $t_0 < 3\ °C$, H is always greater than zero in the range of water temperature which can be operated on multi-connected due to the increase of Q and the decrease of Q_a.

4.2 Variation of Q, Q_a, and H_max with t_0

When t_0 is constant and the heat recovery system operates under t_{wb}, the energy efficiency ratio of the system reaches the maximum $EER_{r\ max}$ and the auxiliary heating quantity also reaches the maximum H_{max}. Q_r and H_{max} of heat supply in waste heat recovery system corresponding to different t_0 are shown in Fig. 4.

There is a strict linear relationship between Q and t_0. Q decreases with the increase of outdoor temperature, while Q_r increases slightly with the increase of t_0. This is because the stable waste heat output of the station is constant while the residual heat offset by ventilation decreases with the increase of t_0. There are the following relationships among them:

$$Q = Q_r + H_{max} \qquad (12)$$

With the increase of t_0, Q decreases and Q_r increases slightly. Thus, H_{max} decreases with the increase of t_0, but the reduction rate is slightly higher than Q.

4.3 Variation of EER_r max and R_r (Q_r's Proportion in Q) with t_0

$EER_{r\ max}$ increases with the increase of t_0. The reason is that when t_0 is high, the multi-connected heating system runs at partial load rate and the heating performance coefficient is higher than the rated working condition. However, the change value of recoverable waste heat does not cause a significant change in the multi-connected load rate of refrigeration, so higher t_0 corresponds to larger $EER_{r\ max}$. $EER_{r\ max}$ reaches a high value of 3.33 at $t_0 = 0\ °C$; even when

$t_0 = -25$ °C, the value is still as high as 3.10. The R_r curve shows that even at the low temperature of $t_0 = -25$ °C, the heating heat of the heat recovery system can still account for 67% of the total heat load.

4.4 Variation of EER$_e$ with t$_w$ and t$_0$

When the auxiliary heat source uses electric energy, the relationships between Q, Q_a, P, H, and total power consumption E are as follows:

$$Q = Q + P + H \tag{13}$$

$$E = P + H \tag{14}$$

EER$_e$ = Q/E is defined to reflect the overall energy efficiency of the system when electric energy is used as an auxiliary heat source. Figure 5 shows the change rule of EER$_e$ with t_w. As shown in the figure, when $t_0 = 5$ °C and $t_w < 24$ °C, the EER$_e$ reaches its maximum and remains constant. The reason is that if t_0 is unchanged, then Q and Q_a are constant. In this condition, EER$_r$ is higher when t_w goes lower (Fig. 2). Thus, less P will be consumed, and the heat supply of the heat recovery system can not meet the heat demand. Therefore, auxiliary electric heating must be used. When $t_w > 24$ °C, EER$_e$ is decreased with the increase of t_w. This is because if t_w deviates from t_{wb}, EER$_r$ is low and P is large, and Q_r can meet all the heating heat demand, even there is surplus. The higher the t_w, the lower the EER$_r$, the larger the P, and the more energy surplus. In actual operation, the system should be avoided under such unfavorable conditions. This may only occur when $t_0 > 3$ °C. When $t_0 < 3$ °C, the auxiliary electric heating needs to be turned on at any t_w, and then, the EER$_e$ will not change with t_w, as shown in the case of $t_0 = -15$ °C in the figure. The variation of EER$_e$ with t_0 is shown in Fig. 6. The higher the t_0, the larger the EER$_e$.

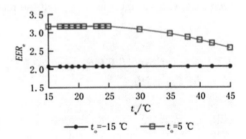

Fig. 5 Changing of EER$_e$ with t_w

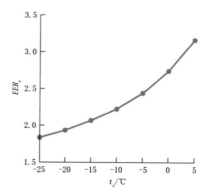

Fig. 6 Relationship between EER_e and t_0

4.5 Prediction of Recovery Potential on Waste Heat

Assuming that the typical non-transfer subway station is located in Harbin, Moscow Helsinki, and Montreal, respectively, and the system operates from 05:00 to 22:00 everyday, the proportion of Q_a, P, and H in the total heating heat in the whole heating season is shown in Fig. 7.

Figure 7 shows that the recovered waste heat accounts for more than half of the total heating heat. For typical non-transfer metro stations, when water-loop multi-connected heat pump system is used to recover waste heat and heat, the recoverable waste heat of each station in a heating season is 575, 674, 676, and 598 MW h in Harbin, Moscow, Helsinki, and Montreal, respectively. If the station is a transfer station, more waste heat can be recovered. Four cities have operated more than 220 underground metro stations, and other cities of the same latitude as Seoul, London, and Milan. It will bring huge energy-saving potential by using water-loop multi-connected heat pump system.

The average EER_e in Harbin, Moscow, Helsinki, and Montreal were calculated to be 2.47, 2.87, 2.87, and 2.7, respectively. The energy efficiency of heat recovery systems in four cities is relatively high with an average EER_r of more than 3 and an average EER_e of more than 2.5.

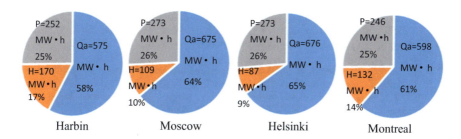

Fig. 7 Proportions of Q_a, P, and H in Q during the whole heating season

5 Conclusion

(1) The equipment room in metro station generates a large amount of stable waste heat all year long. It is feasible to reuse it as a kind of heat source for the station heating.
(2) Water-loop multi-connected heat pump system is the preferred scheme to recover waste heat from equipment room in metro station for the station heating. It has the following advantages: (1) less equipment, small size, flexible installation space; (2) energy of any area in the station could be recovered or distributed relying on the water loop to convey heat; (3) only refrigerant tube enters the equipment room avoiding water leakage.
(3) The waste heat of about 575, 675, 676, and 598 MW h could be recovered from the sample stations, respectively, in Harbin, Moscow, Helsinki, and Montreal in one heating season, using the water-loop multi-connected heat pump system. The regulation of room configuration, room function, waste heat generation, and heating load is basically the same in subway stations of different cities, although the specific conditions of metro stations vary greatly. Therefore, the results of this research could be applied universally in different metro stations.

Acknowledgements National Natural Fund Project (Number 51408457).
China Scholarship Council (Number 201807835013).

References

1. Gilbey, M., et al.: The potential for heat recovery from London underground stations and tuunels. CIBSE Technical Symposium, Leicester UK (2011)
2. Guan, X.Y., et al.: CFD simulation of a novel ventilation system of subway station in Harbin. In: Proceedings Building Simulation, 432–436 2007
3. Toki, Y.: District heating and cooling system utilizing the waste heat from the subway in the city of Sapporo. In: Proceedings Industrial Electronics Society, IECON'88, Singapore, 1000–1005 1988
4. Kojima, S.: Road heating by subway waste heat recovery heat pump. Heat pump **1990**, 781–784 (1990)
5. Vasilyev, G.P., et al.: Technical and economic aspects of using heat pump systems for heating and cooling of the Moscow subway's facilities. Appl. Mech. Mater. **664**, 254–259 (2014)
6. Ninikas, K., et al.: Heat recovery from air in underground transport tunnels. Renew. Energy **96**(10), 843–849 (2016)
7. Gao, H., et al.: Analysis of pipe spacing for capillary heat exchanger in metro tunnel. J. Qingdao Technol. Univ. **37**(5), 106–109 (2016)
8. Chai, Y.J., et al.: Modularly design for waste heat recovery system in subway based on air source heat pump. Procedia Eng. **205**(10), 273–280 (2017)

9. Yao, Y., et al.: Design of water-loop heat pump air conditioning system, pp. 32–33. Chemical Industry Press, Beijing (2011)
10. Wang, W., et al.: Study on the optimum water supply temperature in the middle loop of a new two-stage coupled heat pump system. Fluid mach. **36**(1), 66–69 (2008)

Parametric Analysis and Exergy Analysis for a New Cogeneration System Based on Ejector Heat Pump

Jiyou Lin, Chenghu Zhang and Yufei Tan

Abstract In this paper, a new cogeneration system is proposed by combining ejector heat pump with organic Rankine cycle, which can output power and domestic hot water simultaneously. The ejector heat pump is used to increase the mass flow rate of the working fluid flowing through the expander, thereby increasing the net power output. The performance of the new cogeneration system is studied by parametric analysis, exergy analysis, and economic analysis. The results show that the net power output is better (6.11%) when the high-temperature evaporation pressure is low and the expander intake pressure is high. There is an optimal condensation pressure that optimizes the investment recovery period. When the new cogeneration system's net power efficiency and exergy efficiency is better, the economics of the system is difficult to improve.

Keywords Cogeneration · Ejector heat pump · Parametric analysis

1 Introduction

Energy-saving technologies have made it possible to solve a large number of primary energy consumption and environmental pollution problems. As an efficient energy-saving technology, the cogeneration system has been widely used in the field of central heating and industrial production. According to the Carnot theorem, the system can obtain better thermal efficiency when the evaporation temperature is higher and the condensation temperature is lower. However, a higher evaporation temperature means that the heat source outlet temperature is higher and the heat recovery capacity is low. A lower condensing temperature means that the cold

---

J. Lin · C. Zhang (✉) · Y. Tan
School of Architecture, Harbin Institute of Technology, Harbin, China
e-mail: chenghu.zhang@163.com

C. Zhang · Y. Tan
Key Laboratory of Cold Region Urban and Rural Human Settlement Environment Science and Technology, Ministry of Industry and Information Technology, Harbin, China

source outlet temperature is lower and difficult to utilize, which is undoubtedly a huge waste of energy [1]. Researchers solve the above problems by combining the power cycle with the heat pump cycle [2]. In the present study, a new cogeneration system based on ejector heat pump is proposed. The new cogeneration system combines the organic Rankine cycle and ejector heat pump and could produce both power output and domestic hot water. In this paper, the performance of the new cogeneration system is studied by parametric analysis and exergy analysis.

2 Cycle Description and Assumptions

A new cogeneration system based on ejector heat pump is proposed, which combines the organic Rankine cycle and ejector heat pump. An ejector heat pump system is installed on the inlet side of expander. The primary fluid is heated by the high-temperature (HT)-heat exchanger to illuminate the secondary fluid heated by the low-temperature (LT) heat exchanger. The mixing fluid with the increased mass flow and reduced pressure is heated by the superheater and then enters the expander to do work. Although the inlet pressure of the expander is reduced, the increase of the total mass flow will make the net power output of the new cogeneration system remain unchanged or even increase. By setting an adaptive heat exchanger, the heat transfer matching problem between the HT-heat exchanger and the LT-heat exchanger is automatically adjusted. The cold source is further heated to a higher temperature in the adaptive heat exchanger, which can be used to produce domestic hot water. The heat source passes through superheater, HT-heat exchanger, adaptive heat exchanger, and LT-heat exchanger, which is beneficial to the cascade utilization of energy and improves the heat recovery capacity. The system schematic diagram and the temperature–entropy diagram of the new cogeneration system are shown in Figs. 1 and 2, respectively.

The main purpose of this paper is to study the influence of main parameters of the new cogeneration system on the system performance. Through the reasonable setting of operating parameters, the balance between net power output, heat recovery capacity, and condensation heat reuse can be achieved. Therefore, it is necessary to ignore the parameters such as pressure loss and pipeline loss of the system [3]. The modified constant pressure-mixing model is used to mathematically model the ejector [4]. And the specific modeling process and solution process can be referred in the references [5]. The values of external operating conditions and main parameters of the new cogeneration system are shown in Table 1. If there is no special parameter setting explanation, the parameters in each working condition of this paper shall prevail in Table 1.

The physical properties of R245fa involved in this paper are calculated by REFPROP 9.1. In actual renovation project, the heat exchanger area and heat exchanger capacity are certain. Therefore, this paper proposes the heat used to heat the secondary fluid vapor in the LT-heat exchanger to characterize the secondary fluid mass flow.

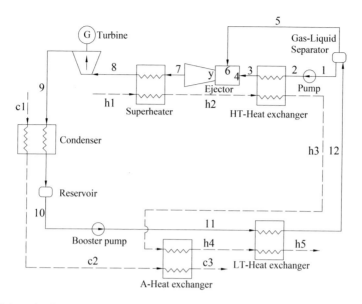

Fig. 1 Schematic diagram of the new cogeneration system

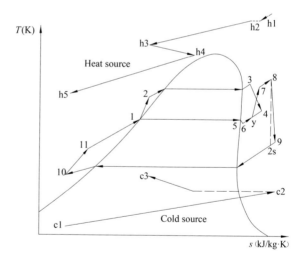

Fig. 2 T-s diagram of the new cogeneration system

$$Q_s = \dot{m}_s(h_5 - h_1) \tag{1}$$

The net power efficiency of the new cogeneration system is as follows.

$$\eta_{net} = \frac{W_{net}}{Q_{sup} + Q_{e,ht} + Q_{e,lt} + Q_a} \tag{2}$$

Table 1 External operating conditions and main parameters of the new cogeneration system

Coefficient	Value
Environment temperature (°C)	20
Turbine isentropic efficiency (%)	85
Pump isentropic efficiency (%)	85
Booster pump isentropic efficiency (%)	90
Nozzle efficiency (%)	95
Diffuser efficiency (%)	85
Superheat of HT-heat exchanger (°C)	5
Superheat of superheater (°C)	5
Supercooling of condenser (°C)	3
Evaporation pressure in HT-heat exchanger (kPa)	1700
Evaporation pressure in LT-heat exchanger (kPa)	800
Expansion pressure (kPa)	1200
Condensation pressure (kPa)	200
Heat source inlet temperature (°C)	130
Heat source mass flow rate (kg/s)	10
Cold source inlet temperature (°C)	15
Cold source outlet temperature (°C)	45
Pinch point temperature difference (°C)	3

Exergy efficiency is defined as the exergy output divided by the exergy input to the new cogeneration system [2]. The exergy input is taken as the available energy change of the heat source. The system heats the cold source from 15 °C to over 45 °C through the condenser and the adaptive heat exchanger. Although the cold source can be used as the heat output of the system, there is still uncertainty. Therefore, the exergy output is the exergy of the net power output only. The exergy efficiency of the new cogeneration system is as follows.

$$\eta_{ex} = \frac{W_{net}}{Ex_{hs}} \tag{3}$$

3 Results

3.1 Parametric Analysis

Under the working conditions described in Table 1, the effects of HT-heat exchanger evaporation pressure and expander intake pressure on system performance are investigated. Figure 3 shows the effect of HT-heat exchanger evaporation pressure and expander intake pressure on the net power generation of the system. The net power output decreases as the HT-heat exchanger evaporation pressure increases and increases as the expander intake pressure increases. When the evaporation pressure is

Fig. 3 Effect of HT-heat exchanger evaporation pressure on the net power output

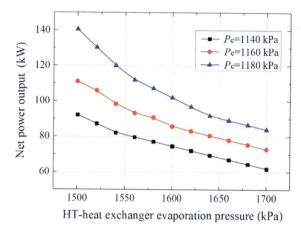

high, it means that more heat is used to heat the primary fluid, and less heat is used to drive the expander to generate electricity. Under the condition that the system condensation pressure is 200 kPa, the higher the expander inlet pressure is, the stronger the power output capacity will be. The net power generation of the system is 141 kW under the condition that the HT-heat exchanger evaporation pressure is 1500 kPa and the expander intake pressure is 1180 kPa.

Figure 4 shows the effect of HT-heat exchanger evaporation pressure and expander intake pressure on the net power efficiency. The higher the HT-heat exchanger evaporation pressure means that there are more heat sources for heating the primary fluid, and the system has a strong entertainment ratio (0.48) and a low thermal efficiency (2.96%). Under the condition that the system condensation pressure is 200 kPa, the higher the expander inlet pressure is, the stronger the net power efficiency will be. The net power efficiency of the new cogeneration system is 6.11% under the condition that the HT-heat exchanger evaporation pressure is 1500 kPa and the expander intake pressure is 1180 kPa.

Fig. 4 Effect of HT-heat exchanger evaporation pressure on the net power efficiency

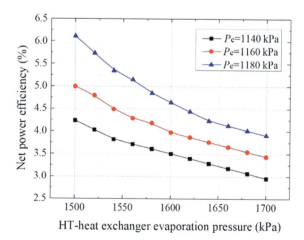

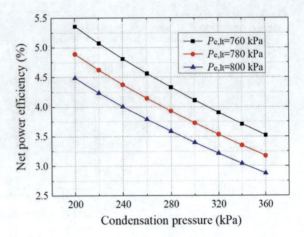

Fig. 5 Effect of condensation pressure on the net power efficiency

Figure 5 shows the effect of condensation pressure and LT-heat exchanger evaporation pressure on the net power efficiency. The net power efficiency decreases significantly as the condensation pressure increases. In the process of increasing the condensation pressure from 200 to 360 kPa, the net power efficiency dropped from 5.36 to 3.52%. For every 20 kPa reduction in LT-heat exchanger evaporation pressure, the net power efficiency can be increased by 0.41%, which is slightly less that of expander intake pressure (0.62%).

The total area of the heat exchanger directly affects the cost of the heat exchanger, and the total cost of the heat exchanger accounts for a large proportion of the total investment of the new cogeneration system. Figure 6 shows the effect of condensation pressure and LT-heat exchanger evaporation pressure on the total heat exchanger area. Excessively increasing the LT-heat exchanger evaporation pressure will result in a significant increase in total heat exchanger area, which is obviously not conductive to the economy of the system. Further reduction of condensation

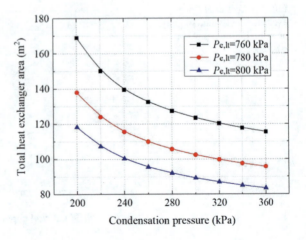

Fig. 6 Effect of condensation pressure on the total heat exchange area

pressure will undoubtedly increase the area of the condenser, which directly leads to the increase of the total heat exchanger area.

3.2 Exergy Analysis

Figure 7 shows the effect of HT-heat exchanger evaporation pressure and expander intake pressure on the exergy efficiency of the new cogeneration system. The exergy efficiency increases as the expander intake pressure increases and increases as the HT-heat exchanger evaporation pressure decreases. This trend is more pronounced at lower HT-heat exchanger evaporation pressures. Figure 8 shows the effect of condensation pressure and LT-heat exchanger evaporation pressure on the

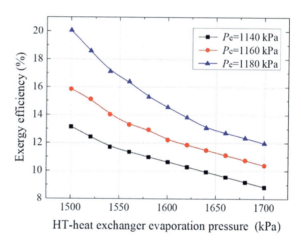

Fig. 7 Effect of HT-heat exchanger evaporation pressure on exergy efficiency

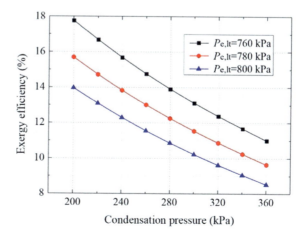

Fig. 8 Effect of condensation pressure on exergy efficiency

Table 2 Main economic parameters of the new cogeneration system

Coefficient	Value	Coefficient	Value
HT-heat exchanger cost (RMB/m^2)	1000	Annual interest rate (%)	6
LT-heat exchanger cost (RMB/m^2)	1000	Annual running time (h)	3600
Superheater cost (RMB/m^2)	1000	Cooling water cost (RMB/t)	4
A heat exchanger cost (RMB/m^2)	750	Net power income (RMB/kWh)	0.5
Condenser cost (RMB/m^2)	800	Hot water income (RMB/(t °C))	0.5
Coal unit price (RMB/t)	400	Supercooling of condenser (°C)	3

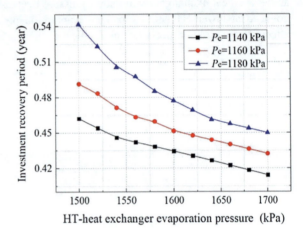

Fig. 9 Effect of HT-heat exchanger evaporation pressure on the investment recovery period

exergy efficiency of the new cogeneration system. The exergy efficiency decreases significantly as the condensation pressure increases. Combined with the foregoing, reducing the condensation pressure as much as possible is an effective way to improve the power generation capacity of the system.

However, the direct discharge of cold source to the environment will result in huge energy waste. Therefore, this paper limits the cold source outlet temperature to above 45 °C to ensure that the cold source can be used as domestic hot water and generate economic value. Based on the values of the main economic parameters described in Table 2, this paper analyzes the dynamic investment payback period of the new cogeneration system. When the cold source is heated above 40 °C, the new cogeneration system can generate economic benefits.

Figure 9 shows that the investment recovery period decreases as the HT-heat exchanger evaporation pressure increases. The ejector has limited ejection capacity when the HT-heat exchanger evaporation pressure is lower, which leads to the limited output capacity of the system. The investment recovery period increases as the expander intake pressure increases. Increasing the expander intake pressure means increasing the net power efficiency and net power output of the system. Obviously, the system's heat output is more economical than the power output.

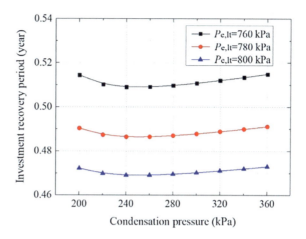

Fig. 10 Effect of condensation pressure on the investment recovery period

Figure 10 illustrates that the investment recovery period decreases as the LT-heat exchanger evaporation pressure increases. However, reducing the LT-heat exchanger evaporation pressure will result in a reduction in the net power efficiency and exergy efficiency of the system. Therefore, there is an optimum LT-heat exchanger evaporation pressure (780 kPa) that optimizes the overall performance of the new cogeneration system. Under the conditions of HT-heat exchanger evaporation pressure of 1700 kPa and expander intake pressure of 1200 kPa, there is an optimal condensation pressure (240 kPa) to optimize the investment recovery period. Lower condensation pressure is not conducive to condenser heat transfer, while higher condensation pressure is not conducive to system power output.

In general, the performance variation of the new cogeneration system is more complicated than the basic ORC. Under the working conditions described in this paper, the economic benefits generated by domestic hot water are greater than that of net power output, which makes it difficult to balance the net power efficiency with the system economy. Since the efficient operation of the ejector requires a certain inlet and outlet pressure condition, the new cogeneration system can only regulate the power output and the domestic hot water output within a limited range. It is still necessary to study the performance limit of the new cogeneration system and the determination of the corresponding working conditions.

4 Conclusions

A new cogeneration system is proposed in this paper, which arranges the ejector heat pump system on the intake side of the expander. It can produce power output and domestic hot water output by using the heat source in cascades. The performance of the new cogeneration system is studied by parametric analysis, exergy analysis, and economic analysis. The system power generation capacity can be improved by increasing the expander intake pressure and reducing the HT-heat exchanger evaporation pressure, but the system economy is not optimal. The variation law of exergy efficiency is almost the same as that of net power efficiency. There is an optimum condensation pressure that makes the new cogeneration system economical optimal.

Acknowledgements The project is supported by the National Key Research and Development Program (Number 2018YFD1100703).

References

1. Lecompte, S., Huisseune, H., Van Den Broek, M., et al.: Review of organic rankine cycle (ORC) architectures for waste heat recovery. Renew. Sustain. Energy Rev. **47**, 448–461 (2015)
2. Dai, Y., Wang, J., Gao, L.: Exergy analysis, parametric analysis and optimization for a novel combined power and ejector refrigeration cycle. Appl. Therm. Eng. **29**(10), 1983–1990 (2009)
3. Wang, J., Dai, Y., Zhang, T., Ma, S.: Parametric analysis for a new combined power and ejector–absorption refrigeration cycle. Energy **34**(10), 1587–1593 (2009)
4. Huang, B.J., Chang, J.M., Wang, C.P., Petrenko, V.A.: A 1-D analysis of ejector performance. Int. J. Refrig **22**(5), 354–364 (1999)
5. Cardemil, J.M., Colle, S.: A general model for evaluation of vapor ejectors performance for application in refrigeration. Energy Convers. Manag. **64**, 79–86 (2012)

Modeling Method of Heat Pump System Based on Recurrent Neural Network

Yin Zheng, Guiqiang Wang, Guohui Feng and Zhiqiang Kang

Abstract Heat pump system is a complex interaction system of multiple mechanical, electrical, and control systems. Traditional modeling methods based on physical laws have large computational complexity and poor precision, which make it not suitable for control strategy optimization. To solve these problems, this paper presents a modeling method based on recurrent neural network (RNN). The network structure and training algorithm were determined according to actual needs. The RNN model was tested and verified on a ground source heat pump system in an office building of a university in Northeast China. The heat pump operation data were continuously monitored and collected, and input into the neural network with three layers. Part of the data set is used for training and the rest is used for testing. The results show that the model has high precision, indicating that this modeling method is effective. This method is considered to be repeatable and can be applied to other heat pump systems.

Keywords Heat pump system · Recurrent neural network · Modeling · RMSprop · Adam

1 Introduction

Heat pump moves thermal energy in the opposite direction of spontaneous heat transfer, by absorbing heat from a cold space and releasing it into a warmer one. A heat pump uses a small amount of external power to accomplish the work of transferring energy from the heat source to the heat sink [1]. Because of these advantages, the heat pump units are widely used in public buildings such as office buildings. There are two types of methods for modeling heat pump units, mathematical methods based on data and physical methods based on characteristic

---

Y. Zheng · G. Wang (✉) · G. Feng · Z. Kang
School of Municipal and Environmental Engineering,
Shenyang Jianzhu University, 110168 Shenyang, China
e-mail: wgq_hit@126.com

parameters. Physical methods require the determination of numerous parameters and equations that are more complex than mathematical methods. The recurrent neural network method used in this paper belongs to one of the mathematical methods.

The artificial neural network relies on the input and output data pairs of the research object. Through learning, a nonlinear mapping describing the input–output relationship of the system is obtained, and the neural network system automatically adjusts its own model [2]. The recurrent neural network is a kind of artificial neural network, which was a method for modeling and predicting nonlinear systems and was first proposed by Samek [3]. In this paper, the complex variable relationships within the heat pump system are fitted to make the model structure simpler and run faster. The recurrent neural network has a ability to fit nonlinear relationships [4], its application has advantages in the modeling of HVAC systems. Therefore, based on the theory of recurrent neural network, this paper studies and builds the model of ground source heat pump system and verifies the reliability of the model by using the model to predict the supply and return water temperature on the demand side of the heat pump system.

2 Methods

2.1 Experimental Data Collection

Starting from the actual demand, when the temperature of the supply and return water on the heat source side is known, the model is established to predict the temperature of the supply and return water on the indoor side. The heat pump unit of an office building in a university in Northeast China was selected as the research object.

The heat pump unit exchanges thermal energy with the deep groundwater pumped by the water pump through the closed-loop system inside the unit through the heat exchanger. Groundwater is drained and injected into the groundwater layer by a pressurized pump (Fig. 1).

The temperature measuring devices were buried under water pipes insulation layer, and the temperature was collected and recorded in real time using a paperless recorder. The recording was recorded every 5 min for a period of 20 days. Get 6,000 pieces of data, 90% of which is used for neural network training, and the remaining 10% is used to test the reliability of neural networks.

Since the sensors of temperature measuring device were not in direct contact with water inside the pipeline, it is obtained by measuring the temperature of the pipe. The presence of heat resistance of air gap can result in inaccurate measurement of the water temperature.

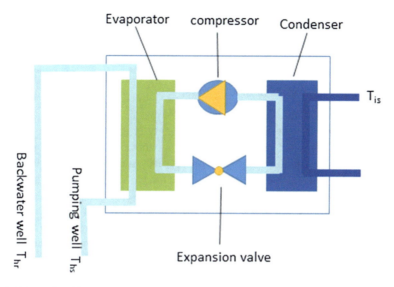

Fig. 1 Schematic of heat pump system

According to the relationship between the heat pump system variables, the heat source side supply water temperature T_{hs} and the heat source side return water temperature T_{hr} are selected as the input of the neural network, and the indoor side supply water temperature T_{is} is selected as the output of the neural network (Fig. 2).

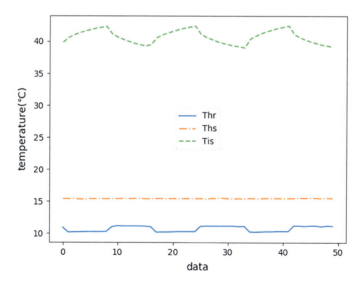

Fig. 2 Test data

2.2 Recurrent Neural Network

A complete recurrent neural network includes the model structure (the number of hidden layers and the number of neurons) and the training algorithm. The specific recurrent neural network is built as follows:

1. Determination of the model structure. Combined with the above study, the number of input neurons is determined to be 1, and the number of output neurons is determined to be 2. When the number of hidden layers is one, the system modeling requirements can be satisfied, and the risk of over-fitting is reduced [5], and so, in this model, a single hidden layer structure was adopted. The determined model structure is shown in Fig. 3.
2. Determination of training algorithm. The training algorithm plays an extremely important role in deep learning process, related to the convergence rate of the results and the accuracy of the results. Representative methods in the recurrent neural network training algorithm are RMSprop method and the Adam method. Among them: RMSprop is an adaptive learning rate method proposed by Geoff Hinton. It only calculates the corresponding average value, so it can alleviate the problem that the learning rate of Adagrad algorithm decreases rapidly. Adam is a first-order optimization algorithm that can replace the traditional stochastic gradient decrease process. It can iteratively update the neural network proportions based on the training data. In the training process, the mean square error (MSE) is used as the loss function to pass back to the hidden layer for network optimization [6].

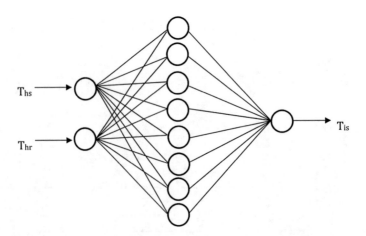

Fig. 3 Structure of neural network for predicting indoor *side* water supply temperature

3 Results and Discussion

3.1 Training Result

The RMSprop method and the Adam method were used to optimize the network model built above. The algorithm was implemented based on the open-source machine learning framework PyTorch. The optimization process of the two algorithms is shown in Fig. 4.

Figure 4 shows the decrease in the MSE value of the RMSporp method and the Adam method. It can be seen from the above figure that the MSE values of both methods decrease rapidly. When the cycle reaches 80 steps, the RMSprop method has reached the optimal MSE value, and the Adam method requires a cycle of 100 steps to achieve the optimal MSE value. In summary, the RMSporp method consumes less time and is more efficient than the Adam method in the training process.

3.2 Application of Network Model on Heat Pump System

When the working conditions are known, the temperature of the indoor side water supply can be predicted by the built network model to provide reference for the optimization control of heat pump systems.

Based on the neural network model built by the RMSporp method, the indoor side water supply temperature is predicted by the reserved 10% data, and the

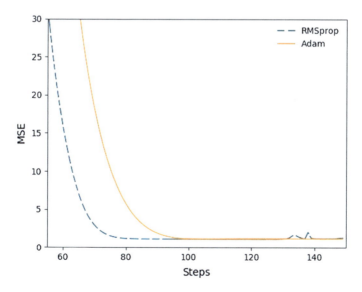

Fig. 4 Optimization process of the RMSporp method and the Adam method

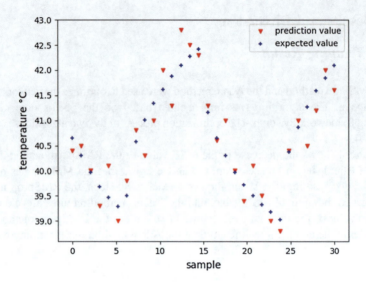

Fig. 5 Comparison of predicted and expected values

expected value and the predicted value are compared, and the predicted values are analyzed by error range.

Figure 5 reflects the predicted temperature and the expected temperature distribution. It can be seen that the two temperature values are similar.

It can be seen from Fig. 6 that the predicted output temperature and the expected output temperature are highly consistent. The error of the two is kept in a small range, indicating that the prediction effect is good.

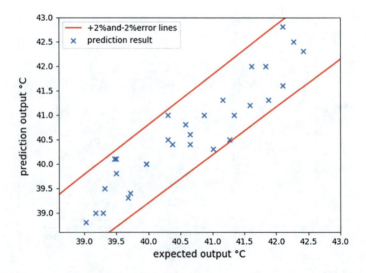

Fig. 6 Error range of prediction results

4 Conclusions

In this paper, a modeling method of heat pump system based on recurrent neural network is studied. The recurrent neural network based on the RMSporp method is applied to the prediction of the water supply temperature in the indoor side of the heat pump, and the error is within 2%. It can be concluded that the model built by this method is effective and considered to be applicable to other heat pump systems.

Acknowledgements The authors wish to acknowledge the support of the Program of National Science and Technology of China during the Thirteenth Five-year Plan (2017YFB0604004-03).

References

1. Robert, A., Ristinen/Jack, J.: Kranshaar energy and the environment, 2nd edn, Wiley, Inc (2006)
2. Chen, W., Zeng, N., et al.: Conductivity prediction and control method of electrode boiler based on artificial neural network. Electron. Technol. Softw. Eng. **01**, 71–72 (2019)
3. Samek, D.: Elman neural networks in model predictive control. In: European Conference on Modelling and Simulation, 577–581 2009
4. Cho, K., et al.: Learning phrase representations using RNN encoder–decoder for statistical machine translation, Empirical methods in natural language processing, pp. 1724–1734 (2014)
5. Yang, Y., Hua, C., et al.: Modeling method of vapor compression refrigeration system based on artificial neural network. J. Armored Force Eng. Inst. **30**(05), 69–72 (2016)
6. Zhang, H.: Research and improvement of optimization algorithm in deep learning. Beijing University of Posts and Telecommunications, China (2018)

Experimental Study on the Influence of Uniformity Liquid Distribution on the Flow Pattern Conversion of Horizontal Tube

Zhennan Qu, Zhixian Ma and Jili Zhang

Abstract This paper experimentally studied the effect of non-uniform liquid distribution on the evolution of the falling film flow mode transition on an array of horizontal tubes. A dedicated experimental bench was designed and built to observe the falling film flow pattern. A smooth copper tube with an outer diameter of 19.05 mm, a tube length of 280 mm, and a tube spacing of 10 mm was selected as the test tube. The experimental results showed that the law of the flow pattern evolution under the non-uniform liquid distribution condition is significantly different from that under the uniform liquid distribution condition: When the mass flow of water gradually increases, the transitional *Re* of the droplet to droplet-column flow pattern conversion is 5.93% lower than that under uniform liquid distribution condition, the transitional *Re* of the droplet-column to column flow pattern conversion is 55.7% higher than that under uniform liquid distribution condition, the transitional *Re* of the column to column-sheet flow pattern conversion is 12.4% higher than that under uniform liquid distribution condition, and the transitional *Re* of column-sheet to sheet flow pattern conversion is 26.2% higher than that under uniform liquid distribution condition. This paper provides a reference for establishing a more accurate condensation heat transfer model of horizontal tube bundle.

Keywords Liquid distributor · Horizontal tube · Flow mode · Falling film Reynolds number

1 Introduction

The falling film flow and heat transfer outside the horizontal tube have important applications in key heat exchange equipment of energy power, petrochemical industry, and so on. Many scholars have found that the falling film flow pattern between tubes has a significant effect on both sensible and latent heat transfer of

Z. Qu · Z. Ma (✉) · J. Zhang
Institute of Building Energy, Dalian University of Technology, Dalian, China
e-mail: mazhixian@dlut.edu.cn

horizontal circular tubes [1–5]. According to the former studies, the flow patterns between the horizontal tubes can be divided into five types, as shown in Fig. 1, namely droplet (D), column(C), sheet(S), droplet-column (DC), and column-sheet (CS).

The evolution law of the falling film flow mode outside the horizontal tube is the basis for further establishing falling film heat transfer model of the horizontal tube and the development of heat exchange equipment. Therefore, several semi-empirical models based on the experimental result were proposed by many scholars, such as Mitrovic [6], Hu [7], and Roques [8]. They all gave a flow mode transition correlation with film Reynolds number (Re) versus the Galileo number (Ga) and gave a mathematical relationship between Re and Ga ($Re = AGa^b$), where A and b are the empirical constants in the formula, as shown in Table 1.

However, all the semi-empirical models were built under the uniform liquid distribution condition. And in a bundle of horizontal tubes of a shell-tube condenser, the condensate of the upper tubes is found not evenly distributed on the lower tubes. To check the validity of these semi-empirical models under this condition, it is necessary to study the effect of the falling film distribution uniformity on the flow mode transitions. Therefore, this paper experimentally studies the effect of non-uniform liquid distribution on the transition of falling film flow pattern in the horizontal tube bundle.

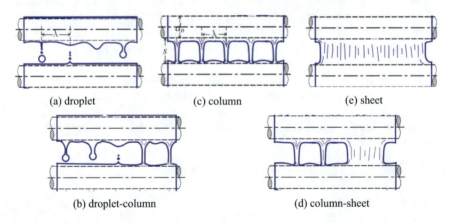

Fig. 1 Schematic diagram of five typical flow patterns in falling film flow

Table 1 Models for judging the transition of falling film flow mode

Author	$D \rightarrow DC$	$DC \rightarrow C$	$C \rightarrow CS$	$CS \rightarrow S$
Mitrovic	$Re = 0.2\ Ga^{1/4}$	$Re = 0.26\ Ga^{1/4}$	$Re = 0.94\ Ga^{1/4}$	$Re = 1.14\ Ga^{1/4}$
Hu	$Re = 0.074\ Ga^{0.302}$	$Re = 0.096\ Ga^{0.301}$	$Re = 1.414\ Ga^{0.233}$	$Re = 1.448\ Ga^{0.236}$
Roques	$Re = 0.0417\ Ga^{0.3278}$	$Re = 0.0683\ Ga^{0.3204}$	$Re = 0.8553\ Ga^{0.2483}$	$Re = 1.068\ Ga^{0.2563}$

2 Experiment

2.1 Experiment System

The schematic of the experimental system is shown in Fig. 2. The integrated system is composed of a high-level water tank, a test section, a backwater tank, a weigh device, valves, and a water pump. The proposed system consists of two subsystems: water system and collection system.

(1) Water system: The working fluid flows to the test section from the high-level water tank, and the flow to the test section is controlled by the valves of the main pipe and the bypass pipe, so that the flow can cover the specified range of $Re \in (10, 1000)$. The working fluid flows through the liquid distributor to a copper tube 1-2 mm below it to make the liquid more evenly distributed, and then flows through the lower test tube into the return water tank. The backwater tank fluid provides a lift through the circulation pump and returns to the high water tank to complete the cycle. The experiment uses the weighing method to test the flow through the test tube within a specified time to obtain the mass flow between the tubes.

(2) Collection system: It includes a data acquisition system and an image acquisition system. In order to measure the image when each flow pattern is changed,

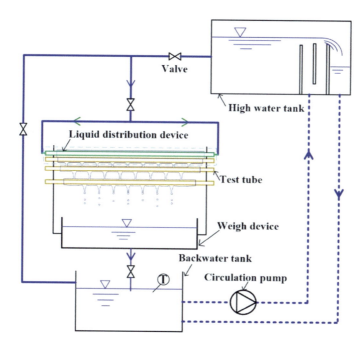

Fig. 2 Schematic of the experimental system

the camera is used for real-time shooting in front of the test section, and the flow mass is measured by the weighing method.

2.2 Test Section

The test section is composed of the liquid distributor and the test tube, wherein the structure of the liquid distributor is special. The liquid distributor consists of three layers of nested stainless steel multi-hole tube, and its specific structure and parameters are shown in Fig. 3 and Table 2, respectively.

Water was used as the working liquid. In Table 3, the properties of the water in this experiment were obtained in accordance with the temperature. The flow rate gradually changed from a low value to a high value and then reduced. Every flow transition was repeated five times to assure an accurate measurement and repeatability of the flow mode transition. The flow modes were recorded by the camera. The test condition was listed in Table 4.

2.3 Uncertainty Analysis

An uncertainty analysis was conducted on the experimental system through the method suggested by the literature [9]. Table 5 presents the relevant physical variables, dimensionless groups, their ranges, and related experimental uncertainties.

Fig. 3 Structure of liquid distributor

Table 2 Parameters of liquid distributor

Distribution tube	Outer diameter d_o/mm	Thickness δ/mm	Length L/mm	Number of holes N	Aperture d_b/mm	Hole distance l_o/mm	Hole direction
DN10	10	1	418	80	1.5	3.5	Down
DN16	16	1	318	56	3	5	Up
DN25	25	1	298	80	1.5	3.5	Down

Table 3 Liquid property

Work fluid	Test temperature /°C	Dynamic viscosity $\mu \times 10^{-6}$/Pa s	Density ρ/kg m^{-3}	Surface tension $\sigma \times 10^{-3}$/N m^{-1}	$Ga^{1/4}$
Water	19.0–21.0	984–1034	998.0–998.4	72.6–72.9	431–452

Table 4 Test condition

Type	d_o/mm	d_i/mm	Tube spacing/mm	Flow rate	Transition mode
Smooth	19.05	16.50	10	Increasing decreasing	Droplet and droplet-column (D-DC) Droplet-column and column (DC-C) Column and column-sheet (C-CS) Column-sheet and sheet (CS-S)

Table 5 Physical variables and dimensionless groups, experimental range, and uncertainty

Physical parameter	Value	Units/uncertainty
Mass flow rate per unit length, Γ	0.03–0.261	kg m^{-1} s^{-1}/± 5%
Liquid density, ρ	998.2	kg m^{-3}/± 0.5%
Liquid dynamic viscosity, μ	1.00×10^{-3}	N s m^{-2}/± 0.5%
Liquid surface tension, σ	7.28×10^{-2}	N m^{-1}/± 0.5%
Gravitational acceleration, g	9.8	m s^2/± 0.5%
Falling film Reynolds number, $Re = 2\Gamma/\mu$	72–570	NA/± 2%
Modified Galileo number, $Ga = \rho\sigma^3/(\mu^4 g)$	3.86×10^{10}	NA/± 3%

3 Results and Discussion

3.1 Experimental Results

Figure 4 illustrates the transitional Re under the liquid distributor of this paper and uniform liquid distribution condition. And the corresponding transitional Re values and falling film flow modes are shown in Table 6 and Fig. 5.

As is shown, for the D-DC, DC-C, C-CS, and CS-S flow mode transitions, the differences between the transitional Re gain under non-uniform conditions and that for the uniform condition are 5.93, 55.7, 12.4, and 26.2%, respectively. For the S-SC, SC-C, C-CD, and CD-D flow mode transitions, the differences are 20.8, 35.0, 7.00, and 38.2%, respectively.

And the hysteresis of falling film flow mode transition is also badly affected by the falling film distribution condition. As the flow rate gradually increased, the

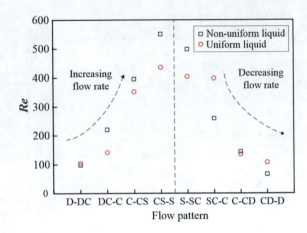

Fig. 4 Comparison of experimental results under uniform and non-uniform liquid distribution conditions

Table 6 Experimental results

Flow pattern	D-DC	DC-C	C-CS	CS-S	S-SC	SC-C	C-CD	CD-D
Flow (kg/s)	0.0137	0.0307	0.0552	0.0768	0.0681	0.0361	0.0202	0.0093
Re	98.30	220.3	395.6	550.6	488.2	258.9	144.4	66.44

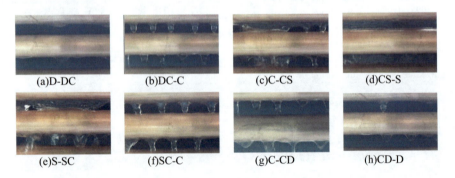

(a)D-DC (b)DC-C (c)C-CS (d)CS-S

(e)S-SC (f)SC-C (g)C-CD (h)CD-D

Fig. 5 Falling film flow modes

transitional Re for DC-C, C-CS, and CS-S modes under non-uniform condition is obviously higher than that under the uniform liquid distribution condition. The observations indicate that the non-uniform liquid distribution condition tends to prevent the DC-C and CS-S modes transition. As the flow rate changes in the opposite direction, the transitional Re of S-SC mode is higher than that under the uniform liquid distribution, but the transitional Re of SC-C mode is obviously lower than that under the uniform liquid distribution. The observations indicate that the non-uniform liquid distribution condition tends to prevent the SC-C mode transition in a horizontal direction.

Fig. 6 Comparison of experimental results with that determined by semi-empirical models

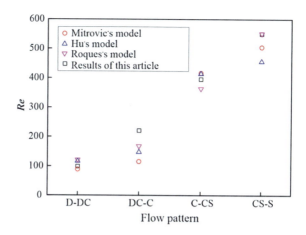

3.2 Comparison with Existing Semi-empirical Models

Figure 6 shows the comparison of the current result of the transitional *Re* and that determined by the semi-empirical models [6–8].

As is shown in Fig. 6, comparing with models of Mitrovic and Hu, the transitional *Re* of C-CS mode is lower, the transitional *Re* of DC-C and CS-S modes is higher. Compared with the semi-empirical model of Roques, it can be seen that the transitional *Re* of D-DC mode is lower, the transitional *Re* of DC-C and C-CS modes is higher, and the transitional *Re* of stable sheet is basically the same. In general, under the condition of non-uniform liquid distribution, it is difficult to form stable flow modes such as column or sheet mode, and the stable flow mode is maintained for a short time.

In particular, it can be seen from comparing with Hu's model [7] that when the flow gradually changes from a low value to a high value, the transitional *Re* of D-DC mode is 15.6% lower than that under uniform liquid distribution condition; the transitional *Re* of DC-C mode is 49.3% higher than that under uniform liquid distribution condition; the transitional *Re* of C-CS mode is 4.47% lower than that under uniform liquid distribution condition; the transitional *Re* of CS-S mode is 20.7% higher than that under uniform liquid distribution condition. This proved that the uniformity of liquid distribution has an effect on the falling film flow mode of horizontal tube.

4 Conclusion

According to the results of this study, the effect of liquid distribution condition on the evolution of the falling film flow mode transition on an array of horizontal tubes is not negligible. For the DC-C and CS-S flow mode transitions, the differences between the transitional *Re* for the non-uniform liquid condition and that for the uniform condition are higher than 20%. For the DC-C flow mode transition, the difference between the transitional *Re* for the non-uniform liquid condition and that for the predicted model achieves 49.3%. The hysteresis of falling film flow mode transition is also badly affected by the falling film distribution condition. The falling film flow mode transitions under non-uniform liquid distribution conditions cannot be well predicted with the classical models established under the uniform liquid distribution condition. The effect of the tube surface structure and tube spacing on the falling film flow mode transition under non-uniform liquid distribution conditions remains to be studied in the future.

Acknowledgements The project is supported by National Natural Science Foundation of China (51606029).

References

1. Kutateladze, S.S., et al.: The influence of condensate flow rate on heat transfer in film condensation of stationary vapour on horizontal tube banks. Int. J. Heat Mass Transf. (1985)
2. Mitrovic, J.: Influence of tube spacing and flow rate on heat transfer from a horizontal tube to a falling liquid film. In: International Heat Transfer Conference, San Francisco, vol. 4, 1949–1956 (1986)
3. Marto, P.J.: Recent progress in enhancing film condensation heat transfer on horizontal tubes. Heat Transf. Eng. **7**, 53–63 (1986)
4. Honda, H., et al.: Film condensation of R-113 on in-line bundles of horizontal finned tubes. ASME J. Heat Transf. **113**, 479–486 (1991)
5. Rogers, J.T., et al.: Turbulent falling film flow and heat transfer on horizontal tubes. In: National Heat Transfer Conference, vol. 12, ASME HTD-vol. 314, 3–12 1995
6. Mitrovic, J., et al.: Fluid dynamics and condensation heating of capillary liquid jets. Int. J. Heat Mass Transf. **38**, 1483–1494 (1995)
7. Hu, X., Jacobi, A.M.: The intertube falling film: part 1-flow characteristics, mode transitions, and hysteresis. ASME J. Heat Transf. **118**, 616–625 (1996)
8. Roques, J.F., Thome, J.R.: Falling film transitions between droplet, column, and sheet flow modes on a vertical array of horizontal 19 fpi and 40 fpi low-finned tubes. Heat Transf. Eng. **24**(6), 40–45 (2003)
9. Chen, J.D., et al.: Falling film mode transitions on horizontal enhanced tubes with two-dimensional integral fins: effect of tube spacing and fin structures. Exp. Therm. Fluid Sci. (2018)

Analysis of the Influence of Atomization Characteristics on Heat Transfer Characteristics of Spray Cooling

Jun Bao, Yu Wang, Xinjie Xu, Xuetao Zhou and Jinxiang Liu

Abstract The development of big data leads to the increasing heat dissipation of chips in data center. As an efficient pattern to remove high heat flux, spray cooling has huge potential for data center cooling. Spray cooling system was established combined with particle image velocimetry (PIV) system. The results show that as the spray diameter decreases, the outlet pressure and outlet velocity of the droplet increase, and the spray cone angle increases. This occasion introduces only minor part of the droplets actually participate in the heat exchange, resulting in a higher velocity and a smaller heat transfer coefficient. It is also inferred that better uniformity of droplets velocity is beneficial for the heat transfer performance.

Keywords Spray cooling · Spray atomization · PIV · Flow pattern distribution · Heat transfer coefficient

1 Introduction

With the sharp increase of the data amount on the Internet, the demand for data centers' processing capacity is getting higher. The augment of server racks power consumption and the improvement of chip integration lead to higher requirements for heat dissipation in data center. The power density of each rack achieved 7 KW in traditional data centers [1], while the cooling capacities could be better between 10 and 15 KW per rack in the new-generation data centers [2]. Air cooling is the most widely used cooling method to remove heat in data centers. Researchers suggested that the maximum heat dissipation capacity of air cooling is about 37 W/cm^2 [3] and it is far less than the heat generation rate of processors which is about 65 W/cm^2 [4]. In order to ensure the reliability and stability of chips, new efficient heat removal technologies need to be adopted urgently.

J. Bao · Y. Wang (✉) · X. Xu · X. Zhou · J. Liu
College of Urban Construction, Nanjing Tech University, Nanjing, China
e-mail: yu-wang@njtech.edu.cn

Plenty of studies concerning the factors affecting spray cooling heat transfer characteristics have been proposed including the coolant flow flux, types of working substance, orifice-to-surface distance, and so on. Experiments were performed by Chen et al. [5] in which the mean droplet velocity had the most significant effect on CHF and the second important factor is the droplet flux. Sozbir et al. [6] studied the atomization heat transfer coefficient under different inlet flow rates in the membrane boiling zone by air atomizing nozzle and found that the heat transfer coefficient was linear with the inlet flow of water. Chen et al. [5] and Oliphant et al. [7] both concluded that surface heat transfer coefficient increases with the mass flux rise of the medium through investigations. Experiments about different type working substance which included FC-72, FC-87 and water on a small area of heat source were conducted by Estes and Mudawar [8], and they found that droplets with a larger diameter and a faster velocity would block its evaporation. Hsieh and Yao [9] carried out a single nozzle spray cooling experiment with pure water at 25 °C and R134a at 14 °C. Conclusions were obtained that the degree of subcooling had little effect on the heat transfer performance of R134a spray cooling, and the cooling effect of R134a was much less than that of water.

Former researchers have been focused on heat transfer coefficient enhancement which could be achieved by enhanced surfaces [10–12] and medium with stronger heat transfer parameters [9, 13] and so on. However, little research was proposed on the heat transfer characteristics of spray cooling through visualization. In this investigation, spray cooling system has been established in combination with PIV system aiming to achieve visualization to investigate the relationship between atomization effects of nozzles with different pore sizes and the heat transfer performance. The results will provide a reference for the selection of nozzles and the optimization of spray effects in spray cooling system design.

2 Equipments and Methods

2.1 Bench Construction

The spray cooling system experimental bench established in this study mainly consists of a spray liquid supply system, a simulated heat source heating system, a data acquisition system, and a particle image velocimetry (PIV) system. The nozzles selected in the experiment are 304 stainless steel solid cone nozzles, and with the outlet diameter of 0.8, 1.2, 1.5, and 2.0 mm, respectively. The experimental medium is water. The surface area of the simulated heat source is a circular shape with a diameter of 24 mm, and the maximum heat that the simulated heat source can achieve is about 1800 W. The distance from the nozzle to the surface of the simulated heat source is 100 mm. Other parameters are listed in Table 1. The spray cooling system is shown in Fig. 1.

Table 1 Spray cooling experimental conditions

Orifice-to-surface distance (mm)	10
Nozzle inlet temperature (°C)	6–12
Working fluid flow (L/h)	20–40
Ambient temperature (°C)	3–15

Fig. 1 Spray cooling system

2.2 PIV System Introduction

A 2D PIV technique was applied in this research. PIV system consists of an illumination laser, a sync controller, an image acquisition board which was built in a computer in advance, a high-speed digital camera, and a computer which was used for recording and calculation. The model types of specific equipment used in this research are shown in Table 2. The schematic diagram of PIV system is shown in Fig. 2.

2.3 Heat Flux Measurement and Data Acquisition Instrument

Four K-type thermocouples were arranged in the vertical direction from the surface of the simulated heat source to the bottom to calculate the surface heat transfer coefficient. The distance to the heating surface was T_1: 17 mm, T_2: 25 mm, T_3: 33 mm, T_4: 41 mm, respectively. The temperature is obtained by an Agilent 34972A data acquisition instrument. According to one-dimensional Fourier heat conduction law, the radial temperature distribution and heat flux of the simulated heat source surface can be determined.

Table 2 PIV system equipment model

Device name	Model type
Illumination laser	Vlite-200
Sync controller	SM-Micropulse725
Digital camera	CLB-B2520 M-SC000

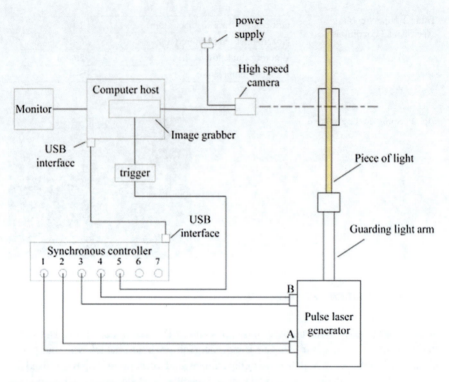

Fig. 2 Operating principle diagram of PIV system

$$q = \lambda \frac{T_i - T_j}{\Delta \delta} \quad (1)$$

where q is heat flux density at the heat source surface, λ is the thermal conductivity of the heat source material copper, T_i, T_j is the temperature measured by thermocouple T_i, T_j, and $\Delta\delta$ is the distance between two adjacent thermocouples.

The temperature of the heat source surface can be calculated by the following formula:

$$T_w = T_1 - \frac{(T_i - T_j)\delta_1}{\Delta \delta} \quad (2)$$

where T_w is the temperature of the heat source surface, and δ_1 is the distance from simulated heat source surface to the thermocouple T_1.

The surface heat transfer coefficient can be calculated as below:

$$H = \frac{q}{T_w - T_{in}} \quad (3)$$

where H is the surface heat transfer coefficient, and T_{in} is the spray substance nozzle inlet temperature.

2.4 Experimental Procedure

The experiments conducted in this paper were divided into two parts. The first part of the experiment is the heat transfer performance experiment of spray cooling. The spray cooling effect under different working conditions could be obtained by altering the nozzle type, heating power, and flow rate of the working fluid. The second part of the experiment is the application of PIV system to take visual graphics of the spray flow field.

3 Results and Discussion

3.1 Effect of Flow Rate Change on Atomization State and Heat Transfer Performance

The heat transfer coefficient of the simulated heat source at the flow rate of 20 and 30 L/h for the nozzle with pore size of 0.8 mm was compared. It can be seen from Fig. 3 that the heat transfer coefficient increases with the increase in input heating power. However, the heat transfer coefficient tends to decrease when the heating power of 200 W/cm^2 is achieved with the flow rate of 20 L/h. This is because with the development of heat flux, the droplets become vapor at the time they fall on the heating surface. The substance droplets are blocked by the vapor, which impairs the heat transfer. With the increase in droplet velocity, the droplets have the ability to overcome the vapor. Therefore, with the flow rate of 30 L/h, the heat transfer is still enhanced when the heat flux of 200 W/cm^2 is achieved.

The velocity distributions of spray from a nozzle with pore size of 0.8 mm at flow rates of 20 and 30 L/h are shown in Figs. 4 and 5, respectively. It can be seen from the two figures that the spray droplet is symmetrically distributed. The droplet velocity at the axis of the cone spray is the greatest, and the droplet velocity from the axis is gradually reduced. It is also found that the spray range of Fig. 5 is larger than that of Fig. 4.

By comparing the surface heat transfer coefficients at the two flow rates, it is found that the heat transfer coefficient at the flow rate of 30 L/h is smaller than the heat transfer coefficient at the flow rate of 20 L/h which is contrary to previous PIV experimental studies. As shown in Figs. 6 and 7, by observing the droplet distribution in the spray chamber, and the comparison of the velocity distributions at the two flow rates in Figs. 4 and 5, it is found that when the flow rate is 20 L/h, the nozzle cone spray with the aperture size of 0.8 mm cannot reach the original

Fig. 3 Comparison of heat transfer coefficients for a nozzle with pore size of 0.8 and flow rates of 20 and 30 L/h

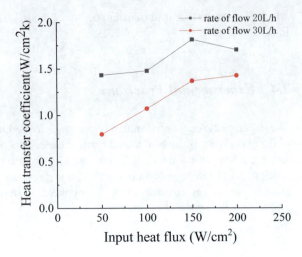

Fig. 4 Flow field distribution with a nozzle aperture of 0.8 mm and a flow rate of 20 L/h

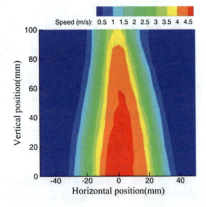

Fig. 5 Flow field distribution with a nozzle aperture of 0.8 mm and a flow rate of 30 L/h

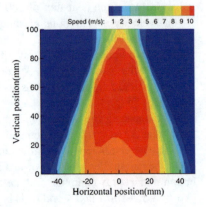

working angel, and the droplets are concentrated on the surface of the heating block to dissipate heat sufficiently. When the flow rate is 30 L/h, the cone spray angle reaches the maximum value, and the droplets sprayed from the nozzle scatter around the edge of the heat dissipation surface, which means part of the droplets are not participated in the heat dissipation of the heating surface, resulting in a large flow rate but a small heat transfer coefficient.

3.2 Effect of Nozzle Pore Size Variation on Atomization State and Heat Transfer Performance

The velocity distribution of the same series of nozzles with aperture size of 1.2, 1.5, and 2.0 mm at a flow rate of 40 L/h and a spray height of 10 cm are shown in Figs. 8, 9 and 10, respectively.

Comparing the three velocity distribution images below, it can be found that as the diameter of the nozzle increases, the droplet velocity of the nozzle ejection decreases. The reason is that when the flow rate is constant, the outlet pressure of the droplet decreases as the diameter of the nozzle increases. As the outlet pressure of the droplet decreases, the droplet velocity also decreases, and the degree of fluid fragmentation becomes smaller so that the initial droplet velocity is lower relatively.

Fig. 6 Droplet distribution in the spray chamber at a flow rate of 20 L/h

Fig. 7 Droplet distribution in the spray chamber at a flow rate of 30 L/h

Fig. 8 Flow field distribution with a nozzle aperture of 1.2 mm and a flow rate of 40 L/h

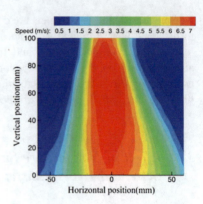

Fig. 9 Flow field distribution with a nozzle aperture of 1.5 mm and a flow rate of 40 L/h

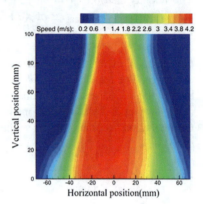

From Fig. 11, the relationship between heat transfer coefficients of droplets ejected from three nozzles with different calibers at a flow rate of 40 L/h can be obtained. The heat transfer coefficient of the heating block surface increases with the augment of the heating power. When the flow rate is constant, the heat exchange effect of the nozzle with the larger diameter is better.

By comparing the velocity maps combined with the droplet distribution state which are shown in Figs. 12, 13 and 14, it is found that although the droplets ejected by the small-caliber nozzle are large in initial velocity, the droplet distribution is not as uniform as that with large-diameter ones, and the edge velocity is weakened causing the surface not completely covered. The droplet velocity of the large nozzle diameter is small, but the velocity distribution is more uniform and can evenly cover the entire surface. Therefore, compared with the velocity values, the velocity uniformity has a major effect on heat transfer.

Fig. 10 Flow field distribution with a nozzle aperture of 2.0 mm and a flow rate of 40 L/h

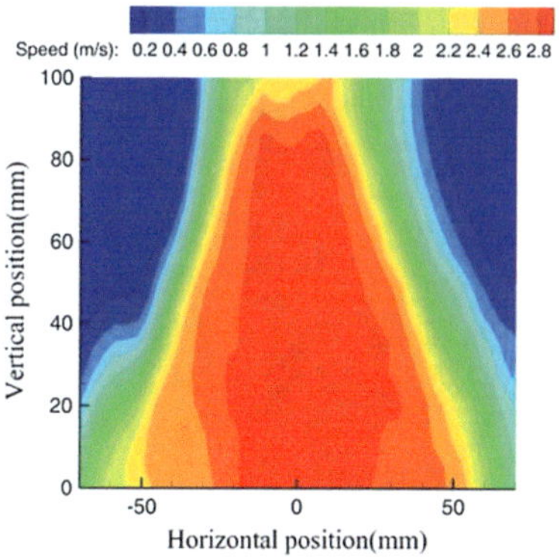

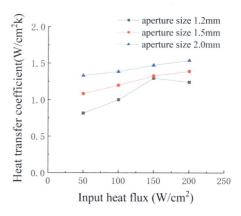

Fig. 11 Comparison of heat transfer coefficients for nozzles with pore size of 1.2, 1.5, and 2.0 mm and flow rate of 40 L/h

Fig. 12 Droplet distribution in the spray chamber of pore size a 1.2 mm nozzle at a flow rate of 40 L/h

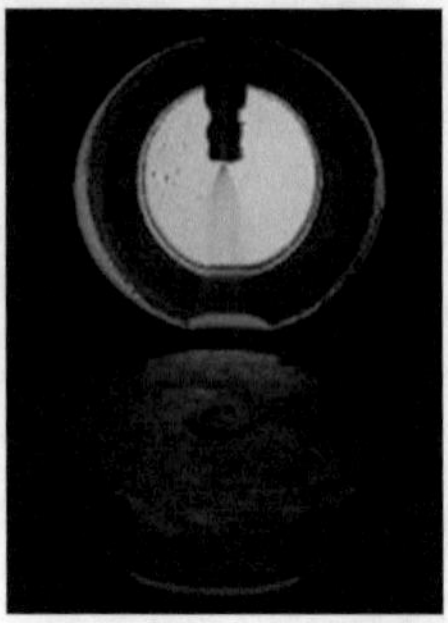

Fig. 13 Droplet distribution in the spray chamber of pore size a 1.5 mm nozzle at a flow rate of 40 L/h

Fig. 14 Droplet distribution in the spray chamber of pore size 2.0 mm nozzle at a flow rate of 40 L/h

4 Conclusion

In this paper, by establishing the spray cooling test bench and PIV system, the velocity distribution of the droplets from different types of nozzles was photographed, and the heat transfer performance corresponding to the velocity distributions was obtained. The relationship between the velocity distribution of the droplets and the heat transfer coefficient was studied.

Conclusions are as follows:

(1) The angle of the spray cone opening has a non-negligible effect on the heat transfer effect of the spray cooling. In some cases, ensuring that the droplets not spraying beyond the surface are more preferred than enhancing the droplet velocity to improve the heat transfer effect.
(2) As the diameter of the nozzle increases, the outlet velocity of the droplet and the angle of the spray vertebral both decrease, the heat transfer effect can be improved.
(3) The better the uniformity of the droplets velocity distribution covering the surface of the simulated heat source, the better the heat transfer of the surface.

Acknowledgements This work is supported by the National Natural Science Foundation of China (Grant No. 51806096) and Natural Science Foundation of the Jiangsu Higher Education Institutions of China (Grant No. 18KJB560007).

References

1. Kant, K.: Data center evolution. Comput. Netw. **53**(17), 2939–2965 (2009)
2. Marcinichen, J.B., Olivier, J.A., Thome, J.R.: On-chip two-phase cooling of datacenters: cooling system and energy recovery evaluation. Appl. Therm. Eng. **41**, 36–51 (2012)
3. Saini, M., Webb, R.L.: Heat rejection limits of air cooled plane fin heat sinks for computer cooling, ITherm 2002. In: Eighth Intersociety Conference on Thermal and Thermomechanical Phenomena in Electronic Systems (Cat. No. 02CH37258), IEEE, 1–8 2002
4. Marcinichen, J.B., Thome, J.R., Michel, B.: Cooling of microprocessors with micro-evaporation: a novel two-phase cooling cycle. Int. J. Refrig. **33**(7), 1264–1276 (2010)
5. Chen, R.-H., Chow, L.C., Navedo, J.E.: Effects of spray characteristics on critical heat flux in subcooled water spray cooling. Int. J. Heat Mass Transf. **45**(19), 4033–4043 (2002)
6. Sozbir, N., Chang, Y., Yao, S.: Heat transfer of impacting water mist on high temperature metal surfaces. J. Heat Transf. **125**(1), 70–74 (2003)
7. Oliphant, K., Webb, B., McQuay, M.: An experimental comparison of liquid jet array and spray impingement cooling in the non-boiling regime. Exp. Thermal Fluid Sci. **18**(1), 1–10 (1998)
8. Estes, K.A., Mudawar, I.: Correlation of Sauter mean diameter and critical heat flux for spray cooling of small surfaces. Int. J. Heat Mass Transf. **38**(16), 2985–2996 (1995)
9. Hsieh, C.-C., Yao, S.-C.: Evaporative heat transfer characteristics of a water spray on micro-structured silicon surfaces. Int. J. Heat Mass Transf. **49**(5–6), 962–974 (2006)
10. Silk, E.A., Kim, J., Kiger, K.: Spray cooling of enhanced surfaces: Impact of structured surface geometry and spray axis inclination. Int. J. Heat Mass Transf. **49**(25–26), 4910–4920 (2006)
11. Bostanci, H., Rini, D.P., Kizito, J.P., Singh, V., Seal, S., Chow, L.C.: High heat flux spray cooling with ammonia: investigation of enhanced surfaces for CHF. Int. J. Heat Mass Transf. **55**(13–14), 3849–3856 (2012)
12. Bostanci, H., Altalidi, S.S., Nasrazadani, S.: Two-phase spray cooling with HFC-134a and HFO-1234yf on practical enhanced surfaces. Appl. Therm. Eng. **131**, 150–158 (2018)
13. Xu, H., Si, C., Shao, S., Tian, C.: Experimental investigation on heat transfer of spray cooling with isobutane (R600a). Int. J. Therm. Sci. **86**, 21–27 (2014)

Research of Personalized Heating by Radiant Floor Panels in Hot Summer–Cold Winter Area

Guoqing Yu, Zhuzheng Diao and Zhaoji Gu

Abstract This paper studies the performance of personalized heating by radiant floor panels on the thermal microenvironment surrounding human subjects in hot summer and cold winter areas in China. Through experiment investigations and numerical simulations, the thermal microenvironment around the human body by heating floors was studied at room temperature of 12, 14, and 16 °C. The following conclusions are obtained:

(1) Thermal sensation of different parts of human body is different under the same environment; the thermal sensation of same part is not same for different subjects.
(2) When the radiant floor panels were placed under the chair and desk, the individual parts of the human body still feel cold at 12 °C; all parts of the human body are close to the heat neutrality except head at 14 °C; all parts of the human body have exceeded the heat neutrality and feet feel slightly warmer at 16 °C.
(3) Numerical simulations found that four heating radiant floor panels have some effect on improving the thermal microenvironment surrounding human body.

Keywords Personalized heating · Radiant floor heating panels · Microenvironment · Numerical simulation

G. Yu (✉) · Z. Diao · Z. Gu
University of Shanghai for Science and Technology, Shanghai, China
e-mail: yuguoqinghvac@163.com

Z. Diao
e-mail: diaozhuzheng@qq.com

Z. Gu
e-mail: 272452987@qq.com

1 Introduction

The hot summer and cold winter areas in China cover about 1.8 million km^2. It is cold in winter, and the indoor air temperature is in the range of 8–15 °C if there is no space heating. Air conditioners are usually used to raise the temperature of the whole room, and the energy consumption is very large. However, a personalized heating system (PHS) can significantly improve the microenvironment around the human body, and the temperature of the room can be appropriately reduced, so that it can save energy for space heating.

Jacob Verhaart et al. calculated and analyzed the use of personalized heating in central heating in Minneapolis (Minnesota, USA); room temperature can be reduced from 21.5 to 18 °C; the maximum energy savings in three months in winter are up to 34% [1]. Veselý M et al. performed a comparative analysis of several different personalized heating devices and control modes. By contrast, it is found that different personalized heating can have a big difference in improving the thermal comfort effect, and some effects are obvious; some effects are limited [2, 3].

Yingdong He and Nianping Li obtained results that indicate that Huotong maintained overall and local comfort in cold environment. With it, 90% acceptable temperature range could be extended to 9 °C [4]. Wilmer Pasut and Hui Zhang et al. evaluated a novel heated/cooled chair's effect on thermal sensation and comfort [5].

This paper studies the effects of personalized heating by radiant floor panels on the thermal microenvironment surrounding human subjects in hot summer and cold winter areas.

2 Methods

2.1 Experimental Environment

In this paper, the experiment was carried out in a room of 6 (m) × 3.4 (m) 3.4 (m). The personalized heating devices of the experiment are four electric heating floor panels (60 × 60 × 2 cm) with 80 W heating power each, which are placed under the chair and desk. The air temperature, the wall temperature, and the surface temperature of the heating floor are tested by thermocouples, and the heating power is monitored by a power meter. The schematic diagram and layout of the laboratory are shown in Fig. 1.

In order to achieve close proximity to the actual working conditions during the experiment, the temperatures are measured during the experiment. The experiment is carried out when the indoor air temperatures are kept at 12, 14, and 16 °C, respectively.

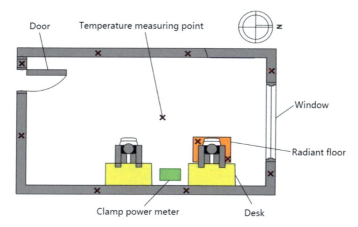

Fig. 1 Room layout

2.2 Experimental Procedures

Experiments were performed at a laboratory on School of Environment and Architectural of the University of Shanghai for Science and Technology and approved by the Ethics Committee of this university. Additionally, informed consent was obtained from all individual participants included in the study and subjects were fully informed about the experimental procedures. There are 10 human subjects taking part in the experiment, their ages are between 20 and 25 years old, and they had good physical condition. Subjects wore long-sleeved warm vests, padded jackets, sweat-absorbent trousers, and sneakers; the clothing resistance is about 1.15 clo [6]. None of the subjects had vigorous exercise before the test, and the diet was normal without taking coffee, alcohol, and any drugs.

The investigation time interval of each subject was 45 min. Before the subjects entered the test room, they waited for 15 min in a waiting room with an indoor air temperature of 18–20 °C and then entered the laboratory for testing. The personalized heating devices are all turned on 20 min in advance; this questionnaire test is divided into two periods: non-personalized heating on station 1 and personalized heating on station 2. After the subject entered the test room, he sat on the station 1 for 10 min. During the period, he could read a book or use a computer. After 10 min, the subject filled out the questionnaire according to the requirements. Then, the subject enters the station 2 and sits for 20 min. During the period, he can also read a book or use a computer, and fill out a questionnaire every 10 min.

During the entire test period, the subjects did not know whether the personalized heating device was turned on or not, thereby eliminating psychological interference and affecting the thermal sensation. After the first subject has been tested, the test room is allowed to wait 15 min to restore the original state, and the second subject enters. The investigation process is shown in Fig. 2.

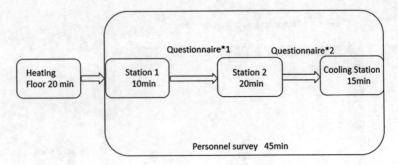

Fig. 2 Investigation process

Table 1 Thermal sensation scoring standard

Scaling	−3	−2	−1	0	1	2	3
Thermal sensation	Cold	Cool	Slightly cool	Neutral	Slightly warm	Warm	Hot

2.3 Questionnaire

Subjects need to fill out a questionnaire during the survey, which includes the subjective responses of local bodies to the surrounding thermal environment. The subjective response of the survey was mainly based on thermal sensation. The surveyed body parts include head, chest, back, buttocks, thighs, calves, and feet.

The thermal sensation evaluations taken in this paper are shown in Table 1. The thermal sensation takes 7 scoring points, from −3 (cold) to 3 (hot) for recording the human thermal sensation votes (TSV) [7].

2.4 Numerical Simulations

This paper uses Airpak to simulate the thermal environment around the human body in a 6 (m) × 3.4 (m) × 3.4 (m) room, the room temperature is 14 °C, and the numbers of radiant floors is 4.

Airpak automatically uses Fluent for calculations. The post-processing has a three-dimensional display function that can display the temperature changes of each section.

3 Results

3.1 Local Thermal Sensation Comparison for Different Parts

In this paper, the local thermal sensations of different parts for different subjects were investigated at 14 °C when there is no heating radiant floor. The results are shown in Fig. 3.

It can be seen from Fig. 3 that the thermal sensations of different parts are different under the same environment. The thermal sensation of the feet is generally lower than other parts, the cold feeling of the head is also obvious, the thermal sensation of the buttocks and thighs is slightly higher than other parts, and the thermal sensation of chest and back is approximately consistent. It can be seen from the range of thermal sensation of each part that the thermal sensation of the same part is different for different subjects, which may be related to the gender, dressing, and living habits of the subjects.

3.2 Local Thermal Sensation Comparison

In this paper, four radiant floor panels with 80 W of each heating power are used as personalized heating devices to study the local thermal sensation compared with no

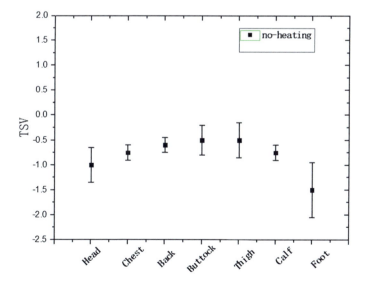

Fig. 3 Local thermal sensation contrast for different parts

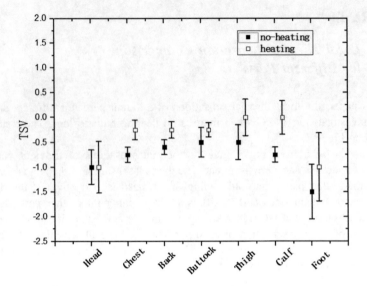

Fig. 4 Thermal sensation comparison at 12 °C

heating condition, when the room temperature is 12, 14, and 16 °C, respectively. The results are shown in Figs. 4, 5 and 6.

From the results of Figs. 4, 5 and 6, it can be seen that heating radiant floors play an important role in improving thermal sensation of human body's local parts compared with no heating radiant floor at 12, 14, and 16 °C, respectively.

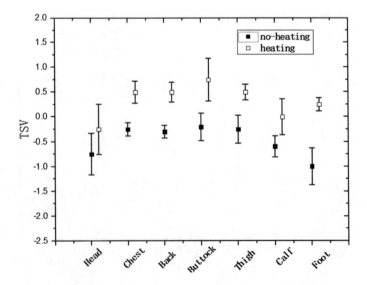

Fig. 5 Thermal sensation comparison at 14 °C

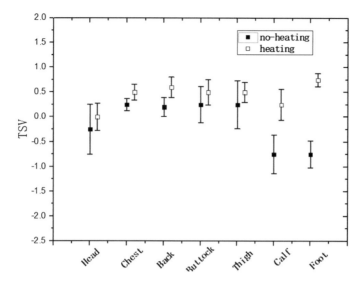

Fig. 6 Thermal sensation comparison at 16 °C

As shown in Fig. 4, the thermal sensation of human body's all parts is lower than the thermal neutrality, when there is non-heating radiant floor at 12 °C. The cold feeling of head, calves, and feet is obvious; when there are heating radiant floor panels, except for the head, thermal sensation of the rest parts has improved, and the thermal sensation of the thighs and calves has reached thermal neutrality. There is no change in the head's thermal sensation. The reason may be that the room temperature is low and the radiant floor heating area cannot reach the head area.

As shown in Fig. 5, when there is no heating radiant floor at 14 °C, the thermal sensation of thighs is close to thermal neutral, but the rest parts still have slightly cold feeling. Feet and head feel a little cool. When there is heating radiant floors, the thermal sensation of human body is improved. However, the head is still lower than thermal neutrality, the calf is thermal neutral, and the rest are slightly higher than thermal neutrality.

As shown in Fig. 6, when there are no heating radiant floors at 16 °C, the thermal sensation of head, calves, and feet is lower than the thermal neutrality, and the other parts are slightly higher than thermal neutrality. When there is heating radiant floors, the thermal sensation of each local part has met the thermal comfort requirement, the head has reached thermal neutrality, the other parts are higher than thermal neutrality, and the feet feels slightly warmer.

3.3 Numerical Simulations

Airpak was used to simulate the effect of the heating radiant floor on the surrounding thermal environment at 14 °C with different floor numbers (0 and 4). This paper adopted the same conditions as the experiment for simulation. The numerical simulation results are shown in Figs. 7 and 8.

It can be seen from Figs. 7 and 8 that the temperature of various body parts is still generally low and is near 14 °C when there is no heating. The area of 20 °C increased significantly around the human body when there are four heating radiant floor panels. It indicates that four radiant heating floors have a significant effect on improving the surrounding thermal environment of human body.

4 Discussions

Due to the limitations of time and conditions, the number of samples of the human thermal sensation survey during the study is small, so the conclusions may have limitations.

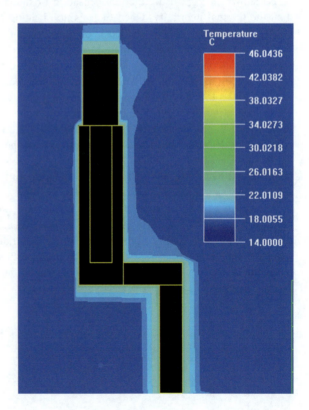

Fig. 7 Non-heating

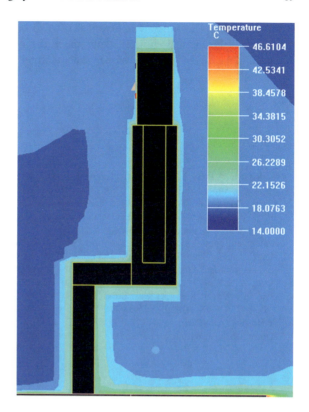

Fig. 8 Four heating floor panels

In the numerical simulation, the human body model is relatively simple and does not consider the human body's breathing, nor does it distinguish between human genders and dressing differences.

5 Conclusions

(1) Thermal sensation of different body parts is different under the same environment; the thermal sensation of same part is not same for different human subjects.
(2) When there are heating radiant floor panels at room temperature of 12 °C, except for head, the thermal sensation of the local parts has improved, and the thermal sensation of the thighs and calves has reached thermal neutrality.
(3) When there is heating radiant floors at 14 °C, the thermal sensation of human body is improved. However, the head is still lower than thermal neutrality, the calves are thermal neutral, and the rest parts are slightly higher than thermal neutrality.

(4) When there is heating radiant floors at 16 °C, the thermal sensation of each part has met the thermal comfort requirement, the head has reached thermal neutrality, the other parts are higher than thermal neutrality, and the feet feels slightly warmer.

References

1. Verhaart, J., Vesely, M., Zeiler, W.: Personal heating: effectiveness and energy use. Build. Res. Inf. **43**(3), 346–354 (2015)
2. Veselý, M., Zeiler, W.: Personalized conditioning and its impact on thermal comfort and energy performance—a review. Renew. Sustain. Energy Rev. **34**(3), 401–408 (2014)
3. Vesely, M., Molenaar, P., Vos, M., Li, R., Zeiler, W.: Personalized heating—comparison of heaters and control modes. Build. Environ. **112**, 223–232 (2017)
4. He, Y., Li, N., Zhou, L., Wang, K., Zhang, W.: Thermal comfort and energy consumption in cold environment with retrofitted Huotong (warm-barrel). Build. Environ. **112**, 285–295 (2017)
5. Pasut, W., Zhang, H., Arens, E., Zhai, Y.: Energy-efficient comfort with a heated/cooled chair: results from human subject tests. Build. Environ. **84**, 10–21 (2015)
6. ASHRAE.: ANSI/ASHRAE Standard 55: thermal environmental conditions for human occupancy. ASHRAE, Atlanta (2017)
7. Zhu, Y.: Built Environment, 3rd edn, China Architecture and Building Press, China (2010)

Passive Thermal Protections of Smoke Exhaust Fans for a High-Temperature Heat Source

Yixiang Huang, Chengqiang Zhi, Qianru Zhang, Wei Ye and Xu Zhang

Abstract Smoke exhaust fans (SEFs) are commonly used to directly remove high-temperature gas exhausted from fire accidents, or industrial applications such as gas turbines and product manufacturing. Most available SEFs used in a confined space can be working properly under specific temperature range for a short period of time. Therefore, thermal protection for the fans is a vital issue to prevent SEFs from failing and causing damages or disasters. In this paper, a linearly five-SEF-array is used to extract exhaust gas from a moving impinging jet at approximately 700 K. Three passive (as opposed to active spraying) measures, i.e., installing a dam board for a fan, changing the angle of inclination of the fans and reducing the running number of fans are studied using the computational fluid dynamics (CFD) method. Three indicators, i.e., gas temperature at the inlet of the fan (T_i), the area of high-temperature zone (A_h) and total space ventilation efficiency of the fans, are assessed by employing at least one of the passive measures. The results show that, first, on the one hand, passive thermal protections are effective tools in decreasing T_i, especially for the fan that is closest to the heat source. On the other hand, A_h will be enlarged with the passive protections, since the exhaust gas is not directly being extracted. Second, changing the angle of inclination of the fans can decrease the temperature of the smoke that enters the SEFs. Third, the smoke

Y. Huang · C. Zhi · Q. Zhang · W. Ye (✉) · X. Zhang (✉)
School of Mechanical Engineering, Tongji University, 201804 Shanghai, China
e-mail: weiye@tongji.edu.cn

X. Zhang
e-mail: xuzhang@tongji.edu.cn

Y. Huang
e-mail: 1830255@tongji.edu.cn

C. Zhi
e-mail: 1810798@tongji.edu.cn

Q. Zhang
e-mail: zhang_qianru@hotmail.com

W. Ye
Key Laboratory of Performance Evolution and Control for Engineering Structures of Ministry of Education, Tongji University, 200092 Shanghai, China

exhaust efficiency can also be significantly lowered by reducing the running number of fans. Last but not least, passive protection control strategy should be up to the location of heat source.

Keywords Exhaust fan · Thermal protection · Heat resource · High-temperature · Impinging jet

1 Introduction

This research is based on an industrial environment that is a long narrow confined space, and there exists a high-temperature impact jet heat source in it. When the heat source is running, an enormous amount of high-temperature gas is exhausted from the heat source with high speed. Then, high-temperature gas may cause damage to the structure of envelope and effect the normal operation of SEFs that usually work properly under the temperature range of 120–280 °C for a short period of time. So, ventilation, a common solution to control indoor thermal environment, should be taken into consideration in order to take a large of heat in time and to ensure safety of SEFs and envelope structure. And it is necessary to take additional cooling measures, e.g., spray cooling, and passive thermal protection measures, i.e., dam board and angle of inclination of the fans, to control the temperature of inclination of fans.

The primary technical researches, which could cope with a challenge of exhausting heating ventilation of a confined cabin, are focused on temperature field distribution in the conditions of tunnel fires. When fire breaks out in a long and narrow space, two main issues must be considered into designing the ventilation system. One is the safety of the people inside the space, where the hot air is of great concern. The second main issue is the endurance of structure under the effect of impacting or high temperature of hot air. So, numerical simulation, multilayer zone model, full-scale experiments and model experiments are employed in smoke spreading under tunnel fires conditions [1–7]. As for ventilating rate determining for a long and confined cabin, ventilation efficiency, a significant indictor, is used in accessing the optimal ventilation rate [8]. And for SEFs optimization problem, SEFs resistance characteristics are combined with spatial coefficient determination by simulation calculation [5]; moreover, improvement fan system efficiency is necessary for ventilation designing industrial application.

As for passive thermal protection measures, they will change pressure distribution of ventilation systems. In addition, passive protection measures increase the flowing resistance of SEFs, and then change the volume flow of SEFs. This affects the performance of the ventilation system. In general, it should be taken into consideration that loss of ventilation performance when taking some passive measures. Details of passive measures, i.e., proper angle of inclination of SEFs and resistance of dem board, are researched by CFD. Moreover, the changing of indicators, i.e., fan inlet temperature (T_i), the area of high-temperature zone (A_h) and ventilation efficiency are assessed by employing one or two passive measures in this paper.

2 Method

2.1 Physical Model

The simulation model is based on an industrial environment. There exists a high-temperature heat source exhausting smoke with high velocity. The ventilation environment is 70 m × 17.4 m × 9.4 m, and symmetrical model is used in this calculation. Five-SEF-array is on the top celling of the symmetrical model, also the five-SEF-array is named fan 1–fan 9 from the front part to the tail part.

As shown in Fig. 1, two different locations of heat source are established in the ventilation space. As for passive thermal protection, inclination of fans is adopted in three fans which is near by the heat source; also, a dem board model is installed below fan 7 and fan 9.

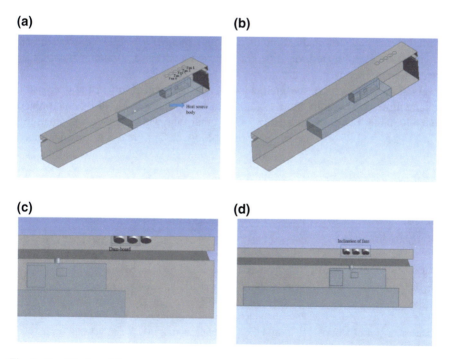

Fig. 1 Simplified models

2.2 Simulation Content and Method

The air supply temperature is 300 K, and ventilation rate of the space is determined by the fan's working condition. Heat resource exhausts high-temperature smoke whose velocity is 70 m/s, and five fans are regarded as pressure outlet whose pressure jump stays constant 280 Pa. Other bounding surfaces are set to adiabatic boundary. Interface between internal environment and surrounding is regarded as free boundary whose total pressure is 0 Pa (Table 1).

The simulations are performed using the commercial software: ANSYS FLUENT 16.0 with realizable k-epsilon turbulence model, because realizable k-epsilon turbulence model can predict the characteristics of jet flow fields [9]. And the standard wall function method employs in simulation. The air density is calculated by incompressible ideal gas model. The pressure velocity coupling adopts the SIMPLE algorithm. The pressure variable adopts the second-order format. The other variables adopt the second-order upwind format. Grid independence of numerical is analyzed; about 6 million unstructured meshes are applied in these case studies.

3 Results

Figures 2 and 3 show that the effect of passive thermal protection changes with the variation of angle of inclination of fans and number of SEFs. Indoor thermal environment is estimated via different indictors such as A_h, volume flow, T_i and ventilation efficiency.

As shown in Fig. (a), the inclination of the fans will result in the decrease of volume flow. And volume flow of fan 1 and fan 5 whose angle of inclination is 0° remains stable, but the volume flow of the other fans whose angle of inclination is 5°, 10°, 15° and 25° decreases remarkably. Flow of fan 7 and fan 9 reduces about

Table 1 Boundary conditions

Zone	Type	Details
Inlet of total space	Pressure inlet	Temperature is 300 K and gauge pressure is set as 0 Pa
Main intake of heat source	Velocity inlet	Calculated by mass balance
Secondary intake of heat source	Velocity inlet	Calculated by mass balance
SEFs	Pressure outlet	Gauge pressure is set as −280 Pa
Exhausting port of heat source	Velocity inlet	Temperature is 700 K and velocity is set as 70 m/s
Other surfaces	wall	Adiabatic boundary

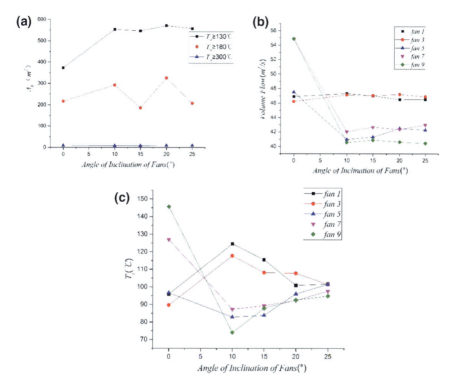

Fig. 2 Gas temperature at the inlet of the fan, the area of high-temperature zone and volume flow as the angle of inclination of fans changes

13 m³/s, and fan 5 decreases about 6 m³/s. In conclusion, the number of angles of inclination of fans influences the volume flow of fans slightly.

As shown in Fig. (b), fan inlet temperature of fan 1 and fan 3 whose inlets incline in a specific angle reduces obviously. However, compared with not taking measure of inclination of fans, the inlet temperature of fan 7 and fan 9 increases about 30 °C. This phenomenon illustrates that the measure will lead to high-temperature air moving and temperature field around the inlets of fan changing. Conclusions can be drawn that the increasement of fans' inlet angle results in that T_i of fans tends to be the same.

Figure (c) illustrates that the measure of inclining inlet angle of fans will result in high-temperature area (A_h) increasement where the temperature is above 130 °C. And A_h whose temperature is over 300 °C remains at about 8–9 m². As for changing of A_h whose temperature is over 180 °C with the angle of inclination of fans, the value reaches in minimum at the angle of 15° and 25°. Taking different factors into consideration, i.e., thermal protection of fans, controlling high-temperature area and resistance conditions, 25° is a proper angle of inclination.

In the part, volume flow, high-temperature area and ventilation efficiency under the condition of different number of fans will be researched. Ventilation efficiency is

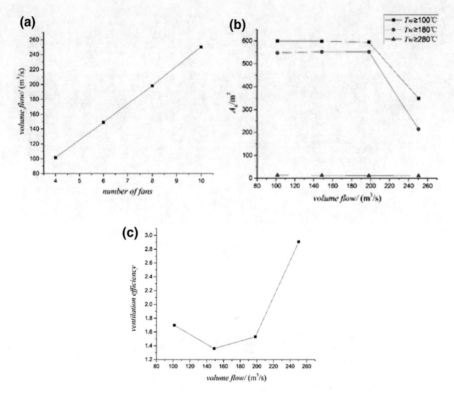

Fig. 3 Volume flow, the area of high-temperature zone and ventilation efficiency as the number of fans changes

introduced in the paper, as a property assessing ventilation effectiveness. Equation of ventilation efficiency using in condition of exhaust heat is below:

$$\eta = \frac{t_o - t_i}{t_a - t_i} \quad (1)$$

where t_o is the exhaust temperature, t_i is the outdoor air temperature, t_a is the average temperature of total space, and η is ventilation efficiency of total space.

As shown in Fig. (a), there exists a liner relationship between the number of fan and volume flow. Also, the results can be drawn from other figures that are parameters such as high-temperature area (A_h) and ventilation efficiency are sensible to the reduction in volume flow or number of fans. When volume flow decreases about 25%, high-temperature area whose wall temperature is above 180 °C and above 280 °C is roughly doubled. So, volume flow is a critical parameter to control A_h.

In this part, two thermal passive protections, from simulation results, dem board installation is contributing to the increasement of fan inlet resistance to flow. When a dem board is installed around the inlets of fan 7 and fan 9, volume flow decreases about 27 m³/s with a constant pressure jump of fans. As the above results are concluding, ventilation efficiency of total space reduces: Because dem board installation not only reduces SEFs inlet temperature, but also decreases total volume flow resulting in total space average temperature increasing. Also, two passive thermal protection measures, i.e., dem board and angle of inclination of the fans, are decreasing volume flow remarkably. But when heat source is located below the inlet of fan 7 and fan 9, it exhausts high-temperature air inflowing inlets of fan 7 and fan 9 resulting in high inlet temperature of fan 7 and fan 9; when two passive thermal protection measures are adopted, inlet temperature of fan 7 and fan 9 decreases by 43–60 °C. But as the moving heat source is moved to another location where heat source is far from original location about 10 m, the effect of decreasing inlet temperature of fan 7 and fan 9 is weakened, and inlet temperature of array of fans tends to be same.

4 Discussion

For common mixing ventilation model, the model assumes that indoor air is mixed to one uniform condition. Under the assumption, indoor air temperature is equal to the outlet temperature. So, indoor air temperature is determined by the following equation.

$$Q = c_p m_v (t_o - t_i) \qquad (2)$$

where c_p is air-specific heat at constant pressure, Pa; m_v is the ventilation mass flow, kg/s; t_o is the indoor air temperature (exhaust temperature), t_i is the supply air temperature, and Q is the heating load unit.

In a word, in this common condition, decrease of ventilation volume flow will result in the increase of indoor air temperature and exhaust temperature while heat source stays constant. In this simulation condition, it is not mixing ventilation model. When not adopted passive thermal protect measures, fans failure or ventilation rate decreasing will lead thermal environment lost control, other fans operation environment worsen. Moreover, when heat source is closed to fans, passive thermal protection measures avoid high-temperature air entering SEFs directly to decrease fans inlet temperature. On the other hand, when heat resource is away from SEFs, different fans inlet temperature tends to be the same; meanwhile, the performance of reducing fans inlet temperature is weakened, and total space ventilation rate decreases lead to ventilation efficiency loss because of passive measures increasing ventilation resistance. In conclusion, passive thermal protection measures are the solutions to protect SEFs when heat resource can exhaust high-temperature smoke influencing fans directly; but during heat resource is far away from fans, ventilation resistance increasing made by passive measures is adverse to eliminate heat. So passive thermal measure, e.g., insulation board, should be designed to be openable which controls according to the location of heat resource.

5 Conclusions

In this paper, CFD method is applied in the investigation about passive thermal protection for SEFs. A_h, T_i and ventilation efficiency are the indicators assessed by employing one or two passive measures. The main conclusions obtained from this study are as follows:

(1) Ventilation flow rate is a critical indictor to control indoor thermal environment; its reduction contributes to the increase of fans inlet temperature and increase of average indoor air temperature.
(2) Passive protection measures are the solutions to avoid high-temperature smoke entering SEFs directly to decrease fans inlet temperature.
(3) Passive protection measures will result in high-temperature area moving and ventilation resistance increasing. It is the disadvantage to total space ventilation system working.
(4) Effect of passive protection measures is related to the location of heat source, and passive protection control strategy should be up to location of heat resource in order to resist negative impact to the ventilation effect on other conditions.

Acknowledgements This research has been supported by the National Natural Science Foundation of China under Grant No. 51878463.

References

1. Zang, J., et al.: Numerical and experimental research on hybrid temperature diffusion for four pairs of symmetric sources and sinks. HV AC **36**(12), 14–17 (2006)
2. Liu, J., et al.: Research on optimization of heat exhaust ventilation for the strong heat disturbance in long and narrow spaces. Build. Energy Environ. **33**(3), 1–4 (2014)
3. Jia, J., et al.: Numerical simulation for temperature distribution during mine fire period. J. Liaoning Technol. Univ. **23**(6), 164–460 (2003)
4. Li, M., et al.: Numerical simulation on temperature field of long narrow confined space fire under negative pressure ventilation condition. Procedia Eng. **52**(1), 272–276 (2013)
5. Xie, Y., et al.: Impedance simulation and experimental analysis of exhaust system in long and narrow space. Build. Energy Environ. **37**(1), 35–39 (2018)
6. Feng, W., et al.: The spatial distribution of toxicant species in fires in a long corridor. J. Univ. Sci. Technol. China **36**(1), 61–64 (2006)
7. Zhao, W., et al.: Analysis of influencing factors on flashover in the long-narrow confined space. Procedia Eng. **62**(11), 250–257 (2013)
8. Chen, X.: Simulation of temperature and smoke distribution of a tunnel fire based on modifications of multi-layer zone model. Tunneling Undergr. Space Technol. **23**, 75–79 (2008)
9. Wang, X.D., et al.: Numerical simulations of imperfect bifurcation of jet in crossflow. Eng. Appl. Comput. Fluid. **6**(4), 595–607 (2012)

Investigation on Polymer Electrolyte Membrane-Based Electrochemical Dehumidification with Photoelectro-Catalyst Anode for Air-Conditioning

Mingming Guo and Ronghui Qi

Abstract Buildings in Southern China require a large amount of energy to handle indoor humidity. As a novel technology, dehumidification with polymer electrolyte membrane (PEM) can remove air moisture using an electric field, without cooling water or absorbents. This study proposed a PEM-based dehumidification element that had simultaneous water electrolysis and photolysis capacities, which could be operated under the joint drive of electricity and solar UV. This element is suitable to be integrated with building envelopes for further energy saving. Several photo-catalysts were prepared, including Pt/TiO_2, IrO_2/TiO_2 and $IrTiO_x$ (IrO_2 : TiO_2 = 2:3), and their photocatalytic properties were measured. Result showed that among these catalysts, under the irradiation of a 300 w xenon lamp, Pt/TiO_2 had the best water photolysis effect, while the durability of Pt under higher current density was poor. The photolysis effect of $IrTiO_x$ was close to that of Pt/TiO_2, and IrO_2 had good catalytic performance and stability in electrolysis. This research helps to improve dehumidification efficiency. More importantly, it attempts a brand-new exploration in the artificial photosynthesis and even power generation relying on humidity difference.

Keywords Electrochemical dehumidification · Water decomposition · Photoelectro-catalyst · Material preparation

M. Guo · R. Qi (✉)
Key Laboratory of Enhanced Heat Transfer and Energy Conservation of Education Ministry, School of Chemistry and Chemical Engineering, South China University of Technology, 510640 Guangzhou, China
e-mail: qirh@scut.edu.cn

1 Introduction

Accurate control of environmental humidity is significant for increasing the comfort level of people's living conditions. Under high-humidity environments, people face problems such as damp houses, mildew, and physical discomfort. Furthermore, equipment and components may suffer from problems of getting moldy, surface corrosion, insulation degradation, or electrical breakdown. In particular, currently, the humidity sensitivity level of machines and production equipment is getting higher. Traditional air handling technology has many limitations such as the possibility of reheating the supply air, low coefficient of performance (COP) [1], insufficient dehumidification capacity and wet surfaces breeding mildew and bacteria. Although, using chemical dehumidification technology by solid or liquid desiccants, some of the drawbacks could be overcome, other practical limitations still exist [2, 3]. Due to the requirement of desiccant regeneration, the system is usually designed to be complex and space consuming. In addition, during dehumidification of the liquid desiccant, desiccant droplets may be carried away by the process air, which may cause corrosion on the metal surface of indoor devices, as well as health problems for the occupants.

As a new type of dehumidification technology, the PEM-based electrochemical dehumidification has the advantages of simple and compact structure, accurate adjustment, low-voltage electric field drive, etc. Qi et al. investigated the dehumidification and energy performance of PEM-based element under various operating conditions [4]. The obtained dehumidification efficiency was $0.6-1.5 \times 10^{-2}$ g/(s V m^2), and energy efficiency was $1.5-2.0 \times 10^{-2}$ g/(J m^2). But the energy conversion efficiency was low, only 20–30%.

Due to its plate and flexible structure, the electrochemical dehumidification is suitable to be integrated with building envelopes, such as windows, to utilize the solar energy. Introducing the solar photocatalysis could make the separation of water vapor during dehumidification more efficient. Besides, driven by an external electric field, recombination of the electron-hole pairs generated in the photocatalytic process can decrease significantly, leading to an increase of separation efficiency, which is supposed to improve the photocatalytic efficiency. Besides, the oxidation and reduction reactions occur, respectively, which reduces the reverse reaction and enhances the reaction efficiency. However, to achieve this purpose, we still face several difficulties. 1) Photocatalytic decomposition of water vapor, especially water vapor in the air, is extremely inefficient. In 1982, K. Domen et al. first carried out an experiment of photocatalytic decomposition of gaseous water on a NiO-SrTiO$_3$ catalyst, with evolution of H$_2$ (4.4×10^{-3} mL h^{-1}) and O$_2$ (2.2×10^{-3} mL h^{-1}) [5]. Investigations on photocatalytic decomposition of water vapor are very limited. 2) Catalysts with both water electrolysis and water photolysis capacities are difficult to find. Electrocatalysts are usually electrically conductive materials, while photocatalysts are usually semiconductors. When preparing catalysts with both water photolysis and water electrolysis performance, the ratio of the two ingredients is difficult to determine.

Herein, an improved PEM-based system was proposed, by introducing an anode-side catalyst with both water electrolysis and water photolysis capacities. The system could dehumidify air under the joint drive of electricity and solar energy. Several photocatalysts were prepared, including Pt/TiO_2, IrO_2/TiO_2, and $IrTiO_x$ ($IrO_2:TiO_2$ = 2:3). Their photocatalytic properties were measured, using an all-glass automatic online trace gas analysis system (Perfect Light-Labsolar 6A). This research helps to enhance the dehumidification efficiency of PEM-based element. More importantly, it is trying to make a new exploration in artificial photosynthesis, and even to generate electricity relying on humidity difference.

2 Methods

2.1 Preparation of the Catalysts

In order to find a photocatalyst with high water decomposition efficiency, three kinds of catalysts, Pt/TiO_2, IrO_2/TiO_2, and $IrTiO_x$ ($IrO_2:TiO_2$ = 2:3) were prepared. The preparations of the catalysts are as follows. Since TiO_2 is inexpensive, stable, and also not easy to cause photo etching, three photocatalysts containing titanium dioxide were selected. TiO_2 (P25), with a particle size of 20 nm, was purchased from Macklin. $IrTiO_x$ ($IrO_2:TiO_2$ = 2:3) was the mixture of IrO_2 (99.9%, mental basis, from Aladdin) and TiO_2 (P25, 20 nm, from Macklin), with (Ir/Ti mole ratio) = 2:3.

Preparation of Pt/TiO_2. Pt/TiO_2 (Pt/Ti mole ratio = 1%) was prepared by the reported method [6]. Five g butyl titanate was dissolved in 50 mL of absolute ethyl alcohol to form solution A. Solution B was prepared by adding 7.5 mL chloroplatinic acid solution to 50 mL of deionized water and adjusting the pH to 1.5 with HNO_3. Then, solution C was formed by adding solution A slowly to solution B. Gel D was obtained after stirring at 30 °C for 24 h. The gel was washed with water and dried at 70 °C for 8 h. Then, the powder was calcined at 400 °C for 1 h.

Preparation of IrO_2/TiO_2. IrO_2/TiO_2 (Ir/Ti mole ratio = 2:3) was prepared by the reported method [7]. An aqueous solution with 30 mL of water and 3 mL of concentrated H_2SO_4 was first prepared. $TiOSO_4$ $0.6H_2SO_4$ $1.3H_2O$ (0.610 g, 2.55 mmol), $IrCl_3$ $3H_2O$ (0.593 g, 1.68 mmol), and $NaNO_3$ (15.0 g, 176 mmol) were then added to the solution under stirring in sequence. After 30 min stirring and mixing, the precursor mixture was dried on a rotary evaporator at 60 °C, 60 mbar. Hereafter, the mixture was calcined in a Muffle furnace first at 150 °C for 2 h, then, after heating to 350 °C, for 1 h. After washing with water and drying in a vacuum oven (150 °C, 16 h), IrO_2/TiO_2 powder was obtained.

Preparation of $IrTiO_x$. $IrTiO_x$ ($IrO_2:TiO_2$ = 2:3) was the mixture of IrO_2 (99.9%, mental basis, from Aladdin) and TiO_2 (P25, 20 nm, from Macklin), with (Ir/Ti mole ratio) = 2:3.

2.2 Test Method of Water Photolysis Performance

To test the water photolysis performance of prepared photocatalyst, an all-glass automatic online trace gas analysis system (Perfect Light-Labsolar 6A) was used to measure the photocatalytic efficiency of the catalysts, as shown in Fig. 1. Before starting the photocatalytic reaction, the reaction apparatus was vacuumed for 30 min. H_2O used for the reaction was degassed by repeated freezing, evacuation, and thaw cycles.

3 Results

3.1 Characterization of the Catalysts

In order to compare the performance differences of the prepared catalysts, the catalysts were characterized by an X-ray diffractometer (BRUKER D8 ADVANCE). X-ray diffraction is a kind of characterization to determine the atomic and molecular structure of a crystal by analyzing the diffraction pattern of the material.

Analyzing the powder pattern of IrO_2/TiO_2 (Ir/Ti mole ratio = 2:3), the oxide phases of anatase TiO_2, rutile TiO_2, and IrO_2 can be noticed (Fig. 2). Apart from IrO_2, the presence of rutile TiO_2 can also be confirmed, due to the asymmetric form of the diffraction peak at 35°. As for the unidentified peak around 23°, it is attributed to SiO_2 dust contamination. Influences of IrO_2 on TiO_2 phase transformation can be determined, according to the detection of both anatase and rutile TiO_2. Such an effect should be reasonable, as IrO_2 also has a rutile structure, and a preferred formation of rutile TiO_2 can be achieved under its presence.

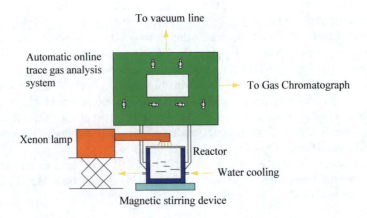

Fig. 1 Schematic view of the experiment apparatus

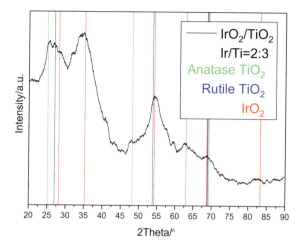

Fig. 2 Powder XRD diffraction patterns of IrO_2/TiO_2 (Ir/Ti mole ratio = 2:3). Green, blue, and red vertical lines refer to the theoretical positions of diffraction peaks for anatase TiO_2, rutile TiO_2, and IrO_2, respectively

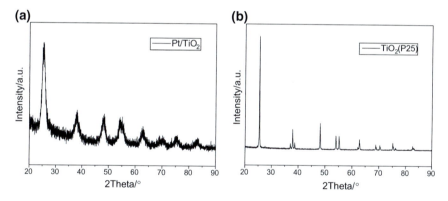

Fig. 3 Powder XRD diffraction patterns of **a** Pt/TiO_2 (Pt/Ti mole ratio = 1%), **b** TiO_2 (P25)

It can be seen from Fig. 3 that the characteristic peaks of the anatase (101), (004), (200), (105), and (204) crystal faces of the sample are all clear (the corresponding 2θ is 25.3°, 37.8°, 48.1°, 54.0°, and 62.7°). The peak diffraction of TiO_2 anatase all appears, no rutile phase diffraction peak is found, and no diffraction peaks of other miscellaneous phases are found, either, indicating that the doping amount of Pt is small, and the distribution on TiO_2 is relatively uniform. It can also be seen from Fig. 3 that the relative intensity of the (101) diffraction peak of the Pt/TiO_2 prepared by the solgel method is weaker than that of the pure TiO_2, meaning that the doping of Pt weakens the lattice signal of TiO_2, leading the lattice to be distorted. The doping of Pt causes new lattice distortion and defect sites, which reduces the lattice order of TiO_2.

3.2 Photocatalytic Decomposition of Liquid Water

In all the experiments, the reaction solution was 100 mL triethanolamine aqueous solution (10 vol %). The amount of each catalyst used was 50 mg. Each mixture was irradiated for 3 h with a 300 w xenon lamp, under magnetic stirring conditions. The results of photocatalytic decomposition of liquid water were showed in Table 1.

From Table 1, it can be determined that the hydrogen production of Pt/TiO_2 (49.73 µmol h^{-1}) was the most, ca. 23 times that of pure TiO_2. The mechanism of this could be as follows. Since Pt and TiO_2 form a heterojunction, the photogenerated charges and holes can be effectively separated. And due to the small hydrogen production overpotential of Pt, the trapped electrons can effectively reduce hydrogen ions to hydrogen. Especially, Fig. 3 shows that the doping of Pt causes new lattice distortion and defect sites, which reduces the lattice order of TiO_2. It has been pointed out that the doping of TiO_2 by metal ions can inhibit the growth of TiO_2 grains [8]. Smaller particle size of the crystalline grain corresponds to a larger specific surface area, thereby facilitating the photocatalytic activity of the photocatalyst. However, in the photoelectro-catalytic reaction, Pt is prone to dissolution, which could cause performance degradation.

Furthermore, comparing the hydrogen production of IrO_2/TiO_2 (15.59 µmol h^{-1}) and $IrTiO_x$ ($IrO_2:TiO_2$ = 2:3) (35.16 µmol h^{-1}), it can be explained that the effective light absorption area of IrO_2/TiO_2 is less than $IrTiO_x$. IrO_2/TiO_2 was prepared by reacting precursors with sodium nitrate at 350 °C and consisted of IrO_2 particles distributed in a matrix of TiO_2 particles. The distribution of IrO_2 in TiO_2 can significantly increase the conductivity of the material, but reducing the effective light absorption area of TiO_2 at the same time. However, $IrTiO_x$ was simply the mixture of IrO_2 and TiO_2. When added to the triethanolamine aqueous solution, IrO_2 and TiO_2 would separate from each other, exerting little influence on the light absorption area of TiO_2. Comparing with the hydrogen production of pure TiO_2 (2.13 µmol h^{-1}), the increase of photocatalytic performance of IrO_2/TiO_2 and $IrTiO_x$ was due to the addition of IrO_2, which might promote the reaction between h^+ and the sacrificial agent (Triethanolamine). Besides, due to the good electrochemical properties of IrO_2, the IrO_2 contained catalysts are promising, for our purpose of preparing photoelectro-catalysts.

From the hydrogen production results, it can be noticed that Pt/TiO_2 had the best water photolysis effect, with a production value of 49.73 µmol h^{-1}. However, the durability of Pt under higher current density was poor. But for $IrTiO_x$ (IrO_2:

Catalyst	H_2 production (µmol h^{-1})
TiO_2 (P25)	2.13
Pt/TiO_2	49.73
IrO_2/TiO_2	15.59
$IrTiO_x$ ($IrO_2:TiO_2$ = 2:3)	35.16

Table 1 Hydrogen production of the catalysts

$TiO_2 = 2:3$), it has good catalytic performance and stability in electrolysis. Meanwhile, the photolysis effect of $IrTiO_x$ was close to that of Pt/TiO_2. Thus, due to the good electrochemical properties of IrO_2, the IrO_2 contained catalysts could be promising, for our purpose of preparing photoelectro-catalysts.

4 Conclusions

Accurate control of indoor humidity is significant and requires a large amount of energy of buildings. Herein, the PEM-based dehumidification system was improved. To utilize the solar energy and reduce energy consumption, catalyst with both water electrolysis and water photolysis performance was introduced to the anode-side. And to find a photocatalyst with high water decomposition efficiency, three kinds of photocatalysts were prepared, including Pt/TiO_2, IrO_2/TiO_2, and $IrTiO_x$ ($IrO_2:TiO_2 = 2:3$), and their photocatalytic properties were measured.

Results showed that among these catalysts, under the irradiation of a 300w xenon lamp, Pt/TiO_2 had the best water photolysis effect, with a hydrogen production of 49.73 $\mu mol\ h^{-1}$, ca. 23 times that of pure TiO_2. The water photolysis effect of $IrTiO_x$ was close to that of Pt/TiO_2. Comparing with pure TiO_2, IrO_2/TiO_2 had better water photolysis effect. However, in the photoelectro-catalytic reaction, Pt is prone to dissolution, which can cause performance degradation. Since IrO_2 has good catalytic performance and stability in electrolysis, for $IrTiO_x$ and IrO_2/TiO_2, the addition of IrO_2 may ensure them with both stable water electrolysis and water photolysis performance. The simultaneous water electrolysis and photolysis performance of gas flow with PEM element will be investigated in our future work. This research helps reduce the energy consumption and improve the efficiency during dehumidification. More importantly, it is trying to make a new exploration in artificial photosynthesis, and even to generate electricity relying on humidity difference.

Acknowledgements The project is supported by the National Natural Science Foundation of China (51876067), the Natural Science Fund for Distinguished Young Scholars of Guangdong Province (2018B030306014). It is also supported by the Science and Technology Planning Project of Guangdong Province: Guangdong-Hong Kong Technology Cooperation Funding Scheme (TCFS), No.2017B050506005.

References

1. Yanjun, H., Amir, K., Farshid, B., Majid, B.: Optimal energy-efficient predictive controllers in automotive air-conditioning/refrigeration systems. Appl. Energy. **184**, 605–618 (2016)
2. Xian, L., Shuai, L., Kok Kiong, T., Qing-Guo, W., Wen-Jian, C., Lihua, X.: Dynamic modeling of a liquid desiccant dehumidifier. Appl. Energy. **180**, 435–445 (2016)

3. Yong-Gao, Y., Baojun, Z., Can, Y., Xiaosong, Z.: A proposed compressed air drying method using pressurized liquid desiccant and experimental verification. Appl. Energy **141**, 80–89 (2014)
4. Ronghui, Q., Dujuan, L., Li-Zhi, Z.: Performance investigation on polymeric electrolyte membrane-based electrochemical air dehumidification system. Appl. Energy **208**, 1174–1183 (2017)
5. Kazunari, D., Shuichi, N., Takaharu, O., Kenzi, T.: Study of the photocatalytic decomposition of water vapor over a NiO-$SrTiO_3$ catalyst. J. Chem. Phys. **86**, 3657–3661 (1984)
6. Vincenzo, V., Lara, M.A., Iervolino, G.: Photocatalytic H2 production from glycerol aqueous solutions over fluorinated Pt-TiO2 with high 001 facet exposure. J. photochem. photobiol. A-Chem. **365**, 52–59 (2018)
7. Emma, O., Dmitry, L., Alexey, F., Frank, K., Jeremy, T., Olha, S., Thomas, J.S., Christophe, C.: A simple one-pot adams method route to conductive high surface area IrO_2-TiO_2 materials. New J. Chem. **40**(2), 1834–1838 (2016)
8. Zhongliang, S., Xiaoxia, Z., Shuhua, Y.: Preparation and photocatalytic activity of TiO_2 nanoparticles co-doped with Fe and La. Particuology **9**(3), 260–264 (2011)

Characteristics of a Natural Heat Transfer Air-Conditioning Terminal Device for Nearly Zero Energy Buildings

Haiwen Shu, Xu Bie, Shan Jiang, Zhiqiang Yang, Yang Zhang, Gao Shu, Hongbin Wang and Guangyu Cao

Abstract As the heating and cooling load of a nearly zero energy building (NZEB) will be greatly reduced, significant energy saving effect can be obtained. In view of the low heating and cooling load and high human thermal comfort level of the NZEB, an air-conditioning terminal device based on natural heat transfer was brought out and studied by the authors. It has the functions of heating, cooling and moisture removal according to the inlet media temperature of the device. Also there is little noise or air disturbance during its operation, thus it is beneficial for the enhancement of human thermal comfort. In the paper, the experimental data of the heating and cooling performance of the device under different operation conditions were collected and analyzed. The calculation models quantifying its heating and cooling capacities were obtained through data regression analysis, and the flow resistance curve of the device was obtained by means of experimental measurement under various flow rates. In addition, comparison was made on the heating and cooling capacities between the device and the radiant floor which also features little noise or air disturbance, and it shows that the heating and cooling capacities of the device are 41.5 and 46.8% higher than the maximum capacities of the radiant floor respectively. The research lays a foundation for the engineering application of the natural heat transfer air-conditioning terminal device.

Keywords Air-conditioning · Terminal device · Natural heat transfer · Nearly zero energy buildings

---

H. Shu (✉) · S. Jiang · Z. Yang · Y. Zhang · G. Shu · H. Wang
School of Civil Engineering, Dalian University of Technology, Dalian, China
e-mail: shwshw313@sina.com

X. Bie
E.N.T. Department, Second Hospital of Dalian Medical University, Dalian, China

G. Cao
Department of Energy and Process, Norwegian University of Science and Technology, Trondheim, Norway

1 First Section

Energy consumption in buildings takes up a large portion in the total energy consumption of society which is more than one third in many countries [1]. In order to decrease the building energy consumption, much research has been done on the nearly zero-energy buildings (NZEB) [2–10]. As the space heating or cooling load can be lowered down significantly by proper load reducing technical measures, air-conditioning terminal devices need not be the same as traditional ones in such kind of buildings. The authors brought out a radiant-convective air-conditioning terminal device with parallel pipes in previous study [11]. As the heat exchange between the terminal device and the ambient environment is carried out by natural convection and radiation which are the common heat transfer methods in the natural world, the air-conditioning terminal device characterized by these two heat transfer methods are called natural heat transfer air-conditioning terminal device here. Such kind of air-conditioning terminal device operates with little noise or air turbulence which is beneficial for the occupants' thermal comfort. What's more, the terminal device can both supply heating in winter and cooling in summer. In other words, it can satisfy both heating and cooling needs of the building all year round. Compared functionally with radiant panels that can also provide heating and cooling for a room quietly with little air disturbance, the terminal device can also undertake the moisture load of the room.

In order to apply the terminal device properly, its actual heating and cooling capacities as well as flow resistance performance should be ascertained. So the emphasis of the paper is to provide the thermal and flow resistance characteristics of the terminal device by experiments under various conditions and compare its thermal performance with the radiant floor.

2 Experiment Setup

The air-conditioning terminal device is shown in Fig. 1. It is made of 9 parallel heat exchange pipes with the spacing of 110 mm between two neighboring pipes, and the dimensions of the device are 2.0 m(H) × 1.0 m(W) × 0.1 m(D). The device can supply heating or cooling according to the inlet fluid temperature, and it allows condensation on its surface in summer since there is a condensate discharge pan under the heat exchange pipes. A test rig of the device is set up (refer to Fig. 2) to examine the heating and cooling performance and water flow resistance characteristics under various experiment conditions.

The experiments were conducted in the experimental chamber in Dalian University of Technology. In the cooling supply mode, filament lamps and electric heaters were selected to simulate the heat source with a humidifier as the moisture source. In the heating mode, outdoor cold air was introduced to form the heating load of the experimental chamber. During various operation conditions of the

Fig. 1 The air-conditioning terminal device

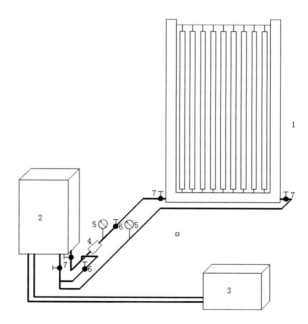

Fig. 2 Test rig of the air-conditioning terminal device *Note* 1. The air-conditioning terminal device; 2. Heating and cooling water producer of the air source heat pump system; 3. Outside unit of the air source heat pump system; 4. High precision turbine flow meter; 5. High precision pressure gauge; 6. Flow control valves; 7. Cut-off valves at inlet and outlet of the device

air-conditioning terminal device, simulation heating or cooling load was adjusted according to the inlet water temperature and flow rate of the device to maintain the indoor air temperature and humidity at a desired level (i.e. 26 °C and 60% in cooling mode and 18–20 °C in heating mode). According to the China national standard 'Test Method for Thermal Performance of Radiant Cooling and Heating Devices' [12], all the experiment data were recorded after the experimental system reached the steady state and the total sampling time should not be less than an hour.

After the measurement and analysis of the data, the heating and cooling capacity per unit projection area of the terminal device will be compared with that of the radiant floor heating and cooling to display the comparative advantage of the terminal device.

3 Test Results and Analysis

3.1 Test Results and Analysis Under Heating Mode

Measured Data of the Device under Heating Mode. The measured data of the terminal device in heating mode under the flow rate of 120 and 240 L/h are listed in Table 1, in which 'the excess temperature' is expressed in Eq. (2).

It can be seen from Table 1 that both the supply water temperature and the flow rate have significant effect on the heating capacity of the device. If the flow rate is kept constant, the higher the temperature of the water supply, the more heat it can supply. And if the supply water temperature is kept constant, the larger the flow rate, the more amount of heat it can supply. For instances, when the water flow rate is kept at 240 L/h, with every 1 °C supply water temperature increase, the amount

Table 1 Measured data in heating mode

Test number	Supply water temperature (°C)	Return water temperature (°C)	Flow rate (L/h)	Indoor reference temperature (°C)	Excess temperature (°C)	Heating capacity (W)
1	51.5	43.8	128.0	18.1	29.4	890.2
2	45.5	38.7	127.2	18.7	23.4	779.2
3	35.9	30.7	125.3	18.1	15.2	585.5
4	40.2	34.5	124.2	20.2	17.2	641.3
5	49.3	45.4	244.9	19.8	27.5	862.1
6	50.5	46.6	240.8	19.9	28.7	852.1
7	48.9	45.2	241.4	18.4	28.6	810.6
8	44.9	41.4	240.6	18.0	25.1	766.3
9	39.6	37.2	240.6	18.0	20.4	519.6
10	40.3	38.0	242.2	20.4	18.7	505.4

of heat the device can supply increases 33.3 W within the supply water temperature range of 40–50 °C. If the supply water temperature is kept at 50 °C, with every 10 L/h supply water flow rate increase, the average heat supply enhancement of the device is about 7.3 W within the water flow rate of 120–240 L/h. From Table 1, the largest amount of heat supply of the device is 890.2 W at test number 1. Though test number 5 has almost the same supply water temperature and even has a higher flow rate of 244.9 L/h, the amount of heat supply is slightly lower than that of test number 1 because the indoor air temperature is raised from 18 °C to nearly 20 °C.

Determination of the heating performance curve of the device. According to the national standard [12], the standard characteristic equation of the heating capacity for the terminal device is shown in Eq. (1).

$$Q_{sh} = K_M \cdot \Delta t^n \qquad (1)$$

In which, Q_{sh} is the heating capacity of the terminal device, W; Δt is the excess temperature, °C, and it is expressed in Eq. (2); K_M and n are constants to be determined.

$$\Delta t = \left| \frac{t_1 + t_2}{2} - t_r \right| \qquad (2)$$

In which, t_1 is the inlet water temperature, °C; t_2 is the outlet water temperature, °C; t_r is the reference indoor air temperature, °C.

In order to obtain the characteristic heating performance curve of the terminal device, the measured data under the water flow rate of 120 and 240 L/h were regressed as shown in Figs. 3 and 4 respectively.

From Fig. 3, the standard characteristic equation of the heating capacity for the terminal device under the water flow rate of 120 L/h is obtained as:

$$Q = 87.51 \, \Delta t^{0.70} \quad (R^2 = 0.9967) \qquad (3)$$

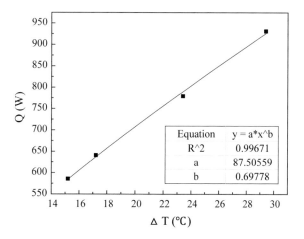

Fig. 3 Heating capacity regression under the flow rate of 120 L/h

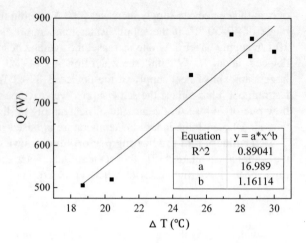

Fig. 4 Heating capacity regression under the flow rate of 240 L/h

From Fig. 4, the standard characteristic equation of the heating capacity for the terminal device under the water flow rate of 240 L/h is obtained as:

$$Q = 16.99 \, \Delta t^{1.16} \quad (R^2 = 0.8904) \tag{4}$$

3.2 Test Results and Analysis Under Cooling Mode

Measured data of the device under cooling mode. The cooling tests of the device under 4 different supply water temperatures were conducted under the water flow rate of 120 and 240 L/h and the results are listed in Table 2 respectively.

It can be seen from Fig. 2 that the supply water temperature and flow rate have significant effect on the cooling capacity of the terminal device. The lower the

Table 2 Measured data in cooling mode

Test number	Supply water temperature (°C)	Return water temperature (°C)	Flow rate (L/h)	Indoor reference temperature (°C)	Excess temperature (°C)	Cooling capacity (W)
1	8.3	13.0	120.2	26.4	15.3	512.6
2	10.9	14.3	118.8	26.7	13.1	395.4
3	12.8	5.9	124.1	26.5	12.1	349.1
4	13.1	16.3	120.4	26.2	11.1	351.8
5	7.7	10.5	235.4	26.2	16.6	598.1
6	12.1	14.0	245.8	26.3	13.0	423.8
7	13.9	15.7	238.6	26.6	11.4	368.1
8	15.6	16.6	243.3	26.3	10.0	242.8

supply water temperature the larger the cooling capacity of the device under a constant flow rate. And the bigger the water flow rate the larger the cooling capacity of the device under the same supply water temperature. For instances, when the cooling water flow rate is kept at 240 L/h, with every 1 °C decrease of the supply water temperature, the cooling capacity increases 45.0 W averagely within the supply cooling water temperature of 8–16 °C. And when the supply cooling water temperature is kept at 8 °C, with every 10 L/h water flow rate increase, the cooling capacity of the device increases 7.3 W averagely within the water flow rate of 120–240 L/h. The largest cooling capacity is 598.1 W within the scope of tests which was at the supply cooling water temperature of 7.7 °C and water flow rate of 235.4 L/h.

Determination of the cooling performance curve of the device. Similar to the heating mode, the standard characteristic curve of the device in cooling mode is:

$$Q_{sc} = K_M \cdot \Delta t^n \tag{5}$$

In which, Q_{sc} is the cooling capacity of the terminal device, W. Other parameters are the same as in the Eq. (1).

So as to obtain the characteristic cooling performance curve of the terminal device, the measured data under the water flow rate of 120 and 240 L/h were regressed as shown in Figs. 5 and 6 respectively.

From Fig. 5, the standard characteristic equation of the cooling capacity for the terminal device under the water flow rate of 120 L/h is obtained as:

$$Q = 13.26 \, \Delta t^{1.33} \quad (R^2 = 0.89776) \tag{6}$$

From Fig. 6, the standard characteristic equation of the cooling capacity for the terminal device under the water flow rate of 240 L/h is obtained as:

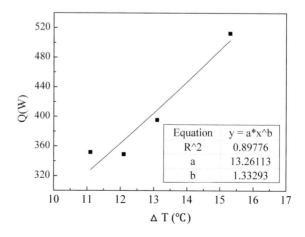

Fig. 5 Cooling capacity regression under the flow rate of 120 L/h

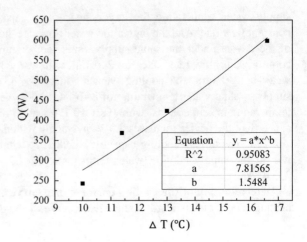

Fig. 6 Cooling capacity regression under the flow rate of 240 L/h

$$Q = 7.82 \, \Delta t^{1.55} \quad (R^2 = 0.95083) \tag{7}$$

4 Discussion

In order to ascertain the relative heat transfer intensity of the natural heat transfer air-conditioning terminal device, radiant floor heating and cooling is selected as the object of comparison as it can also achieve space heating and cooling by radiation and natural convection heat transfer. Normally speaking, the two kinds of air-conditioning terminals should be compared under the same operation conditions which are the same supply water temperature and flow rate. However, so as to prevent thermal discomfort and condensation on the floor surface, the surface temperature of the floor heating and cooling should not exceed 29 °C [13] in heating mode and must not lower than 18 °C in cooling mode. The standard conditions of the device are selected as 50 and 13 °C for the supply water temperature in heating and cooling mode, and 240 L/h for the water flow rate in both modes. Thus the heat transfer intensity per unit projection area of the device is 65.8 and 41.5% higher than of the floor heating and cooling respectively.

5 Conclusions

An air-conditioning terminal device that can be used both in space heating and cooling in buildings featuring low heating and cooling load is brought out and studied. Compared with similar air-conditioning terminals such as radiant floor heating and cooling, the device allows moisture condensation on its surface, that is, it can undertake both sensible and latent cooling load of a room. In the paper, emphasis is given on its heating and cooling capacities under different operation conditions as well as its water flow resistance characteristics. Following conclusions are reached in the study.

(1) Within the experiment scope, the largest heating capacity of the device is 890.2 W (when supply water temperature is 51.5 °C, and the measured flow rate is 128.0 L/h), and the largest cooling capacity of the device is 598.1 W (when the supply cooling water temperature is 7.7 °C and water flow rate is 235.4 L/h).

(2) Both the supply water temperature and the flow rate have significant impact on the heating and cooling capacity of the device. For the convenience of application of the air-conditioning terminal device, the heating and cooling capacity calculation models of the device were obtained through data regression under various supply water temperatures and flow rates. Meanwhile the water flow resistance characteristics of the device is regressed too and it complies well with the law of quadratic power.

(3) In comparison with the radiant floor heating and cooling from the heat transfer intensity per unit projection area, the air-conditioning terminal device studied in the paper can provide 41.5 and 46.8% more heating and cooling respectively in the corresponding standard operation conditions than the maximum capacities of floor heating and cooling.

Acknowledgements This work is supported by China National 13th Five-Year Plan of Key Research and Development Program "The technical system and key technologies development of nearly zero-energy buildings" (2017YFC0702600).

References

1. Li, H., et al.: Operation performance analysis of ground source heat pump system in certain nearly zero energy building based on actual measurement data. Build. Sci. **31**(6), 124–130 (2015)
2. Weißenberger, M., et al.: The convergence of life cycle assessment and nearly zero-energy buildings: the case of Germany. Energy Build. **76**(6), 551–557 (2014)
3. Lindberg, K.B., et al.: Cost-optimal energy system design in zero energy buildings with resulting grid impact: a case study of a German multi-family house. Energy Buil. **127**(9), 830–845 (2016)
4. Ali-Toudert, F., Weidhaus, J.: Numerical assessment and optimization of a low-energy residential building for Mediterranean and Saharan climates using a pilot project. Renew. Energy **101**(2), 327–346 (2017)
5. Tian, Z., et al.: Investigations of nearly (net) zero energy residential buildings in Beijing. Procedia Engineering. **121**, 1051–1057 (9th International Symposium on Heating, Ventilation and Air Conditioning, ISHVAC 2015 Joint with the 3rd International Conference on Building Energy and Environment, COBEE 2015) (2015)
6. Dall'O', G., et al.: An Italian pilot project for zero energy buildings: towards a quality-driven approach. Renew. Energy. **50**(2), 840–846 (2013)

7. Congedo, P.M., et al.: Cost-optimal design for nearly zero energy office buildings located in warm climates. Energy **91**(11), 967–982 (2015)
8. Ferrara, M., et al.: Energy systems in cost-optimized design of nearly zero-energy buildings. Autom. Constr. **70**(10), 109–127 (2016)
9. Pikas, E., et al.: Cost optimal and nearly zero energy building solutions for office buildings. Energy Build. **74**(5), 30–42 (2014)
10. Cellura, M., et al.: Different energy balances for the redesign of nearly net zero energy buildings: an Italian case study. Renew. Sustain. Energy Rev. **45**(5), 100–112 (2015)
11. Haiwen, S., et al.: Cooling performance test and analysis of a radiant–convective air-conditioning terminal device with parallel pipes. Sci. Technol. Built Environ. **23**(3), 405–412 (2017)
12. JG/T 403: Ministry of housing and urban-rural construction of the people's Republic of China (MHURCPRC). Test methods for thermal performance of radiant cooling and heating unit, Standards Press of China, Beijing, China (2013)
13. JGJ 142: Ministry of housing and urban-rural construction of the people's Republic of China (MHURCPRC). Technical specification for radiant heating and cooling. China Architecture and Building Press, Beijing, China (2012)

Study on Independent Air Conditioning System Based on the Temperature and Humidity of Rotary Dehumidifier and Heat Pipe

Jiangbo Li, Haiwei Ji and Liu Chen

Abstract For those areas with higher relative humidity in summer, the traditional air-conditioning system is difficult to meet the humidity requirements. A temperature and humidity independent control air-conditioning system based on rotary dehumidifier and heat pipe is proposed. The research on the working principle and thermodynamic calculation method of the system is carried out, and an office building in Shenzhen is taken as an example for thermodynamic calculation. Thermodynamic calculations were performed for Shenzhen, Beijing, Changsha, Wuhan Xi'an, and Guangzhou. The results indicate that the system can meet the requirements of indoor air supply better. Thermodynamic calculation indicates that the temperature and humidity independent control air-conditioning system based on rotary dehumidifier and heat pipe has the advantages of large dehumidification capacity and low energy consumption.

Keywords Rotary dehumidifier · Heat pipe · Air-conditioning

1 Introduction

In China, the annual humidity is generally higher in hot summer and cold winter and hot summer and warm winter areas, while the traditional air-conditioning system is mainly based on temperature control, which makes many air-conditioning systems fail to meet the need of humidity control in high-humidity season. And it is difficult to obtain better indoor air quality [1]. It is generally believed in HVAC that the

---

J. Li · H. Ji · L. Chen (✉)
Xi'an University of Science and Technology, Shaanxi, China
e-mail: chenliu@xust.edu.cn

J. Li
e-mail: 18829344581@163.com

H. Ji
e-mail: 415543043@qq.com

independent control of temperature and humidity can meet the requirements. The temperature and humidity independent control system mainly consists of two independent systems: latent heat load system and sensible heat load system, which have significant energy-saving potential compared with conventional air-conditioning system [2].

Jiang Yi et al. analyzed the annual operation energy consumption of air-conditioning system with the help of subitem metering meter data. Compared with the same kind of office buildings, the temperature and humidity independent control air-conditioning system can save energy about 35% [3]. John et al. tested the performance of temperature and humidity independent control air-conditioning system and conventional air-conditioning system in two supermarkets. The energy-saving efficiency is about 13% [4]. Chen et al. proposed a temperature and humidity independent control method using dual cooling sources with different temperatures (DCSTHIC), compared with the traditional air-conditioning system, when the heat–humidity ratio is from 8000 to 15,000 kJ/kg. The COP of the system increases from 3.1 to 17.5% [5, 6].

Rotary dehumidification is gradually widely used due to its advantages of large dehumidification capacity and low environmental pollution [7, 8]. Heat pipe is a new type of heat transfer element, which has the advantages of high thermal conductivity, simple structure, reliable operation, and uniform temperature. It can be widely used in heat transfer, heat flux conversion, and heat control [9].

In this paper, a new air-conditioning system is proposed to realize the independent control of temperature and humidity by making use of the high efficiency dehumidification performance of rotary dehumidification and the good heat transfer performance of heat pipe. The system can be applied in areas with higher humidity.

2 System Scheme and Working Principle

Figure 1 shows the schematic diagram of an air-conditioning system based on temperature and humidity independent control of a rotary dehumidifier and a heat pipe. The system consists of a treatment air system and a regeneration air system. The working principle of the system is that the outdoor air to be treated is heated and dehumidified by the rotary dehumidifier under the action of the fan; then, it is cooled in the evaporation section of the heat pipe. At last, it is transported to the indoor through the air supply pipe. The regeneration energy consumption of the rotary dehumidifier is supplied by the condensation section of the heat pipe, and the auxiliary electric heater can provide the needed heat when the heat of the heat pipe is insufficient. The regenerated air treatment process is as follows: Under the action of the fan, the regenerated air is filtered and dedusted first in the filter, and then the regenerated air is heated by the condensation section of the heat pipe. The air is further heated by the auxiliary electric heater and then conveyed to the rotary dehumidifier for dehumidification. Finally, it is discharged to the outside.

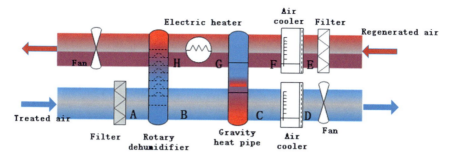

Fig. 1 Schematic diagram of temperature and humidity independent control system

3 Thermal Calculation of the System

As shown in Fig. 1, the air-conditioning system based on the independent control of temperature and humidity of the rotary dehumidifier and heat pipe mainly includes the rotary dehumidifier and heat pipe. Firstly, the mathematical model of the rotary dehumidifier and heat pipe is established, and finally, the thermodynamic calculation model of the whole system is established [10].

3.1 Thermodynamic Calculation of the Rotary Dehumidifier

There are many models for thermodynamic calculation of rotary dehumidifier. The 54 model proposed by Beccali is adopted in the paper, and the formula for calculating the outlet parameters of rotary dehumidifier is as follows:

$$\phi_2 = 0.9428\phi_3 + 0.0572\phi_1 \tag{1}$$

$$i_2 = 0.1313 i_3 + 0.8688 i_1 \tag{2}$$

where ϕ_1 and ϕ_2 are the relative humidity in the initial and final states of treated air, respectively, (%); ϕ_3 is the relative humidity of regenerated air, (%); i_1 and i_2 are the enthalpy values of the initial and final states of treated air, kJ/kg; i_3 is the enthalpy value of regenerated air, kJ/kg.

3.2 Thermal Calculation of Heat Pipe

The heat pipe used in this system belongs to the air–air heat exchanger. Therefore, the design and calculation of heat pipe heat exchanger is basically the same as that of the conventional interwall heat exchanger. The design formulae for heat pipes are as follows:

$$F = \frac{Q}{K \cdot \Delta t_m} \tag{3}$$

$$n = \frac{F}{S_w L_e^h} \tag{4}$$

$$E = \frac{1 - \exp\{-NTU[1-R]\}}{1 - R \cdot \exp[-NTU(1-R)]} \tag{5}$$

$$R = (q_m c)_{\min}/(q_m c)_{\max} \tag{6}$$

where F is the heat transfer area, m²; Q is the heat transfer capacity of the heat pipe, kw; K is the heat transfer coefficient, w/(m² °C); Δt_m is the heat transfer temperature difference, °C; n is the total number of heat pipes; S_w is the heat transfer area per unit length of the heat pipe, m²; L_e^h is the effective length of the evaporation section of the heat pipe, m; E is the heat exchanger efficiency; q_m is the mass flow rate, kg/s; c is the heat capacity, J/(kg/°C); NTU is the number of transfer unit [11–13].

3.3 Performance Indicators of the System

The cooling load of the system. The cooling load Q provided by the system is calculated [14].

$$Q = G(i_A - i_D) \tag{7}$$

where i_A, i_D is the enthalpy value of the state of the treated air, kJ/kg.

Dehumidification of the system. The dehumidification capacity W of the rotary dehumidifier to the treated air is the dehumidification capacity of the whole system.

$$W = G(d_A - d_B) \tag{8}$$

where d_A and d_B are the humidity ratios in the initial and final states of treated air of the system, g/kg.

Thermodynamic coefficient of the system TCOP. TCOP is defined as the ratio of cooling load Q to regenerative heat consumption.

$$\text{TCOP} = \frac{Q}{Q_r} \tag{9}$$

The formula for calculating regenerative heat consumption is

$$Q_r = cG_r(t_{r1} - t_{r2}) \quad (10)$$

where G_r is the flow rate of regenerated air, kg/s; t_{r1} and t_{r2} are the dry bulb temperatures in the initial and final states of treated air of the auxiliary electric heater.

4 Case Analysis

Taking an office building in Shenzhen as an example, carrying out the thermodynamic calculation of temperature and humidity independent control air-conditioning system based on the rotary dehumidifier and heat pipe for a unit, the outdoor gas-phase parameters of Shenzhen in summer are as follows: atmospheric pressure is 110.262 kPa; air parameters of air intake to be treated are as follows: dry bulb temperature is 29.3 °C, relative humidity is 79%, and humidity ratio is 20.3 g/kg. The treatment air duct and regeneration air duct are designed to be 500 mm 500 mm. According to the heat transfer performance requirements of heat pipe, design wind speed is 2 m/s; then, the air volume of treated air and regenerated air is 1800 m³/h. According to the size of the air duct, the net length of the upwind side of the heat pipe is 1.1 m, the net height is 0.5 m, and the width of the intermediate baffle is 0.1 m. The type II DGL-38 × 1.8 aluminum sheet tube of a manufacturer is selected as shell for the heat pipe, and the diameter of the tube is 16 mm, the outer diameter is 38 mm, and the heat transfer area per unit length is 0.923 m²/m. The ethyl ether with lower boiling point temperature is selected as the working medium. The number of heat pipes in each row is 11, the total of number of rows is 23, and the total number of heat pipes is 256. After calculation, the parameters of each air state point are obtained. The calculation results of every state point are shown in Table 1.

It can be seen from Table 1 that in the entire process, the cooling and dehumidification treatment of the treated air is realized. Temperature has been reduced by 7.3 °C, and humidity ratio has been decreased by 8.2 g/kg. Relative humidity

Table 1 Air state parameters

State points	T (°C)	d (g/kg)	ϕ (%)	i (kJ/kg)
A	29.3	20.3	79	82.5
B	58.5	10.8	9.2	87.5
C	28.5	10.8	43.8	56.2
D	22	13.1	78	56.2
E	25	13	65	58.5
F	20	14.8	95	58.5
G	50	14.9	18.8	89.2
H	80	15.9	5	120.6

has been reduced by 20.5%, and enthalpy has been decreased by 26.3 kJ/kg. The air treated by the cooling system can meet the requirement of indoor air supply. The refrigeration capacity Q of the system is 16.96 kw, the dehumidification capacity W is 5.48 g/s, the regeneration heat consumption Q_r provided by the auxiliary electric heater is 19.4 kw, and the thermal coefficient TCOP of the system is 0.87.

Thermodynamic calculation was performed on Beijing, Xi'an, Changsha, Wuhan, and Guangzhou, respectively. The calculation results are shown in Table 2 and the TCOP is shown in Fig. 2.

Through the thermal calculation of six cities, the results suggest that the system has good cooling and dehumidifying ability. On the condition of the same energy consumption, the temperature drop can reach 5 °C, the humidity can reach 8.5 g/kg, and TCOP is close to 1.

Table 2 Air treatment parameters of different cities

City	Parameter						
	ΔT (°C)	Δd (g/kg)	Δi (kJ/kg)	Q (kw)	W (g/s)	Q_r (kw)	TCOP
Shenzhen	7.3	8.2	26.3	16.96	5.48	19.4	0.87
Beijing	4.4	9.3	27.47	17.72	6	19.4	0.91
Xi'an	5.7	9.4	29.3	18.9	6.06	19.4	0.97
Changsha	4.5	9.6	29.7	19.16	6.19	19.4	0.987
Wuhan	4.5	9.6	29.4	18.96	6.19	19.4	0.977
Guangzhou	4.8	8.4	23.8	15.4	5.42	19.4	0.8

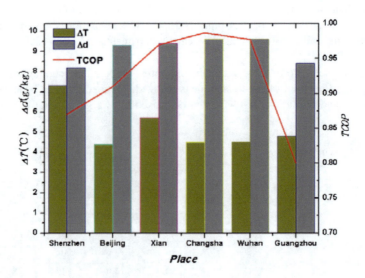

Fig. 2 Comparison of air treatment results among different cities

5 Conclusion

1. An independent temperature and humidity control air-conditioning system based on a rotary dehumidifier and a heat pipe is proposed, and an air handling system and a regenerated air system are included in it. Treated air is cooled and dehumidified by the rotary dehumidifier and the heat pipe. The heat required for regenerated air is provided by the heat pipe, and the insufficient part is provided by the auxiliary electric heater; thus, the regeneration energy consumption can be saved.
2. Deep well cooling system based on rotary dehumidification and heat pipe cooling and heating utilization utilizes the efficient dehumidification ability of rotary dehumidifier to achieve better dehumidification effect and solve the problem of traditional air-conditioning dehumidification in high-humidity areas.
3. Through the thermal calculation of six cities, it shows that the system has a good temperature and humidity treatment ability, which can better meet the requirements of air treatment. In addition, the system has a higher thermal coefficient and has the advantage of low energy consumption.
4. Exhaust air is heated by heat pipe and then passes through the regeneration area of the rotary dehumidifier, the temperature of the air discharged to the outside is higher, and heat recovery can be carried out to achieve the effect of energy-saving.

Acknowledgements This study was supported by the National Natural Science Foundations of China (No.51404191, 51504182, 51674188, 51874229) and Shaanxi Innovative Talents Cultivate Program-New-star Plan of Science and Technology (No.2018KJXX-083). We would like to thank our committee members, sponsors, reviewers, authors, and other participants for support to ISHVAC 2019.

References

1. Gu, J., et al.: Study on operation strategy of ground source and air source heat pump combined air-conditioning system in hot summer and cold winter regions. Refrig. Technol. **37**(05), 41–47 (2017)
2. Xu, Z., et al.: Analysis of energy conservation of temperature and humidity independent control system. HVAC **37**(6), 129–132 (2007)
3. Zhang, T., et al.: Application performance analysis of temperature and humidity independent control air-conditioning system. Arch. Sci. **26**(10), 146–150 (2010)
4. Zhang, H.Q., et al.: Performance comparison of temperature and humidity independent control air conditioning system and conventional air conditioning system. HVAC **41**(01), 48–52 (2011)
5. Chen, T.T., et al.: Applicability and energy efficiency of temperature and humidity independent control systems based on dual cooling sources. Energy Amp Build. 121 (2016)
6. Duan, L.F.: Energy consumption simulation and analysis of temperature and humidity independent control air conditioning system. Beijing University of Architecture (2015)

7. Ge, T.S.: Theoretical and experimental study of two-stage dehumidification air conditioning with runner. Shanghai Jiaotong University (2008)
8. Wang, G.F.: Research on rotary dehumidification air conditioning system. Guangzhou University (2007)
9. Zhuang, H.: Heat pipe technology and engineering application. Chemical Industry Press, Beijing (2000)
10. Chen, L., et al.: Deep well adsorption and cooling mechanism based on heat storage backfill. J. Coal **43**(02), 483–489 (2008)
11. Huang, W.Y.: Design Basis of heat pipe and heat pipe heat exchanger. China Railway Press, Beijing (1995)
12. Zhao, R.E., et al.: Design of gravity heat pipe. Ironmaking **04**, 14–20 (1985)
13. Zheng, J.J:. Experimental study on performance of two-stage evaporative cooling air conditioning system with heat recovery heat pipe. Xi'an Engineering University (2007)
14. Yang, Z., et al.: Comparative study on different ways of condensing reheat air. Refrig. Technol. **37**(01), 67–72 (2017)

The Influence Factors and Performance Optimization Analysis of the Total Heat Exchanger

Jiafang Song, Shuang Liang, Xiangquan Meng and Shuhui Liu

Abstract The energy-saving problem of air-conditioning system has attracted much attention during these years. Total heat exchanger has become one of the effective means to deal with the contradiction between improving air quality and energy saving of air-conditioning system. In this paper, the influence of air flow rate, moisture content difference, and temperature difference on heat exchanger under different operating conditions are analyzed and optimized. Based on the optimization results, we draw some conclusions. The air flow rate has a great influence on the efficiency of total heat exchanger, and the efficiency of heat exchanger decreases linearly with the increase of air flow rate. The difference in moisture content has little effect on the efficiency of the total heat exchanger. Temperature difference has a big influence on heat exchange efficiency and the efficiency of moisture. As the temperature difference increases, the sensible efficiency and the thermal efficiency increase.

Keywords The total heat exchanger · Air flow rate · Moisture content difference · Temperature difference

1 Introduction

With the rapid development of China's economy, energy issues have received increasing attention. According to statistics, building energy consumption accounts for about 30% of the national energy consumption, of which 50–60% is the air-conditioning ventilation energy consumption, and the fresh air consumption in air conditioning ventilation accounts for 30–40% [1]. Since the humidity load in the fresh air often accounts for a large proportion, people need precise control of indoor

---

J. Song · S. Liang (✉) · X. Meng · S. Liu
School of Mechanical Engineering, Tiangong University, Tianjin 300387, China
e-mail: 1831035247@tiangong.edu.cn

J. Song
e-mail: songjiafang@tiangong.edu.cn

humidity as well as indoor air quality. Therefore, the most used energy recovery form is the total heat exchanger. The actual recovery efficiency of the total heat exchangers in China is about 50%, which only meets the minimum efficiency requirement of the national standard GB/T21087-2007(GB/T21087-2017, [2]. It is far from the developed countries, so it is necessary to conduct research on the total heat exchanger.

Many scholars carried out some researches on the total heat exchanger: Fuhuan found that the air flow rate has a negative correlation effect on the heat and humidity exchange efficiency [3]. Ougatla et al. experimentally studied the loss of internal pressure in plate-fin heat exchanger and the law of fluid temperature and humidity [4]. The results showed that the increase of heat transfer area can greatly reduce the heat transfer loss, but at the same time bring about the increase of pressure loss. Michael proposed a simple mathematical model to calculate the plate-type sensible heat exchanger. The model described the relationship between the airflow distribution and the ratio of the fluid capacity, such as the total heat transfer coefficient and the heat exchange efficiency on both sides [5]. Yuhua analyzed the sensible heat and latent heat recovery efficiency and related performance of cross-counter flow and cross-counter flow plate total heat exchangers by establishing an integrated two-room environment testing platform [6]. The study also obtained the conclusions that the heat transfer efficiency decreases with the decrease of the moisture content and the temperature difference with the increase of the air flow rate, and increases with the increase of the temperature difference.

2 Methods

In actual operation, the efficiency of total heat exchanger is affected by different indoor and outdoor air parameters. Under the premise of fully considering the law of heat and moisture transfer on both sides of the thermal and moisture permeable membrane, we applied a thermal and moisture coupling calculation model based on FLUENT to simulate the total heat exchanger. Under the influence of the inlet angle of the air on the windward side and the geometry of the heat exchanger core, we analyze the total heat exchanger from the aspects of air volume, moisture content difference, and temperature difference, in order to develop a total heat exchanger with high efficiency and comprehensive performance.

2.1 Establishment of Physical Model

The 3D model entity was based on the literature [7]; we established the physical model of the pure cross-flow and cross-counter-flow plate-type total heat exchanger. The structure is shown in Fig. 1. The relevant data are shown in Table 1.

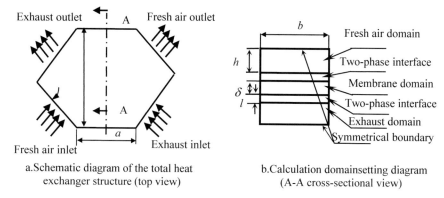

Fig. 1 Schematic diagram of total heat exchanger structure and calculation domain setting

Table 1 3D modeling-specific data

α (°)	b (mm)	h (mm)	a (mm)	δ (mm)	l (mm)
90	230	1.5	$100 \leq a \leq 1500$	0.05	0.01

Firstly, when it is determined that the air condition parameters of the fresh air and the exhaust air are unchanged, that is, the temperature difference (TD) is determined to be 16 °C and the moisture content difference (MCD) is 3 g/kg, the air flow rate (AFR) is set to 50, 100, 150, 200, and 250 m³/h. For these cases, the sensible heat and latent heat efficiency of the heat exchanger under different air flow rate conditions are calculated.

Secondly, the moisture content difference between fresh air and exhaust air is changed. Humidity difference is determined to five conditions of 1, 3, 5, 7, 9 (g/kg) and keep the air flow rate at 150 m³/h and the temperature difference at 16 °C. Constantly, the sensible heat and latent heat efficiency of the heat exchanger are calculated.

Finally, the temperature between fresh air and exhaust air is changed. The air flow rate is 150 m³/h, and the moisture content difference is 3 g/kg. The temperature difference is adjusted to 8, 12, 16, 20, and 24 (°C), and the total heat exchange is calculated. The sensible heat and latent heat efficiency of the heat exchanger are calculated.

2.2 Computational Method

For the above-mentioned simulated statistical data, multiple linear regression fitting is performed by MATLAB programming method to determine the temperature difference between outdoor air and indoor air. The purpose of changing moisture content difference and air flow rate is to find out its influence on sensible heat efficiency and enthalpy exchange efficiency of total heat exchanger. The temperature efficiency influence equation is calculated firstly.

Structure of air flow rate (x_1), temperature difference (x_2), moisture content difference (x_3), heat exchange efficiency (y):

$$x_1 = [50, 100, 150, 200, 250, 150, 150, 150, 150, 150, 150, 150, 150, 150, 150];$$
$$x_2 = [16, 16, 16, 16, 16, 24, 20, 16, 12, 8, 16, 16, 16, 16, 16];$$
$$x_3 = [3, 3, 3, 3, 3, 3, 3, 3, 3, 3, 9, 7, 5, 3, 1];$$
$$y_2 = [0.906, 0.864, 0.822, 0.784, 0.75, 0.827, 0.825, 0.822, 0.817,$$
$$0.807, 0.831, 0.828, 0.826, 0.822, 0.82];$$

temp = linspace(1, 1, 15);

$x = [\text{temp}', x_1', x_2'];$

[b, bint, r, rint, stats]

The regression results as follow:

b = 0.916575 −0.00078 0.0012 0.001425

bint = 0.9037 0.9295 0.0006 0.0018 0.0004 0.00248 −0.0008 −0.0007

r_i = 0.0052 0.0024 − 0.0005 0.0008 0.0059 − 0.0051 − 0.00225

−0.00045 −0.00065 −0.005850 −0.00015 0.0007 −0.00045 0.0004

stats = 0.9915 428.1440 0.0000

In the calculation results, each regression coefficient b_i and the correlation coefficient r_i are within the allowable range. Correlation coefficient $R^2 = 0.9915$, $F = 428.1440 > F0.05(4, 15)$. The linear regression is very significant by the significance statistic F value, $p = 0.00 < 0.05$; at least one regression coefficient in the regression equation of the regression equation is not significantly zero.

The temperature exchange efficiency formula is

$$\eta_x = 0.9166 + 0.0012X + 0.0014Y - 0.00078Z \tag{1}$$

where

η_x Temperature exchange efficiency;
X   TD (°C);
Y   MCD (g/kg);
Z   AFR (m³/h).

Calculate the efficiency equation of enthalpy.

Structure of air flow rate (x_1), temperature difference (x_2), moisture content difference (x_3), and enthalpy exchange efficiency (y):

$x_1 = [50, 100, 150, 200, 250, 150, 150, 150, 150, 150, 150, 150, 150, 150, 150]$;
$x_2 = [16, 16, 16, 16, 16, 24, 20, 16, 12, 8, 16, 16, 16, 16, 16]$;
$x_3 = [3, 3, 3, 3, 3, 3, 3, 3, 3, 3, 9, 7, 5, 3, 1]$;
$y_2 = [0.893, 0.847, 0.804, 0.76, 0.723, 0.815, 0.809, 0.804,$
$\quad 0.784, 0.755, 0.813, 0.812, 0.808, 0.804, 0.798]$;
temp = linspace(1, 1, 15);
$x = [\text{temp}', x_1', x_2']$;
[b, bint, r, rint, stats] = regress[y, x];
The regression results as follow:
b = 0.8633 0.0036 0.0024 −0.00085
bint = 0.8336 0.8930 0.0022 0.0050 0.0001 0.00480 −0.0009 −0.0007
r_i = 0.00725 0.00395 0.00365 0.00235 0.00805 −0.01435 −0.00585
$\quad$ 0.00365 −0.00185 −0.01635 −0.0016 −0.00215 0.0029 −0.00365 0.0024
Stats = 0.9667 106.4671 0.0000

In the calculation results, each regression coefficient b_i and the correlation coefficient r_i are within the allowable range. Correlation coefficient $R^2 = 0.9667$, $F = 106.4671 > F0.05(4, 15)$. The linear regression is very significant by the significance statistic F value, $p = 0.00 < 0.05$. It shows that the regression equation has a significant regression effect, and at least one regression coefficient in the equation is not significantly zero.

The efficiency equation of enthalpy:

$$\eta_h = 0.8633 + 0.0036X + 0.0024Y - 0.00085Z \quad (2)$$

where

η_h The efficiency of enthalpy;
X   TD (°C);
Y   MCD (g/kg);
Z   AFR (m³/h).

3 Results and Discussion

When the temperature difference and the moisture content difference are constant with the change of the air flow rate, the calculated efficiency is shown in the following Table 2 and Fig. 2.

The effect of air flow rate on the efficiency of the total heat exchanger is shown in Fig. 2. The total thermal efficiency decreases linearly with the increase of air flow rate. At low air flow rate of 50 m³/h, the sensible heat efficiency reaches the highest,

Table 2 Efficiency change with AFR

TD (°C)	MCD (g/Kg)	AFR (m³/h)	η_x	η_h
16	3	50	0.906	0.893
16	3	100	0.864	0.847
16	3	150	0.822	0.804
16	3	200	0.784	0.760
16	3	250	0.750	0.723

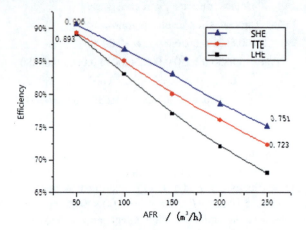

Fig. 2 Relationship between total heat exchanger efficiency and AFR. *Notes* SHE = sensible heat efficiency, TTE = total thermal efficiency, LHE = latent heat efficiency

which is 90.6%, and total thermal efficiency is 89.3%. When the high air flow rate is 250 m³/h, the lowest sensible heat efficiency is 75.1%, while the total thermal efficiency is 72.3%. The decrease in temperature efficiency is 15.5%, and the reduction in latent heat efficiency is 17%, which shows that the air flow rate has a greater impact on the latent heat exchange efficiency of the total heat exchanger. The curve of the moisture exchange efficiency is also shown in Fig. 2. With the increase of the air flow rate, the trend of the moisture exchange efficiency is particularly obvious. It can be seen that the influence of the air flow rate on the moisture exchange efficiency is the largest.

When the temperature difference and air volume are constant, the efficiency of the total heat exchanger is calculated by changing the moisture content difference, and the results are shown in Table 3 and Fig. 3.

The difference in moisture content has little effect on the sensible heat efficiency of the total heat exchanger. However, with the increase of the moisture content difference, the total thermal efficiency and the latent heat efficiency tend to increase linearly. It is obvious that the increase of the total thermal efficiency is mainly caused by the increase of the moisture exchange efficiency. As shown in Fig. 3, when the moisture content difference increases from 1 to 8 g/kg, the latent heat efficiency increases by about 10%, and the total thermal efficiency increases by 2%. The main reason for the above situation is that the increase of the moisture content difference of the fresh air exhaust inlet leads to an increase in the humidity potential

Table 3 Efficiency change with MCD

TD (°C)	MCD (g/Kg)	AFR (m³/h)	η_x	η_h
16	9	150	0.831	0.813
16	7	150	0.828	0.812
16	5	150	0.826	0.808
16	3	150	0.822	0.804
16	1	150	0.820	0.798

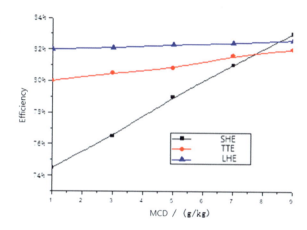

Fig. 3 Relationship between total heat exchanger efficiency and MCD

of the two boundary gases of the hot and humid exchange membrane, thereby increasing the mass diffusion coefficient of water vapor in the exchange paper. The mass transfer process is strengthened.

When the moisture content difference and the air flow rate are constant with the change of temperature difference, the calculated efficiency is as shown in the following Table 4 and Fig. 4.

The influence of temperature difference on heat exchange efficiency and moisture change efficiency is obvious. The latent heat efficiency increases linearly with the increase of temperature difference between fresh air and exhaust, and the variation range is large, reaching 10%. The sensible heat efficiency and the total thermal efficiency increase with the increase of the temperature difference, and the logarithmic function increases. When the temperature difference increases

Table 4 Efficiency change with TD

TD (°C)	MCD (g/Kg)	AFR (m³/h)	η_x	η_h
24	3	150	0.827	0.815
20	3	150	0.825	0.809
16	3	150	0.822	0.804
12	3	150	0.817	0.784
8	3	150	0.807	0.755

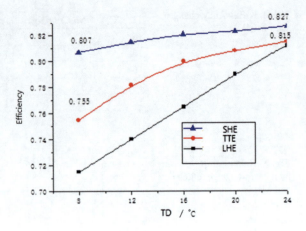

Fig. 4 Relationship between total heat exchanger efficiency and TD

from 8 to 24 °C, the temperature exchange efficiency increases by 2%, and the total thermal efficiency increases by 6%.

4 Conclusions

In this paper, the factors affecting the efficiency of total heat exchanger under different working conditions are studied and analyzed. We drew some conclusions as follows:

(1) At low air flow rate of 50 m³/h, the highest sensible heat efficiency is 90.6%, and the total thermal efficiency is 89.3%; at 250 m³/h, the sensible heat efficiency reaches a minimum of 75.1%, while the total thermal efficiency is at least 72.3%. The decrease of temperature efficiency is 15.5%, and the reduction of enthalpy efficiency is 17%. The air flow rate has a greater influence on the enthalpy exchange efficiency of the optimized core.
(2) When the moisture content difference increases by 8 g/kg, the latent heat efficiency increases by about 10%, and the total thermal efficiency increases by 2%.
(3) The sensible heat efficiency and the total thermal efficiency increase with the increase of the temperature difference, and the logarithmic function increases. When the temperature difference increases from 8 to 24 °C, the temperature exchange efficiency increases by 2%, and the total thermal efficiency increases by 6%.

Acknowledgements The project is supported by National Science Foundation of Tianjin (17JCTPJC48200).

References

1. Zhao, W.: Study on heat transfer and flow in cross triangular corrugated plate flow channel. South China University of Technology, Guangzhou, China (2009)
2. GB/T21087-2017. Indoor air quality. Standards Press of China, China (2017)
3. Fuhuan, W.: Study on performance of plate-fin air total heat exchanger. Donghua University, Shang hai, China, (2004)
4. Ogulata, R.T., et al.: Experiments and entropy generation minimization analysis cross-flow heat exchanger. Int. J. Heat Mass Transf. **41**(2), 373–381 (1997)
5. Michael, W.: Simulation model air-to-air plate heat exchanger. Energy **13**(32), 49–54 (1998)
6. Yuhua, W.: Research on heat recovery performance of air-air plate heat exchanger. Harbin Institute of Technology, Harbin, China (2009)
7. Jinji, X.: ANSYS 13.0 workbench numerical simulation analysis. Water and Power Press, Beijing, China (2012)

On-Site Operation Performance Investigation of Nocturnal Cooling Radiator for Heat Radiation

Yi Man, Shuo Li, Xinyu Zhang, Guoxin Jiang and Tiantian Du

Abstract Nocturnal cooling radiation is one of the effective natural passive cooling technologies by electromagnetic infrared radiation exchange between terrestrial surfaces and the sky. For promoting the application of nocturnal cooling radiation in the active cooling system, a novel nocturnal cooling radiator with simple panel structure and low initial investment is designed in this study. In order to investigate the heat radiation performance of the nocturnal cooling radiator, a practical test rig of nocturnal cooling radiation system with thermometer sensors and flow rate sensors is assembled horizontally against on the roof of a science and technology building located in Jinan, China. Then, the outlet water temperature, cooling capacity, and the exterior wall temperature distribution of the nocturnal cooling radiator for different operation conditions are measured and recorded. According to the practical investigation, the heat dissipating capacity of properly designed and controlled nocturnal cooling radiator is proved to be considerable for merging with various active cooling systems.

Keywords On-site operation test · Nocturnal cooling radiator · Performance investigation

---

Y. Man · S. Li · X. Zhang · G. Jiang · T. Du
School of Thermal Engineering, Shandong Jianzhu University, Jinan, China
e-mail: 937542611@qq.com

X. Zhang
e-mail: 727254368@qq.com

G. Jiang
e-mail: 724848723@qq.com

T. Du
e-mail: 18766110679@163.com

Y. Man (✉)
Shandong Key Laboratory of Renewable Energy Technologies for Buildings,
Shandong Jianzhu University, Jinan, China
e-mail: manyilaura@163.com

1 Introduction

It is well known that nocturnal cooling radiation (NCR) is one of the effective natural passive cooling technologies caused by infrared radiation between the terrestrial surfaces and the sky [1]. The sky far away from atmosphere is a great free radiation heat sink, whose temperature is about 4 K [2]. Since the temperatures of the skyward terrestrial surfaces are always higher than 4 K, these surfaces experience heat loss by infrared radiation to sky. Although the atmosphere forms a screen between terrestrial surfaces and space to prevent the excessive heat loss of earth, the screen function is weak for radiation between infrared wavelengths of 8–13 μm, which is called "atmospheric window." The nocturnal cooling radiation is enhanced if the terrestrial surface has a high emissivity in the wavelength region matched to the "atmospheric window." Then, the terrestrial surface can effectively utilize the space as free heat sink and reach the low equilibrium temperature.

The nocturnal cooling radiation phenomenon was first explored by Arago [3] for dwellings climatization since nineteenth century. Following the energy crisis of the 1970s, the nocturnal cooling radiation technology for buildings received increasing attentions from researchers. Both flat-plate solar collectors and other specially designed radiators were investigated for nocturnal cooling mainly in five topics: the performance investigation of the passive cooling and heating systems in which flat-plate solar collectors were used [4], the simulation model and experimental results of the flat-plate solar collectors as the nocturnal cooling radiator (NCR) [5], the performance of a passive cooling system with other specially designed radiators [6], and the simulation model and experimental results of the other specially designed radiators [7–9]. From the literature review, the existing researches mainly concentrated on the performances investigation of passive cooling systems mainly consisting of nocturnal cooling radiators and their accessories.

In fact, application of the nocturnal cooling radiation technology for normal heat pump applications in active cooling systems is inherently limited. First, the cooling capacity of NCR system cannot match the building cooling loads. Second, the cooling capacity of the NCR is significantly affected by relative humidity of ambient air and cloud cover conditions of a nocturnal sky. However, its inherent feature determined that the NCR can be used as supplemental heat rejecter and to be activated under ideal meteorological conditions. Therefore, it is more suitable to be utilized as efficient supplemental heat sink to assist other stable heat sinks in an active cooling system.

For promoting the application of nocturnal cooling radiation in the practical active cooling system, a novel nocturnal cooling radiator with simple panel structure and low initial investment is designed in this study. The structure of nocturnal cooling radiator's main body is the stainless steel flat tube coated with high-emissivity coating material. The two ends of the stainless steel flat tube are connected by connecting boxes. Then, the circulating water flow inside the stainless steel flat tube can transfer the heat into the external wall of tube and emit the heat into the sky by heat radiation during night.

In order to investigate the actual heat radiation performance of the nocturnal cooling radiator, a practical test rig of nocturnal cooling radiation system with thermometer sensors and flow rate sensors is assembled horizontally against on the roof of a science and technology building located in Jinan. Then, the outlet water temperature, cooling capacity, and the exterior wall temperature distribution of the nocturnal cooling radiator for different mass flow rates, inlet water temperatures, running time, and weather conditions are measured and recorded. Based on the practical test data, the influencing factors about the operation performance of the nocturnal cooling radiator are analyzed and compared. According to the practical operation performance investigation, the heat dissipating capacity of proper designed and controlled nocturnal cooling radiator is proved to be considerable for merging with various active cooling systems.

2 The NCR Test Rig

The NCR test rig is designed and installed in the science and technology building on the campus of the Shandong Jianzhu University, Jinan, Shandong Province of China. The NCR test rig mainly consists of five components, i.e., the NCR, the circulating water pumps, the electrically heated constant temperature water tank, the plate heat exchanger, and data acquisition instruments. The NCR is installed on the roof of the building, and other auxiliary equipments are installed on the floor inside the building. The schematic diagram of the NCR test rig is shown in Fig. 1.

2.1 The NCR

According to Ferrer et al. [10], the NCR with simple metal surface is inexpensive and is easy for installation and maintenance, but it has comparative nocturnal cooling performance compared with other radiators with complex configurations. Therefore, a simple rigid radiator as shown in Fig. 1 is selected in this study. For beneficial in exhausting the excess air, convenient for manufacture, and cheap for initial cost, the stainless steel flat tube with width of 10 cm and thickness of 1 cm is selected. The stainless steel flat tube is selected also because it has a high emissivity in the "atmospheric window" and it is economical as well. For enhancing the infrared radiation, the stainless steel flat tubes are coated with high-emissivity coating material.

For the NCR test rig, 12 stainless steel flat tubes are bonded along the width direction, along the length direction, these stainless steel flat tubes are welded with the stainless steel water separator. The width and the length of the NCR are 1 m and 5 m, respectively. The circulation water flows inside these stainless steel flat tubes. The schematic diagram and on-site photograph of the NRC are shown in Figs. 2 and 3. The two NCRs can be connected in parallel or in series in the NCR test rig.

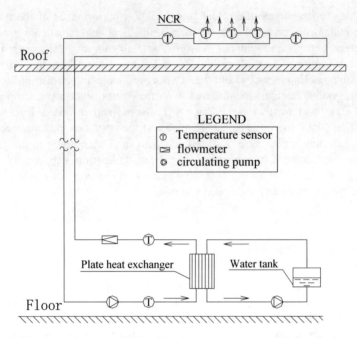

Fig. 1 Schematic diagram of the NRC test rig

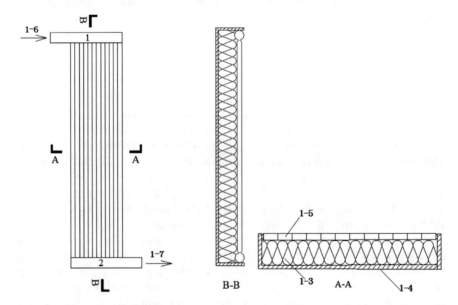

Fig. 2 Schematic diagram of the NRC

Fig. 3 On-site photograph of the NRC

2.2 The Auxiliary Equipments

The galvanized steel pipes with diameter of DN40 are selected to connect the NCR, the circulating water pumps (as shown in Fig. 4a), the electrically heated constant temperature water tank (as shown in Fig. 4b), and the plate heat exchanger (as shown in Fig. 4c).

The data acquisition as well as control instruments and the auxiliary equipments installed in the indoor of the science and technology building are shown in Fig. 5.

Then, the operation data of the NRC test rig such as the temperature distribution of the NRC surface, the temperature of circulating water, the temperature and the relative humidity of outdoor air, the circulating water flow rate of system, the operation powers of the circulating pump and the electric heater inside water tank can be measured and recorded in the data acquisition, as shown in Fig. 6.

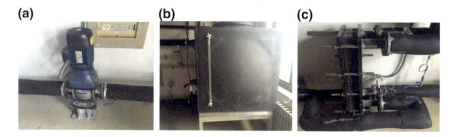

Fig. 4 Auxiliary equipments of the NRC test rig

Fig. 5 Indoor equipments of the NRC test rig

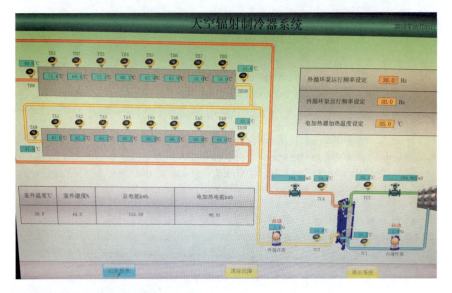

Fig. 6 Operation data presentation of the NRC test rig

3 Test Results

The nocturnal radiation cooling test is carried out during the summer of 2018. The circulating pump is activated from 19:00 to 6:00, and the water circulates from the plate heat exchanger and the NCR. The circulating water absorbs the heat from the plate heat exchanger and releases the heat into the night sky. The circulation water temperature during one operation cycle is plotted and shown in Fig. 7.

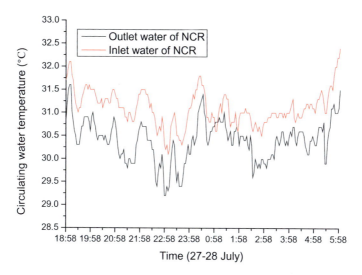

Fig. 7 Circulating water temperature of the NRC

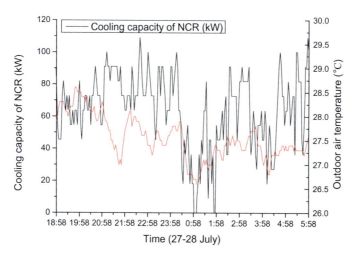

Fig. 8 Cooling capacity of the NRC

As shown in Fig. 7, the average temperature difference between the outlet and the inlet water of the NRC is about 0.72 °C for the water flow rate of 21.67 kg/s. Then, the cooling performance of the NRC is calculated and plotted in Fig. 8.

It should be noticed in Fig. 8 that the transient dysfunction of the thermal resistance temperature sensors at moments of 1:01, 1:04, 1:52, and 1:55 was responsible for the measurement failures of the NRC's cooling capacity, and the

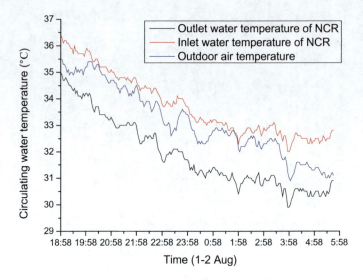

Fig. 9 Relationship between the circulating water and the outdoor air temperature

corresponding bad pixels were eliminated in the following analysis. Under the outdoor air condition as shown in Fig. 8, the average cooling capacity of the NCR from 19:00 to 6:00 is 65.41 kW, and the total cooling load that can be afforded by the NCR is 784.92 kWh for one night. Therefore, the properly controlled NCR can provide considerable cooling capacity for merging with various active cooling systems.

The test data demonstrate that the change trend of the cooling capacity is identical with the outdoor air. As shown in Fig. 9, when the electric heater inside the water tank is turned off, both the outlet and inlet water temperatures of the NCR decrease along with the decline in the outdoor air temperature. When the inlet water temperature of the NCR is higher than the outdoor air temperature, the cooling capacity of the NCR is magnified. Therefore, for the NCR installed in the circumstance with low outdoor air temperature at night, measures should be carried out to enhance the convective heat transfer between the outdoor air and the outside surface of the NCR.

4 Conclusions

For investigating the actual heat radiation performance of the nocturnal cooling radiator, a practical test rig of nocturnal cooling radiation system with thermometer sensors and flow rate sensors is assembled horizontally against on the roof of a science and technology building located on campus of the Shandong Jianzhu University. Then, the outlet water temperature, cooling capacity, and the exterior wall temperature distribution of the nocturnal cooling radiator for different mass flow rates, inlet water temperatures, running times, and weather conditions are measured and recorded.

The nocturnal radiation cooling test is carried out during the summer of 2018. When the circulating pump is activated from 19:00 of the 27 July to 6:00 of the 28 July, the average temperature difference between the outlet and the inlet water of the NCR is measured to be 0.72 °C for the water flow rate of 21.67 kg/s. The average cooling capacity of the NCR is 65.41 kW, and the total cooling load that can be afforded by the NCR is 784.92 kWh for one night. Therefore, the properly controlled NCR can provide considerable cooling capacity for merging with various active cooling systems based on the practical test data.

Acknowledgements The project is supported jointly by a grant from National Natural Science Foundation of China (Number 51808321) and a grant from Shandong Province Natural Science Foundation (Number BS2015NJ016).

References

1. Catalanotti, S., et al.: The radiative cooling of selective surfaces. Sol. Energy **17**(83), 97–112 (1975)
2. Bliss, R.: Atmospheric radiation near the surface of the ground: a summary for engineers. Sol. Energy **1961**(5), 103–115 (1961)
3. Arago, F.: Annuaire du Bureau des Longitudes pour I'an. reprinted in Oeuvres Completes de Francois Arago. 87–95 (1826)
4. Eicker, U., Dalibard, A.: Photovoltaic-thermal collectors for night radiative cooling of buildings. Sol. Energy **85**(7), 1322–1335 (2011)
5. Sikula, O., Vojkuvkova, P.: Hybrid roof panels for night cooling and solar energy utilization in buildings. Energy Procedia. 177–183 (2015)
6. Hollick, J.: Nocturnal radiation cooling tests. Energy Procedia **30**(1), 930–936 (2012)
7. Matsuta, M., Terada, S.: Solar heating and radiative cooling using a solar collector-sky radiator with a spectrally selective surface. Sol. Energy **39**(3), 183–186 (1987)
8. Mostrel, M., Givoni, B.: Windscreens in radiant cooling. Passive Sol J **14**, 229–238 (1982)
9. Harrison, A., Walton, M.: Radiative cooling of TiO_2 white paint. Sol. Energy **20**, 185–188 (1978)
10. Ferrer, J., et al.: Modeling and experimental analysis of three radio convective panel for night cooling. Energy Build. 37–48 (2015)

Simulative Investigation of Ground-Coupled Heat Pump System with Spiral Coil Energy Piles

Yi Man, Xinyu Zhang, Shuo Li, Tiantian Du and Guoxin Jiang

Abstract Combining the heat exchanger and building foundation pile as the energy pile can eliminate the drilling expense and land area requirement of the conventional borehole heat exchanger. For the spiral coil energy pile, the circulation coil pipe is intertwined tightly in spiral shape against the reinforcing steel of a pile. The distinct advantage of spiral coil energy pile is it can offer higher heat transfer efficiency, reduce pipe connection complexity, prevent air chocking and decrease the thermal short-circuit between circulating pipes compared with other configurations. In order to investigate the operation performance of ground-coupled heat pump system with spiral coil energy piles, analytical simulation models of spiral coil energy piles and the entire heat pump system are established. Temperature responses of the circulating water entering/effusing the spiral coil energy piles and the heat pump unit to the short time step building heating/cooling loads are simulated based on the established analytical models. Then, the heat exchange capacity of the spiral coil energy piles is investigated, and the operation performance of the entire system is analysed. The economic superiority of the ground-coupled heat pump system with spiral coil energy piles is found to be obvious compared with traditional air conditioning modes.

Y. Man · X. Zhang · S. Li · T. Du · G. Jiang
School of Thermal Engineering, Shandong Jianzhu University, Jinan, China
e-mail: 727254368@qq.com

S. Li
e-mail: 937542611@qq.com

T. Du
e-mail: 18766110679@163.com

G. Jiang
e-mail: 724848723@qq.com

Y. Man (✉)
Shandong Key Laboratory of Renewable Energy Technologies for Buildings, Shandong Jianzhu University, Jinan, China
e-mail: manyilaura@163.com

Keywords Energy pile · Spiral coil · Ground-coupled heat pump · Simulative investigation

1 Introduction

It is well known that, the high initial cost and land area requirement to install the borehole ground heat exchanger (GHE) remain the major obstacles of the ground-coupled heat pump technology [1]. Based on previous research, the energy pile which combining the heat exchanger and building foundation pile can eliminate the deficiencies of borehole GHE [2–5]. In order to enlarge the inside heat transfer area as well as heat transfer efficiency, and to reduce pipe connection complexity as well as prevent the air chocking of energy pile, this study utilizes the spiral coil energy pile, which intertwined the circulation coil pipe tightly in spiral shape against the reinforcing steel of a pile, as shown in Fig. 1.

For the ground-coupled heat pump system with spiral coil energy piles, heat is extracted from or injected into the ground by circulating water between the heat pump unit and the energy piles, as shown in Fig. 2. It is crucial to calculate the heat transfer of the energy piles and the heat pump unit for simulating the performance of the whole ground-coupled heat pump system.

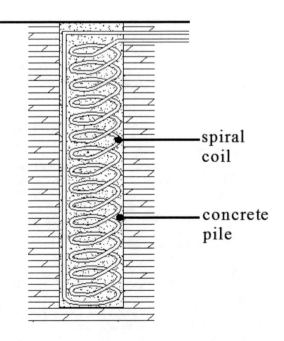

Fig. 1 Schematic diagram of the spiral coil energy pile

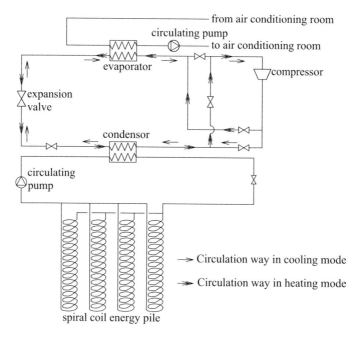

Fig. 2 Ground-coupled heat pump system with spiral coil energy pile

2 Model Establishment of Energy Pile

Compared with borehole GHE, the proposed spiral coil energy pile possesses thicker diameter and shorter depth. As shown in Fig. 3, the buried coil inside spiral coil energy pile can be simplified into spiral heat source. Then, the spiral heat source model can be established by taking the three-dimensional geometrical characteristic of the spiral pile into account.

2.1 Temperature Response to the Spiral Heat Source

According to the Green's function theory, the temperature response at point (r, φ, z) to an instantaneous point heat source with intensity of ρc, located at (r', φ', z') and activated at the instant τ' can be expressed as:

$$G(r, \varphi, z, \tau; r', \varphi', z', \tau') = \frac{1}{8[\pi a(\tau - \tau')]^{3/2}} \cdot \exp\left[-\frac{(r\cos\varphi - r'\cos\varphi')^2 + (r\sin\varphi - r'\sin\varphi')^2 + (z - z')^2}{4a(\tau - \tau')}\right] \tag{1}$$

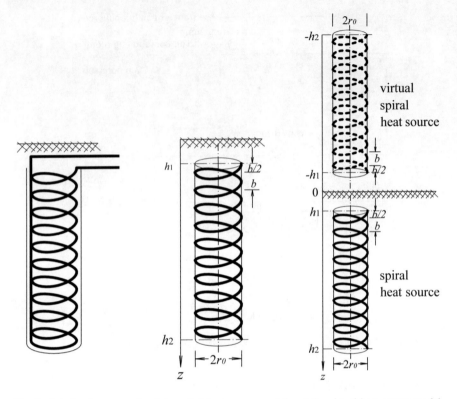

Fig. 3 Spiral coil energy pile, finite spiral heat source model and the virtual heat source model

As shown in Fig. 2. The spiral heat source and heat sink can be approximated as the sum of numerous point heat sources and heat sinks. Then, the temperature response of the medium around energy pile can be deduced based on the Green's function theory and the superposition method:

$$\theta_{f,\text{spiral}} = \frac{q_l b}{2\pi \rho c} \int_0^\tau d\tau' \left[\int_{2\pi h_1/b}^{2\pi h_2/b} G(z' = b\varphi'/2\pi) d\varphi' - \int_{2\pi h_1/b}^{2\pi h_2/b} G(z' = -b\varphi'/2\pi) d\varphi' d\varphi \right]$$

$$= \frac{q_l b}{16\pi \rho c} \int_0^\tau \frac{d\tau'}{[\pi a(\tau - \tau')]^{3/2}} \cdot \exp\left[-\frac{r^2 + r_0^2}{4a(\tau - \tau')}\right]$$

$$\cdot \int_{2\pi h_1/b}^{2\pi h_2/b} \exp\left[\frac{2rr_0 \cos(\varphi - \varphi')}{4a(\tau - \tau')}\right] \left\{ \exp\left[-\frac{(z - b\varphi'/2\pi)^2}{4a(\tau - \tau')}\right] - \exp\left[-\frac{(z + b\varphi'/2\pi)^2}{4a(\tau - \tau')}\right] \right\} d\varphi'$$

(2)

2.2 Temperature Response of Pipe Wall

The spiral heat source can be approximated as located at the centre of coil pipe, and the pipe wall located at r_p away from the spiral heat source, as shown in Fig. 4. Then, the temperature response of pipe wall at moment τ can be deduced in Eq. (3) based on the short time step pulse heat currents q_{l_i}.

$$\begin{aligned}\theta_{\text{ring,pile}} &= \frac{1}{k}\sum_{i=1}^{\infty}(q_{l_i}-q_{l_{i-1}})\cdot p(\tau-\tau_{i-1}) \\ &= \frac{1}{k}\left[\sum_{i=1}^{\infty}q_{l_i}\cdot p(\tau-\tau_{i-1}) - \sum_{i=1}^{\infty}q_{l_{i-1}}\cdot p(\tau-\tau_{i-1})\right] \\ &= \frac{1}{k}\left[\sum_{i=1}^{\infty}q_{l_i}\cdot p(\tau-\tau_{i-1}) - q_{l_0}\cdot p(\tau-\tau_0) - \sum_{j=1}^{\infty}q_{l_j}\cdot p(\tau-\tau_j)\right] \\ &= \frac{1}{k}\sum_{i=1}^{\infty}q_{l_i}\cdot[p(\tau-\tau_{i-1})-p(\tau-\tau_i)] = \frac{1}{k}\sum_{i=1}^{\infty}q_{l_i}\cdot q(\tau-\tau_{i-1}) \end{aligned} \quad (3)$$

2.3 Temperature Response of Circulating Water

Compared with the heat transfer characteristics of ground outside the pile, the heat transfer of circulating water inside the spiral coil energy pile can be approximated as a steady-state process. Then, the entering and effusing fluid temperatures of the spiral coil energy pile can be determined by heat currents and heat transfer resistances:

$$\begin{cases} T'_f = T_p + \dfrac{q_l\cdot b\cdot R_{rp}}{\sqrt{(2\pi\cdot r_0)^2+b^2}} + \dfrac{q_l(h_2-h_1)}{2M\cdot C_p} \\ T''_f = T_p + \dfrac{q_l\cdot b\cdot R_{rp}}{\sqrt{(2\pi\cdot r_0)^2+b^2}} - \dfrac{q_l(h_2-h_1)}{2M\cdot C_p} \end{cases} \quad (4)$$

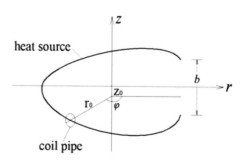

Fig. 4 Spiral coil pipe buried in pile GHE

3 Model Establishment of Heat Pump Unit

Energy consumption of the ground-coupled heat pump system is major affected by the operation performance of the heat pump unit. For simulating the heat pump's performance based on variable building air conditioning loads, it is more feasible to fit the functions of its COP and effusing fluid temperature versus entering fluid temperature by utilizing the least square method and the 2th power polynomial curves. Operation parameters of the selected heat pump unit in this study are plotted in Fig. 5.

4 Results

4.1 Field Temperature Response to the Sample Spiral Coil Energy Pile

According to the normal pile configurations in the practical engineering, a sample spiral coil energy pile with $r_0 = 0.4$ m, $h_1 = 2$ m, $h_2 = 22$ m, $b = 0.4$ m, $r_{pi} = 20$ mm and $r_p = 32$ mm buried in the soil with undisturbed temperature of 12.5 °C is selected in this study. The fluid flow velocity inside coil pipe is set to be 0.5 m/s. By simulation, the temperature fields covering the sample energy pile and its surrounding soil at different operation times are described in Fig. 6. As shown, the temperature rise fluctuates considerably in the vicinity to the spiral coil. The axial heat conduction influence is limited to the two ends of energy pile in relatively short time periods, and penetrate deeper for longer-term operation.

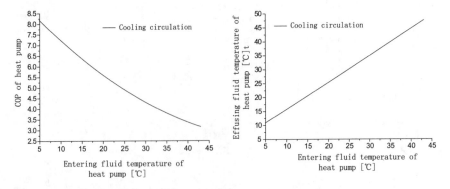

Fig. 5 Operation parameters of the selected heat pump unit

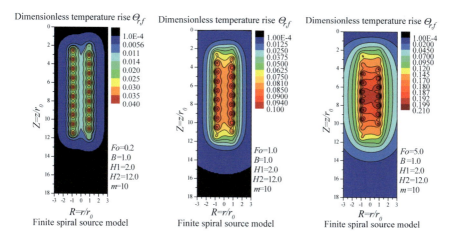

Fig. 6 Temperature response to the spiral coil energy pile

4.2 Operation Parameters of the Sample Spiral Coil Energy Pile

The hourly heat transfer loads afforded by the sample spiral coil energy pile is shown in Fig. 7, and its hourly operation parameters are simulated and plotted in Fig. 8. Based on simulation results, the sample energy pile possesses heat exchange capacity of about 212 W/m. For the sample energy pile with depth of 20 m, it can afford about 100 m^2 building air conditioning areas.

4.3 Performances of the Heat Pump Unit

As shown in Fig. 8, the highest outlet water temperature of the spiral coil energy pile in the ground-coupled heat pump system with spiral coil energy piles for cooling provision during summer is 28.4 °C, and the lowest outlet water temperature of the spiral coil energy pile in the ground-coupled heat pump system for heating provision during winter is 3.8 °C. Compared with the building ambient air, the spiral coil energy pile can provide the heat pump with the heat sink in lower temperature for cooling provision and the heat source in higher temperature for heating provision. Therefore, the heat pump unit of the ground-coupled heat pump system with spiral coil energy piles can obtain high COP value and low operation energy consumption. The operation COP value of the heat pump unit during one year operation is simulated and plotted in Fig. 9.

According to the operation simulation results of the ground-coupled heat pump system with spiral coil energy piles, the average COP of the heat pump unit during one year operation is 4.2, and the total energy consumption of the heat pump unit

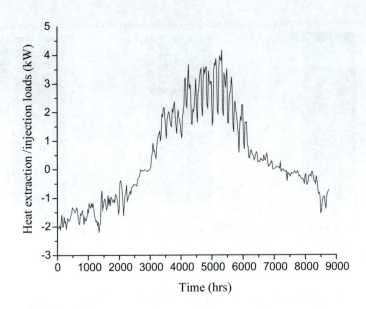

Fig. 7 Hourly heat transfer loads afforded by the sample spiral coil energy pile

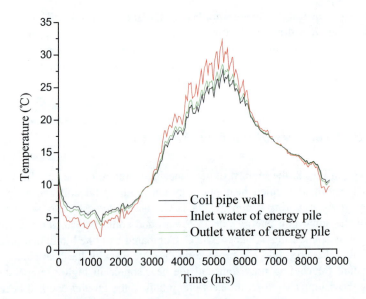

Fig. 8 Hourly operation parameters of sample spiral coil energy pile

for heating and cooling provision during the whole year's operation is about 27.2 kWh/m^2. Taking the energy consumption of the circulating water pumps and the fan coil units into account, total cost of the ground-coupled heat pump system

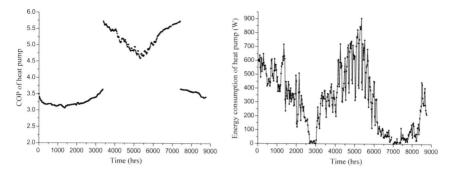

Fig. 9 COP and energy consumption of the heat pump unit

with spiral coil energy piles for the whole year's heating and cooling provision to building is as low as 22.5 RMB Yuan/m^2. The economic superiority of the ground-coupled heat pump system with spiral coil energy piles is obvious compared with traditional air conditioning modes.

5 Conclusions

In order to investigate the operation performance of the ground-coupled heat pump system with spiral coils energy piles, this study simulated the temperature responses of the spiral heat source, the coil pipe wall and the circulating water entering/effusing the spiral coil energy pile to the hourly heat transfer loads based on the established analytical model. Then, the operation performance of the spiral coil energy pile and the whole ground-coupled heat pump system is investigated.

According to the simulative operation parameters of the ground-coupled heat pump system with spiral coil energy piles, the highest outlet water temperature of the spiral coil energy pile for cooling provision during summer is 28.4 °C, and the lowest outlet water temperature of the spiral coil energy pile for heating provision during winter is 3.8 °C. The heat exchange capacity of the sample spiral coil energy pile is about 212 W/m. The average COP of the heat pump unit in one year's operation is 4.2, and the total energy consumption of the heat pump unit for heating and cooling provision during the whole year's operation is about 27.2 kWh/m^2.

The economic superiority of the ground-coupled heat pump system with spiral coil energy piles compared with traditional air conditioning modes is found to be obvious in this study. For further research, the optimal design parameters of the ground-coupled heat pump system with spiral coil energy piles should be discussed.

Acknowledgements The project is supported jointly by a grant from National Natural Science Foundation of China (Number 51808321) and a grant from Shandong Province Natural Science Foundation (Number BS2015NJ016).

References

1. Bose, J.E., et al.: Design/data manual for closed-loop ground coupled heat pump systems. Oklahoma State University for ASHRAE (1985)
2. Mehrizi, A., et al.: Energy pile foundation simulation for different configurations of ground source heat exchanger. Int. Commun. Heat Mass Transfer **70**, 105–114 (2016)
3. Luo, J., et al.: Thermo-economic analysis of four different types of ground heat exchangers in energy piles. Appl. Therm. Eng. **108**, 11–19 (2016)
4. Huerta, L., Krarti, M.: Foundation heat transfer analysis for buildings with thermal piles. Energy Convers. Manag. **89**, 449–457 (2015)
5. Loveridge, F., Powrie, W.: 2D thermal resistance of pile heat exchangers. Geothermics **50**, 122–135 (2014)

A Developed CRMC Design Method and Numerical Modeling for the Ejector Component in the Steam Jet Heat Pumps

Chenghu Zhang, Yaping Li and Jianli Zhang

Abstract Steam jet heat pump is one kind of high energy efficiency heating supplying device. In this system, ejector is an essential component determining the system operating performance. Our work features the advanced ejector design with developed constant rate of momentum change (CRMC) method based on the real properties of working fluid, and the flow friction loss is also considered, which results in a higher thermal compression effect more accurately. With the validated mathematical models, we compare the performances of the designed ejectors using different design methods. The results show a promoted ejector entrainment ratio up to 0.59 with the CRMC method. Furthermore, the variations of the designed entrainment ratio with the change of provided ejector working conditions are analyzed. At last, for the purposed to verify the correct and advantages of the CRMC model, the numerical modeling for operating condition with the specific construction CRMC ejector is also conducted. A more stable and more efficient performance is shown in the modeling results.

Keywords Ejector component · Developed ejector design model · Constant rate of momentum change method · Numerical modeling

---

Y. Li · J. Zhang
Heilongjiang Provincial Key Laboratory of Building Energy Efficiency and Utilization, Harbin 150090, China

C. Zhang · Y. Li (✉) · J. Zhang
School of Architecture, Harbin Institute of Technology, Harbin 150090, China
e-mail: 14B927067@hit.edu.cn

C. Zhang
Key Laboratory of Cold Region Urban and Rural Human Settlement Environment Science and Technology, Ministry of Industry and Information Technology, Harbin 150090, China

1 Introduction

Steam jet heat pump cycle, or called ejector heat pump, is driven by thermal energy and is considered as an alternative to the absorption for its simple construction, low cost, long lifetime, flexible capacity, no-chemical corrosion and chemical reaction [1]. The literature published recently on steam jet refrigeration cycles covers comprehensive research results concerning thermodynamic characteristics, construction configuration, performance prediction and optimization and the working fluid selection. The operating temperatures in each system component are essential parameters influencing the cycle performance [2–10]. Ejectors are key components in each steam jet refrigeration system. In the ERS cycle, ejector is installed for providing the entrainment, mixing and compression function. Many theoretical and modeling researches are existing to promote the ejector design process and performance. It is important to note that the dimensional construction of ejector would greatly affect the device operation. However, the CRMC model proposed by Eames [11] is less than perfect since in their model assumptions, the mixing process of primary flow and secondary flow is done at the inlet of mixing-diffuser section, which deviates from the reality and causes the unsatisfactory ejector performance in practice. In this paper, the developed constant rate of momentum change model for the ejector design is proposed. Based on this advanced CRMC design theory, the mathematical model with real working fluid properties and flow friction loss is built in our work. The conclusion of this work can be a guideline for the primary design of ejector in the practical engineering.

2 Methods

2.1 Developed CRMC Model for Ejector Component

The developed CRMC model for the ejector design is presented based on the real properties of the working fluids and the friction loss factor. The schematic of the developed CRMC model is shown as Fig. 1. Compared with the traditional CRMC

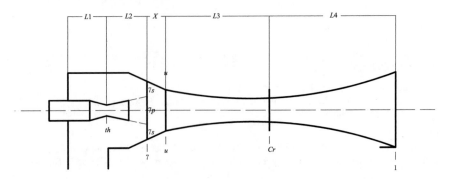

Fig. 1 Schematic diagram of the CRMC ejector model

model, a conical zone (X) is added behind the primary nozzle exit where entrainment takes place. To explain the details of this model and simplify the design process, the assumptions are listed as follows:

The adiabatic expansion process of primary flow occurs in the Laval nozzle. Started from the inlet section of the nozzle, with the pressure decrease of the primary flow, the flow velocity is continuously increasing. At the throat of the Laval nozzle, the cross-section area equals the minimum value, and the primary flow is chocked.

For the primary flow, the thermal states at the inlet and nozzle throat obey the energy conversation theory, expressed as Eq. (1).

$$h_{g,pi} = h_{p,t} + 0.5 V_{p,t}^2 \tag{1}$$

At the nozzle throat, the mass flow rate per unit area is formulated as Eq. (2).

$$\dot{m}_p / A_t = V_{p,t} / v_{p,t} \tag{2}$$

The adiabatic expansion process of the secondary flow between the secondary flow inlet and the aerodynamic throat in 7 s-7 s cross section also can be described using the similar equations. It can be expressed as Eqs. (3) and (4).

$$h_{e,si} = h_{s,7} + 0.5 V_{7,s}^2 \tag{3}$$

$$\dot{m}_s / A_{s,7} = V_{s,7} / v_{s,7} \tag{4}$$

According to the CPM model, at the 7-7 cross section, the pressure relation between the primary and secondary flow is shown as Eq. (5).

$$P_{mix} = P_{7s} = P_{7p} \tag{5}$$

The mass, energy and momentum conservation relation should be obeyed in the control volume for the constant pressure mixing process, which is expressed as Eqs. (6) and (7).

$$(\dot{m}_p + \dot{m}_s) \cdot v_u = V_u \cdot A_u \tag{6}$$

$$h_{p,pi} + \omega \cdot h_{s,pi} = (1 + \omega) \cdot (h_u + 0.5 V_u^2) \tag{7}$$

The normal shock would occur and intensify the irreversibility of the steam jet process. With the same inlet condition and entrainment ratio, the outlet pressure of the mixed flow would be lower. Adopting the CRMC method to design and calculate the process in mixing section and diffuser (mixing-diffusing section), the shock phenomena can be eliminated [12].

The model of CRMC is written as Eq. (8).

$$\frac{d\dot{M}_o}{dx} = (\dot{m}_p + \dot{m}_s)\frac{dV}{dx} = C \qquad (8)$$

On the other aspect, in the mixing-diffusing chamber, the energy balance relationship between any cross section and the inlet is expressed as Eq. (9).

$$h_u + 0.5V_u^2 = h_{x,i} + 0.5V_{x,i}^2 \qquad (9)$$

According to the definition of isentropic efficiency, the enthalpy relation is formulated as Eq. (10).

$$h_{x,i} = h_u + \left(h'_{x,i} - h_u\right)/\eta_{\text{hed}} \qquad (10)$$

Above all, Eqs. (1)–(10) describe the thermodynamic process in the ejector based on the developed CRMC model. Since the additional conical zone is added, the total length of the novel model should be larger than the original one. It extends the space for the mixing process before the CRMC criterion applied, which would lead to a better performance in practical operation.

3 Results and Discussion

With the identical preconditions, the results calculated with CRMC model and the normal CPM model presented in are compared in Fig. 2–4.

Figure 2 shows the change of cross section radius along the x-axis. From this figure, note that compared with the normal CPM model, CRMC model defines the linear velocity variation in the ejector chamber, and therefore, the radius variation tendency of this model is nonlinear. Moreover, attribute to the constant rate of momentum change in the mixing-diffusing section, the static pressure in this

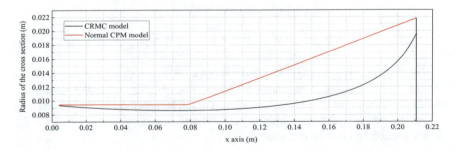

Fig. 2 Comparison between the dimension size of mixing-diffuser section based on CPM-CRMC model and normal CPM model

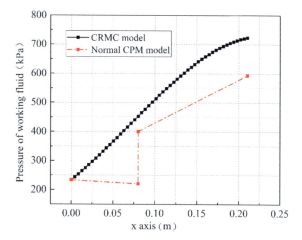

Fig. 3 Comparison between the static pressure of mixing-diffuser section based on CRMC model and normal CPM model

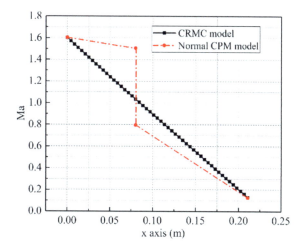

Fig. 4 Comparison between the Mach number of mixing-diffuser section based on CRMC model and normal CPM model

chamber is increasing gently instead of a sudden jump in the normal CPM model shown as Fig. 3. This advanced modification results in a higher outlet pressure of the ejector with the identical inlet conditions. In addition, in Fig. 4, with CRMC model, the Mach number in the mixing-diffusing section is also changing gently from supersonic to subsonic state.

Set the specific operating conditions of the ejector shown in Table 1, with the normal CPM model, normal CRMC model and the developed CRMC model, the dimension sizes of different ejectors can be designed. Therefore, the comparison of the ejector performance could be analyzed by CFD method.

With other boundary conditions constant, the pressure condition of the primary flow varies from 550 to 710 kPa, the entrainment ratios of different ejector constructions are calculated by numerical modeling, and the results are shown in Fig. 5.

Table 1 Thermodynamics condition for ejector modeling

Primary flow inlet temperature (°C)	Primary flow inlet pressure (kPa)	Secondary flow inlet temperature (°C)	Secondary flow inlet pressure (kPa)	Background pressure (kPa)
97	591	36	104.76	203.20

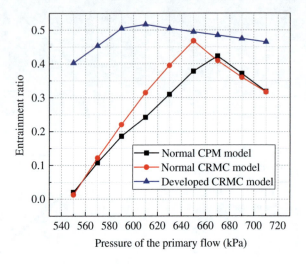

Fig. 5 Entrainment ratio of ejectors under different primary flow pressures

From this figure, it can be seen that regardless of the ejector construction, with the increase of the primary flow pressure, the entrainment ratios grow firstly and then decrease. The pressure of the mixed flow can be seen as the background pressure of the Laval nozzle; therefore, when the primary flow pressure of the ejector is too low, the normal shock will occur in the nozzle. It will result in a speed decrease of the primary flow to the subsonic state before the nozzle outlet. For this reason, the primary flow cannot entrain the secondary flow normally with low speed and high pressure. With the growth of the inlet pressure of the primary flow, the flow at the nozzle outlet would be accelerated to supersonic state, and the entraining capacity of the primary flow is increasing, which lead to a growing entrainment ratio of the ejector. However, when the primary flow pressure is too high, although the entraining capacity is increasing, the primary and secondary flow will not mix completely, and the pressure growth process in the mixing-diffusing section could also be affected. It results in a decrease of the ejector entrainment ratio. On the other aspect, from the variation tendency in Fig. 4, among the three different ejector constructions, under the variable primary flow conditions, the entrainment ratio of the developed CRMC ejector changes gently. The thermodynamic parameters in this ejector model are more stable.

Set the background pressure as 203 kPa, the pressure and temperature of the primary flow as 590 kPa and 97 °C. When the pressure of the secondary flow varies from 84.46 to 204 kPa, with the identical CFD models, the performance parameters

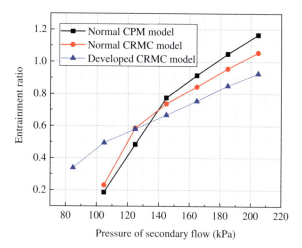

Fig. 6 Entrainment ratio of ejectors under different secondary flow pressures

of these ejectors are calculated, and the variation tendencies are depicted in Fig. 6. The tendencies in this figure show the increasing trend of the ejector entrainment ratios with the growth of secondary flow pressure. That is because with the sufficient primary pressure, a higher secondary flow pressure will lead to a large difference between the primary flow and secondary flow at the outlet of the Laval nozzle and a higher entraining ability of the primary flow in the suction section. Hence, the ejector entrainment ratio would increase. Comparing the performances of the three different construction ejectors, when the pressure of the secondary flow is 84.46 kPa, the entrainment phenomena in the normal CRMC ejector and normal CPM ejector will disappear. This is because at the outlet of the mixing section, the pressure of the primary flow is still higher than the pressure of the secondary flow, and the counterflow will occur in this section. However, since the better mixing effect in the developed CRMC ejector, this kind of ejector still functions normally.

With other boundary conditions constant, the ejector background pressure changes from 163 to 223 kPa, adopting the constructed numerical models, and the variation tendencies of three different ejectors can be shown as Fig. 7. When the ejector background pressure is lower, with increase of this value, the entrainment ratio is remaining constant basically, and when the background pressure grows to a certain value, the entrainment ratio decreases to zero rapidly. The reason for this tendency is that the critical background pressure exists in any kind of the ejectors. When the background pressure is lower than this value, the double chocking condition takes place in the Laval nozzle and mixing-diffusing section. Under this condition, the flow rate of the primary flow and the mixed flow will not change with the variation of the background pressure, and the entrainment ratio keeps constant. When the value of background pressure is higher than the critical value, the chocking condition in the mixing-diffusing section will disappear. In this situation, the secondary flow flowrate is affected badly by the value of background pressure, and therefore, the entrainment ratio is decreasing. With the constantly increase of

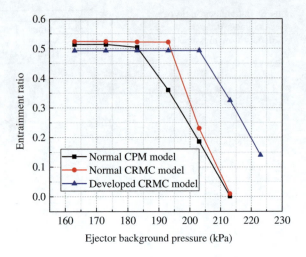

Fig. 7 Entrainment ratio of ejectors under different background pressures

the background pressure, the chocking condition in the Laval nozzle is also vanishing. The capacity of entrainment of the primary flow disappears, and the entrainment ratio of the ejector reduces to zero. Comparing these three ejectors, the critical background pressure of the developed CRMC model is higher than the others, and the decrease slope of the entrainment ratio is smaller, which indicates a more stable operating performance.

4 Conclusion

In this study, a developed CRMC ejector design model is proposed to enhance the operating performance of the ejector refrigeration system. The mathematical model featuring the developed CRMC model for the ejector component based on the properties of real working fluid and friction loss is constructed and verified. Different with the former CPM model for ejector design, the developed CRMC model results in an improved operating condition of the ejector without irreversibility normal shock. By calculation, the outlet pressure obtained from CRMC model is higher than that from normal CPM model by 20%. Furthermore, the variations of the designed entrainment ratio with the change of provided ejector working conditions are analyzed. At last, for the purpose to verify the correct and advantages of the CRMC model, the numerical modeling for operating condition with the specific construction CRMC ejector is also conducted. A more stable and more efficient performance is shown in the modeling results.

However, this work is theoretical, the normal shock in the ejector occurs inevitably when the fluid transfer from supersonic to subsonic, which would still weaken the ejector performance in the system, and the further study with practical experiment would be conducted in our successive research. The results and conclusions of this work can be a guidance for primary design and operating regulation for the SJHE system in the practical engineering.

Acknowledgements This work has been supported by "the Fundamental Research Funds for the Central Universities" (Grant No. HIT. NSRIF. 2017056).

References

1. Chen, J., Zhu, K., Huang, Y., et al.: Evaluation of the ejector refrigeration system with environmentally friendly working fluids from energy, conventional exergy and advanced exergy perspectives. Energy Convers. Manag. 1208–1224 (2017)
2. Chen, X., Omer, S., Worall, M., et al.: Recent developments in ejector refrigeration technologies. Renew. Sustain. Energy Rev. 629–651 (2013)
3. Aidoun, Z., Ouzzane, M.: The effect of operating conditions on the performance of a supersonic ejector for refrigeration. Int. J. Refrig. Rev. Int. Du Froid **27**(8), 974–984 (2004)
4. Diaconu, B.M.: Energy analysis of a solar-assisted ejector cycle air conditioning system with low temperature thermal energy storage. Renew. Energy **37**(1), 266–276 (2012)
5. Selvaraju, A., Mani, A.: Experimental investigation on R134a vapour ejector refrigeration system. Int. J. Refrig **29**(7), 1160–1166 (2006)
6. Sankarlal, T., Mani, A.: Experimental investigations on ejector refrigeration system with ammonia. Renew. Energy **32**(8), 1403–1413 (2007)
7. Besagni, G., Mereu, R., Inzoli, F., et al.: Ejector refrigeration: a comprehensive review. Renew. Sustain. Energy Rev. 373–407 (2016)
8. He, S., Li, Y., Wang, R.Z., et al.: Progress of mathematical modeling on ejectors. Renew. Sustain. Energy Rev. **13**(8), 1760–1780 (2009)
9. Pianthong, K., Seehanam, W., Behnia, M., et al.: Investigation and improvement of ejector refrigeration system using computational fluid dynamics technique. Energy Convers. Manag. **48**(9), 2556–2564 (2007)
10. Eames, I.W.: A new prescription for the design of supersonic jet-pumps: the constant rate of momentum change method. Appl. Therm. Eng. **22**(2), 121–131 (2002)
11. Mazzelli, F., Milazzo, A.: Performance analysis of a supersonic ejector cycle working with R245fa. Int. J. Refrig. Rev. Int. Du Froid. 79–92 (2015)
12. Khennich, M., Galanis, N., Sorin, M., et al.: Effects of design conditions and irreversibilities on the dimensions of ejectors in refrigeration systems. Appl. Energy. 1020–1031 (2016)

Working Fluid Selection and Thermodynamic Performance of the Steam Jet Large-Temperature-Drop Heat Exchange System

Chenghu Zhang, Yaping Li and Jianli Zhang

Abstract Steam jet large-temperature-drop heat exchange system (SJHE) is presented to improve the heat exchange effect between exothermic and endothermic media in the heat exchange process. In this paper, with the higher efficient CRMC ejector, the mathematical model of the SJHE system is constructed. We compared the application scopes of two system constructions with 10 different working fluids and select the more feasible system construction as the optimal system. Based on the real properties of 10 environment-friendly dry pure fluid, R141b is screened as the best cycle working fluid for the system, using which the exothermic medium can be reduced to 33.7 °C in the district heating situation (primary water supply temperature: 130 °C, and the temperature lift range of secondary water is 45–60 °C). Moreover, the thermodynamic characteristics of the SJHE system are evaluated to achieve a better understanding of the thermal performance of this new heat exchanger system.

Keywords Large-temperature-drop heat exchange system · System construction comparison · Working fluid selection · Thermodynamic evaluation

1 Introduction

For the purpose to control the air/water pollution and ensure environmental safety, as the thermal power and the nuclear plants are always set far away from the urban areas. It results in a large scale of heating networks [1]. Therefore, the long-distance

---

Y. Li (✉) · J. Zhang
Heilongjiang Provincial Key Laboratory of Building Energy Efficiency and Utilization, Harbin 150090, China
e-mail: 14B927067@hit.edu.cn

C. Zhang · Y. Li · J. Zhang
School of Architecture, Harbin Institute of Technology, Harbin 150090, China

C. Zhang
Key Laboratory of Cold Region Urban and Rural Human Settlement Environment Science and Technology, Ministry of Industry and Information Technology, Harbin 150090, China

heating supply technology has been applied. However, several problems are still existing [2]. One of them is the ineffectiveness due to the high thermal loss and pump work consumption during the heat media transportation [3]. To address this issue, solutions are presented. The large-temperature-drop heat supply technology is a meaningful and effective way to reduce the thermal dissipation and pump power in the heat transmission process [4]. Easily note that $T_{\text{pri,in}} > T_{\text{sec,out}}$ and $T_{\text{pri,out}} > T_{\text{sec,in}}$, which would cause system failure in the heat substation with normal heat exchangers according to the second law of thermodynamics.

Attempts are conducted to extend the heat exchange capacity in the heat network substation. The concepts of the absorption heat exchanger are proposed by Li and Fu [5], and a new type of the district heating system is designed to promote the heating capacity and energy efficiency of the cogeneration plants. Additionally, for development, Wang and Jiang [6] have optimized the construction design of AHE, and the multistage vertical absorption system is proposed by them. According to the former researches [7–9], any type of heat-driven heat pump cycle including absorption heat pump cycles, adsorption heat pump cycle, and the steam jet heat pump cycle shares the similar theoretical thermodynamic model as multi-heat-reservoirs refrigeration cycle. Hence, the large-temperature-drop heat exchanger also can be constructed by other kinds of heat-driven heat pump cycles. Steam jet heat pump cycle, or called ejector heat pump, is driven by thermal energy and is considered as an alternative to the absorption for its simple construction, low cost, long lifetime, flexible capacity, no-chemical corrosion, and chemical reaction [10].

However, in most of the research work, the working fluid in the ejector is seen as ideal gas and the friction loss between the fluid and the chamber wall is neglected. However, the characteristics of the working fluid would not only influence the ejector performance but also the heat exchange effect in each heat exchanger chamber in the whole system. A successful design for the novel thermal systems should consider the better system construction, better component efficiency, proper system working fluid, and higher system operating performance.

In our research, thermodynamic evaluation model of the steam jet large-temperature-drop heat exchange system is constructed, with this model and the thermos-physical parameters data, the system application scope with different system configurations and the working fluid are studied in the second section.

2 Methods

2.1 SJHE System Model

Basic system construction and operation process
Two basic constructions of the steam jet large-temperature-drop heat exchange system are shown in Figs. 1 and 2. With the operating process shown in these two figures, the mathematical models can be built.

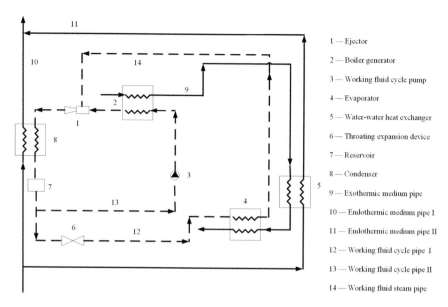

Fig. 1 Basic schematic diagram of steam jet high-temperature-drop heat exchange system I

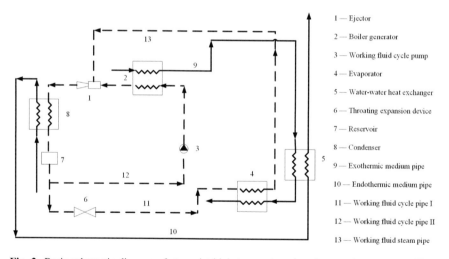

Fig. 2 Basic schematic diagram of steam jet high-temperature-drop heat exchange system II

System mathematical description

Before the construction of the model, the assumptions are made as follows:

- The system operates at steady state, and the flow is 1D. Heat leakage and pressure drop in the connecting pipelines are neglected.

- In the pipes of exothermic and endothermic media, the fluid inside is considered saturated.
- The working fluid outlet states of the boiler generator and evaporator are both superheated and saturated, and the state of the working fluid liquid leaving condenser is sub-cooled or saturated.

System conservation relation

For the heat exchange system II, the system energy conservation relation is shown in Eq. (1). For the heat exchange system I, the system energy conservation relation will be complete with Eqs. (2) and (3).

$$G_{exo} \cdot (h_{11} - h_{12}) = G_{endo} \cdot (h_{22} - h_{21}) \tag{1}$$

$$G_{endo,1} \cdot (h_{21'} - h_{21}) + G_{endo,2} \cdot (h_{21''} - h_{21}) = G_{endo} \cdot (h_{22} - h_{21}) \tag{2}$$

$$G_{endo} = G_{endo,1} + G_{endo,2} \tag{3}$$

Ejector

The process inside the ejector is the mixing process between the two flows with different pressures; hence, it is expressed in Eq. (4). The mass relationship between the two flows is written in Eq. (5).

$$h_{g,go} \cdot \dot{m}_p + h_{g,eo} \cdot \dot{m}_s = (\dot{m}_p + \dot{m}_s) \cdot h_{g,ci} \tag{4}$$

$$\omega \cdot \dot{m}_p = \dot{m}_s \tag{5}$$

Heat exchanger chambers

The energy balance relation in each heat exchanger chamber is shown in Eq. (6).

$$\sum h_{in,i} \cdot \dot{m}_{in,i} = \sum h_{out,j} \cdot \dot{m}_{out,j} \tag{6}$$

Throttling expansion device

Working fluid flowing through the throttling expansion device is an isenthalpic process; hence, it can be written in Eq. (7).

$$h_{l,ei} = h_{l,co} \tag{7}$$

3 Results and Discussion

3.1 System Application Scope and Working Fluid Selection

Different working fluids have different thermos-physics and chemical characteristics. In this section, based on the application scope, the working fluid selection of the large-temperature-drop heat exchange system will be conducted.

The comparing results are shown in Figs. 3 and 4. In these figures, the inlet temperature of exothermic medium $t_{11} = 130\ °C$, and the mass flow rate of the exothermic medium $G_{exo} = 1.0\ kg/s$; the inlet temperature of endothermic medium $t_{21} = 45\ °C$, and the temperature lift of the endothermic medium through the whole system is 15 °C. In the boiler generator, the temperature t_{g1} of the exothermic medium at the end of the boiler process is varying from 95 to 120 °C.

For the system type I shown in Fig. 3, with the increase of boiling temperature, the system outlet temperatures of the exothermic medium calculated with 10 working fluids both decreasing first and then increasing, and the minimum outlet temperatures exist. It results in an increasing first and then decreasing from variation tendency of the temperature lift capacity of the heat pump; therefore, under the constant mass flow rate, the outlet temperature of the exothermic medium decreases first and then increases. It also can be found that, in proper preconditions, the outlet temperature of exothermic medium t_{12} is lower than the system inlet temperature of the endothermic medium t_{21}; thus, the feasibility of the large-temperature-drop heat exchange process is verified. Furthermore, the temperature demand can be achieved by adopting system type I. Comparing the temperature variation tendencies with different working fluids, the minimum applicable temperature with R141b and R236fa is lower than the other fluids, reaching 33.7 and 33.6 °C, separately.

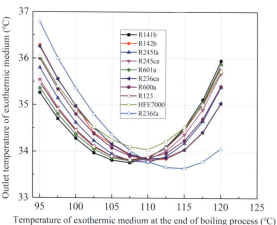

Fig. 3 Lowest outlet temperature variation tendency of the exothermic medium in JHES type I

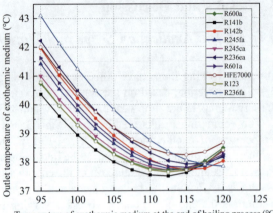

Fig. 4 Lowest outlet temperature variation tendency of the exothermic medium in JHES type II

Meanwhile, as to the system type II shown in Fig. 4, within the value range of the boiling temperature, the lowest outlet temperatures of exothermic medium are decreasing first and then growing slightly. This phenomenon indicates a narrow application scope of system type II. For this type of system, R141b is the most suitable working fluid with the lowest system outlet temperature of 37.5 °C.

By comparison, system type I is the better configuration of steam jet large-temperature-drop heat exchange system, and the R141b is selected as the optimal working fluid of the internal cycle in the steam jet large-temperature-drop heat exchange system.

3.2 Thermodynamic Performance Evaluation

For the purpose to evaluate the thermodynamic performance of the steam jet large-temperature-drop heat exchange system, serval evaluation indexes are presented.

(1) **System heat exchanged per unit area**

Therefore, in order to evaluate the heat exchange effects of different system types with unitive criteria, in this section, the system heat exchanged per unit area is adopted as expressed in Eq. (8):

$$q_{QA} = Q_{total}/A_{total} \tag{8}$$

(2) System heat exchanged distribution ratio

Defining the heat exchanged distribution ratio β_r as the ratio between the heats exchanged in heat pump part and the water–water heat exchanger part, it can be written in Eq. (9).

$$\beta_r = Q_{JHP}/Q_{tw} \tag{9}$$

The variations of the evaluating indexes in the SJHE system are obtained in Figs. 5–8.

Figure 5 illustrates that with the variable boiling temperature, the changing tendency of the heat exchanged per unit area of the steam jet large-temperature-drop heat exchange system. From this figure, under the identical exothermic medium temperature at the outlet of boiler generator, the values of system heat exchanged per unit area q_{QA} are growing continuously. The variation characteristics of heat exchanged distribution ratio between the heat pump and the water–water heat exchanger are shown in Fig. 6. From this figure, note that with the increase of t_g, the heat exchanged distribution ratio is increasing slowly. Moreover, when the temperature of $t_{w,go}$ is set at a higher level, the heat exchanged distribution ratio will also have a higher value. With proper preconditions, the value of β_r is larger than 1.

Figure 7 expresses the system heat exchanged per unit area variation tendency with the change of evaporating temperature under different condensing temperatures. With the increasing trend of evaporating temperature, under a specific condensing temperature, the demand heat transfer area for exchanging identical amount of heat is decreasing. It will result in the growth of the q_{QA} value. Hence, the higher

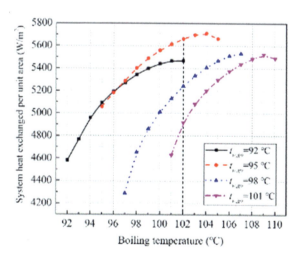

Fig. 5 Variation of the heat exchanged per unit area with the change of boiling temperature

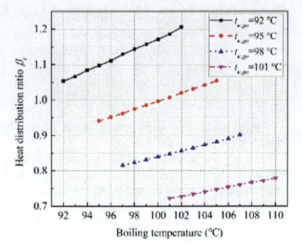

Fig. 6 Variation of the system heat exchanged distribution ratio with the change of boiling temperature

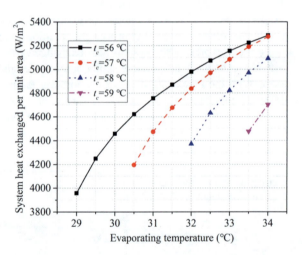

Fig. 7 Variation of the heat exchanged per unit area with the change of evaporating temperature under different condensing temperatures

condensing temperature leads to a narrower value range of the evaporating temperature. Figure 8 shows the proportion of the heat exchanged between the steam jet heat pump component and the water–water heat exchanger under different condensing temperatures. From this figure, note that with the identical heat released in the boiler generator, the increasing temperature in the evaporator will lead the entrainment ratio to increase.

Fig. 8 Variation of the system heat exchanged distribution ratio with the change of evaporating temperature under different condensing temperatures

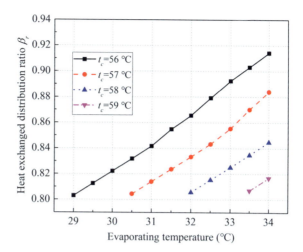

4 Conclusion

The main conclusion of this paper can be summarized as follows:

1. By analyzing the system application scopes with 10 pure working fluid with dry, safety, and environment-friendly properties at the boiling temperatures ranging from 92 to 117 °C, the system type with steam jet heat pump and water–water heat exchanger paralleling is found to have a wider applicable scope than the series connecting form. After comparison, R141b is chosen as the optimal working fluid for the SJHE system.
2. In the district heating situation, with the 130 °C primary water supply and secondary water temperature lift range of 45–60 °C, the primary return water can be reduced to 33.7 °C.
3. From the results, it can be concluded that increase of the boiling or evaporating temperatures and decrease of the condensing temperature would improve the system heat exchange effect. It also leads to the enhancement of the heat pump cycle performance and heat exchanged in the heat pump component relatively. However, with this tendency, the performance of the steam jet heat pump will be deviating the ideal operating situations.

The results and conclusions of this work can be a guidance for primary design and operating regulation for the SJHE system in the practical engineering.

Acknowledgements This work has been supported by "the Fundamental Research Funds for the Central Universities" (Grant No. HIT. NSRIF. 2017056).

References

1. Hirsch, P., Duzinkiewicz, K., Grochowski, M., et al.: Two-phase optimizing approach to design assessments of long distance heat transportation for CHP systems. Appl. Energy. 164–176 (2016)
2. Lin, P., Wang, R.Z., Xia, Z.Z., et al.: Experimental investigation on heat transportation over long distance by ammonia–water absorption cycle. Energy Convers. Manag. **50**(9), 2331–2339 (2009)
3. Danielewicz, J., Śniechowska, B., Sayegh, M.A., et al.: Three-dimensional numerical model of heat losses from district heating network pre-insulated pipes buried in the ground. Energy. 172–184 (2016)
4. Li, Y., Fu, L., Zhang, S., et al.: A new type of district heating method with co-generation based on absorption heat exchange (co-ah cycle). Energy Convers. Manag. **52**(2), 1200–1207 (2011)
5. Wang, S., Xie, X., Jiang, Y., et al.: Optimization design of the large temperature lift/drop multi-stage vertical absorption temperature transformer based on entransy dissipation method. Energy. 712–721 (2014)
6. Chen, J., Yan, Z.: Equivalent combined systems of three-heat-source heat pumps. J. Chem. Phys. **90**(9), 4951–4955 (1989)
7. Zhang, C., Li, Y. Thermodynamic performance of cycle combined large temperature drop heat exchange process: theoretical models and advanced process. Energy. 1–18 (2018)
8. Chen, J., Zhu, K., Huang, Y., et al.: Evaluation of the ejector refrigeration system with environmentally friendly working fluids from energy, conventional exergy and advanced exergy perspectives. Energy Convers. Manag. 1208–1224 (2017)
9. Sun, D.: Comparative study of the performance of an ejector refrigeration cycle operating with various refrigerants. Energy Convers. Manag. **40**(8), 873–884 (1999)
10. Diaconu, B.M.: Energy analysis of a solar-assisted ejector cycle air conditioning system with low temperature thermal energy storage. Renew. Energy **37**(1), 266–276 (2012)

Optimization Method of Reducing Return Water Temperature of Primary Heating Circuit

Haiyan Wang, Yanling Wang, Fang Wang, Xin Xu and Shuai Gao

Abstract How to effectively reduce the return water temperature and increase the temperature difference between supply and return water of the primary heating circuit has become the main concern for many heating companies in order to meet growing heating demands. This paper presents the optimized operation method of plate heat exchangers in the heat exchange station for district heating to improve heat supply capacity of the primary heating circuit. By analyzing operational problems of existing heat exchange stations during several heating seasons in a heating company, appropriate optimization objectives were put forward. Then a couple of basic optimization modules were proposed with theoretical analysis under different application situations. Moreover, schematic diagram of the whole optimization combination selection based on optimization modules within the heat exchange station was given. Reconstruction of several heat exchange stations was conducted and tested during 2016–2017 heating season according to the above theoretical research results.

Keywords District heating · Optimization module · Return water temperature of the primary heating circuit · Plate heat exchanger

1 Introduction

The main methods to reduce the return water temperature of primary circuit for district heating so far are as follows: (1) to increase the heat transfer area of the plate heat exchanger in heat exchange station; (2) to increase the heat dissipation area at

H. Wang · Y. Wang · F. Wang (✉)
Key Laboratory of Cold Region Urban and Rural Human Settlement Environment Science and Technology, Ministry of Industry and Information Technology, Harbin, China

School of Architecture, Harbin Institute of Technology, Harbin, China
e-mail: wfang2004@126.com

X. Xu · S. Gao
Harbin Investment Co., Ltd Heating Company, Harbin, China

© Springer Nature Singapore Pte Ltd. 2020
Z. Wang et al. (eds.), *Proceedings of the 11th International Symposium on Heating, Ventilation and Air Conditioning (ISHVAC 2019)*, Environmental Science and Engineering, https://doi.org/10.1007/978-981-13-9524-6_17

user side; (3) to adopt the absorption heat exchange unit. Although methods (1) and (2) can effectively reduce the return water temperature of the primary circuit, the increase of heat transfer area is costly, which is not linear with the decrease of return water temperature [1]. However, the effect of decreasing return water temperature will be less and less obvious with the increase of area. Method (3) needs large area of land and costs expensively. Therefore, methods (1) and (2) cannot be improved without restriction, and the cost and benefit of reducing the return water temperature should be considered comprehensively.

A large number of operation data of heat exchange stations in existing district heating system show that the actual operating parameters of the heat exchange station are different from the design condition with the large flow rate ratio and the low heat transfer coefficient of plate heat exchangers [2]. Therefore, according to the operation parameters of the heat exchanger unit, optimizing the plate exchanger's combination mode in parallel according to the local conditions, the problem of excessive primary and secondary side flow ratio of the plate exchanger can be alleviated to a certain extent, and the heat transfer performance of the plate exchanger can be improved [3]. Meanwhile, operation mode of plate heat exchangers can be optimized by using the combination of the plate heat exchangers in series or parallel connection in the heat exchange station to improve heat supply capacity of the primary circuit.

In this paper, several optimization modules of plate heat exchanger in the heat exchange station are put forward. The overall optimization scheme selection flow of heat exchange station is given, and the theoretical analysis and calculation of heat transfer optimization are carried out. Aiming at the specific heat exchange station, this paper provides the optimization scheme of heat transfer transformation in the aspects of equipment and operation of the heat exchange station so as to enhance the heat supply capacity of the primary circuit of the existing heating network.

2 Methods

The determination of the total return water temperature of primary heating circuit for each optimization objectives depends on the lower return water temperature of secondary circuit of heat exchanger in each heat exchange station. The temperature difference is better when it is smaller. We confirm the difference between the two temperatures is about 5 °C according to operation experience. Heat supply for the various users of the secondary circuit unchanged is necessary for optimization. Optimization of heat transfer calculation steps are as follows: Firstly, calculate the heat balance and total heat load of the exchanger in parallel before optimization on basis of operation data; secondly, calculate the supply return water temperature of the primary and secondary side of each optimization exchangers before and after optimization; thirdly, calculate the heat transfer area of the high-temperature and low-temperature heat exchangers after the optimization of the series or parallel connection; at last, consider whether to modify optimization result regarding the change of heat transfer coefficient and the existing equipment usability.

The optimization of plate heat exchanger in heat exchange station can be explained by three basic optimization modules. Basic calculation equations of heat exchanger are as follows:

$$Q = KF\Delta T_m = G_1\rho C_p(T_{1g} - T_{1h}) = G_2\rho C_p(T_{2g} - T_{2h}) \quad (1)$$

$$\Delta T_m = \frac{(\Delta T_g - \Delta T_h)}{\ln(\Delta T_g/\Delta T_h)} \quad (2)$$

$$\Delta T_g = T_{1g} - T_{2g} \quad \Delta T_h = T_{1h} - T_{2h} \quad (3)$$

where Q the heat transfer load of the heat exchanger, the input heat of the primary network side is equal to the output heat of the secondary network side (disregarding heat loss of heat exchanger), W, K the heat transfer coefficient of the heat exchanger, W/m² K, F the heat transfer area of heat exchanger, m², ΔT_m the logarithmic average water temperature difference of heat exchanger, °C, T_{1g} water supply temperature of primary side, °C, T_{2g} water supply temperature of secondary side, °C, T_{1h} the temperature of return water at primary side, °C, T_{2h} the temperature of return water at secondary side, °C, G_1 the volume flow at primary side of the heat exchanger, m³/h, G_2 the volume flow of secondary side of the heat exchanger, m³/h, ρ the density of hot water, kg/m³, C_p the constant pressure specific heat of hot water, J/kg K.

Assume that C_p and ρ does not change during the optimization. In the district heating system, there are many heat exchangers operating in the same heat exchange station. T_{1ag} is primary side supply water temperature of "a" heat exchanger, and T'_{2bh} is the secondary side return temperature of "b" heat exchanger after optimization.

2.1 Optimization Module 1-Single Exchanger in Parallel (Increase the Heat Transfer Area of the Plate Heat Exchanger)

It is assumed that the return water temperature can reduced to T'_{1h}; according to the optimization objective, the return water temperature and the flow rate of secondary side remain unchanged.

The heat transfer temperature difference after optimization

$$\Delta T_h' = T_{1h}' - T_{2h} \quad \Delta T_m' = \frac{(\Delta T_g - \Delta T_h')}{\ln(\Delta T_g/\Delta T_h')}$$

The change of heat transfer area for single heat exchanger

$$\Delta F = F' - F = (K \cdot \Delta T_m)/(K' \cdot \Delta T_m') \cdot F - F \qquad (4)$$

Primary circuit flowrate change

$$\Delta G = G_1 - G_1' = \frac{Q}{\rho C_p(T_{1g} - T_{1h})} - \frac{Q}{\rho C_p(T_{1g} - T_{1h}')} \qquad (5)$$

Therefore, in order to reduce the return water temperature of the primary circuit, the plate heat exchanger needs to change ΔF.

2.2 Optimization Module 2-Two Exchangers in Series

According to the inlet and outlet water temperature of the secondary side of the heat exchanger, the exchanger with low return water temperature in the secondary side is used as a low-temperature exchanger b, and the other is high-temperature exchanger "a" as shown in Fig. 1. Temperature for exchanger a:

$$\Delta T_{ag}' = T_{1ag} - T_{2ag} \quad \Delta T_{ah}' = T_{1ah}' - T_{2ah}$$
$$\Delta T_{am}' = \frac{(\Delta T_{ag} - \Delta T_{ah}')}{\ln(\Delta T_{ag}/\Delta T_{ah}')} \quad \Delta T_{bm}' = \frac{(\Delta T_{bg}' - \Delta T_{bh}')}{\ln(\Delta T_{bg}'/\Delta T_{bh}')}$$

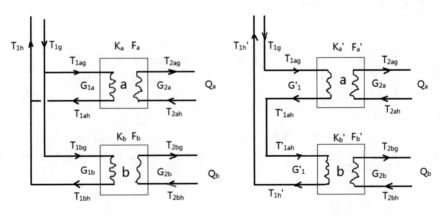

Fig. 1 Two exchangers in series

The change of heat transfer area for two heat exchangers

$$\Delta F = F' - F = F'_a + F'_b - F_a - F_b = \frac{Q_a}{K'_a \Delta T'_{am}} + \frac{Q_b}{K'_b \Delta T'_{bm}} - F_a - F_b \quad (6)$$

Primary circuit flowrate change

$$\Delta G = (G_{1a} + G_{1b}) - G'_1 = \frac{Q_a}{\rho C_p (T_{1ag} - T_{1ah})} + \frac{Q_b}{\rho C_p (T_{1bg} - T_{1bh})} - G'_1 \quad (7)$$

2.3 Optimization Module 3-Two Exchangers in Parallel then in Series with the Third One

According to the outlet water temperature of the secondary side of each heat exchanger, they are called high-temperature heat exchanger a, high-temperature heat exchanger c, and low-temperature heat exchanger b as shown in Fig. 2. Heat exchanger with lower outlet water temperature of the secondary side is called b, and heat exchanger with higher outlet water temperature of the secondary side is called c with better heat transfer coefficient. The premise is that the mixing primary side return water temperature of the exchangers a and c is lower than T_{1bg}, which is in series with the exchangers a and c after the optimization, and T'_{1h} is calculated.

If the T_{1ah} and T_{1ch} are both lower than T_{1bg}, the area of exchangers a and c should be reduced or the original heat transfer area should be kept unchanged. If the T_{1ah} and T_{1ch} are both higher than T_{1bg}, then exchanger c with lower return water temperature

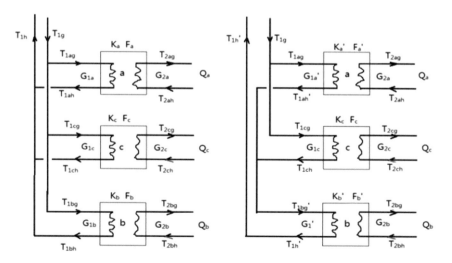

Fig. 2 Two exchangers in parallel and then in series with the third one

will remain unchanged, and the exchanger a will increase the heat transfer area and reduce the return water temperature. The purpose is to make the mixing water temperature for return water of exchangers a and c at the primary side equal to or lower than T'_{1bh}, required by the optimization supply water of exchangers b. If the T_{1ah} is higher and T_{1ch} is lower than T_{1bg}, the exchanger with low return water temperature will not change as exchanger c.

Temperature and flowrate
when

$$T'_{1bh} = T'_{1h} \quad G'_1 = \frac{Q}{\rho C_p(T_{1ag}-T'_{1h})}, \quad T'_{1bg} = T'_{1h} + \frac{Q_b}{\rho C_p G'_1}, \quad G'_{1c} = G_{1c} = \frac{Q_c}{\rho C_p(T_{1cg}-T_{1ch})}$$

The change of heat transfer area for three heat exchangers

$$\Delta F = F' - F = F'_a + F'_b + F'_c - F_a - F_b - F_c$$
$$= \frac{Q_a}{K'_a \Delta T'_{am}} - F_a + \frac{Q_b}{K'_b \Delta T'_{bm}} - F_b + \frac{Q_c}{K'_c \Delta T'_{cm}} - F_c \qquad (8)$$

Primary circuit flowrate change

$$\Delta G = (G_{1a} + G_{1b} + G_{1c}) - G'_1$$
$$= \frac{Q_a}{\rho c_p(T_{1ag} - T_{1ah})} + \frac{Q_b}{\rho c_p(T_{1bg} - T_{1bh})} + \frac{Q_c}{\rho c_p(T_{1cg} - T_{1ch})} + \frac{Q}{\rho C_p(T_{1ag} - T_{1h'})} \qquad (9)$$

2.4 Optimization of Plate Heat Exchangers Combination in Heat Exchange Station

The flowchart of the optimization scheme of heat exchange station is shown in Fig. 3. The relevant calculation programs needed for the theoretical calculation of optimization of heat exchanger can be compiled according to the optimization diagram, and the optimal optimization scheme can be determined on the basis of the overall optimization index β.

The overall optimization index β proposed here is a comprehensive evaluation index for the overall cost and income for optimization. The smaller the overall optimization index is, the change of the heat transfer area needed to reduce the per unit primary circuit flow, the better. In the process of optimization calculation, the variation of heat transfer coefficient ratio before and after optimization can be reasonably analyzed and set up, and the change value of overall heat transfer area after optimization of heat exchangers can be obtained. To find a combination, scheme is to obtain the minimum optimization index of heat transfer station $\beta = \Delta F/\Delta G$ for primary circuit flowrate reduction ΔG. On the basis of the theoretical calculation, the primary side series–parallel combination of heat exchangers in heat exchange station is carried out, and the related heat transfer area ΔF is

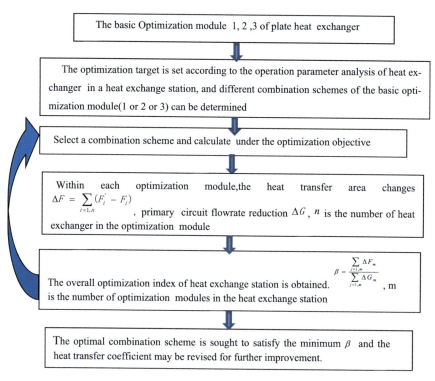

Fig. 3 Diagram of optimization combination scheme in heat exchange station

adjusted with integration and pipeline reconstruction. According to the overall optimization index β, the optimal transformation scheme of the heat exchange station is selected.

3 Results

There are four heat exchangers in the heat exchange station. Optimization combination schemes based on optimization modules within the heat exchange station were given. The results of optimization combination calculation of optimization module 1 and module 2 are as shown in Table 1. Although the optimization modules used in Schemes 1 and 2 are the same, the optimized heat exchangers are different.

The optimization scheme 1 is selected as the theoretical reconstruction scheme as the minimum optimization index. The area of the high-temperature heat exchanger a in practical transformation scheme remains unchanged in fact. The return water of exchanger a at primary side and primary supply water of heat exchange station are mixed to meet the heat demand at the entrance to exchanger b, thus controlling the supply water temperature of secondary side by adjust the

Table 1 Comparison of combination of the optimization modules

Optimization module combination scheme	Optimization module	ΔG (t/h)	ΔF (m^2)	β (m^2/t)
1	2	1.70	−0.30	−0.18
2	2	1.70	0.47	0.28
3	1	1.70	6.48	3.81

Table 2 Comparison of main operating data before and after optimization

Heat exchange station	Characteristics	Primary water supply temperature(°C)	Primary circuit flowrate (t/h)	ΔF (m^2)	Primary return water temperature (°C)
Operation mode 1	Original parallel	103.50	14.12	/	48.20
Operation mode 2	Practical optimization	102.06	12.96	5.46	38.33
Operation mode 3	Theoretical optimization	103.50	12.55	−0.30	39.50

mixing quantity to satisfy the secondary circuit heat demand. So ΔF is 5.6 m^2, not −0.30 m^2. Under the condition that the outdoor meteorological parameters and the water supply temperature of the primary side for the heat exchange station are close to each other, the original parallel operation mode (mode 1) and the actual optimized parallel operation mode (mode 2) are carried out based on the historical data and the field test data. The comparison and analysis of the heat exchange station parameters under the practical series operation mode (mode 2) and the theoretical optimization reconstruction mode (mode 3) were conducted and tested to verify the theoretical calculation results and whether the expected optimization objectives are achieved during 2016–2017 heating season according to the above theoretical research results (Table 2).

4 Discussion

The optimization scheme 1 is selected as the theoretical reconstruction scheme according to the optimization index. The optimization scheme 3 can reduce the return water temperature of the primary circuit in the heat exchanger station to a certain extent, and the method is relatively simple. However, the optimization scheme 3 needs more heat transfer area and reduces the overall heat transfer coefficient of the heat exchanger, which is not advisable for the heat exchanger with low heat transfer efficiency.

Plate heat exchangers in series or parallel optimization combination modules are adopted according to the operation parameters of heat exchange station and local conditions. The optimization scheme 1 and scheme 2, combination of different heat exchangers, will have different β. The problem of excessive primary and secondary side flow ratio can be alleviated to some extent; the heat transfer performance of heat exchanger can be improved; the optimization objectives of reducing the return water temperature of the primary circuit can be achieved; and the heating capacity can be promoted.

The transformation of the existing heat exchange station can adopt the optimization combination of the above three basic optimization modules and provide different combination schemes. Choose the best scheme according to the overall optimization index β. From the view of overall operation effect of heat exchange station, mode 2 and mode 3 can both achieve the optimization objectives by significantly reducing return water temperature of primary circuit, but they are different in essence: Mode 2 has more investment in reconstruction, but it has better user adjustment; mode 3 has less investment and better heat transfer performance, but the adjustment performance of the user is poor.

By comparing operation data before and after optimization, it may be concluded that the proposed optimization modules for plate heat exchanger station are feasible and profitable; the heat supply potentials of primary heating circuit have been improved greatly without much change with pipes and equipments. Theoretical analysis shows that heat transfer performance of existing plate heat exchanger can be improved to some extent.

References

1. Zhang, H.: Energy efficient design of heating system with plate heat exchangers. HV&AC **42**(4), 113–116 (2012)
2. Zheng, R., et al.: Analysis of relative heat transfer coefficient of plate heat exchangers under variable flow conditions. HV&AC **40**(10), 85–88 (2010)
3. Gao, X., et al.: Influence of fluid velocity on overall heat transfer coefficient of titanium plate heat exchanger. Gas Heat **34**(10), 13–15 (2014)

Simulation Study on Nominal Heat Extraction of Deep Borehole Heat Exchanger

Tiantian Du, Yi Man, Guoxin Jiang, Xinyu Zhang and Shuo Li

Abstract In order to study the actual heat extraction performance of non-interference heat transfer to deep geothermal energy, the nominal heat extraction of casing-type deep borehole heat exchanger was simulated by numerical simulation method. The results show the distribution of geological conditions in each soil layer under the location of the deep borehole pipe heat exchanger affects the nominal heat extraction of the heat exchanger. By adjusting the parameters of buried pipes and reasonably selecting the location of buried pipes and other optimization measures, the casing-type deep buried pipe heat exchanger can achieve considerable nominal heat extraction and realize continuous heating in winter together with the heat pump unit.

Keywords Deep borehole heat exchanger · Numerical simulation · Nominal heat extraction · Geological distribution of stratum layer

1 Introduction

Geothermal energy, as a clean and pollution-free new energy source, has been widely used at present. Geothermal energy includes shallow geothermal energy and deep geothermal energy [1, 2]. When using shallow geothermal energy, the shallow geothermal heat source heat pump system has the problem of soil thermal imbalance in some areas, resulting in the performance of the heat pump decaying year by year. In addition, deep geothermal energy mainly refers to the thermal energy resources contained in the strata below 200 m and below 3000 m underground [3]. U-tube heat exchanger and casing heat exchanger can be used when using deep

T. Du · Y. Man (✉) · G. Jiang · X. Zhang · S. Li
School of Thermal Engineering, Shandong Jianzhu University, Jinan 250101, China
e-mail: manyilaura@163.com

T. Du
e-mail: 18766110679@163.com

G. Jiang
e-mail: 724848723@qq.com

geothermal energy. Since the heat resistance of the casing-type ground heat exchanger is smaller than that of the U-tube heat exchanger and its heat exchange capacity per unit length is greater than that of the U-tube heat exchanger, the casing-type ground heat exchanger has a better heat exchange performance, and under the same heat exchange condition, the area occupied by the casing-type ground heat exchanger's buried tubes is smaller than that of the U-tube heat exchanger [4]. Therefore, in this paper, the heat exchange between solid rock and soil and circulating liquid inside the pipe is realized by using the casing-type underground heat exchanger. In order to analyze the exploitation capacity of the deep borehole heat exchanger to the deep geothermal energy, this paper calculates and analyzes the nominal heat extraction capacity of the deep buried heat exchanger under different physical parameters of multi-layer soil by means of numerical simulation.

2 Design Overview of Deep Borehole Heat Exchange

As shown in Fig. 1, the casing heat exchanger that can be used to collect deep geothermal energy flows in the inner and outer pipes with water flowing in and out. Backfill material is filled between the outer pipe and the rock and soil layer, ignoring the influence of groundwater flow. According to the geometric characteristics of the casing-type buried pipe heat exchanger, the computational grid division of the efficient simulation model of the deep borehole heat exchanger established by the finite difference method [5] is shown in Fig. 2.

3 Simulation of Nominal Heat Extraction of Underground Heat Exchanger

3.1 Definition of Nominal Heat Extraction

In order to facilitate the parameter analysis and engineering design of the deep borehole heat exchanger, this study uses "nominal heat extraction" to measure the

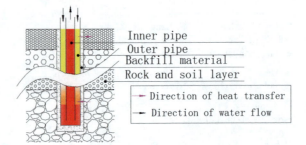

Fig. 1 Heat transfer in DBHE

Fig. 2 Schematic diagram of grid division in thermal numerical simulation

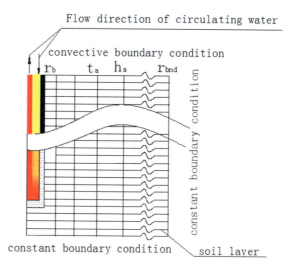

heat extraction performance of the deep borehole heat exchanger. Nominal heat extraction is defined as the maximum heat extraction that can be provided by the deep borehole heat exchanger under the following specific conditions [6]:

(1) Heat extraction is constant within three months (90 days).
(2) The inlet water temperature of the deep borehole heat exchanger system shall not be less than 5 °C during the heat extraction period.
(3) The initial temperature distribution in the rock and soil is determined, and the rock and soil layer is not disturbed by the deep borehole heat exchanger at the beginning of heat extraction.

3.2 Calculation of Nominal Heat Extraction

According to the numerical simulation model, the corresponding computer program can be compiled in Fortran language. After inputting the geological parameters (including earth heat flux, thermal conductivity, volumetric heat capacity, annual average atmospheric temperature, surface convection heat transfer coefficient, etc.), the geometric dimensions of casing-type deep borehole heat exchanger, the thermal properties of the buried pipe and backfilling materials, the calculation area and discretization settings, the circulating water flow, etc., the maximum heat quantity that can be extracted by casing-type deep borehole heat exchanger under the condition of multi-layer soil physical parameters can be obtained through design and calculation, that is, the nominal heat quantity.

The deep borehole heat exchanger is drilled deeply and usually crosses several soil layers with different physical properties. In order to analyze the influence of the geological structure distribution of soil layers on the nominal heat extraction of heat exchangers, this paper takes the buried pipe heat exchangers buried in the soil consisting of four layers of soil layers with different thermal conductivity and depth of 500 m as an example to simulate, and the physical parameters. The thermal conductivities of the four layers of soil are 0.98 W/m K, 1.7865 W/m K, 1.565 W/m K, and 3.585 W/m K, respectively.

In this study, the geometric dimensions of the casing-type deep borehole heat exchanger are: the depth of drilling and buried is 2000 m, the diameter of drilling hole is 326.55 mm, the outer diameter of the heat pipe exchanger is 244.5 mm, the inner diameter is 228.66 mm, the outer diameter of the inner pipe is 110 mm, and the inner diameter of the inner pipe is 85.4 mm.

The physical properties of the ground heat exchanger studied in this paper are: the thermal conductivity of backfill material is 1.5 W/(mK), the volume specific heat capacity of backfill material is 2.2×106 J/(m^3K), the thermal conductivity of outer tube is 60.5 W/(mK), and the volume specific heat capacity of outer tube is 3.4×106 J/(m^3K). The thermal conductivity of the inner tube is 0.41 W/(mK), and the volume specific heat capacity of the inner tube is 1.2×106 J/(m^3K).

The circulating medium in the ground heat exchanger is water, the circulating water flow is 5 kg/s, the surface convection heat transfer coefficient is 15 W/(m^2K), and the annual average temperature of the local atmosphere is 12.2 °C.

4 Results

Under winter conditions, it was found in this study that the thickness of each stratum layer and its position in the depth direction had a great influence on the nominal heat extraction of the heat exchanger when the heat exchanger in the deep borehole heat exchanger appeared in the soil composed of various stratum layers. Therefore, this paper takes the above-mentioned soil composed of four stratum layers as an example and calculates the nominal heat extraction of the ground heat exchanger by changing the different depths of each layer of soil, and the calculation results are shown in Table 1.

As given in Table 1, if the other conditions of the deep borehole heat exchanger are the same, even if the entire stratum where the heat exchanger is located has the same average thermal conductivity, the nominal heat extraction of the obtained deep borehole heat exchanger is also different due to the different depth conditions of soil distribution with different physical properties.

For the above-mentioned deep borehole heat exchanger buried in the soil consisting of four layers of soil layers with a thickness of 500 m, respectively, the calculated nominal heat extraction quantity changes with the thermal conductivity of each layer under the assumption that only the thermal conductivity of one stratum layer is changed and the thermal conductivities of the other stratum layers are unchanged, as shown in Fig. 3.

Table 1 Nominal heat extraction from heat exchangers with different soil layers at different depths

	Thermal conductivity of layer N formation (W/m K)					
Layer 1	0.98	0.98	0.98	0.98	0.98	0.98
Layer 2	1.7865	1.7865	1.565	1.565	3.585	3.585
Layer 3	1.565	3.585	1.7865	3.585	1.7865	1.565
Layer 4	3.585	1.565	3.585	1.7865	1.565	1.7865
Nominal heat extraction Q (KW)	183.3008	167.3340	185.6445	172.9981	155.6641	158.5449

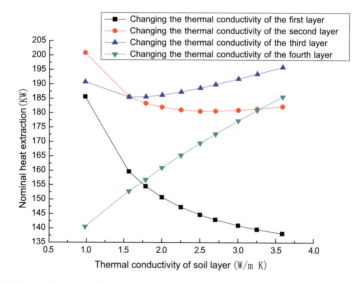

Fig. 3 Variation of nominal heat extraction with thermal conductivity of each soil layer

As shown in Fig. 3, when the thermal conductivity of the first stratum layer closest to the earth's surface gradually increases, the nominal heat extraction of the deep borehole heat exchanger gradually decreases, but the decreasing trend will be slower and slower. This is because the lower the temperature of the first stratum layer, the greater its thermal conductivity, and the greater the heat transfer from circulating water in the outer tube of the deep borehole heat exchanger to the first stratum layer, so the nominal heat transfer decreases with the increase in the thermal conductivity of the shallow stratum layer. The heat transfer resistance between circulating water and soil layer is mainly affected by the thermal conductivity of backfill material in the borehole, in addition to the thermal conductivity of the stratum layer. Therefore, when the thermal conductivity of the first stratum layer continues to increase, the thermal conductivity of backfill material will have an increasing impact on the thermal resistance, which slows down the increase in the heat dissipation of circulating water to the surrounding stratum layer, showing that the nominal heat extraction will decrease with the increase in the thermal conductivity.

When the thermal conductivity of the second and third stratum layers is changed, the nominal heat extraction of the ground heat exchanger will decrease at first. When the thermal conductivity of the second stratum layer is 3 W/(mK), the nominal heat extraction will increase with the increase in the thermal conductivity, and when the thermal conductivity of the third stratum layer is about 1.8 W/(mK), the nominal heat extraction will increase with the increase in the thermal conductivity. This is because when the thermal conductivity of the second and third stratum layers is small, increasing the thermal conductivity will enhance the heat transfer of the water inside the outer pipe to the stratum layer with lower ambient temperature, so the nominal heat extraction will decrease with the increase in the thermal conductivity. When the thermal conductivity increases to a certain extent, the thermal conductivity of the stratum layer increases with the constant heat flow of the earth, the vertical temperature difference in the stratum layer decreases, and the average temperature of the surrounding stratum layer increases. Therefore, the heat transfer from the water in the outer pipe to the surrounding stratum layer gradually decreases. When the temperature of the surrounding stratum layer is higher than the circulating water temperature in the outer pipe, the circulating water will get heat from the surrounding soil layer, and the nominal heat transfer from the heat exchanger will increase with the increase in the thermal conductivity of the stratum layer.

When the thermal conductivity of the fourth stratum layer is changed, the nominal heat extraction of the buried heat exchanger in the deep stratum layer increases with the increase in the thermal conductivity. The burial depth of the fourth stratum layer of soil stratum layer is the largest, so the temperature is the highest and the heat exchange intensity with circulating water inside the outer tube of the heat exchanger is the largest. At the same time, it can be seen that due to the large temperature difference between the stratum layer and the circulating water, the change of thermal conductivity of the shallowest and deepest stratum layer l has a greater impact on the nominal heat extraction of the heat exchanger.

5 Conclusion

Through the numerical simulation, it is found that under the heating condition, the distribution of geological conditions in each stratum layer under the location of the deep borehole heat exchanger also affects the nominal heat extraction of the heat exchanger: the smaller the thermal conductivity of the shallow stratum layer, the greater its nominal heat extraction; the greater the thermal conductivity of the deep stratum layer, the greater the nominal heat extraction. When selecting the drilling site for the deep borehole heat exchanger, the influence of geological distribution of different soil layers on its nominal heat extraction should be fully considered. For the shallow stratum layer with higher thermal conductivity and lower temperature, backfill material with lower thermal conductivity or heat preservation of the outer pipe of the shallow pipe well can be adopted to reduce the heat dissipation of circulating water in the outer pipe to the surrounding soil, thus effectively increasing the nominal heat extraction of the borehole heat exchanger. If there is a low-temperature aquifer in the shallow stratum layer, corresponding well sealing measures should be taken to avoid convection heat dissipation from circulating water inside the outer pipe of the heat exchanger to groundwater. By adjusting the parameters of buried pipes and reasonably selecting the location of buried pipes and other optimization measures, buried in common soil geological conditions, the 2000-m-deep casing-type buried pipe heat exchanger can take up about 150 KW of nominal heat and can provide continuous heating in winter with a heating area of about 3500 m^2.

Acknowledgements The work presented in this paper is supported jointly by a grant from National Natural Science Foundation of China (No. 51808321) and a grant from Shandong Province Natural Science Foundation (No. BS2015NJ016).

References

1. Ma, H.Q., Long, W.D.: Heat balance of ground source heat pump system. HVAC **39**(1), 102–106 (2009)
2. Spider, J.D., Gehlin, S.E.A.: Thermal response testing for ground source heat pump systems-an historical review. Renew. Sustain. Energy Rev. **50,** 1125–1137 (2015)
3. Fang, C.H., Su, F.R.: Technical Progress and development countermeasures of geothermal exploration and development in middle and deep layers. J. Shandong Inst. Commer. Technol. **17**(1), 89–93 (2017)
4. Hu, Y.N., Li, C.H.: Experimental study on heat transfer performance of casing ground heat exchanger. HVAC **41**(9), 100–105 (2011)
5. Wang, D., Hu, S., Gao, Z., et al.: Parameter analysis of the performance of medium and deep casing ground heat exchangers. Dist. Heat. **3,** 1–7 (2018)
6. Fang, L., Diao, N.R., Shao, Z.K.: A computationally efficient numerical model for heat transfer simulation of deep borehole heat exchangers. Energy Build. 167, 79–88 (2018)

Numerical Study of Data Center Composite Cooling System

Xuetao Zhou, Xiaolei Yuan, Jinxiang Liu, Risto Kosonen and Xinjie Xu

Abstract During recent years, the energy consumption of data centers (DCs) has doubled every four years, and DCs accounted for about 3% of the global electricity consumption in 2010. In order to reduce the specific energy consumption, methods for improving the thermal performance of data center have been proposed. In this paper, a composite cooling system combining a heat pipe technology and overhead floor return air technology was introduced. In this study, two novel concepts (i.e., no server upper heat pipe exchanger and server upper heat pipe exchanger) were developed and compared using the Airpak 3.0 software packages. Simulation results show that the new system proposed can effectively improve the thermal environment in high-density data center. Simultaneously, the energy consumption of air conditioning and operating cost of the data center can be significantly reduced. Racks had more uniform temperature distribution, and the number of rack hot spots has been significantly reduced.

Keywords Data center · Heat pipe · Composite cooling system · Energy conservation · Local hot spots

1 Introduction

With the advent of the information age and the rapid development of emerging industries such as the Internet of Things, data has exploded. As the most important infrastructure supporting the rapid development of cloud computing and big data industries, the value of data centers (DCs) is self-evident [1]. DCs are facilities used to house computer systems and related components. It typically includes redundant

---

X. Zhou · X. Yuan · J. Liu (✉) · R. Kosonen · X. Xu
College of Urban Construction, Nanjing Tech University, Nanjing, China
e-mail: jxliu@njtech.edu.cn
URL: https://mail.njtech.edu.cn

R. Kosonen
Department of Mechanical Engineering, Aalto University, Espoo, Finland

or backup components and infrastructure for power supply, data communications connections, environmental controls (e.g., air conditioning, fire suppression), and various security devices [2]. A large data center is an industrial-scale operation using as much electricity as a small town [3]. In recent years, due to the increased integration and operation speed of electronic computer products, the heat dissipation per unit area has increased significantly in the operation and maintenance process of the data center. Meanwhile, the specific heat load of the equipment room has increased significantly, and the energy consumption problems of data centers are increasingly prominent [4]. The number of DCs has increased rapidly from virtually nothing ten years ago to consuming about 3% of the global electricity supply and accounting for about 2% of total greenhouse gas emissions. The global energy data center consumption was estimated 416 TWh per year, which was significantly higher than the UK's total consumption, the UK of about 300 TWh [5].

The data center must be cooled throughout the year to maintain the proper temperature of the IT equipment. Figure 1 shows the energy consumption breakdown in a traditional data center cooled by computer room air conditioning (CRAC) system. Cooling energy consumption typically accounts for 40% of the total energy consumption in a data center. According to a survey conducted by research firm ResearchandMarkets.com [2], the global DC cooling market is expected to exceed $8 billion by 2023, with a compound annual growth rate of approximately 6% from 2017 to 2023. In addition, 30–40% of the energy consumption of traditional data centers is used to cool racks and servers [6]. To ensure long-term reliable operation of IT equipment and the efficiency of IT equipment, it is necessary to use effective cooling technology to improve the thermal environment within the data center.

Some high-efficiency cooling technologies are provided for DC energy conservation. For example, Yuan et al. [7] proposed to set a flexible baffle (FB) on the front door of the rack, which is an innovative optimization method of the airflow distribution that makes the air temperature over the racks more uniform and improves the thermal environment of the data center. In addition, the method of

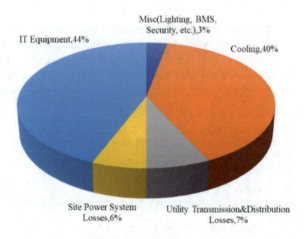

Fig. 1 Example of a data center energy breakdown

using liquid cooling [8] can solve the problem of local hot spots efficiently, but as the data center needs to operate in a highly security environment and water leakage may be dangerous, there are challenges to utilize liquid cooling technology.

In this paper, a new composite cooling system that combines heat pipe technology and overhead floor return air technology was introduced to effectively remove heat and further to minimize local hot spots within the IT server. This method can greatly improve the performance of high heat density data centers.

2 Methods

2.1 Heat Pipe Technology

Heat pipe technology, as an effective heat transfer technology, has been widely applied to electronic and electrical equipment with high specific heat flux cases, providing efficient and reliable heat dissipation. The use of efficient heat pipe technology in the data center cooling system provides a feasible method to reduce required cooling power and energy consumption of the data center [9].

As a working principle, the heat pipe exchanger uses the phase change of the internal working fluid to transfer heat and has higher heat transfer efficiency than the single-phase heat transfer of the heat transfer heat sink member. The operating principle of the heat pipe is shown in Fig. 2. In this paper, the evaporation section of the heat pipe system is installed above the servers and the condensation section is connected with the outside environment, so the excess heat from the server is transmitted to the atmosphere.

2.2 Analyzed Data Center

This study analyzes the actual running data center in Nanjing, China. The specifications of the physical model and the distribution of the rack are provided based on the case study data center. Figure 3 shows the layout of the data center. The size

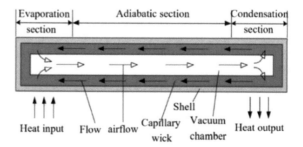

Fig. 2 Operating principle of heat pipe

Fig. 3 Data center layout

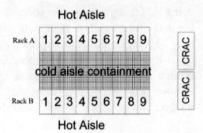

of the data center is 11.0 m (L) × 8 m (W) × 4 m (H). The data center has altogether 18 server racks, which divided into two rows (rack A and rack B) with 27 three-row perforated tiles (0.6 × 0.6 m^2 with 45% open area) in the middle. Two room air conditioning (CRAC) units located on one side of the data center are used to supply cooling air. The cold air supplied by CRAC enters the data center through the floor vents and reaches the underfloor ventilation system. The data center contains a 0.45 m raised floor plenum. The air supplied rises to the cold aisle containment between rack A and rack B and further distributed to the servers in the racks. The cold air is heated by the IT loads of the servers and then enters the hot aisle behind the two rows of racks. Finally, hot air returns to the CRAC through the hot aisle.

Higher temperature of the rack hot spots and more inefficient airflow distribution in the data center, the greater possibility of damage will be caused to the servers. By reducing rack hot spot temperatures, it helps to optimize operating conditions and reduces energy consumption in the CRACs. According to the information provided by the data center technicians, the power density of rack A is much larger than that of rack B. The rack A7 had high power density whose figures both exceeded 10 kW per rack, which is much larger than heat load of other racks. Therefore, this study focuses on thermal environment analysis and heat dissipation optimization for the rack A7.

2.3 Model of Data Center

In this study, a numerical model is established in the CFD software Airpak 3.0. This model is built based on the actual geometry of the data center (Table 1). In Fig. 4, there is presented the three-dimensional model of the data center. The initial supply air temperature and air speed for the simulated CRAC were, respectively, set as 22 °C and 5.33 m/s. The power consumption of each rack is also set according to the actual value, and the airflow organization of the data center in the simulation is also the same as the actual one. The air supply of the CRACs and the total airflow required by the server fans to operate were equal. Besides one, the effects of low wave radiation were ignored.

Table 1 Dimension of the data center

Items	Length/m	Width/m	Height/m
Cold aisle containment	5.4	1.8	2.2
Perforated tiles	0.6	0.6	N/A
CRACs	1.8	0.8	2.25
Rack	2.2	0.6	1.2

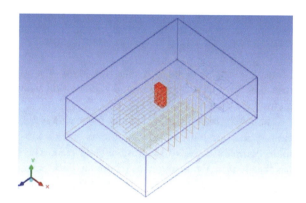

Fig. 4 Three-dimensional (3D) model of an analyzed data center

The challenges on turbulence modeling and numerical errors also require serious consideration in CFD accuracy. The correct governing equations and boundary conditions and appropriate numerical algorithms must be carefully selected [10].

2.4 Governing Equations and Turbulence Model

According to Song [11] and Abanto et al. [12], considering data center size, server rack, and flow conditions, turbulent mixed convection is the flow state of the data center airflow model. The governing equations of fluid flow represent mathematical statements of the conservation laws of physics. The airflow considered in this study has very low velocity, and the change in pressure is very small as well. Therefore, the flow is assumed to be incompressible. The equations for momentum, energy, and continuity conservation can be written as follows:

Momentum equation:

$$\frac{\partial}{\partial t}(\vec{u}) + \vec{u} \cdot \nabla \vec{u} = -\nabla p + \nabla \cdot \left(\bar{\bar{T}}\right) + \vec{g} \qquad (1)$$

where p is the static pressure, T is the stress tensor, and $\vec{g}$ is the gravitational body force.

The stress tensor T is given by:

$$\bar{\bar{T}} = v_{\text{eff}} \cdot \nabla u \tag{2}$$

Energy conservation equation:

$$\frac{\partial}{\partial t}(h) + \nabla \cdot (h\vec{u}) = \nabla \cdot [k_{\text{eff}} \nabla T] + S \tag{3}$$

Continuity equation:

$$\nabla \cdot \vec{u} = 0 \tag{4}$$

where v_{eff} and k_{eff} are the effective fluid viscosity and thermal conductivity, respectively; $\vec{g}$ is the gravitational acceleration vector, and S represents the volumetric heat generation.

The k-ε turbulence model is the most widely used model because of its applicability to wide-ranging flow problems and its lower computational demand than more complex models that are available [13]. The standard k-ε turbulence model has been used by many researchers in data center calculation models and has achieved good results [11, 14–19]. In this paper, the k-ε two-equation model is used to solve the three-dimensional turbulent flow and heat transfer problems. The model requires just a few assumptions and has broad applicability.

2.5 Boundary Conditions Setting

(1) Assuming that the wall of the data center is a steady-state isothermal boundary.
(2) The open area of perforated tile is 45%.
(3) The supply air temperature is 22 °C.
(4) The open area of the rack is 65%, and the excess heat of a single rack is 2 kW.
(5) The convergence criteria were set as follows: The residuals for both the velocity in the directions of X, Y, and Z and the continuity reached 10^{-3}, and that for energy was set at 10^{-6} [15, 16].

2.6 Grid Generation

Grid generation is a key factor to the success of CFD calculations [15]. The grid should meet the following criteria: The change in density in the grid domain should be based on the variables, and the change in elements in the entire solution domain should be smooth [20]. The unstructured grid can eliminate the structural constraints of the nodes in the structural mesh and deal with the boundaries better.

Therefore, the mesh type was Hexa unstructured and the computational domain included about 1.2 million meshes.

3 Results and Discussion

In this section, simulation results of the analyzed cases for data center are presented. The simulation was divided into two cases (case 1: no heat pipe exchanger, case 2: server upper heat pipe exchanger). The air temperature of the data center in each scenario was simulated and the influence of the heat pipe exchanger on the temperature distribution was analyzed and discussed.

The influence of airflow rate and airspeed has a significant effect on the temperature distribution inside racks. To analyze the distribution of air temperature field, the plane parallel to the Z-axis is intercepted in two cases, and the air temperature profiles are, respectively, shown in Fig. 5. It shows clearly that the temperature on the backside of the server racks is highest, and there is a large amount of excess heat accumulation at the back of the rack A7. In addition, the temperatures of the upper part of the servers are higher than the temperatures of the lower part of the servers.

Figure 6a shows that the rack A7 had not uniform temperature distribution with a maximum temperature difference of 9.8 °C. Besides, the highest temperatures of the rack A7 without heat pipe exchangers were approximately 37.5 °C and the average temperature on the back of the rack is 23.6 °C. Figures 5 (case 2) and 6b show the improvement of the egress air temperature distribution by applying heat pipe exchangers. In each case, the temperatures of nine measuring points for racks A7 were used to describe the exhaust temperature distribution of the corresponding racks. Figure 7 shows a comparison of the temperatures at nine simulation points in the rack A7 without and with the heat pipe exchangers. With the heat pipe exchangers, the temperatures at each location decreased and the maximum temperature drop of the rack hot spot was about 5 °C. Without the heat pipe exchanger, the temperature of each simulated locations is slightly higher than with the heat

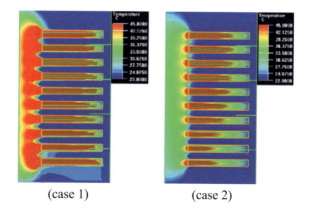

Fig. 5 Temperature distributions of the rack A7 without (case 1) and with (case 2) heat pipe (the plane parallel to the Z-axis)

Fig. 6 Egress air temperature of the rack A7 without (case 1) and with (case 2) heat pipe

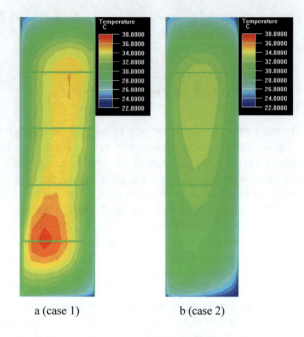

a (case 1)　　　　　　b (case 2)

Fig. 7 Temperatures at nine points in the rack A7 without and with heat pipe exchangers

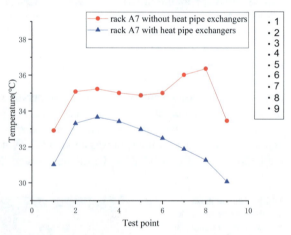

pipe. The rack hot spots with the heat pipe exchangers reduced significantly. Besides, the overall temperature on the back of the cabinet also dropped to 23.1 °C, the rack hot spots were weakened effectively, and the thermal environment in rack A7 was improved significantly.

4 Conclusions

In this paper, the concept of heat pipe exchanger for a DC cooling is numerically simulated and compared with the solution without heat pipe. In the analysis, airflow and temperature distributions are analyzed with the previous concepts and the following conclusions are drawn:

(1) The temperature on the backside of the server racks is highest, and there is a large amount of excess heat accumulation at the back of the rack. The temperatures of the upper part of the servers are higher than that of the lower part of the servers.
(2) The cooling coil of the heat pipe exchanger is closer to the heat source, directly transfers the heat from the local hot spots to the outside environment.
(3) The heat pipe exchangers have effectively improved the thermal environment of the DC; rack A7 had more uniform temperature distribution, and the number of rack hot spots has been significantly reduced.
(4) With the heat pipe exchangers, the temperatures at each location decreased and the maximum temperature drop of the rack hot spot was about 5 °C. At the same time, this composite cooling system ensures the safe and stable operation of the equipment in the data center.

References

1. Zhang, Q., Cheng, L., Boutaba, R.: Cloud computing: state-of-the-art and research challenges. J. Internet Serv. Appl. **1**, 7–18 (2010)
2. Nadjahi, C., Louahlia, H., Lemasson, S.: A review of thermal management and innovative cooling strategies for data center. Sustain. Comput.: Inform. Syst. **19**, 14–28 (2018)
3. Mittal, S.: Power management techniques for data centers: a survey, arXiv preprint arXiv 1404.6681 (2014)
4. Rahman, M.N., Esmailpour, A.: A hybrid data center architecture for big data. Big Data Res. **3**, 29–40 (2016)
5. Dayarathna, M., Wen, Y., Fan, R.: Data center energy consumption modeling: a survey. IEEE Commun. Surv. Tutor. **18**, 732–794 (2016)
6. Ham, S.W., Park, J.S., Jeong, J.W.: Optimum supply air temperature ranges of various air-side economizers in a modular data center. Appl. Therm. Eng. **77**, 163–179 (2014)
7. Yuan, X.L., Wang, Y., Liu, J.X., Xu, X.J., Yuan, X.H.: Experimental and numerical study of airflow distribution optimisation in high-density data centre with flexible baffles. Build. Environ. **140**, 128–139 (2018)
8. Ouchi, M., Abe, Y., Fukagaya, M.: New thermal management systems for data centers. J. Therm. Sci. Eng. Appl. **4**, 031005 (2012)
9. Singh, R., Mochizuki, M., Mashiko, K., Nguyen, T.: Heat pipe based cold energy storage systems for datacenter energy conservation. Energy **36**, 2802–2811 (2011)
10. Li, Y., Nielsen, P.V.: CFD and ventilation research. Indoor Air **21**, 442–453 (2011)
11. Song, Z.: Thermal performance of a contained data center with fan-assisted perforations. Appl. Therm. Eng. **102**, 1175–1184 (2016)

12. Abanto, J., Barrero, D., Reggio, M., Ozell, B.: Airflow modelling in a computer room. Build. Environ. **39**, 1393–1402 (2004)
13. Tennekes, H., Lumley, J.L., Lumley, J.L.: A first course in turbulence, MIT Press (1972)
14. Almoli, A., Thompson, A., Kapur, N., Summers, J., Thompson, H., Hannah, G.: Computational fluid dynamic investigation of liquid rack cooling in data centers. Appl. Energy **89**, 150–155 (2011)
15. Ni, J., Jin, B.W., Zhang, B., Wang, X.W.: Simulation of thermal distribution and airflow for efficient energy consumption in a small data centers. Sustainability **9**, 664 (2017)
16. Alkharabsheh, S., Fernandes, J., Gebrehiwot, B.: A brief overview of recent developments in thermal management in data centers. J. Electron. Packag. Trans. Asme. **137** (2015)
17. Cruz, E., Joshi, Y.: Coupled inviscid-viscous solution method for bounded domains: application to data-center thermal management. Int. J. Heat Mass Transf. **85**, 181–194 (2015)
18. Bazdidi-Tehrani, F., Shahmir, A., Haghparast-Kashani, A., Agonafer, D., Ghose, K., Joshi, J., Sammakia, B.: Numerical analysis of a single row of coolant jets injected into a heated crossflow. J. Comput. Appl. Math. **168**, 53–63 (2004)
19. Wang, S.J., Mujumdar, A.: Flow and mixing characteristics of multiple and multi-set opposing jets. Chem. Eng. Process. **46**, 703–712 (2007)
20. Zhang, K., Zhang, X.S., Li, S.H., Wang, G.: Numerical study on the thermal environment of UFAD system with solar chimney for the data center. Energy Procedia **48**, 1047–1054 (2014)

Influence of Air Supply Outlet on Displacement Ventilation System for Relic Preservation Area in Archaeology Museum

Juan Li, Xilian Luo, Bin Chang and Zhaolin Gu

Abstract Displacement ventilation (DV) system is an energy-saving strategy for buildings with large-space layout. However, the application of this system to the relic preservation area in archaeology museum is facing challenges since there no internal heat sources existed in the occupied area. The primary factors that affect the thermal stratification would be the arrangement of the supply air outlet and supplied air parameters. In this research, three experimental tests were carried out to evaluate the influence of the supply air outlet on the system performance. The experimental results show that the vertical distribution of the supply air outlet has a negligible effect on the temperature and relative humidity (RH) environment parameters in the inner area of the burial pit, but has a great influence on the upper region. That is to say, for the area with slightly higher cultural relic, the location of the supply air outlet should not be too low.

Keywords Relic preservation · Archaeology museum · Displacement ventilation system · Air supply outlet

1 Information

Displacement ventilation is a kind of ventilation mode which relies on the thermal buoyancy generated by indoor heat sources as a power to realize indoor air ventilation. Its special thermal stratification structure divides the room into the lower clean area and the upper dirty area, which could effectively improve the air quality of the occupied area and save nearly 33% energy consumption compared with the traditional mixed ventilation system [1]. The thermal stratification characteristics of

---

J. Li · X. Luo (✉) · B. Chang · Z. Gu
School of Human Settlements and Civil Engineering, Xi'an Jiaotong University, Xi'an, China

Shaanxi Environmental and Building Energy Conservation Engineering Technology Research Center, Xi'an Jiaotong University, Xi'an, China
e-mail: xlluo@mail.xjtu.edu.cn

Table 1 Summary of the test cases

Cases	The operational model	Time
1	Full opening of the supply air outlet	2016/07/31–2016/08/01
2	Lower half of the supply air outlet	2016/07/20–2016/07/21
3	Upper half of the supply air outlet	2016/07/23–2016/07/24

the DV system are affected by many aspects, such as characteristics of the heat source, the arrangement of the air return outlet, and indoor obstacles, among which the heat sources are considered as the most important factors. Many researches have been conducted to evaluate the relationship between the features of the thermal stratification features and the temperature, jet velocity, dispersion features and fixed height of the heat source [2, 3].

DV system has been widely used in various buildings. In our previous research, it has been used to control the preservation area for relic in archaeology museum, a stable preservation environment was achieved when the air exchange rate per hour reaches to 11.9 [4]. However, the thermal stratification features of the DV system for relic area should be considered in a differential way since there is no internal heat source at the preservation area and occupied area is opened without ceiling. While all existing theoretical models for DV system are based on the precondition of internal heat source and closed room [5, 6]. The stratification in the room with DV system mainly depends on driving force. In the case when buoyant force is inadequate, the inertia force of air inlet tends to be the primary factor to push supplied air to spread horizontally on cultural relic area. Meanwhile, the suction of air outlet will produce an upward mechanical extraction to promote airflow, in the vertical direction to form the DV. Lin and Wu [7] used salt bath technology to simulate the displacement flow of mechanical extraction. The results showed that the stability of spatial stratification largely depends on the ratio of inertia force inflow to suction force at the outlet. The inflow inertia force has a significant effect on the middle stratification (Table 1).

In this research, an experimental investigation was conducted to evaluate the influence of the air supply outlet on the performance of the DV system for relic preservation area. A series of tests by varying the height of the air supply outlet were carried out in a simulated exhibition hall of archaeology museum.

2 Experimental System and Method

China is rich in earthen relic, and many of them are excavated and found in pits, such as the unearthed pottery figurine in Emperor Qin's Terracotta Warriors and Horses Museum and the Hanyang Mausoleum Museum. By referring to the funerary pits in these archaeology museums, an experimental hall for environmental control of cultural relic was set up in our previous research [8]. A local DV system was designed and installed in the burial pit.

2.1 Experimental System

As shown in Fig. 1a, the air supply outlet of the DV system is fixed at the bottom of the burial pit, while the air return outlet is installed at the junction of the cultural relic area and the tourist area. The whole experimental system is comprised of a refrigeration unit, an indoor air handling system and a supply and return air system as shown in Fig. 1b. The refrigeration unit stabilizes the water supply temperature and reduces the fluctuation of the air supply parameters according to the set parameters. The air handling system mainly regulates the temperature and RH of the air supply and purifies the air supply. The air supply and return system consist of a plenum chamber, a return air outlet and a circulating air duct. The surface of the back wall of the supply static pressure box is processed into a supply orifice plate with a diameter of 2 mm. The opening rate is 25%, mainly to improve the uniformity of air supply.

2.2 Arrangement of the Measuring Points

In order to evaluate the control performance of the DV system, the air temperature, RH and velocity were monitored and recorded during the experiment. Figure 2 shows the arrangement of measuring points. T1–T5 are temperature and RH sensor. The heights for T1–T5 from the bottom of the pit are 0.2 m, 0.5 m, 1.3 m, 1.9 m and 2.6 m, respectively. The measuring instruments for the temperature and RH are

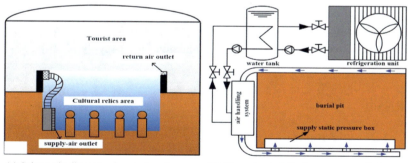

(a) Schematic diagram of control system;   (b) Flow chart of displacement ventilation system

Fig. 1 Displacement ventilation control system

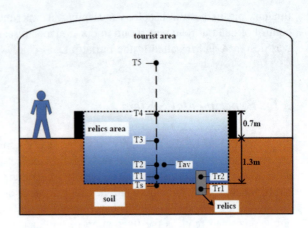

Fig. 2 Distribution of measuring points

TR-72Ui thermal recorders. The measuring accuracy is 0.3 °C for temperature, +5% for RH. TaV is the measuring point of the breeze velocity. The height from the bottom of the pit is 0.35 m. The measuring instrument is Sweden Swema 03 breeze anemometer. The measuring accuracy is 0.03 m/s. In the figure, the simulated cultural relic are pottery square bricks with a length of 0.6 m. Half of them are buried in soil, half are exposed to the air, Tr1 and Tr2 are temperature sensors embedded in pottery square bricks, and the distance from the surface is 0.25 m. Ts is the soil temperature measurement point and the burial depth is 0.05 m. The temperature sensor for ceramic tile and soil is copper–constantan thermocouple with a measurement accuracy of 0.3 °C.

2.3 Experimental Mode

Three experimental modes, i.e. full opening of the supply air outlet, upper half of the supply air outlet and lower half of the supply air outlet, were performed during July–August 2016. The air exchange rate was sustained to be 11.9 times, and the air supply temperature was 20–21 °C for all the tests [4]. At the same time, considering the boundary between dry-crack disease and mildew disease caused by RH, 80% is chosen as the appropriate value of relative humidity for air supply [9, 10].

3 Results

3.1 Outdoor Climate Conditions

During the experiment, the average outdoor temperature ranged from 30.0 to 31.0 °C and the fluctuation of day and night temperature was greater than 10 °C. During the opening period of the museum in the daytime (9 a.m. to 6 p.m.), the average outdoor ambient temperatures are 33.5, 34.5 and 34.9 °C, respectively, which will adversely affect the proper preservation of cultural relic. Therefore, it is necessary to take environmental control measures. During the experiment period for case 2, there was a short rainfall, which resulted in the average temperature and minimum temperature of outdoor climate slightly lower than the other two working conditions.

3.2 Temperature and Humidity Distribution in Experimental Exhibition Hall

Figure 3 shows the vertical temperature distribution at the centre of the burial pit. The average temperatures (T1–T4) under the three cases are 21.6, 21.5 and 21.5 °C, respectively. It indicates that the influence of the vertical position of the air supply outlet on the temperature in the burial pit can be neglected. However, by comparing the fluctuations at these measuring points, it is found that a relative large difference existed at T4 for these cases. In case 2, the temperature fluctuation at T4 is significantly greater than that in the other two cases. This is mainly due to the location of T4 at the interface between the burial area and the tourist area, which is greatly influenced by the upper natural ventilation zone. At the same time, case 2 is the lower part of the air supply, and the distance between the air supply position and T4 is larger than the other two cases, so the influence of the displacement ventilation control system on the measuring point T4 is less than the other two cases. According to the distribution of temperature, the temperature difference in the relic area mainly concentrates in the upper pit area. Therefore, for areas with high cultural relic, the layout height of air supply outlets should not be too low.

Figure 4 shows the distribution of the RH in the experimental hall under cases 1–3. It is found that the statistical data and distribution of RH in burial pits under three cases are very close, and the difference is very small. The RH in the cultural relic area is higher at night and lower during the day. This is mainly because the return air outlet is located at the interface of the cultural relic area and the tourist area. The return air will suck up certain indoor air. In order to control the relative humidity, the humidifier of the system has not been opened in this study.

3.3 Air Velocity Distribution in Cultural relic Preservation Area

The wind speed distributions of case 1 and case 2 are very close. The main reason is that the air supply outlets under these cases are close to the ground. Although the effective area of the air supply outlet under case 2 is smaller than that under case 1,

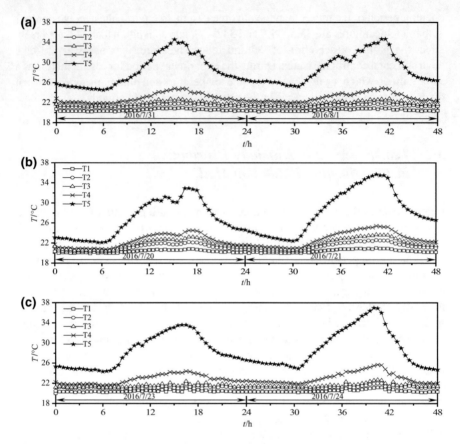

Fig. 3 Temperature distribution in the experimental hall. **a** case 1; **b** case 2; **c** case 3

the effect of orifice air supply outlet on the air velocity distribution at the bottom of the burial pit can be neglected because of its good air diffusion characteristics. The maximum and average wind speeds in case 3 are obviously higher than those in the other two cases, mainly because of the higher arrangement of the air supply outlet. After the cold air leaves the air outlet, it will sink first, which leads to a significant increase in the wind speed at the bottom of the burial pit. In fact, the average wind speed is less than 0.1 m/s. Under this weak air movement, the evaporation of water at the interface of earthen sites will not be significantly enhanced.

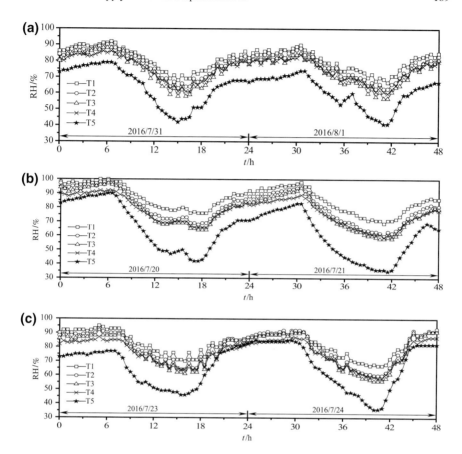

Fig. 4 Relative humidity distribution in the experimental hall. **a** case 1; **b** case 2; **c** case 3

3.4 Balance Analysis of Soil–Air Coupled Preservation Environment

Table 2 shows the temperature difference between soil and air surrounding the cultural relic for three cases. The soil temperature measurement points are located at the depth of 0.05 m, and the air temperature measurement points are located at the height of 0.2 cm at the soil–air interface at the bottom of the burial pit. The interfacial temperature difference corresponding to the three experimental

Table 2 Temperature difference between soil and air around cultural relic

Cases (°C)	1	2	3
Max ΔT	1.24	1.17	1.47
Min ΔT	0.17	0.02	0.18
Mean ΔT	0.67	0.48	0.60

Table 3 Temperature difference between the cultural relic body

Cases (°C)	1	2	3
Max ΔT	1.17	1.17	1.05
Min ΔT	0.39	0.27	0.0
Mean ΔT	0.81	0.74	0.58

conditions is also very close, and the average temperature difference is 0.67, 0.48 and 0.60 °C, respectively. Among them, the temperature difference of the environment interface in case 2 is the smallest, which is mainly because the air supply of case 2 is concentrated in the lower part of the tuyere near the bottom of the funeral pit, and the cooling effect near the soil interface is better than that of case 1 and working condition case 3.

Table 3 shows the statistical distribution of temperature difference between the part buried in soil (Tr1) and the part exposed to air (Tr2). Under the three cases, the average temperature difference of different parts of the cultural relic body is 0.81, 0.74 and 0.58 °C, respectively. There are some differences, but the differences are very small. Among them, the temperature difference of case 1 is the largest, and that of case 3 is the smallest. This is mainly because the cultural relic are located in the centre of the burial pit, and the flow of low-temperature air supply varies with the location of the air supply outlet. Because the air supply of case 2 is closer to the bottom of pit, the temperature difference of case 2 is less than that of case 1. In case 3, the main reason is that the cold air sinks after leaving the tuyere and flows into the cultural relic exposed to the air, which results in a lower temperature at the site, thus the average temperature difference is the smallest.

4 Conclusions

In this paper, the distribution characteristics of temperature, RH, wind speed and soil–air coupling performance of the exhibition hall under three different supply air outlet height modes are studied through experiments. The results show that:

(1) In the case when there is no internal heat source in the occupied area, the height of the supply air outlet has only a minor effect on the average temperature and RH in the cultural relic area, especially the temperature fluctuation value in the upper part of the burial pit. Therefore, in reality, the location of the air supply outlet should not be too low in the areas with high cultural relic.
(2) Under the three cases, the average wind speed in the burial pit is less than 0.1 m/s, and the coupling performance between soil and air is similar, which has little influence on cultural relic. However, relatively high tuyere conditions will lead to the sinking of cold air, accelerate the airflow in the cultural relic area, weaken the cooling effect on the soil interface and reduce the balance of the cultural relic environment.

Acknowledgements This study was supported by the Key Scientific Research Innovation Team Project of Shaanxi Province (under grant No.2016KCT-14), the Shaanxi Provincial Natural Science Foundation (2018JM5091) and the Fundamental Research Funds for the Central Universities (under grant No. zrzd2017003).

References

1. Dunham Inc. HVAC comparison study at 225 bush street San Francisco, Minneapolis, MN, USA. (1999)
2. Wang, X., et al.: Analysis of the influence of heat source characteristics on the thermal stratification height of displacement ventilation. Build. Therm. Energy Air Cond. **24**(3), 72–75 (2005)
3. Park, H.J., Holland, D.: The effect of location of a convective heat source on displacement ventilation: CFD study. Build. Environ. **36**(7), 883–889 (2001)
4. Luo, X., Gu, Z., et al.: Experimental study of a local ventilation strategy to protect semi-exposed relics in a site museum. Energy Build. **159**, 558–571 (2018)
5. Manins, P.C.: Turbulent buoyant convection from a source in a confined region. J. Fluid Mech. **37**(4), 51–80 (1969)
6. Linden, P.F., Lane-Serff, G.F., Smeed, D.A.: Emptying filling boxes: the fluid mechanics of natural ventilation. J. Fluid Mech. **212**(2), 309–335 (1990)
7. Lin, Y.J.P., Wu, J.Y.: A study on density stratification by mechanical extraction displacement ventilation. Int. J. Heat Mass Transf. **110**, 447–459 (2017)
8. Zhang, Z., et al.: In-situ conservation of cultural relics under multi-media coupling environment. J. Xi'an Jiaotong Univ. **11**, 150–156 (2016)
9. Luo, X., Zhaolin, Gu, et al.: Desiccation cracking of earthen sites in archaeology museum—a viewpoint of chemical potential difference of water content. Indoor Built Environ. **24**, 147–152 (2015)
10. Viitanen, H., Vinha, J., Salminen, K., et al.: Moisture and bio-deterioration risk of building materials and structures. J. Build. Phys. **33**(3), 201–224 (2010)

A Study on Polymer Electrolyte Membrane (PEM)-Based Electrolytic Air Dehumidification for Sub-Zero Environment

Tao Li and Ronghui Qi

Abstract Electrolytic dehumidification with polymer electrolyte membrane (PEM) is a compact and effective independent humidity control method for refrigerators. To explore the performance of PEM-based dehumidification under sub-zero conditions, experiments were conducted at low temperatures of −10–0 °C. Results showed that this device could operate stably at temperatures above −8.5 °C. The moisture removal rate was 0.05–0.08×10^{-9} g/s, and the energy efficiency was 0.64–0.68×10^{-2} g/(J m^2), respectively. It has also been found that the moisture removal rate and energy efficiency decreased with the decrease of absolute humidity in the air. Furthermore, the performance under sub-zero conditions was relatively low, which the dehumidification rate at −8.5 °C was only 1/6 of that at 25 °C. The possible reason is that water inside the PEM may be frost, hindering the proton transmission below 0 °C. Besides, the electrical conductivity also reduces due to the low temperature. This study proved that PEM-based dehumidification is a promising alternative for independent humidity control under low-temperature conditions.

Keywords Electrolytic dehumidification · Refrigerator · Sub-zero environment · Independent humidity control · Experiments

---

T. Li · R. Qi (✉)
Key Laboratory of Enhanced Heat Transfer and Energy Conservation of Education Ministry, School of Chemistry and Chemical Engineering, South China University of Technology, Guangzhou 510640, China
e-mail: qirh@scut.edu.cn

Nomenclature

m	Mass flow rate
t	Temperature
RH	Relative humidity
ω	Moisture content

Subscripts

a	Air
A	Anode
In	Inlet
P	Power
w	Water
C	Cathode
Out	Outlet
A_{area}	Membrane area

1 Introduction

In recent years, with the improvement of quality of life, people's storage requirements for foods become higher and higher. For example, dry goods, tea, and some precious herbs need to be stored in an area with a low RH (relative humidity) of 45% RH or less, while the storage humidity of cakes or breads should be kept at about 60% RH. Thus, many companies have developed refrigerators with classified storage. These refrigerators perform well in the temperature control, but not good enough in the classified humidity control. Many dehumidification technologies, such as condensation ones, cannot operate under sub-zero temperatures. Desiccant absorption could be operating at low temperature, but it usually has a big volume. As the humidity control in current refrigerators is usually achieved by changing the operating strategies of evaporators, compressors, and fans [1], the energy efficiency was low and the pipeline was complicated [2]. Furthermore, the space inside the refrigerators is very limited, which increases the difficulty of installing dehumidification equipment.

Electrolytic dehumidification with electrolyte membrane is an innovative independent humidity control technology and driven by low voltage electricity. In 2000, Iwahara et al. first proposed the use of solid electrolyte material SrCeO ceramics for electrical dehumidification and verified its feasibility under DC voltage [3]. In 2005, it was found by Onda et al. that the electroosmotic effect could not be ignored [4]. In 2009, the possibility of dehumidification by electricity at room temperature (20–40 °C) was verified [5]. The V-I characteristics were tested, the influence of air temperature on system performance was analyzed, and an empirical formula was proposed by Sakuma et al. [6]. Previous research of our research group found that under 3 V electric field, the air outlet humidity could be reduced from 90 to 30%. And the dehumidification energy efficiency was 2.0–3.0 × 10^{-2} g/(J m^2), which was competitive with other dehumidification methods [7].

Previous research found that as the moisture transport is driven by the electric field rather than the difference of temperature or water vapor partial pressure, this system has the potential to be operated efficiently under low-temperature conditions. In this paper, the operation performance of PEM-based electrolytic dehumidification was investigated experimentally, under a sub-zero environment with the temperature range of −10–0 °C. The dehumidification device was constructed by a PEM, a porous electrode with a catalytic layer and a diffusion layer on both sides, and two outermost gas passages. During experiments, the voltage and air humidity were 3 V and 55–65% RH, respectively, and the flow rate was 1.9 × 10^{-2} g/s. Nafion 117 was used as an electrolyte membrane. The humidity, temperature, and current were recorded. This study investigated the performance of the device operation under low-temperature conditions(<0 °C) and proofed its feasibility in the application of refrigerators.

2 Methods

In this study, a PEM-based electrolytic experimental system was developed. The PEM component was composed of electrolytic membrane, porous Pt cathode, and IrO$_2$ anode with catalytic layers composed of noble metal particles, and carbon fiber papers as diffusion layers on both sides. The gas passage, whose thickness was 2 mm, was on the outermost side. When a DC voltage was applied, the reactions was shown in Eqs. (1) and (2) occurred. At the anode side, water vapor in the air was decomposed on the catalytic layer, producing H$^+$ protons. The H$^+$ protons were driven toward the cathode side through the proton exchange membrane and reacted with oxygen to form water at the cathode side. At the same time, some water molecules were carried from the anode side to the cathode side by H$^+$ protons. This is called the electroosmotic effect. Then the water at cathode side would be taken away by the sweep gas. Since the humidity at the cathode side was greater than that of anode side, a negative concentration gradient was formed, which caused some moisture moving back toward the anode.

$$\text{Anode}: \quad H_2O \rightarrow 2H^+ + 2e^{-1} + 0.5O_2 \tag{1}$$

$$\text{Cathode}: \quad 2H^+ + 2e^{-1} + 0.5O_2 \rightarrow H_2O \tag{2}$$

In order to study the dehumidification performance under low-temperature conditions, the whole system with the measurement devices was placed into a constant temperature and humidity chamber. The temperature and air humidity inside could be set. Temperature and humidity setting error was ±1 °C and ±3% RH, respectively.

Four 100-mL-three-necked flasks with temperature and humidity sensors were used to record the temperature and humidity. Temperature and humidity recording error were ±0.5 °C and ±3% RH at 25 °C, respectively. Several microporous flowmeters were placed at the anode inlet and cathode inlet to record the gas flow rate. The voltage was applied by a DC power, and the current was recorded by a multimeter, with the measurement error of ±2 mV and ±1 mA, respectively. The measurement error of energy efficiency is $\pm 2.8 \times 10^{-2}$ mg/(J m^2). The data were collected through Agilent, and the schematic diagram of the experimental apparatus was shown in Fig. 1.

During the experiments, the moisture removal rate of the electrolyte membrane could be calculated by Eq. (3). The dehumidification efficiency could be calculated by Eq. (4).

$$m_{\text{removal}} = m_{aA}(\omega_{aA,\text{out}} - \omega_{aA,\text{in}}) \tag{3}$$

$$\eta_{\text{energy}} = m_{\text{removal}}/(P \times A_{\text{area}}) \tag{4}$$

where m is mass flow rate, ω is moisture content, A_{area} is membrane area, P is power, A is anode, η is energy efficiency, and a is air.

Experimental procedures are as follows. (a) The air pump was opened for 1 h before the experiment. (b) The temperature of the chamber was set to 0 °C.

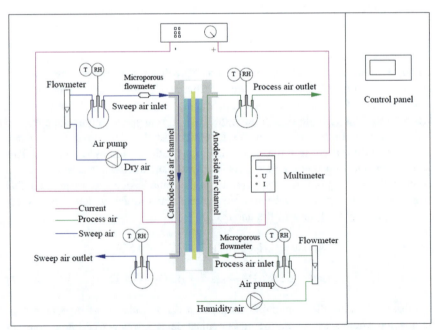

Fig. 1 Schematics of test rig

Table 1 Operation conditions of PEM dehumidification

Electric field (V)	Air inlet of anode side			Air inlet of cathode side		
	m_a (g/s)	t (°C)	RH (%)	m_a (g/s)	t (°C)	RH (%)
3	0.0189	−8.5–0	57 ± 5	0.0189	−9–0	60 ± 5

(c) After the temperature was stable, the DC power was turned on. The power supply was set to 3 V. (d) After the device reached a steady state, the operational data were recorded by the Agilent data logger every 5 s. (e) Every test was conducted for about 20 min. Then reset the temperature and start the steps from the beginning. The experimental conditions were shown in Table 1.

3 Results and Discussion

In the experiment, the temperature dropped at a rate of −5.1 °C/h from 0 to −8.5 °C and the humidity fluctuated around 60% RH. Unlike the proton exchange membrane fuel cell, which had a disadvantage that it cannot be started under low-temperature conditions due to lack of water, the PEM-based dehumidification could decompose water by applying voltage and did not require a large amount of moisture to start. So the dehumidification device could be started and stabilized under lower temperature conditions.

From Fig. 2, the moisture removal rate and current decreased with the decrease of temperature. When the temperature was −0.8 °C, the current and the moisture removal rate were 0.269 A and 0.0747×10^{-4} g/s, respectively. When the temperature dropped to −4.8 °C, the current was 0.217 A, and the moisture removal rate was 0.0582×10^{-4} g/s. The moisture removal rate decreased more than the current. In addition, the decline trend of current and moisture removal rate with time was not steady. As shown in Fig. 2a, the current curve first dropped and then

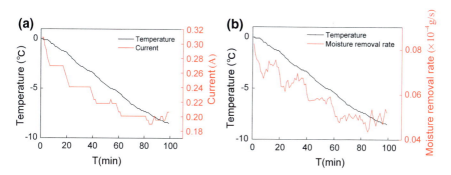

Fig. 2 Variation of dehumidification performance with time: **a** temperature and current and **b** moisture removal rate and temperature

stabilized. After a period of stability, the curve of current would decrease and then stabilize again. Simultaneously, as shown in Fig. 2b that the curve of moisture removal rate first increased and then decreased. After this, the curve would increase and then decrease again. The possible reason was that the water storage capacity of the PEM increased with the decrease of the current. In the experiment, the current was phased down. So the water in the membrane would increase first and then decrease with the decrease of temperature.

As shown in Fig. 3a because of the change of water in PEM, the moisture removal rate decreased with the decrease of absolute humidity. It would rise first and then decrease. It also could be seen from Fig. 3b that the current curve was substantially consistent with the absolute humidity curve. Under low-temperature conditions, the absolute humidity had a greater influence on the current and the moisture removal rate. Changing the temperature actually was changing the absolute humidity.

The energy efficiency of PEM electrical dehumidification was shown in Fig. 4. In the experiment, the observed stability efficiency was $0.64–0.78 \times 10^{-2}$ g/(J m^2),

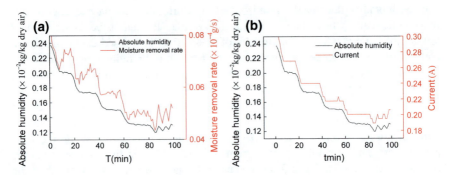

Fig. 3 Variation of dehumidification performance with time: **a** moisture removal rate and absolute humidity and **b** absolute humidity and current

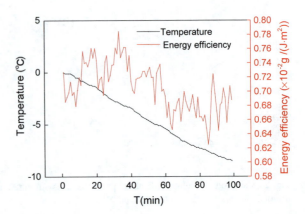

Fig. 4 Energy efficiency and temperature with time

i.e., 0.1–0.13 MJ/kg of electricity demand. It was competitive with current dehumidifiers under low temperatures. For example, Chen et al. investigated a dehumidification process with cascading desiccant wheels with sub-zero dew point and obtained the lowest electricity demand of 0.6 MJ/kg [8]. Besides, the electricity demand of electroosmotic dehumidification and thermoelectric dehumidifier was around 0.09 MJ/kg, while their performance was measured under normal temperature range about 25 °C [9].

From previous experiments, when the temperature was 25 °C, the moisture removal rate was 0.3×10^{-4} g/s. In this experiment, when the temperature was −8.5 °C, the moisture removal rate was 0.05×10^{-4} g/s. The moisture removal rate at −8.5 °C is only 1/6 of that at 25 °C. Under low-temperature conditions, especially below 0 °C, some of the water inside the PEM may frost, which would reduce the internal moisture content of the PEM. The current and dehumidification decreased with the decrease of absolute humidity, because the conductivity of the membrane would reduce due to lack of water. Furthermore, the formed frost in the diffusion layer would hinder the mass transfer during dehumidification. And the catalyst performance was also reduced because of the low temperature. But the device could run stably. Therefore, it was feasible to apply the method to the refrigerator for low-temperature-independent dehumidification.

4 Conclusion

Humidity control under low-temperature conditions has great significance in the food preservation of refrigerators. The existing dehumidification technology has many problems such as unable to operate under low temperature (<0 °C) and large device volume. As the moisture transport is driven by electric field rather than the difference of temperature or pressure, the electrolytic dehumidification has potential to be operated efficiently under sub-zero environments.

In this paper, the feasibility of PEM-based electrolytic dehumidification device was experimentally investigated under low-temperature conditions, and the factors affecting the performance were analyzed. Experiments were conducted at low temperatures of −10–0 °C. Results showed that this device could operate stably at temperatures above −8.5 °C. The moisture removal rate was 0.05–0.08×10^{-9} g/s. And the energy efficiency was 0.64–0.68×10^{-2} g/(J m^2). As it showed that the moisture removal rate and energy efficiency decreased with the decrease of absolute humidity in the air. In addition, the dehumidification rate at −8.5 °C was only 1/6 of that at 25 °C, because the water inside the PEM might frost below 0 °C, which hindering the proton transmission. Furthermore, the electrical conductivity could be reduced due to the low temperature. This study showed that PEM-based dehumidification device can run stably under low-temperature conditions. Therefore, PEM-based dehumidification technology can be a promising alternative for small-scale dehumidification under low temperature for refrigerators.

Acknowledgements The project is supported by the National Key Research and Development Program (No. 2016YFB0901404), National Natural Science Foundation of China (51876067), the National Science Fund for Distinguished Young Scholars of Guangdong Province (2018B030306014). It is also supported by the Science and Technology Planning Project of Guangdong Province: Guangdong–Hong Kong Technology Cooperation Funding Scheme (TCFS), No.2017B050506005.

References

1. Cheng, et al.: Household frost-free appliances humidity control principle and practical application (in Chinese), J. Appl. Sci. Technol. **04**, 56–59 (2017)
2. Liu, et al.: Performance improvement in air dehumidifiers: from ideal to actual (in Chinese). Chin. Sci. Bull. **60**, 2631–2639 (2015)
3. Iwahara, et al.: Electrochemical dehumidification using proton conducting ceramics. Solid State Ionics **136–137**, 133–138 (2000)
4. Onda, et al.: Polymer electrolyte dehumidifying cell and its application to air condi-tioners. J. Appl. Electro-chem. **152**(12), A2369–A2375 (2005)
5. Sakuma, et al.: Water transfer simulation of an electrolytic de-humidifier. J. Appl. Electro-chem. **39**(6), 815–825 (2009)
6. Sakuma, et al.: Estimation of dehumidifying performance of solid polymer electrolytic dehumidifier for practical application. J. Appl. Electro-chem. **40**(12), 2153–2160 (2010)
7. Qi, et al.: Experimental study on electrolytic dehumidifier with polymer electrolyte membrane for air-conditioning systems. Energy Procedia. **142**, 1908–1913 (2017)
8. Chen, et al.: A dehumidification process with cascading desiccant wheels to produce air with dew point below 0 °C. Appl. Therm. Eng. **148**, 78–86 (2019)
9. Li, Y.Y.Yan: Solid desiccant dehumidification techniques inspired from natural electroosmosis phenomena. J. Bionic Eng. **8**, 90–97 (2011)

Experimental Investigation on an Evaporative Cooling System for the Environmental Control in Archaeology Museum

Bin Chang, Xilian Luo, Yanqian Shen, Juan Li and Zhaolin Gu

Abstract High energy consumption of the heating, ventilation and air-conditioning (HVAC) system and unsuitable preservation environment for earthen relics are common problems to the environmental control in archaeology museums, which are the main reasons that many archaeology museums in China do not equip HVAC system and are suffering deteriorations of dry cracking and salt enrichment. In this paper, an environmental control system, which employs evaporative cooling and ultrasonic humidification units to control the local environment for the relics preservation area and provide a "supersaturated" humidity (RH) for suppressing earthen cracking, is proposed. An experimental system was built to validate the feasibility of the control strategy. Experimental tests of the system were performed and the results show that the system could create a relative stable preservation for the relics, and the RH of the preservation area is high enough to replenish water to the earthen relics. It provides a high-efficiency and energy-saving environmental control strategy for historic site in archaeology museum.

Keywords Evaporative cooling · Earthen site, environmental control · Energy-saving

1 Introduction

The museums for cultural relics are mainly categorized into two types of indoor-display museum and archaeology museum according to the different displays of cultural relics [1, 2]. Amongst, archaeology museum is constructed at the

B. Chang · X. Luo (✉) · Y. Shen · J. Li · Z. Gu
School of Human Settlements and Civil Engineering, Xi'an Jiaotong University, Xi'an 710049, China

Shaanxi Energy Environmental & Building Energy Conservation Engineering Technology Research Center, Xi'an Jiaotong University, Xi'an 710049, China
e-mail: xlluo@xjtu.edu.cn

place where the immovable historical sites were found to protect the relics from the deteriorations due to natural weathering. As an ancient civilized country, China has a long history of thousands of years, there are more than 770,000 immovable historical sites have been discovered and over 150 archaeology museums have been built in China. The long-term preservation of cultural relics requires a suitable indoor temperature and RH environment. The environmental control for the collections in museums is becoming a concerning issue and HVAC system is therefore widely equipped in museums to upgrade the preservation environment [3]. Considering that the environmental control system in museums always has to be operated continually during all the year and 24 h per day, an energy-efficient HVAC system is of very important to provide a stable preservation environment for the relics preservation area.

The historical sites are usually preserved in soil–air coupled environment. Dry cracking and salt enrichment are widely appeared to the earthen site in archaeology museum. In order to satisfy the unique requirement of historical site on preservation environment, many efforts have been made in our previous researches, local environment control systems with high RH have been proposed to lower the energy consumption of the HVAC system and restrain the drying rate of the earthen site [2, 4], and considerable progresses in improving the preservation environment have been obtained. However, the moisture transfer from the earthen site to the air is a one-way procedure such that high RH is incapable of completely suppressing the deteriorations of dry cracking and salt enrichment. In this research, a novel control strategy, which employed evaporative cooling and ultrasonic humidification to treat the supplied air of the HVAC system to a "supersaturated state", was proposed to provide a steady preservation for the historical sites in archaeology museum. The evaporative cooling system uses of water as working fluids could significantly reduce the energy consumption of the HVAC system, which has been widely used in the building of workshop, subway platform and office [5, 6]. An experimental system of the proposed strategy was set up, and then experimental tests were carried out to evaluate the performance of the system.

2 Methods

2.1 Moisture Transfer and Water Replenishing for Earthen Site

Dry cracking of earthen site and salt enrichment on relics surface are mainly caused by water evaporation from the soil environment to the air environment. The moisture migration is essentially a phase change from the liquid state in the soil into vapour state in the air, and the chemical potential for the water migration process could be given as [4]:

Experimental Investigation on an Evaporative ...

$$\Delta\mu = \mu_l - \mu_v = -2.41 - 0.002 \times \theta^{-1.75} - \frac{RT}{M_w} \times \ln(RH) \quad (1)$$

where μ_v and μ_l are the chemical potentials for water vapour and liquid water, respectively, θ (kg/kg) is the free water content of topsoil, M_w (kg/mol) is the water molecular mass, R (J/k·mol) is the universal gas constant, T (K) and RH (%) are the temperature and relative humidity of water vapour. When $\Delta\mu > 0$, the migration of free water in topsoil into atmospheric environment occurs; when $\Delta\mu < 0$, the reverse process occurs spontaneously; when $\Delta\mu = 0$, the water exchange equilibrium is achieved. Luo et al. have analysed the migration process and pointed out that $\Delta\mu$ is always greater than 0 only if RH equals to 100%.

In order to meet the demand of RH = 100% for earthen relics, a novel control strategy by integrating dew point indirect evaporative cooling and direct evaporative cooling was proposed in this research.

2.2 Experiment Set-up

An experimental hall with a funerary pit to simulate the exhibition hall of archaeology museum has been constructed in our previous researches [7–9]. Based on this experimental platform, the proposed evaporative cooling system was designed and installed in the experimental funerary pit. The schematic diagram of the evaporative cooling system is shown in Fig. 1.

The fresh air successively flows through the blower unit, the synthetic evaporative cooler, the surface cooler and the ultrasonic humidifier in turn and finally flows into the relic preservation area. The surface cooler will not operate only if the preservation area requires a precision control. In this research, the system is used to implement a relaxed control for the relics preservation area such that the surface cooler is not used. The synthetic evaporative cooler is the most important component of the system, as shown in Fig. 2, the primary air flows into the dry channel, part of them traverse through the orifices in the wall of the duct and entrance into wet channel. The supplied air in wet channel has a direct contact with water droplets and will be cooled by water evaporation. As a result, the product air is cooled along the dry channel by transfer sensible heat to adjacent wet channel. Then, the cooled air flows into the direct evaporative cooling and is treated to saturated state (RH = 100%) by directly contacting with the sprayed water.

In order to evaluate the performance of the system, the dry bulb temperature and RH in relics preservation area and air handle unit were measured by the TR-72Ui thermal recorders. The measuring accuracies of the recorder are 0.3 °C for temperature, +5% for RH. The parameters were monitored at the locations shown in Table 1. At the same time, two soil samples, i.e. S1 and S2, are used to record soil moisture content.

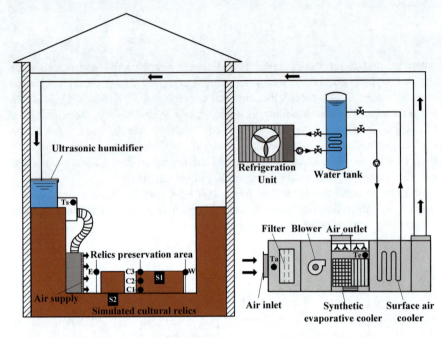

Fig. 1 System diagram of the evaporation cooling control system

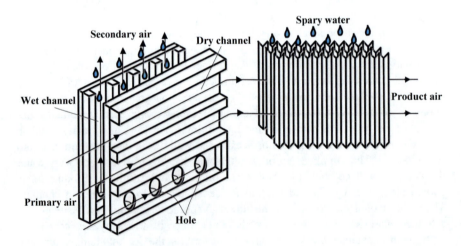

Fig. 2 Schematic of synthetic evaporative cooler

Two experimental tests were conducted in this research, as listed in Table 2. Amongst, case 1 was tested for the condition of natural ventilation and case 2 for the evaporative cooling system; meanwhile, the ultrasonic humidifier was opened as well. The air exchange rate in case 2 is 15 times per hour.

Table 1 Specification of the different monitoring positions

Points	Record unit	Location
C1	Temperature and RH	0.2 m from the bottom of the pit (centre position)
C2	Temperature and RH	0.5 m from the bottom of the pit (centre position)
C3	Temperature and RH	0.7 m from the bottom of the pit (centre position)
E	Temperature and RH	0.7 m from the bottom of the pit (eastern position)
W	Temperature and RH	0.7 m from the bottom of the pit (western position)
Ta	Temperature and RH	Air inlet position
Te	Temperature and RH	Evaporative cooling outlet position
Ts	Temperature and RH	Ultrasonic humidifier outlet position
S1	Gram	Soil sample1 (on the top)
S2	Gram	Soil sample2 (on the bottom)

Table 2 Experiment operation condition

Case	Evaporative cooler	Surface cooler	Humidifier	Air change per hour
1	○	○	○	0 ac/h
2	●	○	●	15 ac/h

Notes "●"equipment switched on; "○"equipment switched off

3 Results and Discussion

3.1 Temperature and RH Distributions at the Preservation Area

The temperature and RH at the central of the funerary are shown in Fig. 3. In case 1, the environmental control system was switched off and the temperature and RH are greatly affected by the outdoor environment, the fluctuations for the temperature and RH are 5.8 °C and 19%, respectively. When the evaporative cooling system

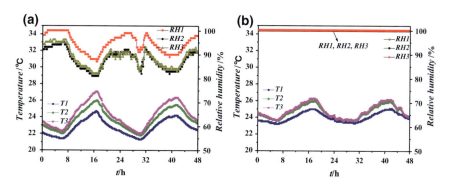

Fig. 3 Vertical distribution of temperature and RH. **a** Case 1 and **b** case 2

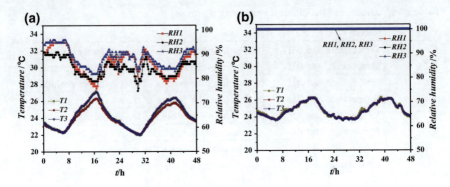

Fig. 4 Horizontal distribution of temperature and RH. **a** Case 1 and **b** case 2

was switched on in case 2, a relative stable preservation environment was created in the funerary pit, the fluctuation of the temperature was reduced to less than 2 °C, which satisfies the specifications (≤ 2 °C) defined in ASHRAE standard [10]. Moreover, the RH for the preservation area were always sustained at saturated state (RH = 100%), which meets the demand for suppressing moisture transfer for earthen relics.

The horizontal profiles of temperature and RH at the height 0.7 m in the preservation area were shown in Fig. 4. Similar to that in vertical direction, there are significant fluctuations in the temperature and RH for case 1, the temperature fluctuating between 21.8 and 27.2 °C, and the RH fluctuating between 75 and 100%. In case 2, the temperature and RH were distributed uniformly at the same height. The RH were kept constant of 100% and the horizontal temperature differences between the three measured points are less than 0.03 °C.

3.2 Water Recovery of the Earthen Site

In order to evaluate the water replenishment of the earthen by the evaporative cooling system, a comparison was made by conducting two additional experiment tests, i.e., the system switched off and on, respectively. Each test lasts 15 days. Two 500 ml beakers, which are filled with soil (the mass water content $\phi = 30\%$) were buried into the bottom and soil column (see Fig. 1), respectively. The weight of the soil samples were recorded regularly at 0:00 am. Figure 5a shows that the weight of soil samples gradually decreased under the natural state, it indicates that the internal water of the soil continuously evaporating into the external environment. Figure 5b shows that the weight of soil samples gradually increased under the control of evaporative cooling, which indicates that the water in soil was replenished.

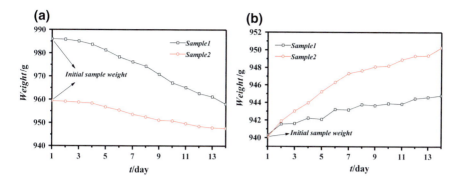

Fig. 5 Weight of soil samples. **a** The system switched off, **b** the system switched on

3.3 Energy Consumption Analysis

The psychometric chart of the air processing procedure for the evaporative cooling system is shown in Fig. 6, where a → b is the process of dew point indirect evaporative cooling and b → c is the process of direct evaporative cooling, the supplied air at point c is saturated (RH = 100%) and was supplied directly into the preservation area.

In the above process, the cooling capacity (Q) for the evaporative cooling procedure can be expressed by Eq. 2 as

$$Q = Q_m \times (h_a - h_c) \tag{2}$$

where Q_m is the mass flow rate, h_a is the enthalpy value of the outdoor air and h_c is the enthalpy value of the air in evaporative cooling outlet, and the values of these variants are listed in Table 3. The average cooling capacity of the system is 1.1 kW, which are provided by evaporative cooling unit.

Fig. 6 Enthalpy–humidity chart of the air-conditioning process

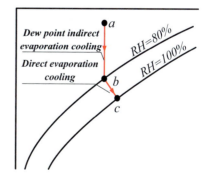

Table 3 Cooling capacity details

h_a (kJ/kg)	h_c (kJ/kg)	Q_m (kg/h)	Q_1 (kW)
75	63	330	1.1

4 Conclusions

In order to provide a stable preservation environment and reduce the energy consumption of the HVAC system, a combined system of evaporative cooling unit and ultrasonic humidification unit is proposed. The main conclusions of the research are as follows:

In the case when the evaporative cooling system was operated, the RH of the cultural relics preservation area were always sustained at 100%, which meets the demand of the high RH environment of the earthen relics. Meanwhile, the temperature fluctuation of the preservation area was reduced to about 2 °C, and the horizontal temperature differences between the three horizontal measuring points are 0.03 °C, which accord with the specifications of recommended preservation environment for museums in ASHRAE.

The evaporative cooling system could effectively recover the water content for earthen sites and suppress the desiccation cracking and salt enrichment of relics sites by replenishing water to their soil environment.

The evaporative cooling system provides an energy-saving strategy for earthen site in archaeology museum. The average cooling capacity of the system is 1.1 kW, which are all bear by the evaporative process.

Acknowledgements This study has been supported by the Key Scientific Research Innovation Team Project of Shaanxi Province (Grant No. 2016KCT-14), Shaanxi Provincial Natural Science Foundation (2018JM5091), the Opening Projects of Beijing Advanced Innovation Center for Future Urban Design, Beijing University of Civil Engineering and Architecture (Grant No. 30) and the Fundamental Research Funds for the Central Universities (Grant No. zrzd2017003).

References

1. Luo, X.L., et al.: Experimental study of a local ventilation strategy to protect semi-exposed relics in a site museum. Energy Build. **159**, 558–571 (2018)
2. Gu, Z.L., et al.: Primitive environment control for preservation of pit relics in archeology museums of China. Environ. Sci. Technol. **47**, 1504–1509 (2013)
3. Liu, S. The museum environment. Sci. Conserv. Archaeol. (2016)
4. Luo, X.L., et al.: Desiccation cracking of earthen sites in archaeology museum—a viewpoint of chemical potential difference of water content. Indoor Built Environ. **24**(2), 147–152 (2015)
5. Fabrizio, A., et al.: Energy saving strategies in air-conditioning for museums. Appl. Therm. Eng. **29**(4), 676–686 (2009)
6. Luo, X.L., et al.: Design of an energy-saving environmental control system for relics preservation in archaeology museum. Energy Procedia **104**, 431–436 (2016)

7. Luo, X.L., et al.: An independent and simultaneous operational mode of air conditioning systems for visitors and relics in archaeology museum. Appl. Therm. Eng. **100**, 911–924 (2016)
8. Luo, X.L., et al.: Efficacy of an air curtain system for local pit environmental control for relic preservation in archaeology museums. Indoor Built Environ. **25**(1) (2016)
9. Luo, X.L., et al.: Environmental control strategies for the in situ preservation of unearthed relics in archaeology museums. Journal of Cultural Heritage **16**(6), 790–797 (2015)
10. ASHRAE Handbook–HVAC applications Chapter 23: Museum, galleries, archives, and libraries. American Society of Heating, Refrigerating and Air-Conditioning Engineers, SI Edition, Atlanta (2015)

Numerical Simulation of a Double-Layer Kang Based on Chimney Effect

Shilin Lei, Bin Chang, Yanqian Shen, Xiaoyu Zhu, Juan Li and Xilian Luo

Abstract The traditional Chinese Kang system is a kind of ancient domestic heating system in China, which is widely used in the northern rural area of China since it has the advantage of heating the room by utilizing waste heat of cooking. However, the traditional Chinese Kang is facing the challenge of dropping out of use due to it heating the room by burning biomaterials or coal directly. In order to reform the existing Kang system, a novel Kang system which employs a capillary radiation mat as heating terminals was proposed in this research. Moreover, a dual heat source of solar energy and air source heat pump was used in the system to replace biomaterials and coal. In order to validate the feasibility of the system, a numerical simulation was performed in this research to obtain the temperature and flow distributed characteristics in the heating room and the results showed that this system could provide a comfortable indoor environment for residents.

Keywords A new kind of Kang · Airflow channel · Clean energy · Numerical simulation

1 Introduction

The traditional Chinese Kang system has been widely used in the northern rural area of China; until now, there are more than 160 million Kang have been built [1]. In recent years, China is vigorously promoting the policy of replacing coal with electricity, such that the traditional Kang system is facing the challenge of abolishment due to it burns coal and biomass to heat the room. The improvement

---

S. Lei · B. Chang · Y. Shen · X. Zhu · J. Li · X. Luo (✉)
School of Human Settlements and Civil Engineering, Xi'an Jiaotong University, Xi'an 710049, China

Shaanxi Energy Environmental and Building Energy Conservation Engineering Technology Research Center, Xi'an Jiaotong University, Xi'an 710049, China
e-mail: xlluo@mail.xjtu.edu.cn

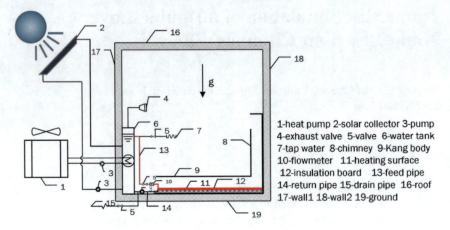

Fig. 1 New kind of double-layer Kang based on chimney effect

scheme of the Kang system is therefore attracting many attentions. Chao et al. proposed a solar energy-based Kang system which employed a paraffin phase change thermal storage unit to implement the continuous operation day and night, a much higher thermal comfort could be obtained since a stable surface temperature was created by the phase change thermal storage unit [2]. In order to investigate heat storage characteristics of the Chinese Kang, Liu et al. conducted a dynamic simulation of a solar energy-based Kang system, the result shown that when the temperature of the supplied water is set to be 40 °C, the system could provide a comfortable indoor environment for the residents [3]. Zhuang et al. carried both experimental and numerical research on the features of the thermal plume of the Chinese Kang system and pointed out that the wasted heating of cooking is inadequate to heating the room and an auxiliary heating source is therefore needed [4].

In this paper, a double-layer Kang by using a capillary radiation mat as heating terminals is proposed (see Fig. 1). A numerical simulation was conducted to evaluate the effects of surface temperature on indoor temperature.

2 Methods

In order to validate the performance of the proposed system, the commercial CFD code of ANSYS was employed to conduct a numerical simulation of the proposed system. A second-order accurate discretization scheme was used to discretize the governing equations. The Boussinesq approximation is used to treat the buoyancy force in the thermal chimney. Before the simulation was performed, a grid independence was validated on four sets of computational grids, and finally, the decomposition of the 307 × 308 was employed.

2.1 The Physical Model

The details of the geometrical model is as follows: (1) The dimensions of the heating room is 3 m × 3 m × 3 m. (2) The thickness of the heating surface, the upper surface is 0.1 m, the length is 2 m, and the width is 1.5 m. (3) The chimney height is 0.5, 1, or 1.5 m. (4) The distance between the upper surface and the heating surface is 0.25 m. (5) The thickness of envelops are 0.3 m.

2.2 Numerical Equations and Calculation Methods

2.2.1 The Governing Equations

2All the variants (velocities, temperature, turbulent kinetic energy, and dissipation) to be solved are denoted by ϕ. The general transport equation for ϕ can be written as:

$$\frac{\partial(\rho\phi)}{\partial t} + \text{div}(\rho \mathbf{U}\phi) = \text{div}\left(\Gamma_\phi \mathbf{grad}\phi\right) + S_\phi \quad (1)$$

where ρ is the density of the fluid, $\mathbf{U}=(u,v)$ is the velocity vector, Γ_ϕ is the generalized diffusion coefficient, and S_ϕ is the source term. If Γ_ϕ, S_ϕ, and ϕ are adequately prescribed, Eq. (1) can express the continuity, momentum, energy, and other scalar equations.

2.2.2 Parameters and Boundary Definition

The Boussinesq assumption was used to treat the buoyant force. The air density is 1.225 kg/m³. Thermal expansion coefficient is β ($\beta = 1/T$), and gravitational acceleration is −9.81 m/s². ρ is the solid density. λ is the thermal conductivity of the solid material. C_p is the specific heat capacity.

The boundary condition for the simulation is defined in Table 1.

Table 1 Boundary conditions

Boundary	Description
Left wall	*Velocity: no slip*, energy: $\partial T/\partial x = 0$
Right wall	*Velocity: no slip*, energy: $T = -3.5$ °C
Bottom wall	*Velocity: no slip*, energy: $\partial T/\partial x = 0$
Ceiling wall	*Velocity: no slip*, energy: $T = -3.5$ °C
Unheated surface of the Kang	*Velocity: no slip*, energy: couple
Heated surface of the Kang	Case 1. *Velocity: no slip*, energy: $T = 35$ °C
	Case 2. *Velocity: no slip*, energy: $T = 40$ °C
	Case 3. *Velocity: no slip*, energy: $T = 45$ °C

Then select simple as the calculation method. Other parameters are unchanged. Run calculation until convergence.

3 Results and Discussion

3.1 The Indoor Temperature and Outlet Velocity in Different Situations

In the case when the airflow channel height is 0.25 m, the heat dissipation of each maintenance structure of the room is given and the heating surface temperature is 40 °C; the higher the chimney, the higher the outlet flow rate and the lower the average indoor temperature. When the height of the chimney is 1.5 m, the airflow velocity at the airflow outlet is the highest (0.405 m/s) and the average indoor temperature is the lowest (14.3 °C). When the chimney height is 0.5 m, the airflow velocity at the airflow outlet is the smallest (0.337 m/s) and the average indoor temperature is the highest (18.53 °C) (Table 2).

When the airflow channel height is 0.25 m, the heat dissipation of each maintenance structure of the room is given and the height of the chimney is 1.0 m; the higher the temperature of the heating surface, the higher the airflow velocity at the air outlet and the higher the average indoor temperature. When the heating surface temperature is 35 °C, the airflow velocity of the outlet is the smallest (0.350 m/s) and the indoor average temperature is the lowest (12.18 °C). When the heating surface temperature is 45 °C, the velocity of the outlet is the largest (0.397 m/s) and the average indoor temperature is the highest (18.14 °C).

3.2 The Streamlines and Temperature Contours

We can see the streamlines and temperature contours when the heating temperature is 40 °C in Figs. 2 and 3. The room is divided into two vortex regions with opposite directions of rotation. In the outer boundary region of the main flow zone, the

Table 2 Indoor temperature and outlet velocity in different situations

The height of the chimney (m)	Heating temperature		Indoor temperature (°C)	Outlet velocity (m/s)
0.5	40 °C		18.53	0.337
1.0	Case 1	35 °C	12.18	0.350
	Case 2	40 °C	14.65	0.360
	Case 3	45 °C	18.14	0.397
1.5	40 °C		14.3	0.405

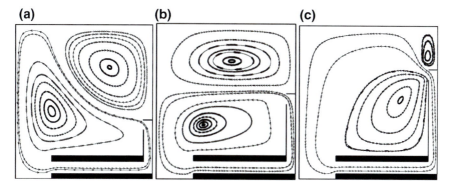

Fig. 2 Streamlines when the heating temperature is 40 °C. **a** The chimney height is 0.5 m, **b** the chimney height is 1 m, **c** the chimney height is 1.5 m

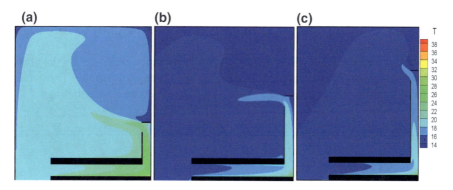

Fig. 3 Temperature contours of the heating temperature are 40 °C. **a** The chimney high is 0.5 m, **b** the chimney high is 1 m, **c** the chimney height is 1.5 m

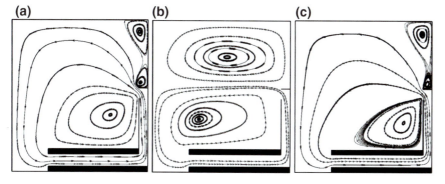

Fig. 4 Streamlines of the chimney high is 1.0 m. **a** The heating temperature is 35 m, **b** the heating temperature is 40 °C, **c** the heating temperature is 45 °C

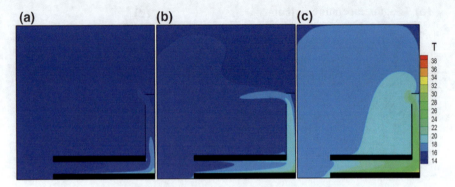

Fig. 5 Temperature contours of the chimney high are 1.0 m. **a** The heating temperature is 35 °C, **b** the heating temperature is 40 °C, **c** the heating temperature is 45 °C

airflow mainly flows from the outlet of the airflow passage to the inlet of the airflow channel. The chimney is higher, and the area occupied by the lower vortex is larger. The temperature is lower.

We can see the streamlines and temperature contours when the height of the chimney is 1.0 m in Figs. 4 and 5. The higher the heating temperature, the smaller the flow radius of the main zone (Excluding the outer area) and the higher the average indoor temperature.

4 Conclusions

The new kind of Kang gets heat energy from the solar collector or the air source heat pump; it is environmental and meets the requirements of coal power modification.

(1) The airflow organization and temperature in the room depend on the height of the chimney and the heating temperature. The room is divided into two vortex regions with opposite directions of rotation, in the outer boundary region of the main fluid zone; the airflow mainly flows from the outlet of the airflow passage to the inlet of the airflow passage. The higher the chimney, the larger the area occupied by the lower vortex and the more uniform the temperature distribution in the room. The shorter the chimney, the smaller the area occupied by the lower vortex, the higher the temperature of the lower vortex and the greater the temperature gradient in the vertical direction. The location of the temperature stratification is consistent with the height of the chimney. The temperature of the airflow at the exit of the chimney is higher. As the airflow flows, the airflow temperature gradually decreases.

(2) In the main heating area, the temperature gradient of P3 area is the largest and the P1 area is the minimal. When the height of the chimney is 1.0 m, the

average temperature is higher; when the heating temperature becomes higher, the heating temperature is lower than 35 °C, and the average temperature is lower than 14 °C.

(3) The air flows along the airflow channel because of the chimney effect, so the heat transfer is enhanced and the temperature is more uniform because of the flow of the air.

Acknowledgements This study has been supported by the Key Scientific Research Innovation Team Project of Shaanxi Province (Grant No. 2016KCT-14), Shaanxi Provincial Natural Science Foundation (2018JM5091), the Opening Projects of Beijing Advanced Innovation Center for Future Urban Design, Beijing University of Civil Engineering and Architecture (Grant No.30), and the Fundamental Research Funds for the Central Universities (Grant No. zrzd2017003).

References

1. Luo, X.L., et al.: Experimental investigation on contacting heating system assisted by air source heat pump in residential buildings. Energy Procedia **152**, 935–940 (2018)
2. Huang, C., et al.: Study on sola-energy phase change storage Kang composited by parrafin and concrate. J. Xi'an Univ. Arch. Technol. **50**(01), 111–116 (2018)
3. Liu, G.S.: Dynamic simulation of thermal process of solar Kang. J. Green Sci. Tecnol. **05**, 307–309 (2013)
4. Zhuang, Z.: Smoke flow and thermal performance of Chinese Kangs. Dalian University of Technology (2009)

Analysis on Control Mechanism and Response Characteristics of Residential Thermostats in District Heating Systems

Baoping Xu, Xi Wang and Yuekang Liu

Abstract In view of large heat inertia and time lag of district heating systems, the terminal thermostat control mechanism and characteristics are examined through theoretical analysis and field investigation. First, the thermal behavior of space heating systems controlled by radiator thermostat valves in multi-family buildings is analyzed and compared with that of the air conditioning system. Then, several district heating systems in Beijing and Tianjin were investigated, where the room variations under different terminal control strategies were monitored. Results indicate that the regulating capacity of terminal valves for room temperature is poor in the short term, but is mainly reflected in historical accumulated influence in the long term. From this point of view, a longer control period is more appropriate for room temperature regulation. In addition, it is not sufficient to rely on terminal control to conquer overheating problems, and it is necessary to implement an effective strategy for combining the source control and terminal control.

Keywords District heating systems · Terminal thermostat control · Heat inertia · Response characteristics · Convection-radiation ratio

1 Introduction

District heating systems in China has gone through the stages of no regulation, manual regulation, self-regulation, and now has entered the intelligent regulation stage. However, the heating overheating problem has not been substantially improved. Fang [1] has investigated typical central heating systems and found that, overheating loss accounts for about 14% of the total heating supply.

There are many ongoing research projects focused on terminal control for district heating system. Some studies focus solely on control algorithm. More recent studies

---

B. Xu (✉) · X. Wang · Y. Liu
School of Energy, Power and Mechanical Engineering, North China Electric Power University, Beijing, China
e-mail: xubp@ncepu.edu.cn

focus on simulation analysis of the building thermal process or actual measurement of the room temperature. Few studies have investigated the regulation characteristics of heating systems from the mechanism level. Therefore, at present, there is a lack of in-depth and clear understanding of room temperature adjustment mechanism and response characteristics in the field of heating, so that most temperature control products cannot provide a scientific temperature control method.

In order to identify the main problems in the terminal regulation of current heating systems, and explore intelligent regulation methods for residential thermostats, this paper preliminarily analyzed the mechanism and response characteristics of the thermostat control in the heating system, by theoretical analysis as well as experimental investigation, and finally put forward some new ideas on the intelligent regulation of the thermostat control for the heating systems.

2 Thermal Process Analysis

2.1 Convection-Radiation Ratio

Compare two kinds of typical terminal heating elements in this study: radiators and fan coils. The differences in heat transfer mechanism between these two terminal forms are indicated in Fig. 1. First, radiators emit heat by convection and radiation, while fan coils immediately affect the thermal state of indoor air by mixing air. For this reason, heat emission from radiators should be divided into two parts-convection heat and radiation heat. However, most of the related researches that analyzing the influence of heating system on building thermal process, always take heat emitting from radiators as a whole item in the dynamic heat transfer equation of the indoor air [2, 3]. It causes the simulated room temperature changes to a certain extent to deviate

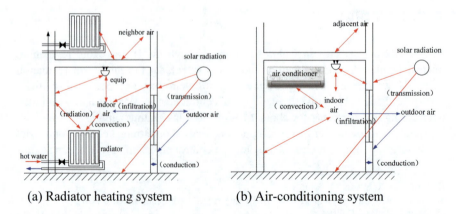

Fig. 1 Thermal process comparison of heating system and air conditioning system

from the actual situation, and much faster than the actual room temperature response rate. As a result, the proposed control strategy that based on above simulation analysis is difficult to achieve a regulatory role in the actual system.

2.2 Heat Transfer Between the Neighborhoods

Second, heat transfer between the neighborhoods is not only caused by the temperature difference but also due to the indirect effects of the radiant heat. For example, radiators heat emitting by radiation influences the outside surface temperature of the partition wall of the adjacent room. Existing researches on thermal dynamic analysis of building heating systems, are often adopting a single-room model [4], or only considering heat transfer caused by the temperature difference between neighborhoods [5], but seldom considering the radiate influences of heating devices between households. It will cover up the real change process, and weaken the heat transfer influence between the neighborhoods during the process.

2.3 Thermal Inertia of Heating System

Furthermore, the thermal inertia of a radiator heating system is greater, making it continuously emit heat at a certain rate even when the heating system is turned off. Therefore, room temperature responding to the terminal regulation of the radiator heating system appears more insensitive.

3 Test Results

Two district heating systems, one in Beijing and another in Tianjin, were investigated during the heating season. The residential thermostats were adjusted under two typical modes, with the main purpose of observing room temperature response to terminal control. Under each situation, choosing a household that has the same house type, floor, and orientation with the regulated room as a reference object, maintaining the state of valves, and measured its room temperature change as a benchmark.

3.1 Room Temperature Response Rate

As shown in Fig. 2, the impact of adjusting valves in a short term (e.g., during one day) on the indoor temperature is very small. When kept the valves closed for 8 h in

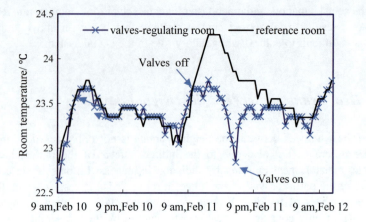

Fig. 2 The measured room temperature change under valves on-off adjustment

Mode I, the maximum temperature difference is less than 1 °C compared with the reference room that maintains the valves opening fully.

In other words, the room temperature reduction caused by valves closing is only about 0.1 °C/h, and the temperature increasing rate when the valves opening again is about 0.5 °C/h. We also have gained similar results in several other tests of district heating systems [6, 7]. While in air conditioning system, the room temperature adjustment rate is about 3 °C/h [8], which is 6–30 times the room temperature regulating speed in heating systems.

3.2 Thermal Inertia Effect in Time Domain

The testing results of Mode II are shown in Fig. 3. When radiator valves of the tested user were turned off, the room temperature decreased from 20.8 to 17.4 °C over 60 h, with the average cooling rate of only 0.06 °C/h. Since then, there is no obvious cooling process, and the fluctuation law of room temperature is consistent with that of the reference room. It can be seen that the thermal inertia effect time domain of indoor heating system and enclosure structure is as long as 60 h, that is, the terminal valve regulation at a certain moment cannot produce an immediate effect, but it needs as long as several days to gradually affect the room temperature.

3.3 Radiator Surface Temperature Variation

When the flow rate transiently becomes zero, the temperature on the radiator surface gradually decreases, as shown in Fig. 4. According to the experimental data,

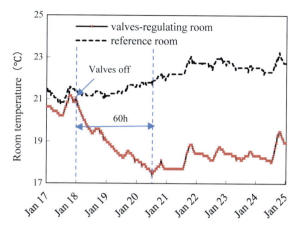

Fig. 3 The measured room temperature decrease when valves closed for a long time

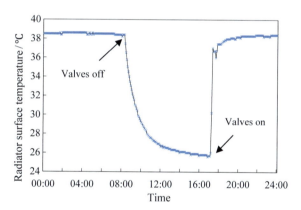

Fig. 4 Measured curve of radiator surface temperature under valves on-off mode

when the valves closed from fully opening, the decrease of the radiator surface average temperature is around 10 °C after 2.5 h. It can be seen that the influence of terminal flow regulation on the change of the radiator surface temperature is very slow and week. In case of taking 30 min as an on-off cycle, even if the whole cycle valves are closed, the radiator surface average temperature drop is not more than 4.5 °C, and the valves regulating effect for radiator heat release is quite limited.

3.4 Disturbance of Solar Radiation and Ventilation

The terminal control has a cumulative effect on room temperature in a long time domain and will not cause a large disturbance of room temperature in a short term. However, when the room temperature of central heating users fluctuates greatly in a short term, it is often caused by solar radiation or windowing ventilation, as shown in Figs. 5 and 6. From the measured room temperature curve, we can see that room

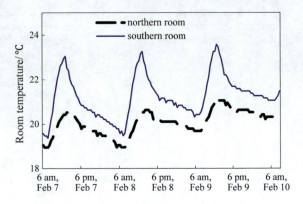

Fig. 5 Measured room temperature under the influence of solar radiation

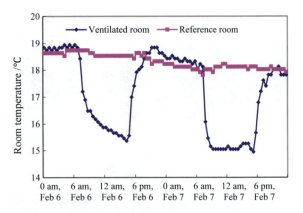

Fig. 6 Measured room temperature fluctuation under the influence of ventilation

temperature rises about 4 °C within 6 h due to solar radiation, and it drops about 2 °C per hour at the beginning of opening the window.

Terminal control plays a role under the premise of basic room temperature, that is, the change range of room temperature in the high-frequency time-domain (such as 1 h) is mostly caused by ventilation, solar radiation, external temperature, etc., and only a small proportion is caused by the change of terminal heat dissipation.

4 Discussions

4.1 The Goal of Terminal Control

At present, most of the related researches on the terminal control of the heating system tries to realize the heating quantity regulation, so as to satisfy the users' demand for heat. In fact, due to the huge thermal inertia of the central heating system and the influence of the coupling between neighboring households, even if there are terminal control valves, it is impossible for the user to realize the dynamic real-time regulation of room temperature.

The achievable goal of terminal control is not to adjust the temperature at the will of any heat user, but to embed certain professional and intelligent algorithms to homogenize the room temperature, and to play a balancing role in restraining local overheating. Secondly, the optimal state of room temperature adjustment is not to precisely control it at a certain point but to make it meet the user's thermal demand and fluctuate within the comfort level [9]. At this point, the thermal inertia of the system is transformed into a factor that can be effectively utilized in the process of heating control optimization, so that under partial load conditions, there are a variety of strategies for room temperature control valves [10, 11]. Therefore, the goal of the terminal control of heating system should be transformed from the precise control of room temperature of a single household to the multivariate optimization, with considering the thermal comfort zone of the user group and other operating parameters (hydraulic conditions, total energy consumption of the system, etc.).

4.2 The Time Scale of Terminal Control

The terminal control is difficult to realize the real-time dynamic adjustment of room temperature. It should adopt the low-frequency regulation mode of quasi-steady-state, supplemented by the compensation action of high-frequency disturbance such as solar radiation and ventilation.

So for determining the time step and adjustment cycle of terminal control, it's better to use the control strategy of dual-frequency domain, with long cycle (for example, a week or more) low-frequency control to determine valve opening or duty ratio value, and short cycle (such as: 1 h or less) high-frequency control to response the solar radiation and ventilation effect.

4.3 Coordinate of Terminal and Source Control

According to simulation results of control effectiveness of TRVs in the district heating system, we find that [12], even if there are thermostats for room temperature control, obvious overheating still occurs, when the supply water temperature is too high. Several actual engineering data also reflect that [1], for the central heating system, it is impossible to completely rely on the terminal control to overcome the overheating problem.

For the room temperature control of the heating system, it is not simply dependent on the terminal control, but need to coordinate with the source regulation, and achieve a decoupling adjustment between the source and the terminals. For the overall overheating of the system in the time distribution, it depends on the supply water temperature regulation. While the overheating of local space or caused by personal thermal comfort can be prevented through the terminal control.

5 Conclusions

At present, most of the studies on the terminal control of the heating system are still in the phase of "phenomenological" analysis of room temperature data, lacking in-depth analysis on the role of terminal regulation and the response characteristics of room temperature. Although there are many studies on air conditioning temperature control, the mechanism of terminal thermostat control of the heating system is different and more complex, which leads to many problems if the air conditioning temperature control algorithm is directly applied to the heating system.

Based on the analysis of the building thermal process and the regulation mechanism of heating systems, this paper made a preliminary theoretical and practical analysis of the response characteristics of room temperature under terminal control. The main conclusions are as follows:

(1) The influence of radiators' radiant heat, the heat transfer between neighbors and the thermal inertia of the system body, make the thermal inertia of the building heating system far greater than that of the air conditioning system. The transient adjustment ability of terminal control to room temperature is very weak (cooling rate is less than 0.1 °C/h during valves closing stage), and its function mainly reflects in the long-term historical accumulation effects (the time domain of thermal inertia effect is about 60 h).
(2) It is impossible for the user to realize the dynamic real-time regulation of room temperature. The achievable goal of terminal control is to homogenize the room temperature and to play a balancing role in restraining local overheating.
(3) It's better to use the control strategy of dual-frequency domain, with the long period low-frequency control to determine valve opening or duty ratio value, and short cycle high-frequency control to response the solar radiation and ventilation effect.
(4) It is necessary to coordinate terminal control with the heat-source regulation and achieve a decoupling adjustment between them.

Acknowledgements This research is supported by the Natural Science Foundation of China (51708210) and the Fundamental Research Funds for the Central Universities (2018 MS023).

References

1. Fang, H., Xia, J.J., Lin, B.R., et al.: Research on current situation and technical route of clean heating in northern cities. Dist. Heat. **1**, 11–18 (2018)
2. Wang, H.: Research and simulation of dynamic performance of heating system and application in electric peak shaving. Master thesis, Shandong University, Shandong (2016)
3. Shao, B., Sun, C.H., Qi, C.Y.: Building thermal characteristic simulation under heating system. Heat. Vent. Air Cond. **47**(1), 124–128 (2017)
4. Gao, X.J.: Research on room temperature control strategy for central heating system. Master thesis, Zhejiang University, Zhejiang (2013)
5. Kang, C.S., Hyun, C.H., Park, M.: Fuzzy logic-based advanced on–off control for thermal comfort in residential buildings. Appl. Energy **155**(6), 270–283 (2015)
6. Xu, B.P., Fu, L., Di, H.F.: Test and analysis on energy saving effects of radiator manual adjustment. Heat. Vent. Air Cond. **37**(11), 82–87 (2007)
7. Xu, B.P., Fu, L., Di, H.F.: Simulation of room thermal dynamics with thermostatic temperature control. J. Tsinghua Univ. (Sci. Technol.) **47**(6), 757–760 (2007)
8. Wang, F.L., Chen, Z.L., Jiang, Y., et al.: Indoor thermal environment control based on thermal sensations. Heat. Vent. Air Cond. **45**(10), 72–75 (2015)
9. Xu, H.B., Duan, M.L., Jin, Q., et al.: Thermal comfort in transient conditions. Chin. J. Ergon. **18**(4), 82–87 (2012)
10. Liu, L.B., Fu, L., Jiang, Y.: A new "wireless on-off control" technique for adjusting and metering household heat in district heating system. Appl. Therm. Eng. **36**(4), 202–209 (2012)
11. Li, Y.M., Xia, J.J., Jiang, Y.: Reducing return water temperature by terminal on-off control. Dist. Heat. **4**, 45–49 (2015)
12. Xu, B.P., Huang, A., Fu, L., et al.: Simulation and analysis on control effectiveness of TRVs in district heating systems. Energy Build. **43**(5), 1169–1174 (2011)

The Study on the Influence of the Containing Ice Ratio on the Flow and Heat Transfer Characteristics of Ice Slurry in Coil Tubes

Changfa Ji, Chenyang Ji, Huan Zhang, Liu Chen and Xiyuan Yu

Abstract The latent heat storage of ice slurry has great potential for application in cold storage air conditioning. However, due to the poor control of the containing ice ratio in the process of transportation and storage, there are phenomena such as stratification, aggregation, and even blockage in pipelines and as a result, such latent heats are still not used on a large scale. The effect of the containing ice ratio on ice flow and heat transfer characteristics of ice slurry in coils has been studied in this paper by numerical simulation method with inlet velocity of 0.1 m/s and pipe outside diameter of 10 mm as the simulation conditions. The results show that containing ice ratio of the ice slurry has a greater influence on the flow and heat transfer of the ice slurry in the coils. To be specific, if the containing ice ratio is too large, the ice particles can form a stagnant layer on top of the pipe, and even ice blocking phenomenon occurs at the bent position of the pipe, which is unfavorable to the flow and heat exchange of ice slurry; when the containing ice ratio of ice slurry is too low, the latent heat released by phase change is not sufficient enough and the heat exchange capacity is rather small. In order to effectively reduce the ice blocking phenomenon, ensure the smooth flow of ice slurry, enhance the heat transfer effect, and reduce the power consumption, according to this study, the ideal containing ice ratio of ice slurry is between for 5–10%.

C. Ji (✉) · C. Ji · H. Zhang · L. Chen · X. Yu
School of Energy Resources, Xi'an University of Science and Technology,
710054 Xi'an, China
e-mail: jicf@xust.edu.cn
URL: http://xust.edu.cn/

C. Ji
e-mail: 7929937422@qq.com

H. Zhang
e-mail: 1097464064@qq.com

L. Chen
e-mail: 745491261@qq.com

X. Yu
e-mail: 1456111240@qq.com

Keywords Heat storage air conditioning · Ice slurry · Heat transfer · Coil

1 Introduction

Cool storage air conditioning technology plays an important role in the peak load shifting of power load. Ice slurry has higher capable of heat transfer. In the regional cooling system, if using the ice slurry with the containing ice ratio of 5%, ice slurry can provide more twice as much cooling capacity as the traditional system's at the same flow rate, and there isn't need to change pumping equipment, distribution network or storage tank for its transportation and storage [1–3]. Since the ice slurry flow is different from the general two-phase flow, it may be stratified in the long pipeline, condensed into ice block through aggregation, then even ice jams. Therefore, the transportation process of ice slurry is an urgent problem to be solved. Many studies have been done on the preparation, transportation and heat transfer of ice slurry at home and abroad.

Hong studied the scraping ice slurry method by combining numerical simulation with experiment [4]. Japanese researcher Yukio Tatsuo carried out the ice slurry preparation experiment by super-cooling method [5]. The preparation method of ice slurry was experimentally studied by researchers at Huazhong University of Science and Technology [6] as well. Qu Kaiyang of Tsinghua University successfully realized the continuous preparation of ice slurry from super-cooled water [7]. Wang Wei of Zhejiang University studied a dual direct contact ice slurry preparation system [8]. Liang Kunfeng of Henan University of Science and Technology borrowed a fluidized bed to swimmingly carry out the preparation experiment of fluid ice [9]. Liu Shengchun of Tianjin Commercial University designed a small air-cooled ice slurry preparation device [10]. Yang Liyuan and others of Huazhong University of Science and Technology conducted numerical simulation of the flow of ice slurry in the three-port valve [11]. Xu Aixiang of Central South University conducted a numerical simulation study on the growth rate of ice layer, thickness change of ice layer and its influencing factors during the indirect ice slurry preparation process [12].

However, in all the above researches being done so far, there are very few studies on the flow characteristics of ice slurry in coil. In this paper, numerical simulation method is used to study the flow and heat transfer characteristics of ice slurry in fan coil. And the paper provides an analysis of the influence of the containing ice ratio on the flow and heat transfer characteristics.

2 Physical Model Establishment, Meshing and Boundary Condition Setting

2.1 Geometric Model and Mesh Generation

This paper has taken the evaporator coil of household split air conditioning cabinet as a research subject. The coil is made of three rows of copper pipes in series. The pipe spacing is 20 mm, the thickness of the pipe wall is 0.35 mm, and the length of the straight pipe is 30 mm. Thick copper sheets with fins of 0.15 mm can increase heat dissipation efficiently. The rib spacing is set to 1 cm, the coil diameter is 10 mm, and the inlet speed is 0.1 m/s. Under the other condition unchanged, the flow of ice slurry with the containing ice ratio (volume percentage) of 1, 5, 10 and 20% in the pipe were numerically simulated. The geometric model of the coil is shown in Fig. 1.

In this simulation, because the ice slurry fluid is viscous, the boundary layer is partitioned firstly, then the quadrilateral unstructured mesh is used to partition the surface mesh, and finally the hexahedral mesh is used to partition the main mesh, including the wedge-shaped unstructured mesh in the appropriate position.

2.2 Boundary Condition Setting

Boundary conditions: In this paper, the inlet of the two-phase flow of ice slurry into the pipeline is set as the velocity inlet for 0.1 m/s, and the inlet temperature of the fluid is set to 274 K; the outlet of the coil is set as the free outlet, the wall of the

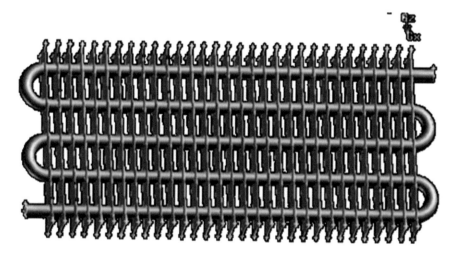

Fig. 1 Pipe diameters of 10 mm coil models

tube is set as the solid wall, and the coefficient of heat transfer is 377 W/(m K). The outer boundary is defined as the heat convection boundary with the fluid temperature of 296 K. The ice-containing rate of the inlet of the ice slurry was set to 1, 5, 10, and 20%. The flow and heat transfer characteristics of ice slurry with different containing ice ratios were compared.

2.3 Basic Assumptions

In the simulation calculation process, the following assumptions are made:

The radial slip of the ice crystal grains is ignored.
The volume change caused by ice particle-phase transformation is neglected.
The ice slurry fluid is regarded as Newtonian fluid.
Ice grains are regarded as uniform flow.
There is no internal heat source in the pipe except the melting latent heat of ice grains.
Only the heat transfer of ice slurry is studied in the coil unit.

3 Simulation Results and Analysis

3.1 Influence of the Containing Ice Ratio on Flow Characteristics

Figures 2, 3, 4 and 5 indicate the flow velocity of the ice slurry in the pipeline under different ice conditions.

In Fig. 2, when the ice content is 1%, the flow characteristics are similar to those of pure water, whether in a straight pipe or a bent pipe. In Fig. 3, when the

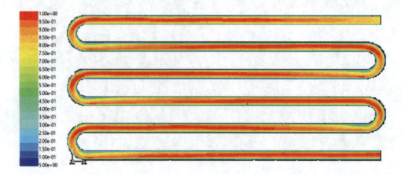

Fig. 2 Distribution of velocity in the pipeline when the containing ice ratio is 1%

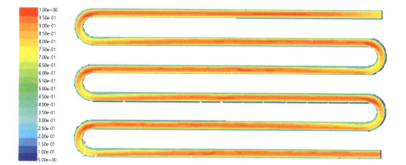

Fig. 3 Distribution of velocity in the pipeline when the containing ice ratio is 5%

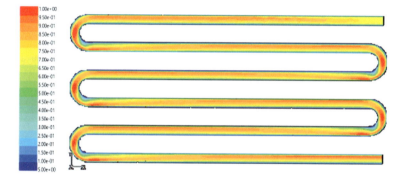

Fig. 4 Distribution of velocity in the pipeline when the containing ice ratio is 10%

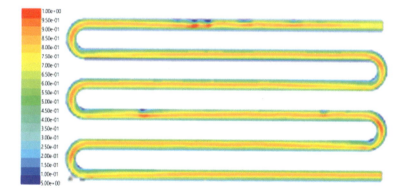

Fig. 5 Distribution of velocity in the pipeline when the containing ice ratio is 20%

containing ice ratio is 5%, the flow velocity in most of the pipeline reaches 0.123 m/s, and the secondary flow phenomenon is obvious at each elbow, but there is no ice jam affecting the flow velocity. In Fig. 4, when the containing ice ratio is 10%, the maximum flow velocity can reach 0.25 m/s; a small part of ice particles gather at the end of the bend, but it has little effect on the flow. In Fig. 5, the containing ice rate is 20%. Because the ice particles with high containing ice rate are gathered at the top of the pipe and detained seriously. There is ice block at the bend, which forms a small local resistance, which changes the flow rate of fluid. The fluidity of ice slurry is not good, and the flow velocity is high in the middle of the elbow.

3.2 Influence of the Containing Ice Ratio on the Distribution of Ice Particles in the Pipeline

Figures 6, 7, 8 and 9 demonstrate the distribution of ice particles in ice-flow tubes with different the containing ice ratio.

It can be seen from Fig. 6, when the ice content is 1%, the melting speed is faster without ice particles accumulating. In Fig. 7, when the ice content is 5%, the unmelted ice particles accumulate at the top of the tube, but the amount is small, and the contact surface of the ice particles with water is large, which is favorable for the phase change of the ice particles to release latent heat and flow. The phenomenon of secondary flow near the elbow is not obvious.

It can be seen from Fig. 8 that when the containing ice ratio is 10%, almost all of the unmelted ice particles accumulate at the top of the pipe, which is consistent with the standard three-layer model of ice flow. After flowing through the elbow, the ice particles will be suspended and accumulated at the top, and there are no ice particles remaining on the outside of the elbow when entering, but as the two-phase flow continues to flow, ice particles accumulate on the outside of the elbow. The vortex

Fig. 6 Ice particles in the pipeline section with the containing ice ratio of 1%

Fig. 7 Ice particles in pipeline section with the containing ice ratio of 5%

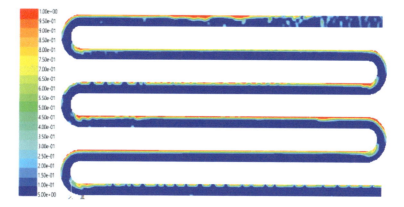

Fig. 8 Ice particles in the pipeline section with the containing ice ratio of 10%

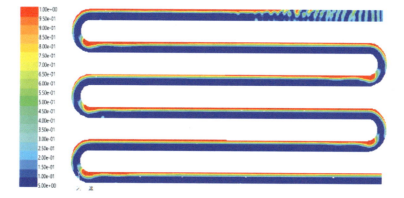

Fig. 9 Ice particles in the pipeline section with the containing ice ratio of 20%

zone is formed inside the elbow, and the secondary flow phenomenon is obvious, and the ice particle retention layer just entering the pipeline is slightly thicker than other parts. As can be seen from Fig. 9, when the ice content of the ice slurry is 20%, the ice particles near the inlet of the pipeline are suspended in the water. Many other parts of the pipeline appear ice particle retention layer, and some parts even appear to be more serious ice blocking phenomenon. Ice particles near the bend are all gathered at the end of the bend, forming larger blocks of ice. Due to the mixing of secondary flow and mainstream, the trajectory of some particles is spiral-shaped. The main aggregation area is below the negative pressure zone of the bending pipe, not only is not conducive to the heat of phase transformation of ice but even blocks the pipeline so that the ice slurry can not flow in the pipeline.

Table 1 shows the import and export ice content and melting ice content of ice slurry with different containing ice ratio.

From Table 1, it can be seen that the melted ice slurry with the containing ice ratio of 5–0% has a higher ice content, which indicates that the heat transfer effect is the best. Before the containing ice ratio increases to 10%, the melting ice content of ice slurry increases with the increase of the containing ice ratio at the entrance, which indicates that the higher the containing ice ratio is at the entrance, the more ice crystals melted and latent heat released during the flow through the coil, and the better efficiency heat transfer will be. When the containing ice ratio exceeds 20%, instead of the increase, melting ice content decreases relatively.

3.3 Effect of the Containing Ice Ratio on Heat Exchange and Pressure Drop

According to the data obtained from the simulation, combined with the calculation formula of ice slurry heat exchange, the heat exchange under different containing ice ratio conditions was calculated.

$$Q_w = WC_{pw}(t_{w2} - t_{w1}) + W(\Delta IPF)L$$

Among them, Q_w represents the heat exchange of ice slurry (kW); W is the flow rate of ice slurry (kg/s); C_{pw} is the specific heat of water at constant pressure

Table 1 Ice content of ice melting with different containing ice ratio

Inlet ice ratio (IPF$_1$) (%)	Flow rate (m/s)	Export ice ratio (IPF$_2$) (%)	Melting ice rate (Δ IPF) (%)
1	0.1	0	1
5	0.1	1.90	3.10
10	0.1	6.02	3.98
20	0.1	16.76	3.24

(KJ/Kg K); T_{W1} and T_{W2} are the inlet and outlet temperatures of ice slurry (K); ΔIPF is the containing ice ratio (%) of ice slurry melting; L is the latent heat of ice melting (335 kJ/Kg).

Figures 10 and 11 show the temperature difference, heat exchange and pressure drop of coil under different containing ice ratio.

It can be seen from Fig. 10 that the temperature difference between the inlet and outlet of the ice slurry does not increase with the increase of the containing ice ratio under given channel diameter and ice slurry's speed. When the containing ice ratio is 1%, the temperature difference between import and export is 2 °C. When the containing ice ratio is 5%, the temperature difference between import and export is 7 °C. When the containing ice ratio is 10%, the temperature difference between import and export is 9 °C. When the containing ice ratio is 20%, the temperature difference between import and export is 8 °C.

It can be seen from Fig. 11 that when the containing ice ratio increases from 1 to 10%, the total heat exchange of ice slurry increases with the increase of the containing ice ratio; when the containing ice ratio exceeds 10%, the total heat exchange of the ice slurry decreases with the increase of the containing ice ratio. This is because the heat exchange of ice slurry is proportional to the flow rate. When the flow rate is constant, the heat exchange rate depends on the melting ice content of ice slurry and the difference in temperature between the inlet and outlet. When the containing ice ratio at the ice slurry inlet is 20%, the phase transformation of ice

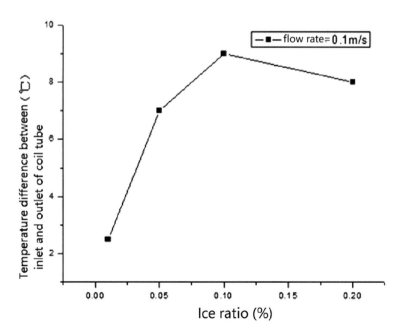

Fig. 10 Temperature difference between the Inlet and outlet of ice at different ice ratio

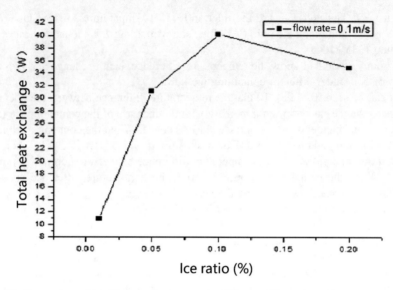

Fig. 11 Comparison of total heat exchange capacity of ice at different ice ratio

grains is poor and when ice particles flow through the pipeline, so the total heat exchange of ice slurry will be reduced.

4 Conclusion

Through the simulation of the heat transfer characteristics of the ice slurry with different ice ratio flowing in the coil tube with an entrance speed of 0.1 m/s and an external diameter of 10 mm, the conclusion is drawn: The ice ratio of the ice slurry has a great influence on the heat transfer of the ice slurry in the coil tube. The ice ratio is too large, and ice particles will form a stagnant layer at the top of the pipe, and even ice blockage will occur at the bend point, which is very unfavorable to the flow and heat transfer of the ice slurry. In order to effectively reduce the phenomenon of ice blocking, ensure the smooth flow of ice pulp, enhance the heat transfer effect, ice ratio should not exceed 20%. However, when the ice ratio of the ice slurry is low, the total number of ice particles is limited, and the latent heat released by the phase transition is too small to obtain the required heat exchange. Therefore, the ideal ice ratio of the ice slurry in this paper is 5–10%.

Acknowledgements We would like to thank our committee members, reviewers, authors, and other participants for support to ISHVAC 2019. This research has been sponsored by the China National Natural Youth Scientific Foundation No. 51404191.

References

1. Li, Y., Liu, S., Rao, Z., et al.: Present state and perspectives of storage and melting as well as flow and heat transfer of ice slurry. Cryog. Supercond. **40**(1), 55–60 (2012)
2. Li, X., Hou, Y., Zhang, X.: Study advance of flow behavior and heat transfer performance of ice slurry. Refrig. Air-Cond. **8**(7), 15–18 (2007)
3. Jan, X.: A study of the flow and heat transfer characteristics of binary ice slurry. Nanjing University of Aeronautics and Astronautics, Nanjing (2008)
4. Kitanovski, A., Poredos, A.: Concentration distribution and viscosity of ice-slurry in heterogeneous flow. Int. J. Refrig. **25**, 827–835 (2002)
5. Domanski, R., Fellah, G.: Thermoeconomic analysis of sensible heat thermal energy storage systems. Appl. Therm. Eng. **18**(8), 693–704 (1998)
6. Xiao, K.: The research on a new dynamic ice making devise. Huazhong University of science and technology, Wuhan (2010)
7. Qu, K.: Factors affecting supercooled water freezing occurrence. Acta Energe. Sols. Sin. **24**(6), 814-821 (2003)
8. Wang, W.: Heat transfer study and entropy generation analysis on contact ice slurry generator. Zhejiang University, Hangzhou (2014)
9. Lang, K., Du, J., Wang, L.: Investigations on flow characteristic of ice slurry in C90 elbow based on numerical simulation method. Cryo. Supercond. **40**(1), 62–68 (2012)
10. Liu, S., Zhu, C., Yang, X.: Experimental analysis on a small ice slurry preparation unit. Cryog. Supercond. **43**(3), 56–60 (2015)
11. Yang, L.: The flow and heat transfer analysis of ice slurry in the pipe network. Huazhong University of science and technology, Wuhan (2012)
12. Xu, A., Liu, Z.: Dynamic characteristics of indirect ice slurry generation system. J. Centl South Univ. Sci. Technol. **46**(2), 710–714 (2015)

Multivariable Linear Regression Model for Online Predictive Control of a Typical Chiller Plant System

Jiaming Wang, Tianyi Zhao and Meng Xu

Abstract Since much attention has been paid to the accuracy of prediction models in chiller plant, the practicality and feasibility of models are often compromised for superior performance. In this paper, a data-driven model and online predictive control strategy are presented for a typical chiller plant to optimize its overall performance and energy consumption from the perspective of real-time application. The model predictive controller is developed based on the multivariable linear regression (MLR) model. Meanwhile, the ordinary least squares (OLS) estimation technique was adopted to identified and update the model coefficients. The input variables of the developed model are the temperature difference of chilled water and the temperature difference of cooling water, which are much easier to obtain in a practical system compared to variables such as water flow and part load ratio, thereby rendering an excellent model capable of easy implementation and duplication. The MLR model-based online predictive control strategy is tested and evaluated in a real chiller plant system. The results of application indicate that the online predictive control strategy can enhance the global COP values by 6.52% on average and reduce electricity consumption by 2.45% daily compared to the local control strategy.

Keywords Regression model · Online application · Predictive control · Chiller plant · Energy efficiency

1 Introduction

The energy consumption of chiller plants accounts for more than 40% of building energy consumption. It is considered to manage the chiller plant with high energy efficiency. The global COP of chiller plant is often used to evaluate the energy

J. Wang · T. Zhao (✉) · M. Xu
Faculty of Infrastructure Engineering, Dalian University of Technology, Dalian, China
e-mail: zhaotianyi@dlut.edu.cn

performance of overall system. Thus, it is significant to operate the chiller plant at an optimal global COP as much as possible [1].

A typical chiller plant consists of different types of components, which makes the characteristics of overall system much more coupled and complicated. And it is challenging to operate the chiller plant appropriately by considering the interactions and characteristics among components. Recently, many studies focused on the optimal solutions for the chiller plant by optimizing control strategies and algorithms [2]. Most of the methods presented in these studies have the capablability in optimal controlling [3–5]. However, the complicated methodologies and executing processes render these significant methods cumbersome to implement online in practical engineering projects. For those studies that considered the feasibility in field applications, their assessments for the methods did not evaluate in a real chiller plant system, only tested on the simulation platform [6].

It is obvious that the researches regarding the practical online application of optimal control in chiller plant are insufficient. Despite the simplified regression model is less capable in illustrating a nonlinear system, its strengths should be taken into consideration while the accuracy is acceptable. It has a superior performance on saving computation time and online implementation.

The aim of this paper is to propose a methodology of modeling and application for the online predictive control of a chiller plant. A multi-input–single-output (MISO) regression model is developed in this study to predict the global COP of chiller plants in real-time. The two input variables are easy to obtain in most field chiller plants; therefore, the model can be easily implemented and duplicated. The output of the model can generally reflect the overall performance of the chiller plant, which makes the model capable of handling complex characteristics of system. Eventually, the model will be applied to a real chiller plant system with acceptable performance.

2 Methods

2.1 Description of the Chiller Plant

Figure 1 illustrates the schematic diagram of the chiller plant, which consisted of chillers, chilled water pumps, cooling water pumps, cooling towers, and water distributor and collector. As shown in Fig. 1, all of the pumps and fans are equipped with a variable frequency drive to ensure that the chiller plant can perform in high-energy efficiency conditions.

The specification of each component of the chiller plant is listed in Table 1. Considering the different operating characteristics between the centrifugal chiller and screw water chiller, we focus only on the centrifugal chiller system herein, which is also the most used chiller system at ordinary times.

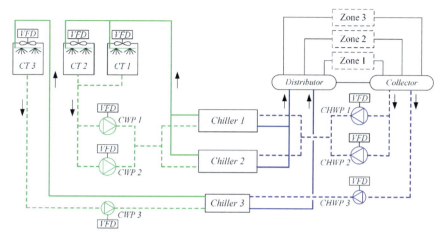

Fig. 1 Schematic of the chiller plant

Table 1 Main components and specifications of the chiller plant

Objective	Type	Specification	Remark
Chillers 1 and 2	Centrifugal chiller	Cooling capacity: 1758 kW; Nominal power: 304 kW	Spared configuration
Chiller 3	Screw water chiller	Cooling capacity: 717 kW; Nominal power: 125 kW	
Chilled water pumps 1 and 2	Variable frequency pump	Lift: 18.5 m; nominal power: 37 kW	Spared configuration
Chilled water pump 3	Variable frequency pump	Lift: 22.0 m; nominal power: 15 kW	
Cooling water pumps 1 and 2	Variable frequency pump	Lift: 23.0 m; nominal power: 37 kW	Spared configuration
Cooling water pump 3	Variable frequency pump	Lift: 22.0 m; nominal power: 15 kW	
Cooling towers 1 and 2	Variable frequency fan	Water flow: 450 m³/h; fan power: 15 kW	Work together
Cooling tower 3	Variable frequency fan	Water flow: 250 m³/h; fan power: 7.5 kW	

The chiller plant was equipped with a supervisory control and data acquisition (SCADA) control system, thereby allowing the operator workstation to monitor the operating status of the chiller plant and provide remote control commands to the actuators in real-time. The global COP was used to evaluate the operating performance of the chiller plant, which can be defined by Eq. (1).

$$\text{COP} = \frac{Q_c}{E_{\text{total}}} = \frac{c_{\text{pw}} m_w (T_{\text{chwr}} - T_{\text{chws}})}{E_{\text{ch}} + E_{\text{cwp}} + E_{\text{chwp}} + E_{\text{ctf}}} \quad (1)$$

where c_{pw} is the specific heat capacity of water (4.19 kJ/(kg °C)), m_w is the mass flow of chilled water (kg/s), Q_c (kW) is the cooling capacity of chiller plant, T_{chwr} and T_{chws} (°C) are the temperatures of chilled water return and supply, respectively, and E_{ch}, E_{cwp}, E_{chwp}, E_{ctf}, and E_{total} (kW) represent the power consumption of chillers, cooling water pumps, chilled water pumps, cooling tower fans, and chiller plant, respectively.

2.2 Multivariable Linear Regression (MLR) and Model Development

Regression analysis has become an extensively used method for statistical data analysis. MLR is an extension of simple linear regression, which involves two or more independent variables and one dependent variable. The MLR model can be described in matrix notation as follows:

$$\boldsymbol{y} = \boldsymbol{X\beta} + \boldsymbol{\varepsilon} \quad (2)$$

where $\boldsymbol{y}$ is an n-dimensional vector of a series of observed values y_i ($i = 1, \ldots, n$) called dependent variables, and n is the number of statistical units of the dataset. X is a matrix of independent variables denoted by a series of ($p + 1$)-dimensional row vectors $\mathbf{x}_i$ ($i = 1, \ldots, n$), where p is the number of independent variables. $\boldsymbol{\beta}$ is a ($p + 1$)-dimensional vector of regression coefficients corresponding to independent variables X_i, and β_0 is called the intercept. ε is an n-dimensional vector of the disturbance term.

In this study, the model is developed with two independent variables ($p = 2$), and the model function can be expressed as follows:

$$y_i = \beta_0 + \beta_1 x_{1i} + \beta_2 x_{2i} + \varepsilon_i \quad (i = 1, 2, \ldots, n) \quad (3)$$

2.3 Model Validation and Model Performance

In this study, the developed prediction model is used for practical applications and also as the basis of predictive control. The requirements for the input variables are easy to obtain and controllable. For a better manipulability of predictive control, the number of variables should not exceed two. Meanwhile, the input variables can reflect the characteristics of the chilled water side and cooling waterside.

Figure 2 shows the Pearson correlation coefficients among the monitored variables of the chiller plant. In addition to the COP, which is the model output, the variables also include the evaporating pressure (p_e), condensing pressure (p_c), flow rate of chilled water (f_{chw}), flow rate of cooling water (f_{cw}), temperature difference of chilled water (ΔT_{chw}), and temperature difference of cooling water (ΔT_{cw}). It can be learnt that the suitable two input variables are ΔT_{cw} and ΔT_{chw}, respectively.

The field operating data are recorded from June 2017 to August 2017. The data log time interval is 5 min. Considering the impacts of system disturbance on real-time data, the mean values of collected data every 15 min are used for model development and validation. Finally, 2092 test datasets are obtained from the collected data. The test data are sorted into two groups by random assignment; the first group with 70% of the test data is used for model development, and the second group with 30% of the test data is used for model validation. The MLR model is developed in R (version 3.4.1), and the modeling results are listed in Table 2.

The accuracy of this model is verified by the validation data that are selected randomly from the datasets. The performance of the developed prediction model is listed in Table 3, the results indicate that the model has a relatively optimistic capability in prediction.

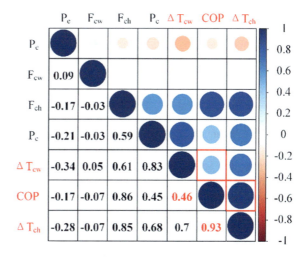

Fig. 2 Plot of Pearson correlation coefficients among variables in chiller plant

Table 2 Results of multivariable linear regression modeling

Coefficients	Estimate	Standard error	t value	P value
ΔT_{chw}	1.2843	0.008278	155.15	<2e−16
ΔT_{cw}	−0.7216	0.014747	−48.93	<2e−16
Intercept	1.1760	0.030469	38.60	<2e−16

Table 3 Performance of MLR model in predicting the COP of the chiller plant

Dataset	R^2 (%)	RMSE	MAE	CV (%)
Training data	95.47	0.3599	0.2470	7.93
Validation data	95.97	0.3328	0.2335	7.42

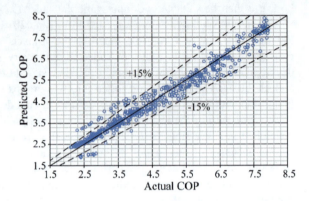

Fig. 3 Scatter plot of predicted COPs against actual COPs

Table 4 Data volume proportions in different ranges of relative error for testing data

Relative error (%)	Data volume (%)
<5	60.29
<10	86.76
<15	92.03
<20	94.10

Figure 3 shows the scatter diagram of the predicted COP values against the actual COP values. The solid line at the catercorner indicates the most desired results, and the two dotted lines beside the catercorner represent the relative error interval within 15%. The data volume proportions for different relative error ranges of validation data are listed in Table 4.

2.4 Online Application of the MLR Model

Figure 4 shows the schematic of the multiple-point predictive method, where the x-direction represents ΔT_{chw}, which is one of the independent variables of the prediction model. The values of ΔT_{chw} are increased from left to right. Similarly, the y-direction represents the other independent variable ΔT_{cw}, and its values are increased from the bottom to the top.

The specific process of the multiple-point predictive method is shown in Fig. 5. The model predictive controller (MPC) integrated with the prediction model built in this study; its inputs were the possible neighboring operating conditions for the next

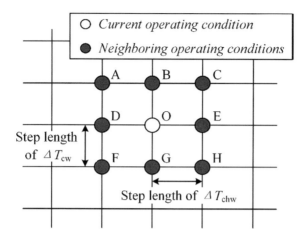

Fig. 4 Schematic of multiple-point predictive method

Fig. 5 Process of multiple-point predictive method

control period, and the outputs were the corresponding predicted COP values of the neighboring operating conditions. Eventually, the corresponding optimized ΔT_{chw} and ΔT_{cw} values of the optimal COP were determined for predictive control strategy.

3 Results and Discussion

3.1 Performance of the Chiller Plant

Figure 6 shows the variation trends of real-time COP in the two typical days. By comparing the real-time COP values between the two control strategies, the COP values of the predictive control strategy are on average 6.52% higher than the values of the conventional control strategy, thus demonstrating the superiority of the predictive control strategy on the energy efficiency of the chiller plant.

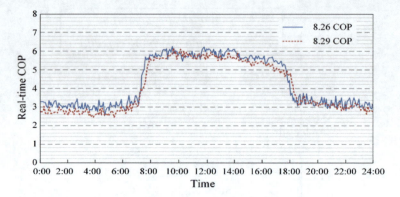

Fig. 6 Comparison of real-time COPs for selected two days

3.2 Energy Consumption of the Chiller Plant

Power consumption values for the selected two days are compared in Fig. 7, in which the x-axis represents the time of the day, the primary y-axis is the power consumption for the chiller and the total, while the power consumption for the pumps and fans are shown on the secondary y-axis.

Compared to the conventional control strategy used in this chiller plant, the daily fractional energy saving of the predictive control strategy can be up to 2.45%. Considering the particularity of the building studied in this research, which imposes strict management regulations on the HVAC air system, the fractional energy saving might be significantly higher in other commercial or institutional buildings.

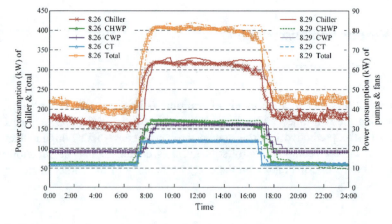

Fig. 7 Comparisons of power consumption for selected two days

4 Conclusions

In this paper, an approach for data-driven modeling and online application was presented to enhance the energy performance of a typical chiller plant. The application results demonstrated that the presented strategy could not only improve by approximately 6.52% on average in the global COP values but also reduced approximately 2.45% of daily electrical energy compared to the original conventional control strategy. Due to the particularity of the building studied in this research, the performance of this approach might be better when applied in other commercial or institutional buildings. Despite the approach proposed herein is based on a typical chiller plant with basic components, the methodology applied could be easily extended to most of the existing chiller plants, which also presents features of practicability and feasibility in integrating the complicated characteristics of the systems. It is significant to apply such a predictive control strategy in a field system online to enhance energy performance in real-time.

Acknowledgements The authors are grateful for the financial support of the National Key Research and Development Project of China No. 2017YFC0704100 (entitled New Generation Intelligent Building Platform Techniques). This work is supported by "The Fundamental Research Funds for the Central Universities" (Grant No. DUT17ZD232), Liaoning Natural Science Foundation Guidance Plan (Grant No. 20180551057), and Dalian High-level Talent Innovation Support Program (Youth Technology Star) (Grant No. 2017RQ099).

References

1. Wei, X., et al.: Modeling and optimization of a chiller plant. Energy **73**, 898–907 (2014)
2. Afroz, Z., et al.: Modeling techniques used in building HVAC control systems: a review. Renew. Sustain. Energy Rev. **83**, 64–84 (2018)
3. Yu, F., et al.: Critique of operating variables importance on chiller energy performance using random forest. Energy Build. **139**, 653–664 (2017)
4. Ma, Z., Wang, S.: An optimal control strategy for complex building central chilled water systems for practical and real-time applications. Build. Environ. **44**(6), 1188–1198 (2009)
5. Wang, S., et al.: An online adaptive optimal control strategy for complex building chilled water systems involving intermediate heat exchangers. Appl. Therm. Eng. **50**(1), 614–628 (2013)
6. Dalibard, A., et al.: Performance improvement of a large chilled-water plant by using simple heat rejection control strategies. Int. J. Refrig. **94**, 1–10 (2018)

Applied Analysis of the Direct Cooling by Cooling Towers in Lanzhou Area

Xiaowei Wang, Wenhen Zhou, Lixin Zhao, Xin Bao, Lu Zheng and Xiaofei Han

Abstract Although the direct cooling to air-conditioning by cooling towers (DCCT) could be practice and energy saving in northwestern area of China due to dry climate, especially it's nearly six months that the outdoor wet-bulb temperature is less than 6 °C in Lanzhou area, and few practical examples can be found. To provide the reference about this, the calculation and analysis about DCCT in Lanzhou area were done to obtain some operating parameters firstly. And then, based on which, a retrofitting work transforming cooling towers attached to the air-conditioning system of a hotel with a large inner zone in Lanzhou to DCCT was carried out, whose energy-saving rate is 84.7% and investment recovery period is about 0.32 years. The paper could be valuable for the practical application of DCCT in Lanzhou area.

Keywords Building energy conservation · Air-conditioning system · Tower direct cooling

Nomenclature

A Area (m^2)
C_P Specific heat of spray water (J kg^{-1} K^{-1})
C_L Specific heat of cooling water (J kg^{-1} K^{-1})
F_o Coil area (m^2)
G_a Air flow rate (kg/h)
h The enthalpy of air (kJ/kg)
h_P Enthalpy of saturated air corresponding to spray water (kJ/kg)
K_o The heat transfer coefficient (kW/m^2 K)
K_m The mass transfer coefficient (kg/m^2 s)

X. Wang · W. Zhou (✉) · X. Han
School of Environmental and Municipal Engineering, Lanzhou Jiaotong University, 730070 Lanzhou, China
e-mail: zwh6888@mail.lzjtu.cn

L. Zhao · X. Bao · L. Zheng
Gansu Institute of Architectural Design and Research Co. Ltd., 730030 Lanzhou, China

L	Water flow rate (kg/h)
Q	Cooling load (kW)
s	The length of heat exchanger coil
t	The temperature of air (°C)
T	The temperature of water (°C)
t_d	The dry-bulb temperature (°C)
t_w	The wet-bulb temperature (°C)

Subscripts

a	Air
i	Inside or inlet
L	Cooling water
o	Outside or outlet
P	Spray water

1 Introduction

In recent years, building energy consumption has increased dramatically. In China, building energy consumption accounts for about 27.6% of total energy consumption, of which air-conditioning energy consumption reaches 40–50% [1]. Therefore, the energy-saving of building HVAC has become particularly important. In general, the air-conditioning system of large public buildings adopts cooling tower to dissipate the heat into environment, and if which was switched to DCCT when condition is allowed in dry areas, a considerable operating cost could be saved. Especially in the northern regions where outdoor air wet-bulb temperatures are relatively low and it's nearly six months that the outdoor wet-bulb temperature is less than 6 °C in Lanzhou area. At present, many scholars have done a lot of research on the principle and application of theoretical analysis and numerical simulation [2]. However, few practical projects can be found, including Lanzhou area.

Lanzhou is located at 36°03′ north latitude and 103°40′ east longitude, whose annual average temperature is 10.3 °C. It is dry all year round, and the air-conditioning cooling load is not very large. Figure 1 shows the outdoor wet-bulb temperature of the typical year in Lanzhou area [3]. The period share whose outdoor wet-bulb temperature is more than 10 °C is only 30–35%. And so, the application of DCCT and its energy-saving potential are worth exploring.

This paper will first obtain the basic parameters for the design and renovation of DCCT in Lanzhou area through theoretical calculations, and then, the transformation work to DCCT of a hotel in Lanzhou is carried out.

Applied Analysis of the Direct Cooling by Cooling ...

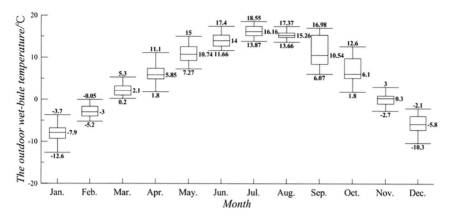

Fig. 1 Outdoor wet-bulb temperature in Lanzhou

2 Theoretical Calculation

2.1 *Principles and Equations*

Considering water quality demand of system and inner zones' cooling requirement of public buildings in winter, a closed wet cooling tower is recommended in this paper. The working mechanism of the closed cooling tower is shown in Fig. 2. The high-temperature water in the cooling coil is cooled by the spray water and air outside the coil. In order to obtain the calculation equations, a heat exchanger coil micro-length ds is selected shown as Fig. 2, and some assumptions are introduced

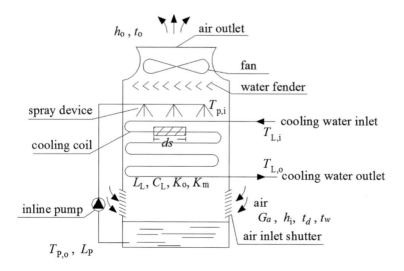

Fig. 2 Working mechanism of closed cooling tower

as follow: (1) The air-mass flow, the spray water flow rate and the cooling water flow rate are regarded as constant; (2) The amount of air actually entering the cooling tower is higher than the theoretical air requirement; (3) Specific heat of water, specific heat of humid air, latent heat of vaporization, convective heat transfer coefficient, convective mass transfer coefficient, etc., are regarded as constant during the cooling process; (4) In the calculation of heat balance, the reduction of the amount of evaporated water is ignored; (5) The temperature of the water film and the water drop are identical, so the thermal resistance of the water is ignored; (6) The saturated water vapor partial pressure and the saturated air enthalpy are linear with the water temperature; (7) Ignoring radiative heat transfer.

Equation (1) shows the heat loss of the cooling water in the micro-element of the heat exchange coil, Eq. (2) shows the heat loss of the spray water in the micro-element of the heat exchange coil and Eq. (3) shows the heat loss of the air in the micro-element of the heat exchange coil.

$$L_L C_L dT = -K_o(T - T_P)dA \tag{1}$$

$$L_P C_P dT_P = K_o(T - TP)dA - K_m(h_P - h)dA \tag{2}$$

$$G_a dh = -K_m(h_P - h)dA \tag{3}$$

where L_L, L_P, G_a are flow rates of cooling water, spray water and air, respectively. C_L, C_P are specific heats of cooling water and spray water, respectively. T, T_P are temperatures of the cooling water and spray water. h, h_P are enthalpy of air and saturated air corresponding to spray water. dA is the heat transfer area of the micro-element. K_o the heat transfer coefficient from cooling water to spray water which is recommended $K_o = 704(1.39 + 0.22T_P)L_P^{1/3}$ by Parker and Treybal [4]. K_m the mass transfer coefficient between the saturated air-water interface and the bulk air which is recommended $K_m = 0.049 G_{ma}^{0.905}$ by Hasan [5].

In order to facilitate the solution of the above equations, Eqs. (1)–(3) are transformed to Eqs. (4)–(6), and then, Eqs. (7) and (8) can be obtained.

$$\frac{dT}{dA} = \frac{K_o}{L_L \cdot C_L} \cdot (T_P - T) \tag{4}$$

$$\frac{dT_P}{dA} = -\frac{K_m}{L_P \cdot C_P} \cdot (h_P - h) + \frac{K_o}{L_P \cdot C_P}(T - T_P) \tag{5}$$

$$\frac{dh}{dA} = \frac{K_m}{G_a} \cdot (h - h_P) \tag{6}$$

$$T_{L,o} = T_P - \frac{1}{\varphi_1 + b_1}\{[b_2(h_o - h_P) + (\varphi_1 + b_1)(T_P - T_{L,i})] \\ e^{F_o \varphi_2} - b_2(h_i - h_P)\} \tag{7}$$

$$T_{L,o} = T_P - \frac{1}{\varphi_2 + b_1} \Big\{ \big[b_2(h_o - h_P) + (\varphi_1 + b_1)(T_P - T_{L,i})\big] \\ e^{F_o \varphi_1} - b_2(h_i - h_P) \Big\} \tag{8}$$

where $a_1 = K_o/(L_L C_L)$, $a_2 = K_m/(L_P C_P)$, $a_3 = K_o/(L_P C_P)$, $a_4 = K_m/G_a$, $b_1 = (a_1 + a_2)$, $b_2 = -a_2$, $b_3 = -ma_3$, $b_4 = ma_2 - a_4$, $\varphi_1 = 1/2\{(-b_1 + b_4) + [(b_1 + b_4)^2 - 4(b_1 b_4 - b_2 b_3)]^{1/2}\}$, $\varphi_2 = 1/2\{(-b_1 + b_4) - [(b_1 + b_4)^2 - 4(b_1 b_4 - b_2 b_3)]^{1/2}\}$.

At the coil inlet, $s = 0$, $T = T_{L,i}$, $T_P = T_{P,i}$, $h = h_o$, $h_P = h_{P,i}$ and at the coil outlet, $s = l$, $T = T_{L,o}$, $TP = T_{P,o} = T_{P,i}$, $h = h_i$, $h_P = h_{P,o} = h_{P,i}$. The calculation process is shown in Fig. 3. It is necessary to assume the cooling water temperature at coil outlet $T_{L,o}$ and the spray water temperature T_P first, and then, two outlet water temperatures (one of which is false) are obtained by Eqs. (7) and (8), respectively. If the difference between assumption value and calculation value is larger than the error controlling range, the cooling water temperature at the coil outlet and the spray water temperature need to be reset and the process is recalculated until the error range controlled is satisfied [6], which is 0.01 in this paper.

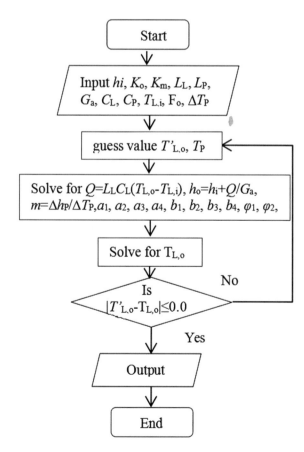

Fig. 3 The flow chart of the calculation

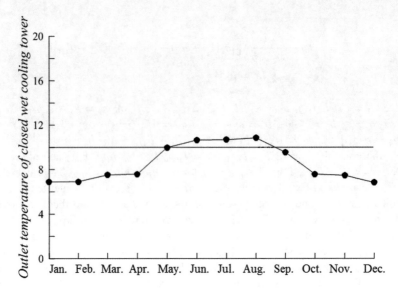

Fig. 4 Outlet temperature of cooling tower

2.2 Results and Analysis

The cooling water temperature at coil inlet is set $T_{L,i}= 13.5$ °C in this paper. The calculation results are shown in Fig. 4. It can be seen that the outlet cooling water temperature is below 10 °C in the transitional seasons and winter, whose wet-bulb temperature is less than 6 °C. Therefore, the wet-bulb temperature of 6 °C is selected as the theoretical switching temperature. It provides a basis for the transform and design of the system of DCCT in Lanzhou.

3 Practice

A hotel with some inner zones in Lanzhou has been investigated, which needs to be cooled throughout the year. In order to reveal the application effect of the DCCT in Lanzhou area, this paper carried out the transformation of the air-conditioning system of the hotel and analyzed its energy-saving effect.

3.1 Project Overview

The total construction area of the hotel is 72516.3 m², of which 60254.8 m² above ground, 12261.5 m² underground, and the total height of the building is 151.8 m.

Table 1 Main equipment and performance parameters

Serial number	Equipment name	Performance parameter	Number
1	Centrifugal chiller	Refrigerating capacity: 1934 kw Input power: 349 kw	2
2	Screw chiller	Refrigerating capacity: 1142 kw Input power: 102×2 kw	2
3	Cooling water pump	$G = 456$ m^3/h, $H = 22$ m, $N = 45$ kw	3
		$G = 270$ m^3/h, $H = 21$ m, $N = 22$ kw	2
4	Chilled water pump	$G = 360$ m^3/h, $H = 13.5$ m, $N = 18.5$ kw	3
		$G = 213$ m^3/h, $H = 13.5$ m, $N = 11$ kw	2
5	Cooling tower	$G = 460$ t/h, $N = 7.5 \times 2$ kw	2
		$G = 270$ m^3/h, $N = 7.5$ kw	1

The air-conditioning system is divided into three zones, of which the podium and the basement are low-zone, the main building is 8–22 floors and the main building is 24–34 floors. In summer, the air-conditioning calculation has a cooling load of 4720 kW, the summer total cooling load index is 78.3 W/m^2, the winter air-conditioning calculates the thermal load to 5250 kW, and the winter total thermal load index is 87.1 W/m^2. The main equipment and parameters are shown in Table 1.

In order to reduce operating costs and saving energy, the hotel air-conditioning system was transformed into DCCT shown in Fig. 5. In April and October, mode 1 (direct cooling to air-conditioning by cooling tower) is set by manual and other times, mode 2 (traditional mode of air-conditioning) is used. The supply water temperature and return water temperature of FCU and the outlet cooling water temperature of the DCCT are set as 10, 15 and 9 °C, respectively. The reconstruction cost of the air-conditioning system is about ¥137 000. After the renovation, the system of DCCT reduces the energy consumption of the refrigeration unit to 74 378.4 kW h, save the operating cost is about ¥5,5738.8 (the commercial electricity price in Lanzhou was ¥0.75/(kW h)).

3.2 Operation and Analysis

In order to obtain the energy-saving potential of DCCT, the direct cooling time of the DCCT in the hotel is counted when wet-bulb temperature is less than 6 °C. The hotel of air-conditioning operation period is from 8:00 to 23:00, and the calculation results are shown in Table 2.

Because the timeshare whose outdoor wet-bulb temperature in September is lower than 6 °C is only 3% based on Table 2, DCCT is not recommended for

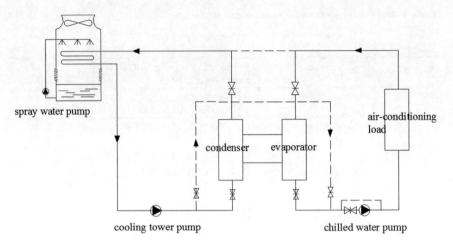

Fig. 5 Cooling tower indirect cooling system

cooling. In January, February, November, and December, the outdoor wet-bulb temperature is almost lower than 6 °C. The period share in March, April, and October whose outdoor wet-bulb temperature was lower than 6 °C accounts for 86.3, 34.4 and 36%, respectively. During January, February, March, April, October, November and December, the total time whose outdoor wet-bulb temperature was below 6 °C is 4197 h. Consider the hotel running time, there are 2672 h for DCCT in these months. Therefore, based on above calculation, The DCCT system can reduce the power consumption of the refrigeration unit to 577 729.73 kW h, which is about ¥43,3297.3, and the energy-saving rate of 84.7%. The recovery period of additional investment is about 0.32 years.

Table 2 Tower cooling time statistics in Lanzhou area

Month	Cumulative time (h) (the wet bulb temperature is less than 6 °C)	Percentage of cooling time (%)	Cumulative time (h) (the wet bulb temperature is less than 6 °C) (8:00–23:00)	Percentage of cooling time (%) (8:00–23:00)
Jan.	744	100	495	99.8
Feb.	672	100	448	100
Mar.	676	90.8	428	86.3
Apr.	338	47	165	34.4
Sep.	61	8.3	14	3
Oct.	322	43.2	179	36
Nov.	701	97.3	461	96
Dec.	744	100	496	100

4 Conclusions

Aiming at decreasing the rapid growth of HVAC energy consumption in public buildings, this paper firstly obtained the relevant parameters of the DCCT in Lanzhou area through calculation. Then, the air-conditioning cooling system of a hotel in Lanzhou was reconstructed and analyzed. The following are conclusions.

(1) The calculation method about DCCT is reasonable and feasible.
(2) Considering some inner zones of public buildings being there, it is reasonable that the outdoor wet-bulb temperature of 6 °C is set as the switching temperature in transitional seasons and winter when DCCT is used in Lanzhou area.
(3) For the hotel, the energy-saving rate of DCCT is 84.7% in January, February, March, April, October, November, and December, and additional investment recovery periods are about 0.32 years.

References

1. Li, L.: An adaptive research for water-side free cooling in Guizhou. Guizhou University, China (2018)
2. Hu, G., Zhang, W.: Energy saving analysis of tower cooling system for a hotel building in Changsha. HVAC **47**(07), 63–66 (2017)
3. Meteorological Information Room of China Meteorological Administration, Tsinghua University. Special meteorological data set for China building thermal environment analysis. China Construction Industry Publishing house (2005)
4. Parker, R., Treybal, E.: The heat and mass transfer characteristics of evaporative coolers. Chem. Eng. Prog. Symp. **57**(32), 139–149 (1961)
5. Hasan, A., Siren, K.: Performance investigation of plain circular and oval tube evaporatively cooled heat exchangers. Appl. Therm. Eng. **24**(5–6), 777–790 (2004)
6. Li, Y.: Closed Cooling Tower for Air-conditioning. China Building Industry Press, China (2008)

Experimental Investigation on Humidity-Sensitive Properties of Polyimide Film for Humidity Sensors

Jianyun Wu, Wenhe Zhou, Xiaowei Wang and Shicheng Li

Abstract Due to excellent properties, polyimide (PI) film is widely used in the capacitive humidity sensor as the humidity-sensitive medium. However, some key properties of this film are uninvestigated and need further research, especially those related to sensor characteristics. So, samples of PI films were self-synthesized in a laboratory. In order to test their effective diffusion coefficients using the self-build test system, films were created from different chemical reagents (PMDA-ODA, BPDA-ODA, BPDA-BAPP and PMDA-BAPP) with four varying thicknesses and four separate concentrations. They were tested in three disparate temperatures, respectively. The effective diffusion coefficients of water molecules in these PI films were 4.25×10^{-13}, 5.49×10^{-13}, 6.34×10^{-13} and 6.95×10^{-13} m^2/s ordered by decreasing thickness, and 6.08×10^{-13}, 5.86×10^{-13}, 4.25×10^{-13} and 3.76×10^{-13} m^2/s ordered by decreasing concentration of PI film. The effective diffusion coefficients of water molecules in these four films were 6.11×10^{-13}m^2/s for PMDA-ODA, 6.38×10^{-13}m^2/s for BPDA-ODA, 6.14×10^{-13}m^2/s for BPDA-BAPP and 5.86×10^{-13}m^2/s for PMDA-BAPP. When the environment temperature rises from 20 to 50 °C, the effective diffusion coefficient rises from 6.38×10^{-13} to 1.67×10^{-12} m^2/s. The results should greatly enhance the properties' improvement of the capacitive humidity sensor.

Keywords Polyimide film · Humidity-sensitive properties · Effective diffusion coefficient

1 Introduction

The capacitive humidity sensor is widely used in many fields, whose characteristics are mainly dependent on the sensing film [1, 2]. Polyimide (PI) film is commonly used as humidity-sensitive material due to its stability and absorbency properties [3].

J. Wu · W. Zhou (✉) · X. Wang · S. Li
School of Environmental and Municipal Engineering, Lanzhou Jiaotong University, Lanzhou 730070, China
e-mail: zwh6888@mail.lzjtu.cn

The research for improving humidity sensors has been ongoing and numerous, but most of that research is focused on the sensor itself. Therefore, there is a glaring vacancy of research on the PI film, an essential component of this system, especially the properties of the PI film related to sensing characteristics. This is a barrier to improve the sensor as a whole. This paper will fill that research void.

2 The Preparation of Polyimide Films

In this paper, two kinds of binary anhydrides and diamines were chosen to form four types of films. Figure 1 shows the fabrication progress of the PI film.

Firstly, the binary anhydride and diamine with same molar mass were put into DMAC. They were stirred and heated to obtain PI acid. The acid with different concentrations can be obtained by changing the mass of binary anhydride and diamine. Then, PI acid was dropped on the glass sheet gripped in the spin coater. By controlling the quantity of PI acid and rotation speed of spin coater, the thickness of PI films can be changed. Finally, put the glass sheets into the electric furnace, controlled and heated. After the process of dehydration cyclization, the PMDA-ODA, BPDA-ODA, BPDA-BAPP and PMDA-BAPP PI films were created. They were in the different thicknesses (25, 20, 13 and 11 μm) and in the varying concentrations (23, 20, 17 and 15%).

3 Test Facility for Effective Diffusion Coefficient

The test system is shown in Fig. 2. The main devices of the test system are shown in Table 1. Before the test, plenty of moist and dry air must be produced from the generators and stored in the gasholders separately. The PI film was then hung on the hook of the electronic scales in the test chamber in order to test its original mass. Refer to Fig. 2, the valve 14 was opened to let the film absorb the moisture from the moist air. When the mass of film reached stabilization, the time of whole process and the film mass were recorded after absorption. In addition, the chamber humidity was recorded before and after the test. Then, the chamber humidity was reduced by inletting the dry air. The process was repeated three times to gain more data. Finally, the effective diffusion coefficient of the water molecule in disparate films can be calculated.

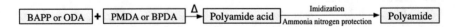

Fig. 1 Flow chart of PI preparation

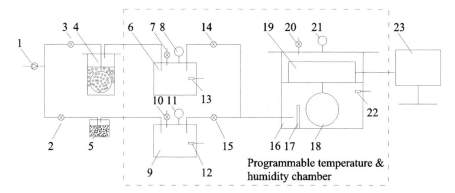

Fig. 2 Schematic diagram of test system on diffusion coefficient. 1—air pump; 2, 3, 7, 10, 14, 15, 20—valve; 4—humidity generator; 5—dryer; 6—gasholder of high humidity gas; 8, 11, 21—pressure gage; 9—gasholder of low humidity gas; 12, 13, 22—temperature and humidity sensor; 16—adsorption chamber; 17—baffle; 18—PI film; 19—electronic scales; 23—computer

Table 1 Equipment and their model and precision

Name	Model	Precision
Electronic scales	HZ-104/55S	1/100000
Oilless air compressor	SQ12	
Temperature and humidity sensor	RS-WS-N01-SMG-7	±2%RH(5–95%RH, 25 °C); ±0.4 °C(25 °C)
Constant temperature humidity chamber	HWHS-225-0	±0.1%RH; ±0.1 °C

4 The Test Data and the Processing

4.1 The Calculation Equations of Effective Diffusion Coefficient

Equation (1) relates the mass flow rate and the partial pressure gradient of vapour [4].

$$g_v = -\delta \cdot \nabla p_v \tag{1}$$

where g_v is the mass flow rate of vapour, kg/s. δ is the effective permeability, kg/(m·s·Pa). p_v is the partial pressure of vapour, Pa.

According to Fick's first law, the one-dimensional instability mass flow rate of water molecules can be expressed as follows when the temperature is constant.

$$g_v = -\rho \cdot D \cdot (\partial w/\partial x) \tag{2}$$

where ρ is the performance density of PI film, kg/m^3. D is the effective diffusion coefficient of water molecules, m^2/s. w is the moisture mass percentage, kg/kg. x is the distance of water molecules diffusion on one direction, m.

Based on the isothermal adsorption, Eq. (2) can be described as follows [5]:

$$g_v = -\rho \cdot D \cdot (\partial w/\partial \varphi) \cdot (\partial \varphi/\partial x) \tag{3}$$

where φ is the environment relative humidity, %RH.

Both partial pressure of vapour in PI film and relative humidity are linear distribution during steady-state diffusion, so g_v can be expressed as follows [6]:

$$g_v = -\rho \cdot D \cdot \left(\frac{\partial w}{\partial \varphi}\right) \cdot \frac{\varphi_2 - \varphi_1}{l} = -\rho \cdot D \cdot \left(\frac{\partial w}{\partial \varphi}\right)\left(\frac{1}{p_{vs}}\right) \cdot \left(\frac{p_{v2} - p_{v1}}{l}\right) \tag{4}$$

where l is the thickness of PI film, m. p_{vs} is the partial pressure of saturated steam in experimental temperature, Pa.

Based on Eqs. (1) and (4), the following equation can be deduced [6]:

$$D = (\delta \cdot P_{vs}/\rho) \cdot [1/(\partial w/\partial \varphi)] \tag{5}$$

According to Eq. (5), it is obviously that the key to obtain effective diffusion coefficient of water molecule is getting effective permeability coefficient and moisture sorption isotherm of water molecule.

The effective permeability of water molecules can be tested according to ASTM Standards [4]; the calculations are as follows:

$$\delta = \xi \cdot (1/2) \tag{6}$$

where ξ is the water molecular permeability; it can be described as Eq. (7):

$$\xi = g/\Delta p_v = g/[p_{vs} \cdot (RH_1 - RH_2)] \tag{7}$$

where Δp_v is the vapour pressure difference, Pa RH_1 and RH_2 are the ambient relative humidity after and before the moist gas is into the chamber, % g is the wet flow density, which can be described as Eq. (8):

$$g = (\Delta m/\Delta t)/A \tag{8}$$

where Δm is the mass change, kg and Δt is the time, s. A is the area of PI film, m^2.

According to Eqs. (6), (7) and (8), the equation of effective permeability coefficient of water molecules can be obtained:

$$\delta = \Delta m \cdot l/[2A \cdot \Delta t \cdot p_{vs} \cdot (RH_1 - RH_2)] \tag{9}$$

Firstly, the partial pressure of vapour and temperature in the film is different from those in the environment. When PI film is exposed in the environment for a certain time, the partial pressure of vapour and temperature in the film matches those in environment. At this time, the equilibrium moisture content, w, is the ratio between water mass that can be separated from the film and the mass of dry PI film. The moisture sorption isotherm $w(\varphi)$ is the curve that linked by equilibrium moisture contents under different relative humidity. According to the ISO 12570 [7], the moisture percentage can be obtained by the following equation:

$$w = (m - m_0)/m_0 \tag{10}$$

where m is the mass of PI film in a certain environment relative humidity, kg. m_0 is the mass of dry PI film, kg.

Peleg's model can describe the expression of w and φ with these parameters [8]:

$$w(\varphi) = k_1 \varphi^{n_1} + k_2 \varphi^{n_2} \tag{11}$$

where k_1, k_2, n_1, n_2 are constants ($n_1 < 1$ and $n_2 > 1$). When k_1 is proximately to k_2, and the relative humidity is low, the first item is much bigger than the second one. While the relative humidity is high, the situation is opposite. Because the relative humidity of tested environment is lower than 50% RH, the Eq. (11) can be simplified as follows:

$$w(\varphi) = k\varphi^{n_1} \tag{12}$$

According to Eqs. (5), (9) and (12), the effective diffusion coefficient, D, can be described as in Eq. (13):

$$D = \Delta m \cdot l / [2A \cdot \Delta t \cdot (RH_1 - RH_2)] \cdot [1/(\partial w/\partial \varphi)] \tag{13}$$

As long as Δm, l, A, Δt, RH_1, RH_2 and $w(\varphi)$ are tested, the effective diffusion coefficient can be obtained.

4.2 Ingredient Influence on Film Humidity-Sensitive Property

Four films were chosen to test in 20 °C. They were shown in Table 3. The areas of them were 0.01 m², and the thicknesses were 0.025 mm. In addition, the concentrations were 20%. The results are listed in Table 2.

Table 2 Permeability coefficients and effective diffusion coefficient of four films

Type	δ(kg/(m·s·Pa))	$w(\varphi)$	$D(m^2/s)$
PMDA-ODA	5.079×10^{-16}	$0.0312\varphi^{0.17898}$	6.11×10^{-13}
BPDA-ODA	7.425×10^{-16}	$0.02073\varphi^{0.32248}$	6.38×10^{-13}
BPDA-BAPP	4.686×10^{-16}	$0.06077\varphi^{0.1079}$	6.14×10^{-13}
PMDA-BAPP	3.569×10^{-16}	$0.06997\varphi^{0.0863}$	5.86×10^{-13}

According to Table 2, four PI films ordered by increasing effective permeability coefficient are PMDA-BAPP, BPDA-BAPP, PMDA-ODA and BPDA-ODA. However, the effective diffusion coefficients of different PI films are similar, and there's no evident regularity. Therefore, the chemical reagents have little influence on humidity-sensitive property of PI film.

4.3 Thickness Influence on Film Humidity-Sensitive Property

The PMDA-BAPP films with different thickness were chosen. The films' areas were 0.01 m^2, and their concentrations were 17%. The test environment was 20 °C. The results are listed in Table 3.

Table 3 shows that both effective permeability coefficients and effective diffusion coefficients increase with the thickness decrease. The thinner the films, the better the humidity-sensitive property of PI film is.

4.4 Concentration Influence on Film Humidity-Sensitive Property

The concentration influence on humidity-sensitive property was discovered by testing PMDA-BAPP films with varying concentrations. The films' areas were 0.01 m^2, and their thicknesses were all 0.0025 mm. The test was run in 20 °C. And results are shown in Table 4.

Table 3 Thickness effect on the coefficients of PI films

Thickness (mm)	δ(kg/(m·s·Pa))	$w(\varphi)$	D(m^2/s)
0.025	2.314×10^{-16}	$0.12907\varphi^{0.06001}$	4.25×10^{-13}
0.02	3.726×10^{-16}	$0.14767\varphi^{0.05758}$	5.49×10^{-13}
0.013	6.264×10^{-16}	$0.04046\varphi^{0.20874}$	6.34×10^{-13}
0.011	9.903×10^{-16}	$0.09978\varphi^{0.13024}$	6.95×10^{-13}

Table 4 Concentration effects on the coefficients

Concentration (%)	δ(kg/(m·s·Pa))	$w(\varphi)$	D(m^2/s)
23	4.391×10^{-16}	$0.05236\varphi^{0.11929}$	6.08×10^{-13}
20	3.569×10^{-16}	$0.06997\varphi^{0.0863}$	5.86×10^{-13}
17	2.314×10^{-16}	$0.12907\varphi^{0.06001}$	4.25×10^{-13}
15	1.924×10^{-16}	$0.05734\varphi^{0.10696}$	3.76×10^{-13}

Table 5 Temperature effects on the coefficients

Temperature (°C)	δ(kg/(m·s·Pa))	$w(\varphi)$	D(m²/s)
20	7.425×10^{-16}	$0.02073\varphi^{0.32248}$	6.38×10^{-13}
35	4.749×10^{-16}	$0.01297\varphi^{0.43392}$	1.00×10^{-12}
50	2.994×10^{-16}	$0.0046\varphi^{0.68938}$	1.67×10^{-12}

Table 4 shows that with the decrement of concentration of PI film, the effective permeability coefficient and diffusion coefficient decrease similarly. Therefore, the higher the concentration, the better the humidity-sensitive property of PI film is, also the better the property of humidity sensor.

4.5 Temperature Influence on Film Humidity-Sensitive Property

The BPDA-ODA films with 0.025 mm thickness and 20% concentration were tested in 20, 35 and 50 °C. The results are shown in Table 5.

According to Table 5, with the increase of temperature, the effective permeability coefficients and effective diffusion coefficients rise simultaneously. Clearly, the temperature has an appreciable impact on effective diffusion coefficient. The higher the temperature, the better the humidity-sensitive property of PI film is.

5 Conclusions

(1) The self-build test system and test method are practical.
(2) The chemical reagents influenced on humidity-sensitive property of PI film is little. The thinner the films, the better the humidity-sensitive property of PI film is. The higher the concentration, the better the humidity-sensitive property of PI film is.
(3) The humidity-sensitive properties of PI films become better when temperature increases.

Acknowledgements This project is supported by the National Natural Science Foundation (Number 51466007).

References

1. Meijuan, C.: A parallel plate capacitive humidity sensor based on PI film and numerical study on its characteristics. Lanzhou jiaotong University (2015)
2. He, X., et al.: Research on dynamic characteristics model of parallel plate capacitive humidity sensor. Instr. Tech. Sensor **4**, 5–26 (2017)

3. Zhou, W., et al.: Numerical study on response time of a parallel plate capacitive polyimide humidity sensor based on microhole upper electrode. J. Micro/Nanolithogr. MEMS MOEMS **16**(3), 034502 (2017)
4. Concepts, R.: E96-93 Standard test methods for water-vapor transmission of materials. Annual Book of Astm Standards (2014)
5. Collet, F., et al.: Water vapor properties of two hemp wools manufactured with different treatments. Constr. Build. Mater. **25**(2), 1079–1085 (2011)
6. Yi, S.: Experimental measurement of the effective water vapor diffusion coefficient of porous building materials by a transient technique toward energy-efficient buildings. Zhejiang University (2016)
7. ISO: ISO 12570: Hygrothermal performance of building materials and products Determination of moisture content by drying at elevated temperature (2014)
8. Peleg, M.: Assessment of a semi-empirical four parameter general Fmodel for sigmoid moisture sorption isotherms. J. Food Proc. Eng. **16**(1), 21–37 (1993)

Research of the Influence of Valve Position on Flow Measurement of Butterfly Valve with Differential Pressure Sensor

Yuanpeng Mu, Zhixian Ma and Mingsheng Liu

Abstract Accurate measurement is the basis of fluid control. Valves as resistance components produce differential pressure which in turn can be used for flow measurement. This paper studies the function among valve position, pressure difference and flow rate of a new designed butterfly valve. In this paper, the flow model of butterfly valve is established based on Bernoulli equation, the discharge coefficient C and the permanent pressure loss ratio φ under different valve opening conditions are studied by CFD simulations. The results show that the discharge coefficient C reaches a stable value with the increase of Reynolds number, and the position of the pressure taps is very important for valve flowmeters. The effect mechanism of valve opening on discharge coefficient C which is a complex and non-linear process needs further study.

Keywords Valve flowmeter · Differential pressure flowmeter · Discharge coefficient · Control valve · Valve position

1 Introduction

Flowmeters and control valves are important components of flow measurement and control in HVAC system. Accurate flowrate measurement is the basis of fluid control. But flowmeters not only increase the investment, but also causes additional pressure loss, especially the inferential type [1]. Valves as resistance components produce differential pressure which in turn can be used for flow measurement [2]. Many valve manufacturers equip the control valves with differential pressure sensors instead of the

Y. Mu · Z. Ma · M. Liu (✉)
Institute of Building Energy, Dalian University of Technology, Dalian, China
e-mail: Liumingsheng@dlut.edu.cn

Y. Mu
e-mail: Ypmu@mail.dlut.edu.cn

Z. Ma
e-mail: Mazhixian@dlut.edu.cn

flowmeters, which can reduce not only the cost but also the pressure loss of pipeline network. Some scholars also use the empirically determined valve characteristic curve to realize flow measurement [3, 4]. However, there are few studies focused on the influence mechanism of valve location on flow measurement, and the flow characteristics of valves are normally obtained by experiments.

The mechanism of valve flowmeter is similar to the orifice or V-cone flowmeter, for which the discharge coefficient C is a very important parameter. Different flowmeters have different stable discharge coefficient ranges, and the orifice flowmeter has the narrowest range compared with other differential pressure types [5, 6]. Generally, differential pressure flowmeters have fixed opening area, but each opening area of the valve is different. Accurate opening angle measurement is the first step to achieve accurate flow measurement of valves. Then, the discharge coefficients at each opening position need to be determined.

Therefore, this paper aims to study the function among valve position, pressure difference and flow rate of a newly designed butterfly valve. The pressure difference mentioned above is referred to the throttling pressure difference which is not the permanent pressure loss of valves and pipes. In this paper, the flow model of butterfly valve is established based on Bernoulli equation, and the influence of opening angle θ on flow measurement is analyzed theoretically. The discharge coefficient C which represents the influence of valve position on streamline is studied by CFD simulations. Finally, the permanent pressure loss ratio at each opening is calculated. Analytical methods proposed in this paper can also be used as a reference for other types of control valves.

2 Methods

The diagrammatic sketch of the butterfly valve with differential pressure sensor is shown in Fig. 1, and the arrows indicate streamline direction.

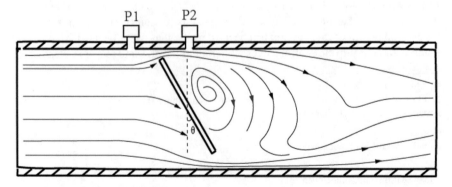

Fig. 1 Butterfly valve with differential pressure sensor

For incompressible fluid, according to Bernoulli equation:

$$\frac{P_1}{\rho} + \frac{1}{2}v_1^2 = \frac{P_2}{\rho} + \frac{1}{2}v_2^2 + \frac{\xi}{2}v_2^2 \tag{1}$$

where P is fluid pressure, ρ fluid density, v fluid velocity, ξ friction coefficient, and subscript 1 represent the cross-section before valve disk, subscript 2 represent the cross-section after valve disk. The position of cross-section is arbitrary.

According to Eq. (1), the volume velocity q_v can be expressed as:

$$q_v = v_2 A_2 = \frac{A_2}{\sqrt{1 + \xi - \left(\frac{A_2}{A_1}\right)^2}} \sqrt{\frac{2}{\rho}(P_1 - P_2)} \tag{2}$$

where A_1 is cross-sectional area of the pipeline, and A_2 cross-sectional area of streamline after valve disk.

As can be seen from Fig. 1, there is always a cross-section area equal to valve opening annulus area A_0, i.e., $A_2 = A_0$. The velocity of the eddy behind the disk is negligible relative to the mainstream velocity. The discharge coefficient C is defined as:

$$C = \sqrt{\frac{P_1 - P_2}{P_1' - P_2'}} = \sqrt{\frac{\Delta P}{\Delta P'}} \tag{3}$$

where $\Delta P' = P_1' - P_2'$ is the actual value of the differential pressure gauge.

The area ratio of valve opening annulus to pipeline:

$$\beta = \frac{A_2}{A_1} = 1 - \alpha \cdot \cos\theta \tag{4}$$

where α is the area ratio of valve disk to pipeline, θ valve opening angle.

The friction loss can be ignored, Eq. (2) can be expressed:

$$q_v = \frac{C \cdot \beta \cdot \pi D^2}{\sqrt{1 - \beta^2}} \sqrt{\frac{\Delta P'}{8\rho}} \tag{5}$$

where D is pipe diameter.

The discharge coefficient C is studied by CFD simulations, and standard k-ε turbulent model is used. Firstly, the correlation between C and Reynolds number is verified. Then, C at each opening is calculated and compared with other differential pressure flowmeters. Finally, the permanent pressure loss ratio at each opening is calculated.

3 Results and Discussion

Pressure contours of $D = 150$ mm, $\theta = 50°$ at $v = 3$ m/s is shown in Fig. 2. The upstream and downstream straight pipes are 10D to avoid the effect of boundary conditions. $P1$ and $P2$ are the pressure taps before and after the valve, respectively. Pin and Pout are inlet and outlet pressure of pipe.

The discharge coefficient C under different diameter conditions ($D = 50, 100, 150$ mm and other structures remain geometrically similar, $\theta = 50°$) were calculated. As can be seen from Fig. 3, C is related to Reynolds number and independent of pipe diameter.

However, the discharge coefficient C under different opening conditions of valve disk is not all the same, and seven opening conditions (10°, 20°, 30°, 40°, 50°, 60°, 70°) were studied, in which pipe diameter $D = 150$ mm, valve disk diameter $D' = 146$ mm, i.e., $\alpha = 0.94$. The results are shown in Fig. 4. The discharge coefficient C reaches a stable value with the increase of Reynolds number. The effect of disk opening on C is a complex and non-linear process, and the maximum difference of stable value of seven opening conditions is about 3.98%.

Comparison of discharge coefficient of valve flowmeter and other type [5] is shown in Fig. 5. Where the discharge coefficient of valve is the modified average value of the seven opening conditions. The characteristic of valve flowmeter is similar to venturi and V-cone flowmeter, which are both better than orifice flowmeter. However, these differential pressure flowmeters have common shortcomings: it is difficult to adapt to small Reynolds number measurement conditions.

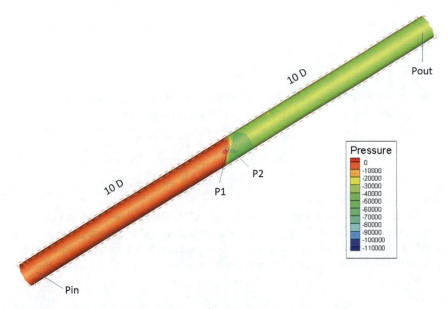

Fig. 2 Pressure contours of $D = 150$ mm, $\theta = 50°$ at $v = 3$ m/s

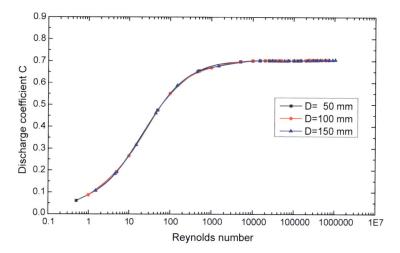

Fig. 3 Discharge coefficient C under different diameter conditions

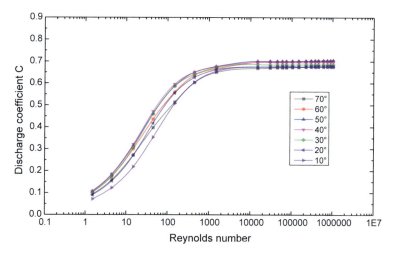

Fig. 4 Discharge coefficient C under different valve opening conditions

When the fluid flows through the valve, part of it will recover and part of it will dissipate. The ratio of permanent pressure loss to total pressure difference is defined:

$$\varphi = \frac{P_{in} - P_{out} - \Delta P_{pipe}}{P_1 - P_2} \qquad (6)$$

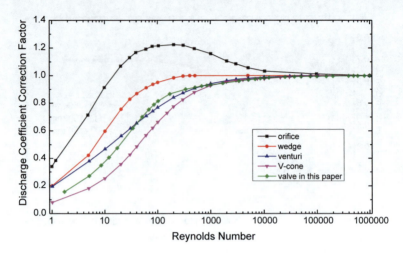

Fig. 5 Comparison of discharge coefficient of several flowmeters

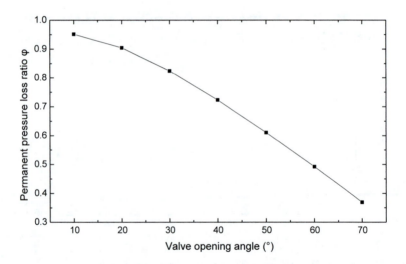

Fig. 6 Permanent pressure loss ratio φ under different valve opening conditions

According to fluid mechanics, φ is independent of the fluid velocity. The results of φ under seven valve opening conditions are shown in Fig. 6, which are calculated at $v = 1.5$ m/s. The larger the valve opening, the smaller the ratio of permanent pressure loss. φ is only 0.37 at 70°. The results can explain that flow measurement with valve characteristic curve which utilize permanent pressure loss cannot achieve high-accuracy.

4 Conclusions

The discharge coefficient C is related to Reynolds number and independent of pipe diameter. Similar to the general differential pressure flowmeters, the discharge coefficient C reaches a stable value with the increase of Reynolds number, and C is about 0.7 in this numerical example.

The effect of valve opening on C is a complex and non-linear process, and the maximum difference of stable value of seven opening conditions is about 3.98%.

The position of the pressure taps is very important for valve flowmeters. With the increase of valve opening, the pressure difference decreases and the permanent pressure loss ratio also decreases, and the accurate of flow measurement which utilize valve characteristic curve will decrease rapidly.

The effect of valve opening on the discharge coefficient C needs further investigation.

References

1. Shaaban, S.: Optimization of orifice meter's energy consumption. Chem. Eng. Res. Des. **92**, 1005–1015 (2014)
2. Atmanand, M.A., Konnur, M.S.: A novel method of using a control valve for measurement and control of flow. IEEE Trans. Instr. Measurement **48**(6), 1224–1226 (1999)
3. Choi, J.: Flow control system design without flow meter sensor. Sens. Actuators A: Phys. **185**, 127–131 (2012)
4. Song, L., Wang, G., Brambley, M.R.: Uncertainty analysis for a virtual flow meter using an air-handling unit chilled water valve. HVAC&R Res. **19**, 335–345 (2013)
5. Hollingshead, C.L., Johnson, M.C., Barfuss, S.L.: Discharge coefficient performance of Venturi, standard concentric orifice plate, V-cone and wedge flow meters at low Reynolds numbers. J. Petrol. Sci. Eng. **78**, 559–566 (2011)
6. Huanga, S., Ma, T., Wang, D.: Study on discharge coefficient of perforated orifices as a new kind of flowmeter. Exp. Therm. Fluid Sci. **46**, 74–83 (2013)

Analysis on Performance of Solar Novel Heat Pipes Radiant Heating System

Yaping Zhang, Yongxin Guo, Pei Wang and Yao Chen

Abstract A kind of novel heat pipe construction was put forward for using solar hot water heating system. An advantage of two-phase heat transfer devices was it can eliminating the entrainment effect because of separated vapor and liquid flow channel. Due to this feature, the novel heat pipe is a perfect product for using the heating systems. But, solar energy utilization is generally restricted by climate factors, leading to overheating in summer and freezing in winter. A new gas–liquid separator type heat pipe heating system based on this problem design is presented. Besides, different types, quantities and area tests were conducted in order to compare the surface temperature of heating concrete block. It indicated that the heat pipe has better thermal conductivity and low thermal resistance. And it can be used to reduce the heating temperature of water, so as to achieve energy-saving purpose. The results show that high-efficiency heat exchanger and rapid thermal response of the copper-water heat pipe heating system. It will be meaningful for improving the thermal efficiency and comfort of the indoor human body, at the same time conducive to safe and efficient operation of the heating system.

Keywords Solar energy · Heat pipe · Radiant heating · Gas–liquid separator

1 Introduction

The commonly used plastic pipes had poor thermal conductivity and require high temperature of water supply for the heating system, making it difficult to directly use low-grade energy such as solar and geothermal energy to heating system. The lower temperature water heating system of plastic system was widely used in all kinds of building heating systems. However, its large thermal resistance and low conductivity coefficient [1] lead to slower thermal response and lower heating efficiency. Moreover, 40–60 °C hot water was offered for the conventional plastic

---

Y. Zhang · Y. Guo (✉) · P. Wang · Y. Chen
Xi'an University of Science and Technology, Xi'an, China
e-mail: 791028985@qq.com

pipe, due to the large hot water circulation flow rate occurring in the plastic heat pipe heating system. This is potentially a problem, because adding bigger building load is risky [2].

Compared with the traditional radiator heating system, the floor radiant heating occupies less indoor space and had a better esthetic level. And the indoor temperature gradients were evenly distributed along the height. The temperature of water supply for heating surface was lower than the traditional heating, which can save about 50% energy.

Heat pipes were widely used in thermal control of satellites, spacecrafts, electronics cooling as well heating systems, while its application in indoor heating system was proposed recently. The generalization of heat pipes is a broad subject covering many applications ranging from heat flux transformation application to isothermal applications. The distance even up to several meters or tens of meters horizontally, and thermal transfer was powered by capillary pumping or gravity. Besides, thermal transfer occurred in small temperature difference [3], these features mean heat pipe had potential to overcome the disadvantages occur in the water plastic coil, e.g., such as the hazard of large water flow rate and low efficiency in terms of energy conversion [4].

Heat pipe had quicker thermal response, smaller thermal resistance and excellent thermal conductivity [5] that enables the transportation of heat while maintaining almost uniform temperature along its heated and cooled sections. However, owing to the influence of unsteady movement of heat pipe vapor flows inside the pipe, vapor fluctuating may appear. In order to prevent vapor plugs of the heat pipe and improve the heat transfer coefficient, the novel heat pipe needs to be redesigned for a better utilization in the floor heating system [6]. The experimental research had been conducted in the floor radiator heating system for the vibrate heat pipe [7] and compared to the plastic pipe [8]. The experimental result had obtained some important parameter relation [9] used in 1000 mm length gravity carbon steel heat pipe for floor radiator heating system. The water resource heat pump integrated heat pipe was used in the floor radiator heating system, and the energy efficiency ratio for the water resource heat pump was concluded by the simulation result [10].

As a kind of clean renewable energy, solar energy had been widely focused nowadays. If we can introduce this outdoor renewable energy into our indoor heating system, it will be a great significance for energy saving and emission reduction [11]. However, in the current application of solar energy, generally, problems such as low heat transfer efficiency, freezing in winter and overheat in summer, meanwhile, the heating effect was restricted by climatic factors, so that the application of solar energy was difficult to popularize. Due to the excellent phase change heat transfer performance of the heat pipe, many scholars' research about heat pipe [12] had been applied in the fields of waste heat recovery, thermal energy extraction and electronic element cooling [13]. And the advantages of heat pipe heat dissipation such as uniform, comfortable and safe had shown in heating system. Zhang et al. proposed a new type of photovoltaic-solar loop heat pipe/heat pump system [14]. Comparative analysis suggested that the new system can realize the efficient conversion between solar energy, heat energy and electric energy for the application indoor heating system. S. Brian et al. had put heat pipes inside the wall and proposed a new type of the passive solar energy utilization technology [8]. This paper proposed a novel heating system that combines solar heating and heat pipe floor radiation. Owing to the floor radiant heating system had more uniform temperature distribution than traditional radiator and required lower heat medium temperature, which had provided the convenience of hot water utilization. The experimental study had demonstrated the feasibility of low-grade direct solar heating technology [15]. The efficiency of collector can be improved by using gas–liquid phase change separation heat pipe instead of the ordinary gravity heat pipe. In order to avoid heat transfer limit and carrying limit of the common heat pipe, the heat absorption efficiency was improved by means of gas–liquid separation. So the utilization of solar heating system during cold climate was guaranteed.

2 Propose the Novel Heat Pipe

A disadvantage in the operation of heat pipe heating system was dry out limitation on the upper side water film of the heat pipe absorber; the reason was the water uplift height limited by insufficient wick capillary force. In order to avoid the "dry limit" phenomenon of the condensation section, a novel heat pipe structure applied to the floor radiant heating system was proposed, as shown in Fig. 1.

The shape of the side of the evaporator of the new heat pipe was an annular cavity. The basic operation is a continuous cycle. The working fluid is located at the bottom of the pipe. Hot water was transferred from the inner annular wall of the evaporator to the outer wall, and the liquid working medium was heated as well, the addition of a heat source allows the liquid pool to evaporate. The annular heat pipe can increase the contact area between the pipe and the hot water. When liquid working medium was boiled and evaporated into gas in the evaporator, gas enters circular flat chambers of the floor support portion due to the pressure difference and releasing the latent heat, and the condensate flows back into the annular cavity of the evaporator under the action of gravity or capillary force, thus the basic operation is a continuous cycle. The proposed heat pipe was closed and contains liquid and gas of the working fluid.

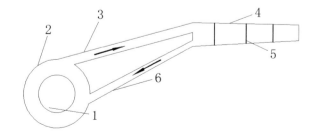

Fig. 1 Heat pipe separated gas and liquid channel.
1—hot water channel;
2—evaporation section;
3—gas medium channel;
4—condensation section;
5—strengthening rib;
6—liquid medium channel

The novel heat pipes were two-phase heat transfer devices separating the vapor and liquid flow channel and eliminating the entrainment limitation effect between them. Due to this advantage, the heat pipe was very suitable for building heating system. It allows the collection of lower quality for reuse. And heat can be transferred to water through a heat exchanger.

(1) The ring chamber of evaporator side acts as a heat exchange surface, which not only increases the heat exchange area, but also allows hot water to flow outside the tube and reducing the local resistance of the heating system.
(2) The liquid phase and the gas phase can flow independently with each other to avoid gas–liquid entrainment and reducing the mutual flow resistance between them.
(3) Increasing the heat exchange area to quickly cool the gas working medium, and ensure efficient reflux of the condensate, and avoiding "dry out" phenomenon of the liquid working medium of the evaporator. And in order to reduce the flatness of the pressure chamber being pressed, the longitudinal reinforcing ribs exert a supporting effect of cyclic annular.
(4) The jet at the outlet of the pipe increases the turbulence of the liquid film at the condensing end, which can destroy the liquid film, so as to enhance the convective heat transfer at the condensation end.

Therefore, the experimental highlights thermal performance of the copper-water heat pipe. All heat pipe parameters are as follows, respectively, the length of the evaporation Section 40 mm, the working fluid water, the diameter 10 mm, the length 300 mm, the wall thickness 0.5 mm, the metal interred powder core 1 mm thick and starting temperature was 20 °C. In the experiment, the optimal refrigerant charge of the heat pipe recommended in literature is 30% [16].

3 The Principle of Heat Pipe Heating System

The hot water flows into the casing pipe connected to the heat pipe, and then the liquid working medium in the evaporation section absorbs heat to evaporate into a gas–liquid two-phase flow. In order to avoid the resistance of the gas–liquid two-phase flow mutual interference, prevent the carrying limit and capillary limit of heat pipe collector as well. The novel gas–liquid separation heat pipe system designed as shown in Fig. 2 realizes the separated flow of gas and liquid in the pipeline, which can improve the heat collection efficiency for the solar energy heating system. When the working medium is cooled, the temperature dropped and returned to the evaporation section of the heat pipe. Thus, a working medium cycle process of heat pipe realized. Since the condensation section that dissipates heat in the room has good isothermality, the comfort of the indoor environment can be maintained.

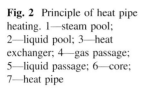
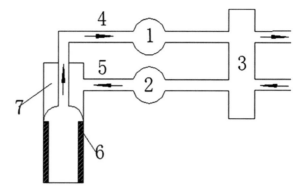

Fig. 2 Principle of heat pipe heating. 1—steam pool; 2—liquid pool; 3—heat exchanger; 4—gas passage; 5—liquid passage; 6—core; 7—heat pipe

4 Experimental Device

The experiment instruments include constant temperature water tank, circulating water pump, heat pipe, temperature inspection instrument, floor heating pipe and concrete block. As Fig. 3 shown the hot water pipe adopts PE-RT, outer diameter 16 mm, inner diameter 12 mm and the temperature range was −40 to 95 °C. Concrete blocks with an area of 30 × 30 m² and 50 × 50 m² were used in the experiment. The post-processing system uses a sensor Pt 100 platinum-thermal resistance and a measurement path for an eight-way temperature patrol.

Four series experiments were designed: concrete temperature characteristics for single heat pipe and double heat pipe; concrete temperature characteristics for different heat sources temperatures; temperature characteristics of concrete blocks heated by double heat pipes with different volumes; the temperature characteristics of 30 × 30 m² concrete blocks heated by heat pipe and floor heating pipe, respectively, for the same heat source. Six measuring points were arranged in concrete block, one measuring point was arranged in water tank, and a measuring

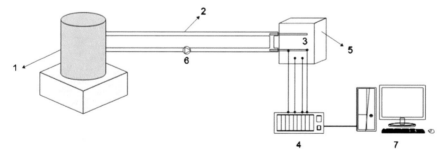

Fig. 3 Installation system. 1—water tank; 2—hot water pipeline; 3—heat pipe; 4—temperature inspection instrument; 5—concert block; 6—water pump; 7—computer

point is arranged in the room. Starting from the water supply of the system, the temperature patrol instrument collected the values until the temperature of each measurement point is stable, and the test duration is 5 h.

5 Results and Analysis

5.1 Thermal Effect of Heat Pipe Heating System

The concrete temperature had a rapid growth trend within 30 min, as shown in Fig. 4. The heat pipe heating system temperature rises by about 5 °C in the first 30 min, while the floor heating pipe temperature rises by only 3 °C. So the heat pipe has the advantages of fast thermal response and high conveying capacity of the heating system. After five hours, the concrete surface temperature was stable. When heated with a heat pipe, the concrete surface temperature was 27 °C. When heated by a common floor heating pipe, the concrete surface temperature is 24 °C. Lower resistance and higher thermal capacity were shown for the heat pipe heating system compared with the traditional heating pipes under the same heat source water supply temperature.

5.2 The Influence of the Number of Heat Pipes

It can be seen from Fig. 5a that when the heat source water supply temperature was lowered by nearly 10 °C, the average temperature of each measuring point of double heat pipe is higher than single. After an hour, the temperature rise of double heat pipe was about 6 °C, and the temperature rise of single heat pipe was only about 3 °C. This shows that the double heat pipe heating temperature rises faster

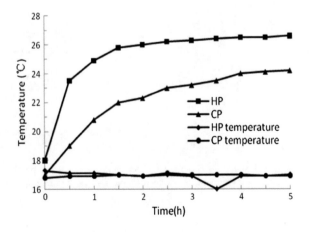

Fig. 4 Temperature change of 30 × 30 m² concrete block under the same heat source

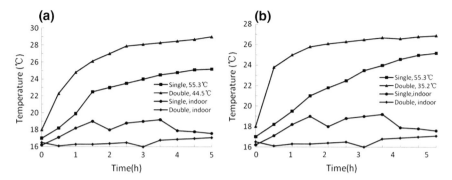

Fig. 5 Temperature change of 30×30 m^2 concrete block

than single. The effect of the different number of heat pipes on the heating of the concrete block is shown in Fig. 5. The heat source temperature is 25 °C with single heat pipe. The heat source temperature of the double heat pipe is 35.2 °C with double heat pipes, and the concrete surface temperature reached 27 °C after for 5 h. Therefore, increasing the heat transfer area is more effective for improving heat transfer performance than increasing the heat source temperature.

5.3 Heat Transfer Effect of Heat Pipe Heating System

A dual heat pipe heating system with heat source temperature of 35.2 °C has faster heating rate than a single heat pipe heating system with heat source temperature of 44.5 °C. After heating for 5 h, temperature of the concrete block heated by different heat source became stable. As shown in Fig. 6, the temperature of the heat source dropped by 9 °C, but the surface temperature of the concrete block was reduced by 2 °C. When heat source temperature was 35.2 °C and steady, the room temperature was 17 °C. The concrete surface temperature is close to 27 °C, which indicates that

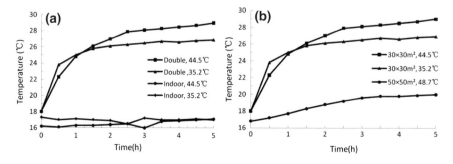

Fig. 6 Temperature change of different volumes concrete block

the heating requirements can meet the heating demand at lower heat source temperatures. The temperature rise of the 50 × 50 cm² concrete block was less than 30 × 30 cm² of the first hour, and the temperature of 50 × 50 cm² concrete surface was 20 °C. This temperature can not meet the indoor heating demand indicating that the double heat pipe cannot transfer enough heat.

6 Conclusions

(1) In gas–liquid separation heat pipe heating system, when water supply temperature of the heat source is same, the temperature of the indoor air heated by the heat pipe is about 3 °C higher than that of the ordinary heating pipe. The results show that the heat pipe heating system has excellent isothermal performance and thermal conductivity.
(2) Experimental tests have found that copper-water heat pipes have faster start-up response and better temperature uniformity. Moreover, the gas–liquid separation heat pipe has the potential of preventing the formation of load limits and capillary limits in the heat pipe heating application and improves the heat dissipation efficiency.
(3) The novel heat pipe heating system has the effect of overcoming the entrapment limit effect. A radiant floor heating system using heat pipes can increase the heat transfer coefficient, reduce the water flow temperature, and reduce energy consumption. Facilitate various waste heat utilization systems such as solar and geothermal energy.

Acknowledgements The project is supported by National Natural Science Foundation Funded Project (51504188).

References

1. Khazaee, I., Hosseini, R., Noie, S.H.: Experimental investigation of effective parameters and correlation of geyser boiling in a two-phase closed thermosyphon. Appl. Therm. Eng. **30**, 406–412 (2010)
2. Tan, B.K., Wong, T.N., Ooi, K.T.: A study of liquid flow in a flat plate heat pipe under localized heating. Int. J. Therm. Sci. **49**, 99–108 (2010)
3. Rassamakin, B., Khairnasov, S., Zaripov, V., et al.: Aluminum heat pipes applied in solar collectors. Solar Energy **94**, 145–154 (2013)
4. Ghoi, J., Jeong, M., Yoo, J., et al.: A new CPU cooler design based on an active cooling heatsink combined with heat pipes. Appl. Therm. Eng. **44**, 50–56 (2012)
5. Suman, Balram, De, Sirshendu, Dasgupta, Sunando: A model of the capillary limit of a micro heat pipe and prediction of the dry-out length. Int. J. Heat Fluid Flow **26**, 495–505 (2005)
6. Gunnasegaran, P., Abdullah, M.Z., Shuaib, N.H.: Influence of nanofluid on heat transfer in a loop heat pipe. Int. J. Heat Mass Transf. **47**, 82–91 (2013)

7. Hobbi, A., Siddiqui, K.: Optimal design of a forced circulation solar water heating system for a residential unit in cold climate using TRNSYS. Solar Energy., vol. 83 (2009)
8. Robinson, B.S., Chmielewski, N.E., Knox-Kelecy, A.: Heating season performance of a full-scale heat pipe assisted solar wall. Sol. Energy **87**, 76–83 (2013)
9. Yang, Weibo, Zhu, Jielian, Shi, Mingheng: Numerical simulation of the performance of a solar-assisted heat pump heating system. Appl. Therm. Eng. **11**, 790–797 (2011)
10. Zhai, X.Q., Wang, R.Z., Dai, Y.J., Wu, J.Y., Ma, Q.: Experience on integration of solar thermal technologies with green buildings. Renew. Energy **33**, 1904–1910 (2008)
11. Li-quan, Yang, Qi, Wang, Shan-shan, Liu: Application and development of solar heating system. J. Agric. Technol. **01**, 169–171 (2016)
12. Ma, H., Yin, L., Shen, X., et al.: Experimental study on heat pipe assisted heat exchanger used for industrial waste heat recovery. Appl. Energy **169**, 177–186 (2016)
13. Zhang, Y., Feng, Q., Yu, X.: Thermal performance analysis of heat pipe used in electronics cooling. Fluid Mach **36**(08), 79–82 (2008)
14. Zhang, X., Zhao, X., Shen, J., et al.: Dynamic performance of a novel solar photovoltaic/loop-heat-pipe heat pump system. Appl. Energy **114**, 335–352 (2014)
15. Experimental investigation on the performance of solar-thermosyphon embedded radiant floor heating system. Acta Energiae Solaris Sinica **29**(6), 637–643 (2008)
16. Zhang, Y., Xie, H., Li, D., et al.: Experimental investigation on heat transfer performance of carbon-steel/water thermosyphon. J. Tianjin Univ. **39**(2), 223–228 (2006)

Numerical Simulation and Optimization of Air–Air Total Heat Exchanger with Plate-Fin

Jiafang Song, Shuhui Liu and Xiangquan Meng

Abstract The air-to-air heat exchanger is a fresh air device that can reduce the energy consumption of air conditioners and improve heat exchange efficiency. In this paper, a mathematical model for heat transfer coupling of air film based on user-defined scalar (UDS) is proposed, and the model calculation is realized by Fluent. The comparison with previous experimental results verifies the accuracy of the calculation model. The influence of core channel height on core efficiency and the influence of inflow angle on heat exchanger performance are simulated when the length of counterflow section is changed. Considering the length of the counterflow section and the setting of the fresh air and exhaust air inflow angle, it is recommended that the length of the counterflow section should be set to 400 mm, and the inlet airflow direction of outdoor air and exhaust air is symmetrically set to 135° or 45° with respect to the symmetry axis of the heat exchanger. The channel height should be set above 2.0 mm.

Keywords The total heat exchanger · The cross-counterflow · Inflow angles · Core channel

1 Introduction

At present, the research of static total heat exchangers mainly focuses on the structure, materials [1, 2], operation control [3, 4] and annual recovery benefit analysis [5–7]. This paper studies the airflow distribution of the total heat exchanger. The performance optimization of the total heat exchanger using the numerical simulation model not only is convenient but also saves time and money.

J. Song · S. Liu (✉) · X. Meng
School of Mechanical Engineering,
Tianjin Polytechnic University, Tianjin 300387, China
e-mail: 15668208498@163.com

J. Song
e-mail: songjiafang@tjpu.edu.cn

This paper draws on a more accurate and simple calculation of the coupled heat and moisture numerical calculation model of the moisture transfer through the user-defined scalar (UDS) equation to explore the performance of the cross-counterflow total heat exchanger. Its accuracy is verified by comparing with previous experimental conclusions. Some conclusions are drawn by comparing the effects of different channel heights on the core. The combined effects of the length of the counterflow section and the inflow angle on the performance of the total heat exchanger are analyzed.

2 Methods

2.1 Establishment of Physical Model

The entity of the model is taken from the literature [8] to establish the physical model of the plate-type total heat exchanger with the pure cross-flow and cross-counterflow. The shape and the structure of the cross-counterflow core are shown in Fig. 1, and the relevant parameters are shown in Table 1.

For the convenience and reliability of modeling, this model is simplified. Because the flow channels are intersected, the horizontal direction cannot be simplified. The vertical direction is to extract a layer of moisture-permeable film and half of the adjacent flow channels, as shown in Fig. 2. The surface facing away from the moisture-permeable membrane is set to a symmetry plane in Fluent. In order to facilitate the independent meshing work and add energy source item in the later work, the heat conductive membrane is cut into three parts and the two-phase domain during the modeling process.

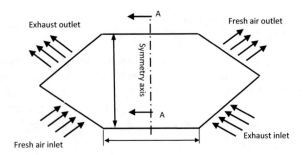

Fig. 1 Schematic diagram of the cross-counterflow structure (top view)

Table 1 Core structure parameters

α (°)	90	a	$100 \text{ mm} \leq a \leq 1500 \text{ mm}$
b (mm)	230	δ (mm)	0.05
h (mm)	1.5	l (mm)	0.01

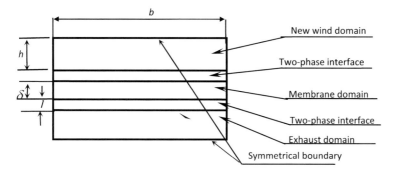

Fig. 2 Schematic diagram of the calculation domain setting (A-A cross-sectional view)

2.2 Mathematical Model

Assumption According to the special case of heat and moisture transfer in the narrow flow path of the total heat exchanger, the assumptions are as follows: (1) The air is incompressible, and the heat and moisture transfer in the fluid domain and the membrane domain is steady; (2) the flow state in the fluid field is laminar flow; (3) the physical property parameters of the air are constant. According to different temperatures, the thermal conductivity of the air is set by piecewise linear settings [8]. The influence of moisture content in fluid field on thermal conductivity is not considered. (4) The phase change heat of adsorption and desorption processes on both sides of the membrane is balanced. (5) Moisture transfer in the film region of the physical model is only along the thickness direction (model y-direction).

Establishment of control equation The equivalent diffusion coefficient D'_{wm} for calculating the diffusion of moisture content in vapor phase is used to replace the water diffusion coefficient in the permeable film when calculating the moisture conduction process in the heat–moisture transfer membrane. To achieve the continuity of moisture transfer at the gas–solid two-phase interface and an analogy with fluid–solid coupling heat transfer. The moisture transfer calculation is achieved by custom scalar equations. Because the membrane material is a porous adsorbent, the following relationship can be obtained from the adsorption theory [9]:

$$\frac{D'_{wm}}{D_{wm}} = \rho_m \rho_a \frac{A\theta_{max}C}{\left(1 - C + \frac{C}{Aw}\right)^2 (Aw)^2} \tag{1}$$

$$A = \frac{e^{5394t}}{10^6} \tag{2}$$

In Formulae (1) and (2), D_{wm} is the water diffusion coefficient in the moisture-permeable membrane, m²/s; ρ_m, ρ_a are the densities of the membrane material and air, respectively, kg/m³; t is the air temperature, °C; θ_{max} is the maximum water absorption of the membrane material, kg/kg; C is the saturation adsorption curve constant; w is the air moisture content, kg/kg.

Heat transfer control equation is as follows [9]:

$$\nabla(\rho_a t U) = \nabla \cdot \left(\frac{K}{c_p} \cdot \nabla t\right) + S_a \quad (3)$$

$$S_a = \pm \frac{L_w \rho_{wm} D'_{wm}}{V_c} \left(\frac{\partial w}{\partial y} A\right) = \pm L_w \rho_{wm} D'_{wm} \left(\frac{\partial w}{\partial yl}\right) \quad (4)$$

In Formulae (3) and (4), U is the velocity vector, m/s; K is the heat transfer coefficient of air, W/(m² K); c_p is specific pressure heat capacity of air, J/(kg K); S_a is an additional source term, caused by the desorption process on both sides of the membrane; L_w is the latent heat of vaporization of water, J/kg; V_c is the calculating unit volume, m³; A is the calculation unit surface vector, m²; l is the two-phase interface height, m.

UDS mass transfer control equation is as follows:

$$\nabla(\rho_a w v) = \nabla \cdot (\Gamma \nabla_w) \quad (5)$$

In formula, v is the air direction thickness direction velocity vector, m/s; Γ is the equivalent diffusion coefficient, kg/(m s), on the air side $\Gamma = \rho_a D_{va}$ (where D_{va} is the mass diffusion coefficient of water vapor in air). At the interface between the membrane and the two phases, the convection term is 0, and the equivalent diffusion coefficient is $\Gamma = \rho_a D'_{wm}$.

2.3 Brief Description of Fluent Internal Setting

Using the double-precision solver, laminar flow model, SIMPLE algorithm, the energy equation source term and the inlet direction control vector are written by VC, and the UDF function is applied to the boundary of the corresponding two-phase interface calculation domain and the windward surface. UDF and UDS are used in reference [10]. The fluid domain uses velocity and outflow conditions. The advantage of this model is that it simplifies the setting of the boundary condition of the interface between the membrane and the air in Fluent, and accurately calculates it by UDS. Table 2 shows the specific physical parameters of air and membrane.

Table 2 Air and membrane property parameters

$\lambda_a/(W/(m\ K))$	Piecewise linear settings	$\lambda_m/(W/(m\ K))$	0.32
$\rho_a/(kg/m^3)$	1.616	$\Theta_{max}/(kg/kg)$	1.52
$D_{va}/(m^2/s)$	2.45×10^{-2}	$\rho_m/(kg/m^3)$	773
Fresh air inlet dry bulb temperature tc/°C	5	$D_{wm}/(m^2/s)$	6.91×10^{-11}
Exhaust air inlet dry bulb temperature th/°C	21	Fresh air inlet moisture content wci/(g/kg)	3.193
C	7.2	Exhaust air inlet moisture content whi/(g/kg)	6.155

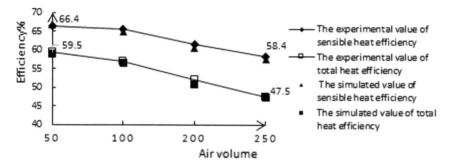

Fig. 3 Comparison of experimental and simulated values of cross-flow heat exchangers with different air volumes

2.4 Verification of the Calculation Model

The cross-flow heat exchanger structure used in this paper is from previous experiments. Under the same simulation conditions as the experimental conditions, the comparison between the obtained simulated values and the experimental values [8] is shown in Fig. 3. It can be seen from Fig. 3 that the maximum error of sensible heat exchange efficiency is 51%, and the maximum error of latent heat exchange efficiency is 3.1%.

3 Results and Discussion

3.1 Analysis of the Effect of Channel Height on Heat Exchanger Performance

The airflow rate is set to 150 m³/h, and the length of counterflow section is 110 mm. The channel height values are set to 1, 1.5, 2, 2.5, 3, 3.5 and 4 mm in

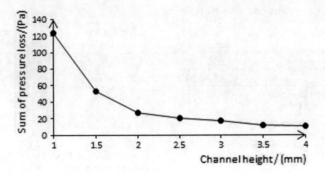

Fig. 4 Effect of channel height on core pressure loss

sequence, and the results were plotted as curves. Figure 4 shows the core efficiency and flow channel pressure loss at different channel heights.

As can be seen from Fig. 4, the core pressure loss decreases with the increase of channel height. When the channel height transits from 1 to 2 mm, the pressure loss decreases obviously, from 124 to 26.8 Pa, and the reduction range of pressure loss is almost 100 Pa. The height of the channel is continuously increased above the height of 2 mm, and the pressure loss value and the channel height are approximately linearly changed. When the channel height is 4 mm, the dual-channel pressure loss is a minimum of 10 Pa. Therefore, the height of the channel is preferably set at a height above 2.0 mm, and the pressure loss caused at this time is no longer obvious.

The relationship between core exchange efficiency and channel height is shown in Fig. 5. With the decrease of channel height, the change of core exchange efficiency and channel height is approximately linearly decreasing. Usually, the height of the core channel is set to 3 mm. This height satisfies the high exchange efficiency (70% sensible heat efficiency and 66% total efficiency), which meets the minimum efficiency requirement of the national standard. It can be seen that the common channel structure of the total heat exchanger is worthy of recognition in the selection of the channel height.

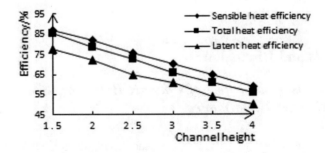

Fig. 5 Effect of single channel height on efficiency

3.2 The Influence of Inflow Angle on Heat Exchanger Performance When the Length of Counterflow Section Changes

In order to comprehensively consider the influence of the inlet angle of the total heat exchanger on the performance, under the circumstance of 200, 300, 400, 500 and 600 mm length of counterflow section, the inflow angles of 45°, 60°, 75°, 90°, 105°, 120° and 135° are simulated, respectively. The effect of different inflow angles on the performance of the total heat exchanger under different lengths of the counterflow section is shown in Fig. 6.

It can be seen from Fig. 6 that when the inflow angle is 45° and 135° the latent heat efficiency and sensible heat efficiency are the largest, while the sum of pressure loss is the smallest. At 90°, sensible heat efficiency and latent heat efficiency are the smallest, and pressure loss the maximum. In order to illustrate the problem easily, the performance curves of three characteristic angles (45°, 135° and 90°) corresponding to different lengths of counterflow section are sorted out, as shown in Fig. 7.

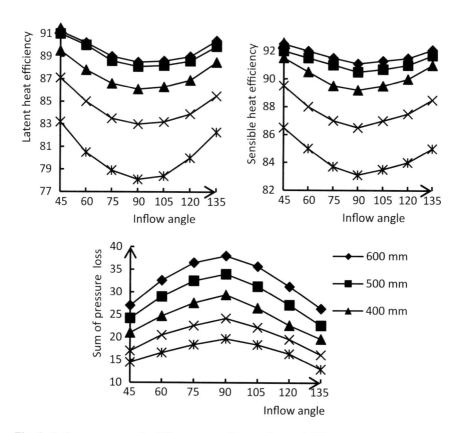

Fig. 6 Performance curves for different counterflow sections and different inflow angles

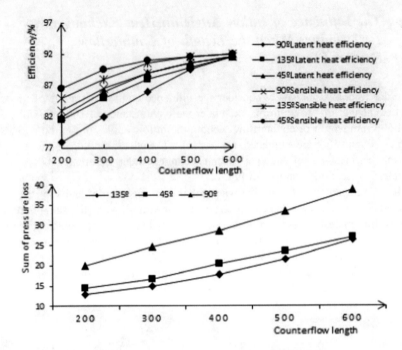

Fig. 7 Curve of the length of the counterflow section and the performance of the heat exchanger under three characteristic angles

It can be seen from Fig. 7 that when the airflow is perpendicular to the windward surface (90°), the effect of the increase of the length of the counterflow section after 400 mm is no longer obvious. The efficiency reaches a maximum at 600 mm, which is consistent with the theoretical solution [8]. When the inlet and outlet airflow angles are simultaneously set to 45° or 135°, the length of the 400-mm counterflow section is sufficient to ensure a large efficiency value. At this time, as the length of the counterflow section increases, the sensible heat efficiency is almost unchanged, and the growth of latent heat efficiency is no longer obvious.

4 Conclusion

By comparing the numerical simulation results with the experimental conclusions, the article can draw the following conclusions:

(1) The efficiency and the sum of pressure loss show a parabolic trend. With the increase of the length of the counterflow section, the simulation results of sensible heat efficiency and latent heat efficiency are in agreement with the theoretical solution results. The sum of pressure loss increases linearly with the increase of the length of the counterflow section.

(2) As the channel height value decreases, the exchange efficiency of the core and the channel height show a nearly linearly decreasing law. Usually, the height of the channel of the core is chosen to be 3 mm, which is sufficient to ensure a high exchange efficiency. When only changing the inflow angle to horizontal or vertical, the fan pressure loss is greatly reduced. The above conclusions provide a reference to practical engineering.

(3) The research in this paper has achieved certain results: The heat exchange efficiency can be significantly improved by changing the air inflow angle. However, due to the constraints of time and existing experimental conditions, this paper did not conduct experiments. In the future research, actual models should be produced and combined with experiments to prove the conclusions of the simulation.

Acknowledgements The research was partially supported by the National Science Foundation of Tianjin (17JCTPJC48200).

References

1. Niu, J.L., Zhang, L.Z.: Membrane-based enthalpy exchanger: material considerations and clarification of moisture resistance. J. Membr. Sci. **189**(2/3), 179–191 (2001)
2. Zhang, L.Z.: Heat and mass transfer in plate-fin sinusoidal passages with vapor permeable wall materials. Int. J. Heat Mass Transf. **51**(3/4), 618–629 (2008)
3. Li, H.: Experimental research on heat recovery ventilator of air conditioning system. Refrig. Air Conditioning **32**(5), 560–562 (2018). (in Chinese)
4. Zhang, L.Z.: Heat and mass transfer in a quasi-counter flow membrane-based total heat exchanger. Int. J. Heat Mass Transf. **53**(23/24), 5478–5486 (2010)
5. Liu, J., Li, W., et al.: Efficiency of energy recovery ventilator with various weathers and its energy saving performance in residential apartment. Energy Build. **42**(1), 43–49 (2010)
6. Wu, J.Y., Wang, R.Z.: Performance of energy recovery ventilator PY with various weather and temperature set-point. Energy Build. **39**(12), 1202–1210 (2007)
7. Nasifa, M., Al-Wakedc, R., et al.: Heat exchanger in HVAC energy recovery systems, systems energy analysis. Energy Build. **42**(10), 1833–1840
8. Wu, W.: Study on heat exchange performance of air-to-air plate heat exchange ventilator. Harbin Institute of Technology 34–36 (2009). (in Chinese)
9. Liu, Y.W., Li, Z.H.: Influence of inlet air flow direction to the performance of the quasi-counter flow total heat exchanger. Build. Energy Effic. **40**(1), 14–17 (2012). (in Chinese)
10. Fluent Inc. Fluent UDF manual. Lebanon: Fluent Inc (2012)

Evaluation of Factors Toward Flow Distribution in the Dividing Manifold Systems with Parallel Pipe Arrays Using the Orthogonal Experiment Design

Wanqing Zhang, Angui Li and Feifei Cao

Abstract Dividing manifold systems have a wide range of applications in the fields of energy transfer and conservation. A uniform flow distribution plays an essential role in industrial processes to improve the efficiency and durability of industrial facilities and equipment. The orthogonal experiment design (OED) was adopted in this study to evaluate the factors that affect flow performance of the dividing manifold systems with parallel pipe arrays (DMS–PPA) under the range of five structural and flow parameters (area ratio (AR), pipe pitch (Δl), height of convex head (h_{head}), roughness factor (K), and inlet Reynolds number (Re_{in})). The non-uniformity coefficient (Φ) and total pressure drop (ΔP_j) were put forward to evaluate flow distribution. The $L_{25}(5^6)$ orthogonal array was selected for the experiment, and the analysis of range (ANORA) and the analysis of variance (ANOVA) are performed. The most significant parameter is identified as AR and Re_{in}, respectively, considering the influence degree on the Φ and ΔP_j. The effect of AR should be further studied for the structural optimization design of the dividing manifold system.

Keywords Flow distribution · Dividing manifold system · Orthogonal experiment design

1 Introduction

A manifold is defined as a flow channel for which fluid enters or leaves through porous sidewalls or lateral pipes under the effect of differential pressure [1]. Dividing manifold systems have a wide range of applications in industrial area.

The design of uniform flow distribution is an essential issue for the performance improvement of the manifold system.

Recently, related researches are mostly focused on the effect of geometrical parameters on the flow distribution [2–4] and optimized the configurations of manifold systems [5–7]. The dividing manifold system with parallel pipe arrays (DMS–PPA) is a commonly used distributing device for industrial application. Considering the significant application value of the DMS–PPA, the flow characteristics need to be investigated for a uniform flow distribution.

Due to many factors affecting the flow distribution in the DMS–PPA, five critical structural parameters and flow condition have been selected based on the design of dividing manifold system. The aim of this study is to analyze the influence degree of each factor on the flow distribution through the orthogonal experiment design. The flow uniformity and pressure drop were used for the evaluation of factors. This study has a guiding role for structural optimization in dividing manifold systems.

2 Methods

2.1 Physical Model

In the water distributing system, the DMS–PPA is a fluid distributing channel, which is composed of cylindrical tank with convex heads, parallel inlet/outlet pipes, and interface fittings. The simplified model of the DMS–PPA with four outlets is shown in Fig. 1, which is a common configuration in engineering application.

Fig. 1 Schematic diagram of the DMS–PPA: 1. Dividing manifold. 2. Convex head. 3. Inlet pipe. 4. Outlet pipes

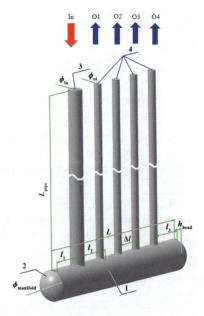

Table 1 Geometric sizes of the DMS–PPA

L	L_{pipe}	Δl	l_1	l_2	l_3	h_{head}	ϕ_{in}	ϕ_{oi}	$\phi_{manifold}$	R_{in}	AR
1910	4800	280	400	350	320	106	150	80	325	0	1.1378

The variable is in millimeters, except for the dimensionless parameter AR

The geometric sizes of the DMS–PPA are shown in Table 1. The sharp-edged inlet/outlet T-junction is adopted. In order to ensure a full development of flow, the branch pipes have been prolonged with a length of 32d/60d at the inlet/outlet pipes.

2.2 Assumptions and CFD Numerical Simulation

In this study, the incompressible, single-phase pure water is adopted as the working fluid. All outlet tubes are of the same diameter and spacing. The fluid has constant physical properties (temperature $T = 368$ K, density $\rho = 961.82$ kg m^{-3}, viscosity $\mu = 2.98 \times 10^{-4}$ kg m^{-1} s^{-1}), and the heat exchange in the system is neglected.

The continuity and momentum equations of the generic form can be expressed as follows for the steady-state simulations:

$$\text{div}(\rho v \varphi) = \text{div}(\Gamma_\varphi \text{grad} \varphi) + S_\varphi \tag{1}$$

where S_φ is generalized source terms, which equals 0 and $-\text{div} p + S_i$, respectively, in the continuity and momentum equations. Furthermore, S_i is 0 for the incompressible fluids with constant viscosity.

ANSYS Fluent is involved for calculating the flow distribution. The realizable k–ε turbulence model and SIMPLE algorithm are adopted in the study. The inlet is set as velocity inlet, and the outlet is pressure outlet. The computational accuracy of numerical method has been validated with a U-type manifold system [8]. The details have been illustrated in the earlier work of authors [9, 10].

3 Orthogonal Experiment Design

An unevenly distributed manifold system has been evaluated using the dimensionless parameter β_i (flow ratio) and Φ (non-uniformity coefficient) [9]. The calculation of β_i can be obtained by $\beta_i = \frac{Q_i}{Q}$, where the Q represents the mass flow rate of branch pipe. The standard deviation of the flow ratio (Φ) is used to assess the degree of unevenness, $\Phi = \sqrt{\frac{\sum_{i=1}^{n}(\beta_i - \bar{\beta})^2}{n}}$. Generally, a smaller value of Φ corresponds to a more uniform flow distribution. Besides, the total pressure drop (ΔP_j) of the DMS–PPA is also an essential indicator to evaluate the flow distribution.

Table 2 Influence factors and level values in orthogonal experiment design

Level	Factor A AR	Factor B Δl (mm)	Factor C h_{head} (mm)	Factor D K (mm)	Factor E Re_{in}
1	0.7511(325-150-65)	$\phi_{oi} + \phi_{o(i+1)} + 80$	0	0.04	0.7×10^5
2	1.1378(325-150-80)	$\phi_{oi} + \phi_{o(i+1)} + 100$	50	0.1	2.1×10^5
3	1.7778(325-150-100)	$\phi_{oi} + \phi_{o(i+1)} + 120$	106	0.25	3.5×10^5
4	2.7778(325-150-125)	$\phi_{oi} + \phi_{o(i+1)} + 140$	162.5	0.5	4.9×10^5
5	4(325-150-150)	$\phi_{oi} + \phi_{o(i+1)} + 160$	180	1	7×10^5

The numerical code in parentheses of Factor A represents the diameters of manifold and inlet/outlet pipe ($\phi_{manifold}$, ϕ_{in}, ϕ_{oi}). The area ratio can be described as $AR = \sum_{i=1}^{n} \frac{\pi \phi_{oi}^2}{4} / \frac{\pi \phi_{in}^2}{4}$

It is calculated according to the most adverse loop, when the direction of flow is from the inlet to the first outlet.

The orthogonal experimental design (OED) is adopted in this research to analyze the influences of various design factors on the flow distribution of the DMS–PPA. The area ratio of inlet/outlet pipe (AR), the pipe pitch (Δl), the height of convex head (h_{head}), the absolute roughness of wall (K), and the Reynolds number of inlet pipe (Re_{in}) are selected as the main influence factors, and there is no interaction among them. Therefore, the $L_{25}(5^6)$ orthogonal array is used to obtain 25 sets of experimental schemes. It considers five physical factors (AR, Δl, h_{head}, K, and Re_{in}) and one additional factor, which is included as error array. Five levels of each factor are listed in Table 2. The multi-factor analysis is carried out by using the non-uniformity coefficient (Φ) and the total pressure drop (ΔP_j) in the DMS–PPA as test indicators. Table 3 shows the results of the OED calculations. The analysis of range (ANORA) and the analysis of variance (ANOVA) are performed to investigate the influence degree of various factors on flow distribution and the significance of each factor.

4 Results and Discussions

4.1 Analysis of Range (ANORA)

Under the orthogonal experiment scheme, the calculation of Ki and R values is obtained in Tables 4 and 5. The Ki value is the sum of the indicator values of the five repetitions for level $i = 1–5$. The $\overline{Ki}$ is obtained by averaging indicator value. The R value is calculated from the difference between maximum and minimum $\overline{Ki}$ value, which can be expressed as $R = \overline{K}_{max} - \overline{K}_{min}$. The greater the R value of a factor, the greater the influence degree of this factor on the test indicator will be.

According to the R value in Table 4, the rank of magnitude is as follows: $R_A > R_D > R_E > R_C > R_B$. The Factor A is the most influential factor for the flow

Table 3 Results of the orthogonal experiment design

No.	Factor A	Factor B	Factor C	Factor D	Factor E	Error	Results	
							Φ	ΔP_j
1	1	1	1	1	1	1	0.01196	647.41
2	1	2	2	2	2	2	0.00847	5663.50
3	1	3	3	3	3	3	0.00674	19,844.61
4	1	4	4	4	4	4	0.00728	46,175.01
5	1	5	5	5	5	5	0.00717	113,073.06
6	2	1	2	3	4	5	0.02497	20,134.02
7	2	2	3	4	5	1	0.01800	46,656.22
8	2	3	4	5	1	2	0.01735	503.62
9	2	4	5	1	2	3	0.03121	2767.27
10	2	5	1	2	3	4	0.02479	8127.88
11	3	1	3	5	2	4	0.05213	3073.46
12	3	2	4	1	3	5	0.07459	5367.72
13	3	3	5	2	4	1	0.07195	10,779.69
14	3	4	1	3	5	2	0.05679	26,139.20
15	3	5	2	4	1	3	0.05748	265.13
16	4	1	4	2	5	3	0.16544	19,194.09
17	4	2	5	3	1	4	0.16487	205.27
18	4	3	1	4	2	5	0.15236	2118.37
19	4	4	2	5	3	1	0.11895	6720.62
20	4	5	3	1	4	2	0.18691	8320.38
21	5	1	5	4	3	2	0.30660	5579.11
22	5	2	1	5	4	3	0.28830	12,385.58
23	5	3	2	1	5	4	0.40016	16,518.47
24	5	4	3	2	1	5	0.34923	185.36
25	5	5	4	3	2	1	0.37139	1708.44

Table 4 Results of the ANORA for the non-uniformity coefficient (Φ)

Index	Factor A (AR)	Factor B (Δl)	Factor C (h_{head})	Factor D (K)	Factor E (Re_{in})	Error
$K1$	0.04162	0.56110	0.53419	0.70482	0.60089	0.59225
$K2$	0.11631	0.55423	0.61004	0.61990	0.61556	0.57612
$K3$	0.31294	0.64856	0.61301	0.62476	0.53166	0.54917
$K4$	0.78853	0.56346	0.63605	0.54171	0.57940	0.64923
$K5$	1.71569	0.64774	0.58180	0.48390	0.64757	0.60832
$\overline{K1}$	0.00832	0.11222	0.10684	0.14096	0.12018	0.11845
$\overline{K2}$	0.02326	0.11085	0.12201	0.12398	0.12311	0.11522
$\overline{K3}$	0.06259	0.12971	0.12260	0.12495	0.10633	0.10984
$\overline{K4}$	0.15771	0.11269	0.12721	0.10834	0.11588	0.12985
$\overline{K5}$	0.34314	0.12955	0.11636	0.09678	0.12951	0.12166
R	0.33482	0.01886	0.02037	0.04418	0.02318	0.02001
Optimal level	A1	B2	C1	D5	E3	/
Ranking	A > D > E > C > B					

Table 5 Results of the ANORA for the total pressure drop (ΔP_j)

Index	Factor A (AR)	Factor B (Δl)	Factor C (h_{head})	Factor D (K)	Factor E (Re_{in})	Error
$K1$	185,403.59	48,628.09	49,418.44	33,621.25	1806.79	66,512.37
$K2$	78,189.00	70,278.28	49,301.73	43,950.52	15,331.04	46,205.80
$K3$	45,625.20	49,764.76	78,080.04	68,031.54	45,639.93	54,456.68
$K4$	36,558.73	81,987.46	72,948.87	100,793.83	97,794.68	74,100.09
$K5$	36,376.96	131,494.89	132,404.39	135,756.35	221,581.05	140,878.53
$\overline{K1}$	370,80.72	9725.62	9883.69	6724.25	361.36	13,302.48
$\overline{K2}$	15,637.80	14,055.66	9860.35	8790.10	3066.21	9241.16
$\overline{K3}$	9125.04	9952.95	15,616.01	13,606.31	9127.99	10,891.34
$\overline{K4}$	7311.75	16,397.49	14,589.77	20,158.77	19,558.94	14,820.02
$\overline{K5}$	7275.39	26,298.98	26,480.88	27,151.27	44,316.21	28,175.71
R	29,805.33	16,573.36	16,620.35	20,427.02	43,954.85	18,934.55
Optimal level	A5	B1	C2	D1	E1	/
Ranking	E > A > D > C > B					

uniformity, and the effect of Factor D is the second most influential, followed by the Factors E and C, while the Factor B has the smallest impact. Consequently, the influence degree of the factors can be ordered as AR (A) > K (D) > Re_{in} (E) > h_{head} (C) > Δl (B). Since the indicator Φ is required to be smaller, the optimal combination scheme of each factor is A1D5E3C1B2. That is, the AR= 0.7511, the K is 1 mm, and the Re_{in} is 3.5×10^5, while the $h_{head} = 0$ mm and the Δl is recommended as $(\phi_{oi} + \phi_{o(i+1)} + 100)$ mm.

Similarly, the results of ANORA for ΔP_j are listed in Table 5. The importance of factors has been ranked as follows: Re_{in} (E) > AR (A) > K (D) > h_{head} (C) > Δl (B). Considering higher ΔP_j would lead to more energy consumption in the DMS–PPA, the lower ΔP_j is preferred. As a result, the optimal combination of factor levels will be E1A5D1C2B1. Namely, under the condition that Re_{in} of 0.7×10^5, the DMS–PPA with AR of 4, K of 0.04 mm, and h_{head} equals 50 mm, as well as $\Delta l = (\phi_{oi} + \phi_{o(i+1)} + 80)$ mm is considered to be the most energy-saving structure.

The relationship between each factor and two indicators is shown in Fig. 2. For the indicator Φ (Fig. 2a), the AR exerts the greatest influence on the flow distribution. With the AR rising, there is an evident increase of Φ which would lead to more severe non-uniformity, while the other factors have a minor influence. The Re_{in} has the largest impact for ΔP_j (Fig. 2b), which followed by the effect of AR. The ΔP_j would sharply climb with Re_{in} rising as the inlet flow increases. All the increasing factors will lead to a pressure rise except for factor AR. In general, the

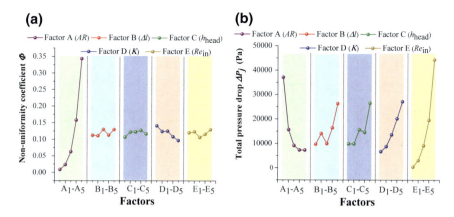

Fig. 2 Comparison of the two test indicators among different levels of each factor: **a** the non-uniformity coefficient (Φ) and **b** the total pressure drop (ΔP_j)

AR is the most influential in both test indicators. Besides, both of the test indicators should be comprehensively considered for the optimized design of the geometric structures.

4.2 Analysis of Variance (ANOVA)

To obtain an accurate quantitative estimation of the importance degree (significance) of each factor, the analysis of variance (ANOVA) is performed in this study. The F statistic is constructed, and the ANOVA with F-test is used to evaluate the significance of various factors on the Φ and ΔP_j. The results are listed in Tables 6 and 7, respectively.

Table 6 Results of the ANOVA for the non-uniformity coefficient (Φ)

Variable	Sum of squares	Degree of freedom	Mean square	F value	Critical $F\alpha$	Significance
Factor A	0.38167	4	0.09542	341.84		**
Factor B	0.00189	4	0.00047	1.69	$F_{0.05}$ (4, 4) = 6.39	
Factor C	0.00122	4	0.00031	1.09		
Factor D	0.00575	4	0.00144	5.15	$F_{0.01}$ (4, 4) = 15.93	
Factor E	0.00150	4	0.00037	1.34		
Error	0.00112	4	0.00028			
Sum	0.39315	24				

** Highly Significant, * Significant

Table 7 Results of the ANOVA for the total pressure drop (ΔP_j)

Variable	Sum of squares	Degree of freedom	Mean square	F value	Critical $F\alpha$	Significance
Factor A	3.20425×10^9	4	8.01062×10^8	2.83	$F_{0.05}$ (4, 4) = 6.39	
Factor B	9.16970×10^8	4	2.29243×10^8	0.81		
Factor C	9.22708×10^8	4	2.30677×10^8	0.82	$F_{0.01}$ (4, 4) = 15.93	
Factor D	1.41425×10^9	4	3.53563×10^8	1.25		
Factor E	6.35500×10^9	4	1.58875×10^9	5.62		
Error	1.13075×10^9	4	2.82686×10^8			
Sum	1.39439×10^{10}	24				

** Highly Significant, * Significant

The F value is the ratio of variance caused by a specific factor to the variance caused by the error term. The critical value of $F\alpha$ is selected through the F-distribution table according to the requirements of degrees of freedom and significant level [11]. Two significant levels, $\alpha = 0.01$ and 0.05, are adopted. When a test F value is greater than $F_{0.01}$, it is regarded as highly significant. While the value is between $F_{0.05}$ and $F_{0.01}$, it is supposed to be significant.

The results indicated that the AR is a highly significant factor for the Φ, with the rest of factors being insignificant. For the indicator ΔP_j, under the confidence level of 0.95, none of the intended factors is significant. Comparing the magnitude of F values, the Re_{in} has a relatively larger influence, and the AR is the second influential factor. In summary, since the AR has a relatively high level of significance for both indicators Φ and ΔP_j, the effect of which on the flow distribution in the DMS–PPA needs to be further studied.

5 Conclusions

In this study, the flow distribution characteristics were analyzed by the orthogonal experiment design (OED) for the dividing manifold system with parallel pipe arrays (DMS–PPA). The influences of various design factors on the non-uniformity coefficient (Φ) and the total pressure drop (ΔP_j) were investigated through the ANORA and ANOVA, respectively. The following conclusions are obtained:

For the indicator Φ, the influence degree of the intended factors can be ranked as, AR (A) > K (D) > Re_{in} (E) > h_{head} (C) > Δl (B). As for the ΔP_j, the importance order of the factors is Re_{in} (E) > AR (A) > K (D) > h_{head} (C) > Δl (B). Among them, the AR has a great influence on both the test indicators.

The AR is a highly significant factor for the Φ. Under the confidence level of 0.95, all the intended factors are not significant for the ΔP_j. Generally speaking, the Re_{in} has a relatively large impact on the ΔP_j, and the AR is the second essential factor.

In summary, there is an opposite tendency between Φ and ΔP_j under the effect of AR and K. Both of the test indicators should be considered for the structural optimization. Since the AR has a relatively high level of significance for both indicators, the effect of which on the flow distribution in the DMS–PPA needs to be further studied.

Acknowledgements The project is supported by the Shaanxi Science and Technology Co-ordination and Innovation Project (No.2016KTCL01-13).

References

1. Bajura, R.A., Jones, E.H.: Flow distribution manifolds. ASME Trans. J. Fluids Eng. **98**(4), 654–665 (1976)
2. Dong, J., et al.: CFD analysis of a novel modular manifold with multistage channels for uniform air distribution in a fuel cell stack. Appl. Therm. Eng. **124**, 286–293 (2017)
3. Yang, H., et al.: Effect of the rectangular exit-port geometry of a distribution manifold on the flow performance. Appl. Therm. Eng. **117**, 481–486 (2017)
4. Lee, S., et al.: A study on the exit flow characteristics determined by the orifice configuration of multi-perforated tubes. J. Mech. Sci. Technol. **26**(9), 2751–2758 (2012)
5. Liu, H.H., et al.: Modeling and design of air-side manifolds and measurement on an industrial 5-kW hydrogen fuel cell stack. Int. J. Hydrogen Energy **42**(30), 19216–19226 (2017)
6. Huang, C.H., et al.: A manifold design problem for a plate-fin microdevice to maximize the flow uniformity of system. Int. J. Heat Mass Transf. **95**, 22–34 (2016)
7. Said, S.A.M., et al.: Reducing the flow mal-distribution in a heat exchanger. Comput. Fluids **107**, 1–10 (2015)
8. Wang, X., Yu, P.: Isothermal flow distribution in header systems. Int. J. Solar Energy **7**(3), 159–169 (1989)
9. Zhang, W., et al.: Effects of geometric structures on flow uniformity and pressure drop in dividing manifold systems with parallel pipe arrays. Int. J. Heat Mass Transf. **127**, 870–881 (2018)
10. Zhang, W., Li, A.: Resistance reduction via guide vane in dividing manifold systems with parallel pipe arrays (DMS–PPA) based on analysis of energy dissipation. Build. Environ. **139**, 189–198 (2018)
11. Winer, B.J.: Statistical Principles in Experimental Design. McGraw-Hill, New York (1962)

Performance Studies of R134a-Dimethylformamide Absorption Refrigeration System for Utilizing Bus Exhaust Gas Based on Aspen Plus Software

Xiao Zhang, Liang Cai, Qiang Zhou and Liping Chen

Abstract In this passage, a steady-state simulation model of air-cooled absorption refrigeration is developed and validated by Aspen Plus software, in which R134a-dimethylformamide (DMF) is used as working pair. The model is originally established under the given conditions, where the temperatures of condensation, evaporation, ambient surroundings, and waste gas in generator are 50, 5, 35, and 550 °C, separately, the mass concentration range of cycle is 0.09, and the refrigeration load is 30 kW. Then, the effects of operating parameters on system performance are investigated. The simulation results present that when the generator heat load is about 65 kW, the COP value is the highest at this time, reaching about 0.385, and the solution circulation ratio is 8.68. It is also found that under most load operating conditions, the value of COP is above 0.35, which proves that R134a-DMF working fluids have great potential for utilizing waste heat in bus.

Keywords Aspen Plus · Heat recovery · Exhaust gas · Absorption refrigeration

1 Introduction

Currently, automotive air conditioning systems have greatly improved the comfort in operation, where the most widely used is the vapor compression refrigeration system, which consumes a tremendous amount of fuel oil. And also commonly, the vehicle engines only have about 30% energy efficiency, and the other approximately two-thirds of the primary energy is rejected to the ambient surroundings [1]. So some effort has been devoted to the heat recovery of exhaust gas. The utilization of the waste heat is generally focused on heating or turbocharging. Then, some

---

X. Zhang · L. Cai (✉) · L. Chen
School of Energy and Environment, Southeast University, Nanjing, China
e-mail: 101008806@seu.edu.cn

Q. Zhou
State Grid Jiangsu Integrated Energy Service Co., Ltd, Nanjing, China

scholars proposed the absorption air conditioning system to replace the conventional systems by utilizing bus exhaust gas.

Koehler et al. [3] proposed the prototype of single-stage absorption system for truck refrigeration using the gas which varied between 440 and 490 °C, the working pair of which is ammonia and water. The results showed that when the ambient temperatures and interior temperatures are separately between 20 and 30, −20 and 0 °C, the COP between 0.23 and 0.3 could be obtained. Horuz [2] investigated the performance of the vapor absorption system utilizing the exhaust gases from a turbocharged diesel engine, among which the Robur Servel ACB-3600 gas-fired VAR system is used for experiments to drive 10 kW rated capacity. The result showed that the refrigeration unit can be powered up successfully to its rated capacity, but a significant reduction in capacity (down to 1 kW) was observed when the engine was run at low speeds. Talbi et al. [4] explored the theoretical performance of four different configurations of a turbocharger diesel engine and the combined absorption refrigeration operating at 35 °C of ambient temperature. It is demonstrated that a pre- and inter-cooled turbocharger engine configuration cycle offers considerable benefits in terms of SFC, efficiency, and output for the diesel cycle performance.

From all of the above literature, most working pairs of the absorption refrigeration systems are focused on the ammonia and water. But the ammonia is poisonous, which will lead to safety problem once leaked. Few researches are performed on the investigation about the performance of the novel refrigerants and their corresponding absorbents. This passage employs R134a-DMF as working pair to study its performance in the absorption cycle.

2 Methods

In this passage, a steady-state simulation model of air-cooled absorption refrigeration is developed and validated by Aspen Plus software [5], in which R134a-dimethylformamide (DMF) is used as a working pair. The models are built to analyze the performance of the absorption unit and the feasibility of working fluids. The establishment of whole models is divided into two parts—one is the appropriate property model of working pair and the other is simulation model of absorption cycle.

2.1 Cycle Description and Working Principles

The schematic representation of the absorption cycle can be shown as in Fig. 1. In the generator, the mixer of the R134a and DMF (12) is heated by the exhaust gas provided by the coach engine. After the process, the vapor–liquid mixer (14) flows into the separator to split into two streams (3) (7) with different phase states, which

Fig. 1 Schematic representation of the absorption cycle

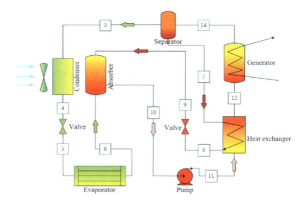

will lead to the vapor stream (3) closing to pure refrigerant state. Then, the refrigerant vapors (3) flow to the air-cooled condenser where they get liquefied. It is noteworthy that in order to save space, condensers and absorbers are placed together and cooled by air. The liquid refrigerant (4) is then circulated to the throttle valve to reduce its pressure and to become the vapor–liquid phase. After that, it is introduced to evaporator to get refrigeration by absorbing the heat of the ambient air. The overheated refrigerant vapor (6) runs into the absorber where the vapor will be mixed by the poor solution (9) coming from the separator. After absorbing, the strong solution (10) is pumped up to generator. During its way, it will transfer heat in the heat exchanger with the weak solution (7) returning to the low-pressure surrounding.

2.2 Property Calculation Model

The establishment of the working fluids calculation model is crucial, especially the selection of method for calculation. Before simulating the absorption cycle in Aspen Plus, to ensure the accuracy of selected method is vital, so the predictions of selected ten property models are compared with the vapor–liquid equilibrium data, then the Aspen Plus Data Regression System facility is used to fit the binary interaction parameter of the methods to the experimental data. The comparisons after adjustment will also be shown in the form of P–X diagrams to choose the best-fitting method.

The selected ten thermodynamic property models in Aspen Plus for the calculation of the properties of the working fluids are presented in Table 1. Firstly, the predictions of selected ten original property models are compared with the vapor–liquid equilibrium (VLE) data reported by Han et al. [6] in the temperatures of 283.15 and 303.15 K. From Figs. 2 and 3, it can be concluded that none of the original property models can calculate the vapor–liquid equilibrium data of R134a and DMF with enough accuracy. In the second step, the Aspen Plus Data

Table 1 Selected ten property models and their deviations

Property models	Full name of the models	Sum of squared errors
BWR-LS	BWR Lee–Starling equation of state	47.2485
BWRS	Benedict–Webb–Rubin–Starling equation of state	62.4539
HYSGLYCO	HYSYS Glycol EOS package	25.7711
HYSPR	HYSYS Peng–Robinson EOS package	22.6031
LK-PLOCK	Lee–Kesler–Plocker equation of state	35.8092
NRTL	Non-random two-liquid	15.505
PENG-ROB	Peng–Robinson equation of state	40.335
SRK	Soave–Redlich–Kwong equation of state	26.8366
WILSON	Wilson	21.7782
RK-ASPEN	Redlich–Kwong–Aspen equation of state	25.7253

Regression System facility is used to fit the binary interaction parameter of the methods to the experimental data. Figures 4 and 5 are the images plotted by the adjusted binary parameters separately in the temperatures of 283.15 and 303.15 K, where it can be noted that all values of the VLE data predicted by the adjusted parameters are now close to the corresponding experimental data. Then, all the experimental data from 263.15 to 303.15 K is used to fit the interaction parameters to select the best property models within the ten methods. By calculation in Aspen Plus, the sum of squared errors for the mentioned ten property models is also summarized in Table 1. It is concluded that the non-random two-liquid (NRTL) model, with fitted parameters, is the most appropriate property method. Figure 6 is the plotted results by the fitted NRTL model and experimental data in the full temperature range.

2.3 Simulation Model of Absorption Cycle

The simulation model of absorption cycle should be based on the establishment of above property models. The model is originally built under the given conditions, where the temperatures of condensation, evaporation, ambient surroundings, and waste gas in generator are 50, 5, 35, and 550 °C, separately, the mass concentration range of cycle is 0.09, and the refrigeration load is 30 kW. The establishment of simulation model is also performed in two steps. Firstly, the simulation modules in Aspen Plus should be connected to integrate the absorption cycle model, where the type of heat exchanger is in design pattern based on the above design conditions. Secondly, the operating pattern of heat exchanger should be converted into simulation pattern based on the following equation of heat transfer characteristics

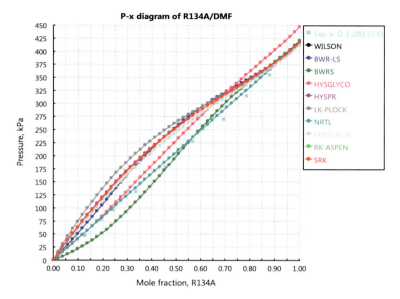

Fig. 2 P–X VLE diagram at $T = 283.15$ K (plotted by the original parameters)

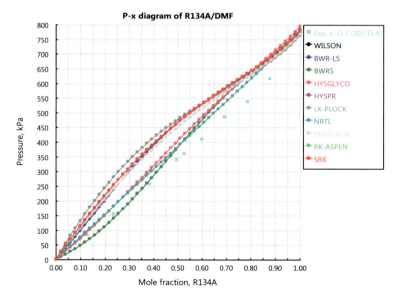

Fig. 3 P–X VLE diagram at $T = 303.15$ K (plotted by the original parameters)

$$Q = (\text{UA}) \cdot \Delta t_{\text{in}} \quad (1)$$

where Q is the heat duty and Δt_{in} is the logarithmic mean temperature difference.

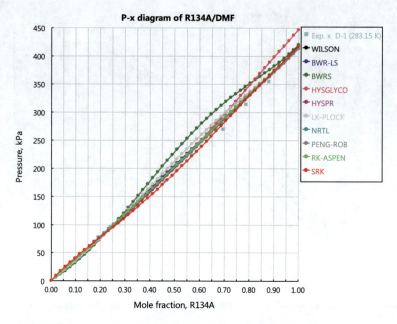

Fig. 4 P–X VLE diagram at $T = 283.15$ K (plotted by the adjusted parameters)

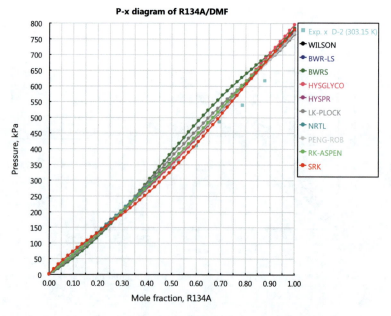

Fig. 5 P–X VLE diagram at $T = 303.15$ K (plotted by the adjusted parameters)

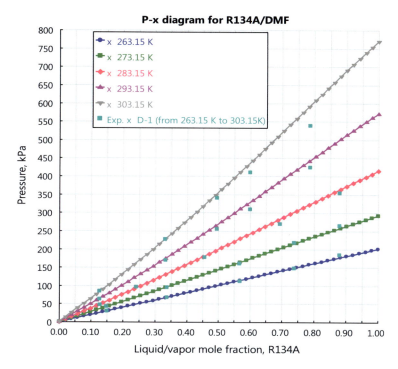

Fig. 6 P–X VLE diagram by fitted NRTL

Table 2 Components and their corresponding modules with input data

Components	Aspen modules	Input data
Evaporator	Heatx (EVAP)	Cold stream outlet temperature increase = 5 °C
Absorber	Mixer and heatx (ABSO and COOLER)	Hot stream outlet vapor fraction = 0
PUMP	PUMP	Discharge pressure = 1.318 MPa Pump efficiency = 1
Heat exchanger	Heatx (HEATX)	Hot stream outlet temperature = 55 °C
Generator	Heatx (GENE)	Cold stream outlet temperature = 128 °C
Separator	Sep2 (SEPA)	The split fraction of the R134a is 0.999 The flow rate of mixed is transferred from stream 6
Condenser	Heatx (COND)	Hot stream outlet degrees subcooling = 1 °C
Valve	Valve (VAL1 and VAL2)	Outlet pressure is 0.349 MPa

Then, the simulation model can be used for analysis.

Table 2 shows the modules corresponding to the components of the cycle as well as the input values in the design pattern. Then, the model in Fig. 7 is running

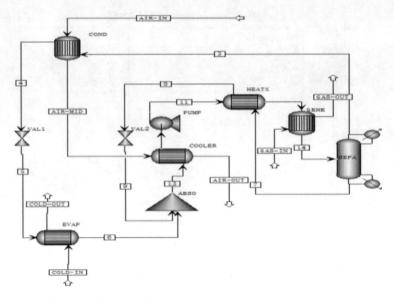

Fig. 7 Simulation model in Aspen Plus

under the above given conditions in design pattern. After the establishment of simulation model in design pattern, the simulation model needs to be converted into simulation pattern according to the above approaches for analysis.

3 Results and Discussion

According to the methods mentioned above, when the appropriate thermodynamic property model is established, the calculated results of the design process in Fig. 7 at given conditions are shown in Table 3. The heat loads of each component are also presented in Table 4. Then, the adjusted model needs to be transformed from the design pattern to simulation pattern by inputting the UA values of heat exchanger. Table 5 shows the UA values of each heat exchanger calculated by the above design data.

After the establishment of simulation model in simulation pattern, the influences of ambient temperature are concluded in Figs. 8 and 9. It can be seen that when the ambient temperature increases, the flow rate and temperature of the refrigerant at the outlet of the generator increase continuously, but the cooling capacity is falling, and the COP also decreases. Especially when the ambient temperature exceeds 36 °C, the COP and cooling capacity drop sharply. The reasons are that when ambient temperature increases, the heat duties of condenser and absorber which can be carried away by the air are decreasing as can be seen from Fig. 8 that will lead to the increase of inlet temperature of generator and so as to the improvement of the

Table 3 Calculated results of the design process

State point	Pressure (MPa)	Temperature (°C)	Vapor fraction	Mass flow (kg/s)	Mass fraction
3	1.318	128	1	0.244	1
4	1.318	49	0	0.244	1
5	0.349	5	0.395	0.244	1
6	0.349	10	1	0.244	1
7	1.318	128	0	1.479	0.364
8	1.318	55	0	1.479	0.364
9	0.349	55	0	1.479	0.364
10	0.349	50	0	1.723	0.454
11	1.318	51.3	0	1.723	0.454
12	1.318	105.8	0	1.723	0.454
13	0.349	56.9	0.046	1.723	0.454
14	1.318	128	0.121	1.723	0.454

Table 4 Heat loads of each component

Block name	Heat loads (kW)
EVAP	30
COOLER	49.61
HEATX	290.147
COND	58.915
GENE	73.073
PUMP	4.704

Table 5 UA values of each heat exchanger

Block name	UA values (J/s K)
EVAP	1336.41496
COOLER	6200
HEATX	26984.6172
COND	2508.52116

flow rate and outlet temperature of the block when given the fixed duty. And when the ambient temperature exceeds 36 °C, the condenser is unable to cool the refrigerant vapor to a subcooled state, as Fig. 9 presents that the outlet refrigerant of condenser goes into vapor–liquid equilibrium state, whose temperature is equal to 50 °C. So the heat duty of condenser decreases sharply, but the situation of absorber is opposite. The reasons are that the vapor fraction of inlet refrigerant increases in evaporator, which will lead to the raise of superheat of the outlet, so as to make the heat load of absorber a little improvement.

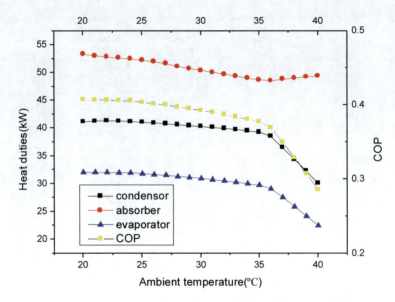

Fig. 8 Heat duties and COP affected by ambient temperature

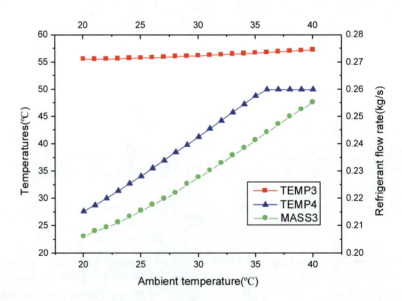

Fig. 9 Stream temperatures and refrigerant flow affected by ambient temperature

Then, the effects of heat duty of generator are also investigated and presented in Figs. 10 and 11. It can be seen that when the heat load increases, the COP, cooling capacity, and other values rise first, but as the load increases to a certain extent, the

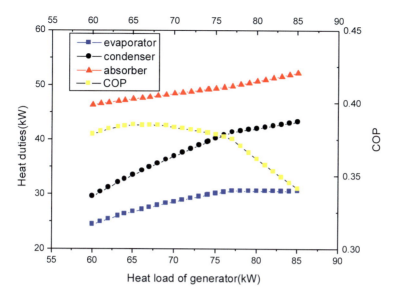

Fig. 10 Heat duties and COP affected by heat load of generator

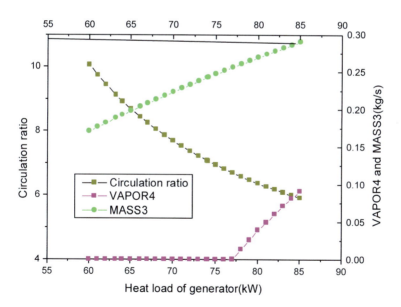

Fig. 11 Parameters of stream affected by heat load of generator

COP will decrease. Especially when it exceeds 76 kW, the COP decreases sharply; however, the cooling capacity will keep steady and other parameters are still growing. The reasons can be found from Fig. 11, as the flow rate of the refrigerant rises with the increase of heat load, which will make more cooling load and more heat transferred by the condenser and absorber. But when the heat load of generator outstrips 76 kW, the vapor fraction of the outlet stream at condenser will get a sharp rise, which will eliminate the influence of excess refrigerant flow and so as to make cooling load a steady state. And also from Fig. 11, the circulation ratio is falling.

From all the charts above, it can also be concluded that when the generator heat load is about 65 kW, the COP value is the highest at this time, reaching about 0.385, and the solution circulation ratio is 8.68.

4 Conclusions

A steady-state simulation model of air-cooled absorption refrigeration is developed and validated by Aspen Plus software, in which R134a-DMF is used as a working pair. The model is originally established under the given conditions. Then, the effects of operating parameters such as ambient temperature, generator load, and solution circulation ratio on system performance are investigated. To get the appropriate property model, based on the experimental VLE data, ten physical methods are selected from software library for data fitting.

The simulation results present that the best operating state is when the generator heat load is about 65 kW, the COP value is the highest at this time, reaching about 0.385, and the solution circulation ratio is 8.68. Through simulation, it is also found that under most load operating conditions, the value of COP is above 0.35. Although the cop is not so high, the cop is stable, especially suitable for occasions that have little effect on load changes, which proves that R134a-DMF working fluids have great potential for utilizing waste heat in bus.

Acknowledgements This work was supported by the National Key Research and Development Program of China (Grant No. 2018YFC0705306) and the Natural Science Foundation of China (Grant No. 6503000103).

References

1. Horuz, I.: An alternative road transport refrigeration. Turk. J. Eng. Environ. Sci. **22**, 211–222 (1998)
2. Koehler, J., et al.: Absorption refrigeration system for mobile applications utilizing exhaust gases. Heat Mass Transf. **32**(5), 333–340 (1997)
3. Horuz, I.: Vapor absorption refrigeration in road transport vehicles. J. Energy Eng. **125**, 48–58 (1999)

4. Talbi, M., Agnew, B.: Energy recovery from diesel engine exhaust gases for performance enhancement and air conditioning. Appl. Therm. Eng. **22**, 693–702 (2002)
5. Mansouri, R., et al.: Modelling and testing the performance of a commercial ammonia/water absorption chiller using Aspen-Plus platform. Energy **93**, 2374–2383 (2015)
6. Han, X., et al.: Solubility of refrigerant 1,1,1,2-tetrafluoroethane in the N,N-dimethyl formamide in the temperature range from (263.15 to 363.15) K. J. Chem. Eng. Data **56**(5), 1821–1826 (2011)

Study on Influence of Outdoor Airflow on Performance of Evaporative Cooling Composite Air-Conditioning System for Data Center

Haotian Wei, Zongwei Han, Chenguang Bai, Qi Fu and Xinwei Meng

Abstract Aiming at the large energy consumption of the air-conditioning system in the data center, an evaporative cooling composite air-conditioning system was proposed. This paper analyzes the impact of outdoor airflow on system performance parameters under different cooling requirements through system modeling and simulation. The results show that the increasing airflow of the outdoor unit increases the energy consumption of the fan, but helps to reduce the power consumption of the compressor and the refrigerant pump. Taking COP value as objective function, the optimum outdoor airflow is obtained. Under the 100, 80, and 60% rated cooling capacity of this system, the optimal outdoor airflow in the heat pipe mode are 2000, 1400, and 800 m^3/h and in the vapor compression mode are 4000, 3600, and 2400 m^3/h, respectively.

Keywords Internet data center · Composite refrigeration technology · Evaporative cooling technology · Outdoor unit airflow

1 Introduction

With the vigorous development of data centers, energy consumption of data centers has attracted worldwide attention. Research shows that data centers with energy consumption of 15 MW cost one million dollars a month in electricity bills [1]. Such huge energy consumption not only brings huge appropriation expenditure to operators, but also aggravates the global environmental pollution.

In order to reduce the PUE value of the data center, scholars mainly research on the cold preparation and the cooling preparation and cold supply. In terms of cooling preparation, the rational use of natural cold source will greatly reduce the energy consumption of the system. Common natural cold source utilization technologies include fresh air cooling technology [2], air-to-air heat exchange technology [3], cooling tower technology [4], refrigerating medium transfer technology [5], and heat pipe technology [6]. These technologies utilize natural cold sources to varying degrees and have better cooling effects, but there are also some problems. Heat pipe heat transfer technology achieves cooling distribution through refrigerant phase change. Compared to air and water as cooling medium, the heat pipe technology has higher efficiency and lower energy consumption, but usually only as an auxiliary technology for energy-saving. In order to make the data center, air-conditioning system can not only make full use of the natural cold source when the ambient temperature is low, but also ensure the cooling effect of the system when the ambient temperature is high. Scholars have combined separate heat pipe heat transfer technology with vapor compression refrigeration technology to study many different forms of heat pipe and vapor compression composite refrigeration system [7].

In terms of cooling supply, closed cold aisle [8] and inter-column air-conditioning [9] can reduce the mixing of hot and cold airflow, improve cooling efficiency, but cannot solve local hot spots. Liquid technology [10] is effective close to the heat source, can achieve precise refrigeration, has a good energy-saving effect, but liquid cooling technology has the safety hazard of liquid leakage; as a mature technology applied to the data center, there is still a distance.

In summary, utilizing natural cold sources, reducing cooling loss, taking terminal unit close to the heat source are effective energy-saving methods. Data center air-conditioning system with reliability, energy-saving, and economy will still need to develop.

2 Data Center Air-Conditioning System

Figure 1 shows an evaporative cooling composite air-conditioning system. This system combines evaporative cooling technology, separate heat pipe heat transfer technology, and vapor compression refrigeration technology to realize the advantage complementation. According to the outdoor environment changes, this system can switch between heat pipe mode and vapor compression mode. In this system, the refrigerant pump enhances the cooling capacity of heat pipe mode. The evaporative condenser extends the running time of heat pipe mode and improves the performance of vapor compression mode. The evaporator placed inside the cabinet realizes on-demand cooling.

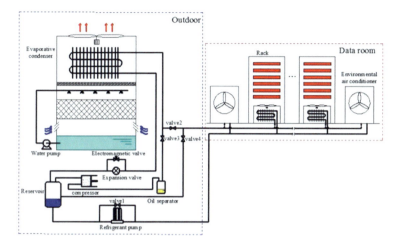

Fig. 1 Schematic diagram of evaporative cooling composite air-conditioning system

3 System Mathematical Model

3.1 Evaporative Condenser Model

Evaporative condenser consists of finned condenser and spray packing. Unlike conventional evaporative condensers, spray water is sprayed directly onto the packing instead of the heat exchanger surface. According to the structural characteristics of this evaporative condenser, the model is divided into two parts—a finned condenser model and a spray packing model.

(1) Finned condenser model

Flow heat exchange equations of refrigerant in the pipe.

$$dQ_r = m_r(h_{ri} - h_{ro}) = \alpha_r dA_i(t_{rm} - t_w) \tag{1}$$

where α_r is the refrigerant-side heat transfer coefficient, W/m^2 °C; in the single-phase zone and the two-phase zone, it is calculated by the heat transfer correlations of Dittus-Boetler and Yu and Koyama, respectively.

Flow heat exchange equation of air out of pipe.

$$dQ_a = m_a(h_{ao} - h_{ai}) = \alpha_a dA_o(t_w - t_{am}) \tag{2}$$

where α_a is the air-side heat transfer coefficient, W/m^2 °C; it is based on the heat transfer correlation obtained by Wu Li.

Two-phase zone pressure drop.

$$\Delta P = 4dLG_r^2 f(\rho_i d_i)^{-1} + G_r^2(\rho_o^{-1} - \rho_i^{-1}) \tag{3}$$

where f is the resistance coefficient.

(2) Spray packing model

The degree of heat and moisture exchange on the surface of spray packing is measured by cooling efficiency.

$$\eta = \frac{t_1 - t_2}{t_1 - t_s} \times 100\% \tag{4}$$

where t_1, t_s, t_2 are the inlet air dry-bulb temperature, wet-bulb temperature, and outlet air dry-bulb temperature, respectively, °C.

The cooling efficiency of spray packing is calculated by empirical formula.

$$\eta = 4.53 t_1^{-0.27} \varphi_1^{-0.19} M_a^{-0.3} M_w^{0.15} \tag{5}$$

where φ_1 is the relative humidity of inlet air, %; M_a is the air quality density, kg/m² s; M_w is the spray density, kg/m² s.

3.2 Finned Evaporator Model

The heat transfer model of finned evaporator is similar to finned condenser. In two-phase zone, the heat transfer coefficient of refrigerant side is calculated by Kandlikar heat exchanger correlation.

3.3 Refrigerant Pump Model

Refrigerant pump head

$$H = -0.29Q^2 - 0.7Q + 5.16 \tag{6}$$

Refrigerant pump power

$$N = \gamma QH/3600 \tag{7}$$

where γ is the refrigerant volumetric weight, N/m³; Q is volume flow rate of refrigerant pump, m³/h;

4 Simulation Results and Analysis

4.1 Influence of Outdoor Airflow on Heat Pipe Performance

Figure 2 shows the variation of the performance parameters of the heat pipe mode with the airflow of the outdoor unit. Under the conditions of the outdoor ambient temperature of 0 °C, relative humidity of 50%, evaporator inlet air temperature of 35 °C, relative humidity of 45%, and cabinet air volume of 800 m^3/h. The condensing temperature and evaporating temperature decrease with the increasing outdoor airflow that because the increasing outdoor airflow increases the air-side heat transfer coefficient of the condenser. When heat exchange area and heat exchange capacity of the condenser are constant, the condensing temperature will decrease with the decreasing heat transfer temperature difference. If the pressure drop of the heat exchanger is not taken into account, the evaporating temperature will be equal to the condensing temperature in heat pipe mode, so the evaporating temperature varies with the condensing temperature. The reduction of the evaporating temperature causes an increase in refrigerant enthalpy difference between inlet and outlet of the evaporator. Therefore, the refrigerant mass flow rate reduces gradually when cooling capacity is constant, and the decreasing refrigerant circulation reduces the total flow loss of the system, so the frequency of the refrigerant pump also reduces. At 100, 80, and 60% rated cooling capacity, the refrigerant pump power consumption decreased by 37.04, 37.79, and 22.96%, respectively. However, the outdoor airflow is not as bigger as better. In order to overcome flow resistance, the fan power consumption increases greatly. Therefore, with the increase of the outdoor unit airflow, the total energy consumption of the system has minimum values which are 494.88, 365.95, and 279.21 W, respectively. At this time, the outdoor unit airflow reaches the optimal value which is 2000, 1400, and 800 m^3/h, respectively, and the corresponding COP values are the largest which are 16.17, 17.49, and 17.20, respectively.

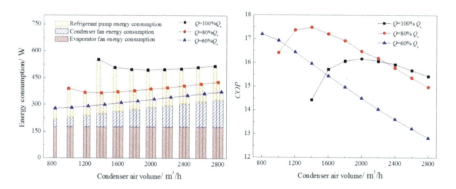

Fig. 2 Influence of outdoor airflow on heat pipe performance

4.2 Influence of Outdoor Airflow on Vapor Compression Performance

Figure 3 shows the variation of the performance parameters of vapor compression mode with the airflow of the outdoor unit, under the conditions of the outdoor ambient temperature of 32 °C, relative humidity of 65%, evaporator inlet air temperature of 35 °C, relative humidity of 45%, and cabinet air volume of 800 m³/h. The condensing temperature decreases with the increasing outdoor airflow, and the evaporating temperature remains constant. Because the increasing outdoor airflow increases the air-side heat transfer coefficient of the condenser and advance the cooling efficiency of the packing that reduced the inlet air temperature of the condenser, so the condensing temperature gradually decreases. However, the increasing outdoor airflow has no effect on the heat transfer performance on the evaporator side. When the cooling capacity is constant, the heat transfer temperature difference will remain constant, so the evaporating temperature does not change, and the refrigerant enthalpy at the inlet and outlet of the evaporator does not change too. It can be understood from Formula (2), the refrigerant mass flow rate will remain constant when the cooling capacity and the evaporator inlet and outlet enthalpy remain unchanged, and because the refrigerant state of the compressor suction port does not change, the reduction in condensing temperature increases the volumetric efficiency of the compressor, so the compressor frequency is slightly reduced. Meanwhile, the reduction of the condensing temperature also reduces the unit power consumption of the compressor, when the refrigerant flow rate is constant, the power consumption of the compressor will gradually reduce which reduces the power consumption of the compressor by 25.90, 25.22, and 23.67%, respectively. However, the increase in the airflow of the outdoor unit will increase the energy consumption of the fan. Therefore, with the increase of the outdoor unit airflow, the total energy consumption of the system has minimum values which are 1.54, 1.11, and 0.78 kW, respectively. At this time, the outdoor unit airflow reaches the optimal values which are 4000,

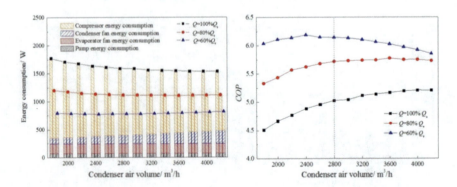

Fig. 3 Influence of outdoor airflow on vapor compression performance

3600, and 2400 m^3/h, respectively, and the corresponding COP values are the largest which are 5.21, 5.78, and 6.19, respectively.

5 Conclusions

This paper studies the influence of outdoor airflow on the system performance parameters under 100, 80, and 60% rated refrigeration capacity. The results show that: In heat pipe mode, evaporating temperature, condensing temperature, refrigeration mass flow rate, refrigerant pump frequency, and power consumption all decrease with the increasing outdoor airflow. Under the three cooling capacities, the optimum outdoor airflow are 2000, 1400, and 800 m^3/h, respectively, and the corresponding maximum COP values are 16.17, 17.49, and 17.20. In vapor compression mode, with the increase of outdoor airflow, evaporation temperature, and refrigerant mass flow rate remain unchanged, condensation temperature, compressor frequency, and compressor power consumption all decrease. Under the three cooling capacities, the optimum outdoor airflow are 4000, 3600, and 2400 m^3/h, respectively, and the corresponding maximum COP values are 5.21, 5.78, and 6.19.

Acknowledgements The authors gratefully acknowledge the support from the Natural Science Foundation of China (grant No. 51778115) and the Fundamental Research Funds for the Central Universities (grant No. N182502043)

References

1. Barroso, L.A., Clidaras, J., Hölzle, U.: The Datacenter as a Computer: An Introduction to the Design of Warehouse-Scale Machines. Morgan & Claypool Publishers (2013)
2. Siriwardana, J., Jayasekara, S., Halgamuge, S.K.: Potential of air-side economizers for data center cooling: a case study for key Australian cities. Appl. Energy **104**, 207–219 (2013)
3. Dunnavant, K.: Indirect air-side economizer cycle data center heat rejection. ASHRAE J. **3**, 44–54 (2011)
4. Gao, T., David, M., Geer, J., et al.: Experimental and numerical dynamic investigation of an energy efficient liquid cooled chiller-less data center test facility. Energy Build. **91**, 83–96 (2015)
5. Wang, D., Chen, K.: Analysis of energy-saving technical scheme for air-conditioning of mobile communication base station. Mech. Electr. Eng. Technol. **39**(4), 46–49 (2010)
6. Zhu, D.D., Yan, D., Li, Z.: Modelling and applications of annual energy-using simulation module of separated heat pipe heat exchanger. Energy Build. **57**(57), 26–33 (2013)
7. Futawatari, N., Tsukimoto, H., Kohata, Y., et al.: Packaged air conditioner incorporating free cooling cycle for data centers. In: Telecommunications Energy Conference. IEEE (2016)

8. Lyu, C., Chen, G., Ye, S., et al.: Enclosed aisle effect on cooling efficiency in small scale data center. Procedia Eng. **205**, 3789–3796 (2017)
9. Nadjahi, C., Louahlia, H., Lemasson, S.: A review of thermal management and innovative cooling strategies for data center. Sustain. Comput. Inf. Syst. S2210537917304067 (2018)
10. Douchet, F., Nortershauser, D., Le Masson, S., et al.: Experimental and numerical study of water-cooled datacom equipment. Appl. Therm. Eng. **84**, 350–359 (2015)

Numerical Simulation of Heat Transfer and Pressure Drop Characteristics of Elliptical Tube Perforated Fins Heat Exchanger

Jiaen Luo, Zhaosong Fang, Lan Tang and Zhimin Zheng

Abstract The heat performance of the elliptical tube perforated fins heat exchanger was investigated with numerical simulation methods in comparison with that of the circular tube plain fins exchanger. Based on the analysis of heat transfer rate, Colburn factor j, pressure drop, friction factor f of the air side of the fin, and the comprehensive performance index $j/f^{1/3}$, the performance of the elliptical tube perforated fin is better than that of the circular tube plain fin. When the air flow velocities range between 1 and 4 m/s, the heat transfer factor j of the elliptical tube perforated fin is higher than that of the circular tube plain fins by 7.1–15.3%, while the pressure drop of the elliptical tube exchanger reduces by 30.1–36.6%, and the comprehensive performance index $j/f^{1/3}$ is higher than that of the circular tube plain fin by mean 18%. The maximum rate is near 31.9%, which indicates that the performance of the elliptical tube perforated fin is better than that of the circular tube plain fin.

Keywords Opening sickle-shaped hole · Numerical simulation · Pressure drop · Enhanced heat transfer · Heat performance

1 Introduction

Finned tube heat exchangers are commonly used in air source heat pump outdoor units, and the improvement of heat transfer and resistance characteristics on the air side helps to improve the overall performance of the heat pump unit [1].

In many technologies of enhanced heat transfer, it is the most common and simple method to enhance the heat transfer performance of heat exchangers by optimizing the distribution of air flow field and temperature gradient field [2]. For the finned tube heat exchanger of air source heat pump, the purpose of the opening is to have superior heat transfer performance under the dry condition than the plain

---

J. Luo · Z. Fang (✉) · L. Tang · Z. Zheng
School of Civil Engineering, Guangzhou University, Guangzhou, China
e-mail: zhaosong0102@126.com

fin, and in the frosting condition, the frost layer is not easy to block the small hole. Excellent enhanced heat transfer characteristics are maintained during the defrost cycle [3]. It is necessary not only to enhance the heat transfer, but also to optimize the flow resistance to enhance the comprehensive heat transfer performance of the finned tube. Usually, the base tube of the heat exchanger is a circular pipe, but it has been found that the profiled tube having an elongated cross section has less air resistance loss. Li and Zhao by numerical simulation concluded that the air flow resistance of the elliptical tube heat exchanger is smaller than that of the circular tube heat exchanger, and the heat transfer coefficient is also increased [4].

In summary, the fins open-hole heat exchanger improves the airflow disturbance and effectively enhances the heat transfer performance, while the elliptical tube heat exchanger can effectively reduce the air flow resistance. In this study, the above advantages are combined. Optimizing of the shape of the base tube and the opening of the fin, the simulation explores the combined effects of the shape of the base tube and the perforated fins on the heat transfer and pressure drop characteristics.

2 Numerical Solution

2.1 Physical Model and Boundary Conditions

Figure 1 illustrates the geometric model of the elliptical tube perforated fin considered in the present work. Figure 2 shows the dimension of the simulation domain of the circular tube plain fin, the elliptical tube plain fins, and the elliptical tube perforated fin.

For other size parameters, the fin thickness is 0.5 mm and space between two fins is 10 mm. To verify that the numerical model is independent of the number of grids, each model was tested by a number of groups of progressively increasing the number of grids. Finally, the grids with 2.15 million cells were used to discretize the computational domain and the air flow field.

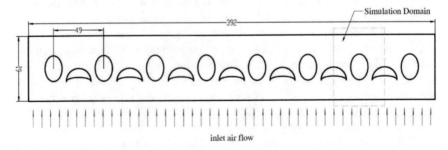

Fig. 1 Schematic of the elliptical tube perforated fin

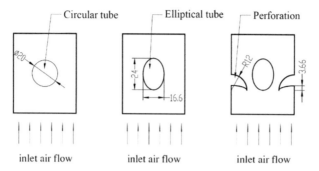

Fig. 2 The dimension of three fin models

Figure 3 shows boundary conditions used in the present study. The models consider only air side, neglecting collar contact resistance and thermal conduction resistance in the tube wall.

2.2 Data Reductions

The heat transfer performance of finned heat exchangers can be evaluated by the heat transfer rate and Colburn factor j at the same mass flow rate and air inlet temperature. The heat transfer rate calculation formulas are as follow:

$$Q = q_m c_{p,c} \left(T_{c,\text{out}} - T_{c,\text{in}} \right) \quad (1)$$

The Colburn factor j (ratio of convection heat transfer per unit duct surface area) to the amount transferable (per unit of cross-sectional flow area) is defined as [5]:

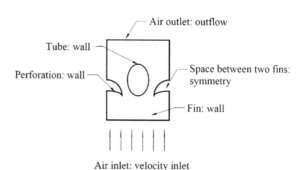

Fig. 3 Computational domain of the elliptical tube perforated fin with boundary conditions

$$j = \frac{Q_c P_r^{2/3}}{A_0 \Delta T \rho u c_p} \quad (2)$$

The friction factor f of the air side of the fin (ratio of wall shear stress to the flow kinetic energy), which is related to pressure drop in tube-and-fins heat exchangers is expressed as:

$$f = \frac{2A_c \Delta P}{A_0 \rho u^2} \quad (3)$$

$$\Delta P = P_{in} - P_{out} \quad (4)$$

To evaluate the comprehensive performance of the fins heat exchanger, the dimensionless factor $j/f^{1/3}$ is defined as the comprehensive performance index [6].

2.3 Model Validation

Experimental data by Su [7] were used to validate the present numerical model, as shown in Fig. 4. Very good agreement is obtained between the experimental and the present CFD results, which are within less than 9% of discrepancy. The main reason is that there is water in the tube exchanging heat with the outside air in the literature [7] while this is not taken into account in the present simulation by assuming the tube surface constant temperature. Also, the present results have been compared with numerical results of Fang [8] obtained using CFD software.

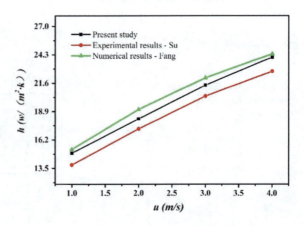

Fig. 4 Comparison of present results with experimental results of Su and CFD results of Fang

3 Results and Discussion

3.1 Temperature Field and Velocity Field

The development of the velocity boundary layer and the thermal boundary layer determines the heat transfer and pressure drop characteristics as the fluid flows through the fins. Based on this theory, the analysis from the velocity field and the temperature field has become one of the main research methods of researchers [9].

The surface temperature distribution of three fin models is shown in Fig. 5. The unit corresponding to the numerical value in the figure is K. The isotherms of the circular tube plain fin and the elliptical tube plain fin are rare, while the isotherms of the elliptical tube perforated fins are denser and the area of the low-temperature isotherms is significantly larger than that of the plain fins. According to the research in the literature [10], in the same working conditions, the air absorbs the heat of the fin by scouring the surface of the fin, so the better the heat transfer effect, the lower the temperature of the surface. Therefore, the surface heat transfer performance of the elliptical tube perforated fin is superior to that of the plain fins.

Figure 6 shows the air velocity distribution on a plane with a vertical distance of 0.05 mm from the fin when the inlet velocity is 4 m/s. Comparing Fig. 6a with b, it can be found that the velocity distribution of elliptical tube plain fin tends to be uniform due to the reduction of the air blocking effect. Moreover, the wake and vortex of the elliptical tube fins are much smaller than the circular tube plain fin. This is the core reason why the flow energy loss of elliptical tube fins is small.

Comparing Fig. 6b with c, perforation tends to increase the uniformity of the overall distribution of the velocity field. The velocity boundary layer is destroyed at the opening, causing the velocity boundary layer to form intermittently, effectively reducing the thickness of the thermal boundary layer on the surface of the fin. Thus, heat transfer of perforation fin is significantly enhanced.

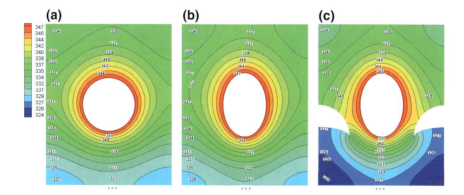

Fig. 5 Temperature distribution of three fin models

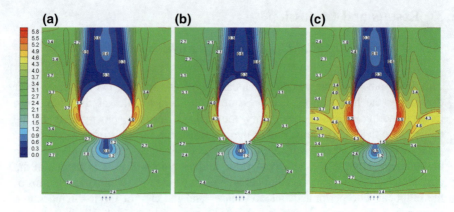

Fig. 6 Velocity distribution of three fin models

3.2 Heat Transfer and Pressure Drop Characteristics

Figures 7 and 8 are comparison diagrams of numerical simulation results of Colburn factor j and heat transfer rate of three kinds of fins.

From Fig. 8, as the inlet velocity increases, the heat transfer rate of the elliptical tube perforated fin also gradually increases. Colburn factor j of the elliptical tube perforated fin is higher than that of the circular tube plain fin by 7.1–15.3%. The fin opening can improve the heat transfer in the tail region of the fin under the premise of retaining the high heat exchange area at the front.

Figures 9 and 10 are comparative diagrams of numerical simulation results of friction factor f and pressure drop of three kinds of fins. Figure 10 shows that the air pressure drop across the three models increases as the air inlet velocity increases, like parabolas. However, the air flow velocities range between 1 and 4 m/s, the

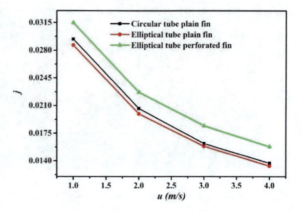

Fig. 7 Comparison of Colburn factor j of three kinds of fins

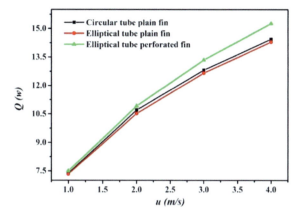

Fig. 8 Comparison of heat transfer rate of three kinds of fins

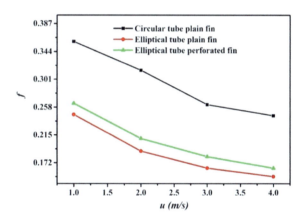

Fig. 9 Comparison of friction factor f of three kinds of fins

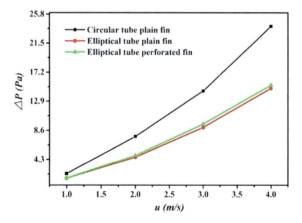

Fig. 10 Comparison of pressure drop of three kinds of fins

pressure drop of the elliptical tube exchanger reduced by 30.1–36.6%. The numerical simulation results are consistent with the experimental results of the literature [4].

The elliptical tube perforated fin has a slightly higher pressure drop than the elliptical tube plain fin. This is because the airflow through the perforation is continuously increased, thereby causing a certain loss of mechanical energy, so that the pressure drop grows faster [1]. In short, the effect of perforation on air resistance is little.

3.3 Comprehensive Performance

The performance of elliptical tube perforated fins is determined by heat transfer and pressure drop characteristics. Given that the improvements in heat transfer are also accompanied by increases in the pressure drops, it is necessary to evaluate the net enhancement in the heat exchanger with the effect of air flow distribution [5]. The index $j/f^{1/3}$ combines the heat transfer and flow resistance factors from the same fan power consumption to compare the heat transferred. Comparison of $j/f^{1/3}$ of three kinds of fins is shown in Fig. 11. Compared with the circular tube plain fin and the elliptical tube plain fin model, the comprehensive performance index of the elliptical tube perforated fin is always higher. The comprehensive performance index $j/f^{1/3}$ is higher than that of the circular tube plain fin by mean 18%. The maximum rate is near to 31.9%, which indicates that the performance of the elliptical tube perforated fin is better than that of the circular tube plain fin and the elliptical tube plain fin.

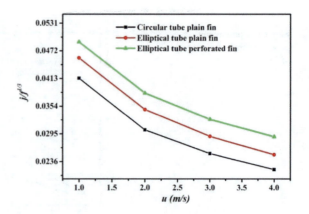

Fig. 11 Comparison of $j/f^{1/3}$ of three kinds of fins

4 Conclusions

In order to reveal the mechanism of heat transfer enhancement of perforated fin, the temperature field and velocity field of the surface of the circular tube plain fin, the elliptical tube plain fin, and the elliptical tube perforated fin were analyzed from the FLUENT software simulation results. After research and analysis, the following conclusions are drawn:

(1) Through the sickle-shaped opening, the disturbance of the fluid is increased while maximally retaining the high heat exchange area at the front of the fin. Moreover, the velocity boundary layer is destroyed at the opening, causing the velocity boundary layer to form intermittently, effectively reducing the thickness of the thermal boundary layer on the surface of the fin.
(2) At the same inlet boundary, heat transfer rate and Colburn factor j of the elliptical tube perforated fin are higher than that of the circular tube plain fin and the elliptical tube plain fin. When the air flow velocities range between 1 and 4 m/s, Colburn factor j of the elliptical tube perforated fin is higher than that of the circular tube plain fins by 7.1–15.3%. The pressure drop of the elliptical tube exchanger reduced by 30.1–36.6%, indicating that the flow resistance characteristics of the elliptical tube fins model are better than those of the circular tube plain fin model.
(3) The comprehensive performance index $j/f^{1/3}$ is higher than that of the circular tube plain fins by mean 18%. The maximum rate is near to 31.9%, which indicates that the comprehensive performance of the elliptical tube perforated fin is better than that of the circular tube plain fins.

Acknowledgements The project is supported by the Opening Funds of State Key Laboratory of Building Safety and Built Environment and National Engineering Research Center of Building Technology (Number BSBE2018-03).

References

1. Case, W.M., London, A.L.: Compact Heat Exchangers. Science Press (1997)
2. Ma, J.Z., et al.: Numerical study on heat and mass transfer of rectangular eccentric corrugated fin-and-oval tube heat exchangers. Therm. Power Gener. **43**(08), 89–93 + 97 (2014)
3. Zhu, R.X., et al.: Heat transfer and pressure drop characteristics of circular hole finned-tube heat exchangers. GAS&HEAT **43**(1), 67–71 (2017)
4. Li, Q.L., Zhao, L.P.: Numerical simulation of flow behaviors and heat transfer characteristic of rectangular finned elliptical tube heat exchanger. J. Fluid Mech. **34**(8) (2006)
5. Yaïci, W., et al.: 3D CFD analysis of the effect of inlet air flow maldistribution on the fluid flow and heat transfer performances of plate-fin-and-tube laminar heat exchangers. Int. J. Heat Mass Transf. **74**, 490–500 (2017)
6. Zhu, J.L., et al.: Numerical simulation and analysis of different louver arrangements of automotive heat exchangers. J. Tianjin Univ. **3**, 244–249 (2013)

7. Su, H.: Experimental Study on Heat Transfer and Flow Resistance of Spoiler Finned Tubes and Numerical Calculation of Fin Efficiency. Chongqing Jianzhu University (1999)
8. Fang, Z.S.: Study on Energy Conservation Performance of Circle Holes Fin-Tube Refrigeration Heat Exchanger. Chongqing University (2008)
9. Lei, Y.G., et al.: An investigation of the airside performance of fin-and tube heat exchangers with delta-winglet. J. Eng. Thermophys. (1999)
10. Wang, H.H., Fang, Z.S.: Numerically simulation of flowing and heat transfer with airflow over holes fins tube. J. Tongji Univ. **37**(7), 969–973 (2009)

Research on Cooling Effect of Data Center Cabinet Based on on-Demand Cooling Concept

Qi Fu, Zongwei Han, Xiaopeng Bi, Xinwei Meng and Haotian Wei

Abstract In order to solve the problems of uneven air supply and mixing of cold and hot airflow in the data center at present. This paper proposes a cabinet with on-demand cooling function, introduces the structure and working principle of it, establishes a simulation experiment platform for simulation calculation, and studies the influence of different factors on the cooling effect. The results show that compared with the traditional cooling method, the cooling effect of this cabinet is better, the average temperature of the servers is reduced by 7.89 K; the cooling effect can be improved by increasing the air intake volume and reducing the air inlet temperature, and the influence of air temperature on the cooling effect is greater than that of air volume: every 2 K increase in inlet air temperature, the minimum intake air volume will be increased by up to 600 m^3/h.

Keywords Cabinet of data center · Cooling on demand · Simulation · Cooling effect

1 Introduction

In recent years, the energy consumption of data centers is increasing, according to statistics, for the power consumption composition of data centers, the power consumption for refrigeration accounts for about 50% of the total power consumption [4]. So the key to energy saving in data centers is to improve the energy efficiency of refrigeration systems. At present, the heat dissipation in most data centers is achieved by blowing cold air through air conditioning system [1]. However, due to the large amount of air supply and the different requirements for the cooling capacity of each part of the data center, the above method may lead to a series of problems such as the mixing of cold and hot airflow and the uneven distribution of cooling capacity. Therefore, at present, many scholars and institutions have done a

Q. Fu · Z. Han (✉) · X. Bi · X. Meng · H. Wei
School of Metallurgy, Northeastern University, 110819 Shenyang, China
e-mail: 2417478333@qq.com; hanzongwei_neu@163.com

lot of research work in the distribution of cooling capacity, which is embodied in the air distribution and the terminal cooling method.

The research on air distribution mainly includes two aspects: evaluating the effect of air distribution [8] and optimizing the air distribution [5]. In the aspect of evaluating the effect of air distribution, Udakeri compared and analyzed the cooling effects of the upper air supply and the underfloor air supply [7]. In the aspect of optimizing the air distribution, separating the cold and hot channels and closing the cold channel can effectively improve the utilization of cold air and reduce airflow disturbance [3]. However, for the cabinets with high heat density and the computer room with uneven calorific value, these methods can neither meet the cooling demand nor quantify the cooling on demand.

Therefore, in order to achieve accurate cooling and improve the efficiency of cooling supply, some scholars have conducted some research on the terminal cooling method. Ouchi proposed liquid cooling methods such as water-cooled backboard technology and cabinet-level water-cooling unit, and the research shows that water cooling can increase cooling effect by 33% compared with air cooling [6]. However, due to the hidden danger of water leakage in this way, its promotion and application are limited.

In addition to the above-mentioned methods that can affect the cooling effect of data center, factors such as air volume, inlet air temperature and calorific value of server can also affect the cooling effect of the cabinet [2].

Based on the above background, considering the problems such as uneven air supply and mixing of hot and cold airflow in the existing data centers, this paper proposes a cabinet structure with on-demand cooling function, simulates and analyzes it by establishing a simulation platform, and studies the influence of different factors on its cooling effect. The research results have certain promoting effects and reference value for improving the cooling efficiency of data center.

2 Working Principle of the Cabinet

Figure 1 shows the working principle of the cabinet. As shown in the figure, the cabinet mainly contains an evaporator, axial-flow fan, and server. The fan is installed at the bottom of the cabinet, blowing air into the cabinet. The servers are arranged vertically in the cabinet. The evaporator is placed between the servers and the fan, cooling the air blown in by the fan. In addition, the evaporator use refrigerant as cooling medium, the refrigerant tube is connected to the refrigeration unit outside the cabinet, and the expansion valve is installed on the refrigerant tube to adjust the refrigerant flow rate according to the thermal value inside the cabinet, and then realize on-demand cooling.

Fig. 1 Working principle of the cabinet

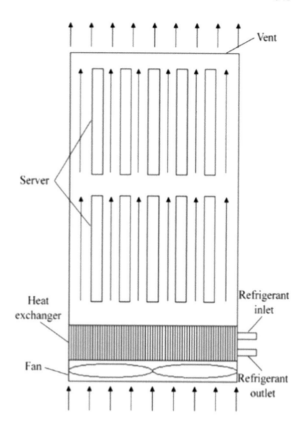

2.1 Other Parameters

The size of the cabinet is 700 mm × 700 mm × 1200 mm. The server is 1U server, the power is 200 W, and the endurance temperature of it is 333 K through the research of predecessors. There are 10 servers in each cabinet, so the total heating power of the cabinet is 2 kW, and the rated refrigeration capacity of the evaporator is 2 kW, which ensures that the servers can achieve good cooling effect under different operating conditions. The cross-sectional area of the evaporator is 700 mm × 700 mm, so that the area of the evaporator swept by air is maximized.

3 Methods

In this paper, the simulation model of the above cabinet and components is established by Airpak software, and the mathematical model of the heat transfer process is established according to the heat conduction differential equation, the convection heat transfer equation, the mass conservation equation, and the momentum conservation equation.

4 Results and Discussion

The purpose of proposing this cabinet is to ensure the normal operation of the server and reduce energy consumption. In the study of this paper, besides taking the maximum temperature, average temperature of the servers and the standard deviation of the server temperature as the evaluation index to analyze the influence of air intake and air inlet temperature on the cooling effect of the cabinet, the variation of minimum air intake required by cabinet with air temperature is also studied. In addition, the cooling effect of this cabinet is compared with that of traditional underfloor air supply.

4.1 Influence of Air Intake

Because the above cabinet places the evaporator inside the cabinet so that the air is cooled after entering the cabinet, the temperature in the computer room only needs to be kept at a comfort value of about 300 K through central air-conditioning. Therefore, we set the range of inlet air temperature at 300 ± 2 K. However, the air volume required for the cabinet will vary with the inlet air temperature. It can be seen from Fig. 2 that when the inlet air volume reaches a certain value, the above evaluation indexes change little with the inlet air volume. In order to reduce the fan energy consumption as much as possible, we set this value as the required air volume under this condition. So when the range of inlet air temperature is 298–302 K, the range of required air volume is 400–1200 m^3/h according to the simulation results in Fig. 2.

In order to discuss the influence of intake air volume on the cooling effect of cabinet, the simulation results in Fig. 2b is taken as an example, it can be concluded that: (1) when the air volume is small, the temperature of the servers decrease with the increase of the air volume. When the air volume is large enough, the increase of the air volume has little effect on the cooling effect; (2) with the increase of air volume, the standard deviation decreases gradually, and the temperature distribution of the servers becomes more uniform.

4.2 Influence of Air Inlet Temperature

In order to further study the influence of the air inlet temperature on the cooling effect, we control the air inlet volume are the same, all of which are 800 m^3/h, and

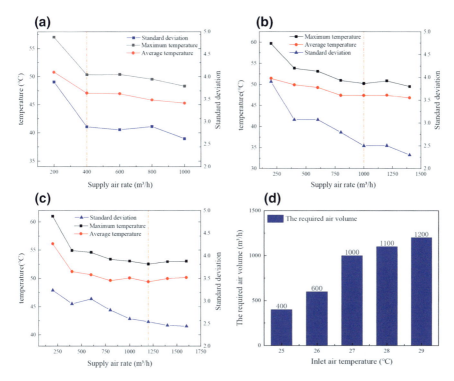

Fig. 2 The change of the required air volume with the air inlet temperature. **a** Air inlet temperature: 298 K; **b** Air inlet temperature: 300 K; **c** Air inlet temperature: 302 K; **d** Required air volume at different air inlet temperatures

the inlet air temperature is 298, 300, and 302 K, respectively. The cooling effect inside the cabinet is simulated as shown in Fig. 3, it can be seen that for every 2 K drop in air inlet temperature, the cooling effect becomes better; however, it has almost no effect on the standard deviation of the temperature value of all servers.

From Fig. 2d, it can be concluded that for every 2 K increase in air inlet temperature, the minimum air intake needs to increase 600 and 200 m³/h, respectively. In summary, the air inlet temperature not only affects the cooling effect inside the cabinet, but also has a greater impact on the minimum air intake required by the cabinet, and then affect the energy consumption of the fan.

4.3 Influence of the Structure of Cabinet

In order to study the cooling effect of this new cabinet, we compare it with the underfloor air supply which is widely used in the data center at present. We take the median value: the air inlet temperature is 300 K, the air intake is 800 m³/h, and

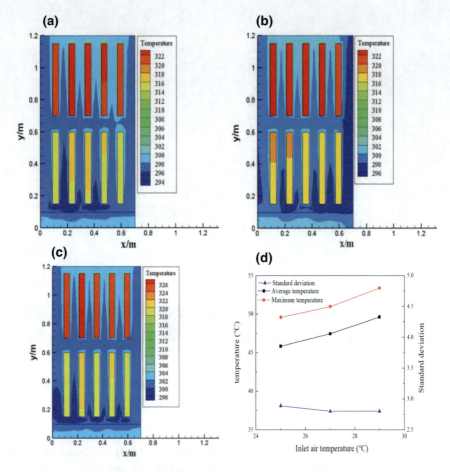

Fig. 3 Variation of cooling effect with air inlet temperature. **a** Air inlet temperature is 298 K; **b** Air inlet temperature is 300 K; **c** The air inlet temperature is 302 K; **d** The carve of cooling effect with the air inlet temperature

other parameters are the same. In addition, the other parameters of the underfloor air supply are shown in Table 1.

According to the above parameters, the velocity field and temperature value of each server are obtained by simulation as shown in Fig. 4. It can be seen from Fig. 4b that there is a phenomenon of hot air reflux in the room and the gas flow velocity of each layer in the cabinet is small and uneven in the underfloor air supply mode. While the servers in the new cabinet are arranged vertically, the temperature of the servers in each layer is more uniform. So the figure shows that the cooling effect of the new cabinet is better compared with the traditional underfloor air supply: the average temperature of the server drops from 328.3 to 320.5 K.

Table 1 Parameters of underfloor air supply model

	Name					
	Room	Vent	Fan 1	Fan 2	Server	Cabinet
Size (mm)	2000 × 1450 × 800	200 × 700	200 × 700	1200 × 700	600 × 450 × 44.45	700 × 700 × 1200
Quantity	1	1	1	1	10	1

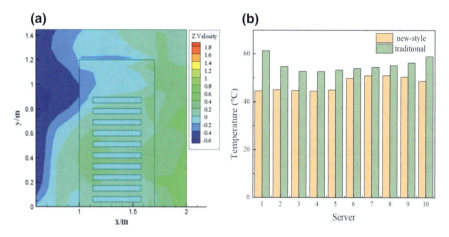

Fig. 4 Cooling effect of two cabinet structures. **a** Velocity distribution of underfloor air supply; **b** Temperature values for each server in two models

5 Conclusions

This paper puts forward a kind of data center cabinet with on-demand cooling function, introduces the concrete structure and working principle of it, and takes it as the research object, simulates and calculates it under different working conditions to study the influence of different factors on its cooling effect. The following conclusions were drawn:

1. The cooling effect of this cabinet is better than that of traditional underfloor air supply.
2. When the air volume is small, the cooling effect will be better with the increase of air volume, but when the air volume reaches a certain value, the increase of air volume has little effect on the cooling effect.
3. The lower the air inlet temperature, the better the cooling effect inside the cabinet, and the smaller the minimum air volume required.

Acknowledgements The authors gratefully acknowledge the support from the Natural Science Foundation of China (grant No.51778115) and the Fundamental Research Funds for the Central Universities (grant No. N182502043).

References

1. ASHRAE.: ASHRAE TC 9.9: Expanded data center classes and usage guidance, ASHRAE, Atlanta (2011)
2. Breen, T.J., et al.: From chip to cooling tower data center modeling: part I influence of server inlet temperature and temperature rise across cabinet. In: Thermal and Thermomechanical Phenomena in Electronic Systems, Las Vegas, America, pp. 1–10 (2010)
3. Chen, J.: Application of closed cold aisle technology to data center room and simulation analysis. Heating Ventilating Air Conditioning **6**, 37–40 (2015)
4. Ebrahimi, K., et al.: A review of data center cooling technology, operating conditions and the corresponding low-grade waste heat recovery opportunities. Renew. Sustain. Energy Rev. **31**, 622–638 (2014)
5. He, Z., et al.: Study of hot air recirculation and thermal management in data centers by using temperature rise distribution. Build. Simul. **9**(5), 541–550 (2016)
6. Ouchi, M., et al.: Liquid cooling network systems for energy conservation in data centers. ASME, Portland, America, pp. 443–449 (2011)
7. Udakeri, R., et al.: Comparison of overhead supply and underfloor supply with rear heat exchanger in high density data center clusters. In: Semiconductor Thermal Measurement and Management Symposium, San Jose, America, pp. 165–172 (2008)
8. Yuan, S., Lu, S.: Simulation of air distribution in data center based on thermal environment evaluation indexes. Heating Ventilating Air Conditioning **46**(1), 66–72 (2016)

Analysis for Vibration Characteristics of Water Pump Piping System

Tiantian Liu and Zhiyong Liu

Abstract The vibration calculation model of the pipeline system is established by the finite element analysis method, and the calculation model is verified by experimental tests. The differences between calculation result and the actual measurement are controlled at about 10%. On this basis, the influence of the support bracket, tee, and elbow on the vibration of the pipeline system is calculated, which provides a reference for the vibration reduction measures of the pipeline system.

Keywords Pipe vibration · Bracket · Elbow · Tee

1 Introduction

The most intuitive manifestation of large stress in a pipe system is the deformation of the pipe, i.e., the generation of vibrational displacement. We can judge whether the force of each pipe section of the pipeline system is reasonable by observing whether the vibration displacement of the pipeline is within a reasonable range and whether it is necessary to modify the local pipe section of the pipeline system.

Long-term vibration beyond the safe range will cause uneven stress on the main equipment, affecting its performance and operation. Moreover, it also destroys the sealing property. If the pipeline contains explosive or toxic gases, it is easy to have a terrible accident [1–3].

Using software, it also can predict the vibration displacement and each mode shape under the operating conditions by vibration analysis, besides the severity of pipeline damage. The brackets and the linking methods can be adjusted according

---

T. Liu · Z. Liu (✉)
College of Environment and Municipal Engineering,
Lanzhou Jiaotong University, 730070 Lanzhou, China
e-mail: 1760184353@qq.com

T. Liu
College of Civil Engineering, Hunan University, 410000 Changsha, China

to the vibration displacement. Thus, it can avoid excessive vibration when the pipeline system is running, causing severe stress, which will damage the piping system equipment seriously [4].

2 Pipe Vibration Principle

2.1 Calculation of Reaction Force R_m Under Extreme Displacement

According to ASME Pressure Pipeline Specification B31.1, Sect. 319.3.1(b), the Extreme Displacement Conditions R_m is calculated by using the highest or lowest metal temperature, whichever produces greater reactive force [5].

The formulation of maximum force calculation is as follows:

$$R_m = R\left(1 - \frac{2C}{3}\right)\frac{E_m}{E_a} \tag{1}$$

where C is cold-spring factor varying from 0 to 1; E_a is reference elastic modulus at 21 °C (70 °F), Pa; E_m is elastic modulus at the highest or lowest metal temperature, Pa; R is E_a_based reaction force range corresponding to the full displacement stress range, N; R_m is instantaneous maximum reaction force calculated at the highest or lowest metal temperature, N.

2.2 Acceleration Calculation of the Ultimate Vibration

The formulation of acceleration calculation is as follows:

$$R_m = ma \tag{2}$$

where m is total unit mass of pipe and fluid, kg; a is acceleration of the pipe system vibration in the corresponding direction under operating conditions, m/s^2.

2.3 Calculation of the Ultimate Vibration Displacement

The formulation of calculating the ultimate vibration displacement is as follows:

$$X = \frac{1}{2}at^2 \tag{3}$$

where X is the vibration displacement of pipe system node in the corresponding direction, m; t the unit time of the node corresponding to the direction vibration under operating conditions, s.

2.4 Calculation of Natural Frequency

The formulation of natural frequency calculation is as follows [6]:

$$f = \alpha \sqrt{\frac{EI}{(M_{ef} + M)l^3}} \quad (4)$$

where α is the frequency coefficient; f is natural frequency of pipe, Hz; l is the pipe length, m; E is elastic modulus of pipe section material at design temperature, Pa; I is inertial moment of section, m^4; M is concentrated mass, kg; M_{ef} is equivalent concentration mass, kg.

3 Brief Introduction of Pipeline System

The pipeline system includes mainly two pumps (the other with a backup.), a number of gate valves, safety valves, check valves, etc., as shown in Fig. 1. (Note: The black part in Fig. 1 is the pump piping system tested.)

4 Simulation Results and Discussion

4.1 Simulation Method

Modeling from the right suction side by software, where set the node number 10 at the origin coordinate. The coordinates of the remaining segment nodes are determined according to their positional relationship with the origin node.

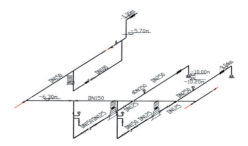

Fig. 1 Schematic diagram of pump piping system

4.2 Simulation Process

Since only one pump is in the starting state during the experiment process, the main pipe of the other pump does not enter the fluid due to the interception of the valve, but its vibration is affected by the side. Therefore, the part of the left side before node 1370 is ignored. The pipe wall thickness specification is set up SCH40, other parameters of the pipe were taken from measurement (the operating temperature $T = 5$ °C, the pump inlet pressure $P_{in} = 0.3$ MPa, and the pump outlet pressure $P_{out} = 1.1$ MPa).

4.3 Boundary Conditions

As shown in Fig. 2, the five constraints of the water pump connection, the end of the pipe (i.e., the node numbers 725, 1370), and the branch part for the node between 1940 and 1990 are fixed brackets. Setting the guide bracket at the constraint of the pipe section passing through the wall (i.e., the part before the node number 2540 and 35). The constraint condition of the remaining pipe sections is the sliding support hanger, which only serves to support and limit the displacement of the support direction, and does not play a role of limiting displacement except the support direction.

4.4 Node Number

Numbering from the suction section of the side, and number the nodes where the vibration displacement such as valves, brackets, tees, pumps, and elbows may be significantly marvelous. Points of interest in the stress isometric are identified by node points. Increase the number as much as possible to ensure that the maximum stress of pipeline occurs at the node. Therefore, it can be accurately expressed, which makes the simulation more realistic [7].

Fig. 2 The structure model of pump pipe

4.5 Mode Shape Analysis

The first-order and second-order modal shapes of the pump piping system are generated by software, thus it can be clearly seen which nodes are more deformed when the pipeline system resonates. Then vibration damping measures can be taken. (The first-order vibration mode shape is shown in Fig. 3, and the second-order vibration mode shape is shown in Fig. 4.)

From Figs. 3 and 4, we can find that:

The closer the spray pump is, the more serious the vibration deformation is. For example, the part between node 70 and 100.

The closer to the end equipment of the pipeline system, the more obvious the deformation of the joint position is. For example, the part between the node 2480 and 2520.

The vibration displacement of the elbow, nozzle, and valve is more significant. The deformation of tee, the valve, and the elbow are larger than connection of other pipe fittings; the deformation of the node 2300 is the most obvious. The more elbows there are, the more obvious the deformation is. For example, the node from 2480 to 2520.

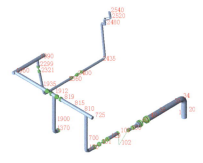

Fig. 3 The 1st pipe mode shape

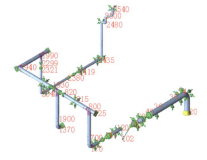

Fig. 4 The 2nd pipe mode shape

5 Test

The experiment measurement used a Pocket Vibrometer Measurement Instruments (type EMT220).

Measuring range: acceleration varying from 0.1 to 199.9, m/s^2; speed varying from 0.01 to 19.99, cm/s; displacement varying from 0.001 to 1.999, mm.

Frequency range: low frequency, 10–1000 Hz; high frequency, 1–15 kHz.

Error range: ±5%.

Using a Vibration meter to measure the vibration displacement of the node when the system was running. When measuring points, select the nodes near the elbow, tee, the water pump, valve, and the size head. The specific situation is shown in Fig. 2. The measurement nodes are 170, 2340, 2420, and 2460.

6 Analysis

6.1 Analysis Between Calculations and Measurements

The differences analysis of vibration displacement between the measurement and the calculation is shown in Table 1. (Note: When measuring data, the vibration displacement value displayed by the measuring instrument does not show the direction. Therefore, in order to facilitate the calculation during the analysis process, the obtained value when the model is subjected to the vibration analysis is also processed in no direction. Take its absolute value.)

As can be seen from Table 1, the maximum error between the calculated and experimental results is 15.38%, the minimum error is 0%; the average error is about 10%.

Table 1 Comparison between calculations and measurements

Node	Direction	Displacement (mm) Maximum	Simulation displacement (mm) Maximum	Error (%)
170	X	0.013	0.011	15.38
	Z	0.180	0.163	9.44
2340	X	0.009	0.008	11.11
	Z	0.760	0.838	10.26
2420	X	0.002	0.002	0.00
	Y	0.075	0.066	12.00
2460	Y	0.530	0.592	11.70
	Z	0.421	0.473	12.35

6.2 Analysis of Error

The main reasons for the vibration displacement error between the actual measurement and the model are as follows:

The error of vibration measuring instrument comes from itself.

When measuring data, the vibrometer probe is not completely perpendicular to the plane of the measurement point.

Although the pipeline system of spray pump is short, the scale is small as well, the pipeline system condition is a little complicated. There are two pumps, the other with a prepared, and some segments are shared with fire piping system. It is difficult to express accurately.

The error between the actual construction and design drawings of the pipeline system cannot be predicted and considered.

7 Optimized Methodology

In order to ensure the piping system to operate more smoothly and safely, the following methodologies are determined to solve the problem. Different models are established by changing the number of brackets, the three-way, and elbow factors. Then the vibration displacement analysis is carried out to observe the damping effect of cases.

7.1 Adjusting the Numbers of Hangers

Hanger is an important part of the pipeline system's structural design. In this process, the purpose of reducing the vibration can be achieved by changing its position and number. This case changes the pipe system to establish model by adjusting the bracket. The vibration displacement analysis is performed to observe the vibration damping effect.

According to Figs. 3 and 4 obtained from the model, as well as the vibration displacement of the operating conditions (as shown in Table 1), The vibration displacement for the nodes from 2338 to 2480 is significantly large. Therefore, the bracket is added to the elbow of the pipe section, i.e., the bracket is added at the nodes from 2440 to 2460, and the detailed information is shown in Fig. 5. Then, the correlation result of the vibration displacement in the original pipeline system operating condition is compared with the optimism model, the calculation result analysis is shown in Table 2.

It can be seen from Table 2 that increasing the bracket in the maximum direction of the vibration displacement for the relevant node, the original vibration displacement can be reduced to 0, the reduction is 100%. Adding a bracket at a

Fig. 5 Optimum pump piping system

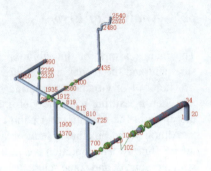

Table 2 Case 1 vibration displacement change local part

Node	Direction	Prototype	Case 1	Reduction (%)
2440	X	−0.048	0	100
	Y	−0.199	0	100
	Z	1.429	0	100
2460	X	−0.376	0	100
	Z	0.473	0	100

certain node will also affect the vibration displacement of other nodes near the position. So adding a bracket is one of the most direct and effective methods.

7.2 Replacing Elbow with Tee

It can be obtained from Figs. 3 and 4, the vibration displacement at the node 2440 is most obvious during operation, therefore the elbow at the node 2440 is changed to a tee. The comparative analysis results between the case 2 and the original pipeline model are shown in Table 3.

It can be found in Table 3, after turning the elbow into a tee, the vibration displacement in the Z direction for the node 2440 is significantly reduced. The amplitude of the vibration in the X direction is reduced by 81.25%, the pipe coordinate Y direction is reduced by 73.77%, and the Z direction is reduced by 99.23%. Thus, the tee can effectively reduce vibration displacement.

Table 3 Case 2 vibration displacement change of local part

Type	X	Y	Z	X-reduction (%)	Y-reduction (%)	Z-reduction (%)
Elbow	−0.048	−0.199	1.429			
Tee	0.009	0.053	0.011	81.25	73.37	99.23

Table 4 Case 3 vibration displacement change of local part

Bends' number	2480 − X	2480 − Z	X-reduction (%)	Z-reduction (%)	2540 − x	2540 − z	X-reduction (%)	Z-reduction (%)
2	−0.332	0.403			0	0		
1	−0.162	0.394	51.2	2.23	0	0	0	0

7.3 Decreasing the Number of Elbows

Reducing the number of bends between the pipe sections for nodes from 2480 to 2540 and turn them into straight pipes. The comparative analysis of the vibration displacement results between the project 3 and the original pipeline model is shown in Table 4.

As indicated in Table 4, reducing the number of elbows can decrease the vibration displacement near nodes, but the amplitude over different coordinate is not same. The gap between them is quite significant. Displacement decreases dramatically. The pipe coordinate X, the decreased amplitude reaches to 51.20%; and the pipe coordinate Z decreases by 2.23%.

8 Conclusions

Throughout the calculation analysis and experimental tests, the following conclusions are as drawn:

First, it can be seen from Table 1 that the maximum error for the calculated and experimental results is 15.38%, the minimum error is 0%, and the average error is about 10%. It illustrates that the establishment and calculation of the model are feasible.

Second, as illustrated by Table 2, increasing the number of constraints can completely reduce the vibration displacement of the corresponding direction. Compared with other cases, adding the corresponding direction of bracket at the corresponding node, it's the most direct and effective methodology to decrease the vibration displacement.

Third, it can be observed in Table 3, turning the elbow into a tee, which can reduce dramatically the vibration displacement. The vibration displacement in different directions is reduced differently, and the maximum reduction reaches to 99.23%.

Fourth, as evident from Table 4 that reducing the number of elbows can reduce displacement at the nearby nodes. Compared with other cases, the reduction in the two directions is quite different, and the vibration displacement is larger in the X-direction. The decrease is by 51.20%.

References

1. Wang, J.: Pipeline vibration and its measurement method. Compressor Technol. **30**(2), 40–41 (1993)
2. Yuan, S.: Dynamic analysis and research of vibration reduction of the fluid pipeline. Department of Chemical Process Machinery of Lanzhou Technology University, Lanzhou, China (2017)
3. Song, W., et al.: On the present status and prospect of pipeline vibration control technology. J. Saf. Environ. **12**(3), 184–188 (2012)
4. Peng, J., Wu, S.: Analysis and improvements for the pipes of air conditioner for transporting and operating process. Power Gener. Air Conditioning **32**(6), 75–78 (2014)
5. ASME.: B31.3: Process Piping Design Code for Pressure Piping. Pittsburgh, Pennsylvania, USA (2014)
6. Dang, X., Chen, S.: Piston Compressor Airflow Pulsation and Pipeline Vibration. Xi'an Jiaotong University Press, Xi'an, China (1984)
7. Wei, W., Li, F.: Several problems in piping stress analysis by using beam element model. Process Equip. Pip. **47**(3), 36–38 (2010)

Analysis on Transient Thermal Behaviors of the Novel Vapor Chamber

Yao Chen, Yaping Zhang, Pei Wang and Yongxin Guo

Abstract A heat pipe substrate module diffused heat by phase change was designed, and transient thermal properties of the vapor chamber were analyzed. The time of the vapor chamber to get steady-state mainly depends on the heat transfer coefficient. As the heat transfer coefficient increases, the time to reach steady state is shorter. Reducing the temperature drop of the vapor chamber core portion can effectively improve thermal performance for the vapor chamber. The transient temperature rise of vapor chamber tube core is smaller than that of the pure copper substrate module and the air heat pipe substrate module with fixed thermal resistance, it is beneficial to overcome the power "swell" and improve thermal shock resistance of the power device. The thermal resistance of the heat sink accounts for more than 70% of the thermal resistance of the entire heat dissipation module. The transient thermal performance of the vapor chamber and the power module integrated packaged can ensure thermal diffusion efficient and smooth operation.

Keywords Vapor chamber · Thermal performance · Transient behaviors · Substrate

---

Y. Chen · Y. Zhang (✉) · P. Wang · Y. Guo
Xi'an University of Science and Technology, 710054 Xi'an, China
e-mail: 603685433@qq.com

Y. Chen
e-mail: 2862986239@qq.com

P. Wang
e-mail: 781019685@qq.com

Y. Guo
e-mail: 791028995@qq.com

1 Introduction

With the development of electrical components toward miniaturization and high power density, how to effectively cope with the heat dissipation issue of electronic component has become the key to ensuring efficient operation of products [1]. It should be noted that the failure rate of electronic equipment caused by high temperature is more than 50% [2]. As the core component of power equipment, the power module is prone to the parasitic inductance of circuit due to the mutual coupling of the electric field, the magnetic field and the thermal field for the integrated power module, resulting in current drift and distortion of the electrical signal [3], so the heat transfer issue become the bottleneck of power module integration technology.

The thermal diffusion device used in the power module is mainly composed of a metal substrate having high thermal conductivity such as a copper plate and an aluminum plate. Although the thermal conductivity of the metal substrate is high, there is still a temperature difference on the bottom surface of the substrate and the radiator efficiency is low. In addition, the main ways of heat dissipation are external heat dissipation such as forced air cooling [4] and water cooling. Nevertheless, these cooling technologies require additional equipment and are extremely limited in application to high heat flux densities.

Compared with other electronic component radiator, heat pipe uses the internal working fluids to conduct phase change heat transfer, which has small thermal resistance and high heat transfer efficiency and has become an essential measure to solve the heat dissipation issue of electronic components [5]. The vapor chamber deformed in the form of a common heat pipe structure achieves high thermal conductivity by reducing thermal resistance [6], meanwhile reduce the module size and quality [7]. If the heat pipe substrate is used instead of the metal substrate, it is expected that the thermal diffusion of the substrate can be greatly enhanced, and the isothermal performance of the vapor chamber is beneficial for reducing the thermal resistance, thereby providing a guarantee for the integrated packaging of the substrate and the electronic component.

At present, the research hotspots of most scholars focus on analyzing the steady-state performance of the heat pipe heat dissipation structure to determine whether the heat source temperature meets the requirements and thus continues to operate in normal conditions [8, 9]. The research on the cooling mode of power module focuses on the flow and heat transfer analysis of cooling system itself while ignoring the combination of cooling system and power devices. There is less research on the transient thermal performance of the entire module when the heat pipe and the power module are packaged. Webb [10] analyzed the performance of an integrated plate heat pipe with a heat spreader at the base. Sauciuc [11] compared the performance of a FPHP with a copper block in the same size. Wei [12] studied experimentally the thermal behavior of a FPHP heat sink and compared its performance with metal base heat sinks. Mariya [13] presented the cooling system with heat pipes embedded in direct bonded copper (DBC) structure, which eliminates the existence of a thermal interface between the device and cooling system. Gao [14], carried out thermal tests on the vapor chamber and copper plates of the same size, verify the excellent thermal conductivity and temperature uniformity of vapor chamber.

Due to small fluctuations in voltage and slight changes in the external environment, the temperature inside the electronic device will be affected, resulting in a sudden change in the thermal stress inside the electronic device. Studying the transient thermal performance of the power module can ensure efficient and smooth operation of the power module, which helps to improve the efficiency and service life of the power module. Therefore, based on the previous research on the steady-state performance of power modules, this paper studies the transient performance of the vapor chamber power modules.

2 Test Device

The purpose of the design is to provide a vapor chamber for thermal diffusion which can fully ensure the mechanical strength of the heat pipe and greatly reduce the return path of the liquid, thereby improving the heat transfer performance of the heat pipe. Designed and developed a small radial vapor chamber for heat dissipation of integrated power electronic module. The vapor chamber uses negative ion water as liquid. Since the power module is in the non-steady state during the start-up and shutdown processes, in addition, the start-up time and characteristics of the power module are crucial for the smooth operation of the electronic components, so the transient performance of the vapor chamber power module is studied here.

The experimental setup is shown in Fig. 1. The experimental equipment includes IGBT constant current control circuit, heater chip, integrated vapor chamber module, axial flow fan, data acquisition system, three-phase adjustable voltage source, thermocouple, power measuring instrument and computer. The IGBT chip is soldered to the surface of the vapor chamber module as the heat source, and the heating power ranges from 0 to 230 W. Axial flow fan is used at the upper end of the rib for forced convection cooling. Multiple thermocouples are arranged to take the substrate temperature, and the ambient temperature is measured using a mercury thermometer. The aluminum finned heat sink is used in conjunction with heat pipe to dissipate heat, and the wind speed is measured by a hot wire anemometer.

3 Temperature Rise at Start and Stop

Since the distance between the evaporation end and the condensation end of the vapor chamber is very small, several main factors affecting the thermal cycle, including the thickness of the vapor chamber, the filling amount of the liquid and

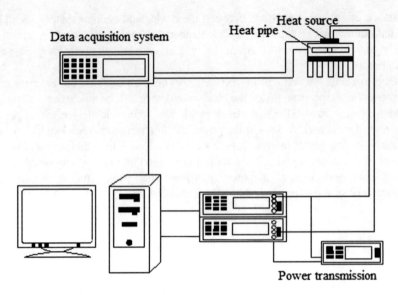

Fig. 1 Experimental facility schematic diagram

the response time of the heating power. The start-up time constant of the vapor chamber is affected by the combination of diffusivity, wall and core thickness and heat flux density. The expression of time constant the response of the vapor chamber obtained by the experiments are as follows:

$$t_{c,up} = 35.2 - 0.0341 h_{conv} + 1.50 \times 10^{-5} h_{conv}^2 \quad (1)$$

$$t_{c,down} = 31.7 - 0.032 h_{conv} + 1.44 \times 10^{-5} h_{conv}^2 \quad (2)$$

$$h_{conv} = \frac{Q}{A_c(T_{st} - T_a)} \quad (3)$$

In the formula: T_{st} is the average temperature of the condensation, K; A_c is the surface area of the condensation, m²; h_{conv} is the heat transfer coefficient of the condensation, W/(m² K), $t_{c,up}$, $t_{c,down}$, are the time constants when the heat pipe starts and closes, respectively.

As can be seen in Fig. 2, the transient thermal performance change is carried out when the vapor chamber is turned on and off at a heating power 200 W. The average temperature at the condensing end is equal to that of the evaporation end, which stands the average temperature of the steam chamber. The time of the vapor chamber to get steady-state mainly depends on heat transfer coefficient. As the heat transfer coefficient increases, the time to reach steady state is shorter. When the thermal load is subtracted, the temperature of all the measuring points eventually approaches a uniform initial temperature. The evaporation end core body is the

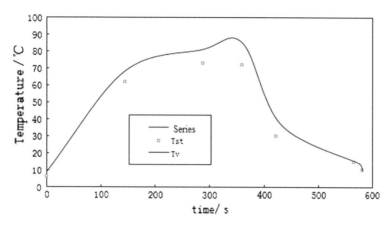

Fig. 2 Transient temperature variation of the heat pipe start and shut

main thermal resistance. Due to the thermal conductivity of the vapor chamber wall surface is large, so that reducing the temperature drop of the vapor chamber core portion can effectively improve thermal performance for the vapor chamber. The working fluid inside the heat pipe is recirculated by the capillary core and which constitutes the main composition of the heat resistance of the heat pipe. Therefore, reducing the temperature gradient of the capillary core can greatly improve the performance of the heat pipe.

4 Thermal Resistance Analysis

Heat transfer experiment was carried out the heating power 200 W under the substrate of the same size, and the heat pipe substrate module with air as the internal medium was added for comparison. The results are shown in Table 1. It is worth noting that the negative ion water vapor chamber substrate can effectively reduce the thermal resistance of the chip, DBC, and solder, and is much smaller than the

Table 1 Thermal model parameters of substrate module (200 W)

Thermal resistance $R/(K/W)$	Negative ion water vapor chamber module	Pure copper module	Air heat pipe module
Radiator thermal resistance	0.313	0.291	0.292
Substrate thermal resistance	0.044	0.045	0.200
Die, solder and DBC thermal resistance	0.120	0.278	0.289
Crust thermal resistance	0.164	0.323	0.489

thermal resistance of the air-filled heat pipe substrate module. The negative ion water in the heat pipe boils to achieve the isothermal temperature in the vapor chamber cavity is better as well, thereby improving the heat dissipation efficiency of tube core, the DBC, and the solder layer. When the vapor chamber substrate module is impacted by a high heat flux density of 200 W/cm², the thermal resistance of the crust is reduced to 0.164 K/W. The transient temperature rise of the tube core is less than the fixed thermal resistance of the pure copper substrate and the air heat pipe substrate module. Therefore, the vapor chamber substrate improves the ability of the integrated module to resist thermal shock. The heat capacity of the negative ion water vapor chamber substrate and the air-filled heat pipe substrate is similar, indicating that the heat storage of the substrate is mainly determined by the heat conductive copper layer.

For the negative ion water vapor chamber substrate module, only the heat resistance of the heat sink is kept constant, and the heat resistance of vapor chamber substrate and the tube core, the solder, and the DBC are all decreased as the heat flow density is increased. Therefore, when the vapor chamber substrate module is impacted by high heat flux density, the thermal resistance of the integrated module is reduced, and the transient temperature rise of the vapor chamber tube core is smaller than the pure copper substrate module and the air heat pipe substrate module with fixed thermal resistance. The long-time required for the chip to reach the same temperature during the start-up of the heat pipe module is beneficial to overcome the power "swell" and improve the thermal shock resistance of the power device. Due to the small thickness of the vapor chamber, it is impossible to take the temperature of the adiabatic section and the steam pressure. Therefore, the thermal resistance calculation formula for measuring the overall performance is given:

$$R_1 = (\overline{T_e} - \overline{T_c})/Q \tag{1}$$

$$R_2 = (\overline{T_c} - T_\infty)/Q \tag{2}$$

In the formula: $\overline{T_e}$ is the average temperature of the evaporation end of the vapor chamber, K; $\overline{T_c}$ is the average temperature of the condensation end, K; T_∞ is the ambient temperature, K; Q is the heating power, W.

Compare the thermal resistance of different heat sinks. The results are shown in Table 2. It can be seen that the thermal resistance of the air heat pipe system and entire thermal resistance were the largest, and the heat dissipation effect is the worst. The gravity heat pipe and the negative ion water vapor chamber heat dissipation system have lower total thermal resistance than other heat dissipation methods, and the heat transfer performance is higher than other devices. The thermal resistance of the gravity heat pipe is higher than that of the negative ion water vapor chamber and the 3 mm copper substrate. Due to the vacuum in the gravity heat pipe does not meet the requirements and a small amount of air is mixed into the steam chamber, which weakened the heat transfer effect. Compared to a 9 mm copper substrate, the

Table 2 Thermal resistance of heat dissipating device

Thermal resistance R/(K/W)	3 mm copper substrate	9 mm copper substrate	Air heat pipe	Gravity heat pipe	Vapor chamber
Heat pipe thermal resistance R_1	0.017	0.034	0.191	0.036	0.027
Radiator thermal resistance R_2	0.648	0.650	0.782	0.562	0.346

gravity heat pipe is roughly equivalent for their thermal resistance, but the weight of the vapor chamber module is reduced by about 50%.

The thermal resistance of the heat sink accounts for more than 70% of the thermal resistance of the entire heat dissipation module. Therefore, for the entire heat dissipation module designed, although the vapor chamber shows the characteristics of strong heat diffusion and low thermal resistance, the large thermal resistance of the heat pipe radiator affects the heat dissipation performance of the whole module. The improved vapor chamber design alone can improve thermal diffusion performance and reduce the thermal resistance of the heat pipe, but it cannot fundamentally improve the performance of the heat dissipation system. The heat pipe only serves as the heat transfer medium. Therefore, optimizing the structure of the heat sink can effectively improve the heat dissipation performance.

5 Conclusions

(1) The time of the vapor chamber to get steady-state mainly depends on heat transfer coefficient. As the heat transfer coefficient increases, the time to reach steady state is shorter. Due to the thermal conductivity of the vapor chamber wall surface is large, so that reducing the temperature drop of the vapor chamber core portion can effectively improve thermal performance for the vapor chamber.

(2) Due to the vapor chamber substrate module is impacted by high heat flux density, the thermal resistance of the integrated module is reduced, and the transient temperature rise of vapor chamber tube core is smaller than that of the pure copper substrate module and the air heat pipe substrate module with fixed thermal resistance. The long-time required for the chip to reach the same temperature during the start-up of the heat pipe module is beneficial to overcome the power "swell" and improve the thermal shock resistance of the power device. The thermal resistance of the heat sink accounts for more than 70% of the thermal resistance of the entire heat dissipation module.

Acknowledgements The project is supported by the National Natural Science Foundation of China (Grant No. 51504188).

References

1. Zhu, K., et al.: Experimental study of energy saving performances in chip cooling by using heat sink with embedded heat pipe. Energy Procedia **105**, 5160–5165 (2017)
2. Jaroslaw, L., et al.: Measurements and simulations of transient characteristics of heat pipes. Microelectron. Reliab. **46**, 109–115 (2006)
3. Kim, K.S., et al.: Heat pipe cooling technology for desktop PC CPU. Appl. Therm. Eng. **23**(9), 1137–1144 (2003)
4. Guo, L.: Development of heat dissipation in electronics components. Cryog. Supercond. **42**(2), 62–66 (2014)
5. Zhang, M., Liu, Z.L., Wang, C.: The integrated design of heat pipe spreader and heat sink. J. Eng. Thermophys. **31**(5), 853–856 (2010)
6. Zhang, L.H., et al.: Thermal characteristic of a novel flat plate heat pipe for hybrid integrated power electronic module. Acat Electronica Sin. **37**(8), 1848–1853 (2009)
7. Chen, T.S., Chen, K.H., Wang, C.: A simplified transient three-dimensional model for estimating the thermal performance of the vapor chambers. Appl. Therm. Eng. **26**, 2087–2094 (2006)
8. Rahman, M.L., et al.: Effect of fin and insert on the performance characteristics of close loop pulsating heat pipe (CLPHP). Procardia Eng. **05**, 129–136 (2015)
9. Tran, T., et al.: Experimental investigation on the feasibility of heat pipe cooling. Appl. Therm. Eng. **63**(2), 551–558 (2014)
10. Take, K., Webb, L.R.: Thermal performance of integrated plate heat pipe with a heat spreader. J. Electron. Packag. **123** (2001)
11. Sauciuc, I., Chrysler, G., Mahajan, R., Prasher, R.: Spreading in the heat sink base: phase change systems or solid metals. IEEE Trans. Compon. Packag. Technol. **25** (2008)
12. Wei, J., Cha, A., Copeland, D.: Measurement of vapor chamber performance. In: IEEE, Semi-Therm Symposium (2013)
13. Ivanova, M., Avenas, Y., et al.: Heat pipe integrated in direct bonded copper technology for cooling of power electronics packaging. IEEE Trans. Power Electron. **21**(6) (2016)
14. Gao, M., Cao, Y.: Flat and U-shaped heat spreaders for high-power electronics. Heat Transfer Eng. **24**(3), 57–65 (2013)

The Role of Cylinder Obstacles before Air Conditioning Filter in the Quenching of Flame Propagation during Their Gas Deflagration Production Process

Lijia Fan, Chenghu Zhang, Jihong Wei and Yufei Tan

Abstract The reticulated polyurethane foam has an extensive application in the ventilation and air conditioning systems as filtration material due to its high filtration velocity, low flow resistance and high filter efficiency. Mostly, the reticulated polyurethane foam is produced from the closed-cell polyurethane foam by the deflagration process of the combustible gas. However, in the actual production process, the quenching of the flame in the porous media of the polyurethane foam always leads to the failure of production process, which reduces the production efficiency and results in economic losses. The thickness of the reticulated foam highly affects the propagation characteristics of the flame due to the effect of the interactions of flame and the reticulated foam. Thus, this paper aims to reveal the effect of the obstacles of the porous media of polyurethane foam on the flame propagation characteristics of gas deflagration. A cylindrical explosion test tank was designed, which has a diameter of 315 mm. The flame propagation characteristics and pressure wave in the presence of reticulated polyurethane foam were numerically investigated. A parameter study of the thickness of the reticulated polyurethane foam on the propagation and quenching of the flame is discussed. It can be concluded that the flame propagation speed decreases when the flame enters into the reticulated polyurethane, it increases as the thickness of the reticulated polyurethane increases, in addition, the final deflagration temperature and the temperature rise rate decrease with the increase of the thickness of the reticulated

L. Fan · C. Zhang (✉) · J. Wei · Y. Tan
School of Architecture, Harbin Institute of Technology, Harbin, China

Key Laboratory of Cold Region Urban and Rural Human Settlement Environment Science and Technology, Ministry of Industry and Information Technology, Harbin, China
e-mail: chenghu.zhang@163.com

L. Fan
e-mail: fanlijia@hit.edu.cn

J. Wei
e-mail: 15704600720@163.com

Y. Tan
e-mail: tanyufei2002@163.com

polyurethane, and the temperature rise rate decreases when the flame enters into the reticulated polyurethane. It is due to the destruction effect of the free radicals and absorption effect the heat of the flame by the reticulated polyurethane. Besides, the deflagration overpressure and the pressure rise rate decrease with the increase of the thickness of the reticulated polyurethane, which is due to the inhibitory effect of deflagration overpressure.

Keywords Reticulated polyurethane foam · Deflagration overpressure · Flame propagation · Gas deflagration · Deflagration temperature

1 Introduction

The reticulated polyurethane foam has an extensive application in the ventilation and air conditioning systems as filtration material due to its high filtration velocity, low flow resistance and high filter efficiency. Mostly, the reticulated polyurethane foam is produced from the closed-cell polyurethane foam by the deflagration process of the combustible gas. However, in the actual production process, the quenching of the flame in the porous media of the polyurethane foam always leads to the failure of production process, which reduces the production efficiency and results in economic losses. The thickness of the reticulated foam highly affects the propagation characteristics of the flame due to the effect of the interactions of flame and the reticulated foam. Thus, this paper aims to reveal the effect of the obstacles of the porous media of polyurethane foam on the flame propagation characteristics of gas deflagration.

The flame propagation characteristics and pressure wave in the presence of various obstacles and porous mediums have been investigated by many researchers. Pang [1] explored the quenching effect of the full-cloth aluminum alloy wire mesh on the deflagration flame, the results showed that, for the hydrogen-air mixture, the aluminum alloy wire mesh not only could not quench the flame but also increased the maximum explosion overpressure, while for the methane-air mixture, the results were just the opposite. Wen [2] made a large vortex simulation on the CH_4–O_2 deflagration flame in the small scale restricted space with obstacles, in which the flame structure, flame propagation velocity, deflagration overpressure and the flame mode were studied. Zhang [3] studied the influence of the parameters of the ring obstacles in the connected container on the gas explosion characteristics. Wan [4] studied the effect of the position of the side-wall opening on the deflagration flame characteristics of methane-air mixture in the end-opening pipeline with obstacles. The results showed that the deflagration effect of explosion venting was significantly enhanced with the reduction of the distance between the sidewall outlet and the ignition point. Wen [5] studied the quenching behavior of the deflagration flame by the porous medium in the square container with different obstacles. The result showed that the interaction between the obstacle in front of the porous medium and the flame would lead to the increase of the flame velocity and overpressure, which finally made the quenching behavior of the porous medium failed. Yu [6] studied the influence of three different hollow shapes in the rectangular obstacles on the propagation of deflagration flame in a semi-closed container, and the results showed that the rectangular hollow shape can generate the maximum turbulence intensity, flame velocity and explosion overpressure. Wang [7] studied the propagation characteristics of methane gas deflagration flame with trapezoid, rectangular and spherical obstacles in the rectangular pipeline. The results showed that the angle of the obstacle in the flow direction and the effective obstacle area had a great influence on the overpressure and propagation characteristics of the deflagration flame.

However, the flame propagation characteristics and pressure wave in the deflagration process in the presence of reticulated polyurethane foam have not been reported before. Thus, the flame propagation characteristics and pressure wave in the presence of reticulated polyurethane foam were numerically investigated. A parameter study of the thickness of the reticulated polyurethane foam of 10 ppi on the propagation of the flame is discussed.

2 Methods

The flame propagation characteristics and pressure wave in the presence of reticulated polyurethane foam were numerically investigated in this paper, and the accuracy of the numerical method has been verified in our previous studies [8].

2.1 The Geometry Model and Meshing

A cylindrical explosion test tank was designed, which has a radius of 157.5 mm, and the geometric model of the tank with the reticulated polyurethane foam was shown in Fig. 1. The ignition point is set at the center of the vessel, in addition, the reticulated polyurethane foam is placed at the distance of 40 mm away from the center of the spark ignition, i.e., the inner radius of the reticulated polyurethane foam $r1$ is 40 mm, and the thickness of the reticulated polyurethane foam is shown in Table 1. The Designmodeler is applied to establish the geometric model, and the ICEM is applied to generate mesh. In addition, to improve the calculation accuracy, the mesh near the location of the ignition point is partially encrypted, and the divided mesh is shown in Fig. 2.

Fig. 1 Sketch of the tank and the obstacles of reticulated polyurethane foam

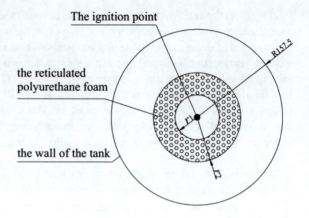

Table 1 Thickness of the reticulated polyurethane foam

Case	Thickness (mm)
1	20
2	30
3	40

Fig. 2 Sketch of the divided mesh

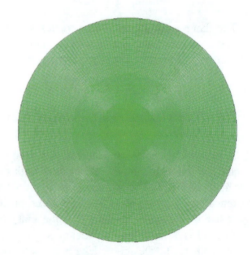

2.2 The Mathematical Model

In this research, the standard model is applied to describe the turbulence properties of the premixed combustion process, and the EBU-Arrhenius is applied to describe the turbulent combustion rate of the premixed combustion model. In addition, the viscous resistance coefficient and inertial resistance coefficient in the porous medium model are shown in Eqs. (1) and (2) [9].

$$\frac{1}{\alpha} = \frac{150(1-\varepsilon)^2}{D_p^2 \varepsilon^3} \tag{1}$$

$$C_2 = \frac{3.5(1-\varepsilon)}{D_p \varepsilon^3} \tag{2}$$

where $\frac{1}{\alpha}$ and C_2 are the viscous resistance coefficient and the inertial resistance coefficient of the porous medium in all directions, ε is the porosity of the porous medium of the reticulated polyurethane foam, and D_p is the average diameter of the particles.

2.3 The Boundary Conditions and the Initial Conditions

The boundary conditions and the initial conditions are set as follows:

(1) The initial temperature of the wall and the fluid domain is 300 K, and the initial pressure of the fluid domain is 101,325 Pa.
(2) The electric spark is used for the ignition, the ignition energy is 5 J, and the ignition cycle is 0.001 s.
(3) The mixture ratio of methane to air is the equivalent ratio.
(4) The parameters of the porous medium are shown in Table 2.

3 Results and Discussion

The flame propagation characteristics and pressure wave in the presence of reticulated polyurethane foam were numerically investigated. A parameter study of the thickness of the reticulated polyurethane foam of 10 ppi on the propagation and quenching of the flame is discussed.

Table 2 Parameter of the reticulated polyurethane foam

Parameter	Value
Pore diameter (mm)	1.27
Coefficient of heat conductivity (W/(m K))	0.030
Density (kg/m^3)	24
Porosity	0.9195

3.1 The Effect of the Thickness of Reticulated Polyurethane on the Position of the Flame Front

Figure 3 shows the position of the flame front of the deflagration process of the premixed mixture of methane and air for the three cases in Table 1, in which the porous obstacles of reticulated polyurethane foams have the same inner radius of 40 mm and different thickness of $r1 = 20, 30$ and 40 mm.

It can be seen that the flame propagation speed decreases when the flame enters into the reticulated polyurethane, and the flame propagation time increases as the thickness of the reticulated polyurethane increases. The reason behind this is that the reticulated polyurethane has the effect to destroy the free radicals of the chemical reactions and absorb the heat of the flame, which results in the decrease of the flame propagation speed, and thus, the thicker the reticulated polyurethane is, the longer the flame propagation time will be.

3.2 The Effect of the Thickness of Reticulated Polyurethane on the Inhibition of Pressure Wave

Figure 4 shows the overpressure-time response of the deflagration process of the premixed mixture of methane and air for the three cases in Table 1, in which the porous obstacles of reticulated polyurethane foams have the same inner radius of 40 mm and different thickness of $r1 = 20, 30$ and 40 mm.

As can be seen in Fig. 4, for all the three cases, the deflagration overpressure increases with time gradually, and finally, it reaches a constant value. In addition, the deflagration overpressure decreases with the increase of the thickness of the reticulated polyurethane. The reason behind this is that thicker the reticulated

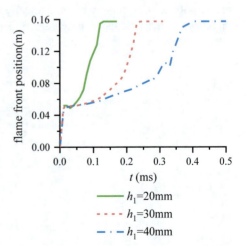

Fig. 3 Effect of the thickness of reticulated polyurethane on the position of the flame front

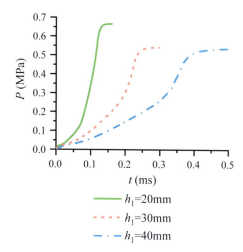

Fig. 4 Effect of the thickness of reticulated polyurethane on the inhibition of pressure wave

polyurethane is, the bigger the volume of the reticulated polyurethane will be; thus, the absorption capacity of the pressure wave is larger, and finally, the inhibitory effect of deflagration overpressure is weaker. Thus, the deflagration pressure rise rate decreases with the increase of the thickness of the reticulated polyurethane for the same reason.

3.3 The Effect of the Thickness of Reticulated Polyurethane on the Temperature Distribution of the Deflagration Process

Figure 5 shows the temperature distribution at the distance of 0.04, 0.06, 0.10 and 0.14 m away from the ignition point for the three cases in Table 1, in which the porous obstacles of reticulated polyurethane foams have the same inner radius of 40 mm and different thickness of $r1$ = 20, 30 and 40 mm.

In Fig. 5b–d, deflagration temperature at the position of x = 0.06, 0.1 and 0.14 mm increases with time gradually, and finally, it trends to a constant value; however, the high temperature at the initial stage in Fig. 5a is caused by the spark ignition.

In addition, the final deflagration temperature decreases with the increase of the thickness of the reticulated polyurethane. In addition, the temperature rise rate decreases when the flame enters into the reticulated polyurethane, and the temperature rise rate decreases as the thickness of the reticulated polyurethane increases. The reason behind this is that the thicker the reticulated polyurethane is, the higher the collision probability between the free radicals of the chemical reactions and the internal structures of the reticulated polyurethane will be, which results in the increase of the quantity of the destroyed free radicals, and finally leads

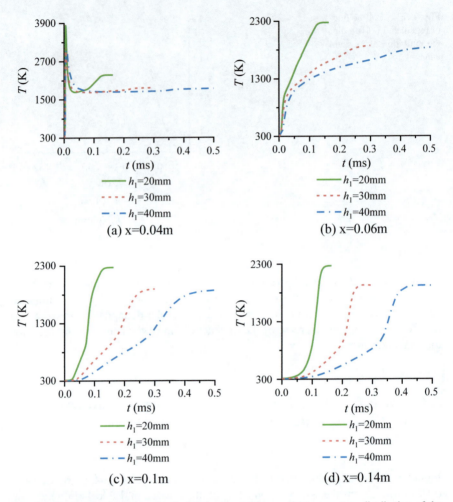

Fig. 5 Effect of the thickness of reticulated polyurethane on the temperature distribution of the deflagration process

to the decrease of the deflagration temperature. In addition, the three-dimensional porous structure of the materials can absorb the heat from burning, which will also result in the decrease of final flame temperature causes that the temperature rise rate decreases when the flame enters into the reticulated polyurethane and decreases as the thickness of the reticulated polyurethane increases.

4 Conclusions

The reticulated polyurethane foam has an extensive application in the ventilation and air conditioning systems as filtration material due to its high filtration velocity, low flow resistance and high filter efficiency. This paper numerically investigated the flame propagation characteristics and pressure wave in the deflagration production process of the presence of reticulated polyurethane foam. A parameter study of the thickness of the reticulated polyurethane foam of 10 ppi on the propagation of the flame is discussed. Here are the main conclusions:

(1) The flame propagation speed decreases when the flame enters into the reticulated polyurethane, and the flame propagation time increases as the thickness of the reticulated polyurethane increases, due to the destruction effect of the free radicals of the chemical reactions and absorption effect the heat of the flame by the reticulated polyurethane.
(2) The deflagration overpressure and the pressure rise rate decrease with the increase of the thickness of the reticulated polyurethane, which is due to the inhibitory effect of deflagration overpressure.
(3) The final deflagration temperature and the temperature rise rate decrease with the increase of the thickness of the reticulated polyurethane. In addition, the temperature rise rate decreases when the flame enters into the reticulated polyurethane. It is also due to the destruction effect of the free radicals of the chemical reactions and absorption effect the heat of the flame by the reticulated polyurethane.

Acknowledgements The project is supported by Natural Science Foundation of Heilongjiang Province of China (Number E2016030).

References

1. Pang, L., Wang, C., Han, M., et al.: A study on the characteristics of the deflagration of hydrogen-air mixture under the effect of a mesh aluminum alloy. J. Hazard. Mater. **299**, 174–180 (2015)
2. Wen, X., Ding, H., Su, T., et al.: Effects of obstacle angle on methane–air deflagration characteristics in a semi-confined chamber. J. Loss Prev. Process Ind. **45**, 210–216 (2017)
3. Zhang, K., Wang, Z., Ni, L., et al.: Effect of one obstacle on methane–air explosion in linked vessels. Process Saf. Environ. Prot. **105**, 217–223 (2017)
4. Wan, S., Yu, M., Zheng, K., et al.: Influence of side venting position on methane/air explosion characteristics in an end-vented duct containing an obstacle. Exp. Thermal Fluid Sci. **92**, 202–210 (2018)
5. Wen, X., Xie, M., Yu, M., et al.: Porous media quenching behaviors of gas deflagration in the presence of obstacles. Exp. Thermal Fluid Sci. **50**, 37–44 (2013)
6. Yu, M., Zheng, K., Chu, T.: Gas explosion flame propagation over various hollow-square obstacles. J. Nat. Gas Sci. Eng. **30**, 221–227 (2016)

7. Wang, L., Si, R., Li, R., et al.: Experimental investigation of the propagation of deflagration flames in a horizontal underground channel containing obstacles. Tunn. Undergr. Space Technol. **78**, 201–214 (2018)
8. Fan, L.: Research on experiment and mechanism of preparation process of reticulated polyurethane foam by detonation method. Harbin Institute of Technology (2015)
9. Chen, Y.: Explosion suppression characteristics of aluminum silicate wool in pipeline. Dalian University of Technology (2012)

An Association Rule-Based Online Data Analysis Method for Improving Building Energy Efficiency

Chaobo Zhang, Yang Zhao and Xuejun Zhang

Abstract Association rule mining has been applied to reveal variable relations from numerous operational data in buildings. However, there is a lack of effective methods to take full advantage of the discovered relations. This study proposes a real-time data analysis method for diagnosing building operational problems based on the discovered relations. In this method, the historical data are explored by the association rule mining to generate raw association rules. The abnormal and normal rules are extracted manually to build a rule base. The rule base is then used to analyze the real-time measurements of the relevant variables in the extracted rules. Operational problems are detected if the measurements of the relevant variables match with an abnormal rule or break all the related normal rules. Evaluations are made using the operational data collected from the chiller plant of a commercial building located in Shenzhen, China. Results show that the proposed method can detect operational problems effectively.

Keywords Data mining · Association rule mining · Expert system · HVAC system · Operational problem diagnosis

1 Introduction

The building sector contributes more than one-third of the final energy consumption worldwide. About 15–30% of the energy used in commercial buildings is wasted because of poor maintenance, performance degradations, and improper control strategies [1]. With the population of building automation systems, massive amounts of building operational data are available, which can reflect the actual operational conditions of buildings. Association rule mining is a useful technology

---

C. Zhang · Y. Zhao (✉) · X. Zhang
Institute of Refrigeration and Cryogenics, Zhejiang University, Hangzhou 310027, China
e-mail: youngzhao@zju.edu.cn

to reveal building operational patterns from the massive amounts of building operational data. Yu et al. discovered energy-inefficient behavior patterns of occupants using the association rule mining from the energy consumption data of residential buildings [2]. Yu et al. also detected energy-inefficient operational patterns and equipment faults of VAV air-conditioning systems using the association rule mining [3]. Cabrera et al. utilized the association rule mining to detect energy wastes from lighting energy consumption data [4]. Xiao and Fan developed a data mining framework based on the association rule mining to detect abnormal operational patterns hidden in the operational data of commercial buildings [5]. Based on this framework, Fan et al. further proposed an association rule mining-based framework to identify non-typical building operational patterns and sensor faults from the operational data of a central chilling system [6].

In previous researches, the mined association rules were only used to discover historical operational patterns of buildings. However, the discovered historical operational patterns are also very useful for evaluating future operations of buildings. This study is inspired by expert systems which can be used to analyze building operational data automatically based on a set of rules. For instance, Peña et al. developed a rule-based expert system to detect energy efficiency anomalies in smart buildings [7]. Seven if–then rules were extracted based on domain knowledge and historical data. Schein et al. also proposed a rule-based fault detection method for air handling units [8]. The rules in these methods were very similar to the association rules. Considering system characteristics in different buildings are usually various, expert systems usually need to be customized for different buildings. It is very hard and time-consuming to design comprehensive rule bases for the customized expert systems.

To combine with the association rule mining technology and the expert systems, this study proposed a generic association rule-based online data analysis method. It can help build customized expert systems for buildings based on their historical operational data. This method includes two stages, i.e., offline training and online analysis. The operational data collected from the chiller plant of a commercial building are used to evaluate the method.

2 Methods

The flowchart of the proposed association rule-based online data analysis method is shown in Fig. 1. It has two stages, i.e., offline training and online analysis. The offline training stage aims to build an association rule base based on historical operational data. The online analysis stage is developed to detect operational problems in real time based on the association rule base.

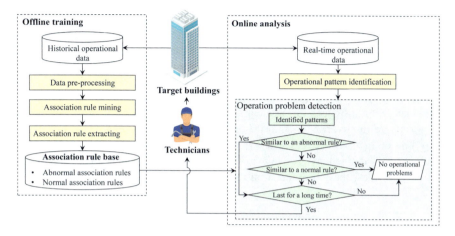

Fig. 1 Flowchart of the proposed association rule-based online data analysis method

2.1 Stage 1: Offline Training

The offline training stage has three steps, i.e., data preprocessing, association rule mining, and association rule extracting.

The step of data preprocessing aims to improve data quality. Missing values are common in building operational data because of signal transmission errors, sensor faults, upper computer crash, etc. In this study, the measurements of a variable which only miss for a short time, i.e., 30 min, are filled in using the linear interpolation algorithm. If the measurements are missed more than 30 min, they are discarded directly. Outliers also commonly exist in building operational data. In this study, the boxplot-based approach is applied to identify outliers preliminarily [5]. In this approach, the interquartile range is defined as the difference between the third quartile (i.e., Q_3) and the first quartile (i.e., Q_1). The lower bound and the upper bound of normal measurements are defined as $Q_1 - 1.5(Q_3 - Q_1)$ and $Q_3 + 1.5(Q_3 - Q_1)$, respectively. If the measurements of a variable are not within the range between the lower bound and the upper bound, they are regarded as outliers. However, if the outlier proportion of a variable is too high (i.e., larger than 1% in this study), the outliers of the variable are re-identified manually. The same strategy for handling missing values is utilized to handle the identified outliers. Data transformation is also important because most association rule mining algorithms cannot process numerical data [5]. In this study, the numerical data are transformed into the categorical data using a width-based binning approach. The width-based binning approach categorizes the numerical data into several intervals which are determined manually. The categorized data are then transformed into a uniform form combined with the variable name and the measurement interval. For instance, all the numerical measurements of the supply chilled water temperature which are between 5 °C and 6 °C can be transformed into a unified form "The supply chilled water temperature is between 5 °C and 6 °C."

The step of association rule mining aims to discover variable association relations in the form of "$A \rightarrow B$." FP-growth is utilized in this study [9]. The amount of mined association rules shows an explosive increase with the increase of the number of the related variables in the association rules. In this study, only the association rules related to two variables are mined. However, the mined association rules are in pairs, which means that at least 50% of the raw association rules are redundant. For instance, the association rule "$A \rightarrow B$" and the association rule "$B \rightarrow A$" are discovered together using the association rule mining. In this study, only one of them which has a larger confidence is left for further analysis. The confidence is defined by Eq. (1). In order to keep the integrality of the mined association rules, the left association rules are not further filtered.

$$\text{confidence}(A \rightarrow B) = \frac{P(A \cup B)}{P(A)} \quad (1)$$

where $P(A \cup B)$ is the probability that A and B coincide in the data set to be analyzed, and $P(A)$ is the probability that A appears in the data set.

In the step of association rule extracting, all the left association rules are analyzed by experts. Only the association rules which can reveal the operational problems or the normal operation patterns are used to build an association rule base. Considering operation patterns of devices are totally different when they are turned on or turned off, the rules in the association rule base are further classified into two types based on the on–off status of the related devices.

2.2 Stage 2: Online Analysis

The online analysis stage has three steps, i.e., operational pattern identification, operational problem detection, and manual operation optimization and maintenance.

In the step of operational pattern identification, the variables in the same rule from the association rule base are regarded as relevant. The real-time measurements of these relevant variables are identified as the operational patterns to be analyzed. For instance, if there is an association rule "The 1# chiller is turned on → The supply chilled water temperature of the 1# chiller is between 5 °C and 6 °C" in the association rule base, the real-time measurements of the on–off status of the 1# chiller and its supply chilled water temperature are extracted continually.

In the step of operational problem detection, the identified real-time operational patterns are compared with the related rules, which have the same variables as the operational patterns, in the association rule base. It needs to be noted that the on–off status of devices related to the identified patterns should be the same as the corresponding rules in the association rule base. If an operational pattern matches with an abnormal association rule or breaks all the related normal association rules, it is detected as potentially abnormal patterns. The operation patterns, whose variable measurements are both within the measurement interval of an association rule related to the variables, are regarded to match with the rule. The operation patterns, whose variable measurements are not both within the measurement interval of an association rule related to the variables, are regarded to break the rule. Considering the interferences caused by sensor signal fluctuations, transient operations and so on, a detected problem is reported until it is steady (i.e., until the detected problem lasts for a given time threshold value).

In the step of manual operation optimization and maintenance, all the potentially abnormal patterns are checked by technicians. If there are operational problems, the technicians can optimize the system operation or maintain the faulty devices timely.

3 Results

3.1 Data Source

The operational data (from December 2015 to April 2018) of the chiller plant of a complex building located in Shenzhen, China, are chosen as the data source in this study. The operational data of the system are sampled at an interval of 10 min. There are 8 chillers, 10 primary chilled water pumps, 14 secondary chilled water pumps, 20 cooling towers, and 10 cooling water pumps in the system. A total of 116 variables are chosen in this study. The chosen variables include the outdoor temperature, the outdoor relative humidity, the chilled water flow, the supply and return chilled water temperatures, the supply and return cooling water temperatures, the chilled water temperature difference, the cooling water temperature difference, the chiller load ratios, the on–off states of devices and valves, the device powers, the pump frequencies, and the numbers of running devices in the same type.

3.2 Offline Training Results

Two-year operational data (from December 2015 to November 2017) are chosen as the historical data to build an association rule base for the system. In the data preprocessing step, a total of 0.01% of the raw data are discarded which are missed more than 30 min. And a total of 7.03% of the raw data are regarded as outliers and discarded. The left data which are numerical are categorized into several intervals. The measurement intervals of each variable are determined manually based on the domain knowledge. For instance, in this study, the measurements of the supply chilled water temperatures are categorized into nine intervals, i.e., [3, 5], (5, 6], (6, 7], (7, 8], (8, 9], (9, 11], (11, 13], (13, 15], and (15, 25]. A total of 101,787 association rules related to two variables are mined in this study. After checking each association rule carefully, two kinds of association rules are extracted to build the association rule base, i.e., normal association rules and abnormal association rules. There are 2871 normal association rules and 1709 abnormal association rules in the association rule base.

Table 1 lists five typical normal rules included in the association rule base. The first, second, and third association rules are related to the running device(s). The first association rule reveals that the supply chilled water temperature of the 7# chiller and the 8# chiller was both between 6 °C and 7 °C when the two chillers were turned on. The second association rule shows that the supply chilled water temperature of the 7# chiller was between 6 °C and 7 °C when the chiller was turned on. The third association rule indicates that the chilled water valve of the 8# chiller was opened when the chiller was turned on. The fourth and fifth association rules are related to the idle device(s). The fourth association rule shows that the supply chilled water temperature of the 5# chiller was between 13 °C and 15 °C when the chiller was turned off. The fifth association rule reveals that the chilled water valve of the 8# chiller was closed when the chiller was turned off.

Table 2 lists four typical abnormal rules included in the association rule base. The first association rule shows that the supply chilled water temperature of the 7# chiller and the 5# chiller was significantly different when the two chillers were both turned on. Actually, the chilled water temperatures of two running chillers should be similar to keep high-efficiency operation. The second association rule reveals that the chilled water temperature difference was too small, i.e., between 2 K and 3 K, when the chillers were running. In practice, the chilled water temperature difference is usually more than 5 K for improving the energy efficiency of chillers. The third association rule indicates that the chilled water valve of the 8# chiller was still opened when the chiller was turned off. It means that the chilled water valve might be stuck. The fourth association rule shows that chilled water pumps were still running when all the chillers were turned off. The operations of chilled water pumps were invalid in this condition, which wasted lots of energy.

Table 1 Examples of the normal rules included in the association rule base

No.	Association rule
The related device(s) is(are) turned on	
1	The supply chilled water temperature of the 7# chiller is between 6 °C and 7 °C → The supply chilled water temperature of the 8# chiller is between 6 °C and 7 °C
2	The supply chilled water temperature of the 7# chiller is between 6 °C and 7 °C → The 7# chiller is turned on
3	The 8# chiller is turned on → The chilled water valve of the 8# chiller is opened
The related device(s) is(are) turned off	
4	The supply chilled water temperature of the 5# chiller is between 13 °C and 15 °C → The 5# chiller is turned off
5	The chilled water valve of the 8# chiller is closed → The 8# chiller is turned off

Table 2 Examples of the abnormal rules included in the association rule base

No.	Association rule
The related device(s) is(are) turned on	
1	The supply chilled water temperature of the 7# chiller is between 3 °C and 5 °C → The supply chilled water temperature of the 5# chiller is between 6 and 7 °C
2	The chilled water temperature difference is between 2 K and 3 K → The number of running chillers is 2
The related device(s) is(are) turned off	
3	The chilled water valve of the 8# chiller is opened → The 8# chiller is turned off
4	The number of running chillers is 0 → The number of running chilled water pumps is 2

3.3 Online Analysis Results

Five-month operational data (from December 2017 to April 2018) are used to simulate the online analysis process. The time threshold value is 30 min in this study. Two detected operational problems are shown in this study as examples including the energy-inefficiency valve control strategy and the total supply chilled water temperature sensor fault.

The first detected abnormal operational pattern is that the chilled water valve of the 7# chiller was closed when the 7# chiller was turned on. This operational problem matched with the abnormal rule "The 7# chiller is turned on → The chilled water valve of the 7# chiller is closed" which was included in the association rule base. As illustrated in Fig. 2, the chilled water temperature in this abnormal operation pattern was significantly lower than the chilled water temperature in the

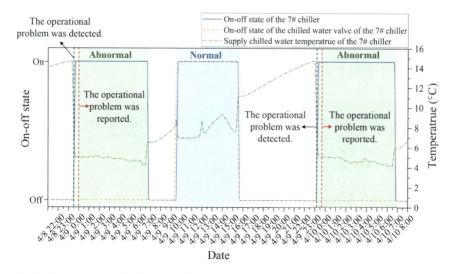

Fig. 2 Detected energy-inefficiency valve control strategy

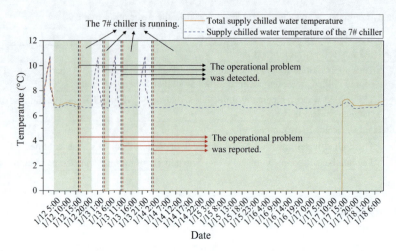

Fig. 3 Detected temperature sensor fault

normal operation pattern. The operational problem was detected immediately when the problem arose. And, it was reported to the technicians 30 min later.

The fault of the total supply chilled water temperature sensor is also detected successfully which had never occurred before. As illustrated in Fig. 3, the supply chilled water temperature of the 7# chiller was between 6 °C and 7 °C when the 7# chiller was running. But, the total supply chilled water temperature was 0 °C when the 7# chiller was running since 16:30, January 12, 2018, which meant that the total supply chilled water temperature sensor might be broken. No abnormal association rules and no normal association rules were similar to this faulty operational pattern. However, this faulty operational pattern was still detected successfully.

4 Conclusions

An association rule-based online data analysis method is proposed in this study to detect the operational problems in buildings. The method has two stages, i.e., offline training and online analysis. The offline training stage aims to build an association rule base using the historical data. The online analysis stage aims to detect the operational problems in real time based on the association rule base. The operational data collected from the chiller plant of a commercial building in Shenzhen, China, are used to evaluate the method.

The offline training stage is evaluated based on the two-year historical operational data of the chiller plant collected from December 2015 to November 2017. A total of 2871 normal association rules and 1709 abnormal association rules are extracted from 101,787 raw association rules to build an association rule base. Then, in the online analysis stage, the association rule base is utilized to detect the operational problems in the chiller plant from December 2017 to April 2018. Two typical detected problems, i.e., the energy-inefficiency valve control strategy and the fault of the total supply chilled water temperature sensor, are described in detail as examples in this study. The results show that the method can detect not only the operational problems which had occurred in the past but also the operational problems which had never occurred before.

Further researches about how to analyze the association rules mined from the historical data more efficiently are encouraged. And, how to transfer the association rule base discovered in one building to other similar buildings is also an interesting research direction in the future.

Acknowledgements This study is supported by the National Nature Science Foundation of China (Number 51706197).

References

1. Katipamula, S., Brambley, M.R.: Review article: methods for fault detection, diagnostics, and prognostics for building systems—a review, part I. HVAC&R Res. **11**, 3–25 (2005)
2. Yu, Z., Haghighat, F., Fung, B.C.M., et al.: A methodology for identifying and improving occupant behavior in residential buildings. Energy **36**, 6596–6608 (2011)
3. Yu, Z., Haghighat, F., Fung, B.C.M., et al.: A novel methodology for knowledge discovery through mining associations between building operational data. Energy Build. **47**, 430–440 (2012)
4. Cabrera, D.F.M., Zareipour, H.: Data association mining for identifying lighting energy waste patterns in educational institutes. Energy Build. **62**, 210–216 (2013)
5. Xiao, F., Fan, C.: Data mining in building automation system for improving building operational performance. Energy Build. **75**, 109–118 (2014)
6. Fan, C., Xiao, F., Yan, C.: A framework for knowledge discovery in massive building automation data and its application in building diagnostics. Autom. Constr. **50**, 81–90 (2015)
7. Peña, M., Biscarri, F., Guerrero, J.I., et al.: Rule-based system to detect energy efficiency anomalies in smart buildings, a data mining approach. Expert Syst. Appl. **56**, 242–255 (2016)
8. Schein, J., Bushby, S.T., Castro, N.S., et al.: A rule-based fault detection method for air handling units. Energy Build. **38**, 1485–1492 (2006)
9. Han, J., Pei, J., Yin, Y.: Mining frequent patterns without candidate generation. In: Proceedings of the 2000 ACM SIGMOD International Conference on Management of Data, pp. 1–12. ACM, New York (2000)

Numerical Study on Two-Phase Flow of Transcritical CO_2 in Ejector

Xu Feng, Zhenying Zhang, Jianjun Yang and Dingzhu Tian

Abstract The transcritical CO_2 two-phase flow in the ejector was investigated numerically by the computational fluid dynamics (CFD) method. The accuracy of the built three-dimensional CFD model was validated by contrasting the simulating results with the available experimental data in the literature. The distribution of pressure, velocity and two-phase volume fraction inside the ejector was analyzed. The effect of the primary nozzle diverging angle on the performance of the ejector was obtained. The results showed that the pressure of the CO_2 is decreased and the velocity is increased after the stream enters the primary nozzle. The velocity at the exit of the primary nozzle reaches a maximum value, which is about 168 m/s. The two streams are then mixed in the mixing chamber; the velocity and the pressure were found to be shocked initially and then tend to be stable. The velocity decreases and the pressure increases gradually in the diffuser section. The vaporization of the transcritical CO_2 was found to be occurred near the throat of the primary nozzle. The optimum diffusion angle of the primary nozzle was found to be about 6°, where the mass entrainment ratio is 0.83.

Keywords Ejector · CO_2 · Numerical simulation · Primary nozzle

1 Introduction

CO_2 is one of the environmentally friendly refrigerants, however, the coefficient of performance (COP) of CO_2 refrigeration cycle is lower than ordinary refrigerant. Therefore, improving the refrigeration cycle and optimizing the components in the cycle to improve the system performance has become a research hotspot. Ejectors can be used to recover the expansion work to increase the system efficiency. In 2014, Zhang et al. [1] thermodynamically showed that the maximum COP of the

X. Feng · Z. Zhang (✉) · J. Yang · D. Tian
School of Architecture and Civil Engineering, North China University of Science and Technology, Tangshan, China
e-mail: zhangzhenying@ncst.edu.cn

ejector expansion transcritical CO_2 refrigeration system is up to 45.1% higher than that of the basic cycle through optimizing the value of suction pressure drop. It is necessary to optimize the ejector internal structure due to the sophisticated flow process of the fluid, the flash and the choked flow in two phases of the ejector. In 2012, Liu et al. [2] found that mass entrainment ratio and COP increased significantly as the primary nozzle throat diameter (D_t) decreased in transcritical CO_2 refrigeration cycle used a controllable ejector through the experiment analysis. Using the ejector with D_t of 1.8 mm, the COP and cooling capacity can be increased, respectively, by approximately 60 and 46%. In 2014, Hu et al. [3] made simulation and experimental analysis of ejector expansion transcritical CO_2 refrigeration cycle, and also found that D_t has a significant influence on the system performance. In 2016, Liu et al. [4] experimentally examined an ejector expansion CO_2 transcritical system with an adjustable ejector and found that total capacity and total COP reach maxima at D_t of 2.0 mm under extreme operating conditions. The entrainment ratio can be increased by over 30% under the proper D_t. In 2016, Wang et al. [5] adopted the constant-pressure mixing model to simulate the ejector, and found that the optimum ejector mixing pressure is a little lower than the entrained fluid's pressure, but far larger than its critical pressure. In 2018, Haida et al. [6] proposed an optimized homogeneous model, and found that the accuracy of the homogeneous model is improved by 5–10% when the pressure of the primary nozzle is greater than 59 bar. In 2018, Zhang et al. [7] used numerical simulation to study the effect of friction on ejector performance. An increase in the roughness level will essentially lead to a decline in the ejector's working performance. In 2017, Zheng and Deng [8] found that the primary nozzle efficiency decreased with the increase of the nozzle exit position. In 2018, Baek and Song [9] used a validated RANS simulation with an evaporation–condensation model found that the optimal ejector design had a maximum entrainment ratio of 0.603, which was 55% higher than the baseline model. In 2018, Taslimitaleghani et al. [10] developed a new thermodynamic model for ejectors and found the irreversibility due to friction in the nozzles and the diffuser is accounted for by polytropic efficiencies. In 2018, Li et al. [11] used the direct photography method to observe the two-phase flow in the primary converging–diverging nozzle of a transcritical CO_2 ejector. It was found that the phase change position moved upward when the primary flow inlet pressure and temperature decreased simultaneously.

However, the effect of primary nozzle diverging angle on the ejector performance has not being deeply discussed. Findings from only few studies are sometimes conflicting. For example, in 2003, Nakagawa and Morimune [12] discovered that the efficiency of R744 two-phase flow nozzle increases with the increase of the diverging angle. However, in 2012, Lawrence and Elbel [13] experimentally found that smaller diverging angles (2.3°) of the primary nozzle yield better performance than larger diverging angles (4.5°) using R134a as the working fluid. In 2017, Wang et al. [14] found that primary mass flow rate and entrainment ratio first rise moderately and then decrease steeply with the increase of convergent portion of the ejector by CFD simulation. In 2018, Baek et al. [15] found that the highest entrainment ratio was achieved with 20 mm of diverging length through the numerical simulation of five diverging lengths of primary nozzle for R134a. In this paper, the transcritical CO_2 two-phase flow in the ejector was investigated numerically by the CFD method. The distribution of pressure, velocity and two-phase volume fraction inside the ejector was analyzed. The effect of the primary nozzle diverging angle on the performance of the ejector was investigated.

2 Mathematical Models

The structure diagram of the two-phase flow ejector is shown in Fig. 1. The initial design is to confirm the size as follows: the entrance diameter of the primary nozzle (D_1), the throat diameter (D_t), the exit diameter of the primary nozzle (D_3), the constant area mixing chamber diameter (D_4), the diffuser exit diameter (D_5), the entrance diameter of the suction chamber (D_6), the constant area mixing chamber length (L_1), the divergent portion length of the primary nozzle (L_2), the diffuser length (L_3), the angle of divergent portion of the primary nozzle (α_1), the angle of convergent portion (α_2), the angle of divergent portion (α_3). All above are basing the design method of the two-phase flow ejector from Tang [16].

The mathematical model can show the relationships between each physical quantity. The calculating can be finished according to three basic laws: continuity equations, momentum equations and energy equations.

The continuity equations for the mixture are:

$$\frac{\partial \rho_m}{\partial t} + \nabla \cdot (\rho_m u_m) = 0 \qquad (1)$$

$$u_m = \frac{\sum_{r=1}^{n} \alpha_r \rho_r u_r}{\rho_m} \qquad (2)$$

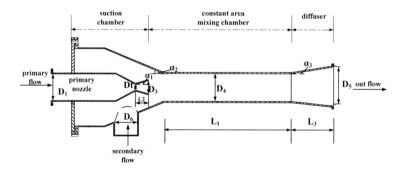

Fig. 1 Structure diagram of two-phase ejector

$$\rho_m = \sum_{r=1}^{n} \alpha_r \rho_r \tag{3}$$

The momentum equations for the mixture are:

$$\frac{\partial}{\partial t}(\rho_m u_m) + \nabla \cdot (\rho_m u_m u_m) = -\nabla p + \nabla \cdot [\mu_m(\nabla u_m + \nabla u_m^T)] + \rho_m g + F \tag{4}$$

$$\mu_m = \sum_{r=1}^{n} \alpha_r \mu_r \tag{5}$$

The energy equations for the mixture take the following form:

$$\frac{\partial}{\partial t}\sum_{r=1}^{n}(\alpha_r \rho_r E_r) + \nabla \cdot [\rho_r u_r(\rho_r E_r + p)] = \nabla \cdot (k_{\text{eff}}\nabla T) + S_E \tag{6}$$

$$E_r = h_r - \frac{p}{\rho_r} + \frac{u_r^2}{2} \tag{7}$$

The standard k-ε turbulence model equation is used as follows:

$$\frac{\partial}{\partial t}(\rho_m k) + \frac{\partial}{\partial x_i}(\rho_m k u_m) = \frac{\partial}{\partial x_j}\left[(\mu_m + \frac{\mu_t}{\sigma_k})\frac{\partial k}{\partial x_j}\right] + \tau_{ij}\frac{\partial u_m}{\partial x_j} - \rho_m \varepsilon \tag{8}$$

$$\frac{\partial}{\partial t}(\rho_m \varepsilon) + \frac{\partial}{\partial x_i}(\rho_m \varepsilon u_m) = \frac{\partial}{\partial x_j}\left[(\mu_m + \frac{\mu_t}{\sigma_\varepsilon})\frac{\partial \varepsilon}{\partial x_j}\right] + C_{1\varepsilon}\frac{\varepsilon}{k}\tau_{ij}\frac{\partial \mu_m}{\partial x_j} - C_{2\varepsilon}\rho_m\frac{\varepsilon^2}{k} \tag{9}$$

where ρ_m is the mixture density (kg/m³), t is time (s), u_m is the mixture velocity, vector (m/s), u_r is the velocity vector of phase r (m/s), a_r is volume fraction, μ_m is the viscosity coefficient of the mixture (Pa s), ∇p is the pressure gradient (Pa/m), g is the gravitational acceleration (m/s²), F is the volume force (N/m³), E_r is the mechanical energy per unit mass (J/kg), T is temperature (K), k_{eff} is the thermal conductivity (W/m K), S_E is the heat source for fluid (J), p is pressure (Pa), h_r is the enthalpy for phase r (J/kg).

3 Results and Discussion

The simulation conditions of the ejector are as follows: (1) The boundary condition of the primary nozzle entrance was set as temperature- and pressure-free inlet. (2) The boundary condition of the secondary nozzle entrance was set as the free entrances of temperature and pressure. (3) The boundary condition of the diffuser exit was set as the pressure outlet. (4) The boundary condition of the ejector wall surface was set as non-slip wall. (5) The two-phase fluid was assumed to be homogeneous. (6) The working fluid is CO_2. (7) The evaporating and condensing temperature is −5 and 40 °C, respectively.

The ejector model validation with the experimental data of Bilir et al. [17] is shown in Fig. 2. Bilir et al. [17] carried out a comparative experiment on the influence of changing the outlet pressure of the condenser on the entrainment ratio. The entrainment ratio calculated by experiment was compared with that calculated by numerical simulation model to verify the accuracy of the calculation process. It was found from Fig. 2 that the maximum relative error between numerical entrainment ratio and experimental data is 9.4%, which is within the acceptable range, indicating that the calculation process is reasonable and credible.

The flow line and the velocity magnitude inside the ejector are shown in Figs. 3 and 4, respectively. It is found that the velocity increases gradually just the working fluid enter the primary nozzle and rises considerably at the throat, then achieves the maximum at the exit of the primary nozzle. The outlet velocity of the mean primary nozzle is about 168 m/s. With the producing of the shock wave, the high-speed liquid–gas fluid from the primary nozzle and the low-speed gas from the suction chamber mixed initially in the convergent section of the mixing chamber. It can be seen that the mean speed of working fluid is about 130 m/s in the convergent section of the mixing chamber. The mixed fluid continues mixing in the constant area of the mixing chamber. The mean velocity in the constant area of the mixing chamber is about 27 m/s. At the end, the velocity of the mixed fluid decreases gradually in the diffuser.

The pressure distribution inside the ejector is shown in Fig. 5. It can be found that after the high-pressure flow enters the primary nozzle, the pressure drops off sharply near the throat and reaches the minimum at the exit of the primary nozzle. After that,

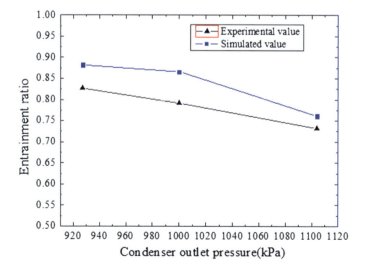

Fig. 2 Model verification

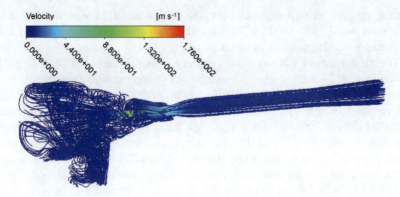

Fig. 3 Three-dimensional flow line in ejector

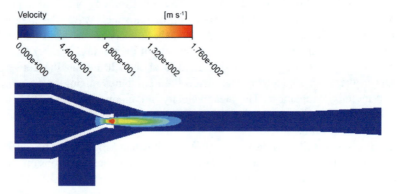

Fig. 4 Velocity–magnitude distribution in ejector

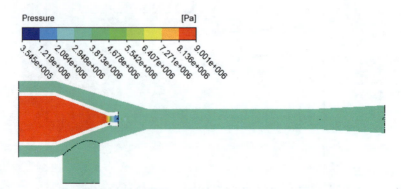

Fig. 5 Pressure distribution in ejector

the pressure shows step change owing to production of the shocking wave during the two streams preliminary mixing in the convergent section of the mixing chamber. And then, the mixed flow's shock wave disappears gradually in the constant area mixing chamber. The pressure rises gradually after the flow enters the diffuser.

The volume fraction of the gas inside the ejector is shown in Fig. 6. It is shown that phase change occurred near the primary nozzle throat, and the CO_2 gas volume fraction achieves 0.2 at the throat. After that the volume fraction increases sharply because of the dropping off of the pressure and achieves the maximum at the exit of the primary nozzle.

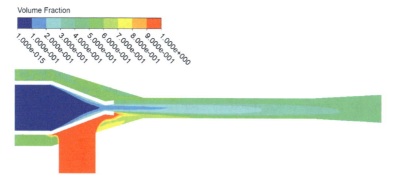

Fig. 6 Gas volume fraction in ejector

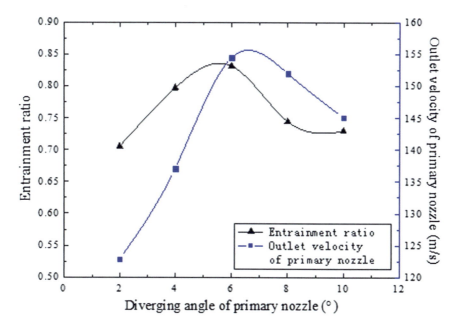

Fig. 7 Effect of diverging angle on entrainment ratio and primary nozzle outlet velocity

The ejector performance is affected by the shape and the size of the primary nozzle. When the outlet diameter of the primary nozzle remains unchanged, the variation of the diverging angle of the primary nozzle will affect the divergent section length of the primary nozzle. When the divergent section length of the primary nozzle is small, the velocity of the fluid cannot be sufficiently increased due to the underexpansion of the fluid in the primary nozzle. As the divergent portion length of the primary nozzle increases, the wall friction increases accordingly. The changes of the ejector entrainment ratio and the outlet velocity of the primary nozzle with the diverging angle of the primary nozzle are shown in Fig. 7. When the throat diameter and the exit diameter of the primary nozzle are constant, the outlet velocity of the primary nozzle rises up at the beginning and drops down with the increase of the diverging angle of the primary nozzle. There is an optimum diverging angle making the transformation efficiency inside the nozzle achieves the maximum. With the diverging angle raising up, the length of divergent section of the primary nozzle becomes shorter and the entrainment ratio raises up at the beginning and drops down after that. There is an optimum diverging angle for the primary nozzle making the entrainment ratio achieves the maximum. The optimum diverging angle is about 6° where the entrainment ratio and the primary nozzle outlet velocity o is about 0.83 and 154 m/s, respectively.

4 Conclusions

The flowing characteristics of the CO_2 in the ejector were investigated numerically, and the following conclusions can be drawn from this paper:

(1) The maximum relative error between numerical entrainment ratio of the model and experimental data is 9.4%, which is within the acceptable range, indicating that the built model in this paper is reasonable and credible.
(2) The velocity increases gradually just the working fluid entering the primary nozzle and rises considerably at the throat, then achieves the maximum at the exit of the primary nozzle. The mean primary nozzle outlet velocity is about 168 m/s. The pressure drops off sharply near the throat after the high-pressure flow enters the primary nozzle. The phase change occurred near the primary nozzle throat, and the CO_2 gas volume fraction achieves 0.2 at the throat.
(3) There is an optimum diverging angle making the transformation efficiency inside the nozzle achieves the maximum. The optimum diverging angle is about 6° where the entrainment ratio and the primary nozzle outlet velocity are about 0.83 and 154 m/s, respectively.

Acknowledgements The project is supported by Scientific and Technological Research Projects of Universities in Hebei Province (ZD2017061), Hebei Province Construction Science and Technology Research Project (2017-131), supported by the Graduate Student Innovation Fund of North China University of Science and Technology (2019S21).

References

1. Zhang, Z., Tian, L.: Effect of suction nozzle pressure drop on the performance of an ejector-expansion transcritical CO_2 refrigeration cycle. Entropy **16**(8), 4309–4321 (2014)
2. Liu, F., Li, Y., Groll, E.A.: Performance enhancement of CO_2 air conditioner with a controllable ejector. Int. J. Refrig. **35**(6), 1604–1616 (2012)
3. Hu, J., Shi, J., Liang, Y., et al.: Numerical and experimental investigation on nozzle parameters for R410A ejector air conditioning system. Int. J. Refrig. **40**, 338–346 (2014)
4. Liu, F., et al.: Comprehensive experimental performance analyses of an ejector expansion transcritical CO_2 system. Appl. Therm. Eng. **98**, 1061–1069 (2016)
5. Wang, F., Li, D.Y., Zhou, Y.: Analysis for the ejector used as expansion valve in vapour compression refrigeration cycle. Appl. Therm. Eng. **96**(5), 576–582 (2016)
6. Haida, M., et al.: Modified homogeneous relaxation model for the R744 transcritical flow in a two-phase ejector. Int. J. Refrig. **85**, 314–333 (2018)
7. Zhang, H., et al.: Influence investigation of friction on supersonic ejector performance. Int. J. Refrig. **85**, 229–239 (2018)
8. Zheng, L., Deng, J.: Research on CO_2, ejector component efficiencies by experiment measurement and distributed-parameter modeling. Energy Convers. Manag. **142**, 244–256 (2017)
9. Baek, S., Song, S.: Numerical study for the design optimization of a two-phase ejector with R134a refrigerant. J. Mech. Sci. Technol. **32**(9), 4231–4236 (2018)
10. Taslimitaleghani, S., et al.: Modeling of two-phase transcritical CO_2 ejectors for on-design and off-design conditions. Int. J. Refrig. **87**, 91–105 (2018)
11. Li, Y., et al.: Visualization of two-phase flow in primary nozzle of a transcritical CO_2 ejector. Energy Convers. Manag. **171**, 729–741 (2018)
12. Nakagawa, M., Morimune, Y.: Subsequent report on nozzle efficiency of two-phase ejector used in carbon dioxide refrigerator. Therm. Sci. Eng. **11**, 51–52 (2003)
13. Lawrence, N., Elbel, S.: Experimental and Analytical Investigation of Automotive Ejector Air-Conditioning Cycles Using Low-Pressure Refrigerants. International Refrigeration and Air Conditioning at Purdue, West Lafayette, USA (2012)
14. Wang, L., et al.: Numerical study on optimization of ejector primary nozzle geometries. Int. J. Refrig. **76**, 219–229 (2017)
15. Baek, S., et al.: Numerical study of high-speed two-phase ejector performance with R134a refrigerant. Int. J. Heat Mass Transf. **126**, 1071–1082 (2018)
16. Tang, B.: Study on Compression-Ejection Refrigeration System and Ejector. Zhengzhou University (2013)
17. Bilir Sag, N., et al.: Energetic and exergetic comparison of basic and ejector expander refrigeration systems operating under the same external conditions and cooling capacities. Energy Convers. Manag. **90**, 184–194 (2015)

3D Numerical Simulation on Flow Field Characteristic Inside the Large-Scale Adjustable Blade Axial-Flow Fan

Lin Wang, Kun Wang, Nini Wang, Suoying He, Yuetao Shi and Ming Gao

Abstract To explore a higher efficiency and lower noise operating condition of adjustable axial-flow fans, a three-dimensional (3D) numerical model is established to study the inside flow field distribution characteristics. Thus, it is indispensable extremely to study the flow field distribution characteristics inside axial-flow fans. In this study, the adjustable blade is installed at seven different angles $\Delta\beta = 0°, \pm 4°, \pm 8°, \pm 12°$ to investigate the effects on the static pressure distribution in an adjustable blade axial-flow fan. The numerical results manifested that with the increase of the installation angle, the high-pressure region of the intermediate flow surface remains unchanged, but that of the top flow surface is movable. The average static pressure of the radial flow surface increases from the blade bottom to top, and the growth rate increases sharply at the highest 30% of blade. This study can provide guidance for the variable angle operation of axial-flow fans and lay a theoretical foundation for the further study of the evolution rules of aerodynamic noise under different blade angles.

Keywords Adjustable blade · Axial-flow fan · Flow field characteristics · Static pressure distribution · Numerical simulation

1 Introduction

By adjusting the installation angle of the rotor blade, an adjustable blade axial-flow fan has inherently diverse merits, including a compact structure, a wide operating scope, and high efficiency under varying-load conditions. Meanwhile, the axial fan has a complicated three-dimensional internal viscous flow, which directly

---

L. Wang · K. Wang · S. He · Y. Shi · M. Gao (✉)
School of Energy and Power Engineering, Shandong University, Jinan, China
e-mail: gm@sdu.edu.cn

N. Wang
Shandong Electric Power Engineering Consulting Institute Corp, LTD, Jinan, China

determines the operating performance of the fan [1], so it is meaningful to investigate the internal flow field under different installation angle conditions.

For the three-dimensional flow field characteristics of axial fans, domestic and foreign scholars mainly conducted research through experimental measurement and numerical simulation. Chow et al. [2] used the laser velocimeter to explore the wake flow evolution of the upstream moving blades between the dynamic and static stages, and its interference with the downstream static blades. Aim at the adjustable blade axial-flow fan with inlet guide blade, Oro et al. [3, 4] studied the flow characteristics, turbulence intensity, and scales of the flow fan through the hot wire anemometer under steady and unsteady conditions at different operating points and the axial gap between dynamic and static impeller, respectively. Chu et al. [5] experimentally investigated the effect of the shape of the pillars at the downstream of the rotor on the static pressure characteristics and internal flow of the fan. With regard to the morphology of the blades, Nho et al. [6] experimentally studied the effect of tip chamfering, Zhang et al. [7] proposed the cavity design method of the blade, and verified the rationality of the optimization design through experiments, which discovered the reduction of unstable flow factors such as secondary flow and reflow in the operation of the fan, thereby improved the static pressure and working efficiency of the fan.

With the continuous upgrading of computer hardware system and the improvement of numerical algorithms, more and more attention has been paid to the numerical simulation of steady and unsteady flow field. Through numerical simulation, Li [8] found the distribution laws of total pressure at the design installation angle. Additionally, Chen et al. [9] investigated the effects of non-uniform tip clearance. In order to improve wake flow and solve the boundary layer separation problem, trailing edge serrated blades [10] and perforated blades [11] have been applied. Tong et al. [12] compared the upright blades and the leading edge wave guide blades, and derived that the wave shape reduces the flow vortices at the leading edge position and the pressure fluctuations in the flow field. Liu et al. [13] studied the influence of the exit angle of the front guide vane on the overall aerodynamic performance. It was found that under the rated working conditions, as the outlet angle increases, the static pressure and efficiency of the fan firstly increase and then decrease.

Based on the above summary, the study on the internal flow field of axial-flow fans mainly focuses on rotor–stator interaction, blade shape, inlet guide blades, tip slotting [14] etc. Furthermore, the research on radial characteristic flow surface is only for a single installation angle. The comparative study of the same radial flow surface under different installation angles has not been involved. Therefore, this paper establishes a three-dimensional geometric model of a large-scale adjustable blade axial-flow fan, and then performs the numerical simulation research for the pressure characteristics of the radial flow surface under varying installation angles. This research can provide significant guidance for the structural optimization design of the axial fan and lay a theoretical foundation for the subsequent study of the aerodynamic noise.

2 Geometric Model

The complete axial fan model consists of mainly four components: a suction inlet, an impeller, an outlet guide blade, and a diffuser, which are shown in Fig. 1, and the specific geometric parameters are given in Table 1.

3 Numerical Model and Calculation Method

The hybrid meshing method is selected in this study due to the complex structure of the axial fan. Specifically, the suction inlet, guide blade, and diffuser are meshed by structured grid and the impeller by unstructured one. In order to eliminate the effect of the grid number on the modeling and ensure the simulation accuracy, the grid independence is checked by the total pressure of outlet using four sets of grids. It can be seen that when the number of grids reaches 5.5 million, the calculated total pressure of the outlet is close to the design value of 4354 Pa. Therefore, it is selected for the following calculations. Comparing the numerical results with design parameters of the axial fan, both the absolute error and relative error (4.1%) are acceptably accurate (Fig. 2).

In this numerical research, the steady flow is simulated by the calculation of Reynolds-averaged Navier–Stokes equations coupled with standard k-epsilon turbulence model. The standard wall function was used to model the flow near the wall. The SIMPLE algorithm is adopted to couple pressure and velocity. The second-order upwind difference scheme is selected to discrete the governing equations.

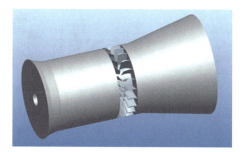

Fig. 1 Flow fan shaft geometric model

Table 1 Axial-flow fan geometry parameters

Parameter	Numerical value
Impeller diameter, mm	2628
Wheel ratio	0.61
Blade tip clearance, mm	4
Rated speed, rpm	990
Number of moving blades	22

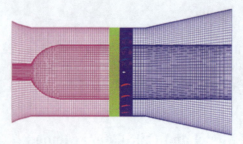

Fig. 2 Axial-flow fan calculation grid

4 Simulation Results and Analysis

As shown in Fig. 3, the installation angle denoted as β is defined as intersection angle between chord at the root blade and the inverse direction of circumferential velocity.

$\beta_0 = 73°$ is the design working condition, expressed as $\Delta\beta = 0°$. This following numerical study majorly focuses on the comparison of the flow field distribution characteristics $\Delta\beta = 0°, \pm 4°, \pm 8°, \pm 12°$.

In order to analyze the radial flow characteristics in the impeller, the annular cascade flow surfaces are selected in the radial direction, as shown in Fig. 4.

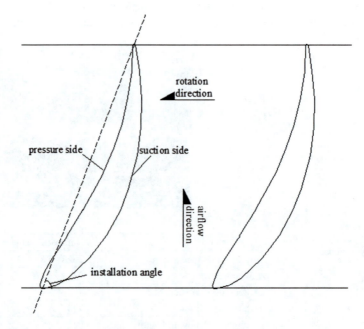

Fig. 3 Blade installation angle diagram

Fig. 4 Schematic diagram of radial cross section

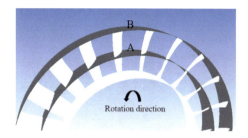

Position A represents the intermediate flow surface (50% of blade height), while position B represents the surface between the blade tip and the casing.

Figure 5 depicts the static pressure distribution on position A. It can be seen that the static pressure in the flow passage decreases from the pressure side to the suction one, at the leading edge of which the high-pressure region and the low-pressure region appear separately. Because of the great pressure difference, it is easy to generate airflow disturbance leading to high turbulence intensity, which is one of the main causes of aerodynamic noise.

As can be seen from the figure, the static pressure value ranges from −10,000 to 5000 Pa. As the installation angle changes from $\Delta\beta = -12°$ to $\Delta\beta = 12°$ with an interval of 4 degrees, the maximum static pressure at the leading edge increases monotonically from 2600 to 5000 Pa. There is a wide spectrum of negative pressure region at the leading edge of the suction side, and the changing amplitude of pressure under different installation angles is relatively small. In addition, a smaller negative pressure region also occurs at the trailing edge of the blade. This is caused by the energy loss caused by the wake effect of the trailing edge. No matter how the working condition changes, the pressure value is always much smaller than that at the trialing edge of the suction side.

Figure 6 shows the static pressure distribution at position B. The pressure gradually decreases from the pressure side to the suction side, which forms adverse pressure gradient in the tip region. The adverse pressure gradient could interrupt the mainstream inducing a secondary flow loss, which is another factor causing flow-induced noise.

The static pressure value of position B lies in the range of −9000 to 6000 Pa at the studied installation angles. With the increase of the installation angle, the high-pressure region at the trailing edge moves toward the leading edge, and the pressure value increases from 3600 to 6000 Pa. The maximum pressure difference appears in the process of adjusting installation angle from 8° to 12°. The negative pressure region near the leading edge does not change that much. In detail, the pressure value decreases from −7133 Pa at $\Delta\beta = -12°$ to −9000 Pa at $\Delta\beta = 4°$ and then increases to −7333 Pa at $\Delta\beta = 12°$. The turning point is nearly $\Delta\beta = 4°$.

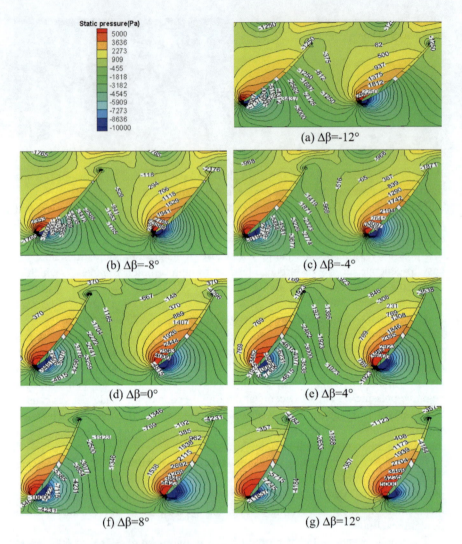

Fig. 5 Static pressure distribution of surface A

In order to thoroughly investigate the variation laws of radial static pressure in the impeller region, 11 annular flow sections are evenly intercepted from the blade root to the top. The parameter R_p, equals to $(r - r_0)/(R_0 - r_0)$, is used to describe the position information on the monitoring surface. Here, r represents the radius of the monitoring surface, r_0 is the blade root radius, and R_0 stands for the outer-shell radius. The variation of static pressure values for varying monitoring surfaces are revealed in Fig. 7.

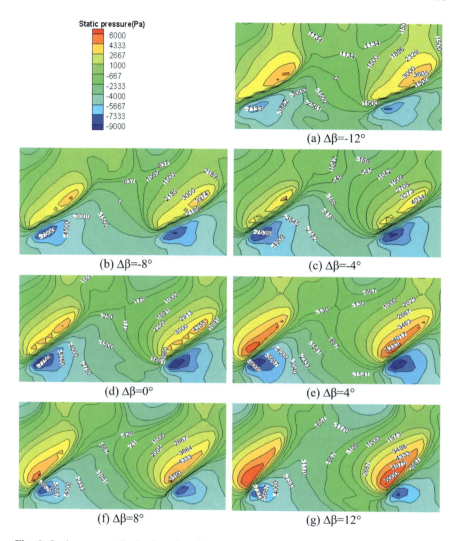

Fig. 6 Static pressure distribution of surface B

It can be seen from Fig. 7 that the static pressure of any installation angles increases with the rise of the blade height, and the growth rate is greatly improved in the highest 30% of blade. The static pressure difference of the same flow surface between two adjacent working conditions decreases with the augment of blade height. With the change of installation angle, the static pressure value increases nonlinearly from the blade root surface to the middle of the blade, and then to the blade top surface.

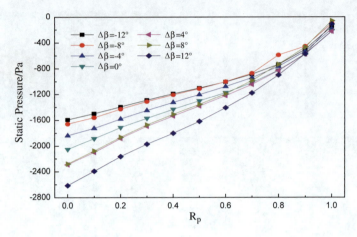

Fig. 7 Impeller radial static pressure diagram

5 Conclusions

In order to reveal the flow characteristics of the axial-flow fan impeller, this study simulates and analyzes the static pressure distributions of the intermediate flow surface and blade top flow surface under different installation angles [15]. Moreover, the changing rules of the static pressure of the radial flow surface along the blade height are studied.

1. With the increase of the installation angle, the high-pressure region of the intermediate flow surface remains unchanged, while that of the top flow surface changes.
2. The average static pressure of the radial flow surface increases with the climb of blade height, and the growth rate increases sharply in the highest 30% of blade. For the same annular flow surface, the static pressure difference between two adjacent working conditions gradually decreases as the blade height rises.
3. The fixed pressure in the flow passage is reduced from the pressure side to the suction side, and the high-pressure region and the low-pressure region are, respectively, present at the leading edges on both sides, which is one of the main causes of aerodynamic noise.
4. The pressure gradually decreases from the pressure side to the suction side, forming a reverse pressure gradient in the tip region, which is one of the main sources of noise caused by the flow.

References

1. Ye, X., et al.: Numerical simulation of internal flow characteristics of axial fan with rear guide vane. Therm. Energy Eng. **24**(02), 163–166 (2009)
2. Chow, Y., et al.: Flow non-uniformities and turbulent "hot spots" due to wake-blade and wake-wake interactions in a multistage turbomachine. J. Turbomach. **124**(4), 553–563 (2002)
3. Oro, J., et al.: Unsteady flow and wake transport in a low speed axial fan with inlet guide vanes. J. Fluids Eng. **129**(8), 1015–1029 (2007)
4. Oro, J., et al.: Numerical simulation of the unsteady stator-rotor interaction in a low-speed axial fan including experimental validation. Int. J. Numer. Meth. Heat Fluid Flow **21**(2), 168–197 (2007)
5. Chu, W., et al.: Simulation and experiment research of aerodynamic performance of small axial fans with struts. J. Therm. Sci. **25**(3), 216–222 (2016)
6. Nho, Y., et al.: Effects of turbine blade tip shape on total pressure loss and secondary flow of a linear turbine cascade. Int. J. Heat Fluid Flow **33**(1), 92–100 (2012)
7. Zhang, L., et al.: Numerical and experimental investigation on aerodynamic performance of small axial flow fan with hollow blade root. J. Therm. Sci. **22**(5), 424–432 (2013)
8. Li, J.: Study on Aerodynamic Characteristics of Installation Angle Anomalies of Axial Flow Ventilation Motor Blades. North China Electric Power University (Hebei) (2009)
9. Chen, Y., et al.: Effects of nonuniform tip clearance on fan performance and flow field. In: ASME Turbo Expo: Power for Land, Sea and Air, Quebec, Canada (2015)
10. Tang, J., et al.: Effects of sinusoidal sawtooth edge on the wake and aerodynamic performance of axial fan. J. Eng. Thermophys. **38**(10), 2145–2150 (2017)
11. Wu, C.: Research on Aerodynamic Noise Characteristics and Noise Reduction Mechanism of Large-Scale Marine Axial Fans. Jiangsu University of Science and Technology (2016)
12. Tong, F., et al.: On the study of wavy leading edge vanes to achieve low fan interaction noise. J. Sound Vib. **419**, 220–226 (2018)
13. Liu, Y., et al.: Effect of inlet guide vanes on the performance of small axial flow fan. J. Therm. Sci. **6**(26), 504–513 (2017)
14. Ye, X., et al.: Numerical study on the effect of tip slotting on the performance of axial fan. Proc. CSEE **35**(03), 652–659 (2015)
15. Li, C., et al.: Noise characteristics of axial flow fan with single blade installation angle anomaly. Proc. CSEE **35**(05), 1183–1192 (2015)

A Study of Wet-Bulb Temperature and Approach Temperature Based Control Strategy of Water-Side Economizer Free-Cooling System for Data Center

Jiajie Li, Zhengwei Li and Hai Wang

Abstract Data centers are specific facilities gathering quantities of servers consuming large energy for which cooling part account two-thirds. Water-side economizer (WSE) system is an effective way to reduce the cooling part consumption. In this study, an ambient wet-bulb temperature (T_{wet}) and approach temperature (T_{ap}) control strategy for WSE system is simulated in five cities in China. The result shows that T_{ap} affects the free-cooling hours and the performance characteristic of chiller and cooling tower. Under most circumstance, the optimal T_{ap} is mainly determined by chiller-cooling tower system performance characteristic while the free-cooling hours are not the dominant factor. However, when the city-ambient T_{wet} decrease, the free-cooling hours is more vital. When city-ambient T_{wet} is low enough, the energy-saving rate is higher when T_{ap} decrease because of more free-cooling hours.

Keywords Water-side economizer · Free-cooling · Control · Data center

1 Introduction

Data centers are specific facilities gathering quantities of servers which work all day through a whole year to compute and storage digital data. The most notable feature of data centers is their high energy consumption which has increased significantly

---

J. Li · Z. Li (✉) · H. Wang
School of Mechanical and Engineering, Tongji University,
201804 Shanghai, China
e-mail: zhengwei_li@tongji.edu.cn

J. Li
e-mail: jiajie_lee@qq.com

H. Wang
e-mail: wanghai@tongji.edu.cn

in recent years. According to Patterson [1], there are three main sources of energy consumption: the IT part includes all the servers and computing units, the cooling part contains all machines involved in the chilled air management, and the power supply system. In general, the cooling part costs more than one-third of the total energy consumption. Thus, how to optimize the cooling system to make it more efficient has become a challenging problem.

As an effective solution for the above problem, free-cooling is widely used in data center nowadays. Data center utilize free-cooling techniques to eliminate or reduce mechanical cooling and improve energy efficiency of cooling system when outdoor air has low temperature or enthalpy. According to Gozcu [2], free-cooling techniques vary with the system. Normally there are three kinds of free-cooling system: direct air-side economizer (ASE), indirect air-side economizer (IASE) and water-side economizer (WSE). In WSE, a heat exchanger is set between condenser water loop and chilled water loop. WSE is favorable in large data centers since it is saved and relatively energy-saving.

Unlike ASE, few control strategies have been proposed for WSE system. Thus, the objective of this paper is to discuss an ambient wet-bulb temperature (T_{wet}) and approach temperature (T_{ap}) based control strategy, that when T_{wet} calculated by T_{ap} reaches the set point the cooling mode change by developing a model-based approach. In this paper, a 3000 kW large data center chilled by a water cooling chiller with WSE is modeled and simulated in five different locations in China on the DYMOLA software. The result shows that T_{ap} will not only affect the free-cooling hours but also the performance characteristic of chiller and cooling tower. Thus, an optimal T_{ap} set-point is found to reach the highest energy saving rate and balance these parameters. Meanwhile, as the ambient T_{wet} varies in different location, the T_{ap} set-point in different cities differ. The optimal T_{ap} is lower when T_{wet} decrease.

2 Methods

To complete the study, models of an ideal data center are established and simulated. The simulated data center is set to locate in Shijiazhuang in Hebei province, China with a constant 3000 kW load. According to ASHRAE TC9 (2011) [3], data center operating environment is defined into different classes. In this study, the inlet air temperature is set to 24 °C and 50%RH. The temperature rise is set to 10 °C to calculate the supply air flow rate. The inlet air is supplied by a blower. Because the load and inlet air conditions are constant, the air flow rate is also constant. Thus, the power consumed by a fixed air flow rate blower is constant which will not change whether free-cooling mode is used. Under this circumstance, this part of energy will not be considered in this study. The simulations and analysis will focus on cooling system.

The cooling system of the simulated data center is water-cooling system. To make the simulations close to reality, the chiller was given the performance parameters of YORK-YK4396 (water-cooling chiller manufactured by YORK). And the supplied chilled water is set to 13 °C, return chilled water 19 °C. The mass flow rate of chilled water can be calculated. A variable speed chilled water pump (CHW pump) is taken from the Wilo products to supply the chilled water. For cooling water, though the return temperature changes with outdoor wet bulb temperature, the difference between supply and return cooling water is set constant to 5 °C. Also, the cooling water mass flow rate can be calculated and a variable-speed cooling water pump (CW pump) is chosen from the Wilo products. The cooling tower with variable speed fan use the York calculation for the rated capacity and rated approach temperature.

To calculate the energy-saving rate, a baseline case is necessary. The cases of free-cooling simulation results will be compared with the baseline result. The baseline case uses normal water-cooling system without WSE. The heat rejection loop only includes condenser and cooling tower. And the chilled water is generated by chiller all around the year.

WSE free-cooling case is shown as in Fig. 1. The biggest difference between baseline and free-cooling case is a 0.8 constant efficiency heat exchanger (HX) added into the cooling plant. Also, 6 valves are set in system to switch the cooling mode when the ambient air condition reaches the switch point. There are three modes in free-cooling system. In normal mode, the HX doesn't work and the chilled water is generated by chiller just like baseline. In partial free-cooling mode, the return water will be pre-cooling by HX and then chilled by chiller. Pre-cooling reduce the load then reduce the energy chiller consuming. In free-cooling mode, the chiller is off. The water is chilled by HX and the heat taken by low temperature ambient air via cooling tower.

Cooling mode switch controlling depends on the temperature of cooling water. Considering the heat exchange efficiency, when $T_{CWR} \leq 12\,°C$, the system can switch to free-cooling mode. When $T_{CWR} > 18\,°C$, the system can switch to normal mode. When $12\,°C < T_{CWR} < 18\,°C$, however, T_{CWR} is not a constant which is decided by ambient wet-bulb temperature (T_{wet}) and approach temperature (T_{ap}). The relationship within CWR temperature T_{CWR}, approach temperature T_{ap} and ambient wet bulb temperature T_{wet} is shown in Eq. (1).

$$T_{CWR} = T_{wet} + T_{ap} \qquad (1)$$

Thus, the direct controlling parameter is T_{wet} and T_{ap}. When T_{wet} reaches the switch point which calculated by Eq. (1), the cooling mode change. Detailed control strategy is shown in Table 1 and Fig. 2.

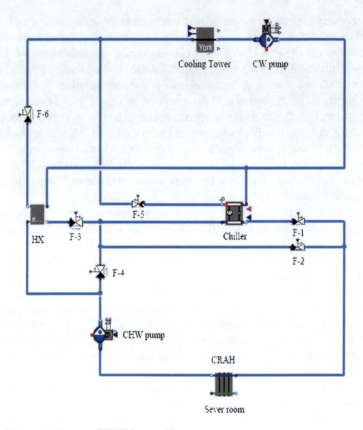

Fig. 1 Schematic diagram of WSE free-cooling system

Table 1 WSE free-cooling system control strategy

Free-cooling	Partial free-cooling	Normal
• $T_{wet} \leq 12\,°C - T_{ap}$ • The chiller is off • F-2, F-3, F-6 is on, F-1, F-4, F-5 is off. As shown in Fig. 2a • Cooling tower fan speed is controlled to maintain the chilled water at 13 °C	• $12\,°C - T_{ap} < T_{wet} \leq 18\,°C - T_{ap}$ • The chiller is on • F-1, F-3, F-5, F-6 is on, F-2, F-4 is off. As shown in Fig. 2b • Cooling tower fan speed is controlled to maintain $T_{CWR} = T_{wet} + T_{ap}$	• $T_{wet} > 18\,°C - T_{ap}$ • The chiller is on • F-1, F-4, F-5 is on, F-2, F-3, F-6 is off. As shown in Fig. 2c • Cooling tower fan speed is controlled to maintain $T_{CWR} = T_{wet} + T_{ap}$

According to Cooling Technology Institute (CTI), T_{ap} should be greater than 3 °C. Because the cooling tower is designed with a 4 °C designed T_{ap}. In order to study the relationship between energy-saving characteristic and this T_{wet} and T_{ap} based control

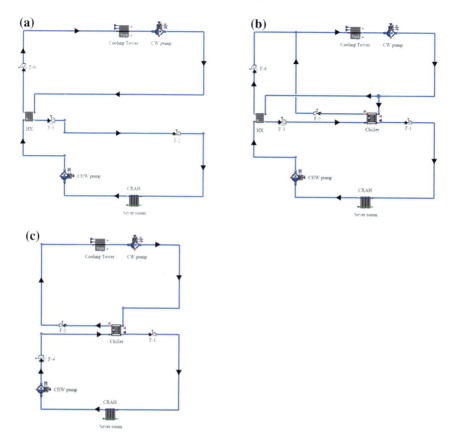

Fig. 2 Schematic diagram of three cooling mode

strategy, the model is simulated in five different cities in China: (From West to East) Xining, Xi'an, Shijiazhuang, Nanjing, and Shanghai with five different T_{ap}: 3, 3.5, 4, 4.5, 5 °C by DYMOLA software.

3 Results

The energy consumption of each part in cooling system is simulated. The results show that WSE free-cooling system can significantly save energy in the modeled climate zone. Average energy-saving rate (The average saving rate of 5 T_{ap} models in the same city) and average free-cooling hours (including free-cooling and partial free-cooling) in five locations are shown in Fig. 3. It is obvious that energy-saving rate depends the time that WSE system works. Meanwhile, we can find that some

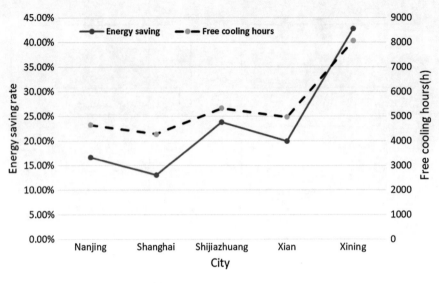

Fig. 3 Average energy saving rate and free-cooling hours in five cities

cites have a tiny difference in free-cooling hours while differ more in energy saving rate like Nanjing (16.65%-4896 h) and Shanghai (13%-4530 h).

To study the energy consumption of each part in cooling system at different T_{ap}, the year energy consumption of the model set in Shijiazhuang is chosen and the result is shown in Fig. 4. It is apparent that the most difference at different T_{ap} is the energy consumed by chiller and cooling tower fan. Chiller consumption increases

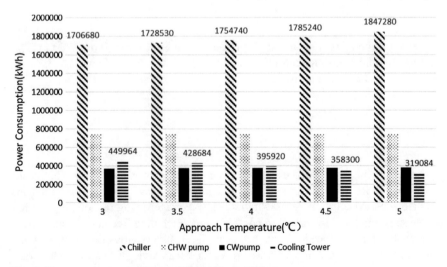

Fig. 4 Year energy consumption at different T_{ap} in Shijiazhuang

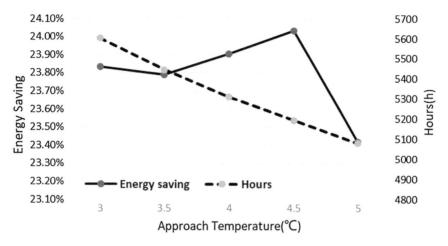

Fig. 5 Year energy saving rate and free-cooling hours at different T_{ap} in Shijiazhuang

when T_{ap} is higher while cooling tower fan consumption decrease. Energy consumed by pumps rarely changes. In Fig. 5, the energy-saving rate and free cooling hours at different T_{ap} in Shijiazhuang are compared. Free cooling hours decrease when T_{ap} grows which means Chiller may cost more power. However, Energy-saving rate peaks when T_{ap} is 4.5 °C with only 5193 h. This result indicates that the free-cooling hours is not the decisive factor for energy saving rate and chiller performance has a great influence.

In Fig. 5, we can find that the energy-saving rate decrease when T_{ap} from 3 to 3.5 °C and then increase. However, in other cities with different T_{wet}, the result change. Year energy-saving rate at different T_{ap} in five cities is shown in Fig. 6. From (a) to (e) is Shanghai, Nanjing, Xi'an, Shijiazhuang and Xining and the average T_{wet} in these five cities decreases in this order. First, it is obvious that the energy-saving rate of WSE system is much higher when T_{wet} is low. In Shanghai the energy-saving rate is about 13% while in Xining is over 40%. In Shanghai and Nanjing, energy-saving rate grows when T_{ap} from 3 to 4.5 °C. However, in Xi'an and Shijiazhuang there is a decreasing part. Different from other cities, energy saving rate declines when T_{ap} rise in Xining whose T_{wet} is the lowest and free-cooling hours is the highest (about 8000 h a year). This result indicates that when the ambient T_{wet} is lower enough, the free-cooling hours are more vital for the WSE system, and the energy-saving rate is higher when T_{ap} is lower. Under most circumstance, the optimal T_{ap} is determined by chiller and cooling tower performed characteristic and the free-cooling hours is a secondary consideration.

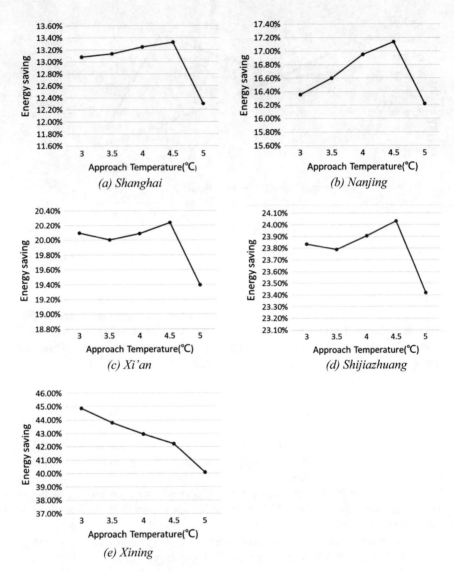

Fig. 6 Year energy-saving rate at different T_{ap} in five cities

4 Discussion

The energy-saving rate will change when T_{ap} vary because the energy-consuming performance of chiller and cooling tower are influenced by T_{ap}, and here is a balance between chiller and cooling tower performance. According to Chia-Wei Liu, at the same T_{wet}, the cooling tower consumes more energy when T_{ap} declines while the chiller consumes less. Thus, there is a best point to reach the highest system performance factor and when T_{wet} is lower this optimal T_{ap} is lower. This could be verified by the above data. But for WSE system, T_{ap} also influent the free-cooling hours. For WSE system in data center, both of these two aspects should be concerned. As shown in Fig. 6, T_{wet} in (a) to (e) decrease, so the influence of the free-cooling hour has more influence, which explains the decline part in (c) and (d). And in (e), free-cooling hour is the major factor for WSE system.

Meanwhile, the simulation shows that WSE system for data center in China has significant energy-saving effectiveness in China, especially in north and west China.

5 Conclusions

In this study, we can find that WSE free-cooling system has a great potential on energy saving, and energy-saving rate is higher in north China and west China with more free-cooling hours. An ambient wet-bulb temperature (T_{wet}) and approach temperature (T_{ap}) based control strategy are studied and the result shows that T_{ap} will not only affect the free-cooling hours but also the performance characteristic of chiller and cooling tower. Under most circumstance in China, the optimal T_{ap} is determined by chiller and cooling tower performance characteristic while the free-cooling hours is a secondary consideration, and the energy-saving rate—T_{ap} curve is not monotonous. In some cities where the ambient T_{wet} is lower enough, the free-cooling hours is more vital for the WSE system, and the energy saving rate is higher when T_{ap} decrease.

References

1. Patterson, M.K., et al: From UPS to silicon an end-to-end evaluation of data center efficiency. In: Digital Enterprise Group Architecture and Planning (2006)
2. Gozcu, O., et al.: Worldwide energy analysis of major free cooling methods for data centers. In: 2017 16th IEEE Intersociety Conference on Thermal and Thermomechanical Phenomena in Electronic Systems (2017)
3. ASHRAE 2011. ASHRAE TC9.9: Thermal Guidelines for Data Processing Environments Expended Data Center Classes and Usage Guidance (2011)

Effect of Adding Ambient Air Before Evaporating on Multi-stage Heat Pump Drying System

Xu Jin, Peng Xu, Baorui Wang, Zhongyan Liu and Wenpeng Hong

Abstract Drying is a high energy consumption process, accounting for more than 12% of total industrial energy consumption. It is therefore urgent to develop newer energy-efficient and saving structures and processes to replace the traditional drying equipment. In this paper, a multi-stage heat pump drying system is proposed. Based on the theoretical simulation, drying energy efficiency was studied by analyzing energy grade and moisture content between the exhaust gas and ambient air. The results show that: SMER increases by up to 12% when winter ambient air is added before evaporating, and moisture condensation has some influence on SMER and water removal rate. Both temperature and relative humidity of ambient air affect the system performance. Therefore, specific control strategies for ambient air ratio during different seasons may solve the low energy recovery rate of exhaust gas, and also improve the efficiency of energy consumption.

Keywords Drying · Multi-stage heat pump · Ambient air · Moisture condensation · Saving energy

---

X. Jin (✉) · P. Xu · B. Wang · Z. Liu · W. Hong
School of Energy and Power Engineering, Northeast Electric Power University, Jilin, China
e-mail: jinxu7708@sina.com

P. Xu
e-mail: 1228983771@qq.com

B. Wang
e-mail: wangbaorui1213@163.com

Z. Liu
e-mail: llzzyy198584@126.com

W. Hong
e-mail: hwp@neepu.edu.cn

1 Introduction

Drying improves product quality in various fields of the economy, and it is the most energy-consuming stage of the processing unit. Presently, nearly 20% of the total fuel is used in the drying process, which accounts for more than 12% of the total industrial energy consumption [1]. Therefore, efficient drying processes and equipment are required to save energy. As a high-efficiency and energy-saving drying technology, heat pump drying can effectively recover the heat discharged from drying exhaust gas and ensure the quality of drying products. For this reason, extensive and in-depth studies have been performed heat pump drying. Xie et al. [2] studied different structures and working principles of heat pump drying and proposed relevant control parameters and calculation formulas. Zhang et al. [3] developed a multi-stage tandem heat pump corn drying system combining heat pipes for cold climate characteristics of Northeast China to prevent high energy consumption and high pollution in this region. The system can increase heat recovery from exhaust gas, and achieve energy-saving as well as emission. In previous studies, the effect of moisture condensation on the systems was ignored, and the moisture content between the exhaust gas and ambient air was not analyzed.

In this paper, a multi-stage heat pump drying system (MHPDS) is proposed, which has both enclosed and semi-open circulation. The drying medium can be dehumidified and heated sectionally. Meanwhile, moisture condensation and drying energy efficiency were studied by analyzing energy grade and moisture content between the exhaust gas and ambient air during different seasons. Supply air temperature and moisture content of the new system can be controlled effectively to improve the performance and extend the service life of the equipment.

2 Structure and Mathematical Simulation

2.1 Introduction the Structure of Multi-stage Heat Pump System

As shown in Fig. 1, MHPDS consists of refrigerant circulation and drying medium circulation. The refrigerant of the first two stages is R134a, and that of the third stage is R22. The drying air circulation consists of material drying section, evaporation dehumidification section, and condensation heating section. The system also contains an auxiliary condenser and an auxiliary electric heater to regulate heat effectively. It has nine air regulating valves and four variable frequency fans, which allow closed and semi-open circulation.

Effect of Adding Ambient Air Before Evaporating ...

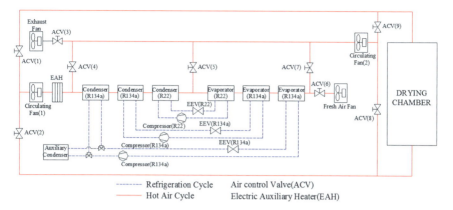

Fig. 1 Structure of multi-stage heat pump system

2.2 Mathematical Models

The performance of MHPDS can be analyzed theoretically by the Engineering Equation Solver (EES). To simplify the calculation, the following assumptions are made: (1) Outlet superheat of evaporator is 5 °C and outlet sub-cooling of condenser is 5 °C; (2) The flow loss of refrigerant is included in the efficiency of heat pump cycles, and the volume coefficient of compressors is 0.9, and the isentropic coefficients are 0.88; (3) The evaporation temperature of each heat pump is 8 °C lower than the air temperature at the outlet of evaporator, and the condensation temperature is 8 °C higher than the air temperature at the outlet of condenser.

In the numerical simulation of drying air, the relationship among mass flow rate of the drying air, volume flow rate, density, and moisture content is expressed. Based on the physical properties of air and principle of mass conservation, the mass flow before and after heat exchanger is equal when drying air passes through heat exchangers.

The conservation of energy drying air cools and dehumidifies through evaporators. The water removal rate of evaporator:

$$G_{m,e,i} h_{e,i} = G_{m,e,i+1} h_{e,i+1} + (\omega_{e,i} - \omega_{e,i+1}) G_{m,e,i} h_{\text{water},T=T_{i+1}} + Q_{\text{eva},i} \quad (1)$$

$$m_{\text{water}} = G_{m,e,i+1} \omega_{e,i+1} - G_{m,e,i} \omega_{e,i} \quad (2)$$

where $G_{m,e,i}$ and $G_{m,e,i+1}$ are dry air mass flow rate before and after evaporator, kg/h, $w_{e,i}$ and $w_{e,i+1}$ are moisture content before and after evaporator, kg/kg; $h_{e,i}$ and $h_{e,i+1}$ are enthalpy before and after evaporator, kJ/kg, $h_{\text{water},T=T_{i+1}}$ is enthalpy of precipitated water, kJ/kg. m_{water} is water removal rate of evaporator, kg/h, $Q_{\text{eva},i}$ is the cooling load of condenser, kW.

When the drying medium passes through condensers, temperature rises and moisture content does not change:

$$G_{m,c,i}h_{c,i} = G_{m,c,i+1}h_{c,i+1} + Q_{con,i} \tag{3}$$

$$\omega_{c,i} = \omega_{c,i+1} \tag{4}$$

where $G_{m,c,i}$ and $G_{m,c,i+1}$ are dry air mass flow rate before and after condenser, kg/h, $w_{c,i}$ and $w_{c,i+1}$ are moisture content of humid air before and after condenser, kg/kg, $h_{c,i}$ and $h_{c,i+1}$ are the ratio, kJ/kg, $Q_{con,i}$ is heat load of condenser, kW.

When drying air mixing with ambient air, mass conservation, and energy conservation. The mixed air also needs to follow energy conservation and mass conservation. When moisture condenses after mixing with air, the condensation of moisture in the air takes away liquid heat, so that enthalpy of the air is slightly reduced and relative humidity is 100%, reference *Air Conditioning Engineering* [4].

2.3 Evaluation Index of Heat Pump Drying System

To evaluate the performance of MHPDS under different working conditions, water removal rate and SMER are selected as evaluation indexes of system performance in this paper [5].

$$m_{\text{water}} = \sum_{i=1}^{3} m_{\text{water},i} \tag{5}$$

$$\text{SMER} = \frac{m_{\text{water}}}{\sum_{i=1}^{3} W_{\text{com},i}} \tag{6}$$

where m_{water} is water removal amount of each stage heat pump, kg, $W_{\text{com},i}$ is the power of each stage heat pump, kW.

3 Results and Discussions

Based on the refrigerant cycle principle and dry media thermodynamic calculation, assuming that drying is performed one hour, the circulation air volume is 3000 m³/h, the temperature of exhaust gas is 40 °C and relative humidity is 80%, and exhaust volume of three compressors is 9.4 m³/h.

Typical environment state parameters of Jilin City were used to draw the meteorological envelope on the psychrometric chart shown in Fig. 2. In winter, temperature range from −28 to 7 °C, and moisture content is between 0.0002 and 0.0027 g/kg, in which the span is small. Therefore, the difference in moisture content is ignored, and the moisture content of 0.0002 g/kg is chosen for calculation. In summer, the span of temperature and relative humidity is large, temperature

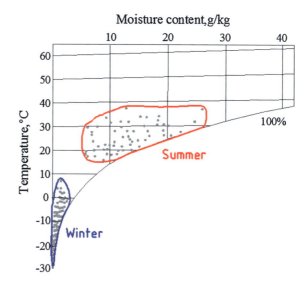

Fig. 2 The typical environmental state parameters of Jilin City

ranges from 18.5 to 36.5 °C, and relative humidity concentrates between 40 and 80%. The effects of these factors when ambient air mixes with exhaust gas cannot be ignored.

As is shown in Fig. 3, the temperature of point A is −28 °C, and moisture content is 0.0002 g/kg. Exhaust gas (point B) mixes with that of point A to point C where moisture condenses in the air, and relative humidity of the mixed air reaches 100% (point D) because it cannot be supersaturated, Then, mixed air enters evaporation zone to cool and dehumidify (from point D to point E), which finally enters condensation zone to be heated (from point E to point F).

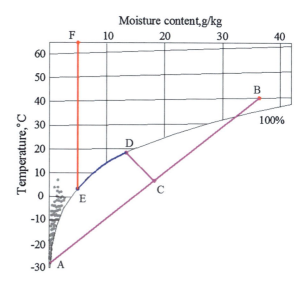

Fig. 3 Mixing process of 50% winter ambient air and 50% exhausting gas

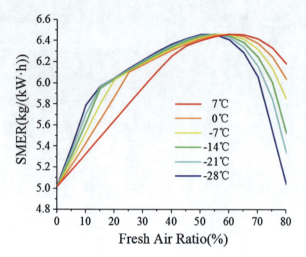

Fig. 4 The effect of different air temperatures on SMER as the fresh air ratio increases

As is shown in Fig. 4, as the amount of fresh air ratio increases, SMER increases at first and then decreases. SMER increases rapidly when no moisture condenses. Moisture condensation influences SMER state. The maximum SMER produced by fresh air of different temperatures is the same, and SMER can increase by up to 12%. When fresh air is −28 °C, the SMER value is maximum at the fresh air ratio is 50%, and when fresh air temperature is 7 °C the SMER value is maximum at the fresh air ratio is 50%. SMER falls rapidly when the fresh air ratio is more than 60%.

As is shown in Fig. 5, as the fresh air ratio increases, the water removal rate increases when no moisture condenses. Different temperatures lead to different extreme value of water removal rates when fresh air ratio is between 10 and 30%. Moisture condensation has a significant effect on water removal rate. As shown in

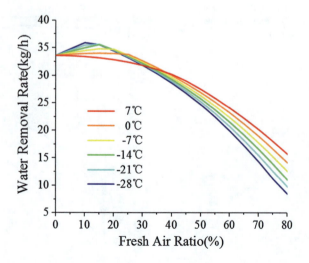

Fig. 5 The effect of different air temperatures on water removal rate as the fresh air ratio increases

Fig. 6 The effect of different fresh air temperatures on supply air temperature as the fresh air ratio increases

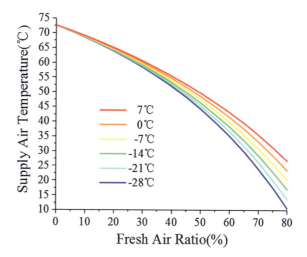

Fig. 6, supply air temperature gradually decreases as the fresh air ratio increases, and the lower the fresh air temperature, the faster the decrease in supply air temperature. When the fresh air ratio is less than 15%, variation in fresh air temperatures has little impact on the supply air temperature. The moisture condensation has no effect on the changes of supply air temperature.

As is shown in Fig. 7. The temperature of point A is 16.5 °C, and moisture content is 0.0058 g/kg. Exhaust air of drying chamber (point B) mixes with point A to point C, enters evaporation zone to release sensible heat and then releases latent heat (from point C to point D), enters the condensation zone to be heated (from point E to point F).

Fig. 7 Mixing process of 50% summer ambient air and 50% exhausting air

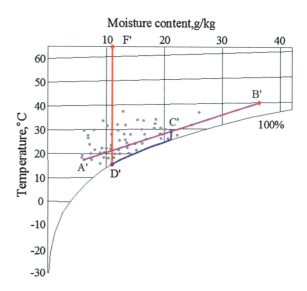

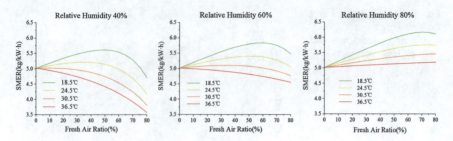

Fig. 8 The influence of summer fresh air on SMER

Figure 8 indicates that when the relative humidity of fresh air is constant, different fresh air temperatures have different effects on SMER as the fresh air ratio increases. When the fresh air temperature is the same, the effects of different relative humidity on SMER are different. When the relative humidity is 80%, regardless of the ambient temperature, SMER shows an increasing trend as the fresh air ratio increases. Both temperature and relative humidity of ambient air affect SMER when summer ambient air is added before evaporating, and the increase in SMER can reach 24%.

4 Conclusions

To improve the energy efficiency of the drying system, a new multi-stage heat pump system was proposed in this paper. The effect of ambient air which enters the system before evaporating was studied. The conclusions are as follows: When circulation air volume is 3000 m³/h, the temperature of exhaust gas is 40 °C and relative humidity is 80%, SMER tends to increase at first and then decrease as the amount of fresh air ratio increases. The maximum SMER reaches 12% when fresh air ratio is between 50 and 60% in summer, and the maximum of SMER can reach 24% when fresh air (temperature is 18.5 °C and relative humidity is 80%) ratio is 60% in winter. Moisture condensation when ambient air is added will have a significant impact on the system. Adding ambient air before evaporating can regulate supply air temperature and improve energy efficiency effectively.

Acknowledgements This work is supported by the Key Technology Research Project of Science and Technology Commission of Jilin province, China (No. 20180201006SF) and the Development and Innovation Project of Science and Technology Commission of Jilin city, China (No. 201750214).

References

1. Jiang, H., Wang, Y., Zhao, L.: Development and new technology of heat pumps at home and abroad. Build. Therm. Vent. Air Condition. **22**(12), 34–42 (2003)
2. Chen, D., Xie, J.: Heat Pump Drying Device, 1st edn. Chemical Industry Press, China (2007)
3. Li, W., Sheng, W., Zhang, Z.: Performance test of heat pipe combined with multi-stage tandem heat pump corn drying system. Trans. Chin. Soc. Agric. Eng. **34**(4), 278–283 (2018)
4. Huang, X.: Air Conditioning Engineering, 1st edn. Machinery Industry Press, China (2006)
5. Lu, Y.: Practical Heating Air Conditioning Design Manual, 1st edn. China Building Industry Press, China (2008)

Study on Moisture Transfer Characteristics of Corn during Drying Process at Low Air Temperature

Xu Jin, Chen Wang, Qingyue Bi, Zhongyan Liu and Wenpeng Hong

Abstract Traditional hot air drying is a high-energy consuming process, and the drying quality is often decreased due to high drying temperature. To improve the drying process and drying efficiency, in this paper, the heterogeneity of corn kernels' shape and physical structure is designed by modeling three-dimensional (3D) corn geometry and using different moisture diffusivity values for different corn components. The model includes evaporation of water inside the corn, the diffusion of water into the surface layer and the surface vaporization in the form of a liquid. In addition, the influence of the pericarp on the mass transfer process was included. Based on the experimental and simulation results, variation of the internal temperature and moisture content of the corn and its internal temperature field and humidity field during the drying process were analyzed. The results show that the main mass transfer resistance during drying comes from pericarp, and the influence of external conditions is not significant. It also shows that moisture diffusivity decreases from soft endosperm, hard endosperm, germ, and pericarp, in that order. The results obtained by the three-dimensional multi-component mathematical model are closer to the actual temperature and moisture content of the drying process, which proved that the model has high practicability.

Keywords Simulation · Heat and mass transfer · Hot air drying · Corn

1 Introduction

The Northeast China (Heilongjiang, Jilin, Liaoning and Inner Mongolia) regions are the main corn-producing areas in China. The corn in this region is harvested when approaching winter, and its moisture is generally about 20–30%. The corn has a

X. Jin (✉) · C. Wang · Q. Bi · Z. Liu · W. Hong
School of Energy and Power Engineering, Northeast Electric Power University, Jilin, China
e-mail: jinxu7708@sina.com

C. Wang
e-mail: wangchen371428@163.com

high moisture content after harvest, which is beneficial to the reproduction of microorganisms. Therefore, it is necessary to quantitatively analyze heat and mass transfer process inside corn. Findings from such studies will improve the drying speed and quality in addition to reducing energy consumption.

Due to the small kernel size, complex geometry, and heterogeneous structure, it is difficult to study the moisture transport in corn using purely experimental methods. Previously, numerical simulation methods were used to predict the temperature and moisture content distribution inside the corn [1–4]. During the drying process, the components of corn are changed, and the moisture diffusion rate is also altered. The interior of the corn can be categorized as a combination of hard endosperm, soft endosperm, and germ. Chen et al. [5] separated each component from the corn kernels and measured its drying curves at controlled conditions. They described the relationship between the diffusion coefficient of different components and the drying temperature and water content, which provided a theoretical basis for the selection of the diffusion coefficients of various components of corn in this paper. Based on previous studies, a three-dimensional multi-component mathematical model for the internal heat and mass transfer of corn during hot air drying was established in this paper. The model is numerically simulated by COMSOL Multiphysics to provide a more accurate temperature and moisture distribution.

2 Drying Model

2.1 Physical Model

Assuming that the corn kernels are structurally symmetrical in width direction, only half of them need to be calculated. The physical model and meshing are shown in Fig. 1. The four parts of the corn kernel are shown in Table 1 [6].

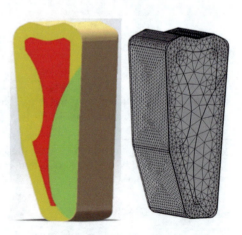

Fig. 1 Corn geometry and mesh generation

Table 1 Four components volume of corn kernel

Components	Pericarp	Germ	Soft endosperm	Hard endosperm
Volume fraction (%)	4	9	21	66

2.2 Mathematical Model

For this study, we make the following assumptions. (1) The internal temperature and moisture of corn are evenly distributed at the beginning of drying process. (2) The changes in shrinkage and shape of the corn are disregarded during the drying process. (3) The various components of corn grain are isotropic and uniform. (4) Water is diffused to the external boundary, and water vaporization occurs on the surface. Then, the temperature governing equation [7] is:

$$\frac{\partial}{\partial x}\left(\lambda \frac{\partial T}{\partial x}\right) + \frac{\partial}{\partial y}\left(\lambda \frac{\partial T}{\partial y}\right) + \frac{\partial}{\partial z}\left(\lambda \frac{\partial T}{\partial z}\right) + q_v = \rho c_p \frac{\partial T}{\partial t} + \rho h_{fg} \frac{\partial M}{\partial t} \quad (1)$$

where x, y, z represent different direction vectors of Cartesian three-dimensional coordinate system; T is the temperature inside the corn, K; λ is the thermal conductivity of corn, W/(m K); q_v is the heat rate of the internal heat source, W/m^3; ρ is the density of corn, kg/m^3; c_p is the specific heat capacity of corn, J/(kg K); h_{fg} is the latent heat of vaporization of corn, J/kg; M is the dry basis moisture content of corn, kg/kg.

Initial condition:

$$t = 0, T = T_0 \quad (2)$$

Boundary condition:

$$-\lambda \left(\frac{\partial T}{\partial x} + \frac{\partial T}{\partial y} + \frac{\partial T}{\partial z}\right) = h_t(T_w - T_a) + \rho\left[h_{fg} + c_v(T - T_w)\right]\frac{V}{A}\frac{\partial M}{\partial t} \quad (3)$$

where T_0 is the initial temperature of corn, K; h_t is the convective heat transfer coefficient, W/(m^2 K); T_a is the temperature of the drying medium, K. T_w is the surface temperature of corn, K; c_v is the specific heat capacity of water, J/(kg K); V is the volume of corn, m^3; A is the surface area of corn, m^2.

The moisture diffusion differential equation [7] is:

$$\frac{\partial M}{\partial t} = \frac{\partial}{\partial x}\left(D_i \frac{\partial M}{\partial x}\right) + \frac{\partial}{\partial y}\left(D_i \frac{\partial M}{\partial y}\right) + \frac{\partial}{\partial z}\left(D_i \frac{\partial M}{\partial z}\right) \quad (4)$$

where x, y, z represent different direction vectors of Cartesian three-dimensional coordinate system; M is the dry basis moisture content of corn, kg/kg; D_i is the effective diffusion coefficient of water for different components of corn, $i = 1, 2, 3, 4$, m^2/s.

Initial condition:

$$t = 0, M = M_0 \tag{5}$$

Boundary condition:

$$-D\left(\frac{\partial M}{\partial x} + \frac{\partial M}{\partial y} + \frac{\partial M}{\partial z}\right) = h_m(M - M_e) \tag{6}$$

where t is the drying time, s; M_0 is the initial moisture content of corn, kg/kg; h_m is the convective mass transfer coefficient, m/s; M_e is the equilibrium moisture content of corn, kg/kg.

2.3 Thermo Physical Properties and Values

The surface heat transfer coefficient is: $h_t = 40$ W/(m^2 K); the density of the corn kernel is uniform at $\rho = \rho_{germ} = \rho_{endosperm} = 1150$ kg/m^3 [8]; The surface mass transfer coefficient is: $h_m = 0.05$ m/s.

The specific heat [9] due to the moisture content is expressed as:

$$C_p = [1.46 + 3.56M/(M+1)] \times 10^3$$

The latent heat of water vaporization [8] is:

$$h_{fg} = 383650 \times [1 + 0.8953 \exp(-12.32M)]$$

The thermal conductivity [7] is:

$$\lambda = \exp\left[-1.74 - 3.56M + 4720(T-273) + 6.48M^2 - 0.00015(T-273)^2 + 0.0672M(T-273)\right]$$

The equilibrium moisture content (d.b.) [7] is:

$$M_e = \left(-\log(1 - RH_{air})/(8.654 \times 10^{-5})/(T_{air} - 73.15 + 49.81)\right)^{1/1.8634}/100$$

The diffusion coefficient of germ (d.b.) [5] is:

$$D_1 = 8.97 \times 10^{-6} \exp(-3020.2/T)M^{1.5}$$

The diffusion coefficient of hard endosperm (d.b.) [5] is:

$$D_2 = 4.08 \times 10^{-6} \exp(-2506.6/T) M^{1.5}$$

The diffusion coefficient of soft endosperm (d.b.) [5] is:

$$D_3 = 9.65 \times 10^{-7} \exp(-1588.1/T) M^{1.5}$$

The diffusion coefficient of pericarp (d.b.) [5] is:

$$D_4 = 3.69 \times 10^{-6} \exp(-3559.1/T) M^{1.5}$$

3 Discussion

In this paper, a mathematical model is numerically solved by the heat transfer module and PDE module in COMSOL Multiphysics. Using the custom control equation in the mathematical model, a simulation of the corn was carried out, and the simulation conditions were the same as the experimental conditions. The grid-independent verification shows that the conventional physical grid is used as the computational grid to meet the accuracy requirements.

Presently, the average temperature and mass average moisture content of corn are conventionally used to compare experimental results and verify the accuracy of the simulation results. The experimental results of drying were carried out using the experimental data of Fortes et al. [9]. The experimental conditions were: hot air temperature of 75 °C, air relative humidity of 4.7%, wind speed of 1.63 m/s, initial moisture content of 0.3 (d.b.), and initial temperature of 25 °C. The experimental results are mainly illustrated by the drying curve.

Figure 2 shows the comparison of the simulation results of corn moisture content and temperature field with experimental results. Single-component model has been previously used in simulations. Given that corn is a multi-component, irregularly shaped object, the results obtained by multi-component models are likely to be closer to the experimental results. Figure 2a illustrates that the three-dimensional multi-component model has higher accuracy, the simulation results obtained by using the 3D multicomponent model are close to the experimental values, the error is within ±5%, while Forte's model error is within ±6% and the max error is ±9%. As seen from Fig. 2a, during the first 7–8 min of the drying process, the average moisture content of the corn rapidly decreased, and then, the moisture content decreased more slowly as the drying process continued. Therefore, this model has high practicability and provides some theoretical guidance for perfecting the drying process and improving drying efficiency.

Figures 3 and 4 show the moisture and temperature distribution in the corn kernel at different times, respectively. Different colors represent the specific values of the moisture content. Figure 5 shows the drying curves (d.b.) for common corn and

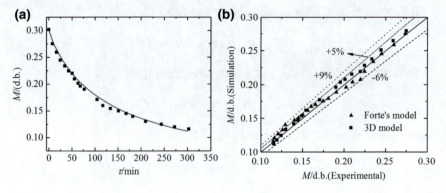

Fig. 2 Comparison of simulation with experimental results

peeled corn, which depicts that the drying rate of peeled corn is significantly higher. When considering the influence of the epidermis, the mass transfer process becomes relatively slow due to the barrier effect of the pericarp, that is, the stronger the resistance, the slower the diffusion and penetration of moisture out of the endoplasm. Therefore, the influence of the pericarp on mass transfer should not be ignored.

Figure 6 shows the change in moisture content of different components under experimental conditions. Figure 6 reveals that the pericarp has the largest dehydration rate in the early stage of drying. Given that the corn pericarp cells are single, evenly arranged, and with a relatively dense structure, which hinders the diffusion of moisture in the endoplasmic part during the drying process. Thus, the main mass transfer resistance during corn drying comes from the pericarp.

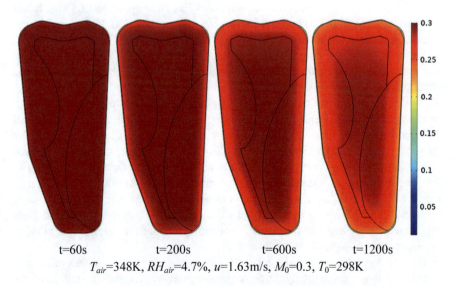

T_{air}=348K, RH_{air}=4.7%, u=1.63m/s, M_0=0.3, T_0=298K

Fig. 3 Slices of moisture distribution in corn cross section at drying temperature

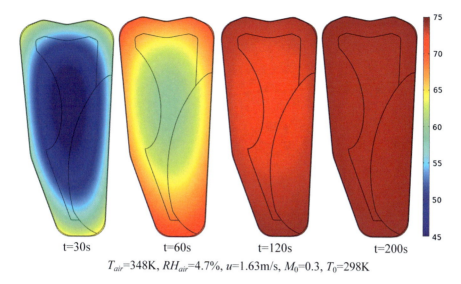

T_{air}=348K, RH_{air}=4.7%, u=1.63m/s, M_0=0.3, T_0=298K

Fig. 4 Slices of temperature distribution in corn cross section at drying temperature

Fig. 5 Drying curves (d.b.) for common corn and peeled corn

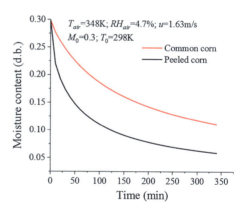

Fig. 6 Drying curves (d.b.) for different components

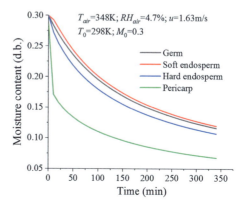

Fig. 7 Drying curves (d.b.) at different air temperatures

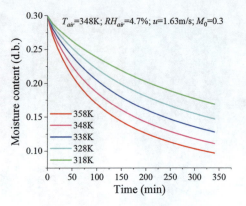

Fig. 8 Drying curves (d.b.) at different air velocities

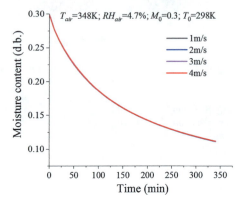

The drying curves (d.b.) at different temperatures and air velocities are shown in Figs. 7 and 8. Temperature significantly affects the mass transfer process, whereas air velocity has barely any effect. It is further revealed that the drying process of corn is determined by internal diffusion, and the external conditions have little effect. The main resistance to drying originates from inside the corn.

4 Conclusions

Compared with the previous one-component models, the results obtained by the three-dimensional multi-component mathematical model are closer to the changes in temperature and moisture content during the actual drying process, which indicates that the model has high practicability. When the drying process of corn is in the falling rate period, the flow rate of 1–4 m/s have no significant effect on the average moisture content of corn. It was found that the main resistance of drying is internal, and the influence of external conditions is not significant. The main mass transfer resistance during corn drying comes from the pericarp.

Acknowledgements This project is supported by the Key technology research Project of Science and Technology Commission of Jilin Province, China (No. 20180201006SF), and the Development and Innovation Project of Science and Technology Commission of Jilin city, China (No. 201750214).

References

1. Jia, C., Cao, C.: Process of two-dimensional heat and mass transfer in corn kernel. J. Beijing Agric. Eng. Univ. **15**(1), 45–51 (1995)
2. Haghighi, K., Irudayaraj, J., Stroshine, R.L.: Grain kernel drying simulation using the finite element method. Trans. ASAE **33**(6), 1957–1965 (1990)
3. Haghighi, K., Segerlind, L.J.: Modeling simultaneous heat and mass transfer in an isotropic sphere—a finite element approach. Trans. ASAE **31**(2), 629–637 (1991)
4. Takhar, P.S., Maier, D.E., Campanella, O.H., Chen, G.: Hybrid mixture theory based moisture transport and stress development in corn kernels during drying: validation and simulation results. J. Food Eng. **106**(4), 275–282 (2011)
5. Chen, G., Maier, D.E., Campanella, O.H., Takhar, P.S.: Modeling of moisture diffusivities for components of yellow-dent corn kernels. J. Cereal Sci. **50**(1), 82–90 (2009)
6. Muthukumarappan, K., Gunasekaran, S.: Moisture diffusivity of corn kernel components during adsorption part III: soft and hard endosperms. Trans. ASAE **37**(4), 1275–1280 (1994)
7. Zhu, W., Cao, C.: Computer Simulation of Agricultural Product Drying Process, 1st edn. China Agricultural Press, Beijing (2001)
8. Zhang, S., Kong, N., Zhu, Y., Zhang, Z., Xu, C.: 3D model-based simulation analysis of energy consumption in hot air drying of corn kernels. Math. Probl. Eng. (2013)
9. Fortes, M., Okos, M.R.: Non-equilibrium thermodynamics approach to heat and mass transfer in corn kernels. Trans. ASAE **24**(3), 761–769 (1981)

The Study on a Dual Evaporating Temperatures-Based Chilled Water System Applied in Small-to-Medium Residential Buildings

Zhao Li, Jianbo Chen, Chunhui Liu, Lei Zhang, Shangqing Yang and Xiaoyu Liu

Abstract In small-to-medium-scaled residential buildings, direct expansion air-conditioning (DX A/C) systems were commonly used in indoor environment control only indoor air temperature. The uncontrolled humidity would cause indoor health and comfort issue. In this study, a dual evaporating temperatures-based chilled water system was proposed and designed for the application in small-to-medium-scaled residential buildings. An experimental prototype was established and tested in an existing psychrometric chamber. With two suction inlets of the compressor and the cooperation of two different electronic expansion valves, two evaporating temperatures were achieved. The lower one was used to generate the low-temperature chilled water, say 4–6 °C, to handle the outdoor air, which could be used to deal with the indoor latent load. On the other hand, the higher one was used to generate high-temperature chilled water of 16–18 °C, to deal with the indoor sensible heat, so as to achieve temperature and humidity independent control. Experiments were carried out under typical operation conditions in Shanghai. The results showed that the proposed system was able to operate stably under different conditions, and the chilled water temperatures could be maintained at required level. Therefore, the system proposed is feasible and would provide a novel solution for the issue of temperature and humidity independent control in small-to-medium-scaled residential buildings.

Keywords Dual evaporating temperatures · Operational characteristics · Chilled water system · Temperature and humidity independent control · Residential building

---

Z. Li (✉) · J. Chen · S. Yang · X. Liu
School of Environment and Architecture, University of Shanghai for Science and Technology, Shanghai 200093, China
e-mail: lizhao_8749@usst.edu.cn

C. Liu · L. Zhang
Shanghai Highly Electrical Appliances Co. Ltd, Shanghai, China

1 Introduction

With the development of economy and society, the proportion of building energy consumption keeps rising rapidly nowadays. Air-conditioning systems are the main energy consumption in buildings [1]. Therefore, the energy saving of air-conditioning systems plays an important role in reducing the building energy consumption. In southern China, air humidity is relatively high all around the year. Many scholars put forward the independent temperature and humidity control system, which can control the air temperature and humidity separately and avoid the reheating energy consumption of the traditional air-conditioning system [2].

Temperature and humidity independent control technology has been developed for a few decades and there were several kinds of solutions. For realizing totally separated control of temperature and humidity, rotated wheel and liquid desiccant were usually adopted. However, additional heat or wasted heat was necessary for regenerating the desiccant, which made this method more applicable in large-scaled buildings. In some other applications, two cold sources were used to generate high- and low-temperature chilled water. But the cost was high and system configuration was complicated. Therefore, most of the current solutions for temperature and humidity independent control were suitable in large-scaled buildings.

At present, the dehumidification of small-to-medium-scaled building is usually condensation dehumidification, but it cannot meet the requirement of varying indoor heat and humidity ratio [3]. Besides, most of the residential buildings nowadays adopt the DX A/C system for indoor temperature control, leaving indoor humidity uncontrolled. The ON/OFF cycle of compressor for controlling temperature would lead to significant fluctuation of the indoor humidity, causing problems of thermal comfort, indoor air quality and energy efficiency. Therefore, a temperature and humidity independent regulating air-conditioning system based on dual evaporating temperatures was proposed in this paper, in which chilled water system was applied. The system design and system test will be introduced in the following sections.

2 System Design

2.1 Object Building

To determine the system configuration, the cooling and heating load of a representative building was simulated. The set points of indoor air states were list in Table 1. The building was located in Shanghai, which is in the climate zone of hot-summer–cold-winter. Additionally, the humidity of the territory of mid-to-down stream of Yangtze River is high all year round. Specially, in the period of late spring–early summer, the weather is of low temperature but extremely high humidity, which is the challenge of humidity control.

Table 1 Indoor settings

Mode	Dry bulb temperature (°C)	Relative humidity (%)	Moisture content (g/kg dry air)
Heating	20	40	5.8
Cooling or dehumidifying	25	55	10.9

Table 2 Building information

Building location	Shanghai	
Area (m^2)	87.3	
Number of floors	1	
Indoor height (m)	3	
Shape coefficient	0.194	
Window–wall area ratio	South	0.38
	North	0.23
Stair	Non-heating and air-conditioning area	
The balcony	Closed and connected with heating and air-conditioning area	

The simulated building is an apartment with two bedrooms and a living room. The basic information of the simulated building model is shown in Table 2.

2.2 *Cooling and Heating Loads Results*

Heating load: According to the simulation results, the maximum heating load of the building was 5167.99 W, occurring on December 23rd. At the time of the peak heating load, the outdoor dry bulb temperature was 1.24 °C, with a moisture content of 1.68 g/kg dry air.

Cooling load: As the load simulation results showed, the maximum cooling load of the building was 9714.25 W, occurring at 18:04 on June 30th, when the outdoor dry bulb temperature was 35.87 °C and moisture content 20.47 g/kg dry air. With the simulated cooling and heating loads, the air-conditioning system can be designed. However, the system we proposed in this study concerns the dehumidification progress, the humidity load should also be specifically considered, as follows.

The summer outdoor parameters for air-conditioning system design in Shanghai are 34.6 °C for the dry-bulb temperature and 28.2 °C for the wet-bulb temperature. The designated outdoor air (DOA) system plus dry fan coil unit (FCU) can be adopted to meet the indoor thermal comfort requirements. The handled outdoor air deals all the indoor humidity load and part of the sensible heat load, while the FCU only deals the indoor sensible load, so as to achieve temperature and humidity independent control. The outdoor air is cooled and dehumidified using low-temperature chilled water, say 4–6 °C, and the return air can be handled by FCU using high-temperature chilled water about 16–18 °C. Therefore, two different evaporating temperatures can be determined accordingly in the actual design process. Since the purpose of this study is to explore the feasibility of this proposed dual evaporating temperatures system, the terminal air handling devices were not considered, and only two plate heat exchangers (PHEs) were used as evaporators to produce chilled water.

With the analysis of air handling process using the psychrometric chart, in order to make outdoor air take away the indoor humidity load, the moisture content d_L of outdoor air supplied into the room should be [4]:

$$d_L = d_N - \frac{W}{\rho G_w} = 9.7 \text{ g/kg} \qquad (1)$$

where d_N is the moisture content of the indoor air, W the humidity load and G_w the outdoor air flow rate. Then, the cooling capacity Q_w required by the humidity control is:

$$Q_w = \frac{\rho G_w (i_w - i_L)}{3600} = 2.5 \text{ kW} \qquad (2)$$

where i_w and i_L are the enthalpy of outdoor air and supply air, respectively. The cooling capacity Q_e required for the temperature control is:

$$Q_e = \frac{\rho (G - G_w)(i_n - i_c)}{3600} = 6.3 \text{ kW} \qquad (3)$$

where G is total air supply. Therefore, considering certain safety margin, the cooling capacity of the low-temperature evaporator is 3 kW, and that of the high-temperature evaporator is 7 kW.

2.3 System Design

In this paper, conceptual dual evaporators chilled water system applied in temperature and humidity separately controlled air-conditioning system is proposed. In this system, the low-temperature plate heat exchanger (LTPHE) provides chilled water at low temperature to deal all the latent load and part of sensible load, and the high-temperature plate heat exchanger (HTPHE) provides chilled water at high temperature to deal the remaining sensible load. The system also includes a

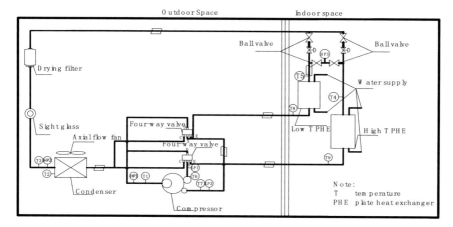

Fig. 1 Schematic diagram of the experimental rig

condenser, a receiver, a compressor with two suction inlet, four-way valves, electronic expansion valves (EEVs) and certain pipeline. The schematic of the system is shown in next section.

In summer condition, the refrigerant from the compressor is divided into two ways, by passing the four-way valves, the refrigerant of two ways enters the condenser together, then passes the receiver, the filter drier, and is divided into two ways, entering two EEVs and two different PHEs. And finally, through four-way valves, the refrigerant is sucked into the compressor with its two suction inlets, completing the refrigeration cycle. In the winter mode, the cycle is reversed using the four-way valves, and the condensing temperatures in two PHEs are the same, the system is a regular heat pump.

3 Introduction of Experiments

3.1 The Experimental Rig

The detailed configuration of the experimental dual evaporating temperatures system is shown in Fig. 1. As shown in Fig. 1, the prototype of the system is built in an existed psychrometric chamber. The chamber has two well-thermally insulated rooms, one simulates the outdoor space and the other the indoor space. As shown, two evaporators were placed in the indoor space, in the future, the terminals can also be installed to carry out related experiments. Several temperature and pressure sampling points were placed as Fig. 1 shown to monitoring the operational state of the system. All the temperatures were tested by RTD PT100, with an accuracy of 0.1 °C. The pressure of the refrigeration system was tested by pressure transducers, with an accuracy of ±0.13% of full-scale reading. All the temperature and pressure values of the system were sampled by the system controller. The water supply for two PHEs adopted the water system of the chamber with its own temperature and flow rate measure and data logging devices. The temperature and humidity in the outdoor and indoor space can be steadily maintained at required level by different air-conditioning systems of the psychrometric chamber. Then, different operational conditions can be achieved.

3.2 Experimental Conditions

Since the HTPHE is aimed to deal the sensible load, chilled water supply temperature should be set higher than the dew point of indoor air. Considering the transportation and the loss through heat transfer, the supply and return water temperatures for HTPHE were set around 18 and 21 °C, respectively. For the LTPHE, the supply water temperature should be lower than the dew point of the indoor air to realize humidity control. Considering the heat transfer loss, the supply and return water temperatures for LTPHE were set around 6 and 11 °C, respectively.

According to the water supply temperatures of two PHEs, the evaporating temperatures can also be determined. The low evaporating temperature should be lower than 4 °C, and high evaporating temperature should be lower than 12 °C. The experiments were carried out at two different outdoor conditions. Set I was the standard condition, the dry-bulb temperature was 35 °C and wet-bulb temperature 24 °C. In Set II, the dry-bulb temperature was 24 °C while wet-bulb temperature was 23 °C, representing the high-humidity condition. During these two experiments, the indoor air states were set at the design point. The sub-cooling of refrigerant was 3 °C. The degrees of superheat at two discharge points of PHEs were set at 5 °C. The experimental results were reported in Sect. 4.

4 Results and Analysis

In Set I, the frequency of the compressor was 55 Hz. The water flow rate for LTPHE was 0.338 and 2.000 m^3/h for the HTPHE. The water supply and return temperatures and evaporating temperatures of two PHEs were monitored and recorded after the system reached the steady state. The water supply and return temperatures of two PHEs are shown in Fig. 2. Evaporating temperatures are shown in Fig. 3.

As shown in Fig. 2, the water return temperature was maintained at 21.0 and 11.0 °C by the water system of the psychrometric chamber. The chilled water supplied by the tested system was at 18.6 and 7.7 °C, respectively. Furthermore, in Fig. 3, the evaporating temperature in HTPHE was stably maintained at 9.7 °C,

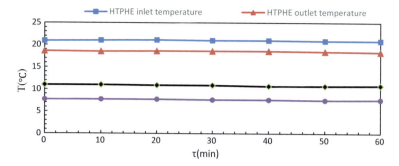

Fig. 2 Water supply and return temperatures in experiment Set I

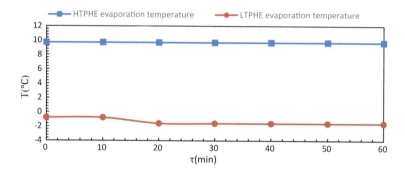

Fig. 3 Evaporating temperatures in experiment Set I

while the one in LTPHE was around −0.8 to −1.6 °C. These temperatures suggested that the system was operated stably at the standard summer condition.

Figures 4 and 5 present the water temperatures and evaporating temperatures of the system in experiment Set II, representing the high-humidity–low-temperature condition. In experiment Set II, only the outdoor air states were changed, all other experimental conditions were remained the same.

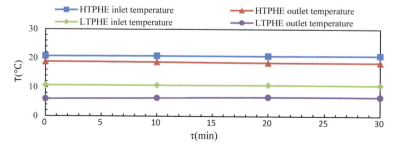

Fig. 4 Water supply and return temperatures in experiment Set II

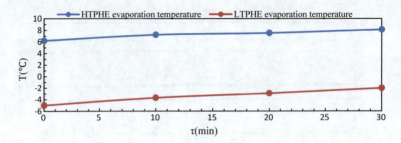

Fig. 5 Evaporating temperatures in experiment Set II

As shown in Fig. 4, water supply temperatures were maintained around 18.7 and 6.3 °C for two PHEs, respectively. And in Fig. 5, although increased slightly, the evaporating temperatures were maintained around 7.5 and −3 °C, respectively. The results of experiment Set II suggest a steady operation of the system under a relatively high-humidity but low-temperature condition.

According to the experimental results, the dual evaporating temperatures-based water system is able to provide chilled water at two different temperatures to realize the independent control of indoor air temperature and humidity. The system was operated stably at both standard operation condition and the high-humidity–low-temperature condition. From the aspect of thermal dynamics of the refrigeration cycle, the rise of the evaporating temperature in the HTPHE may result in potential energy saving. Although the operation of the system was stable in the experiment, it should be noted that the evaporating temperature in the LTPHE is relatively low and even below 0 °C, which is not a suitable operation condition for a chilled water system. However, the purpose of this paper is to prove the feasibility of the dual evaporating temperatures-based chilled water system. As to the detailed application of this system, a well-developed control strategy over the EEV and compressor is indispensable to maintain the evaporating temperatures at required level.

5 Conclusion

Through this experimental study, the following conclusions can be drawn according to the results and the analysis.

The system proposed can operate stably and reliably under both high-humidity and standard operation conditions, and the outlet water temperature of the PHEs with high and low temperature can basically meet the requirements. The refrigerating capacity of high- and low-temperature evaporator can be distributed according to the demand of dehumidification capacity and cooling capacity. Therefore, satisfying independent control of temperature and humidity can be expected with the application of this system, indicating that the proposed system is feasible.

In both experiment condition, the evaporating temperatures of LTPHE were relatively low and even below 0 °C, suggesting that a well-developed control strategy regulating EEVs and the compressor to maintain the evaporating temperature at required level is indispensable in the future application of this system on independent control of temperature and humidity. Furthermore, a more detailed operational characteristics study of this system is required in the future, including the energy performance, the dynamic distribution of cooling capacity, before this system can be actually applied.

References

1. Ma, J.: Optimal Operation of Central Air-Conditioning System. Xi'an University of Science and Technology, Xi'an (2006)
2. Liu, J., et al.: Experimental and theoretical study on a novel double evaporating temperature chiller applied in THICS using R32/R236fa. Int. J. Refrig. **75**, 343–351 (2017)
3. Jiang, Y., et al.: Solution type air conditioning and its application. HVAC **34**(11), 88–97 (2004)
4. Zhang, X., et al.: Heat Transfer Science, 5th edn. China Building Industry Press, Beijing (2007)

Study on the Inter-Stage Release Characteristics of Two-Stage Compression Heat Pump System

Xu Jin, Zhe Wu, Zhongyan Liu and Wenpeng Hong

Abstract The intermediate injection combined with two-stage compression technology can improve the low-temperature adaptability of air source heat pump, but the problem of "frosting" is still to be improved. Considered the two characteristics of variable capacity and inter-stage release, the technology of inter-stage hot gas bypass defrosting for two-stage compression is presented. In addition, a two-stage compression dynamic coupling model with inter-stage release characteristics is established. Based on the simulation model, the influences of matching relations between inter-stage release parameters and compressor capacity, the dynamic characteristics of inter-stage parameters on system performance were analyzed in the absence of injection. It aims to provide a reliable guiding significance for solving the frosting problems encountered in the promotion of two-stage compression heat pump heating technology. Based on the simulation results, there is an optimum COP when the release ratio m_{rat} = 0.1–0.3. When the release ratio m_{rat} = 0–0.6, the inter-stage release process can reduce the system heating capacity Q_{hot} by 7–37%, which can increase the system heating performance coefficient COP_{hot} by 7–11%.

Keywords Heat pump · Two-stage compression · Inter-stage release · Inter-stage injection · Dynamic coupling

1 Introduction

Air source heat pump can greatly reduce the consumption of fossil energy and greenhouse gas emissions [1]. The "Three-North" region of China is gradually promoting air source heat pump heating technology [2]. It has been used on a large scale in the southern part of China, but it is rarely used in cold regions of the north. This is because the air source heat pump has the following disadvantages in the low-temperature environment: the heating capacity does not match the heating load, the heating coefficient COP_{hot} decreases, the compressor exhaust temperature rises, and the evaporator surface is frosted. To improve the performance of air source heat pump in low-temperature environment. In recent years, scholars have conducted many researches. Navarro et al. [3] built a (quasi) two-stage compression test bench. Comparing the system performance of the injection and non-injection, it is pointed out that the injection increases the system Q_{hot} and COP by 20 and 10%. Jiang et al. [4] simulated and experimentally studied the variable capacity two-stage compression heat pump system. It is concluded that the optimal intermediate pressure for actual operation is 1.2–1, 56 times of the geometric mean pressure when the condensing conditions are fixed. Jin et al. [5] pointed out that the two-stage compression is not a continuous compression process. The intermediate pressure has pulsating characteristics. The system heating has a linear growth trend and the COP has the optimal value with the increase of the intermediate pressure. Based on the above research on the two-stage compression injection technology, it can be seen that various scholars have compared the performance of the two-stage compression system with injection characteristics from various aspects. However, the moisture deposited in the air will frost on the outdoor machine fins when the outdoor heat exchanger surface temperature is below the air dew point temperature and below 0 °C. As the frost layer increases, the heat production and heating performance gradually decrease. It is necessary to remove the frost layer on the fins when the performance of the heat pump is reduced to a certain extent. The most widely used defrosting methods are reverse cycle defrosting and hot gas bypass defrosting [6]. However, the study of the two defrosting methods is limited to the single-stage air source heat pump system, the frosting and defrosting performances of the two-stage compression air source heat pump system need further study. The hot gas bypass defrosting has the following characteristics: low noise, small fluctuations in indoor temperature, and no blowing cold feeling. Therefore, this paper designs and builds experiment bench for the discharge of heat between the stages (hot gas bypass). It is a two-stage compression heat pump system (SC-IC) with intercooler as the inter-stage structure.

2 Mathematical Model

Based on the assumptions of the model and the structural characteristics of the rotary compressor, with the reference points of crank angles θ_{cf}, θ_{vo}, θ_{vc} relative to the initial compression of compressor and the opening or closing of the exhaust valve, the model divided the suction and compression process of compressor cylinder into four phases. i means the characteristic parameter of exhaust valve opening, j means the characteristic parameter of operational phase, and the specific meaning is shown in Table 1. According to the dividing method for the operational phase of cylinder, being aimed at the intermediate control volume, the dynamic compressor coupling process was divided into eight modes.

As is shown in Table 2, there are totally eight controlling volumes in this system, and the compressor is dynamically coupled. Equations should be established for each control volume in each micro-time step. Based on the continuity equation and the first law of thermodynamics, the mass and temperature of the refrigerant in the control volume can be calculated by Eqs. (1) and (2).

$$\frac{dm}{d\Gamma} = \sum \frac{dm_{in}}{d\Gamma} - \sum \frac{dm_{out}}{d\Gamma} \qquad (1)$$

$$\frac{dT}{d\Gamma} = \frac{1}{mC}\left[-T\left(\frac{\partial P}{\partial T}\right)_v \left(\frac{dV}{d\Gamma} - v\frac{dm}{d\Gamma}\right) + \sum \frac{dm_{in}}{d\Gamma}(h_{in} - h_{out}) + \frac{dQ}{d\Gamma}\right] \qquad (2)$$

Table 1 Meaning of characteristic parameters for division of operational modes

Operating phase	Operating range of crankshaft angle	i value	j value
1	$0 \le \theta < \theta_{cf}$	0	1
2	$\theta_{cf} \le \theta < \theta_{vo}$	0	2
3	$\theta_{vo} \le \theta < \theta_{vc}$	1	3
4	$\theta_{vc} \le \theta < 2\pi$	0	4

Table 2 Operational modes of dynamic low- and high-stage compressor coupling process

Mode	Low compressor running of $1^{\#}$ cylinder	Low compressor running of $2^{\#}$ cylinder	High compressor running of the cylinder
1	$i = 0; j = 1, 2, 4$	$i = 0; j = 1, 2, 4$	$i = 0; j = 1, 2, 4$
2	$i = 1; j = 3$	$i = 0; j = 1, 2, 4$	$i = 0; j = 1, 2, 4$
3	$i = 0; j = 1, 2, 4$	$i = 1; j = 3$	$i = 0; j = 1, 2, 4$
4	$i = 1; j = 3$	$i = 1; j = 3$	$i = 0; j = 1, 2, 4$
5	$i = 0; j = 1, 2, 4$	$i = 0; j = 1, 2, 4$	$i = 1; j = 3$
6	$i = 1; j = 3$	$i = 0; j = 1, 2, 4$	$i = 1; j = 3$
7	$i = 0; j = 1, 2, 4$	$i = 1; j = 3$	$i = 1; j = 3$
8	$i = 1; j = 3$	$i = 1; j = 3$	$i = 1; j = 3$

Table 3 Baseline test conditions

Operating conditions	Parameter values	Operating conditions	Number values
Evaporation temperature	−20 (°C)	Low-stage compressor frequency	50 (Hz)
Condensing temperature	40 (°C)	High-stage compressor frequency	50 (Hz)
Artificial environment normal temperature room temperature	15 (°C)	Compressor housing surface wind speed	2.5 (m s^{-1})

3 Results

In order to reflect both the degree of release quality in the intermediate release process and the relationship between release volume and the refrigerant mass flow through the evaporator, the relative parameters of release process and the impact of compression performance were analyzed in this paper. In addition, in order to make the results more general, the relative release mass flow rate was used to analyze the data. It should be pointed out that m_{rat} is the ratio of the mass flow of the intermediate release to the evaporator, $m_{rat} = m_{pump}/m_{eva}$. The specific operating parameters are shown in Table 3.

3.1 Mass Flow Rate of Each Mass in the Circulation Loop

In order to better understand the release ratio and the relative size of the mass cycle of each part of the system. First, based on the simulation results of the baseline test conditions, the variation rules of the release amount m_{pump}, the mass flow rate of the refrigerant in the evaporator $m_{eva,}$ and the mass flow rate of the refrigerant in the condenser m_{con} with the relative release amount (release ratio) m_{rat} were analyzed, as shown in Fig. 1.

As can be seen from Fig. 1, as m_{rat} increases, the m_{pump} increases significantly and the rate of increase gradually decreases. The m_{con} and m_{eva} show a trend of decrease with the increase of m_{rat}, which are close to a linear change. The main reason is that the mass circulation flow of the low pressure compressor increases due to the decrease of the suction superheat, so the mass flow through the evaporator decreases.

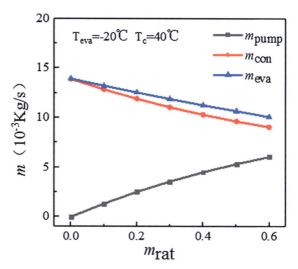

Fig. 1 Variations of m_{eva}, m_{con}, and m_{pump} with m_{rat}

3.2 Influence of Different Release Characteristics on Intermediate Pressure

Figure 2 shows the variation of the intermediate pressure P_m with m_{rat} under different operating conditions. P_m had a trend of rapid decrease first and then slow decrease with the increase of m_{rat}. The reason is that the variation trend of m_{pump} is nonlinear increasing, and a slow variation trend also exists. It can be concluded that the smaller the evaporation temperature, the condensation temperature, and the gas transmission ratio, the lower the P_m value, and the influence of the evaporation temperature and the gas transmission ratio on the intermediate pressure P_m is more significant.

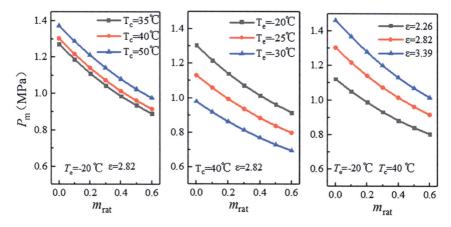

Fig. 2 Variations of P_m with m_{rat}

3.3 Influence of Different Release Characteristics on System Heat Capacity Q_{hot}

It can be seen from Fig. 3 that for different T_e, T_c, and ε working conditions, the system heat capacity Q_{hot} has a tendency to gradually decrease with the increase of the relative injection amount m_{rat}, and this trend tends to be gentle. This change is mainly due to two factors of closely related system Q_{hot}; the condensation side refrigerant circulation amount m_{con} and the high compressor exhaust temperature T_{comH} are formed with different variations of the m_{rat}. The T_{comH} substantially maintained a relatively stable value when the m_{rat} gradually increases. The variation of m_{con} plays a decisive role in Q_{hot}. It is seen in Fig. 1 that the refrigerant side circulation amount m_{con} of the condensing side is significantly decreased with the increase of the m_{rat}, so that the system heat capacity Q_{hot} and the refrigerant circulation amount m_{con} have the same tendency to change with the m_{rat}. In general, it can be concluded that increasing the evaporation temperature T_e, decreasing the condensation T_c, and increasing the gas transmission ratio can increase the heating capacity of the system. However, in terms of the degree of influence on the system heat capacity, the evaporation temperature has a greater influence on the system heat capacity. For example, when T_e is increased by 5 °C and T_c is decreased by 5 °C, the former system heat capacity Q_{hot} will increase by about 300–500 W, which is about five times of the latter variation. Therefore, it is known that when the defrosting is released between the two-stage compression stages, the condensing temperature is appropriately increased to improve the indoor comfort, and the influence on the heat generation of the system is not large. The release process between stages can reduce the system Q_{hot} by 7–37% (about 160–1120 W).

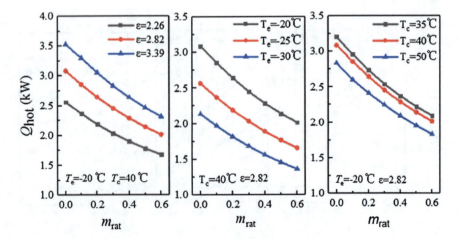

Fig. 3 Variation of Q_{hot} with m_{rat} in variable working condition

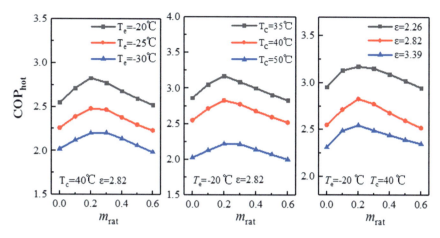

Fig. 4 Variation of COP_{hot} with m_{rat} in variable working condition

3.4 Influence of Different Release Characteristics on System Heat Capacity COP_{hot}

Figure 4 shows the variation of the system heating coefficient COP_{hot} with relative release amount. For the system COP_{hot}, there is mainly due to two factors including Q_{hot} and P_{com}. When the m_{rat} is less than 0.2, the rate of power decay is greater than the rate of heat decay. Therefore, the COP_{hot} of the system gradually increases with the increase of m_{rat}, which can increase the COP_{hot} of the system by 7–11% (the COP_{hot} value increases by about 0.18–0.3); when m_{rat} is more than 0.2, the rate of power decay is less than the rate of heat decay. Therefore, the COP_{hot} of the system gradually decreases with the increase of m_{rat}. The optimal COP_{hot} of the system appears near $m_{rat} = 0.2$. And the optimal COP_{hot} increases as T_e increases and T_c decreases. Under the baseline conditions, when $m_{rat} = 0.2$, the venting amount $m_{pump} = 2.512$ g/s, the system's heating capacity $Q_{hot} = 2.643$ kW, and the system heating coefficient $COP_{hot} = 2.827$. This performance index further verifies the reliability of the two-stage compression inter-stage defrosting system and has guiding significance for the future research of two-stage compression bypass defrosting of hot gas.

4 Conclusion

The influence of evaporation temperature on the intermediate pressure and system heat capacity is obviously more than that of condensation temperature. It is pointed out that with the increase of m_{rat}, the mass flow rate of the refrigerant in the evaporator and condenser, intermediate pressure, and heating capacity is decreasing. For the system Q_{hot} and COP_{hot}, with the increase of m_{rat}, the system heat capacity Q_{hot} decreases, and the system heating coefficient COP_{hot} first increases and then decreases. In addition, the release process can reduce system Q_{hot} by 7–37% (about 160–1120 W) and increase system COP_{hot} by 7–11% (about 0.18–0.3).

Acknowledgements The project is supported by the Science and Technology Project of Educational Commission of Jilin Province, China (No. JJKH20180097KJ), the Development and Innovation Project of Science and Technology Commission of Jilin City, China (No. 201750214), and the Key Technology Research Project of Science and Technology Commission of Jilin Province, China (No. 20180201006SF).

References

1. Wu, X., Xing, Z., He, Z.: Performance evaluation of a capacity-regulated high temperature heat pump for waste heat recovery in dyeing industry. Appl. Therm. Eng. **93**, 1193–1201 (2016)
2. Le, H., Li, H., Jiang, Y.: Using air source heat pump air heater (ASHP-AH) for rural space heating and power peak load shifting. Energy Procedia **122**, 631–636 (2017)
3. Navarro, E., Redón, A., Gonzálvez-Macia, J., Martinez-Galvan, I.O., Coberán, J.M.: Characterization of a vapor injection scroll compressor as a function of low, intermediate and high pressures and temperature conditions. Int. J. Refrig. **36**(7), 1821–1829 (2013)
4. Jiang, S., Wang, S., Wang, F.: Simulation and optimization of heating performance of two-stage compressed air source heat pump. China Sci. Pap. **12**(01), 24–31 (2017)
5. Jin, X., Wang, S., Zhang, T.: Intermediate pressure of two-stage compression system under different conditions based on compressor coupling model. Int. J. Refrig. **35**(4), 827–840 (2012)
6. Shi, W.: A general model for two-stage vapor compression heat pump systems. J. Refrig. **51**, 88–102 (2015)

Theoretical and Experimental Analysis of a New Evaporative Condenser

Yaxiu Gu, Yang Zou, Song Pan, Junwei Wang and Guangdong Liu

Abstract In this paper, a new evaporative condenser with an annular elliptic finned tube heat exchanger was studied and some of its main performance parameters were simulated. For comparison, the actual operating parameters of the round finned tube evaporative condensation air-conditioning system using in a subway station were also measured and simulated. This paper provides a reference for related researches on the actual operation of subway stations. Meanwhile, the heat transfer performance of the annular elliptic finned tube heat exchanger was compared with the round finned tube heat exchanger by numerical simulation. The results showed that the annular elliptic finned tube heat exchanger had a smaller air resistance and a larger heat transfer coefficient. Besides, the new annular elliptic finned tube proposed in this paper could increase the COP of the air-conditioning system by roughly 5–17%, and the energy consumption is reduced by about 8–22%.

Keywords Evaporative condenser · Theoretical analysis · Compared analysis · Annular elliptic finned tube · Practical measured

1 Introduction

The evaporative condenser is a combination of air-cooled condenser and cooling tower. It is researched that the condensing temperature of the air-cooled condenser is greatly affected by the surrounding environment and is about 5–10 °C higher than the evaporative condenser [1, 2]. Water-cooled condenser has a lower condensing temperature than evaporative condenser, but it needs more cooling water. As a result,

---

Y. Gu (✉) · Y. Zou · J. Wang · G. Liu
Department of Building Environment and Energy Engineering,
Chang'an University, Xi'an 710061, China
e-mail: guyaxiu@foxmail.com

S. Pan
Beijing Key Laboratory of Green Built Environment and Energy Efficient Technology,
Beijing University of Technology, Beijing 100124, China

the water pump in water-cooled condenser consumes more energy [3, 4]. Many studies showed that evaporative condensation air-conditioning system can save energy by 11–70% compared to air-cooled or water-cooled air-conditioning units [5].

The heat transfer enhancement and energy saving of evaporative condenser have been widely investigated in recent years. Yu and Chan [6] studied the performance of a residential evaporative condensation air conditioner with various strategies for staging evaporative condenser fans under different operating conditions. Hao et al. [7] changed the cooling pad's thickness of an evaporative condenser to research the heat transfer enhancement and energy saving. Salah et al. [8, 9] investigated a new rotating disk evaporative condenser under a wide range of option parameters, and they found that the heat removal capacity of this evaporative condenser was 13 times higher than the air-cooling condenser. Vrachopoulos et al. [10] presented an evaporative condenser which was installed with a cooling-water basin and a system of water-drop cloud, and the results showed that it could save roughly 58% energy in comparison with the air-cooled condenser. Hwang et al. [11] put forward a new evaporative condenser which was installed with a tube immersing in a water tub. Yang et al. [12] developed a water mist evaporative condenser in a subtropical climate. It was found that the COP of the air-conditioning system could be improved by around 18.6%.

Many researchers have carried out experiments and numerical simulation studies of evaporative cooling system [13, 14]. However, the research on evaporative condenser composed with new tubes is scarcely reported and researched. This paper introduced an evaporative condenser which is installed with a new annular elliptic finned tube heat exchanger. Besides, to verify the accuracy of the simulation results, the simulation results were verified with the measured data testing in the actual evaporative condensation air-conditioning system in a subway station.

2 Methods

2.1 Actual System Synopsis

The air-conditioning system measured in this paper is located in a subway station in Beijing, China. This subway station is an underground two-story island station, and its total length is 249.6 m, standard section width is 21.3 m, and main building area is 19,184 m^2. This station has part A and part B, which are jointly responsible for the indoor loads of this station. Part A offers the cooling capacity for the large system (including the station hall and the platform) and covers the small system's loads of the station office area and the equipment room at the same time. The large system and the small system operate independently. The measured data used in this paper is mainly derived from the operating test results of the refrigeration system at part A.

The system consists of several single decentralized systems comprising an evaporator, an evaporative condenser, and a compressor unit. The evaporator was placed in the air supply channel, and the evaporative condenser was placed in the air exhaust channel. The design value of the cooling capacity of the evaporative condenser is 1150 kW, and the cooling air is mainly from the exhaust air in the subway station. When the air volume is insufficient, it will be assisted by fresh air.

2.2 Actual Measurement Methods and Contents

This paper selected one week per year with the highest temperature in Beijing from July to August in 2016–2018 to measure the air-conditioning system. The test contents mainly included the relevant environmental parameters of the station hall and the platform of this subway station. Besides, this paper also tested the main operating parameters of the air-conditioning system such as the airspeed, air temperature, air relative humidity, the compressor power, the pump power, the fan power, and so on. This test used the automatic recording instruments to continuously monitor the above parameters of the air-conditioning system.

When testing the airspeed of the evaporative condenser, 25 survey points were uniformly distributed in the cross section of the wind channel, and the average value of the testing data was taken as a standard operating value.

2.3 Modeling

In the actual tested evaporative condensation air-conditioning system, the heat exchanger used in the evaporative condenser is the round finned tube heat exchanger. The schematic diagram of the round finned tube flow cross-sectional structure is shown in Fig. 1a. The length of the round tube is $l_1 = 1.42$ m, the diameter is $d_r = 10$ mm, the tube is arranged in a row, and the distance between two adjacent tube centers is 30 mm. The new annular elliptic finned tube heat exchanger is shown in Fig. 1b. The heat exchanger is constituted by two tubes: the inner tube and the outer tube. The outer tube is an elliptical tube, and the inner tube is a round tube. Because the refrigerant is flowing in the annular space between the two tubes, in order to keep the same cross-sectional areas as the round finned tube heat exchanger, the size of the new heat exchanger was computed as follows: length $l_2 = 1.42$ m, inner round tube diameter $d_e = 8$ mm. As for the outer elliptic tube, its long axis length is $2a = 16$ mm, and its short axis length is $2b = 12$ mm. And the distance between two adjacent centers of the annular elliptic tube is 35 mm, and the cross-sectional schematic diagram of the annular elliptic finned tube heat exchanger is shown in Fig. 1b.

In this paper, the heat transfer and flow conditions inside and outside the tubes were separately modeled as shown in Fig. 2 and calculated.

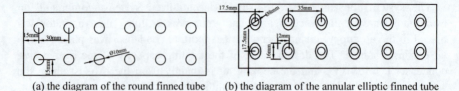

(a) the diagram of the round finned tube   (b) the diagram of the annular elliptic finned tube

Fig. 1 Cross-sectional schematic diagram

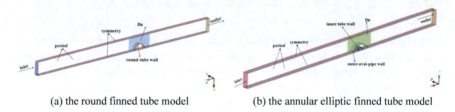

(a) the round finned tube model   (b) the annular elliptic finned tube model

Fig. 2 3-D model of two types of tube round tube

This paper took a single complete heat exchange tube as the calculation unit, in order to avoid the interference of the inlet effect and export effect. The length of the model was 1.42 m the same as the real tube.

2.4 Equations and Parameters Setting

In this paper, the evaporation condensation model in the mixed multiphase flow model of FLUENT software was used to simulate the phase transition between vapor and liquid phases. The principle of the model is to apply the Lee model for calculating the phase transition. The governing equation for mass transfer (evaporation condensation) between liquid and vapor phases is:

$$\frac{\partial}{\partial t}(\alpha_v \rho_v) + \nabla \cdot \left(\alpha_v \rho_v \vec{V}_v\right) = \dot{m}_{lv} - \dot{m}_{vl} \quad (1)$$

Where α_v is vapor volume fraction, ρ_v is vapor density, $\vec{V}_v$ is vapor speed, $\dot{m}$ is quality flow, the subscript 'lv' is from liquid to vapor, and 'vl' is from vapor to liquid.

The mass transfer of evaporation condensation model is determined by the Formulae (2) and (3), which is mainly determined by the relationship between the vapor and liquid phases' temperature and the saturation temperature. The specific calculation process is as follows.

When the liquid phase temperature is higher than the saturation temperature, the mass transfer direction is from the liquid phase to the vapor phase, which is the evaporation process:

$$\dot{m}_{lv} = \xi_{coef} \times \alpha_l \rho_l \left(\frac{T_l - T_{sat}}{T_{sat}} \right) \quad (2)$$

where ξ_{coef} is evaporation and condensation coefficient, T is temperature, the subscript 'l' is liquid, and 'sat' is saturated.

When the vapor phase temperature is higher than the saturation temperature, the mass transfer direction is from the vapor phase to the liquid phase, and that is the condensation process:

$$\dot{m}_{vl} = \xi_{coef} \times \alpha_v \rho_v \left(\frac{T_{sat} - T_v}{T_{sat}} \right) \quad (3)$$

According to the Hertz–Knudsen equation:

$$F = \beta \sqrt{\frac{M}{2\pi RT_{sat}}} L \left(\frac{\rho_v \rho_l}{\rho_l - \rho_v} \right) \frac{(T^* - T_{sat})}{T_{sat}} \quad (4)$$

β is defined with the accommodation coefficient and the physical characteristics of the vapor. T^* stands for the vapor partial pressure at the interface on the vapor side. Provided all vapor bubbles have the same diameter, and the interfacial area density is $A_i = \frac{6\alpha_v \alpha_l}{d_b}$, the phase source term (kg/(s m³)) should be:

$$FA_i = \frac{6}{d_b} \beta \sqrt{\frac{M}{2\pi RT_{sat}}} L \left(\frac{\rho_v \alpha_v}{\rho_l - \rho_v} \right) \left[\rho_l \alpha_l \frac{(T^* - T_{sat})}{T_{sat}} \right] \quad (5)$$

According to the above analysis, the ξ_{coef} is defined as

$$\xi_{coef} = \frac{6}{d_b} \beta \sqrt{\frac{M}{2\pi RT_{sat}}} L \left(\frac{\rho_v \alpha_v}{\rho_l - \rho_v} \right) \quad (6)$$

where d_b is the vapor bubble diameter, m.

3 Discussion

In order to study the heat transfer and flow performance of the annular elliptic finned tube heat exchanger, this paper simulated the annular elliptic finned tube and the traditional round finned tube heat exchangers on the basis of the measured data.

The measured values would be compared with the simulated values under the same conditions, such as the refrigerant outlet temperature, condensation temperature, heat transfer coefficient, heat dissipation, and external pressure of the heat exchanger. In this paper, 2268 sets of continuously recorded data were arranged in order to refrigerant inlet temperature, and 1 °C was taken as a calculation node to divide all data into groups; for example, when the refrigerant inlet temperature ranged from 39.5 to 40.5 °C, the eligible data would be counted as a group, and the average value of this group would be the valid data for the simulation calculation.

3.1 Refrigerant Temperature Measured and Simulation Results

As shown in Fig. 3, this paper used FLUENT software to simulate the condensation temperature value of the refrigerant in the condensing tube of the annular elliptic finned tube heat exchanger and the round finned tube heat exchanger under the same conditions. According to the actual measurement situation, the mass flow rate of cooling spray water was 0.2 kg/s, and the temperature range of cooling spray water was from 27.21 to 33.24 °C. In order to ensure the accuracy of the simulation value, the measured value and simulation value of the condensation temperature were compared and analyzed. The error between the FLUENT simulation results and the measured results of the condensation temperature of the round finned tube heat exchanger was between 2.45 and 11.55%.

According to the simulation results, due to both inner sides and outer sides of the annular elliptic finned tube heat exchanger were flown by the cooling water,

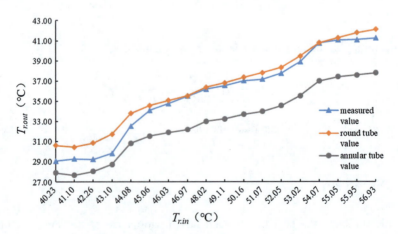

Fig. 3 Comparison of condensation temperature under the same conditions

the condensing temperature of the annular elliptic finned tube heat exchanger was obviously lower than the round finned tube. And with the increasing temperature of cooling water, the advantage of annular elliptic finned tube heat exchanger would be more outstanding. As a result, there would be two sides cooling the refrigerant flowing in the annular tube, that is, under the same area of cross section, the annular elliptic finned tube heat exchanger presented in this paper has about 52.8% area more than the round finned tube heat exchanger. The calculation results showed that when the cooling water temperature ranged from 27.21 to 34.24 °C, the condensing temperature of annular elliptic finned tube heat exchanger was 8.45–10.02% lower than the round finned tube heat exchanger. According to the above literature review [15], the condensing temperature decreases 1 °C, and the energy consumption of the refrigeration system will decrease around 4%; meanwhile, the COP of the refrigeration system will increase about 3%. By the calculation results, the new annular elliptic finned tube investigated in this paper could increase the COP of the air-conditioning system by roughly 5–17%, and the energy consumption could reduce by about 8–22%.

3.2 Heat Transfer Capacity and Air Resistance

In this paper, the influence of inlet air velocity on the heat transfer capacity of the evaporative condenser was studied, as shown in Fig. 4, it could be known that when the inlet air velocity was less than 1.52 m/s, the evaporation heat exchanger would decrease with an increase of the air velocity. Due to the increase of the inlet air velocity, the contact time between cooling air and the liquid film would decrease, so the heat transfer capacity at the gas-liquid interface would be insufficient. When the inlet air velocity was higher than 1.52 m/s, the increase of the air velocity would promote the heat transfer capacity of the evaporative condenser. When the air velocity was higher than 2.88 m/s, with the air velocity increasing the increasing speed of the heat transfer capacity would gradually slow down.

According to other research results [15, 16], when the air velocity exceeded 5 m/s, the continually increasing of the air velocity would cause heat transfer capacity reducing again. This is because the larger the inlet air velocity becomes, the larger the amount of cooling air would flow through the heat exchanger per unit time, as a result, the more latent heat exchange of vaporization would be carried out by the air. When the flowing direction of the air was the same as the spray direction of the spray water, the liquid film on the surface of the heat exchanger would be thinner with the larger the air velocity. Within a certain range, the heat transfer resistance would be reduced and the heat transfer coefficient would be increased. But when the air velocity was too large, the liquid film formed by the spray water would be blown off and forms the dry spots on the surface of the heat exchanger at the same time; this would not benefit the heat exchange.

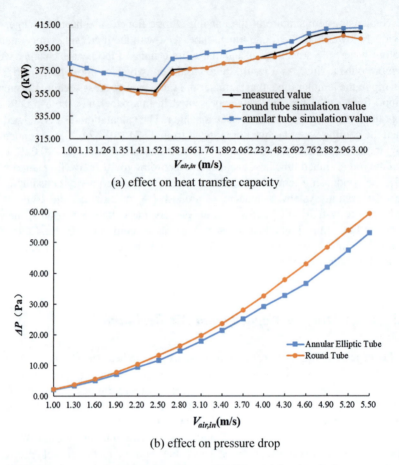

Fig. 4 Effect on heat transfer and pressure drop under different inlet air velocity

This paper also compared the air resistance of the round finned tube and the annular elliptic finned tube heat exchanger under the same air velocity, and the air pressure drop between the entrance and the exit of the heat exchanger would be taken as the judgment criteria of the air resistance. As shown in Fig. 4b, the new annular elliptic finned tube heat exchanger proposed in this paper had a larger cross-sectional area, so the air pressure drop was lower than that of the round finned tube at the same air velocity. With the inlet air velocity becoming larger, this advantage would be more obvious. It showed that when the inlet air velocity was between 1.0 and 5.5 m/s, the average pressure drop of the annular elliptic finned tube heat exchanger was about 11.74% lower than that of the round finned tube heat exchanger.

4 Conclusions

This paper firstly conducted the measurement of the actual evaporative condensation air-conditioning system, got the operating condition of the traditional round finned tube evaporative condenser in actual engineering. Based on the measured data, the models of the traditional round finned tube heat exchanger and the new annular elliptic finned tube heat exchanger were established for simulation calculation. Through the actual measurement combined with the simulation, the parameters such as condensation temperature and refrigerant temperature were analyzed in detail. The results could be concluded as follows:

(1) Under the same spray water temperature, the condensation temperature of the new annular elliptic finned tube heat exchanger was 8.45–10.02% lower than that of the traditional round tube, and the lower spray water temperature would enhance the heat transfer performance of the new heat exchanger.
(2) When the inlet air velocity was between 1.0 and 5.5 m/s, the air resistance of the annular elliptic finned tube heat exchanger was smaller than the round tube. The calculation results showed that the average pressure drop of the annular elliptic finned tube heat exchanger was about 11.74% lower than that of the round tube.

Acknowledgements The project is supported by Natural Science Basic Research Plan in Shaanxi Province of China (Number 2018JM5084), Shaanxi Provincial Key Program of Science and Technology Innovative Research Groups (Number 2016KCT-22), and the Fundamental Research Funds for the Central Universities, CHD300102289203.

Permissions Appropriate permissions from responsible authorities were obtained for study in air-conditioning system measurement in the underground two-story island subway station in Beijing.

References

1. Thu, H.T.M., Sato, H.: Proposal of an eco-friendly high-performance air-conditioning system part 1. Possibility of improving existing air-conditioning system by an evapo-transpiration condenser. Int. J. Refrig. **36**(6), 1589–1595 (2013)
2. Thu, H.T.M., Sato, H.: Proposal of an eco-friendly high-performance air-conditioning system part 2. Application of evapo-transpiration condenser to residential air-conditioning system. Int. J. Refrig. **36**(6), 1596–1601 (2013)
3. Sreejith, K., Sushmitha, S., Vipin, D.: Experimental investigation of a household refrigerator using air-cooled and water-cooled condenser. Int. J. Eng. Sci. **4**(6), 13–17 (2014)
4. Chen, H., Lee, W.L., Yik, F.W.: Applying water-cooled air conditioners in residential buildings in Hong Kong. Energy Convers. Manag. **49**(6), 1416–1423 (2008)
5. Hajidavalloo, E.: Application of evaporative cooling on the condenser of window air conditioner. Appl. Therm. Eng. **27**(11–12), 1937–1943 (2007)
6. Yu, F.W., Chan, K.T.: Improved condenser design and condenser-fan operation for air-cooled chillers. Appl. Energy **83**(6), 628–648 (2006)

7. Hao, X.L., Zhu, C.Z., Lin, Y.L., et al.: Optimizing the pad thickness of evaporative air-cooled chiller for maximum, energy saving. Energy Build. **61**, 146–152 (2013)
8. Salah, H., Youssef, M.A.: Theoretical model for a rotating disk evaporative condenser used in a split air conditioner. J. Eng. Technol. **32**, 1–20 (2013)
9. Nasr, M.M., Salah, M.H.: Experimental and theoretical investigation of an innovative evaporative condenser for residential refrigerator. Renew. Energy **34**(11), 2447–2454 (2009)
10. Vrachopoulos, M.G., Filios, A.E., Kotsiovelos, G.T., et al.: Incorporated evaporative condenser. Appl. Therm. Eng. **27**(5–6), 823–828 (2007)
11. Hwang, Y.H., Radermacher, R., Kopko, W.: An experimental evaluation of a residential-sized evaporatively cooled condenser. Int. J. Refrig. **24**(3), 238–249 (2001)
12. Yang, J., Chan, K.T., Wu, X., et al.: Performance enhancement of air-cooled chillers with water mist: experimental and analytical investigation. Appl. Therm. Eng. **40**, 114–120 (2012)
13. Ertunc, H.M., Hosoz, M.: Artificial neural network analysis of a refrigeration system with an evaporative condenser. Appl. Therm. Eng. **26**(5–6), 627–635 (2006)
14. Islam, M.R., Jahangeer, K.A., Chua, K.J.: Experimental and numerical study of an evaporatively-cooled condenser of air-conditioning systems. Energy **87**(1), 390–399 (2015)
15. Chow, T.T., Lin, Z., Yang, X.Y.: Placement of condensing units of split-type air conditioners at low-rise residences. Appl. Therm. Eng. **22**(13), 1431–1444 (2002)
16. Anica, I., Rodica, D., Liviu, D., et al.: Experimental investigation of heat and mass transfer in evaporative condenser. Energy Procedia **112**, 150–157 (2017)

Studying the Performance of an Indirect Evaporative Pre-cooling System in Humid Tropical Climates

Xin Cui, Le Sun, Weichao Yan, Sicong Zhang, Liwen Jin and Xiangzhao Meng

Abstract The application of evaporative cooling technique is getting more and more attention. The aim of the work is to propose a hybrid air-conditioning system. The hybrid system employs an indirect evaporative heat exchanger (IEHX) as a pre-cooling unit that is operated in tandem with conventional air handling unit. The present work has developed a numerical model by considering the pre-cooling effect of the IEHX. The IEHX is able to adopt the room exhaust air as its working air. In addition, the mathematical formulation for the conventional cooling coil has been developed to study the influence of the pre-cooling effect on the chilled water temperature. The calculated results have demonstrated the ability of the evaporative pre-cooling unit to cool and dehumidify the ambient air under humid tropical climates. The chilled water supply temperature can be also raised due the pre-cooling process. Consequently, an improvement on the coefficient of performance for the chiller is achieved. The hybrid system is able to obtain a potential energy saving as a result of the pre-cooling effect and the enhanced efficiency.

Keywords Indirect evaporative cooling · Mathematical model · Heat and mass transfer · Air-conditioning

1 Introduction

Plate type heat exchangers are commonly used for indirect evaporative cooling systems [1]. For a typical indirect evaporative heat exchanger (IEHX), the product air usually flows along the alternatively arranged dry channels, while the working air acts as the heat sink due to the evaporation process in wet channels [2].

X. Cui (✉) · W. Yan · S. Zhang · L. Jin · X. Meng
Institute of Building Environment and Sustainable Technology, School of Human Settlements and Civil Engineering, Xi'an Jiaotong University, Xi'an 710049, Shaanxi, China
e-mail: cuixin@xjtu.edu.cn

L. Sun
Xi'an Aerospace Propulsion Test Technology Institute, Xi'an 710100, China

Since the air temperature is reduced by vaporizing water, the IEHX is generally suitable for hot and dry areas [3]. However, a single IEHX is often insufficient to maintain comfort indoor thermal conditions for building in humid tropical climates [4]. Figure 1 schematically shows the design for the hybrid indirect evaporative pre-cooling system. The ambient air is first treated through an IEHX. The exhaust air from the indoor is adopted as the working air in order to promote the cooling performance of the IEHX. Thereafter, the ambient air is further conditioned by the conventional mechanical vapor compression unit before supplying to the building.

The pre-cooling IEHX is associated with a complicated air treatment process. In humid tropical climates, the ambient air generally has a higher temperature and a higher humidity ratio compared with the indoor air conditions [5]. As a result, the working air temperature in the pre-cooling IEHX for this hybrid system may be lower than the dewpoint temperature of the ambient air [6]. In other words, the intake ambient air in the pre-cooling unit has a possibility to condense water in the product channel. The aim of the work is to evaluate the energy performance of the hybrid system with pre-cooling air treatment process.

2 Methods

The mathematical formulation for the IEHX has been established to study the pre-cooling performance. The key governing equations are expressed as follows.

$$\frac{\partial u_a}{\partial x} + \frac{\partial v_a}{\partial y} = 0 \qquad (1)$$

$$u_a \frac{\partial u_a}{\partial x} + v_a \frac{\partial u_a}{\partial y} = -\frac{1}{\rho_a}\frac{dp}{dx} + v_a \frac{\partial^2 u_a}{\partial y^2} \qquad (2)$$

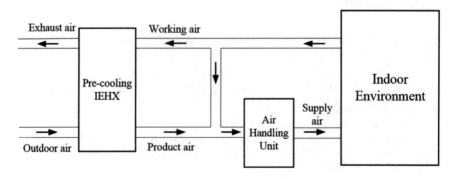

Fig. 1 Schematic of the hybrid indirect evaporative pre-cooling system

$$\frac{\partial}{\partial x}(u_a T_a) + \frac{\partial}{\partial y}(v_a T_a) = \alpha_a \frac{\partial^2 T_a}{\partial y^2} \qquad (3)$$

$$u_a \frac{\partial c_a}{\partial x} + v_a \frac{\partial c_a}{\partial y} = D_a \frac{\partial^2 c_a}{\partial y^2} \qquad (4)$$

The interfacial boundary condition at the working channel surface is given as:

$$-k_w \frac{dT_w}{dy} = -k_a \frac{dT_a}{dy} + M_{H_2O} h_{fg} D_a \frac{\partial c_a}{\partial y} \qquad (5)$$

After the treatment of the pre-cooling IEHX, the product air is further conditioned through the cooling coil in the conventional AHU. A mathematical model was also developed for the cooling coil as follows.

The total heat transfer for the chilled water and the air

$$\Delta Q_{(i,j)} = m_w c_{pw} \left(T_{w(i,j+1)} - T_{w(i,j)} \right) \qquad (6)$$

$$\Delta Q_{(i,j)} = m_a \left(i_{a(i,j)} - i_{a(i+1,j)} \right) \qquad (7)$$

By considering the convective heat transfer, we obtain:

$$\Delta Q_{(i,j)} = h_i \Delta A_i \left(T_{s,m(i,j)} - T_{w,m(i,j)} \right) \qquad (8)$$

$$\Delta Q_{(i,j)} = h_o \Delta A_o \eta_s \left(T_{a,m(i,j)} - T_{s,m(i,j)} \right) \\ + h_{fg} h_m \Delta A_o \eta_s \left(\omega_{a,m(i,j)} - \omega_{s,m(i,j)} \right) \qquad (9)$$

The calculation of the water/air temperature for the next grid is written as

$$T_{w(i,j+1)} = T_{w(i,j)} - \frac{\Delta Q_{(i,j)}}{m_w c_{pw}} \qquad (10)$$

$$T_{a(i+1,j)} = \frac{1 - \frac{\Delta \text{NTU}_o}{2}}{1 + \frac{\Delta \text{NTU}_o}{2}} \cdot T_{a(i,j)} \\ + \frac{\Delta \text{NTU}_o}{1 + \frac{\Delta \text{NTU}_o}{2}} \cdot T_{s,m(i,j)} \qquad (11)$$

3 Results

The computational model is first validated against experimental data. Thereafter, the model is employed to theoretically study the performance of the proposed hybrid system operating under humid tropical climates.

3.1 Validation

Firstly, the validation of the model is carried out based on the experimental data of an IEHX. The simulation has been conducted under the experimental conditions [7]. The calculated outlet air temperature is compared with the experimental measurement as illustrated in Fig. 2. It is observed from figure that the developed model shows a good agreement compared with the experimental data with a maximum discrepancy around 5%.

Secondly, the mathematical model for the cooling coil is validated to demonstrate the capability. The validation relied on the experimental data acquired from literature [8]. Table 1 compares the calculated outlet air temperature with experimental data. The comparison has demonstrated the accuracy of the model to predict the air cooling performance.

3.2 Pre-cooled Air Treatment Profiles

For the pre-cooling IEHX, the validated model has been employed to predict the air stream conditions profiles. The intake air condition is assumed as a typical ambient air condition in humid tropical climates ($T_{a,\,in}$ = 35 °C, and $RH_{a,\,in}$ = 80%). The working air in the IEHX utilizes the room exhaust air ($T_{a,\,in}$ = 25 °C, and $RH_{a,\,in}$ = 50%).

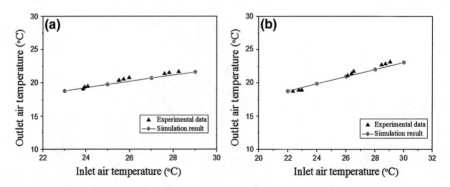

Fig. 2 Validation of the model for a counter-flow IEHX, **a** the intake air flow rate is 4.5 L/s; **b** the intake air flow rate is 6.0 L/s

Table 1 Validation of the model for the cooling coil

Test	Coil row	$T_{a,\,in}$ (°C)	$RH_{a,\,in}$ (%)	m_a (kg/s)	$T_{a,\,out}$ Experiment	Model
1	Four-row	28.68	55.60	1.16	17.44	17.58
2	Four-row	25.99	70.22	1.20	17.17	16.91
3	Four-row	25.99	54.13	1.20	15.27	15.38
5	Eight-row	23.94	70.80	1.13	13.24	14.46
4	Eight-row	24.59	55.02	1.27	15.83	16.43
5	Eight-row	27.29	51.19	1.04	13.47	14.23

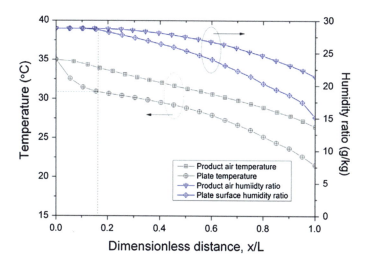

Fig. 3 Temperature and humidity ratio profiles in the pre-cooling IEHX

Figure 3 shows the profiles of the product air in terms of the temperature and humidity ratio along the flow passages. It can be inferred from figure that the product air temperature decreases along the flow direction. The product air humidity ratio is kept constant when the interface temperature of the plate is higher than its dewpoint temperature. Once the plate temperature is reduced to below dewpoint temperature, the product air humidity ratio will decrease accordingly. In other words, the condensation process occurs in this region of the product channel.

3.3 Impact of the Pre-cooling on the Chiller

Figure 4 illustrates the variation of the product air condition through the pre-cooling IEHX. A larger temperature change of the product air is obtained by

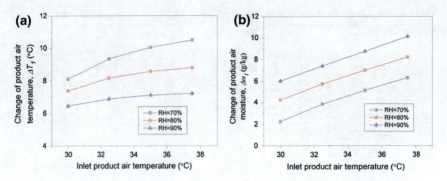

Fig. 4 Variation of the product air condition through the pre-cooling IEHX. **a** Change of temperature; **b** change of humidity ratio

increasing the intake air temperature under a specific relative humidity. In addition, the change of product air moisture shows a similar trend. The reduction of the humidity ratio demonstrates the possibility of vapor condensation. In general, the pre-cooling IEHX shows the ability to cool and dehumidify the ambient air.

Figure 5 illustrates an example of the calculated air treatment conditions for the hybrid system. On the psychrometric chart, the point O represents the selected outdoor air condition ($T = 33$ °C, RH = 80%), and the point R shows the assumed indoor air condition ($T = 24$ °C, RH = 60%).

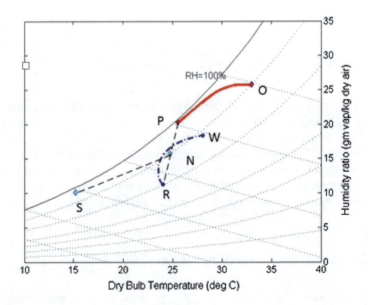

Fig. 5 Description of air treatment conditions on psychrometric chart for the hybrid indirect evaporative pre-cooling system

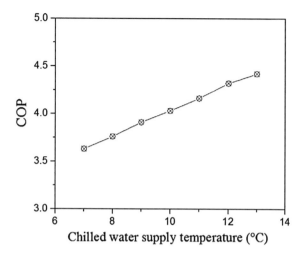

Fig. 6 Influence of the chilled water temperature on the COP of the chiller

In Fig. 5, the outdoor air (O) and the exhaust air (R) are first treated in the IEHX. The product air in the IEHX is cooled and dehumidified as the condition is varied from point O to point P. The exhaust air absorbs heat and moist in the working channel resulting in a final condition at point W. The product air and the return air are then mixed to point N. In addition, a higher chilled water supply temperature (10 °C) can be employed in the hybrid system for achieving a similar supply air condition at point S compared with the conventional air handling unit.

Figure 6 shows the impact of the chilled water temperature on the chiller's coefficient of performance (COP) [5]. It is observed that the chiller's efficiency can be improved by increasing the supply temperature of the chilled water [9]. For example, the average COP is ranged from 3.6 to 4.4 due to the increase in chilled water temperature. The enhanced chiller's performance can potentially reduce the energy consumption.

4 Conclusions

The present work has introduced a hybrid air-conditioning system by using an IEHX as a pre-cooling unit. Mathematical model has been established for both the IEHX and the cooling coil. The pre-cooling IEHX employs the room exhaust in the working channel to enhance the cooling performance for the ambient intake air. The temperature and humidity distribution along the passages have been evaluated for the IEHX. Simulation results have illustrated the capability of the IEHX to cool and dehumidify the intake air. An energy-saving potential can be achieved due to the application of the energy-efficient IEHX, reduced cooling load, and improved chiller's performance.

Acknowledgements The project is supported by the Fundamental Research Funds for the Central Universities (xjj2018074), and China Postdoctoral Science Foundation (2018M631153).

References

1. Chua, K.J., Chou, S.K., Yang, W.M., Yan, J.: Achieving better energy-efficient air conditioning—a review of technologies and strategies. Appl. Energy **104**, 87–104 (2013)
2. Duan, Z., Zhan, C., Zhang, X., Mustafa, M., Zhao, X., Alimohammadisagvand, B., et al.: Indirect evaporative cooling: past, present and future potentials. Renew. Sustain. Energy Rev. **16**, 6823–6850 (2012)
3. Xu, P., Ma, X., Zhao, X., Fancey, K.S.: Experimental investigation on performance of fabrics for indirect evaporative cooling applications. Build. Environ. **110**, 104–114 (2016)
4. Woods, J., Kozubal, E.: A desiccant-enhanced evaporative air conditioner: numerical model and experiments. Energy Convers. Manag. **65**, 208–220 (2013)
5. Cui, X., Mohan, B., Islam, M.R., Chua, K.J.: Investigating the energy performance of an air treatment incorporated cooling system for hot and humid climate. Energy Build. **151**, 217–227 (2017)
6. Chen, Y., Yang, H., Luo, Y.: Indirect evaporative cooler considering condensation from primary air: model development and parameter analysis. Build. Environ. **95**, 330–345 (2016)
7. Cui, X., Islam, M.R., Mohan, B., Chua, K.J.: Developing a performance correlation for counter-flow regenerative indirect evaporative heat exchangers with experimental validation. Appl. Therm. Eng. **108**, 774–784 (2016)
8. Zhou, X., Braun, J.: A Simplified Dynamic Model for Chilled-Water Cooling and Dehumidifying Coils—Part 2: experimental validation (RP-1194). HVACR Res. **13**, 805–817 (2007)
9. Thu, K., Saththasivam, J., Saha, B.B., Chua, K.J., Srinivasa Murthy, S., Ng, K.C.: Experimental investigation of a mechanical vapour compression chiller at elevated chilled water temperatures. Appl. Therm. Eng. **123**, 226–233 (2017)

Preheating Strategy of Intermittent Heating for Public Buildings in Cold Areas of China

Chunhua Sun, Jiali Chen, Yuan Liang and Shanshan Cao

Abstract Intermittent heating is commonly applied and considered as energy efficient in public buildings in cold areas considering their occupation schedules. Rational preheating time is of significance to guarantee indoor thermal comfort, facility safety and energy conservation in intermittent operation strategy. In this paper, indoor temperature and energy consumption under continuous and intermittent heating conditions of a typical office building in Shijiazhuang City, China, are, respectively, studied using DeST-C software. The influence of building thermal parameters and outdoor temperature on preheating time is also studied. The results show that indoor temperature fluctuates periodically within a larger margin in a day under intermittent heating while it is relatively stable under continuous heating. The lowest indoor temperature of energy-efficient buildings and ordinary ones is as low as 11 and 9 °C, respectively, under intermittent heating during the coldest days in winter. The preheating time for energy-efficient buildings is 2, 2 and 1 h; for non-energy-efficient buildings, it is 3, 3 and 1 h during the early, peak and final heating period. The energy-saving rate of intermittent heating compared with the continuous heating mode is above 20% for both energy-efficient buildings and ordinary ones in cold area.

Keywords Public buildings · Intermittent heating · Preheating time · DeST-C · Indoor temperature · Energy consumption

---

C. Sun · J. Chen · Y. Liang · S. Cao (✉)
School of Energy and Environment, Hebei University of Technology, Tianjin 300401, China
e-mail: css_2005@126.com

C. Sun
e-mail: sunchunhua@163.com

J. Chen
e-mail: 15732676469@163.com

Y. Liang
e-mail: 806011147@qq.com

1 Introduction

Intermittent heating is studied a lot and commonly known as energy saving in commercial buildings. E. Wang studied heat consumption of a public building in hot summer and cold winter climate zone and founded that intermittent heating operation saved most energy in the studied three-heating strategy [1]. B. Xu, et al. founded that intermittent heating could save 20% energy in weekly operation while maintaining the indoor thermal comfort [2]. Z. Wang, et al. compared two types of intermittent heating and founded that intermittent heating could improve indoor thermal comfort and reduce heat consumption in hot summer and cold winter climate zone [3]. A. A. Badran et al. using HAP software to study the energy consumption in Jordan founded that intermittent 14 h operation could achieve the same indoor thermal comfort and saving energy [4]. N. Cardinale and P. Stefanizzi compared the heat consumption of continuous operation and intermittent operation and founded that lower energy is needed under intermittent heating in short time [5]. Z. Wang, et al. conducted field test of intermittent heating in Cambridgeshire; it can achieve thermal comfort under intermittent heating [6]. D. Pupeikis, et al. founded that reasonable preheating time should be chosen considering different thermal inertia [7].

Preheating time is of great significance to heat saving potential of intermittent heating. However, in practical operation, it is usually decided by the rule of thumb with rational analysis. In this paper, the preheating strategy of office building in cold climate will be studied to supply theoretical support to operating staff. The rest of this paper is organized as follows: Simulation model of an office building in Shijiazhuang City, China, is established using commercial software DeST-C, set relative parameters; decide preheating time by considering different outdoor temperature and building types, and the corresponding heat consumption and heat saving rate are studied.

2 Methods

2.1 Simulation Software

The commercial software DeST-C (Building environment design simulation toolkit for commercial building) can analyze building and environment management, according to Chen and Deng [8]. In this section, the physical model of an office building in Shijiazhuang City, China, is built firstly using DeST-C. The architectural plan of standard floor is shown in Fig. 1. This office is a four floor building, with total height of 17.4 m and area of 4448.79 m^2. The shape coefficient of this building is 0.24. The office time is 7:00 am–6:00 pm. The simulation model setup in the DeST-C software is shown in Fig. 2.

Fig. 1 Architectural plan

Fig. 2 Architectural model in DeST-C software

2.2 Internal Disturbance and Ventilation Rates Setting

Internal heat source of the studied office building includes staffs, lights, equipment, etc. The internal disturbance factors are set as follows. Working day: 1 for 7:00–11:00, 0.3 for 12:00, 1 for 13:00–16:00, 0.5 for 17:00–18:00. Off day: 0.3 for 7:00–11:00, 0.1 for 12:00, 0.3 for 13:00–16:00, 0.1 for 17:00–18:00.

The ventilation rate is set as 0.8 per hour and 1 per hour for energy-efficient building (EEB) and non-energy-efficient building (NEEB), respectively, according to Lu [9].

2.3 Building Thermal Parameters

In order to compare the influence of building types on the preheating time and energy-saving potential of intermittent heating systems, the main structure and heat transfer coefficients of the two kinds of buildings are given in Table 1.

Table 1 Heat transfer coefficient of envelope in EEB and NEEB

Envelope	EEB (W/(m² K))	NEEB (W/(m² K))
Exterior wall	0.433	0.929
Roof	0.399	0.603
Floor	0.408	0.684

3 Results and Discussion

3.1 Indoor Temperature

In this paper, the whole heating period is divided into three parts according to the variation of outdoor temperature, marked as heating period 1–3, representing the beginning, coldest and ending period, respectively. December 7–December 11, January 15–January 19 and February 20–February 24 are selected as typical days for period 1–3, respectively, and outdoor temperatures during the three periods are shown in Fig. 3.

Indoor temperature of a certain office in standard floor is studied. Under continuous heating, indoor temperature during the three-heating period are relatively steady around 19.26, 19.12, 19.32 °C for NEEB, 19.2, 19.14, 19.46 °C for EEB.

Figure 4 is the indoor temperature while preheating 3 h during the three-heating period in NEEB. The lowest indoor temperature is 13.2, 11.2 and 12.9 °C during the three periods. The lowest indoor temperature is much higher than 5 °C, so it indicates that there is no pipe frozen risk when reasonable preheating time of intermittent heating is chosen in Shijiazhuang City.

Figure 5 shows the indoor temperature of NEEB and EEB while preheating 2 h during heating period 2. It can be seen that under the same outdoor temperature and preheating time, the lowest indoor temperature is quite different. The building type should be taken into consideration when deciding the preheating time.

3.2 Decision of Preheating Time

It is time consuming to simulate the preheating time in daily operation. This section tries to select the typical day of heating period 1–3 and decide their preheating time.

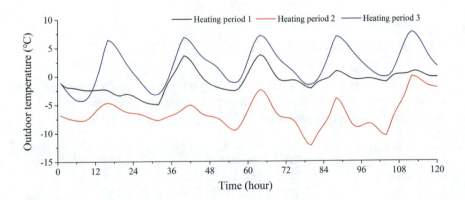

Fig. 3 Outdoor temperature during simulation periods

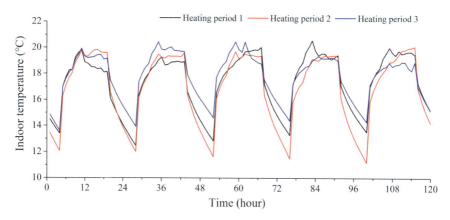

Fig. 4 Indoor temperature in NEEB while preheating 3 h

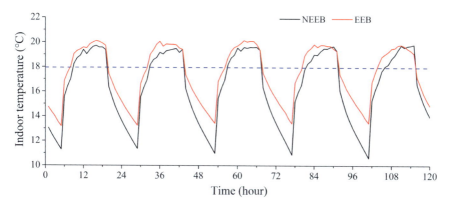

Fig. 5 Indoor temperature of NEEB and EEB preheat 2 h during heating period 2

December 7, January 15 and February 21 were selected, respectively. Their indoor temperature is simulated using the DeST-C model established, in order to maintain indoor temperature at 7:00 am not lower than 18 °C. The results are shown in Table 2. Preheating time for NEEB is 3, 3 and 2 h, while 2, 2 and 1 h for EEB.

Applying the preheating time in Table 2 in corresponding heating period, simulate the indoor temperature at 7:00 am. While indoor temperature at 7:00 am is higher than 18 °C, its indoor temperature meets the requirements. The overall satisfied days and rates are shown in Table 3. About 80% satisfied rates can be reached when using the preheating time in Table 2 except heating period for NEEB. This will reduce the simulation cost in practical operation.

Table 2 Preheating time of typical day for the three-heating periods

Typical day	Heating period	Preheating time of NEEB (hour)	Preheating time of EEB (hour)
Dec. 7	1	3	2
Jan. 15	2	3	2
Feb. 21	3	2	1

Table 3 Heating quality under studied preheating time

Heating period	Duration (day)	NEEB		EEB	
		Indoor temperature satisfied time (day)	Satisfied rate (%)	Indoor temperature satisfied time (day)	Satisfied rate (%)
1	31	29	93.55	31	100
2	62	45	72.58	50	80.65
3	28	22	78.57	25	89.29

3.3 Heat Consumption Analysis

Taking January 15 for example (shown in Figs. 6 and 7), the heat load of EEB and NEEB under continuous and intermittent heating reaches its maximum of 71.9, 147.2, 89.1 and 172.9 W/m^2 at 6:00 am. Although the maximum heat load of intermittent heating in two types of building is higher than continuous heating, the average heating load is reduced.

The average heat consumption during the whole heating season is shown in Fig. 8. The energy-saving rates of intermittent heating in NEEB are 22.76%, with a reduction from 47.7 to 36.8 W/m^2, while in EEB is 28.7 W/m^2, which is 9.5 W/m^2 smaller than continuous heating, with a 22.76% energy-saving rate. Whether the building is NEEB or EEB, the energy-saving rate of intermittent heating is above 20% in cold climate zone.

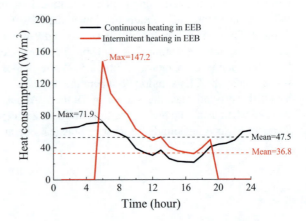

Fig. 6 Heat consumption of EEB

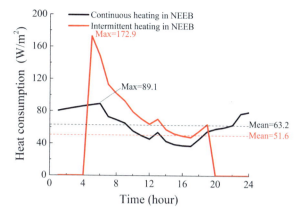

Fig. 7 Heat consumption of NEEB

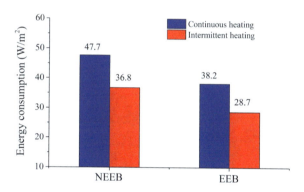

Fig. 8 Average heat consumption of EEB and NEEB during overall heating period

4 Conclusions

This paper studied the indoor temperature and heat consumption of a public building in Shijiazhuang under continuous and intermittent heating using the DeST-C software. The main conclusions are as follows:

Indoor temperature under intermittent heating fluctuates periodically within a large margin each day, and the indoor temperature may drop as low as 11 and 9 °C in EEB and NEEB, respectively.

Preheating time of intermittent heating varies with outdoor temperature and building types. The optimal preheating time for EEB in the three-heating periods is 2, 2 and 1 h, respectively, while it is 3, 3, and 2 h of NEEB.

Although the maximum heat load of intermittent heating is larger than continuous heating, the average heat load is much smaller. The energy-saving rate of intermittent heating is above 20% for both type buildings in a cold climate zone.

Acknowledgements This project is supported by National 13th Five-Year Science and Technology Support Program (2016YFC0700707) and Natural Science Foundation of Hebei Province, China (E2015202063).

References

1. Wang, E.: Research on energy consumption and operation strategy of different heating system in intermittent operating condition. Master Degree's Thesis of Harbin Institute of Technology (in Chinese) (2016)
2. Xu, B., et al.: An intermittent heating strategy by predicting warm-up time for office buildings in Beijing. Energy Build. **155**, 35–42 (2017)
3. Wang, Z., et al.: A model to compare convective and radiant heating systems for intermittent space heating. Appl. Energy **215**, 211–226 (2018)
4. Badran, A.A., et al.: Comparative study of continuous versus intermittent heating for local residential building: case studies in Jordan. Energy Convers. Manag. **65**, 709–714 (2013)
5. Cardinale, N., Stefanizzi, P.: Heating-energy consumption in different plant operating conditions. Energy Build. **24**(3), 231–235 (1996)
6. Wang, Z., et al.: Modeling and measurement study on an intermittent heating system of a residence in Cambridgeshire. Build. Environ. **92**, 380–386 (2015)
7. Pupeikis, D., et al.: Required additional heating power of building during intermitted heating. J. Civ. Eng. Manag. **16**(1), 141–148 (2010)
8. Chen, F., Deng, Y.: Building Environment Design Simulation Toolkit DeST. J. HVAC **29**(4), 58–63 (1999)
9. Lu, Y.: Practical HV & AC Design Handbook. China Architecture & Building Press, Beijing (2008)

On-Site Performance Investigation of the Existing Ground Source Heat Pump Systems in Residential Buildings in Cold Area

Lixia He, Han Du, Peng Gao, Zhuangzhuang Zheng and Ping Cui

Abstract In this paper, three typical ground source heat pump (GSHP) projects applied to residential buildings in cold area in China are investigated. According to the on-site tested data, such as temperatures and flow rates of the circulating water on the user side and ground side and the power consumption, the operating performance of the whole system is evaluated. The results of the on-site test show that some equipment capacity of the residential GSHP systems investigated in this study was inappropriate in the actual operation conditions. The main problem of the systems is the improper design with "large flow rate and small temperature difference." In general, the investigated systems are in an operation mode with high energy consumption, which means they have great energy-saving potential. In addition, compared with the systems applied to pure residential buildings, those employed in mixed residential and commercial buildings can efficiently reduce the thermal imbalance of the underground. Finally, some possible energy-saving retrofit measures have been suggested according to the results.

Keywords GSHP · On-site investigation · Residential buildings · Cold area

1 Introduction

Ground source heat pump (GSHP) technology has been widely used in recent years [1, 2]. However, some critical problems caused by the improper design, construction, and commissioning have been exposed during the operation years and

L. He · H. Du · P. Cui (✉)
School of Thermal Engineering, Shandong Jianzhu University, Jinan 250101, China
e-mail: sdcuiping@sdjzu.edu.cn

P. Gao
Shandong Yateer Group Co., Ltd, Jinan 250101, China

Z. Zheng
First Exploration Team of Coal Geology Bureau of Shandong Province, Zaozhuang 277100, China

needed to be solved especially in cold area where the GSHP system operation may cause a severe problem [3, 4]. Therefore, it is necessary to investigate the current situation and propose the diagnostic evaluation methods for the existing GSHP systems, especially the systems used in residential buildings in cold area. The existing research was focused on the theoretical calculation, optimal design, and feasibility study [5, 6]. Few studies have been conducted on the on-site investigation and diagnosis for the existing GSHP systems [7].

In this study, three typical GSHP systems applied to residential buildings in Shandong Province are selected to investigate the operating performance. According to the on-site tested data, some serious problems are detected and the feasible solutions to these problems are proposed to improve the system performance. The results can provide a reference and technical support for the GSHP system applications in cold area in China.

2 Project Overview

Three typical residential GSHP systems, named project A, B, and C, are selected for the on-site investigation. Projects A and B are used for providing heating and cooling only for residential buildings. Project C is used for both residential and commercial buildings. The basic parameters about the three systems are listed in Table 1. The parameters tested include the indoor air temperature and humidity, the temperature and flow rate of the circulating water of the heat pump unit, and the power consumption of the system. The main equipment parameters of the three equipment rooms are shown in Table 2.

3 Test Plan

The existing GSHP systems in the three projects were not equipped with the data measuring instruments. Firstly, a set of data collection system was designed and installed on the GSHP plant to collect the operating data, as shown in Fig. 1.

The operating parameters of the GSHP system with the buried pipe are measured. The parameters of the measuring instruments are shown in Table 3.

Table 1 Basic information sheet of the test project

	Location	Floor area (m^2)	Number of boreholes	Borehole depth (m)	Total drilling depth (m)	Buried pipe parameters
A	Jinan	156,000	1000	120	120,000	Double U-tube
B	Zaozhuang	13,256	140	100	14,000	Single U-tube
C	Jining	52,000	444	100	44,400	Single U-tube

Table 2 Main equipment parameters of the equipment room

	Equipment	Design parameter	Quantity
A	User-side water pump	Flow rate: 300 m^3/h; power: 45 kW	3
		Flow rate: 139 m^3/h; power: 22 kW	1
	Ground-side water pump	Flow rate: 400 m^3/h; power: 55 kW	3
		Flow rate: 206 m^3/h; power: 30 kW	1
	Heat pump	Cooling: capacity of 1664 kW; power of 259 kW Heating: capacity of 1737 kW; power of 343 kW	3
B	User-side water pump	Flow rate: 100 m^3/h; head: 32 m; power: 15 kW	2
	Ground-side water pump	Flow rate: 40 m^3/h; head: 30 m; power: 5 kW	6
	Heat pump	Cooling capacity: 452 kW, cooling power: 92 kW Heating capacity: 358 kW; heating power: 108.1 kW	2
C	User-side water pump	Flow rate: 245 m^3/h; power: 37 kW	2
	Ground-side water pump	Flow rate: 320 m^3/h; power: 45 kW	2
	Heat pump	Cooling: capacity of 1350 kW; power of 243 kW; Heating: capacity of 1336 kW; power of 309 kW	2

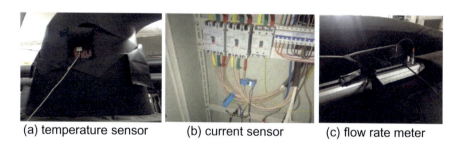

(a) temperature sensor   (b) current sensor   (c) flow rate meter

Fig. 1 On-site measurement pictures

4 Test Data Processing and Results

The three GSHP systems were measured during the heating period of December 25, 2017, to March 15, 2018, and the cooling period of June 12, 2018–June 21, 2018. Based on the measured data, the main performance parameters are analyzed.

The average indoor air temperature in the selected rooms during the cooling period was calculated to be 26.81, 26.56, and 26.63 °C, and the average indoor relative humidity during the test period was 52.85, 56.39, and 57.36%, respectively.

Table 3 Performance of the measuring instrument

Measuring instrument	Measurement range	Accuracy
External ultrasonic flowmeter	0–3×10^4 m^3/h	RE $\pm$ 1%
Temperature sensor	0–100 °C	$\pm$0.2 °C
Current sensor	0–800 A	Accuracy level 0.5
Temperature and humidity loggers	30–60 °C, 0–100%	$\pm$0.3 °C, $\pm$3%

Table 4 Comparisons of measured and rated flow rates

	Flow rate (m^3/h)		Mode		Flow rate (m^3/h)		Mode	
			Cooling	Heating			Cooling	Heating
A	Buried tube side	Measured	334	282	User side	Measured	299	221
		Rated	326	288		Rated	286	298
B		Measured	212	152		Measured	141	100
		Rated	156	74.1		Rated	125	61.4
C		Measured	677	417		Measured	472	455.5
		Rated	625	362.4		Rated	415	395

According to national standard of GB50736-2012, it can be concluded that the indoor thermal and humid environment of the three residential quarters is acceptable.

4.1 Flow Rate Measurement Data Processing and Analysis

The rated flow rate can be calculated according to the rated heating capacity, rated cooling capacity, and design temperature difference (5 °C on the user side and 4 °C on the buried pipe side). Table 4 compares the measured and the rated flow rates under heating and cooling modes.

It can be seen from Table 4 that only the flow rates of the water pumps of project A are close to the rated values. However, the measured flow rates of the water pumps in projects B and C are significantly greater than their rated values obtained from the designed condition. Obviously, this causes a large flow rate with a small temperature difference between supply and return water which leads to the increase of power consumption and decreases of the system efficiency accordingly.

4.2 Circulating Water Temperatures

It can be seen from Figs. 2, 3, 4, 5, 6, and 7 that the average inlet and outlet water temperatures in both user side and buried pipe side of projects are relatively stable during the test period. The groundwater temperatures are relatively higher in heating mode and lower in cooling mode compared to the normal designed temperatures. However, all the temperature differences between the inlet and outlet water are extremely small, except the cooling mode in project C.

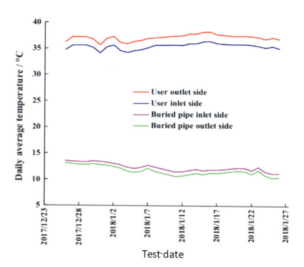

Fig. 2 Inlet and outlet water temperatures of buried pipe side and user side under heating condition in project A

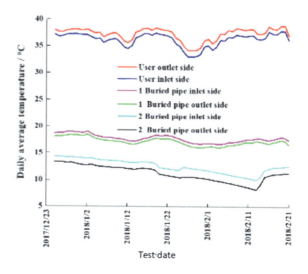

Fig. 3 Inlet and outlet water temperatures of buried pipe side and user side under heating condition in project B

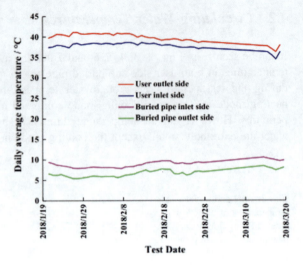

Fig. 4 Inlet and outlet water temperatures of buried pipe side and user side under heating condition of project C

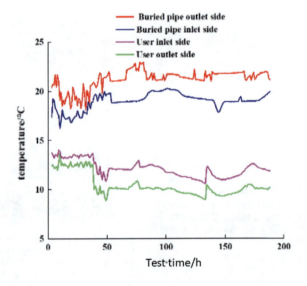

Fig. 5 Inlet and outlet water temperatures of buried pipe side and user side under cooling condition of project A

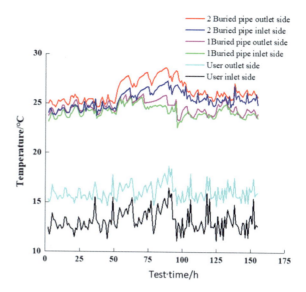

Fig. 6 Inlet and outlet water temperatures of buried pipe side and user side under cooling condition of project B

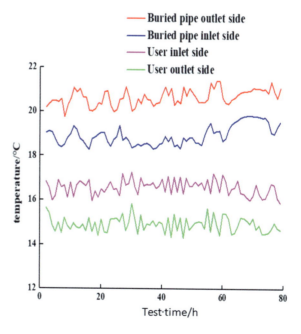

Fig. 7 Inlet and outlet water temperatures of buried pipe side and user side under cooling condition of project C

4.3 System Performance Analysis

The coefficient of the performance (COP) of each system can be calculated according to the measured temperature, flow rate and power consumption, and the average COP of the GSHP systems during the test period are shown in Table 5.

It can be detected from Table 5 that the COP of project B is quite lower compared to the other two projects, which is even lower than the value of air source heat pumps. The main reason is the improper operating mode with large flow rates and small temperature difference, which causes a large percentage of power consumption of water pump.

4.4 Thermal Analysis of Underground

The total energy and power consumption by the system during the entire heating season and cooling season can be estimated based on the measured data and the degree-day method. The total heat release and extraction to/from the ground can also be predicted based on the average COP. The calculation results are shown in Table 6.

It can be seen from Table 6 that the GSHP air-conditioning systems in residential buildings in cold area cannot achieve a thermal balance in the underground

Table 5 Average COP of the systems

	Project A		Project B		Project C	
Operation mode	Cooling	Heating	Cooling	Heating	Cooling	Heating
System cop	3.85	3.0	2.46	2.2	4.2	3.1

Table 6 Estimated accumulated energy and power consumption of the systems

Energy (MWh)	Project A	Project B	Project C
Accumulated energy consumption on the user side in summer	910.3	361.8	750.1
Power consumption in summer	279.1	137.5	192
Accumulated energy consumption on the user side in winter	2505.9	523	1666.7
Power consumption in winter	828.2	281	645.6
Accumulated heat release in the ground in summer	1189.4	400	904.1
Accumulated heat extraction from the ground in winter	1627.3	622	1021.2
Thermal imbalance in the ground (%)	27.0%	35.6%	11.5%

after a whole heating and cooling circle. Obviously, the system which provides heating and cooling for residential and commercial buildings may have a better thermal balance in the ground, such as project C.

5 Conclusions

This study takes three typical existing GSHP systems installed in Shandong Province as an investigation example to measure the real operation performance of the GSHP systems. A set of data collection system was designed and installed on each GSHP plant to collect the operating data. The system performance and the thermal balance of the underground have been carried out on the basis of the measured operating data during a heating and cooling season. The following several problems of the existing systems have been found according to the diagnosis results:

(1) The annual underground thermal imbalance is quite serious. This is caused by the reason that the heating time is much longer and the heating load is significantly lager compared to the cooling season for the residential buildings. (2) Most majority circulating pumps were oversized, which leads to a large fluid flow and small temperature difference in summer. (3) Lack of the necessary automatic monitoring equipment and improper management strategies also caused a high energy consumption and dangerous operation state.

As for project B, according to the measured data, the overall operating COP of the GSHP systems was quite low, only 2.46 in cooling and 2.2 in heating case.

Some possible energy-saving retrofit measures have been suggested according to the simulation results, as follows:

(1) To increase the air source heat pump as an auxiliary heating source to solve the problem of thermal imbalance in the underground, the optimal operation strategy of the hybrid system has been suggested;
(2) The existing circulating pumps are suggested to optimize to further reduce the transmission energy consumption. The on-site management system should be standardized.
(3) The reasonable capacity of the boreholes should be designed based on the cooling load and the excess heating load should be taken by other supplemental heating source.

Acknowledgements The project is supported by Shandong Province key research and development project (Project No. 2017GGX40117) and Shandong Natural Science Foundation (Project No. ZR2017MEE037).

References

1. Yang, H., Cui, P., Fang, Z.: Vertical-borehole ground-coupled heat pumps: a review of models and systems. Appl. Energy **87**(1), 16–27 (2010)
2. Li, M., Lai, A.C.K.: Review of analytical models for heat transfer by vertical ground heat exchangers (GHEs): a perspective of time and space scales. Appl. Energy **151**, 178–191 (2015)
3. Liu, Z., Xu, W., Qian, C., et al.: Investigation on the feasibility and performance of ground source heat pump (GSHP) in three cities in cold climate zone. China. Renew. Energy 2015: S0960148115300422
4. Ruiz-Calvo, F., Cervera-Vázquez, J., Montagud, C., et al.: Reference data sets for validating and analyzing GSHP systems based on an eleven-year operation period. Geothermics **64**, 538–550 (2016)
5. Xia, L., Ma, Z., et al.: Experimental investigation and control optimization of a ground source heat pump system. Appl. Therm. Eng. **127**, 70–80 (2017)
6. Garber, D., Choudhary, R., et al.: Based lifetime costs assessment of a ground source heat pump (GSHP) system design: methodology and case study. Build. Environ. **60**, 66–80 (2013)
7. Chu, G., Wang, Y., Chu, M.: Measurement and analysis of a GSHP system operation in winter. Procedia Eng. **146**, 573–578 (2016)

Analysis of Heat and Mass Exchange Performance of Enthalpy Recovery Wheel

Hong Fan and Liu Chen

Abstract Enthalpy recovery wheel is a high-efficiency heat recovery device with fresh air and return air, which can recycle both the obvious and latent heat from return air at the same time, and reduced energy consumption for handling fresh air. In this paper, a mathematical model of enthalpy recovery wheel was established, taking into account the air side and adsorbent side. The heat and mass transfer equation of the enthalpy recovery wheel through the coupled heat and mass transfer equations on the air side and the adsorbent side was established. The COMSOL Multiphysics which multiphysics coupling software is used to simulate the effect of fresh air inlet temperature, air inlet humidity, and face velocity on the enthalpy recovery wheel under summer conditions, revealing the heat and mass exchange performance of the enthalpy recovery wheel. This provides a reference for the optimal operation of the enthalpy recovery wheel.

Keywords Enthalpy recovery wheel · Fresh air · Heat recovery efficiency · Porous media

Nomenclature

c Water vapor concentration, g/m^3
c_p Constant pressure specific heat capacity, J/(kg k)
c_l Liquid water concentration on the side of the adsorbent, g/m^3
C_{pa} Specific heat of air, J/(kg K)
C_{cp} Adsorption, liquid phase and gas combined with specific heat, J/(kg K)
d_z Adsorbent layer thickness, m
d_e Hydraulic diameter, m
G_p Exhaust volume, kg/s

H. Fan · L. Chen (✉)
School of Energy, Xi'an University of Science and Technology, Xi'an 710054, China
e-mail: chenliu@xust.edu.cn

H. Fan
e-mail: 1065546892@qq.com

h	Heat transfer coefficient, W/(m² K)
K	Mass transfer rate, 1/s
K_a	Air–steam diffusion coefficient, m²/s
K_{dv}	Air–Steam effective diffusion coefficient of gas phase, m²/s
K_{dl}	Liquid water diffusion coefficient on the side of the adsorbent, m²/s
M	Dehumidification (parsing) volume, g/(m³ s)
q_{st}	Heat of sorption, J/kg
Q_t	On apparent heat exchange, kw
Q_d	Latent heat exchange, kw
W	Extent of adsorption kg adsorbate/kg adsorbent
λ_{cp}	Adsorbent, liquid phase and gas phase combined with thermal
t	Time, s
T	Temperature, K
u	Face velocity, m/s
x	Axial coordinate, m
y	Axial coordinate, m
Y	Moisture content, g/kg
T	Temperature, K

Greek letters

γ	Latent heat of vaporization, kJ/kg
λ	Thermal Conductivity, W/(m K)
ρ_{da}	Gas phase density on the adsorbent side, kg/m³
ρ	Density, kg/m³
ρ_{cp}	Adsorbent, liquid phase and gas phase bonding density, kg/m³

Subscripts

a	Air
d	Adsorbent
f	Fresh air
r	Return air
0	Initial state
1	Import
2	Export

1 Introduction

Desiccant wheels have two major applications: air dehumidification [1, 2] and enthalpy recovery [3, 4]. For dehumidification wheels, process air is dried after it flows through the wheel, which rotates constantly between the process air and a hot regenerative air stream. However, enthalpy recovery wheels are used to recover energy by transferring sensible and latent heat between supply air and exhaust air. Due to different operating conditions, heat and moisture transfer behaves quite differently in the wheels [5]. The enthalpy recovery wheel handles a large volume of air, flexible in layout, high heat recovery efficiency, and easy to clean. It is widely used as a heat recovery component in large air handling units. Enthalpy recovery wheel uses the wheel as the only heat exchange core to maintain 10–25 r/min speed rotation, and its performance determines the efficiency of the entire heat recovery system [6].

Pan and Gang [7] analyzed the working characteristics and energy-saving effect of the enthalpy recovery wheel. Nóbrega and Brum [8] established a mathematical model for the enthalpy wheel, an effectiveness number of thermal units (NTU) analysis is carried out. La [9] compared the effects of the changes of wheel thickness, rotation speed, and face velocity on the enthalpy recovery wheel efficiency when silica gel and lithium chloride were used as moisture absorbents. Horton [10] established a one-dimensional transient heat and mass transfer model and analyzed the performance of enthalpy recovery wheels both with and without purge air.

In this paper, the physical model of the enthalpy recovery wheel was established, taking into account the air side and adsorbent side. A coupled heat and mass transfer equation for the air and adsorbent side was established. The porous media saturated heat and mass transfer model is applied to the adsorbent side. Numerical simulation was carried out by using COMSOL Multiphysics, investigated the effect of the temperature of process air, humidity of process air, and face velocity on the performance of enthalpy recovery wheel under typical summer conditions.

2 Model of Enthalpy Recovery Wheel

Simulation study on the enthalpy recovery wheel with silica gel as adsorption material and rotational speed as 10 r/min, and the shape of the honeycomb channel is sinusoidal. The enthalpy recovery wheel is composed of the air and the adsorbent side, in which the adsorbent side is a porous medium composed of solid skeleton and pores, and the pores contain liquid water and gaseous steam. Silica gel remains solid during adsorption and desorption, and the adsorption process is generally physical adsorption [11].

2.1 Teat and Mass Transfer Model

For the air side and the adsorbent side of the enthalpy recovery wheel, establish the two-dimensional geometric model shown in Fig. 1 and assume that:

(1) The diffusivity of steam and air are assumed to be constant.
(2) The inlet air conditions are uniform in space.

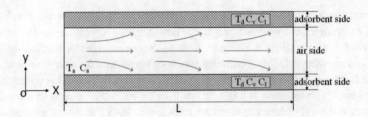

Fig. 1 Physical model of single-channel for enthalpy recovery wheel

(3) The adsorbent material is isotropic.
(4) All honeycombed channels in the wheel are identical and evenly distributed over the entire wheel.
(5) The analytical heat of silica gel is approximately equal to the adsorption heat.
(6) The adsorption potential energy of the pore surface of the adsorption material is negligible.
(7) The physical parameters of steam and liquid water are constant.
(8) The heat dissipation of the wheel shell is negligible.
(9) The effect of centrifugal force is negligible

Based on the above assumptions, the mathematical model of enthalpy recovery wheel is established:

(1) Energy balance differential equation on the air side:

$$\frac{\partial(\rho_a C_{pa} T_a)}{\partial t} - \frac{\partial}{\partial x}\left(\lambda_a \frac{\partial T_a}{\partial x}\right) - \frac{\partial}{\partial y}\left(\lambda_a \frac{\partial T_a}{\partial y}\right) = \frac{1}{d_e} h(T_d - T_a) \quad (1)$$

(2) Mass balance differential equation on the air side:

$$\frac{\partial c_a}{\partial t} - \frac{\partial}{\partial x}\left(K_a \frac{\partial c_a}{\partial x}\right) - \frac{\partial}{\partial y}\left(K_a \frac{\partial c_a}{\partial y}\right) = K(c_v - c_a) \quad (2)$$

(3) Energy balance differential equation on the adsorbent side:

$$\frac{\partial(\rho_{cp} C_{cp} T_d)}{\partial t} - \frac{\partial}{\partial x}\left(\lambda_{cp} \frac{\partial T_d}{\partial x}\right) - \frac{\partial}{\partial y}\left(\lambda_{cp} \frac{\partial T_d}{\partial y}\right) = \frac{1}{d_z} h(T_a - T_d) + Mq_{st} \quad (3)$$

(4) Gas phase mass balance differential equation on the adsorbent side:

$$\frac{\partial c_v}{\partial t} - \frac{\partial}{\partial x}\left(K_{dv} \frac{\partial c_v}{\partial x}\right) - \frac{\partial}{\partial y}\left(K_{dv} \frac{\partial c_v}{\partial y}\right) = K(c_v - c_a) + M \quad (4)$$

(5) Liquid phase mass balance differential equation on the adsorbent side:

$$\frac{\partial c_l}{\partial t} - \frac{\partial}{\partial x}\left(K_{dl}\frac{\partial c_l}{\partial x}\right) - \frac{\partial}{\partial y}\left(K_{dl}\frac{\partial c_l}{\partial y}\right) = M \qquad (5)$$

The initial conditions for the adsorbent and air are:

$$\begin{cases} W = W_0 \\ T_d = T_{do} \end{cases} \qquad (6)$$

$$\begin{cases} T_a = T_{a0} \\ Y_a = Y_{a0} \end{cases} \qquad (7)$$

The temperature and humidity boundary conditions for the air are:

$$T_a = \begin{cases} T_f \\ T_r \end{cases} \qquad (8)$$

$$Y_a = \begin{cases} Y_f \\ Y_r \end{cases} \qquad (9)$$

The above governing Eqs. (1)–(5), initial conditions (6)–(7) and boundary conditions (8)–(9), constitute a complete mathematical model of the enthalpy recovery wheel.

2.2 Performance Indexes

(1) Apparent heat exchange:

$$Q_t = G_p c_p (T_1 - T_2) \qquad (10)$$

(2) Latent heat exchange:

$$Q_d = G_p \gamma (Y_1 - Y_2) \qquad (11)$$

3 Model Parameters

The predefined interface [12, 13] of COMSOL Multiphysics is used to simulate. The standard inlet parameters and parameter changes during simulation are shown in Table 1, and the defined parameters in the model are shown in Table 2.

Table 1 Standard inlet parameters and parameter changes in simulation

Inlet parameter	Standard condition	Variation range
Fresh air temperature °C	35	28–37
Fresh air relative humidity %	50	40–70
Face velocity m/s	2	1–3
Return air temperature °C	24	–
Return air relative humidity %	50	–

4 Results and Discussion

When the return air parameters are constant, the face velocity, inlet temperature, and relative humidity of fresh air are, respectively, changed to simulate the change of the performance of the wheel in one turn (6 s).

(1) Face velocity

When other conditions are under standard conditions, the face velocity is changed from 1 to 3 m/s, simulating the effect of face velocity on the performance of the enthalpy recovery wheel.

Figure 2a, b, c shows the single-channel temperature distribution of fresh air when the face velocity increases from 1 to 3 m/s, and it can be observed that the temperature range near the wall surface of the micro-channel is large. With the increase of face velocity, only the temperature near the wall of the micro-channel changes and the amplitude is small. The analysis shows that the increase of the face velocity shortens the time that the fresh air stays in the micro-channel and shortens the heat transfer time between fresh air and adsorbent.

As shown in Figs. 3 and 4, Q_t and Q_d decrease with an increasing face velocity. The analysis shows that the increase of face velocity makes the time of air stay in the wheel shorter, and the heat and moisture exchange is not sufficient. In addition, the hygroscopic ability of adsorbent decreases with the increase in face velocity, which deteriorates the performance of the adsorbent and weakens the hygroscopic ability to fresh air. Therefore, Q_t and Q_d are gradually reduced. In practical application, under the condition of ensuring airflow, a larger wheel should be chosen to increase the airflow channel area and reduce the face velocity.

(2) Fresh air temperature

Temperature of fresh air increases from 28 to 37 °C when other operating conditions are in standard conditions, simulating the influence of fresh air temperature on the performance of the enthalpy recovery wheel.

As shown in Fig. 5, Q_t increases with temperature of fresh air, and the rise of fresh air temperature increases the temperature difference between fresh air and adsorbent and strengthens the heat transfer of the fresh air through the wheel, and so, Q_t increases. As shown in Fig. 6, the increase of air temperature causes a slight reduction of Q_d, The surface temperature of absorbent rises with the fresh air

Table 2 Model parameters and constants

Description	Value
Channel length m	0.15
Thickness of the wheel m	0.05
Adsorbent porosity	0.7
Adsorbent permeability m^2	1.28×10^{-14}
Adsorbent thermal conductivity w/(m·k)	0.175
Adsorbent heat capacity J/(kg·k)	921
Adsorbent density kg/m^3	1201
Ambient pressure pa	101,325

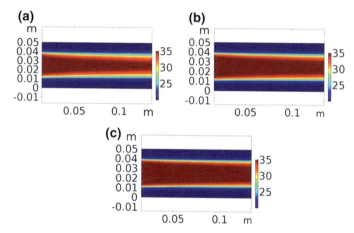

Fig. 2 Single-channel air temperature distribution **a** u = 1 m/s; **b** u = 2 m/s; **c** u = 3 m/s

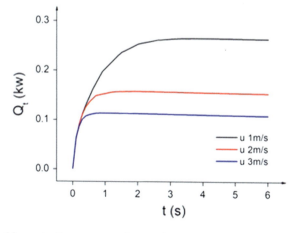

Fig. 3 Effect of face velocity on apparent heat exchange

Fig. 4 Effect of face velocity on latent heat exchange

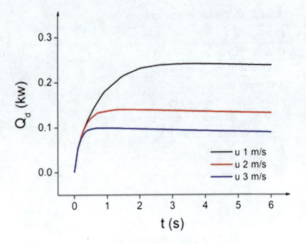

Fig. 5 Effect of fresh air temperature on apparent heat exchange

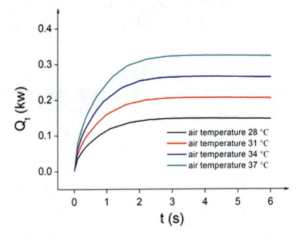

temperature, which causes the water vapor partial pressure difference between the fresh air and the adsorbent to decrease, and the mass transfer driving potential difference is reduced, which weakens the moisture transfer of the fresh air through the wheel, so Q_d decreases.

(3) Fresh air relative humidity

Relative humidity is increased by 65% from 50% (the corresponding moisture content increased from 17.7 to 23.21 g/kg) when other operating conditions are in standard conditions, simulating the influence of fresh air relative humidity on the performance of the enthalpy recovery wheel.

As shown in Fig. 7, Q_t remains basically unchanged with the increase of the relative humidity, and the fresh air humidity has no influence on the sensible heat exchange. As shown in Fig. 8, Q_d increase with inlet humidity, When fresh air is

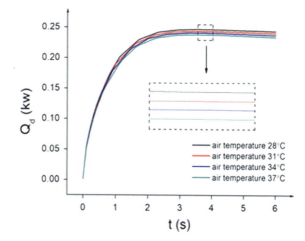

Fig. 6 Effect of fresh air temperature on latent heat exchange

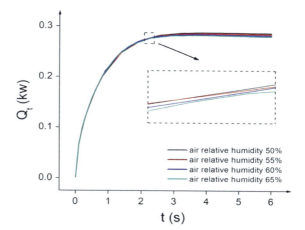

Fig. 7 Effect of relative humidity on apparent heat exchange

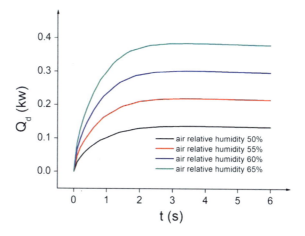

Fig. 8 Effect of relative humidity on latent heat exchange

more humid, a higher difference of vapor partial pressure between fresh air and the surface of the adsorbent side, which enhances the moisture transfer, so Q_d increases.

5 Conclusion

In this paper, the heat and mass transfer equations are established for the air side and the adsorbent side of the single channel of the enthalpy recovery wheel, and the numerical simulation is solved using COMSOL Multiphysics. The temperature distribution characteristics of the enthalpy recovery wheel are studied, and the influence of face velocity, fresh air temperature, and fresh air humidity on the performance parameters of the enthalpy recovery wheel is analyzed. By simulation results, it obtains the following conclusions: (1) The temperature distribution on the fresh air side of the enthalpy recovery wheel indicates that the temperature range near the wall is large, and the heat exchange mainly occurs near the wall surface. (2) Q_t and Q_d gradually decrease with an increasing face velocity. Therefore, in practical applications, the face velocity should not be too large in the case of ensuring airflow. (3) Q_t gradually increases and Q_d decreases slightly with the increase of fresh air temperature. (4) remains basically unchanged and the Q_d increases with the increase of relative humidity of fresh air.

Acknowledgements This work is supported by the Natural Science Foundation of China (NSFC) (Number51176104).

References

1. Kodama, et al.: Experimental study of optimal operation for a honeycomb adsorber operated with thermal swing. J. Chem. Eng. Jpn. **26**(5), 530–535 (1993)
2. Tauscher, R., et al.: Transport processes in narrow channels with application to rotary exchangers'. Heat Mass Transf. **35**(2), 123 (1999)
3. Besant, R.W., Simonson, C.J.: Heat and moisture transfer in energy wheels during sorption, condensation, and frosting conditions, J. Heat Transf.: Trans. ASME **120**(3), 699–708 (1998). Contribution, F.: Title. In: 9th International Proceedings on Proceedings, pp. 1–2. Publisher, Location (2010)
4. Kassai, M.: Experimental investigation on the effectiveness of sorption energy recovery wheel in ventilation system. Exp. Heat Transf. **31**(2), 106–120 (2018)
5. Zhang, L.Z., Niu, J.L.: Performance comparisons of desiccant wheels for air dehumidification and enthalpy recover. Appl. Therm. Eng. **22**(12) (2002)
6. Fang, J.H., et al.: Mathematical model performance analysis with variable condition of enthalpy recovery wheel. J. Shanghai Jiaotong Univ. **48**(6), 809–815 (2014)
7. Pan S.H, Gang, Z.: Character analysis on the running wheels of air ventilator with total thermal recovery. Refrig. Air Cond. Electr. Power Mach. **28**(5), 76–78 (2009)
8. Nóbrega, C.E.L., Brum, N.C.L.: Modeling and simulation of heat and enthalpy recovery wheels. Energy **34**(12), 2063–2068 (2009)

9. La, D., et al.: Numerical simulation of characteristics of enthalpy recovery wheel with different desiccants. J. Chem. Ind. Eng. **59**(S2), 64–69 (2008)
10. Horton, W.T., et al.: Modeling analysis of an enthalpy recovery wheel with purge air. Int. J. Heat Mass Transf. **55**(17–18), 4665—4672 (2012)
11. Rouquerol, etc., Adsorption by powders and porous solids: principles, methodology and applications, Academic Press (2013)
12. Datta, A.K.: Porous media approaches to studying simultaneous heat and mass transfer in food processes. II: property data and representative results. J. Food Eng. **80** (2013)
13. Datta, A.K.: Porous media approaches to studying simultaneous heat and mass transfer in food processes. I: problem formulations. J. Food Eng. **80** (2007)

A New Two-Stage Compression Refrigeration System with Primary Throttling Intermediate Complete Cooling for Defrosting

Chaohui Xuan, Yongan Yang and Ruishen Li

Abstract A new type of two-stage compression refrigeration system with primary throttling intermediate complete cooling is proposed in this paper. The defrost cycle of this system is still a two-stage compression cycle, and the refrigeration system can provide two evaporation temperatures at medium temperature or low temperature. Studies have shown that R410A is best suited for this system; increasing the evaporation temperature of 2 °C can increase COP by 4.3%; lowering the condensation temperature of 2 °C can increase COP by 6.4%; the COP of the system decreases with the increase of cooling capacity of the low-temperature stage but increase with the increase of the cooling capacity of the medium-temperature stage and the range of the system COP when the cooling capacity changes is obtained.

Keywords Two-stage compression · Defrosting · COP · System design

1 Introduction

As a low-temperature component of the refrigeration system, evaporator is easy to cause frosting [1–5] and lower the efficiency of system [6, 7]. Therefore, one hand of energy saving of the cold storage is to control of defrosting [8]. In order to solve the problem of evaporator frosting, scholars at home and abroad have mainly done research on the physical properties of frost, the technique of frost suppression, and the method of defrosting in the process of frosting [9–13].

At present, the effective defrosting method in the two-stage compression refrigeration system, that is, the pipeline connecting the inlet and outlet of the evaporator, is divided into a cooling branch and a defrosting branch, and the

---

C. Xuan · Y. Yang (✉) · R. Li
Tianjin Key Laboratory of Refrigeration Technology, Tianjin University of Commerce, Tianjin, China
e-mail: yyan@tjcu.edu.cn

C. Xuan
e-mail: xuanchaohui@163.com

© Springer Nature Singapore Pte Ltd. 2020
Z. Wang et al. (eds.), *Proceedings of the 11th International Symposium on Heating, Ventilation and Air Conditioning (ISHVAC 2019)*, Environmental Science and Engineering, https://doi.org/10.1007/978-981-13-9524-6_53

evaporator inlet defrosting pipe is connected to the compressor, and the evaporator outlet is connected to the gas–liquid separator. Since the method is switched from a bipolar compression refrigeration cycle system to a single-stage compression heat pump cycle during defrosting, the working pressure difference of the compressor participating in the unipolar cycle is extremely increased, causing damage to the compressor. Moreover, the temperature of the working medium of the cycle which enters the evaporator defrosting is low, as well as the defrosting efficiency.

To solve the above problems, this paper proposes a new two-stage compression refrigeration system with primary throttling intermediate complete cooling and selects different working conditions for thermal calculation.

2 System Overview

While considering the defrosting problem, the new system takes into account the needs of different warehouse temperatures in actual projects.

Figure 1 is a schematic diagram of this system. The system includes high-pressure compressors 1-2, low-pressure stage compressor 1-1, four-way reversing valve 2-1, throttle valve 4-2, low-temperature evaporator 6-1, medium-temperature evaporator 6-2, two-way valve 8, check valves 7-1, 7-2, and 7-3, condenser 5, a throttle valve 4-1, and an intercooler 3.

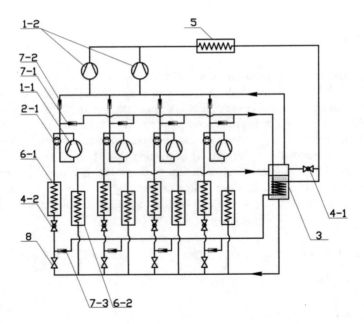

Fig. 1 System schematic

2.1 Thermodynamic Process of Refrigeration Cycle

The low-pressure stage compressor 1-1 draws low-pressure steam, compresses the steam into medium-pressure superheated steam and then presses the steam into the intercooler 3 for cooling. The high-pressure compressor draws medium-pressure saturated steam from the intercooler 3; then, the vapor condensed within the condenser 5 after being compressed. The high-pressure working fluid from the condenser 5 is divided into two parts, one part of which is throttled and depressurized by the throttle valve 4-1, becomes medium-pressure wet steam and enters the intercooler 3, and the other part of the high-pressure working fluid flows through coil in the intercooler 3 is cooled to a supercooled liquid. The medium-pressure liquid in the intercooler 3 partially evaporates and cools the medium-pressure superheated vapor and the high-pressure liquid fluid. The medium-pressure saturated liquid fluid from the intercooler 3 is evaporated in the intermediate-temperature evaporator 6-2, and the gaseous fluid from the intermediate-temperature evaporator 6-2 is returned to the intercooler 3. The supercooled liquid fluid from the coil in the intercooler 3 is throttled to low-pressure wet steam and then evaporated in the low-temperature evaporator 6-1. The low-pressure vapor from the low-temperature evaporator 6-1 is returned to the suction inlet of the low-pressure stage compressor 1-1 to complete the refrigeration cycle.

2.2 Thermodynamic Process of Defrosting Cycle

Based on the refrigeration cycle thermodynamic process, the defrosting cycle thermodynamic process is as follows: low-pressure stage compressor 1-1 sucks medium-pressure saturated vapor from intercooler 3 and compresses it into high-pressure superheated vapor. After throttling, the high-pressure superheated steam is discharged into the low-temperature evaporator 6-1 to condense and defrost the low-temperature evaporator 6-1. After condensation, the high-pressure liquid refrigerant changes to medium-pressure wet vapor and mixes with medium-pressure liquid coming out from intercooler 3 to form wet vapor. The wet vapor enters the medium-temperature evaporator 6-2. The refrigerant from each medium-temperature evaporator mixes and enters the intercooler 3 to complete the defrosting cycle.

2.3 P-H Schematic Diagram of the System

Figure 2 is the P-H diagram of the system. 1-2 represents the compressor compression process of low-pressure stage. 2-3 represents the cooling process of the exhaust gas from the compressor of the low-pressure stage in the intercooler.

Fig. 2 System P-H schematic

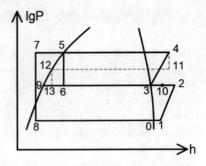

3-4 represents the compressor compression process of high-pressure stage. 4-5 represents the condensation process of the refrigerant. 5-6 represents the throttling process of a portion of the refrigerant entering the intercooler in the throttling valve. 6-3 represents the evaporation process of the refrigerant in the intercooler after throttling. 5-7 represents the subcooling process of the refrigerant entering the coils of the intercooler. 7-8 represents the throttling process of refrigerants flowing out of the intercooler coil in the throttle valve. 8-0 represents the evaporation process of refrigerant in the low-temperature evaporator. 9-10 represents the evaporation process in the medium-temperature evaporator. 10-3 represents the cooling process of superheated steam from a medium-temperature evaporator in an intermediate cooler. 4-11 represents the throttling process before defrosting. 11-12 represents the defrosting process of the evaporator in the low-temperature stage. 12-13 represents the throttling process of refrigerants from the evaporator of the low-temperature stage in the throttle valve.

3 Thermal Analysis

To facilitate thermal calculation, the following assumptions are made:

(1) The compression process is an adiabatic non-isentropic process;
(2) The refrigerant has no heat loss and pressure drop in the cycle;
(3) The enthalpy remains the same before and after throttling;
(4) The system runs in a stable state

Mass flow rate of refrigerant in low-pressure stage (Kg/s)

$$G_d = \frac{Q_0}{h_0 - h_8} \tag{1}$$

Mass flow rate of refrigerant in medium-pressure stage (Kg/s)

$$G_z = \frac{Q_z}{h_{10} - h_9} \tag{2}$$

Mass flow rate of refrigerant in high-pressure stage (Kg/s)
The energy conservation equation of the intercooler is presented:

$$G_g(h_3 - h_6) = G_d(h_2 - h_7) + G_z(h_{10} - h_9)$$

The equations above facilitate:

$$G_g = \frac{G_d(h_2 - h_7) + G_z(h_{10} - h_9)}{(h_3 - h_6)} \tag{3}$$

Intermediate pressure (MPa)

$$P_m = \sqrt{P_k * P_0} \tag{4}$$

Shaft power of compressor (KW)

$$P_e = \frac{G_d(h_2 - h_1)}{\eta_{id}\eta_m} + \frac{G_g(h_4 - h_3)}{\eta_{ig}\eta_m} \tag{5}$$

Coefficient of Performance

$$COP = \frac{Q_0 + Q_z}{P_e} = \frac{(Q_0 + Q_z)\eta_m}{\frac{G_d(h_2-h_1)}{\eta_{ig}} + \frac{G_d(h_4-h_3)}{\eta_{id}}} \tag{6}$$

In the formula, P_k is condensation pressure, kPa; P_0 is evaporation pressure, kPa; Q_0 is refrigeration capacity of low-pressure stage, kW; Q_z is refrigeration capacity of medium-pressure stage, kW; η_{id} is indicative efficiency of low-pressure stage, 0.85; η_{ig} is indicative efficiency of high-pressure stage, 0.85; η_m is mechanical efficiency of compressor, 0.95; T_0 is evaporation temperature, K; T_m is intermediate temperature, K; $h_0, h_1, h_2, h_3, h_4, h_6, h_7, h_8, h_9$, and h_{10} correspond to the enthalpy of each point in Fig. 2, KJ/kg.

4 Results and Analysis

4.1 Operating Parameters of the System with Different Refrigerants

Select the following parameters as the system runtime status: $T_k = 40\,°C$, $T_0 = -45\,°C$, $Q_0 = 6\,kW$, $Q_z = 4\,kW$, $T_a = 35\,°C$, $T_{s0} = -35\,°C$, and $T_{sz} = T_m + 5\,°C$. Ammonia, R410A, R404A, R134a, R502, and R407C are simulated by EES software. P_0, P_k, P_m, T_{sz}, and other operating parameters are given in Table 1.

When ammonia is used as refrigerant, the COP of the system is the largest which is 2.02. In refrigerant mass flow, ammonia is about 10% of the mass flow of other refrigerants. But the evaporation pressure of ammonia is 54.47 Pa, which is less than atmospheric pressure. The system is easy to mix with air when it is running, resulting in poor heat transfer performance of heat exchanger, thereby reducing the suction pressure of compressor and increasing the exhaust pressure. This problem also exists in R134a and R407C. In contrast, R410A is suitable for refrigerant in new system.

According to the design principle that the difference of heat transfer temperature between evaporator and storage temperature is 5 °C, the temperature of medium-temperature cold storage can reach −5 °C except R407C, which meets the temperature requirement of general precooling storage or fruit and vegetable storage. It is proved that the assumption of adding medium-temperature evaporator to the new system to produce two evaporation temperatures, medium temperature and low temperature, meets both theoretical requirements and engineering practice.

4.2 Effect of Evaporation and Condensation Temperatures on System Performance

R410A was selected as refrigerant, set $Q_0 = 6\,kW$ and $Q_z = 4\,kW$. T_0 ranges from −50 to −40 °C, and T_w ranges from 35 to 45 °C. Assume $T_a = T_w - 5\,°C$, $T_{s0} = T_0 + 5\,°C$, and $T_{sz} = T_m + 5\,°C$. EES software is used to simulate.

Table 1 System operating parameters under different refrigerants

Refrigerant	P_0 (kPa)	P_m (kPa)	P_k (kPa)	T_{sz} (°C)	G_d (Kg/s)	G_z (Kg/s)	G_g (Kg/s)	COP
Ammonia	54.47	291.10	1555.00	−5	0.0058	0.0031	0.0108	2.020
R410A	139.60	580.80	2416.00	−6	0.0409	0.0167	0.0774	1.718
R404A	105.10	439.00	1833.00	−5	0.0688	0.0224	0.1229	1.475
R134a	39.15	199.60	1017.00	−5	0.0493	0.0190	0.0867	1.771
R502	103.30	416.10	1676.00	−5	0.0717	0.0255	0.1276	1.617
R407C	66.09	338.80	1737.00	−3	0.0474	0.0176	0.0854	1.501

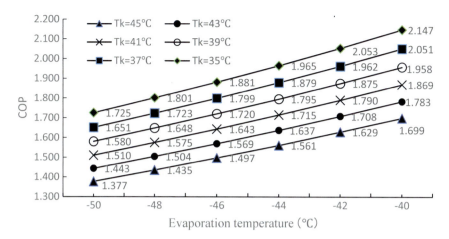

Fig. 3 COP at different evaporation and condensation temperatures

Figure 3 shows that the COP of the new system increases with the increase of evaporation temperature and the decrease of condensation temperature, but the growth rate is different. At the same condensation temperature, COP increases by about 4.3% for each 2 °C higher evaporation temperature, while at the same evaporation temperature, COP increases by about 6.4% for each 2 °C lower condensation temperature.

4.3 Effect of Refrigeration Capacity on COP of the System

R410A was selected as refrigerant, set $T_k = 40$ °C, $T_0 = -45$ °C, $T_a = 35$ °C, $T_{s0} = -45$ °C, and $T_{sz} = T_m + 5$ °C. The parameters under different Q_0 and Q_z were simulated by EES software.

Figures 4 and 5 show that COP of the new system shows an upward trend with the increase of the refrigeration capacity of the intermediate-temperature stage and a downward trend with the increase of the refrigeration capacity of the low-temperature stage. This is because when the refrigeration capacity of intermediate-temperature stage is 0, the new system is the same as the ordinary one-throttle intermediate complete cooling system. When the intermediate refrigeration capacity exists, the system has a part of additional refrigeration capacity, and the effect on the system is only due to the increase of a part of the high-pressure stage flow rate. When the low-temperature refrigeration capacity is zero, the new system is equivalent to a single-stage refrigeration cycle operating under condensation pressure and intermediate pressure. With the increase of low refrigeration capacity, the system is closer to a two-stage compression refrigeration cycle, and the COP decreases. Therefore, the COP of the system ranges from 1.323 to 3.110 under the simulated conditions.

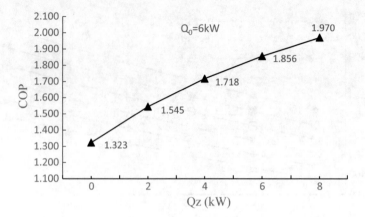

Fig. 4 COP at different refrigeration capacities of intermediate-temperature stage

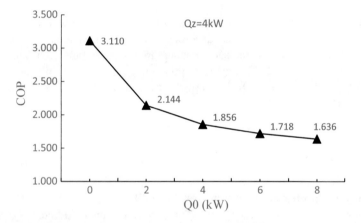

Fig. 5 COP at different refrigeration capacities of low-temperature stage

5 Conclusion

A new two-stage compression refrigeration system with primary throttling intermediate complete cooling is proposed. The thermodynamic model of the new system under different working conditions is established and calculated. The conclusion is drawn as follows:

(1) R410A is recommended as the refrigerant of the new system;
(2) Reducing condensation temperature and increasing evaporation temperature can both increase COP, but the effect of reducing condensation temperature is more obvious;

(3) The lower the condensation temperature is, the more obvious the improvement of system performance by increasing the evaporation temperature; the higher the evaporation temperature is, the more obvious the improvement of system performance by decreasing the condensation temperature.
(4) COP increases with the increase of refrigeration capacity of medium-temperature stage and decrease of refrigeration capacity of low-temperature stage. Change the refrigeration capacity under the same working conditions. The range of COP is from COP when the refrigeration capacity of medium temperature stage is 0 to COP when the refrigeration capacity of low temperature stage is 0.

References

1. Feng, H., et al.: Study on flow control to improve the defrosting performance of liquid refrigerants. Low Temp. Supercond. **45**(5), 70–73 (92) (2017)
2. Dong, Y.: Experimental study on frost formation on ultra-low temperature surfaces. Beijing University of Technology, Beijing (2016)
3. Huang, D., et al.: Effects of fin types on frosting characteristics of heat pump air conditioning. J. Refrig. **3**(2), 12–17 (2012)
4. Amer, M., et al.: Review of defrosting methods. Renew. Sustain. Energy Rev. **73**, 53–74 (2017)
5. Yu, C., et al.: Study on frosting and defrosting performance of centralized air supply cold storage. Low Temp. Supercond. **46**(05), 88–92 (2018)
6. Li, Z., et al.: Comparative analysis of hot gas defrosting and electrothermal defrosting in cold storage. Refrig. Air Cond. **25**(06), 577–579 (2011)
7. Kandula, M.: Frost growth and densification flow over flat surfaces. Int. J. Heat Mass Transf. **54**(15), 3719–3731 (2011)
8. Kim, D., et al.: Frosting model for predicting microscopical and local frost behaviors on a cold plate. Int. J. Heat Mass Transf. **82**, 135–142 (2015)
9. Jung, H., et al.: An experimental study on performance improvement for an air source heat pump by alternate defrosting of outdoor heat exchange. Int. J. Air Cond. Refrig. **22**(03), 1450017 (2014)
10. Kim, K., et al.: Local frosting behavior of a plated-fin and tube heat exchanger according to the refrigerant flow direction and surface treatment. Int. J. Heat Mass Transf. **64**, 751–758 (2013)
11. Su, S., et al.: Application of refrigerant flow control method in defrosting of air-cooled heat pump. Refrig. Air Cond. **25**(3), 213–215 (2011)
12. Han, Z., et al.: New system and experimental study of air source heat pump energy storage hot gas defrosting. J. Harbin Univ. Technol. **6**(6), 901–903 (2007)
13. Li, X.: Refrigeration Principle and Equipment. Machinery Industry Press (2006)

Prediction of Pressure Drop in Adsorption Filter Using Friction Factor Correlations for Packed Bed

Ruiyan Zhang, Zhenhai Li, Lingjie Zeng and Fei Wang

Abstract Adsorption filters in air cleaners are usually produced as honeycomb structure filled with granular adsorbent, which can be seen as numerous parallel arranged fixed beds. In order to investigate if the friction factor correlations for packed bed can be used in adsorption filter, experimental study was conducted using adsorption filters filled with spherical, columnar and crushed activated carbon from 2 to 20 mesh. The range of void fraction of adsorption filters is 0.29–0.92, and the range of particle Reynolds number is 8–1072. Comparing the experimental data with the calculation results using 12 published friction factor correlations, results show that large differences exist between the experimental data and the calculated values. Considering such large differences, friction factor correlations for packed bed are likely not suitable for prediction of pressure drop in adsorption filter.

Keywords Pressure drop · Adsorption filter · Friction factor

R. Zhang · Z. Li (✉) · L. Zeng · F. Wang
School of Mechanical Engineering, Tongji University, Shanghai, China
e-mail: Lizhenhaioffice@163.com

R. Zhang
e-mail: 1610290@tongji.edu.cn

L. Zeng
e-mail: zenglingjie1990@foxmail.com

F. Wang
School of Environment and Architecture, University of Shanghai for Science and Technology, Shanghai, China
e-mail: fidle_cn@163.com

1 Introduction

Adsorption is currently considered to be one of the most effective method for removing gaseous contaminations such as volatile organic compounds (VOCs) from air in buildings and protecting the persons indoor [1]. As a kind of technical application of adsorption, adsorption filters are commonly used in portable air cleaners and air-handling units in air-conditioning system.

Not like the industrial waste gas treatment equipment designed to remove as much waste gas as possible before air is released, a considerable portion of adsorption filters for indoor air cleaning are designed to remove contaminants slightly but continuously from the circulating air. Therefore, they have significantly lower resistance and lower removing efficiency compared with the former [2]. For selection of an adsorption filter, its resistance and efficiency are important information, which obviously correlate with its friction factor [3, 4] and are fundamentally determined by its structure (such as the size and shape of the filled granular adsorbent and the degree of fullness).

The correlations of friction factor for fixed bed are usually adopted for calculating the pressure drop in adsorption equipment with the structure similar with a fixed bed, namely a flow path filled with granular matter. The pressure drop in packed bed can be correlated in terms with the dimensionless friction factor, as

$$\Delta p = \frac{Lf\rho_f v^2}{d_p} \qquad (1)$$

where Δp is pressure drop, f is dimensionless particle friction factor, ρ_f is fluid density, v is superficial velocity and d_p is particle diameter.

Most of published equations for calculating pressure drop in fixed bed can be easily converted into the form of correlations of friction factor with Reynolds number Re and several structural parameters, as shown in Table 1.

The Reynold number is defined as

$$\mathrm{Re} = \frac{\rho_f v d_p}{\mu} \qquad (18)$$

where μ is fluid viscosity.

Although extensive research has been carried out on pressure drop calculation of fixed bed, the structural characteristics of most adsorption filters make them more extreme cases compared with the applicable range of these correlations. Firstly, the diameter ratio between the flow channel units and the adsorbent particles D/d_p are relatively small, making the low resistance area near the wall, namely the wall effect, could not be neglected. Secondly, the granules filled in adsorption filters are usually irregular in shape and size, producing greater randomness in the way of stacking. Thirdly, the Reynolds number of adsorption filters at their normal working

Table 1 Friction factor correlations for the main range of Re in adsorption filters

Source	Equation	
Chilton and Colburn [5]	$Re < 40 \ f = 805/Re$	(2)
	$Re > 40 \ f = 38/Re^{0.15}$	(3)
Carman [6]	$f = \frac{180(1-\varepsilon)^2}{\varepsilon^3 Re} + \frac{2.87(1-\varepsilon)^{1.1}}{\varepsilon^3 Re^{0.1}}$	(4)
Rose [7]	$f = 1000 Re^{-1} + 60 Re^{-0.5} + 12$	(5)
Rose and Rizk [8]	$f = 1000 Re^{-1} + 125 Re^{-0.5} + 14$	(6)
Ergun [9]	$f = \frac{150(1-\varepsilon)^2}{\varepsilon^3 Re} + \frac{1.75(1-\varepsilon)}{\varepsilon^3}$	(7)
Hicks [10]	$f = \frac{6.8(1-\varepsilon)^{1.2}}{\varepsilon^3 Re^{0.2}}$	(8)
Brauer [11]	$f = \frac{160(1-\varepsilon)^2}{\varepsilon^3 Re} + \frac{3.1(1-\varepsilon)^{1.1}}{\varepsilon^3 Re^{0.1}}$	(9)
Tallmadge [12]	$f = \frac{150(1-\varepsilon)^2}{\varepsilon^3 Re} + \frac{4.2(1-\varepsilon)^{\frac{7}{6}}}{\varepsilon^3 Re^{\frac{1}{6}}}$	(10)
Kuerten, reported by Watanabe [13]	$f = \left[\frac{25(1-\varepsilon)^2}{4\varepsilon^3}\right](21 Re^{-1} + 6 Re^{-0.5} + 0.28)$	(11)
Macdonald et al. [14]	Smooth[a] particles $f = \frac{180(1-\varepsilon)^2}{\varepsilon^3 Re} + \frac{1.8(1-\varepsilon)}{\varepsilon^3}$	(12)
	rough particles $f = \frac{180(1-\varepsilon)^2}{\varepsilon^3 Re} + \frac{4(1-\varepsilon)}{\varepsilon^3}$	(13)
Eisfeld and Schnitzlein [15]	$f = \frac{155(1-\varepsilon)^2}{\varepsilon^3 Re} A^2 + \frac{1-\varepsilon}{\varepsilon^3}\frac{A}{B}$	(14)
	$A = 1 + \frac{2}{3(D/d_p)(1-\varepsilon)}$	(15)
	$B = \left[1.42\left(\frac{d_p}{D}\right)^2 + 0.83\right]^2$	(16)
Montillet et al. [16]	$f = a\left(\frac{1-\varepsilon}{\varepsilon^3}\right)\left[\frac{D}{d_p}\right]^{0.2}(1000 Re^{-1} + 60 Re^{-0.5} + 12)$	(17)
	$a = 0.061$ for $\varepsilon < 0.4$	
	$a = 0.05$ for $\varepsilon \geq 0.4$	

[a]Surface elevation changes are about two orders of magnitude smaller than the local channel diameter
ε is void fraction of adsorption filter, D is diameter of a unit in adsorption filter

flow rate is relatively small while the void ratio can be really large in some cases. So, there is no clear published evidence that which equation of the fixed bed friction factor is the best for adsorption filters.

In order to solve the above-mentioned problem, in this paper, the resistance curves of the adsorption filters with typical skeletal structure were experimentally measured in the range of their common working flow rates, with emphasis on changing thickness, particle shape and size distribution, to cover as many common adsorbents as possible. The results of above experiment were compared with the predicted results using various friction factor equations, and the overall deviation by those equations was investigated, to find if there is one best suited for adsorption filters.

2 Experimental Set up and Procedure

Since the roughness differenced by material is not included in all of the above friction factor formulas, activated carbon was selected as the representative adsorbent in this experiment because of its pervasiveness, and it has the largest variety of shapes, containing most of the common shapes of other adsorbents. The physical properties of activated carbon used are shown in Table 2.

The approximation of diameter using sphericity and screen size was introduced in the experiment, because it is difficult to accurately measure the precise geometry of mass of activated carbon particles. Particles of different screen size d_{scr} are obtained by sorting them using sieves in Tyler standard. Then, the particle diameter d_p can be converted by the following formulas [18].

For activated carbon balls and coal-crushed activated carbon,

$$d_p \cong \psi d_{scr} \tag{19}$$

For columnar activated carbon and coconut shell activated carbon,

$$d_p \cong \psi^2 d_{scr} \tag{20}$$

The void fraction ε of the adsorption filter is calculated by the following formula,

$$\varepsilon = \frac{V_{carbon}}{V_{filter}} \tag{21}$$

where V_{carbon} is the total volume of the filled activated carbon particles, and V_{filter} is the volume of filter.

The tested adsorption filters have two different thickness, 1.5 and 0.8 cm, which are within the usual thickness range. The volume of filter is obtained by filling 200–300 mesh sand in the adsorption filter and measuring the volume of filled sand. The total volume of the filled activated carbon particles was measured by volume displacement technique using water. In order to eliminate the error caused by water adsorption, the activated carbon was pre-soaked to saturated, and then the excess water on the surface was removed by tissue paper before measuring.

The experimental apparatus is shown in Fig. 1. The adsorption filters were vertically fixed in the air duct at the direction perpendicular to the wind speed.

Table 2 Physical properties of packing materials

Sorts of activated carbon	Tyler standard screen size (mm)	Sphericity ψ [17]
Activated carbon balls	3.3–6.7	1
Columnar activated carbon	4.7–2.4	0.65–0.75
Coal-crushed activated carbon	3.3–0.83	0.81
Coconut shell activated carbon	4.7–2.4	0.55–0.7

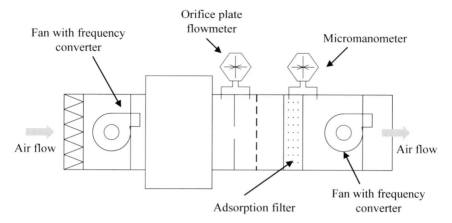

Fig. 1 Schematic of pressure drop experiment

The fans connected with frequency converter and the orifice plate flowmeter worked together to obtain a certain volume flow of air, making wind speed for each of the tested adsorption filter ranges from 0.1 to 2 meters per second. The pressure drop across the adsorbent filter was measured by a micromanometer.

3 Results and Discussion

3.1 Comparison Between the Experimental Data and the Calculation Results

The measured value and the calculated value were compared in 558 different cases with diverse combination in shape and size of particles (spherical, columnar and crushed activated carbon from 2 to 20 mesh), thick of filters (1.5 and 0.8 cm), void fraction (0.29–0.92) and air velocity (0.1 to 2 m/s with Reynolds number ranges from 8 to 1072), for the purpose of containing most of practical common cases. The measured value is the actual measured pressure loss in Pa to a certain filter at a certain wind speed, while the calculated value is the values of pressure loss in Pa calculated, respectively, according to the 12 formulas in Table 1, using the measured parameters d_p, L, ε, and the Reynolds number converted from the velocity of corresponding experiment.

Since the measured resistance values range from 1 to 491 Pa, which is largely differenced with each other, the measured and calculated values are compared with relative errors but not the absolute errors. The measured values are regarded as true values.

In Fig. 2, the overall accuracy of these friction factor correlations is compared. The relative errors for all cases of each formula are divided into five levels:

(1). The relative errors are greater than one order of magnitude (the calculated value differs from the measured value by more than ten times), and the calculated values are greater than the measured values, (2). The relative errors are greater than 30%, but less than one order of magnitude, and the calculated values are greater than the measured values, (3). The relative errors are less than 30%, (4). The relative errors are greater than 30%, but less than one order of magnitude, and the predicted values are less than the measured value, and (5). The relative errors are greater than one order of magnitude, and the calculated values are greater than the measured values. The proportion of the relative errors of each level in all prediction cases are marked in dark red, dark yellow, green, light yellow and pink in Fig. 2, respectively.

Results show that big difference exists between the experimental data and the calculated values. Friction factor correlations of Chilton and Colburn, Carman, Ergun, Hicks, Brauer, Tallmadge, Macdonald and Eisfeld and Schnitzlein can predict the order of magnitude of pressure drop in adsorption filter in more than 90% of all 558 cases. However, the ratio when the relative errors are less than 30% is very small, only about 10%, or even less. And only the correlation of Eisfeld and Schnitzlein has achieved the order of magnitude accuracy in all cases.

The difference between predicted values is compared in Fig. 3. For each case, the largest predicted value is represented by a solid yellow line, the smallest predicted value is represented by a yellow dashed line, and the corresponding measured value is represented as a black line. The data sets are sorted by the measured values.

It can be seen in Fig. 3, large differences exist among the calculated values of different friction factor correlations, which can be two orders of magnitude in the

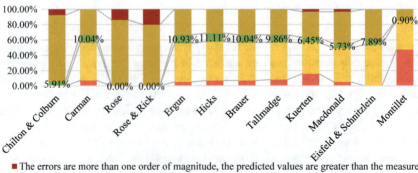

■ The errors are more than one order of magnitude, the predicted values are greater than the measured value

■ The relative errors are greater than 30%, the predicted values are greater than the measured value

■ The relative errors are less than 30%

■ The relative errors are greater than 30%, the predicted values are less than the measured value

■ The errors are more than one order of magnitude, the predicted values are less than the measured value

Fig. 2 Overall comparison of relative errors of different friction factor correlations

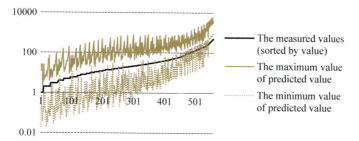

Fig. 3 Measured pressure loss and the range of predicted pressure loss

area of small pressure loss. However, in most cases where the measured pressure loss is lower than 100 Pa, the measured values lie in the middle of the band of the predicted values. When the pressure loss is greater than 100 Pa, the measured values tend to be smaller than the predicted values. This phenomenon may be due to the enhancement of the wall effect at high air velocity, considering the diameter ratio in the adsorption filter is usually smaller than the fixed bed, while the wall effect is ignored or underestimated in friction factor correlations.

3.2 Accuracy and Common Applicability Range Parameters

Most of the friction factor correlations have their announced scope of application, usually confined by the parameters of Re, ε and D/d_p, which are also the most important parameters in the formulas. The prediction accuracy may change, as the values of these three parameters change. In order to clearly figure out the relationship in adsorption filters, the Pearson correlation coefficients between the relative errors and each parameter values are calculated and compared in Fig. 4.

The Pearson correlation coefficients of Reynolds numbers and relative errors are all negative, and the absolute value is less than 0.4, which means for all correlations

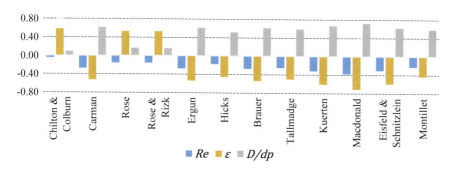

Fig. 4 Pearson correlation coefficients between the relative errors and Re, ε and D/d_p

the larger the Reynolds number is, the lower is the prediction accuracy, but the relationship between the two is not significant.

On the contrary, the Pearson correlation coefficients of diameter ratios and relative errors are positive, and the absolute values are relatively lager except for correlations of Chilton and Colbum, Rose and Rose and Rizk, which means the larger the diameter ratio is, the higher is the prediction accuracy, and for most correlations it has more significant influence than of Reynolds number.

In a similar way, the accuracy of friction factor correlations of Chilton and Colburn, Rose and Rose and Rizk increase when the void fraction increases, while the accuracy of correlations of Carman, Ergun, Hicks, Brauer, Tallmadge, Kuerten, Macdonald, Eisfeld and Schnitzlein and Montillet increase when the void fraction decreases. However, considering the prediction accuracy of the first three correlations are relatively low among all correlations, the improvement of the prediction accuracy seems to be limited when using them by large void fraction.

4 Conclusions

This work experimentally investigates the pressure drop in different adsorption filters and evaluates the applicability of 12 friction factor correlations in prediction of the pressure drop. And the scope of application of each correlation is briefly discussed.

For all tests, the differences between the calculated values and the measured values are mostly in an order of magnitude. But only in less than 10% cases, the relative errors are less than 30%. Such accuracy is not sufficient for engineering application. And the potential improvement of accuracy is small, even if the friction factor correlations are selected according to the operating conditions.

Therefore, in general, the friction coefficient correlations of the packed bed are likely to be unsuitable for predicting the pressure drop in adsorption filters. It is necessary to establish a friction factor correlation and corresponding resistance prediction formula more suitable for adsorption filters.

References

1. Zhang, Y., et al.: Can commonly-used fan-driven air cleaning technologies improve indoor air quality? A Lit. Rev. Atmos. Environ. **45**(26), 4329–4343 (2011)
2. Chen, W., et al.: Performance of air cleaners for removing multiple volatile organic compounds in indoor air. ASHRAE Trans. **111**(1), 1101–1114 (2005)
3. Pei, J., Zhang, J.: Modeling of sorbent-based gas filters: development, verification and experimental validation. Build. Simul. Tsinghua Press. **3**(1), 75–86 (2010)
4. Mugge, J., et al.: Measuring and modelling gas adsorption kinetics in single porous particles. Chem. Eng. Sci. **56**(18), 5351–5360 (2001)

5. Chilton, T.H., Colburn, A.P.: II—pressure drop in packed tubes. Ind. Eng. Chem. **23**(8), 913–919 (1931)
6. Carman, P.C.: Fluid flow through granular beds. Trans. Lond. Inst. Chem. Eng. **15**, 150–166 (1937)
7. Rose, H.E.: On the resistance coefficient—Reynolds number relationship for fluid flow through a bed of granular material. Proc. Inst. Mech. Eng. **153**(1), 154–168 (1945)
8. Rose, H.E., Rizk, A.M.A.: Further researches in fluid flow through beds of granular material. Proc. Inst. Mech. Eng. **160**(1), 493–511 (1949)
9. Ergun, S.: Fluid flow through packed columns. Chem. Eng. Prog. **48**, 89–94 (1952)
10. Hicks, R.E.: Pressure drop in packed beds of spheres. Ind. Eng. Chem. Fundam. **9**(3), 500–502 (1970)
11. Brauer, H.: Grundlagen der Einphasen- und Mehrphasenströmungen. Vol. 2. Sauerländer AG, Aarau, Switzerland (1971)
12. Tallmadge, J.A.: Packed bed pressure drop—an extension to higher Reynolds numbers. AIChE J. **16**(6), 1092–1093 (1970)
13. Watanabe, H.: Drag coefficient and voidage function on fluid—flow through granular packed-beds. Int. J. Eng. Fluid Mech. **2**(1), 93–108 (1989)
14. Macdonald, I.F., et al.: Flow through porous media—the Ergun equation revisited. Ind. Eng. Chem. Fundam. **18**(3), 199–208 (1979)
15. Eisfeld, B., Schnitzlein, K.: The influence of confining walls on the pressure drop in packed beds. Chem. Eng. Sci. **56**(14), 4321–4329 (2001)
16. Montillet, A., et al.: About a correlating equation for predicting pressure drops through packed beds of spheres in a large range of Reynolds numbers. Chem. Eng. Process. **46**(4), 329–333 (2007)
17. Brown, G.G.: Unit operations. Wiley, New York (1950)
18. Levenspiel, O.: Engineering flow and heat exchange. Springer (2014)

Evaluations and Optimizations on Practical Performance of the Heat Pump Integrated with Heat-Source Tower in a Residential Area in Changsha, China

Fenglin Zhang, Nianping Li, Haijiao Cui, Jikang Jia, Meng Wang and Meiyao Lu

Abstract The heat-source tower (HST) integrated heat pump system has more potential in energy saving than air-source heat pump system in the Yangtze River region. However, the heat pump unit and HST had mismatch on operation in field tests. And the actual maximum load of the HST integrated heat pump system was generally smaller than its design load, which will lead to higher energy consumption, and therefore it is important to optimize the system with the climate characteristics of the Yangtze River region. The HST integrated heat pump system in a certain residential area in Changsha, China was taken for an example to be monitored in real time. And the practical energy performance and typical issues of the system were analyzed. The results showed that the system failed to achieve a high COP for its low part-load rate (PLR) with oversized capacity design and unreasonable control strategies. Based on the analysis, some suggestions were made in this paper, which gave solutions to the heating and cooling of residential buildings in the Yangtze River region.

Keywords Heat-source tower integrated heat pump · Residential buildings · Operation optimization · Practical energy performance · Control strategies

1 Introduction

The heat-source tower (HST) integrated heat pump system is a new energy-saving air-source heat pump system (ASHPS), which can prevent the unit from freezing under low temperature and high humidity [1]. It has been well promoted and used for heating and cooling in the Yangtze River region, 32.7% more efficient than ASHPS [2].

F. Zhang · N. Li (✉) · H. Cui · J. Jia · M. Wang · M. Lu
College of Civil Engineering, Hunan University, Changsha 410082, China
e-mail: linianping@126.com

Previous researches have mainly studied the applicability of the HST integrated heat pump system. The study of Li et al. [3] compared the economic analysis of HST integrated heat pump system and ASHPS in an office building model. The results showed that HST integrated heat pump system had a lower initial and annual operating cost. When it came to the system performance of the HST integrated heat pump system, a majority of research mainly aimed to study the winter performance ignoring the summer performance, which had a greater energy-saving potential (23.1%) [2]. Besides, the study of Gong [4] demonstrated that cooling load distribution of HVAC system in residential buildings was quite different from that of public buildings. It had large discreteness and low load ratio. As the units will not always operate under design load value in actual operation, strategies need to be made to increase the part-load rate (PLR) of heat pumps to improve the energy efficiency. Especially, for the heat pump unit with screw compressor, its efficiency was greatly influenced by PLR and will have better performance with increasing PLR [5]. Besides, the integrated part-load value (IPLV) is generally used to evaluate the annual energy consumption levels [6]. Based on the load characteristic of residential areas, this paper introduced the field test results on one residential area in Changsha. According to the field test data, some typical problems were found in HST integrated heat pump systems about the heat pump unit configuration, water pump selection and the unreasonable control strategies. Therefore, some main improvement measures were submitted.

2 Methods

2.1 System Description and Analytical Method

The project was a residential area in Changsha, China with a total construction area of 67,731 m^2. The HST integrated heat pump system was adopted. The HST located in the northwest corner of the residential area and the equipment room was in the basement. The design cooling, heating and hot water load were 3584 kW, 3060 kW and 342 kW, respectively. The system mainly consisted of two screw heat pump units (1#, 2#) with heat recovery and one maglev heat pump unit (3#). Each heat pump unit was equipped with two chilled water pumps (one for standby) and two cooling water pumps (one for standby). The specifications of the major components were presented in Table 1. The circulating medium of the HST system in summer was water while in winter was antifreeze solution ($CaCl_2$). In winter, the HST directly extracted heat from air. Heat pump unit 1# and 3# supplied air conditioning hot water and unit 2# supplied domestic hot water. In summer, all heat pump units were used for cooling, while only unit 2# supplied domestic hot water at the same time for its heat recovery.

The real-time operation data was obtained by field data acquisition system. The temperature and flow rate of both supply and return water in user side, heat source

Table 1 Specification of the major components in the system

Component	Number	Parameter
Heat pump unit 1#	1	Rated cooling capacity: 2140 kW Cooling power consumption: 376 kW Rated heating capacity: 1497 kW Heating power consumption: 397 kW
Heat pump unit 2#	1	Rated cooling capacity: 513 kW Cooling power consumption: 110 W Rated heating capacity: 343 kW Heating power consumption: 102 kW
Heat pump unit 3#	1	Rated cooling capacity: 950 kW Cooling power consumption: 178 W Rated heating capacity: 1134 kW Heating power consumption: 248 kW
Heat-source tower 1#	3	Flow rate: 350 m^3/h Fan power consumption: 11 kW
Heat-source tower 2#	1	Flow rate: 250 m^3/h Fan power consumption: 7.5 kW
Chilled water pump 1#	2	H = 42 m H$_2$O, Q = 370 m^3/h
Chilled water pump 2#	2	H = 42 m H$_2$O, Q = 110 m^3/h
Chilled water pump 3#	2	H = 42 m H$_2$O, Q = 220 m^3/h
Cooling water pump 1#	2	H = 40 m H$_2$O, Q = 440 m^3/h
Cooling water pump 2#	2	H = 40 m H$_2$O, Q = 130 m^3/h
Cooling water pump 3#	2	H = 40 m H$_2$O, Q = 220 m^3/h

side and heat recovery side were monitored. Moreover, measurements were all taken for the electricity consumption of the heat pump units, water pumps and HST. According to the field tests data and energy performance index, the practical energy performance of the HST integrated heat pump system can be analyzed.

2.2 Practical Energy Performance of Heat Pumps in Cooling and Heating Season

(1) System energy consumption in cooling season

According to the measured data from November 15, 2017 to November 15, 2018, it was found that the cooling time of this project was from June 10, 2018 to September 19, 2018 and the total energy consumption was 497,385 kWh. Figure 1 depicted the statistic of energy consumption. Heat pump units consumed energy the most (70%). Among them, unit 3# accounted for 50% for its longest running time, while unit 2# accounted for 15% for its intermittent operation (worked for 30 min and stopped for one hour). Unit 1# worked only on weekends with the least running time (5%).

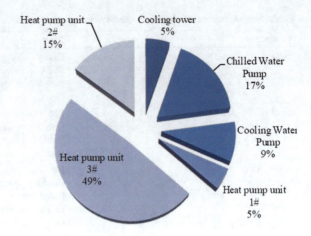

Fig. 1 Cooling energy consumption

The power consumption of cooling water pump accounted for only 9% and cooling tower power 5%, which were lower than normal level. It was due to the heat recovery system that only unit 3# was connected to the cooling tower in summer. In addition, the selected head (H) of the cooling water pump (40 m H_2O) was oversized leading to a low efficiency of water pumps. And the power consumption of chilled water pump accounted for 17% higher than the normal level. It was concluded from the analysis that the selected H (42 m H_2O) of the chilled water pump was too high. Besides, each unit was equipped with two pumps (one for standby), without considering the mutual standby between the equipments.

(2) **Practical performance of heat pump unit under PLR in cooling season**

The cooling load time frequency (CLTF) refers to the ratio of the actual operating hours under a certain load rate to the total operating hours of the HVAC system during the cooling season. The partial load rate (PLR = Q_e/Q_0) is defined as the ratio of the actual capacity to the rated capacity [7], where Q_e is the rated capacity, Q_0 is the actual capacity. For the unit studied in this paper, the discrete points and fitting curves of COP under various PLR can be obtained through the calculation of the measured data under typical conditions. The CLTF and COP under various PLR of heat pump units were shown in Figs. 2, 3 and 4, respectively. For unit 1#, the highest CLTF of 26.01% was achieved when PLR was 60% but COP was only 4.1. For unit 2#, the highest COP was achieved when PLR was 80%, but CLTF was only 5.98%. For unit 3#, it ran over 40% time at low PLR and COP. All units failed to work in high PLR resulting in lower IPLV.

(3) **COP of the system in typical summer days**

Figure 5 depicted the power consumption and system COP calculated from the field tests in four typical summer days (i.e., 11th August–14th August) in 2018. It was apparent to find that COP fluctuated between 2.5 and 3.5 with the mean value of 3.03, much lower than the normal value. Moreover, the power consumption of

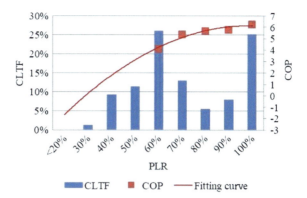

Fig. 2 CLTF and COP under PLR of unit 1#

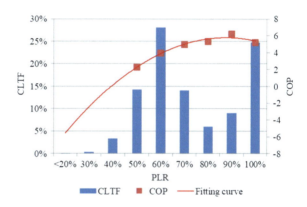

Fig. 3 CLTF and COP under PLR of unit 2#

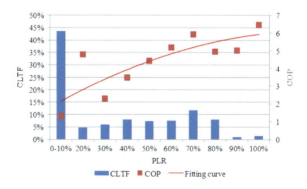

Fig. 4 CLTF and COP under PLR of unit 3#

heat pump unit reached the minimum during 0:00–9:00 and the maximum during 18:00–23:00. The value showed periodic changes affected by load characteristics of residential buildings.

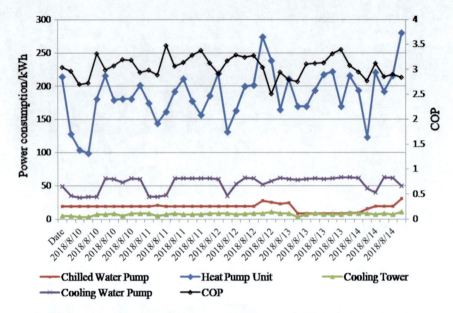

Fig. 5 Power consumption and COP of the system with the variation of times in summer

(4) **Cooling temperature in a typical summer day**

Figure 6 analyzed the variation of temperature of the supply chilled water (T_1), return chilled water (T_2), evaporation temperature (T_0) of unit 2#, ambient temperature (T_a) and chilled water flow rate (G) in a typical summer day (11th August). Results showed that: T_1 fluctuated in the range of 6.79–12.46 °C, with the average temperature difference (Δt_1) between T_2 and T_1 of 2.31 °C design code recommends "small G and large Δt_1," while in this case Δt_1 was only about 1.41–2.90 °C, inevitably leading to a large G and more energy consumption of distribution system. And T_0 fluctuated greatly, due to the frequent start and stop of the unit and the lack of control strategies. It reduced to −5.2 °C at 12:05 and increased to 16.4 °C.

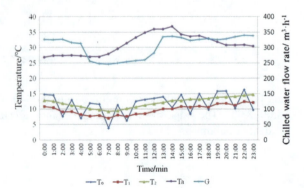

Fig. 6 Cooling temperatures and chilled water flow rate in typical summer day

(5) **Heating energy and the partial load rate operation of unit 1# for heating**

According to the field tests result, it was found that the heating time was from November 20, 2017 to February 25, 2018. The average daily COP of the system was 2.34. Figure 7 described the heating energy of the system. It can be seen that heat pump unit 1# consumed more than half of the system energy consumption (69%) providing air conditioning heat water. Unit 2# was used for domestic hot water, basically running at full load, accounting for 8%, and unit 3# basically not worked, accounting for merely 3%. Therefore, this paper focused on the analysis of the operation of unit 1# in winter. The heating load time frequency (HLTF) and COP under various PLR of heat pump unit 1# were shown in Fig. 8. It was clear that unit 1# spent most of the time working under 90–60% PLR, but COP was only 3.93 and 3.35, respectively. So it would be better to work in high-efficiency area (PLR = 70–100%).

(6) **Heating temperatures on typical heating day**

Figure 9 depicted the temperature difference between supply and return water in heat source side (Δt_2). The supply water temperature reached the lowest value of

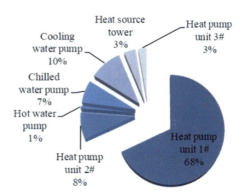

Fig. 7 Heating energy consumption

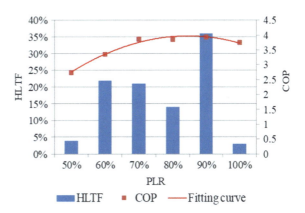

Fig. 8 HLTF and COP under PLR of unit1#

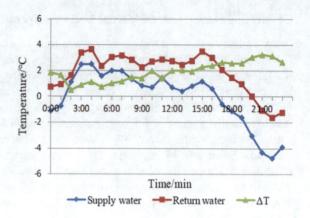

Fig. 9 Δt in heat source side

−4.76 °C at 22:00, and Δt_2 fluctuated between 0.5 and 3.3 °C. The average temperature difference was 1.92 °C, which was much lower than the design Δt_2 of 3 °C.

3 Results—Summary and Evaluations

According to the diagnosis of the field test results and analysis of the current technical specification (Chinese Std CECS 362: 2104, Chinese Std GB 50736-2012), the following problems were found.

3.1 Heat Pump Unit Configuration

(1) The practical cooling load was only 2609 kW, much lower than the installed capacity of 3603 kW. The oversized installed capacity made it difficult to ensure each unit to work in high PLR during practical process.
(2) The models of units were different. The installed capacity of unit 1# accounted for 59.3%, while unit 2# 14.2% and unit 3# 26.3%, thus leading to the poor mutual backup, low operational reliability and high maintenance cost.

3.2 Water Pump Configuration

(1) The excessive H led to the phenomenon that the practical working position deviated from the high-efficient position. Therefore, the pump worked with low efficiency and poor adjustable performance most of the time.

(2) The water pumps did not set separately in user side for cooling and heating seasons.
(3) The minimum of Δt between supply and return water was 1.41 °C for chilled water and 1.92 °C for cooling water. The system operated under "small Δt and large G."
(4) Each unit was equipped with two water pumps (one for standby), which resulted in complicated pipes and high cost. Without considering about the mutual backup between pumps of the same type, it was not conducive to system control.

3.3 Unreasonable Control Strategy

(1) Cooling water outlet temperature was high of 35 °C (21:30–22:10, August 14, 2017). Lack of interlocking control (between cooling towers, cooling water pumps and heat pump units) and inefficiency group control of the cooling tower module were important driving factors of that. In addition, the cooling tower fan operated with a fixed frequency, unable to adjust to changes in load accordingly.
(2) Inefficient control between the load side water pumps and the heat pump units contributed to the widely fluctuated T_0 from 3.8 to 16.4 °C (August 11, 2018), which had an impact on the reliability for system operation.

Table 2 Specification of the major components in the system

Component	Number	Parameter
Heat pump unit 1#, 2# (standard)	2	Rated cooling capacity: 1125 kW Cooling power consumption: 225 kW Rated heating capacity: 814 kW Heating power consumption: 245 kW
Heat pump unit 3# with two condensers (with heat recovery)	1	Rated cooling capacity: 464 kW Cooling power consumption: 142 W Rated heating capacity: 398 kW Heating power consumption: 118 kW
Heat-source tower 1# (with two speed fan)	5	Flow rate: 100 m^3/h Fan power consumption: 11/5.5 kW
Chilled water pump	3	H = 32 m H$_2$O, Q = 250 m^3/h; (two for use and one for standby)
Cooling water pump 1#	3	H = 24 m H$_2$O, Q = 300 m^3/h; (two for use and one for standby)
Cooling water pump 2#	2	H = 24 m H$_2$O, Q = 138 m^3/h; (one for use and one for standby)
Hot water pump	2	H = 16 m H$_2$O, Q = 112 m^3/h; (one for use and one for standby)

4 Discussion—Optimization and Solutions

Based on the summary and analysis of typical issues of the system, some solutions were put forward as follows to improve COP.

Firstly, reasonable configuration of units should be adopted. Three HST heat pump units were selected. Two were standard units and each unit had two compressors with high safety and good load adaptability (12.5–100%). The other was a total heat recovery unit with two condensers which can implement cooling, heating and domestic hot water. The total cooling and heating capacity were 2813–2026 kW. The specifications of the components were presented in Table 2. Secondly, reasonable configuration of water pump should be taken. In the optimized system, unit 1# and 2# were equipped with three water pumps (one for standby) and unit 3# was equipped with two water pumps (one for standby). The optimized selection was shown in Table 2. Thirdly, it should strengthen the joint adjustment and group control. For example, quality and quantity regulation can be achieved by climate compensation control in winter, which can realize on-demand heating. Then the equipment should be started and stopped properly with t_a. Besides variable frequency fan should be adopted in HST and interlocking control (HST fan-cooling water pump-chilled water pump-heat pump unit) would be taken.

5 Conclusions

1. The average COP of the heat pump unit 1# was 2.34 in winter and 6.45 in summer. The HST can be treated as an efficient cooling tower,
2. HST heat pump system can realize cooling, heating and domestic hot water supply, which improves the equipment utilization rate and reduces the initial cost.
3. Excessive capacities of units led to long-time running in PLR with low COP.
4. High H and unreasonable configuration of water pumps result in high energy consumption.
5. The system failed to realize the group control according to the actual demand with the trend prediction, resulting in a poor matching performance of devices.

Acknowledgements This work was financially supported by the China National Key R & D Program "Solutions to heating and cooling of buildings in the Yangtze River region" (Grant No.2016YFC0700305) and the Natural Science Foundation of China (No. 51878255).

References

1. Cui, H., Li, N., Peng, J., Cheng, J., Li, S.: Study on the dynamic and thermal performances of a reversibly used cooling tower with upward spraying. Energy **96**, 268–277 (2016)
2. Huang, S., Zuo, W., Lu, H., Liang, C., Zhang, X.: Performance comparison of a heating tower heat pump and an air-source heat pump: a comprehensive modeling and simulation study. Energy Convers. Manag. **180**, 1039–1054 (2019)
3. Li, N., Zhang, D., Cheng, J., He, Z., Chen, Q.: Economic analysis of heat pump air conditioning system of heat-source tower. J. Shenzhen Univ. Sci. Eng. **32**, 404–410 (2015)
4. Gong, Y.: Research on partial load distribution of air conditioning for residential buildings. HVAC **35**, 91–94 (2005)
5. Deng, J., Wei, Q., Liang, M., He, S., Zhang, H.: Does heat pumps perform energy efficiently as we expected: field tests and evaluations on various kinds of heat pump systems for space heating. Energy Build. **182**, 172–186 (2019)
6. Jing, J.: ShiMinqi: influence of lPLV and COP on chiller's annual energy consumption. Refrig. Air Cond. **12**, 89–92 (2012)
7. Yan, X., Meng, Q., Ren, Q., Tong, W.: Optimal load sharing strategy for multiple-chiller systems and its simulation. HVAC **37**, 18–21 (2007)

Research on the Effect of Solid Particle Diameter on the Performance of Solid-Liquid Centrifugal Pump

Kuanbing CaoZhu and Changfa Ji

Abstract In order to do the research on the effect of solid particle diameter on the performance of solid-liquid centrifugal pump, based on MRF method and by using Eulerian model, standard k-e model combined with the algorithm—Phase Coupled SIMPLE— carries out constant numerical simulation of internal two-phase flow of the solid-liquid centrifugal pump, mainly analyzes external characteristic and volume fraction and velocity distribution of sand particles on the impeller surface. The result suggests that with the increase in the solid particles' diameter, the head and efficiency of the solid-liquid two-phase flow centrifugal pump both gradually decrease. No matter how much diameter it is, volume fraction and velocity of sand particles reach the maximum on the surface of blade, and it indicates that abrasion mainly happens on the blade surface of the solid-liquid centrifugal pump. Centrifugal pump can perform better when the solid particle diameter is 0.003 mm or 0.0035 mm because the head and efficiency of solid-liquid centrifugal pump are higher, and centrifugal pump is less affected by abrasion.

Keywords Centrifugal pump · Solid-liquid two-phase flow · Solid particle diameter

1 Introduction

Solid-liquid two-phase flow centrifugal pump is widely used in all sectors of industry and agriculture in China. Compared with single-phase flow, two-phase flow is more complex [1, 2]. Ye et al. [3] proposed that due to the complexity of the solid-liquid two-phase flow, the solid-liquid two-phase flow centrifugal pump has

---

K. CaoZhu (✉) · C. Ji (✉)
The School of Energy Resource, Xi'an University of Science and Technology, Shaanxi 710054, China
e-mail: 2786459033@qq.com

C. Ji
e-mail: jicf@xust.edu.cn

more obvious defects in performance, noise and life. Besides the complexity of solid-liquid two-phase flow, abrasion can also influence the performance of centrifugal pump, and many experts have made the analysis of it. For instance, Liu et al. [4] analyzed the abrasion characteristic of the blade of single stage double suction centrifugal pump with sand-filled water, concluding that the middle and rear part of the lower pressure side of blade is more affected by abrasion. When analyzing the abrasion of flow components by using particle model and nonuniform phase model, Wang et al. [5] reached the conclusion that abrasion is most serious near the end of suction surface of the impeller blade.

Besides abrasion, several scholars analyze other factors of influencing the performance of solid-liquid centrifugal pump through numerical simulation. For example, Liao et al. [6] carried out three-dimensional unsteady numerical simulation of the whole flow passage for centrifugal pump without splitter blade and centrifugal pump with three splitter blades, proposing that there is an obvious increase in the head and efficiency of solid-liquid centrifugal pump after adding splitter blade, which shows that the performance of solid-liquid centrifugal pump is improved. Li et al. [7] carried out the research on the unsteady flow characteristic in centrifugal pump by means of particle image velocimetry technique; Noon and Kim [8] proposed that with the increase of slurry particle concentration, the pressure head and efficiency ratio both decrease through the analysis of impact abrasion of slurry flow on the centrifugal pump shell.

In this paper, the effect of solid particle diameter on the performance of solid-liquid centrifugal pump is mainly analyzed. Besides it, by changing the particle diameter in the process of numerical simulation, the proper particle diameter will be determined to make sure the effect on the performance of centrifugal pump is lessened, which lays a foundation for the optimum design of centrifugal pump. The model that will be used in this paper is shown in Fig. 1.

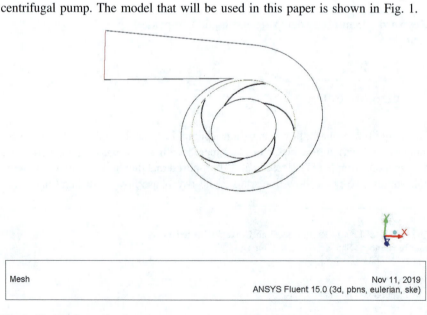

Fig. 1 Model of centrifugal pump

2 Analysis of Numerical Simulation

2.1 Basic Assumption

Because the diameter of sand particles is smaller, the water that contains sand can be treated as a continuous medium. That is, two-fluid model is used for solid-liquid two-phase flow [9]. Assumptions are (1) internal flow of the solid-liquid centrifugal pump is turbulent and steady flow; (2) physical parameters of two phases are constant; (3) interaction between the two phases exists; (4) the volume fraction of solid phase at the inlet is constant.

2.2 Determination of Calculation Parameters

The main design parameters of the solid-liquid centrifugal pump studied in this paper are rotational velocity of the impeller: 2800 r/min, inlet diameter of the impeller: 140 mm, outlet diameter of the impeller: 220 mm, the number of blades: 5 and the width of exit: 30 mm. When the rated flow rate of centrifugal pump is known, the computational formula of inlet velocity is $v = Q/A = 4Q/\pi d^2$ [10]. Here, the inlet flow velocity is 2 m/s, and inlet diameter of the impeller is 140 mm. The rated flow rate of the solid-liquid centrifugal pump can be calculated by the formula $Q = v*A$, which is 110.8 m³/h. The density of liquid phase (water) is 998.2 kg/m³, and the density of solid phase (quartz sand dry sand) is 1500 kg/m³.

2.3 Numerical Calculation Method

In this paper, Phase Coupled SIMPLE algorithm is used as pressure-velocity coupling method, and the method MRF is chosen. Impeller zone is set as rotating coordinate system, and volute zone is set as stationary coordinate system [11]. Turbulent standard *k-e* model and Eulerian model are adopted here. Schiller–Naumann is used as the interaction between water and sand grain. Second-order upwind scheme is put to use for momentum, turbulent kinetic energy and turbulent dissipation rate for the higher solution accuracy. The absolute criterion of residual error is 0.001 for every quantity.

2.4 Boundary Condition Settings

At the inlet of the impeller, the velocity of water phase and sand grain phase is set as 2 m/s, and the inlet boundary condition is velocity inlet condition. The volume fraction of sand grain at the inlet is 5%. For pressure and velocity are unknown at the outlet, "outflow" boundary condition is used as the outlet boundary condition [11]. Interface between impeller and volute is set as dynamic-static interface. Water phase and sand grain phase both exist in the rotating area of the impeller.

3 Analysis of Calculation Results

3.1 Contours of Volume Fraction in different particle diameters

When the particle diameter is 0.005 mm, 0.0045 mm, 0.0035 mm, 0.003 mm, respectively, cloud pictures of volume fraction of solid phase in the impeller zone are shown in Fig. 2 as follows:

It can be seen from Fig. 2 that the volume fraction of sand particles reaches the maximum on the part zone of the blade surface. And the volume fraction of quartz sand dry sand on the blade surface is smaller when the diameter of sand particles is 0.003 mm or 0.0035 mm, and thus, abrasion on the blade surface is less serious.

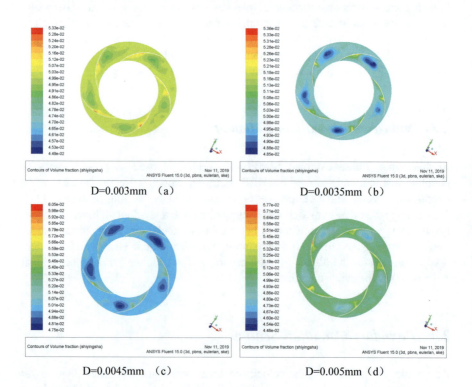

Fig. 2 Cloud pictures of volume fraction of sand particle

It can be seen from Fig. 2 that the volume fraction of sand particles reaches the maximum on the part zone of the blade surface. And the volume fraction of quartz sand dry sand on the blade surface is smaller when the diameter of sand particles is 0.003 mm or 0.0035 mm, and thus, abrasion on the blade surface is less serious.

3.2 Velocity Vector Charts

When the particle diameter of sand grains is 0.005, 0.0045, 0.0035, 0.003 mm, velocity vector charts of the impeller surface are shown in Fig. 3 as follows:

It can be seen from Fig. 3 that minor increase in the particle diameter of sand grains exerts a little effect on the velocity distribution of sand grains. Velocity magnitude reaches the maximum on the blade surface. With the increase in the particle diameter, the velocity of sand grains on the blade surface gradually increases.

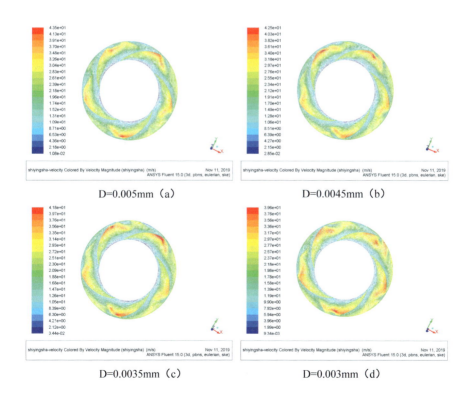

Fig. 3 Velocity vector charts

3.3 Analysis of External Characteristics

As the two factors that affect the performance of the solid-liquid centrifugal pump, head and efficiency of the solid-liquid centrifugal pump need to be determined.

The formulas for calculating head and efficiency are as follows:

$$H = (P_{0out} - P_{0in})/\rho g \tag{1}$$

where H is head, P_{0in} is the total pressure at the inlet of impeller, P_{0out} is the total pressure at the outlet of volute [12].

$$\eta = \rho g Q H / 3600 P \tag{2}$$

$$P = M * w \tag{3}$$

where ρ is the density of water, M is torque of impeller to z-axis, Q is mass flow rate of centrifugal pump, H is head of centrifugal pump.

The head of solid-liquid centrifugal pump can be determined after calculating the total inlet and outlet pressure of mixture through the fluent software.

When the volume fraction of solid phase is kept unchanged at 5%, the head variation curve and the efficiency variation curve are as follows:

As can be seen in Figs. 4 and 5, when the volume fraction of sand particles is 5%, head and efficiency of solid-liquid centrifugal pump both gradually decrease with the increase of particle diameter of sand particles. The more the particle diameter increases, the faster head and efficiency of centrifugal pump drop.

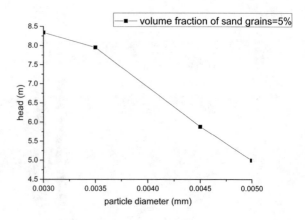

Fig. 4 Head variation curve

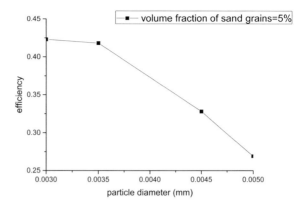

Fig. 5 Efficiency variation curve

4 Conclusions

(1) It can be seen from cloud pictures of volume fraction of sand grain and velocity vector charts that the volume fraction of sand grains is higher on the blade surface.
(2) When the particle diameter is 0.003 mm or 0.0035 mm, head and efficiency of the solid-liquid centrifugal pump are higher, thus choosing the sand particles whose diameter is 0.003 mm or 0.0035 mm is more suitable.
(3) No matter how much the particle diameter of sand grains is, volume fraction and velocity of sand particles both reach the maximum on the blade surface of the impeller, and it is affected by abrasion more easily.

Acknowledgements This paper has been supported by China National Natural Youth Science Foundation Project (No. 51404191). We would like to thank the committee members, sponsors, reviewers, authors and other participants for the support to ISHVAC 2019.

References

1. Xie, Z.B., et al.: Effect of blade inclination angle on flow field and pulsation characteristics of centrifugal pump. J. Shanghai Univ. Technol. **39**(5), 430–437 (2017)
2. Xiang, J.L., et al.: Study on unsteady flow characteristics in solid-liquid two-phase centrifugal pump. Mech. Electr. Eng. **31**(6), 702–706 (2014)
3. Ye, Q., et al.: Unsteady numerical simulation of three-dimensional flow field in solid-liquid two-phase flow centrifugal pump. China Rural Water Hydropower **10**, 76–79 (2014)
4. Liu, Z.L., et al.: Analysis of blade wear characteristics of single-stage double-suction centrifugal pump with sediment-laden water. J. Lanzhou Univ. Technol. **40**(4), 56–61 (2014)
5. Wang, J,Q., et al.: Numerical simulation of internal flow field and wear characteristics of solid-liquid two-phase flow centrifugal pump. J. Agric. Mach. **44**(11), 53–60 (2013)
6. Liao, X., et al.: Effect of splitter blade on solid-liquid two-phase flow in centrifugal pump. Yangtze River **48**(10), 79–82 (2017)

7. Li, W., et al.: PIV experiment of the unsteady flow field in mixed-flow pump under part loading condition. Exp. Therm. Fluid Sci. 191–199 (2017)
8. Noon, A.A., Kim, M.H.: Erosion wear on centrifugal pump casing due to slurry flow. Wear **364**(10), 103-111 (2016)
9. Li, Y.S., Wei, J.F.: Numerical simulation of the motion of solid-liquid two-phase flow with small size in swirl pump. Fluid Mach. **38**(7), 20–23 (2010)
10. Guo, R.N., et al.: Numerical simulation of internal flow field of centrifugal pump based on Fluent. World Sci. Technol. R&D **33**(2), 185–188 (2011)
11. Hu, R,X., Kang, S.T.: Fluent 16.0 flow field analysis from introduction to proficiency. Mechanical Industry Press, Beijing (2016)
12. Wang, K.J., et al.: Numerical research on the influence of fine sediment on the working characteristics of centrifugal pump. China Rural Water Hydropower **5,** 129–132 (2013)

Experimental Study on Flow Maldistribution and Performance of Carbon Dioxide Microchannel Evaporator

Jing Lv, Guo Li, Tang fuyi Xu and Chenxi Hu

Abstract The effects of three inlet parameters on flow distribution and performance of a microchannel evaporator are investigated experimentally. The three research factors are evaporation pressure, inlet mass flow rate and dryness. The configuration is made of brazed aluminum microchannel flat tubes with multi-louver fin structure using CO_2 as the refrigerant. Those 19 parallel flat tubes are divided into 9 intervals, and 72 temperature measurement points are set. The unevenness standard deviation is adapted to evaluate the unevenness of each interval. It is found that mass flow rate has the greatest influence on the flow maldistribution for a CO_2 microchannel evaporator, followed by inlet dryness and evaporation temperature. The position of the inlet header has a great influence on the flow maldistribution of the heat exchanger and its performance. For a CO_2 microchannel evaporator, the unevenness of flow maldistribution is reduced when increasing the inlet dryness or decreasing the mass flow rate. Moreover, the inlet dryness has much greater impact on performance than the mass flow rate and evaporation temperature. The CO_2 microchannel evaporator is more apt to uneven flow maldistribution when the load is large, which will seriously affect its performance under extreme conditions.

Keywords Microchannel evaporator · Flow maldistribution · Heat transfer distribution · CO_2

J. Lv (✉) · G. Li · T. Xu · C. Hu
School of Environment and Architecture, University of Shanghai for Science and Technology, Shanghai, China
e-mail: lvjing810@163.com

G. Li
e-mail: usst13liguo@163.com

1 Introduction

Microchannel evaporator with extruded aluminum channels and folded louvered fins is shown in Fig. 1. Based on numerous studies, using CO_2 as refrigerant can significantly reduce the size of a system due to its good heat transfer and high area to volume ratio [1, 2]. However, maldistribution of the refrigerant is a challenge in this type of heat exchangers, especially for evaporators where the entering fluid is usually in two-phase condition [3]. Choi et al. [4] founded that uneven air distribution reduced evaporator capacity by 8.7%, while uneven refrigerant distribution caused evaporator capacity to decay by 30%. Therefore, "dry steaming" and "excessive liquid supply" caused by uniform distribution in each flat tube will seriously affect the performance of microchannel evaporator, as shown by Agostini et al. [5].

Several experimental researches have been conducted to study the liquid separation and heat transfer characteristics of carbon dioxide microchannel evaporators. Yoon et al. [6] founded the main reasons for the uneven distribution was the uneven distribution of inlet and outlet pressure lose due to uneven flow resistance in the channel. Pettersen et al. [7] founded an interesting phenomenon that the convective heat transfer coefficient would be larger if the mass flow rate was larger in a high refrigerant dryness condition. But in a low dryness condition, there was no significant influence. Subsequently, Cheng et al. [8] discovered that the critical dryness of carbon dioxide was generally between 0.5 and 0.7, which was much lower than that of R22 with a critical dryness usually between 0.8 and 0.9. Recently, Oh and Son [9] founded that the CO_2 convective heat transfer coefficient increased before

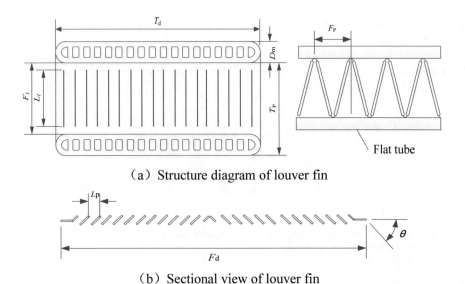

(a) Structure diagram of louver fin

(b) Sectional view of louver fin

Fig. 1 Diagram of louver fin of microchannel evaporator

the occurrence of dry up phenomenon. But after that, the convective heat transfer coefficient was gradually reduced with an increased evaporation temperature. It can be seen that the heat transfer performance degradation caused by the uneven liquid separation is more severe than that of the conventional heat exchanger because of a two-phase flow and a lower critical dryness in microchannel evaporator. Therefore, a more detailed study of maldistribution of CO_2 in microchannel evaporators is interesting and necessary.

A carbon dioxide microchannel evaporator and a whole relatively refrigeration system were designed. The mass flow rate, evaporation temperature and inlet dryness were analyzed to figure out their impact on the liquid distribution unevenness and performance of CO_2 microchannel evaporator, providing a theoretical basis for the improvement and optimization.

2 Microchannel Evaporator

The refrigerant flows into the inlet header from a connecting pipe and then into the microchannels of the flat tubes in different height. The experimental evaporator is composed of 19 parallel flat tubes, each of which has 6 microchannels with equivalent diameter of 1.096 mm. Table 1 shows its main structural parameters. Figure 2 shows the artificial partition of the 19 parallel flat tubes, which are divided into 9 intervals, and 72 temperature measurement points are set. The temperature distribution field of the microchannel evaporator was initially obtained by using a thermal imager.

Table 1 Main structural parameters of microchannel evaporator

Upwind surface width	Upwind surface height	Air direction depth	Evaporator volume	Heat exchange area		Equivalent diameter
				Air side	Refrigerant side	
L_y (mm)	W_y (mm)	T_d (mm)	V_c (cm^3)	A_a (m^2)	A_r (m^2)	d (mm)
810	350	25	7087.5	9.464	2.28	1.096

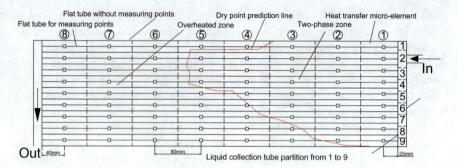

Fig. 2 Collector interval distribution

3 Experimental System

An experimental equipment system including CO_2 microchannel evaporator was set up (see Fig. 3). The conditions of the evaporator side were measured in the Psychometric Room where several temperature and pressure sensors are installed to measure the inlet and outlet temperature and pressure of CO_2. Thermocouples were fixed on the surface of the evaporator in order to measure tube wall temperature. The air side state parameters were measured by using temperature and humidity

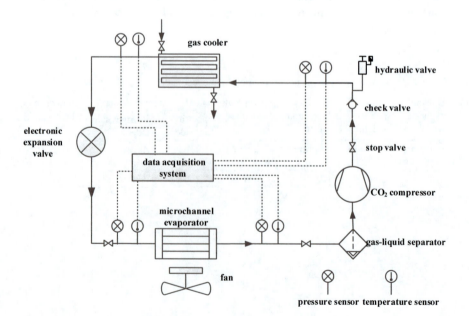

Fig. 3 Experiment system diagram of CO_2 microchannel evaporator

4 Calculations and Analysis

The air side of the evaporator maintained at a constant air velocity of 0.5 m/s, a temperature of 35 °C and a relative humidity of 40% during the experiment. Firstly, the heat transfer micro-element segments were divided according to the temperature distribution and drypoint prediction line as shown in Fig. 2. It was supposed that the wall temperature of one micro-element is approximately the same. And then, the heat exchange between the two-phase region and the overheat region in each interval could be calculated by using the heat balance equation. Also, the mass flow rate of each flat tube can be obtained by selecting a appropriate correlation formula based on mass balance equation. Finally, the inlet dryness of each flat tube was obtained according to the position of the drypoint and the length of the two-phase area. The calculated value of one flat tube represented the entire interval, which was easier for next analysis. The unevenness standard deviation S_V [10] was adapted to evaluate the flow unevenness of each interval.

$$S_V = \sqrt{\frac{1}{(n-1)} \sum_{i=1}^{n} \left(\frac{V_i}{\overline{V}} - 1\right)^2} \quad (1)$$

where V represent q, x and Q, q mass flow, x inlet dryness, Q heat transfer amount, n is equal to 9 and S_V unevenness.

The smaller the value of S_V, the more uniform the distribution of refrigerant in each flat tube.

4.1 Impact of Carbon Dioxide Mass Flow Rate

Test experiments were carried out under the condition that the inlet pressure of the evaporator was 4.5 MPa, the dryness was about 0.86 and the temperature was 10 °C. The carbon dioxide mass flow rate was controlled by the opening degree of an electronic expansion valve, which changed from 3.76, 6.79 to 10.8 g/s.

Figure 4a shows that the temperature distribution of the microchannel is mainly 11–34 °C which is below the critical temperature, indicating that the working state is subcritical when flow rate equals 10.8 g/s. The location where the temperature rises sharply is called drypoint, which before is the two-phase zone and followed by the overheated zone. The highest measured heat flux reaches 545.47 J/(m$^2 \cdot$ s).

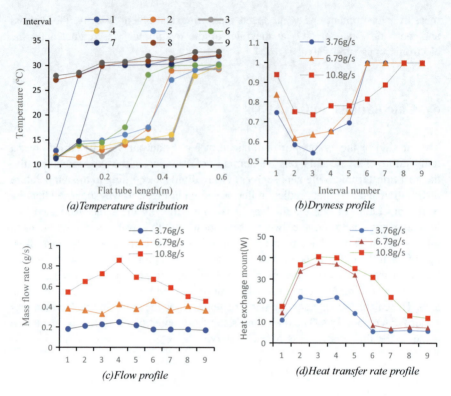

Fig. 4 Selected parameters distribution as a function of three different inlet CO_2 mass flow rate

So the heat transfer rate is 528.3 W, which is 30.0% (380.5 W) more than case of 6.79 g/s and 54.4% (240.7 W) of 3.76 g/s.

Intervals 8 and 9 are both located in the overheated zone because the inlet dryness of two intervals equal to 1 as seen in Fig. 4b. While intervals 3 and 4 have longest two-phase zone length, which is about 0.5 m. Their inlet dryness is lower as seen in Fig. 4b and more liquid separation as seen in Fig. 4c, resulting in the maximum heat exchange as seen in Fig. 4d. Intervals 3 and 4 are both close to the refrigerant inlet header, indicating that the position of the inlet header has a great influence on the flow maldistribution of the heat exchanger and its performance. An obvious conclusion is that the larger the flow rate in the interval, the more the liquid quality and the greater the boiling heat exchange, which is consistent with the findings of Ducoulombie and Colasson [11]. Finally, when the mass flow rate is larger, the more uniform the flat tube inlet dryness distribution, but the more uneven flow maldistribution.

4.2 Impact of Evaporator Temperature

The evaporation temperatures were controlled by adjusting the operating pressure from 10.2 °C, 14.4 °C to 18.2 °C, respectively, under the condition that the evaporator total inlet mass flow rate was 10.72 g/s and the inlet dryness was 0.8.

Figure 5 shows that the flow rate and the heat transfer distribution are both high on the left side and low on the right side. Additionally, the mass flow rate is mainly in the range of 0.4–0.8 g/s in each interval. Usually, a higher total heat transfer is expected for a lower evaporation temperature. However, higher evaporator temperature causes a higher heat transfer from interval 6 to 9, which may be due to the more refrigerant flow in the flat tube as seen in Fig. 5a. The evaporation temperature has a great influence on the heat exchanger distribution of the evaporator but a little influence on the flow maldistribution according to preliminary analysis.

4.3 Impact of Carbon Dioxide Inlet Dryness

Under the condition of evaporator inlet pressure of 4.5 MPa, mass flow rate of 6.15 g/s and inlet temperature of 10 °C, the inlet dryness was controlled by changing the cooling amount on the air cooler side to 0.555, 0.895 and 0.956, respectively.

Figure 6b shows that the interval 5 has the highest heat flux density of 661.38 J/(m² · s) when the inlet dryness is 0.555. Its total heat exchange capacity of the evaporator is 586.5 W, which is more than 54.9% of the case of 0.895 and 65.2% of the case of 0.956. It is obvious that the inlet dryness is much greater impact on heat transfer coefficients than the evaporation temperature in Fig. 5b and the flow rate in Fig. 4d. An interesting phenomenon is that the smaller the inlet dryness, the lower the flow peak will move to the right when the inlet dryness is

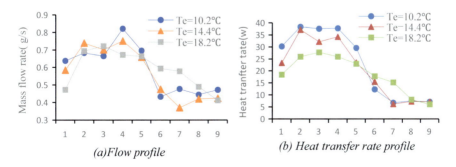

Fig. 5 Flow and heat transfer rate profile under three different evaporator temperature

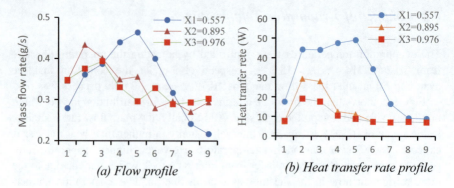

Fig. 6 Flow and heat transfer rate profile under three different inlet dryness

Table 2 Mass flow rate, inlet dryness and heat transfer unevenness

Parameter		S_x	S_q	S_Q
Inlet mass flow rate (g/s)	3.560	0.244	0.102	0.604
	6.590	0.205	0.144	0.583
	10.80	0.122	0.195	0.452
Evaporator temperature (°C)	10.20	0.126	0.234	0.621
	14.40	0.103	0.263	0.593
	18.20	0.054	0.185	0.420
Inlet dryness	0.555	0.555	0.252	0.695
	0.895	0.128	0.165	0.563
	0.956	0.034	0.115	0.459

below 0.9, as shown in Fig. 6a. That is the same direction of gravity. That means the carbon dioxide in the liquid phase is more susceptible to gravity duo to lower dryness.

4.4 Analysis of Unevenness

The flow rate, inlet dryness and heat transfer degree of unevenness were calculated according to the experimental data and formula Eq. (1) (Table 2).

The dryness distribution and heat transfer are relatively uniform while the flow maldistribution is rather uneven when the CO_2 mass flow is larger. The greater the inlet dryness, the more uniform the heat transfer and liquid distribution, but the heat transfer efficiency of the evaporator will decrease. When the evaporation temperature changes, the unevenness of dryness and flow distribution is not much different. Therefore, it has little effect on the dryness and flow distribution. However, the higher the evaporation temperature, the more uniform the heat transfer distribution, but the heat exchange of the whole evaporator will be reduced.

5 Conclusions

A CO_2 microchannel evaporator with 72 temperature measurement points had been designed in order to study three possible factors affecting uneven flow separation and its performance. The experimental data were obtained by the control variable method. The finite element models established by the lumped parameter method were used to solve the distribution, which were mass rate flow, dryness and heat transfer over 9 intervals of evaporator. The main conclusions were as follows:

(1) Mass flow rate had the greatest influence on the flow maldistribution for a CO_2 microchannel evaporator, followed by inlet dryness and evaporation temperature among the three factors.
(2) The unevenness of flow maldistribution was reduced when increasing the inlet dryness or decreasing the mass flow rate for a CO_2 microchannel evaporator.
(3) The inlet dryness had much greater impact on performance of CO_2 microchannel evaporator than the mass flow rate or evaporation temperature.
(4) The CO_2 microchannel evaporator was more apt to uneven flow maldistribution when the load was large that would seriously affect its performance in extreme conditions.

References

1. Khan, M.G., Fartaj, A.: A review on micro-channel heat exchangers and potential applications. Int. J. Energy Res. **35**(5), 553–582 (2011)
2. Brix, W., et al.: Modelling distribution of evaporating CO_2 in parallel minichannels. Int. J. Refrig **33**(6), 1086–1094 (2010)
3. Kandlikar, S.G.:A roadmap for implementing minichannels in refrigeration and air-conditioning systems current status and future directions. Heat Transf. Eng. **28**, 953–985 (2005)
4. Choi, J.M., et al. Effects of nonuniform refrigerant and air flow distribution on finned tube evaporator performance. In: International Congress of Refrigeration, pp. 1–8, Washington, D. C., USA (2003)
5. Agostinni, B., et al.: Friction factor and heat transfer coefficient of R134a liquid flow in mini-channels. Appl. Therm. Eng. **22**(16), 1821–1834 (2002)
6. Yoon, S.H., et al.: Characteristics of evaporative heat transfer and pressure drop of carbon dioxide and correlation development. Int. J. Refrig **25**(2), 111–119 (2004)
7. Pettersen, J.: TWo-phase flow pattern, heat transfer, and pressure drop in microchannel vaporization of CO_2. ASHRAE Trans. (Symp.) 109 (1), 523–532 (2003)
8. Cheng, L., et al.: New flow boiling heat transfer model and flow pattern map for carbon dioxide evaporating inside horizontal tubes. Int. J. Heat Mass Transf. **49**(21–22), 4082–4094 (2006)
9. Oh, H.K., Son, C.H.: Flow boiling heat transfer and pressure drop characteristics of CO_2 in horizontal tube of 4.55 mm inner diameter. Appl. Therm. Eng. **31**(2), 163–152 (2011)

10. Habib, M.A., et al.: Evaluation of flow maldistribution in air-cooled heat exchangers. Comput. Fluids **8**(3), 655–690 (2009)
11. Ducoulombie, M., Colasson, S.: Carbon dioxide flow boiling in a single microchannel-partII: heat transfer. Exp. Thermal Fluid Sci. **35**, 595–611 (2011)

Capture of High-Viscosity Particles: Utilizing Swirling Flow in the Multi-Layer Square Chamber

Leqi Tong and Jun Gao

Abstract Cooking oil fume (COF) contains a variety of high-viscosity particles with multiple diameters. For protecting the exhaust fan in range hood, these particles should be captured before they reach the fan blade. However, current particle filter component in range hood possesses either low capture efficiency or high flow resistance. This paper proposed a novel structure with multi-layer square chamber for the capture of high-viscosity particles in COF. The centrifugal effect of the swirling flow that generated in the chamber made the particles gradually moving towards the inner walls, thus increasing the possibility of particle trap. Orthogonal experiment based on CFD simulation was involved to explore the effect of the flow and structure factors. Discrete phase model (DPM) was chosen to track the trajectory of the particles moving in the chamber. The results indicated that with proper design, the capture efficiency of particles above 5 μm reached 95.0%, and particles above 3 μm reached 86.4%, while the pressure loss is just 32.5 Pa under the designed flow rate, which showed much higher capture efficiency and more acceptable flow resistance than those existing filter components.

Keywords Particle capture · High-viscosity liquid particles · Swirling flow · Multi-layer square chamber

1 Introduction

Cooking oil fume (COF) in Chinese kitchen contains a variety of high-viscosity liquid particles with multiple diameters, mainly 1–10 μm [1]. Chronic exposure to these massive particles constitutes health hazard to human. Measurements showed

---

L. Tong · J. Gao (✉)
Institute of HVAC Engineering, School of Mechanical Engineering,
Tongji University, Shanghai, China
e-mail: gaojun-hvac@tongji.edu.cn

L. Tong
e-mail: 1730202@tongji.edu.cn

that the cooker's vital capacity declined, while peak expiratory flow rose significantly after the cooking process, which means the lung function was harmed by COF to some extent [2]. On the other hand, indoor airborne polycyclic aromatic hydrocarbons (PAH) with severely carcinogenic risk [3] that generated while cooking is mostly attached to the COF particles. Therefore, it is necessary and significant to improve the ventilation condition for removing these particles. Range hood is mostly used in the domestic kitchen as the local exhaust machine. Experimental study showed that it is beneficial for improving the indoor air quality (IAQ) of the kitchen during [4] and after cooking [5]. Many scholars were devoted to optimizing the exhaust efficiency of the range hood. The effects of exhaust flow rate, particle size and burner position were investigated [6]. Experimental results showed that air curtain assisted range hood could be more efficient at lower flow rates [7].

Except for improving the exhaust effectiveness, COF particles treatment is also considerable. For protecting the exhaust fan in the range hood and thus prolonging its life length, those high-viscosity liquid particles in COF should be captured before they reach the fan blade. Existing approach for COF particle capture in range hood is utilizing metal filter components with different types such as wire netting or grid. These components showed either low capture efficiency or high flow resistance. This paper proposed a novel structure with multi-layer square chamber for the capture of the high-viscosity particles in COF. The performance of the multi-layer component with different parameters was investigated. Orthogonal experiment based on CFD simulation was involved to explore the effect of the flow and structure factors. Discrete phase model (DPM) was chosen to track the trajectory of the particles moving in the chamber.

2 Methods

2.1 Development of the Multi-layer Component

The component consists of multiple layers with many square chambers. The outlet of the chamber in this layer is the inlet of that in the next layer so that the airflow can pass through sequentially. Figure 1 shows its overall installation in the range hood and the detailed structure. The face velocity keeps perpendicular to the inlet of each chamber. Therefore, the inlet airflow can be attached to the inner wall and form the swirling flow in the square chamber of each layer. In this process, the high-viscosity particles in COF can gradually move towards the inner walls of the chamber due to the centrifugal effect of the swirling flow and then be trapped.

The key indexes to evaluate the performance of a particle filter are the capture efficiency, flow resistance and particle containing capability. The component can be cleaned regularly and reused so that the particle containing capability is not considered. The rest two performance indexes are influenced by the flow and structure

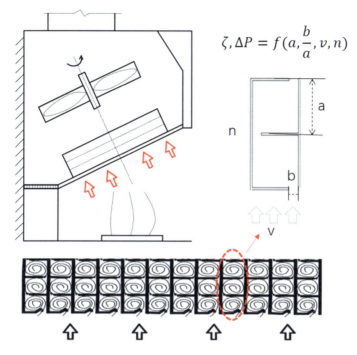

Fig. 1 Detailed structure and the operating unit of the component

factors. The structure factors of this component are the length of the square a, the layer number n and the proportion of the inlet length b/a. The outlet length is kept equal to the inlet in each chamber. The flow factor is namely the face velocity v.

2.2 CFD Simulation Setup

Orthogonal experiment based on CFD simulation was adopted to investigate the impact of flow and structure factors on the performance of the component and find the optimal structure. Each factor picked four appropriate levels. The length of the square was chosen from 20 to 50 mm, and the layer number was picked from 1 to 4, which are constrained by the geometry size of the range hood. The face velocity was picked from 0.5 to 1.25 m/s, considering the proper exhaust airflow rate and size of range hood. At the designed exhaust rate as 600 m³/h, the rated face velocity of the normal range hood is around 0.5 m/s. The proportion of the inlet was chosen from 0.10 to 0.25. Totally, 16 cases were simulated.

The standard k-ε turbulence model combined with the QUICK difference scheme and SIMPLE algorithm for coupling pressure and velocity was found to be adequate and hence used for the present work. The pressure-based segregated solver and structured mesh were chosen. Discrete phase model (DPM) was chosen to track the trajectory of the particles moving in the chamber. The inlet of the first layer was set as velocity inlet, and the set values of velocity were calculated from the face velocity. The outlet of the last layer was set as pressure outlet, and the gauge pressure value was set as 0 pa. For the DPM boundary setting, the inner walls were set as trap, while the velocity inlet and pressure outlet were set as escape. Besides, the inlet of the first layer was set as a face injection. Ten kinds of particles with the diameter from 1 to 10 μm were injected.

3 Results

Figure 2 shows the simulation results of case 10. As we can see in Fig. 2a, swirling flow generated in the chambers of each layer. Massive airflow was circling around the inner wall of the chamber. The high-viscosity particles moved with the airflow and circled in the chamber, as is shown in Fig. 2c. The swirling flow prolonged the residence time of the particles in the chamber and increased the possibility of particles capture. As is shown in Fig. 2b, the majority of the particles were trapped by the inner walls of the first and second chambers. While at the outlet of the last layer, the particle concentration was greatly dropped. The simulation results of other cases were similar to that described above.

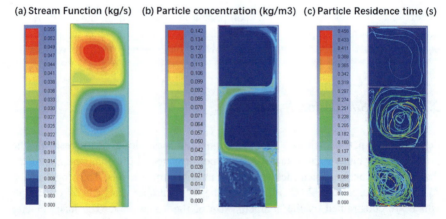

Fig. 2 Simulation result of case 10: **a** stream function distribution; **b** particle concentration distribution; **c** particle traces coloured by particle residence time

3.1 Capture Efficiency

The capture efficiency of particles is related to the structure and flow factors, as well as the particle diameters. Particles in COF with the diameter from 1 to 10 μm were all investigated. The capture efficiency of the particles with single diameter d was calculated by the equation below.

$$\eta_d = 1 - N_{ed}/N_{sd} \tag{1}$$

where N_{ed} is the number of the particles that escaped from the outlet, N_{sd} is the number of the particles that released from the inlet.

Figure 3 shows the capture efficiency of the cases. It indicated that the number of layers greatly affected the particle trap process. The opponent with more layers possesses higher capture efficiency. Particles with larger diameter were much easier to be captured. As is shown in Fig. 3, the capture efficiency of that with four layers illustrated almost 100% on the particles with 10 μm diameter and above 90% on the particles above 5 μm. The opponent with three layers also showed good performance on particles trap, while that with one layer showed unsatisfied capture efficiency.

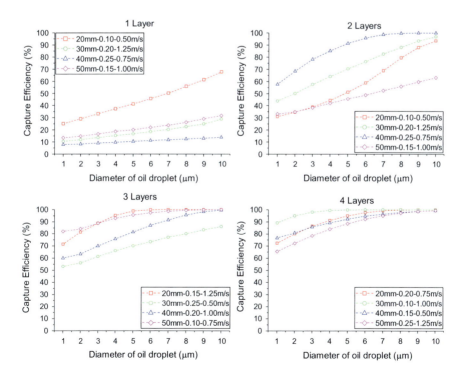

Fig. 3 Simulation results of capture efficiency

On the other hand, the proportion of the inlet is found also important influence factors. It is concluded that smaller inlet proportion donated higher capture efficiency. However, the inlet proportion can not be shrunk unlimitedly for the reason that extremely small inlet could give rise to manufacture difficulty. The capture efficiency seemed not sensitive to the side length of square chamber. In other words, small chamber presented almost the same performance as the large ones. It is more advantageous to utilize smaller chamber for saving installation room, which also needed higher manufacture requirements. Therefore, proper combination of these parameters is significant to obtain good performance on particle trap. Besides, the flow factor also presented little influence on the capture efficiency so that lower face velocity should be adopted in design stage for lower flow resistance. Nevertheless, the face velocity is constrained by the exhaust airflow rate and the size of the range hood.

3.2 Flow Resistance

Except for the capture efficiency, the flow resistance also needed consideration. It is not economical to cost much high fan energy consumption for pursuing little improvement on capture efficiency. The flow resistance is the difference value of the total pressure between the inlet and outlet of the opponent. The resistance coefficient is defined by the equation below.

$$\xi = 2\Delta P / \rho v^2 \quad (2)$$

where ΔP is the flow resistance, ρ is the density of the air, v is the face velocity of the opponent.

Figure 4 showed the results of the flow resistance and resistance coefficient of the cases. It presented that more layers and smaller inlet length proportion caused higher flow resistance coefficient. The opponent with just one layer merely has flow resistance, compared to that with more layers. In those opponents with the same number of layers, resistance coefficient also showed different owing to different inlet length proportion and the side length of the square chamber.

As is shown in the figure below, the flow resistance of the opponent with more layers could be cut down with appropriate choice on other structure factors. Therefore, it is necessary to conduct factor analysis and pick the reasonable combination of those influence factors so that high capture efficiency as well as low flow resistance can be both realized.

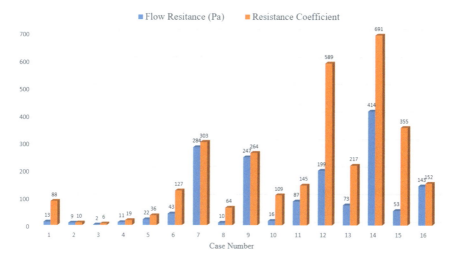

Fig. 4 Flow resistance and the resistance coefficient

4 Discussions

As is listed in Fig. 3, four levels were picked for each influence factor. The average values of each factor under the certain level were calculated and showed in Fig. 5. The capture efficiency of 3 μm particle was chosen to present the overall capture efficiency for the COF particles now that 3 μm is the peak diameter in COF.

As is shown in Fig. 5a, the factor A (the side length of square chamber) and C (the face velocity) were seemed to have little impact on the capture efficiency. The factor D (the number of layers) was the vital parameter and showed the most important influence on the particle trap process. The factor C (the proportion of inlet length) also played a non-ignorable role. Therefore, it could be concluded that more layers and smaller inlet proportion are advantageous for improving the particle capture efficiency.

As is shown in Fig. 5b, the factor A showed less effect on the flow resistance than the other three factors. $A2$ (the side length of square at 30 mm) seemed to present highest flow resistance than other three levels. Higher inlet length proportion and less layers brought lower flow resistance, which showed similar trend to the capture efficiency. $B3$ (the inlet length proportion at 0.2) seemed to be enough to decrease the flow resistance.

Considering both the requirements of higher particle capture efficiency and lower fan energy consumption, the optimal combination of $A1$ (the side length of square at 20 mm), $B3$ (the inlet length proportion at 0.2), $D4$ (4 layers) was finally determined. At the designed exhaust flow rate for the range hood as 600 m^3/h, the face velocity is usually 0.5 m/s ($C1$).

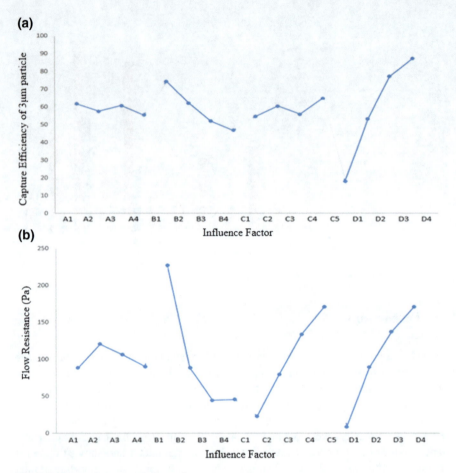

Fig. 5 Influence factors analysis result

5 Conclusions

This paper proposed a novel structure with multi-layer square chamber for the capture of high-viscosity particles in COF. The centrifugal effect of the swirling flow that generated in the chamber made the particles gradually moving towards the inner walls, thus increasing the possibility of particle trap. With optimal design on its structure, the capture efficiency of the proposed opponent on particles above 5 μm reached 95.0%, and particles above 3 μm reached 86.4%, while the pressure loss is just 32.5 Pa under the designed flow rate as 600 m^3/h, which showed much higher capture efficiency and more acceptable flow resistance than other existing filter opponents.

References

1. Gao, J., Cao, C., et al.: Volume-based size distribution of accumulation and coarse particles (PM0.1–10) from cooking fume during oil heating. Build. Environ. **59**, 575–580 (2013)
2. Du, B., Gao, J., et al.: Particle exposure level and potential health risks of domestic Chinese cooking. Build. Environ. **123**, 564–574 (2017)
3. Singh, A., Chandrasekharan Nair, K., et al.: Assessing hazardous risks of indoor airborne polycyclic aromatic hydrocarbons in the kitchen and its association with lung functions and urinary PAH metabolites in kitchen workers. Clin. Chim. Acta **452**, 204–213 (2016)
4. Singer, B.C., Pass, R.Z., et al.: Pollutant concentrations and emission rates from natural gas cooking burners without and with range hood exhaust in nine California homes. Build. Environ. **122**, 215–229 (2017)
5. Dobbin, N.A., Sun, L., et al.: The benefit of kitchen exhaust fan use after cooking—an experimental assessment. Build. Environ. **135**, 286–296 (2018)
6. Rim, D., Wallace, L., et al.: Reduction of exposure to ultrafine particles by kitchen exhaust hoods: the effects of exhaust flow rates, particle size, and burner position. Sci. Total Environ. **432**, 350–356 (2012)
7. Claeys, B., Laverge, J., et al.: Performance testing of air curtains in residential range hoods. Procedia Eng. **121**, 199–202 (2015)

Experimental Study on Vertical Temperature Profiles under Two Forms of Airflow Organization in Large Space during the Heating Season

Chenlu Shi, Xin Wang, Gang Li, Hongkuo Li, Minglei Shao, Xin Jiang and Bingyan Song

Abstract Stratified air-conditioning systems are widely used in large space. The vertical temperature gradient is an important index to evaluate characteristics of the indoor thermal environment, and reasonable airflow organization in a large space is closely related to heating efficiency. The researches of the cooling season behavior of the system are various, but few researches on airflow organization for heating season have been carried out. This paper takes a large space building as the experimental object in winter, comparing the indoor vertical air temperature distribution, the temperature distribution of the working area, and the temperature distribution of the non-air-conditioned area of the two forms of airflow under different outdoor air temperatures and different air supply volumes. Furthermore, the heating load and heating efficiency are also compared based on experimental data. It is concluded that the vertical temperature gradient of the nozzle supply is slightly smaller than that of the columnar supply, and the indoor temperature of the nozzle is higher. In general, the heating effect of the nozzle supply is better than the columnar supply.

Keywords Large space · Airflow organization · Vertical temperature profiles · Heating efficiency

1 Introduction

The structure of large space buildings is becoming more and more complex, and the span is gradually increasing. They are widely used in public buildings such as theaters, stadiums, airport terminals, and industrial buildings. In winter, the heat of air

---

C. Shi · X. Wang (✉) · M. Shao · X. Jiang · B. Song
School of Environment and Architecture, University of Shanghai for Science and Technology, Shanghai, China
e-mail: wangxinshiyun@126.com

G. Li · H. Li
Technology Center of Tong Yuan Design Group Co., Ltd., Shanghai, China

supply is consumed in the upper part of the large space building, which makes it difficult to obtain sufficient heat in the lower working area. Said et al. [1] tested the thermal environment of the hangar in the heating season and found that the energy consumption increased by 4.8% as the temperature difference between the roof and the floor increased every 1 °C. In summer, temperature stratification can reduce energy consumption, while in winter the lower the temperature difference between the upper and lower is, the lower the energy consumption is. In large space, the vertical temperature gradient in a stratified air-conditioned room is less than that of a full air-conditioned room [2]. Wang et al. [3] point out that higher air temperature and air velocity can ensure thermal comfort, which need to increase the heating load in winter. However, increasing heating load reduced heating efficiency and yet increased energy consumption [4]. If the stratified air-conditioning airflow is well designed in winter, the obvious vertical temperature stratification will not occur, and Gorton and Bagheri [5] and Lai and Wang [6] both measured the thermal environment of the large space with nozzle air-conditioning in the heating season. It is found that the temperature difference between the air-conditioned area and the non-air-conditioned area is small, and no obvious temperature gradient is formed. Therefore, reasonable airflow organization in a large space is closely related to heating efficiency.

The airflow organization is directly related to the heating effect of the stratified air conditioner. Whether the temperature distribution in the working area is uniform, the effect of stratified air-conditioning and the purpose of energy saving are largely determined by reasonable airflow organization [7]. The main airflow organization of the stratified air conditioner is the nozzle-side supply and the columnar down supply. Indoor vertical air temperature in large space is one of the important parameters to characterize the airflow characteristics of stratified air-conditioning. The study of vertical temperature and airflow organization is the main content of basic research.

This paper will take the numerical control area of the Engineering Training Center of University of Shanghai for Science and Technology (USST) as the research object. Compare the indoor vertical air temperature distribution, the temperature distribution of the working area, and the temperature distribution of the non-air-conditioned area of the two forms of airflow under different outdoor air temperatures and different air supply volumes. Furthermore, the heating load and heating efficiency are also compared based on experimental data.

2 Experiment

2.1 Experimental Base Overview

The experimental object is shown in Fig. 1. The area of the air-conditioned area is about 500 m^2, of which the east–west length is 18.3 m and the north–south length is 28 m.

Fig. 1 Real map of the engineering training center of USST

Fig. 2 Nozzle air supply inlet, columnar air supply inlet, and return air inlet

The nozzle air supply inlet, the columnar air supply inlet, and the return air inlet are shown in Fig. 2. The aperture of each nozzle is 373 mm, and the spacing between the two nozzles is 1.5 m. The experimental nozzle height in this paper is 5.5 m. The return air inlet with a length of 3 m and a width of 2 m is arranged below the nozzle side, the lowest point of which is 0.6 m from the ground. The columnar air supply inlets are located on both sides of the north and south wall. There are four air supply inlets on both sides, and they are arranged symmetrically. The air volume of each inlet is 3500 m^3/h, the height and the diameter of which are, respectively, 1.5 m and 1.0 m.

2.2 Experimental Measuring Points and Instruments

In the experimental area, the uniformity of the measurement points was fully considered when the measuring points were arranged, and finally a total of nine basic measuring points were set. There were seven thermocouple temperature measuring points on the vertical fixed suspensions on the three measuring lines A, B, and C, and eight thermocouples were arranged in the vertical fixed suspension on the six measuring lines of D, E, F, I, J, and K. Each measurement line started from 3 m from the ground, and the measurement points were arranged upward at a

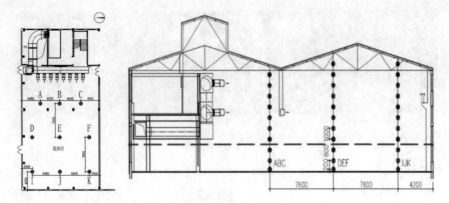

Fig. 3 Plan and elevation of indoor air-measuring points

distance of 1 m in the vertical direction. The positions from bottom to top are 3 m, 4 m, 5 m, 6 m, 7 m, 8 m, 9 m, 10 m, respectively, and no points were arranged at 10 m of A, B, C. The moving points were set below 3 m, and the measuring points in the vertical direction were, respectively, 2.0 m, 1.1 m, and 0.3 m. A total of 96 measuring points are shown in Fig. 3. The above fixed measuring points were collected by the PT1000 acquisition module and directly fed back to the computer through the network. For the moving measuring points, the required data were obtained by hanging the temperature recorder and manual reading.

The main instruments used in the experiment are shown in Table 1.

2.3 Experimental Conditions

The experimental conditions are shown in Table 2. During the experiment, the machine was turned on at 8:00 every day, and the measurement was started at 9:00. The data were recorded every half hour until 17:00.

Table 1 Experimental main instrument parameters

Measuring parameters	Instrument names	Resolution	Measuring range	Precision
Outdoor temperature and humidity	Outdoor weather station	0.1 °C	−40–60 °C	±0.2 °C
		0.1%	0–100%	±2.5%
Indoor air temperature below 3 m	Testo 174 logger	0.1 °C	0–50 °C	±0.1 °C
Vertical measuring point temperature sensor above 3 m	Four-wire platinum resistance PT100	–	–	A

Table 2 Experimental conditions

Conditions	Outdoor air temperature (°C)	Supply air temperature (°C)	Return air temperature (°C)	Supply air volume (m³/h)	Remarks
CASE 1	9.5	32.1	21.2	20,822	Nozzle air supply
CASE 2	6.2	32.4	19.8	20,831	
CASE 3	3.2	31	18.9	20,840	
CASE 4	7.4	31.1	18.8	23,217	
CASE 5	5.9	36.8	15.1	12,488	
CASE 6	10.8	33.1	20.5	20,080	Columnar air supply
CASE 7	7.6	31	18	21,010	
CASE 8	3.6	31.8	18.3	20,645	
CASE 9	8.9	37.8	19.3	12,823	

3 Experiment

The experimental data of two forms of airflow organization conditions under different outdoor temperatures and air supply volumes in large space were arranged. The effects of indoor vertical air temperature distribution and the working area temperature under different conditions were analyzed. Based on the calculated heat supply and thermal efficiency, the heating effects were compared.

3.1 Vertical Temperature Distribution

Under the condition of air volume of 21,000 m³/h, the influence of different outdoor temperatures on vertical temperature distribution of nozzle and columnar air supply was studied, the distribution law of which was analyzed. Besides, the experimental results of different air supply volumes were compared under the conditions of similar outdoor temperature to study the vertical temperature distribution.

From the nozzle conditions, it is observed that the temperature below 2 m rises slowly with the increase of the height, while the temperature rises sharply at 2–3 m. At 3–5 m, due to the temperature rise and fall caused by the air supply at 5.5 m, a temperature inflection point was generated. It can be seen from Fig. 4b that the temperature change trend below 3 m is similar to that of the nozzle. Under the nozzle conditions, the air volume is larger, but the indoor temperature is lower. When the air volume is much smaller, not only is the overall temperature lower, the temperature of the working area also cannot meet the thermal comfort requirement. Under the columnar conditions, when the air volume is small, increasing the supply air temperature can raise the temperature of the working area, but the increase of the supply air temperature will result in waste of energy. Therefore, it is most appropriate to use the conditions of 21,000 m³/h as the reference conditions. Also, the conclusion has reference significance for determining the air supply volume in air-conditioning design.

Taking the air volume of 21,000 m³/h as the reference working condition, the indoor vertical temperature distribution of the nozzle and the columnar supply is compared.

It can be seen from Fig. 5 that although the outdoor temperature under the nozzle condition is lower, the indoor vertical temperature is generally higher. Comparing the temperature of the working area, the nozzle supply is also superior to the columnar supply.

In winter, the smaller the vertical temperature gradient is, the higher the heating efficiency is. The reduction of the excessive vertical temperature gradient is beneficial to decrease the heating load of the air conditioner and improve the comfort of the indoor thermal environment. It can be seen from the figure that although the temperature of the non-air-conditioned area of the nozzle is higher, the temperature

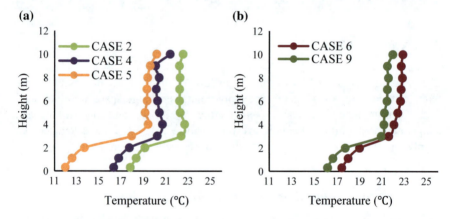

Fig. 4 Relationship between indoor air temperature and supply air volume for: **a** nozzle air supply and **b** columnar air supply

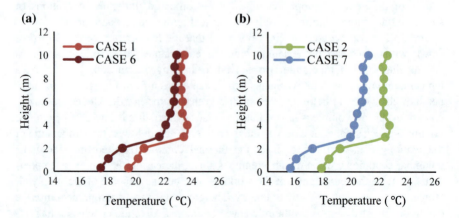

Fig. 5 Vertical temperature comparison between spout and delivery for: **a** 10 °C and **b** 7 °C

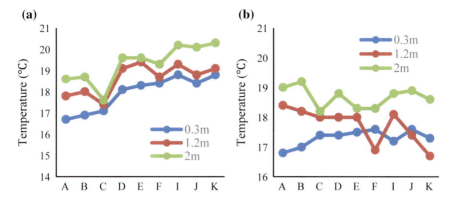

Fig. 6 Comparison of the working area temperature for: **a** CASE 1 and **b** CASE 6

gradient is significantly smaller, which shows that the heating efficiency of the nozzle supply is better than that of the columnar supply.

In order to compare the temperature distribution of the working area of the nozzle and the columnar airflow organization, the temperature of the measuring points below 2 m of CASE 1, CASE 2, CASE 6, and CASE 7 was selected for comparison (Fig. 6).

For the nozzle conditions, the temperature at ABC is significantly lower than that of the other measuring points. It can be seen from the jet trajectory that the jet is almost impossible to reach the area within 5 m from the nozzles, and the three measuring points of ABC are about 4 m away from the nozzles, so the temperature is lower than the others. The temperature at C is the lowest, because the C is closest to the door, and during the experiment period, it is the students' class time. Frequent entry and exit often lead to cold wind intrusion and infiltration. Observing the columnar conditions, the overall temperature distribution is relatively uniform, which has a great relationship with the uniform arrangement of the columnar inlets.

3.2 Heat Supply and Thermal Efficiency

The airflow organization is directly related to the heating effect of the stratified air conditioner. Only comparing the heat supply of the two airflow organizations cannot explain which type of air supply is more energy saving, so the outdoor temperature, the working temperature, the temperature of the air-conditioning zone, and the temperature of the non-air-conditioned zone are considered. It is more appropriate to use the concept of thermal efficiency for comparison.

$$\eta = \frac{X}{q} \tag{1}$$

Table 3 Calculation results of heat supply and heat efficiency

Conditions	Outdoor air temperature	Supply air volume	Heat supply	η_1	η_2	η_3	η_4
CASE 1	9.5	20,822	142.8	0.0873	0.0786	0.0077	0.0162
CASE 2	6.2	20,831	166.6	0.0876	0.0788	0.0080	0.0165
CASE 3	3.2	20,840	162.0	0.0982	0.0870	0.0089	0.0196
CASE 4	7.4	23,217	173.3	0.0652	0.0587	0.0075	0.0127
CASE 5	5.9	12,488	163.4	0.0661	0.0520	0.0162	0.0287
CASE 6	10.8	20,080	155.2	0.0494	0.0494	0.0232	0.0234
CASE 7	7.6	21,010	161.7	0.0563	0.0563	0.0247	0.0252
CASE 8	3.6	20,645	177.5	0.0668	0.0668	0.0202	0.0201
CASE 9	8.9	12,823	157.6	0.0567	0.0567	0.0230	0.0231

where η is thermal efficiency, X_1 is indoor and outdoor temperature difference, X_2 is working area and outdoor temperature difference, X_3 is temperature difference between non-air-conditioned area and air-conditioned area, and X_4 is temperature difference between non-air-conditioned area and working area.

The experimental calculation results are shown in Table 3. In order to facilitate the comparison, the working area takes the average temperature below 2 m.

Based on the above studies, a baseline condition of 21,000 m^3/h (CASE 1–3 and CASE 6–8) was selected for comparison. It can be seen from the above formula that when the heat supply is the same, the larger the X_1 and X_2 are, the larger the η_1 and η_2 are. In summer, the greater the vertical temperature gradient of indoor air is, the better the heating effect is, which is opposite in winter. Therefore, from the formula, when the heat supply is same, the smaller the X_3 and X_4 are, the larger the η_3 and η_4 are. It can be seen from the data in Table 3 that the η_1 and η_2 of the nozzle conditions are larger than those of the columnar conditions. At the same time, the η_3 and η_4 of the nozzle conditions are smaller, so it can be stated that the heating effect of the nozzle supply is better than that of the columnar supply.

4 Conclusions

This paper takes the numerical control area of the Engineering Training Center of University of Shanghai for Science and Technology as the research object and analyzes the influence of different outdoor temperatures and different air supply volumes on the vertical temperature distribution of large space under the two forms of airflow organization. Also, the thermal efficiency was compared. The research results are as follows:

- Through the actual measurement and analysis, the indoor vertical temperature distribution is obtained under two forms of airflow organization in winter. Under the nozzle supply conditions, the vertical temperature distribution presents the trend of slowly increase, rapid increase, decrease and remaining unchanged. Under the columnar supply conditions, the temperature below 3 m is same as the nozzle, and then the temperature rises with a slower trend.
- The vertical temperature gradient of the nozzle supply is slightly smaller than that of the columnar supply. Under the same conditions, the indoor temperature of the nozzle is higher than that of the columnar, including the temperature of the working area, which indicates that the nozzle supply is better than the columnar supply under the winter condition.
- According to the measured data, the thermal efficiency is calculated. Comparing the nozzle supply and the columnar supply conditions, it can be concluded that the heating effect of the nozzle supply is better.

In summary, in winter the nozzle-side supply is superior to the columnar down supply in terms of thermal comfort and energy saving. The research in this paper has reference value for the selection of airflow organization in the design of large space in heating seasons.

Acknowledgements This work was supported by the Natural Science Foundation of Shanghai (Grant No. 16ZR1423200).

References

1. Saïd, M.N.A., MacDonald, R.A., Durrant, G.C.: Measurement of thermal stratification in large single-cell buildings. Energy Build. **24**(2), 105–115 (1996)
2. Xu, L., Weng, P.: Numerical analysis of air distribution in a stratified air conditioning room. J. Shanghai Univ. (Nat. Sci.) **8**(5), 447–451 (2002)
3. Wang, L., Du, Z., Zhang, J., et al.: Study on the thermal comfort characteristics under the vent with supplying air jets and cross-flows coupling in subway stations. Energy Build. **131**, 113–122 (2016)
4. Yang, H., Cao, B., Zhu, Y.: Study on the effects of chair heating in cold indoor environments from the perspective of local thermal sensation. Energy Build. **180**, 16–28 (2018)
5. Bagheri, H.M., Gorton, R.L.: Performance characteristics of a system designed for stratified cooling operating during the heating season. ASHRAE Trans. **93**, 367–381 (1987)
6. Lai, H., Wang, Y.: Measurement and analysis of air conditioning heating in high space of Chongqing in winter. Refrig. Air Cond. (Sichuan) **26**(5), 514–520 (2012)
7. Li, X., Gao, J., Xu, S.: Numerical simulation and discussion of stratified airflow in sideways of large space buildings. Build. Heat Vent. Air Cond. **23**(2), 64–66 (2004)

Experimental Study of Electroosmotic Effect in Composite Desiccant

Shanshan Cai, Xu Luo, Xing Zhou, Wanyin Huang, Xu Li and Jiajun Ji

Abstract Electroosmotic technology is widely used in dehydration applications, such as soil consolidation and sludge dewatering. It has the potential to be used as a dehumidification tool combined with solid desiccant—proposed as a novel dehumidification fin. In this study, composite desiccants are prepared by mixing macroporous silica gel with liquid desiccant and placed in the environmental chamber to determine the adsorption capacity. Both the type and the concentration of the liquid desiccant are compared on the amount of moisture adsorbed in the composite materials. Further, the selected composite samples are installed between two humidity chambers to test the dehumidification rate under the electroosmotic effect. The main factors, such as the size of particles, the intensity of electric field, and the density of material, on the dehumidification rate are compared and discussed in detail.

Keywords Electroosmotic · EOF · Composite desiccant · Adsorption · Dehumidification rate

1 Introduction

Humidity is a key parameter in the air quality which is closely related to human life and energy consumption in buildings [1]. Condensation dehumidification, solid adsorption, and liquid absorption are three main types of dehumidification methods considered in the air-conditioning systems. Condensation dehumidification is a typical procedure of coupled temperature and humidity control. Solid adsorption and liquid absorption are commonly used in the applications of independent control or weakly coupled control of temperature and humidity, which improve the overall efficiency and decrease the energy consumption of air-conditioning systems.

S. Cai (✉) · X. Luo · X. Zhou · W. Huang · X. Li · J. Ji
School of Energy and Power Engineering,
Huazhong University of Science and Technology, Wuhan, China
e-mail: shanshc@hust.edu.cn

However, the additional dehumidification devices may increase the system complexity [2–4]. Electroosmotic effect is an electrochemical method that widely used in the applications such as electroosmotic pumps [5], soil consolidation [6], sludge dewatering [7], walls protection [8], and desiccant regeneration. The method of electroosmotic regeneration removes moisture in the desiccant when the solid desiccant is saturated with moisture adsorbed from humid air [9–11]. This method has the advantages of low regeneration energy consumption and simple system structure [12]. In the previous research work, an electroosmotic dehumidification fin, which can be installed next to a traditional heat exchanger or a dehumidification unit, is proposed to utilize the potential of electroosmotic effect in the porous hygroscopic materials. The efficiency of the dehumidification fin is closely related to the adsorption ability of the solid desiccant. In the current literature, the commonly used solid desiccants are active alumina, silica gel, and zeolite [13], while the traditional liquid desiccants are lithium chloride, lithium bromide, calcium chloride and tri-ethylene glycol [14]. In order to enhance the ability of moisture adsorption, researchers [15–17] prepare different types of composite desiccants, of which the adsorption rate is higher than that of pure solid desiccants. By analyzing the regeneration of solid desiccants, researchers [9, 12, 18] have observed the obvious electroosmotic effect in single-solid desiccants (silica gel and zeolite) with appropriate water content and DC voltage. However, the studies of electroosmotic effect on composite desiccant are few. In this study, the electroosmotic effect is experimentally studied on the composite hygroscopic materials. The factors that affect the electroosmotic flow, which is a key related to the dehumidification rate of the novel fin, are further analyzed and discussed.

2 Theoretical Description of a Novel Dehumidification Fin

Electroosmotic effect refers to the phenomenon that the liquid in a solid material migrates directionally in an electric field. Take the silica gel as an example. When the porous hygroscopic material adsorbs moisture from humid air, it will form a negative ion layer on the surface of the solid. These negative ions attract the cations from the aqueous solution in the pores and form a positive ion layer on the surface, which is called "stern layer," as shown in Fig. 1a. The adjacent region with ions, which has the ability to migrate in diffusive distributions is named as "diffuse layer." These two layers form the "electric double layer" (EDL). When an electric field is applied (Fig. 1b), the ions in the diffuse layer drive surrounding water molecules moving together from one electrode to the other under the effects of Coulomb and viscous forces, thus forming the electroosmotic flow (EOF) [19]. The structure of the dehumidification fin, which utilizes the electroosmotic effect, is indicated in Fig. 1c [20]. The fin is composed of two thin sheets of solid desiccant, an inner cavity for water drainage and the electrodes. When air passes through the dehumidification fin, the hygroscopic sheets adsorb water vapor and decrease the humidity in the air. When the hygroscopic sheets are about to reach saturation, a

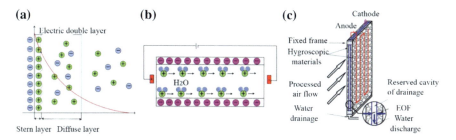

Fig. 1 Schematics demonstrated **a** the "electric double layer" (EDL), **b** the electroosmotic flow (EOF), and **c** a dehumidification fin

low voltage is applied to the electrodes and water migrates from one electrode to the other, which is next to the internal cavity for water drainage. The direction and the rate of electroosmotic flow highly depend on the type and physical properties of the adsorbent, as well as conditions of electric field. In order to better design the dehumidification fin, it is required fundamental studies on the main factors that affect adsorption capacity and dehumidification rate of composite desiccants.

3 Experimental Methods

The experimental methods include the fabrication of composite desiccants and the test of dehumidification rate.

3.1 The Fabrication of Composite Desiccant

Macroporous silica gel is a common porous hygroscopic material with high values of specific surface area, porosity, and hygroscopicity. Therefore, in this study, macroporous silica gel is selected as the base for the composite materials. The procedure for preparing composite materials is summarized as follows (see Fig. 2):

(1) Molding: Mix the powder of macroporous silica gel (0–0.05, 0.05–0.1, and 0.1–0.2 mm) with curing agent (PVA) to form the solid sheet.
(2) Drying: Dry the solid sheets in the oven at 90 °C to remove liquid water. The dry mass is recorded as m_1.
(3) Immersion: Immerse the fabricated sheets in three types of liquid for one hour to fill the pores with additional liquid desiccant. The three types of liquid desiccants are selected as calcium chloride ($CaCl_2$) solution, lithium chloride (LiCl) solution, and tri-ethylene glycol (TEG) solution. Based on multiple trials, the concentrations are selected as 30%.

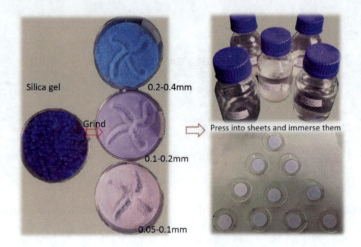

Fig. 2 Preparation of test samples

(4) Preconditioning: Rinse and dry the sheets and then humidify the sample in an environment chamber (80% RH, 15.2 °C) to observe the adsorption procedure. The mass of the sample after drying and during humidification is recorded as m_2 and m_3, respectively.

3.2 The Measurement of Dehumidification Rate

The test apparatus to measure the dehumidification rate is indicated in Fig. 3. The test sample is located between the high and the low humidity chambers. The initial moisture content of the test samples is maintained around similar level by adsorption procedure, which can be calculated as $w_2 = 1 - m_2/m_3$. The upper and lower surfaces of the test samples are tightly attached to two electrodes. The relative humidity of the high humidity chamber is around 98% (at 15.6 °C), and the humidity is maintained by using saturated solution of potassium sulfate. The relative humidity of the low humidity chamber is around 30% (at 14.3 °C), and the humidity is maintained by using the solid desiccant of macroporous silica gel. The variations of current, temperatures, and humidity in the chambers are recorded by a multimeter, thermocouples, and humidity sensors, respectively. The dehumidification rate is estimated by comparing the increasing rates of relative humidity in the chamber with applied voltage and the one without.

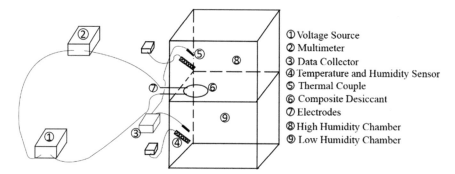

Fig. 3 Schematic of test apparatus

4 Experimental Methods

Both the results of adsorption capacity and dehumidification rate and the impact factors are discussed in the following sections.

4.1 Adsorption Capacity

The total time length for moisture adsorption procedure is 232 h. The description of the test samples is listed in Table 1, and the test results are indicated in Fig. 4.

Figure 4a compares the adsorption curve among four test samples composited with different liquid desiccants. Results indicate that for the pure macroporous silica gel, the saturation amount is 0.26 g/g; for the samples with lithium chloride, calcium chloride, and tri-ethylene glycol, the adsorption amount is 0.664 g/g, 0.358 g/g, and 0.163 g/g, respectively, after 232 h continuous test. By the end of the adsorption test, the adsorption capacity of composite materials still keeps increasing and the capacity of M30 LiCl and M30 $CaCl_2$ is 2.55 and 1.38 times higher than

Table 1 Description of composite samples preconditioned in the environment chamber

Number[a]	Samples (diameter of 34 mm, height of 6.2 mm)						
	M30 LiCl	M30 $CaCl_2$	M30 TEG	L30 LiCl	S30 LiCl	M20 LiCl	M40 LiCl
Liquid	LiCl	$CaCl_2$	TEG	LiCl	LiCl	LiCl	LiCl
Particle size (mm)	0.1–0.2	0.1–0.2	0.1–0.2	0.2–0.4	0.05–0.1	0.1–0.2	0.1–0.2
Concentration (%)	30	30	30	30	30	20	40
w_1(%)[b]	21.60	15.98	11.88	23.84	17.79	17.75	24.59

[a]"S," "M," and "L" represent the samples fabricated from powders of small, medium, and large size particles
[b]w_1 is the content of inorganic salts in the composite desiccants is $w_1 = (m_2 - m_1)/m_2$

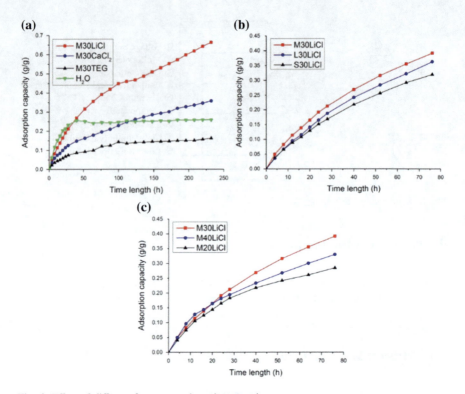

Fig. 4 Effect of different factors on adsorption capacity

that of the pure macroporous silica gel. However, during the initial stage, the moisture adsorption is less than the pure solid desiccant. The possible reason is explained as follows: Composite desiccant includes both physical adsorption and chemical adsorption. Compared with pure macroporous silica gel, the adsorption heat of composite desiccant is more and cannot be released in time, which slows down the absorption of water vapor from air [21]. However, the inorganic salts absorb moisture continuously; therefore, the saturation level increases and the adsorption period extends. For the sample M30 TEG, the final adsorption capacity is even lower than that of pure macroporous silica gel due to the formation of additional substance on the exterior surface. Among the three composite samples, M30 LiCl behaves better than the other two samples (M30 $CaCl_2$ and M30 TEG) and is selected for further investigation. Figure 4b compares the effect of particle size of macroporous silica gel on the moisture adsorption of composite desiccants. In the samples with small particle sizes, which related to the pore sizes, less amount of lithium chloride is filled in than in the samples with medium and large particle sizes (see Table 1). Therefore, both M30 LiCl and L30 LiCl indicate higher adsorption capacity than S30 LiCl does. Due to the fact that the lithium chloride filled in the pores or adhered to the surface may reduce the region of physical adsorption, the final adsorption capacity is also affected. Therefore, compared to the sample of L30 LiCl, which is filled with more salts, the sample of M30 LiCl has a higher adsorption capacity. Figure 4c compares the effect of solution concentration by varying the value from 20 to 40%. It is observed that the sample of M20 LiCl has the lowest adsorption capacity and the sample of M30 LiCl has the highest. As like the findings in the particle sizes, a balance is required between physical and chemical adsorption to derive a maximum adsorption capacity.

4.2 Dehumidification Rate

Based on the findings in the adsorption test, samples with 30% LiCl solution are selected in the dehumidification test. Test samples prepared at different particle sizes and densities are tested under different values of electric field intensity. The parameters of test samples and the dehumidification rate are provided in Table 2 and Fig. 5.

Figure 5a compares the effect of particle size (related to the pore sizes) on the dehumidification rate. Different from the findings in the pure test samples [20], which observes high dehumidification rate in the samples with small particle sizes; in this study, the samples with medium and large particle sizes indicate overall high values of dehumidification rate. As explained in the previous section, with more amount of lithium chloride filled in the samples of medium (P2) and large pores (P3), the conductivity of ions increases and enhances the EOF rate to some extent. In addition, the medium and large pores in samples P2 and P3 may lead to less flow resistance when compared to the small pores in the sample PED, and less flow resistance benefits the EOF rate. Figure 5b compares the effect of electric field intensity on the dehumidification rate. The electric field intensity is controlled by varying the applied voltage between the two electrodes. Generally, within the given range, it is observed that the dehumidification rate increases with the field intensity. Based on the theory of EOF, it is acknowledged that the ions in EDL are mainly driven by the Coulomb force and are migrated in a directional way. Therefore, the

Table 2 Description of test samples

	Samples (diameter of 34 mm, height of 6.2 mm)						
Number	PED[a]	P2	P3	E1	E3	D1	D3
Particle size (mm)	0.05–0.1	0.1–0.2	0.2–0.4	0.05–0.1	0.05–0.1	0.05–0.1	0.05–0.1
Intensity (V/mm)	6.45	6.45	6.45	4.84	8.06	6.45	6.45
Density (g/cm^3)	0.85	0.85	0.85	0.85	0.85	0.81	0.89
w_1 (%)	17.2	21.4	23.5	17.2	17.2	19.6	16.1
Miscellaneous	Liquid desiccant: 30% LiCl; initial moisture content w_2 (%): 21.4						

[a]"P," "E," and "D" represent "particle size," "electric field intensity," and "density"

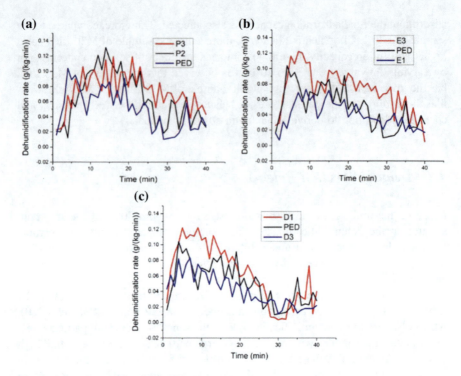

Fig. 5 Comparison of the dehumidification rate

higher the value of electric field intensity is, the higher the migration rate of ions will be. However, there is an appropriate range of electric field intensity. Although high intensity may benefit the movement of ions, it may also lead to serious deformation of the composite desiccant, which can destroy test samples. Figure 5c compares the impact of density on the dehumidification rate. It is observed that the samples of low density lead to high dehumidification rate. The possible reason is that density will affect the number of flow paths and the overall flow resistance (caused by the tortuosity of the flow path). In the low-density sample D3, there are more flow paths and the overall flow resistance is smaller. Besides, the increase of lithium chloride content in samples can also increase the conductivity of aqueous solution, and ultimately improve the dehumidification rate.

5 Conclusions

In order to evaluate the potential of composite desiccants on the application of a novel dehumidification fin, which utilizes the electroosmotic effect, the factors affecting the adsorption capacity and dehumidification rate are studied experimentally. The impact factors related to the adsorption capacity are the particle size of macroporous silica gel, the type, and concentration of liquid desiccant. The impact factors related to the dehumidification rate are the size of particles, the intensity of electric field, and the density of material. The main conclusions are summarized as follows:

(1) Among the three liquid desiccants ($CaCl_2$, LiCl and TEG), both $CaCl_2$ and LiCl increase the adsorption capacity of composite desiccants by 1.38 and 2.55 times when compared to that of pure macroporous silica gel. By varying the concentration of LiCl from 20 to 40%, the optimal concentration is around 30%. The particle size of macroporous silica gel affects the filling of salt solution, and the optimal particle size is around 0.1–0.2 mm.

(2) It is generally observed that within the given range, the dehumidification rate in the composite desiccant is linearly related to the electric field intensity (with the maximum EOF rate appeared around 8.06 V/mm). The samples with medium (0.1–0.2 mm) and large particle (0.2–0.4 mm) sizes indicate a better dehumidification rate than the samples in small particle size (0.05–0.1 mm) do. Material density represents the number and complexity of flow paths and the dehumidification rate in the low-density materials behaves better than that in high-density materials.

Acknowledgements The authors would like to thank the National Natural Science Foundation of China (Grant No. 51706078) and the Natural Science Foundation of Hubei Province in China (Grant No. 2017CFB131) for funding and supporting this work.

References

1. Zhang, M.J., Qin, M.H., Chen, Z.: Study on impact of composite phase change moisture-controlled materials on indoor thermal and humid environment. Build. Sci. **32**(12), 72–79 (2016)
2. Huang, S.F., Zhang, X.S., Xu, Y.: Study on direct air dehumidification method based on electro-osmosis. J. Eng. Thermophys. **37**(12), 2532–2535 (2016)
3. Wang, Q.: Dehumidification in air conditioning system. J. Guangdong Univ. Petrochem. Technol. **23**(4), 63–67 (2013)
4. Lu, D.W., Shao, S.Q.: Development progress of new type high efficient dehumidification technologies. Refrig. Air Cond. **18**(8), 97–101 (2017)
5. Chen, N.X., Guan, Y.F., Ma, J.P.: Controlling electro-osmotic flow in electro-osmotic pump. Chin. J. Anal. Chem. **21**(5), 619–623 (2003)
6. Liu, F.Y., Wang, Y.J., Wang, J.: Electro-osmotic consolidation property of layered soft soil containing soluble salt. J. Highw. Transp. **29**(5), 19–25 (2016)
7. Ma, D.G., Qian, J.J., Zhu, H.M., et al.: Drying characteristics of electro-osmosis dewatered sludge. Environ. Technol. **37**(23), 3046–3054 (2016)
8. Bertolini, L., Coppola, L., Gastaldi, M., et al.: Electro-osmotic transport in porous construction materials and dehumidification of masonry. Constr. Build. Mater. **23**(1), 254–263 (2009)

9. Zhang, G.Y., Tian, C.Q., Shao, S.Q., et al.: Investigation on electro-osmotic regeneration of solid desiccants. Contemp. Chem. Ind. **42**(8), 1043–1046 (2013)
10. Qi, R.H., Tian, C.Q., Shao, S.Q.: Electro-osmotic regeneration for solid desiccants. J. Chem. Ind. Eng. **61**(3), 642–647 (2010)
11. Zhang, G.Y., Shao, S.Q., Lou, X.M., et al.: Investigation on the adsorption mechanism and electro-osmosis regeneration of common solid desiccants. J. Refrig. **35**(1), 8–13 (2014)
12. Duan, D.X., Liu, X.Y., He, J., et al.: Orthogonal regression experiment based on electro-osmosis regeneration of macroporous silica gel. ShanDong Chem. Ind. **47**(8), 43–52 (2018)
13. Liu, L., He, Z.H., Chen, J.C., et al.: Development on solid composite desiccants for desiccant cooling systems. Adv. New Renew. Enengy. **5**(5), 377–385 (2017)
14. Li, M., Yu, X.M., Tian, J.: Analysis of liquid desiccant air conditioning system. Build. Energy Effic. **45**(3), 40–43 (2017)
15. Yuan, H.C.: Experimental study on performance of a new silica gel/calcium chloride composite desiccant. Dalian Maritime University, Dalian (2014)
16. An, B.C.: Experimental study on adsorption performance of silica gel/calcium chloride composite desiccant. Dalian Maritime University, Dalian (2013)
17. Chen, F.K., Ma, Y.: Study on adsorption properties of composite desiccant based on silica gel and calcium chloride. J. Ezhou Univ. **24**(3), 100–102 (2017)
18. Zhang, Y.Q.: Research on air conditioning dehumidification method based on electro-osmosis. Beijing University of Technology, Beijing (2007)
19. Rao, W., Li, D.G.: Dynamics research on ion adsorption of electric double layer. J. Nav. Univ. Eng. **21**(4), 108–112 (2009)
20. Huang, W.Y., Zhou, X., Zhang, B.X., et al.: Experimental study on the impact factors of electro-osmotic dehumidification. In: The 9th Asian Conference on Refrigeration and Air-conditioning, Sapporo, Japan (2018)
21. Wang, L.B., Piao, X.B., Ma, W.B., et al.: Study on preparation and properties of calcium chloride/silica gel composite desiccant for waste heat refrigeration. J. Eng. Therm. Energy Power **27**(3), 366–371 (2012)

District Heating System Load Prediction Using Machine Learning Method

Meng Jia, Chunhua Sun, Shanshan Cao and Chengying Qi

Abstract Accurate prediction of heating load can help improve operational efficiency of district heating systems (DHSs). The selection of feature variables is of great significance to prediction performance. Most existing methods only use the meteorological data and historical thermal demand data. In this study, correlation analysis method is employed to analyze predominant variables affecting prediction accuracy. The correlation of supply/return temperature, outdoor temperature, and historical load data were examined. The obtained results were used to select minimal input variables subset so as to avoid multiple input variables. The extreme learning machine (ELM) was used to predict the energy consumption of the next 6, 12, and 24 h. The approach was adopted to predict heating load of a DHS in Changchun, China. Historical heating load data were proved to be the most essential prediction inputs. The results show that the root-mean-square error predicted by the ELM model can reach 4.1%.

Keywords District heating system · Heating load prediction · Correlation analysis · Extreme learning machine

---

M. Jia · C. Sun · S. Cao (✉) · C. Qi
School of Energy and Environment, Hebei University of Technology, Tianjin 300401, China
e-mail: css_2005@126.com

M. Jia
e-mail: hegongdajiameng@163.com

C. Sun
e-mail: sunchunhua@163.com

C. Qi
e-mail: qicy@163.com

1 Introduction

Accurate prediction of the short-term heat load is a prerequisite for efficient and stable operation of district heating system (DHS). Most existing thermal load prediction methods considered limited influencing factors, like meteorological and historical parameters, and the prediction accuracy is unstable [1–6]. These models usually reflect a smooth linear relationship between load and weather variables, which is of great nonlinearity and complexity actually [1]. A. Kusiak et al. used weather forecast data to predict steam load [2]. Nicolas Perez-Mora et al. used historical heat demand data to predict and manage DHS loads [3]. E. Dotzauer took weather forecasting and social component modeling into account [4]. H. A. Nielsen et al. obtained a regression equation between meteorological parameters (i.e., outdoor temperature, solar radiation, relative humidity, and wind speed) and building heat consumption [5]. O. Yetemen et al. found that the monsoon circulation has some influence on the long-term energy consumption prediction [6].

With the continuous development of machine learning theory, nonlinear prediction methods have been successfully applied in the field of load forecasting. Huang et al. [7] developed extreme learning machine (ELM), which is an evolutionary neural network method with good generalization ability. Sajjadi et al. established a DHS thermal load prediction model by using ELM method, revealing the robustness of this method, [8].

This paper studied the correlation of historical heating load, historical secondary supply/return temperature, and outdoor temperature. The selected input variables were used to predict heat load for the next 6, 12, and 24 h using ELM method. The proposed method was applied and analyzed in a DHS in Changchun, China.

2 Data Preprocessing

2.1 Data Outlier Elimination

Test values with coarse errors are called outliers, which are undesirable and should be removed from the measured data [9]. PauTa criterion is commonly used to judge the gross error, whose basic idea is that any error beyond triple standard deviation limit is considered to be gross error rather than random error.

When using the PauTa criterion to judge and eliminate outliers, the average value $\overline{X}$ and residual error $V_i = X_i - \overline{X}$ of the independent measurement column $X_i (i = 1, 2, 3, …, n)$ should be calculated first. The standard deviation S of the measurement column is calculated. If the residual error V_d of a measured value X_d satisfies $V_d > 3S$, it is considered that X_d is an outlier needs to be rejected.

2.2 Correlation Analysis

The selection of the characteristic variables plays a crucial role in the thermal load prediction model. Through correlation analysis, the relative factors that have a great influence on load can be taken as the input factors of the prediction model to improve accuracy. In this study, the correlation coefficient method was used to analyze the correlation between two variables. r can be calculated by Eq. (1):

$$r = \frac{\sum_{i=1}^{n}(X_i - \overline{X})(Y_i - \overline{Y})}{\sqrt{\sum_{i=1}^{n}(X_i - \overline{X})^2}\sqrt{\sum_{i=1}^{n}(Y_i - \overline{Y})^2}} \tag{1}$$

where X and Y represent the two variables. The r is between $[-1, 1]$. A positive value of r indicates a positive correlation, vice versa. The greater the absolute value of r, the stronger the correlation.

3 Prediction Methods

3.1 Extreme Learning Machine (ELM)

ELM refers to an artificial neural network model that is developed with the improvements on single-hidden layer feedforward networks (SLFNs) [10], as shown in Fig. 1.

For M arbitrary samples (x_i, t_i), in which $x_i=[x_{i1}, x_{i2}, \ldots, x_{in}]^T \in R^n$ and $t_i = [t_{i1}, t_{i2}, \ldots, t_{in}]^T \in R^m$. The number of single-hidden layer nodes is $\tilde{N}$, the standard SLFNs model with an activation function $g(x)$ is as follows:

$$\sum_{i=1}^{\tilde{N}} \beta_i g_i(x_j) = \sum_{i=1}^{\tilde{N}} \beta_i g_i(a_i \cdot x_j + b_i), j = 1, \ldots, N \tag{2}$$

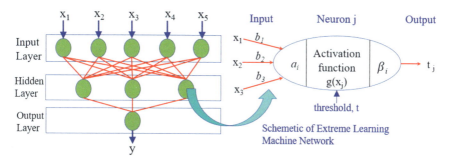

Fig. 1 Schematic of ELM network

where $a_i = [a_{i1}, a_{i2}, \ldots, a_{im}]^T$ is the weight vector that connects the ith hidden layer node; b_i is the threshold of ith hidden layer nodes; $\beta_i = [\beta_{i1}, \beta_{i2}, \ldots, \beta_{im}]^T$ is the output weight vector connecting ith hidden layer nodes; $a_i \cdot x_j$ represents the inner product of a_i and x_j.

The ELM model can approach the output value t_j of N training samples with zero error, and we get:

$$\sum_{i=1}^{\tilde{N}} \beta_i g_i (a_i \cdot x_j + b_i) = t_j, j = 1, \ldots, N \tag{3}$$

Equation (4) is written in the matrix form as follows:

$$\beta H = T \tag{4}$$

where H is the hidden layer output matrix of the network; the ith column represents the output vector of the ith hidden layer node associated with the input $x_1, x_2, \ldots, x_N$, and the jth row represents the implicit layer output vector associated with the input. The hidden layer matrix day is a deterministic matrix, so training SLFNs is equivalently converted to a least-squares solution, so that $\beta H = T$, which is expressed as follows:

$$\hat{\beta} = \min_{\beta} \left\| T\left(a_i, \ldots, a_{\tilde{N}}, b_i, \ldots, b_{\tilde{N}}\right) \beta - T \right\| \tag{5}$$

Equation (6) can be expressed as follows:

$$\hat{\beta} = H^+ T \tag{6}$$

where H^+ is the molar generalized inverse matrix of the hidden layer output matrix.

3.2 Prediction Model Performance Evaluation Criteria

The mean absolute percentage error (MAPE) and root-mean-square error (RMSE) are used to evaluate the performance of the thermal load prediction model, which are relative and absolute indicators, respectively. They can be calculated by Eq. (7):

$$\begin{cases} \text{MAPE} = \frac{1}{n} \sum_{t=1}^{n} \left| \frac{\text{observed}_t - \text{predicted}_t}{\text{observed}_t} \right| \times 100\% \\ \text{RMSE} = \sqrt{\frac{1}{n} \sum_{t=1}^{n} (\text{observed}_t - \text{predicted}_t)^2} \end{cases} \tag{7}$$

where observed$_t$ is actual heat load and predicted$_t$ is the predicted heat load.

4 Results and Discussion

In order to verify the feasibility and effectiveness of the proposed prediction algorithm, filed test of a DHS station in Changchun City was conducted from October 21 to December 7, 2018. Outdoor temperature t_w, supply temperature t_g, return temperature t_h, and heating load q were collected every 10 min, and a total of 6840 data were collected, as shown in Fig. 2. It can be seen that t_g and t_h are relatively stable. t_w and q fluctuate more severely, which may have a certain impact on the later prediction accuracy. The measured variables were averaged every 6, 12, and 24 h, to study different timescale heat load predictions.

4.1 Correlation Analysis

The measured factors were normalized and then calculate the correlation coefficient with heat consumption according to Eq. (1), and the results are shown in Tables 1, 2, and 3.

As shown in Table 1, when the heat consumption prediction period is 6, 12, and 24 h, the historical heat consumption and the historical secondary return temperature have a strong correlation with the heating load. The correlation coefficient of historical heat consumption, historical secondary return temperature, and heating load reached the maximum when the prediction period is 12 h.

When the prediction period is 6h, 12h, 24h, the correlation coefficient between heating load and outdoor temperature is -0.485, -0.523, -0.561, respectively. Although the correlation between outdoor temperature and heating load is weak, it is the key factor in updating the heating load prediction model. Finally, we use historical heating load, secondary return temperature, and outdoor temperature as the heating load variables with prediction periods of 6, 12, and 24 h.

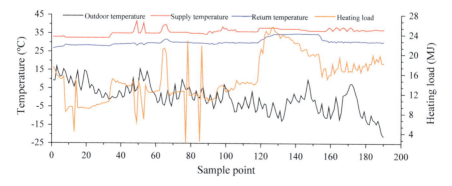

Fig. 2 Measured data

Table 1 Correlation coefficient of 6-h averaged heat load and measured data

Variable	$t_{w,i}^{(6h)}$	$q_{i-1}^{(6h)}$	$q_{i-2}^{(6h)}$	$q_{i-3}^{(6h)}$	$t_{g,i-1}^{(6h)}$	$t_{g,i-2}^{(6h)}$	$t_{g,i-3}^{(6h)}$	$t_{h,i-1}^{(6h)}$	$t_{h,i-2}^{(6h)}$	$t_{h,i-3}^{(6h)}$
$r^{(6h)}$	−0.485	0.806	0.804	0.778	0.652	0.643	0.610	0.767	0.736	0.712

Table 2 Correlation coefficient of 12-h averaged heat load and measured data

Variable	$t_{w,i}^{(12h)}$	$q_{i-1}^{(12h)}$	$q_{i-2}^{(12h)}$	$q_{i-3}^{(12h)}$	$t_{g,i-1}^{(12h)}$	$t_{g,i-2}^{(12h)}$	$t_{g,i-3}^{(12h)}$	$t_{h,i-1}^{(12h)}$	$t_{h,i-2}^{(12h)}$	$t_{h,i-3}^{(12h)}$
$r^{(12h)}$	−0.523	0.923	0.850	0.781	0.687	0.643	0.640	0.812	0.760	0.710

Table 3 Correlation coefficient of 24-h averaged heat load and measured data

Variable	$t_{w,i}^{(24h)}$	$q_{i-1}^{(24h)}$	$q_{i-2}^{(24h)}$	$q_{i-3}^{(24h)}$	$t_{g,i-1}^{(24h)}$	$t_{g,i-2}^{(24h)}$	$t_{g,i-3}^{(24h)}$	$t_{h,i-1}^{(24h)}$	$t_{h,i-2}^{(24h)}$	$t_{h,i-3}^{(24h)}$
$r^{(24h)}$	−0.561	0.875	0.759	0.613	0.664	0.640	0.638	0.777	0.687	0.611

4.2 Prediction Analysis

The data sets are divided into two categories by setting the number of test sets: the number of training sets = 7:3. As the ELM method is used to predict the heating load of the periods of 6, 12, and 24 h, the results are shown in Figs. 3, 4, and 5, respectively. It can be seen that when the predicted period of heating load is 6, 12, and 24 h, the corresponding MAPE values are 4.1, 6.8, and 9.3%. The corresponding MSE value is 0.941, 1.459, and 2.063. Comparing the prediction results, it is found that the heating load prediction model has the best degree of agreement in 6 h, the 12-h result is the second, and the 24-h fitting degree is the worst.

When the predicted period of heating load is 6 h, the trend of the predicted load curve is similar to the actual load trend. At 1–20 and 35–40 sample points, the

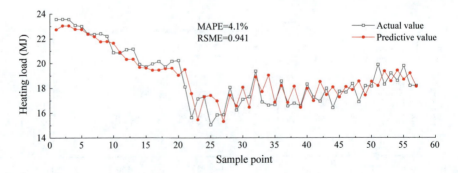

Fig. 3 Next 6-h heating load prediction results

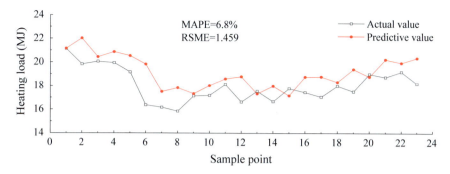

Fig. 4 Next 12-h heating load prediction results

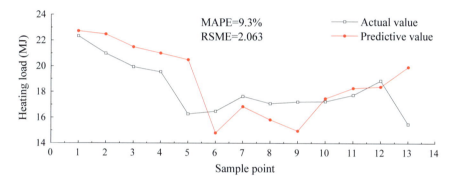

Fig. 5 Next 24-h heating load prediction results

predicted value is closer to the true value. The prediction results show that the ELM method has effectiveness in the application of short-term heating load prediction research.

With the extension of prediction time, the accuracy of heating load prediction decreases gradually. The main reason may be that the collected data samples are located in the early stage of heating, the heating load fluctuates greatly, and the collected heating load and other data are insufficient.

5 Conclusions

In this paper, the method of ELM heating load prediction is studied and verified in a heating network in Changchun. Through the establishment of ELM prediction model, the following conclusions can be drawn:

(1) Studying the influence of different characteristic variables on heat load prediction, the MAPE values of predicted future heating loads at 6 and 12 h are 4.1 and 6.8%. It is proved that the optimized feature set model has good prediction performance.
(2) In this study, the accuracy of the future 24-h heating load prediction is lower than the heat load forecast for the future 12 and 6 h, and its improvement measures need to be further researched.

References

1. Islam, S.M., et al.: Forecasting monthly electric load and energy for a fast growing utility using an artificial neural network. Electr. Power Syst. Res. **34**(1), 1–9 (1995)
2. Kusiak, A., et al.: A data-driven approach for steam load prediction in buildings. Appl. Energy **87**(3), 925–933 (2010)
3. Perez-Mora, N., et al.: DHC load management using demand forecast. Energy Procedia **91**, 557–566 (2016)
4. Dotzauer, E.: Simple model for prediction of loads in district-heating systems. Appl. Energy **73**(3–4), 277–284 (2002)
5. Nielsen, H.A., et al.: Modelling the heat consumption in district heating systems using a grey-box approach. Energy Build. **38**(1), 63–71 (2006)
6. Yetemen, O., et al.: Climatic parameters and evaluation of energy consumption of the Afyon geothermal district heating system, Afyon, Turkey. Renew. Energy **34**(3), 706–710 (2009)
7. Huang, G.B., et al.: Extreme learning machine: theory and applications. Neurocomputing **70**(1–3), 489–501 (2006)
8. Sajjadi, S., et al.: Extreme learning machine for prediction of heat load in district heating systems. Energy Build. **122**, 222–227 (2016)
9. Zhang, M., Yuan, H.: PauTa criteria and data outlier elimination. J. Zhengzhou Univ. (Eng. Sci.) **1**, 87–91 (1997)
10. Bilhan, O., et al.: The evaluation of the effect of nappe breakers on the discharge capacity of trapezoidal labyrinth weirs by ELM and SVR approaches. Flow Meas. Instrum. **64**, 71–82 (2018)

A Case Study on Existing Building HVAC System Optimization of a Five-Star Hotel in Shanghai

WeiFeng Zhu, Zhuling Zheng, Mengyuan Liu and Guangwei Deng

Abstract Due to the complex interaction of components and the improper O and M in existing building HVAC system, a large amount of HVAC systems is not able to operate as their design condition. In order to improve the HVAC energy performance, the common methods are renovation and commissioning, and these days, commission is getting attention due to its efficiency and low cost. This paper aims to illustrate the existing building optimal method and its effects. A five-star hotel is used as example, and after field study, several optimal methods were applied, including improving performance of main chiller by applying a boost pump to the inlet pipe of evaporator, adjusting chilled water hydraulic balance and shutting down the chilled water bypass, adding free-cooling system, adjusting the pump frequency and air-conditioning terminal. The result shows that after the whole optimization, from the theoretical analyzation, the system can save about 319,200 kWh per year, according to the energy bill, the final energy saving is 359,000 kWh per year, and performance of building system and indoor thermal comfort are significantly improved.

Keywords Existing building · HVAC system · Commission · Optimization

1 Introduction

In order to improve the HVAC energy performance, the conventional mode is renovation and commissioning [1]. However, due to the normal renovation will replace the core component of the whole system, the owners may not accept the energy saving renovation mode due to the cost and complex process, which leads to the deadlock in the energy saving work of some projects and hinders the deepening and expansion of the energy saving work of existing public buildings. Another method is commissioning. Commission implements a quality-oriented optimal process for adjusting and optimizing the equipment operation status, making the equipment running in the optimal state point, improving the performance of the device itself and comfort with a few changes of the existing system. Commission has shown a very cost-effective way to improve productivity and optimize operational costs. A study reported a one-to four-year payback with energy saving of 13–16% [2].

This paper will introduce the whole work content and on-demand optimization of the adaptation project and the energy saving and comfort effect through the existing building commissioning of the HVAC system of a five-star hotel.

2 Project Description

The five-star hotel is located in Shanghai which was completed in 2000. The original air-conditioning system of the hotel used three large screw chillers. In early 2017, after the energy saving reconstruction, three small magnet bearing centrifugal chillers were used to replace one of the screw chillers, from design, the new chillers can basically meet the demand, and the screw machine is only used in hot summer with extremely high outdoor temperature. The primary pump operates at a constant frequency. The secondary pump can be operated at a variable frequency according to the variation of the building load. Cooling water pump and cooling tower fan are running at constant frequency. The guest rooms of the hotel are equipped with FCU, and the public areas are equipped with fixed air volume system (Table 1); (Fig. 1).

3 Operation Diagnosis

The performance of the whole station was tested before the commission. The result is shown as follows, the outdoor temperature of the day is 33.6 °C, and RH is 61.3% (Table 2).

We conducted in-depth detection and analysis of the whole system to diagnose the operation problems. The main reasons were determined as follows:

Table 1 Device parameters of the hotel after the energy saving reconstruction

Device name	Quantity	Performance parameters
Screw chiller	2	Cooling Capacity: 1050 kW, Input power: 210 kW, COP: 5.0
Magnet bearing centrifugal chiller	3	Cooling Capacity: 528 kW, Input power: 96.9 kW, COP: 5.45
Chilled water primary pump	4	Flow rate: 180 m^3/h, Lift: 12 m, Input power: 7.5 kW
Chilled water secondary pump	3	Flow rate: 270 m^3/h, Lift: 20 m, Input power: 22 kW
Cooling water pump	4	Flow rate: 320 m^3/h, Lift: 25 m, Input power: 30 kW
Cooling tower	3	Flow rate: 300 m^3/h, Input power: 7.5 kW

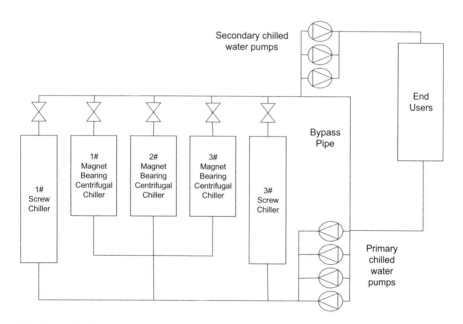

Fig. 1 Hotel chilled water system

Magnetic suspension chillers showed poor performance. Before the commission, chiller water flow rate was 63 m^3/h, which is lower than rated value 90 m^3/h. It will reduce the heat exchange efficiency of the chilled water side of the evaporator and cause the reduction of COP of the whole unit. Meanwhile, in hot summer, the low primary chilled water flow rate will cause large temperature difference between supply and return water temperature. Also, it causes instability of magnetic suspension chillers.

Screw chiller cooling water bypass. In daily operation, staff mainly operates three maglev units in summer. 3# screw machine is added in few high outdoor temperature conditions. However, when the 3# screw chiller is turned off, the

Table 2 Device parameters of the hotel after the energy saving reconstruction

Devices	No.	Power (kW)	Supply/return water temperature (°C)	Chilled water flow rate (m³/h)	Cooling capacity (kW)
Magnet bearing centrifugal Chiller	1#	54.8	6.2/10.1	61.6	279.4
	2#	57.8	6.1/10.1	63.4	294.9
	3#	56.2	6.1/10.1	64.0	297.7
Chilled water primary pump	3#, 4#	14.9	/	189	/
Chilled water secondary pump	1#, 2#	35.2	/	445	/
Cooling water pump	1#, 4#	59.2	/	/	/
Cooling tower	1#, 2#, 3#	21.5	28.6/30.5	/	/
Comprehensive operation efficiency of refrigeration system		2.9			
Magnetic suspension chiller COP		5.2			

cooling water pipeline valve is still open. According to the test, half of the cooling water of the system is bypassed from the screw unit pipe, and the cooling water flow rate of the magnetic suspension unit is low with the average flow rate of 89 m³/h, lower than the rated flow rate which is 107 m³/h.

Flow rate of the secondary chilled water bypass pipe was too high. In order to master the operation status of the secondary water system, the total flow rate and the temperature of water supply and return were monitored for a long time. In summer, two secondary chilled water pumps were operated under high frequency. The maximum flow was up to 462 m³/h, and the average flow was about 420 m³/h. The supply-return water temperature difference between is between 1.3 and 3° centigrade which is far below the design temperature difference of 5 degree centigrade.

Meanwhile, we found that the secondary air-conditioning water system is a serious hydraulic imbalance. For the first floor, due to heavy cooling load, chilled water temperature difference was 3.4–5.1 °C, and meanwhile, the temperature difference of high area was very small. For instance, the temperature difference of 9th floor is only about 1.4–2.4 °C which cannot meet the demands. In hot summer, in order to ensure the air condition demand in those areas, operators can only increase the flow rate by increasing the secondary pump frequency. Meanwhile, the rated flow of 3 magnet chillers is 270 m³/h, and the frequency of primary pumps are fixed, the actual flow of magnet bearing chiller is only 190 m³/h, and the residual chilled water flow to bypass pipe, which resulted in great energy consumption and waste.

Some indoor areas have poor environmental comfort. Terminal equipment, piping and control valves were checked. It was measured in the field test that the lobby the air supply rate is only 9000 m³/h, which was far lower than the rated

21,000 m³/h. After detailed inspection, we found three main reasons. First, some air supply outlets in lobby were not connected to air duct, so existing air supply rate could not meet the demand of staffs and customers in front reception area. Secondly, severe scaling problems appeared in AHU which resulted to low air inlet volume. Third, lobby air distribution was poor. The air supply outlet blows to the curtain wall directly, so the supply airflow was directly heated by the curtain wall resulting in poor air-conditioning performance in the lobby area. The side door of the lobby is always open, so unorganized fresh air increased the cooling load and relative humidity of the lobby, which seriously affecting indoor comfort.

4 Commission and Post-evaluation

According to the evaluation and diagnosis of the hotel air-conditioning system, some appropriate solutions are selected for the commission in order to deal with the problem.

4.1 Add a Chilled Water Booster Pump to Magnetic Suspension Chiller Unit

In view of the problem that the chilled water flow rate of the maglev unit is relatively low, a booster pump was installed according to the evaporator resistance and flow demand of the unit, which was specially used to increase the water supply pressure of the maglev unit and increase the chilled water flow rate to the rated flow rate (Table 3).

After booster pump installation, total chilled water flow of the unit was 263 m³/h, all chillers' chilled water flow has reached the rated flow 90 m³/h, which improved the running stability of the magnetic levitation unit, and significantly reduced the use of screw machine. Test proofed that its operation efficiency can be improved to 5.6 which is much higher than the existing screw machine. For the same operation condition, the COP increased by 36.6%. According to the operating record, during 70 days from middle of July to August, the total energy consumption of the screw chiller is 350,000 kWh, and after adding the boost pump, the maglev chiller can reach the total energy saving for about 93,800 kWh.

For the booster pump, its rated power is 22 kW, but from the test, its operation power is 15. 9 kW. For 24 h for 70 days, the total energy consumption of booster pump is 26,700 kWh. Thus, the actual energy saving of the chiller plant after adding the booster pump is 67,100 kWh.

Table 3 Rated parameters of booster pump

Device name	Flow (m³/h)	Lift (m)	Power (kW)	Frequency
Booster pump	322	14	22	Constant

4.2 Turn off the Water Bypass

Turn off the cooling water bypass of the screw machine. Screw chiller is only used in hot summer, for the most of time, it keeps in shutdown state, but 3# screw chiller cooling water bypass keeps open. In the commission, the cooling water bypass of #3 screw machine was turned off, and the average cooling water flow rate of the maglev unit increased from 89.2 to 111.4 m^3/h which reached the rated flow requirement. The total flow of cooling water decreased from 537.7 to 334.2 m^3/h, and the corresponding actual operating power decreased significantly. According to 120 days of system operation from June to September in summer, 24 h of operation per day, the energy saving of cooling water pump is calculated to be 510,000 kWh

When the flux of maglev cooling water increases, both the heat exchange efficiency of the condenser and operating efficiency of the chiller increase. The refrigeration power consumption of the original screw machine in summer is 971,000 kWh. According to the efficiency of the main engine, the refrigeration power consumption of the magnetic suspension main engine (COP = 5.6 h) in summer is 719,900 kWh. After the cooling water flow is increased, the power consumption of the magnetic suspension main engine is 663,500 kWh, which can save 47,400 kWh (Table 4).

Turn off the bypass of both primary side and secondary side chilled water pipe. The field test showed that the secondary water flow is much larger than the primary water flow, a large amount of chilled water bypass. By adjusting the hydraulic balance of the secondary water system and the operation frequency of the secondary pump, the total flow rate of the secondary water was significantly reduced. In addition, a stop valve was installed on the bypass pipe to shut off the bypass in order to achieve the balance between the primary water and the secondary water. After turning off the chilled water bypass, the energy saving effects can be achieved as follows:

Before the adjustment of the operating frequency of the secondary pump, two pumps are operated with a total power of 37.6 kw and a flow rate of 445 m^3/h at a frequency of 47 Hz. After commission, two pumps are operated with a total power of 20 kW and a flow rate of 330 m^3/h at a frequency of 35 Hz. According to

Table 4 Cooling water bypass data

Device name	Operation status in summer	Rated flow (m^3/h)	Measured flow (m^3/h)	
			Before commission	After commission
1# Magnetic suspension chiller	On	107.5	90.4	117.1
2# Magnetic suspension chiller	On	107.5	82.0	105.1
3# Magnetic suspension chiller	On	107.5	95.3	112.0
1# Screw chiller	Off	216.7	0	0
2# Screw chiller	Off	216.7	270	0
Total			537.7	334.2

120 days of air-conditioning operation from June to September in summer, the calculated energy saving is about 507,000 kWh/year.

When there is no mixing of primary water and bypass water, the outlet chilled water temperature of the magnetic levitation unit is equal to the secondary water supply temperature. Before the commission, supply water temperature chillers are 6 °C in summer, and after the commission, the supply water temperature of chillers could rise to 7.5 °C. Furthermore, the temperature could be set to 8 °C in transition season, so the optimal supply water setpoint is shown in Table 5. By increasing maglev units supply water temperature from 6 to 7.5 °C, the maglev chiller COP could rise to 6.5. Based on the summer cooling energy consumption of the maglev units, which is 663,500 kWh, the improvement of the chill water temperature will result the energy saving for about 51,000 kWh.

4.3 Install Free-Cooling System

In the commission, a free-cooling system is applied to the HVAC system. During the transition season and at night, when the wet-bulb temperature of outdoor air is relatively low, the low-temperature backwater of the cooling tower is directly used through heat exchange with the chilled water of the air conditioner, so as to reduce the running time of the chiller and save energy. The system can switch to free cooling mode when the outdoor wet-bulb temperature is below 16 °C.

4.4 Air-Conditioning Terminal Adjustment

In view of the poor comfort of some indoor areas, the project adopted the following measures:

Hydraulic balance commissioning of secondary water system. Install pressure reducing valves in the water supply pipes of hotel rooms and other areas with exceeding chilled water flows. Equipped the hotel room FCU water supply pipes with orifice plate, restriction and balance the room running supply water flow, make the temperature difference between supply water and return water between 4 and 5 °C. Then, installed booster pumps in the chilled water supply pipes

Table 5 Supply water temperature setting of the chiller

Working condition	Outdoor temperature (°C)	Supply chilled water set value
1	Above 32	7
2	27–31	7.5
3	22–26	8
4	17–21	9

of lobby and other areas whose AHUs with high cooling demand to increase their chilled water flow and increase the cooling capacity. Last, maintained and added electric valves of each branch pipe of the chilled water supply system. When the air conditioner in special areas such as banquet turned off, cut off the waterway to prevent waste of cooling capacity caused by irrational use of air conditioning terminal, reduce heat loss of pipes, reduce total secondary water flow and reduce power consumption of water pumps.

AHU rectification. By cleaning the cooling coil in the AHU, air blocking effect caused by the dirt can be solved. After cleaning, the air supply volume of the AHU increased from 9000 to 12,000 m^3/h, and the indoor comfort was improved.

Improvement of indoor air distribution. Redecorated the lobby, adjusted the air supply direction, connected the air outlets which were not connected with the air ducts improve the air distribution of in the lobby and finally improve the indoor comfort level.

5 Results

After the commissioning of HVAC systems in the hotel, under the similar weather conditions, the average operating efficiency of the HVAC system was 3.7 which is higher than the 2.9 before the commission. The comprehensive energy efficiency of the chiller plant was improved by 27.6% after the commission (Tables 6, 7).

Table 6 Operation data after the commission

Devices	No.	Power (kW)	Supply/return water temperature (°C)	Chilled water flow Rate(m^3/h)	Cooling capacity (kW)
Magnetic suspension chiller	1#	43.8	7.6/10.4	86	280.1
	2#	46.4	7.5/10.4	88	296.8
	3#	46.2	7.5/10.4	89	300.2
Chilled water primary pump	3#, 4#	11.9	/	263	/
Chilled water secondary pump	1#, 2#	20.6	/	/	/
Booster pump	1#	15.9	/	263	/
Cooling water pump	1#, 4#	41.5	/	/	/
Cooling tower	1#, 2#, 3#	10.8	29.1/31.8	/	/
Comprehensive operation efficiency of refrigeration system		3.7			
Magnetic suspension chiller COP		6.5			

Note The secondary chilled water pump and cooling tower are operated at 35 Hz, and other pumps are operating with motor in constant frequency

Table 7 Commission project total investment and energy saving

Commission project	Investment/yuan	Energy saving/kWh
Add a chilled water booster pump	2,000	67,100
Turn off the bypass (cooling water)	/	98,400
Turn off the bypass (chill water)	/	101,700
Others (hydraulic balance commission in, etc.)	/	/
Add free cooling	180,000	52,000
Total	182,000	319,200

According to the theoretical calculation, the total energy saving is about 319,200 kWh. And actually, according to the energy bill, after commission, the total energy saving is about 359,000 kWh.

Acknowledgements The authors gratefully appreciate the support from the National 13th Five-year Science and Technology Support Project of China (Grant No. 2017YFC0704207).

References

1. Tsinghua University building energy conservation research center: China building energy conservation annual development research report. Tsinghua University, China (2016)
2. Mills, E.: Building commissioning: a golden opportunity for reducing energy costs and greenhouse gas emissions in the United States. Energy Effic. **4**(2), 145–173 (2011)

Research on Heat and Moisture Transfer Characteristics of Soil in Unsaturated and Saturated Condition with Soil Stratification under Vertical Borehole Ground Heat Exchanger Operation

Yao Wang and Songqing Wang

Abstract In this paper, the heat and moisture transfer characteristics of soil in unsaturated and saturated condition with soil stratification under vertical borehole ground heat exchanger operation were analyzed by studying the theory of heat and moisture migration. A three-dimensional unsteady heat transfer model of the ground heat exchanger was established which further coupled the soil-layered structure and considered the effects of vertical moisture migration and seepage. The vertical migration of the moisture which has different initial saturation was considered and compared with the case that only considered horizontal moisture migration at the layered soil. Based on the heat-moisture-coupled heat transfer model under two conditions, the soil temperature field and the moisture field were compared and analyzed. The heat and moisture transfer law in layered soil boundary was analyzed, and its influence on the heat transfer performance of the ground heat exchanger was summarized.

Keywords Soil stratification · Unsaturated soil and saturated soil · Temperature and moisture field · Heat transfer performance

1 Introduction

Ground source heat pump (GSHP) takes advantage of shallow ground temperature energy and with good stability and environmental protection. At present, the studies of GSHP are centralized on increasing the heat exchange efficiency of the ground

Y. Wang · S. Wang (✉)
School of Civil Engineering, Northeast Forestry University, Harbin, China
e-mail: wsqnefu@163.com

Y. Wang
e-mail: yaowang_hvac@126.com

heat exchanger, by means of improving the model of the ground heat exchanger, determining rationally the optimum size, pipe arrangement, and operation mode. In the process of soil formation, the soil was affected by environmental and geological conditions and showed the characteristics of horizontal stratification. Vertical ground heat exchangers usually span different geological layers. One part is located in the unsaturated soil zone; there is a complex heat and mass transfer process in which heat transfer and moisture migration are coupled together by temperature gradient and soil water potential. The other part is located below the groundwater level; there is a heat transfer process in which heat conduction and convection heat transfer are coupled together by temperature gradient and hydraulic gradient [1]. A new temperature response function of the underground pipe group which was composed of U-shaped ground heat exchanger and energy pile was proposed to analyze the change of soil temperature field and the energy efficiency of ground source heat pump [2]. The heat transfer process of the ground heat exchanger was simulated by the mixed time-step method which could effectively improve the calculation efficiency [3]. The heat transfer process of the ground heat exchanger was studied by numerical simulation method, and the influence of heat flux from the tube wall was considered [4]. The effect of layered soil with different physical properties on the heat transfer model of vertical ground heat exchangers was studied. But, the influence of moisture migration on the performance of ground heat exchangers was not considered [5]. A model was established to study the effect of heat and moisture migration on the thermal diffusion of the ground heat exchanger in unsaturated soil [6]. It is pointed out that the model calculation results had large deviations when the ground heat exchangers were all located in saturated soil or partially in saturated soil layers. But, it did not consider the effect of heat and moisture migration of unsaturated soil on the heat transfer process [7]. The influence of atmospheric environment and the groundwater seepage on the heat transfer was considered between the ground heat exchangers. However, the effect of temperature change of circulating fluid and soil stratification on the heat transfer performance of the ground heat exchanger was not considered [8]. In summary, the influence of the horizontally layered soil which simultaneously across the unsaturated and saturated soil on the heat transfer of the ground heat exchanger should be considered in order to accurately describe the heat transfer performance of the ground heat exchanger under actual conditions.

2 Theories and Methods

2.1 *Heat and Moisture Migration Equation*

If neglected the temperature change of circulating liquid in the buried heat exchanger with depth, a large deviation would be obtained from the calculation result. Therefore, the internal turbulence model of the heat exchanger adopted a two-equation model. Since the circulating water which flows in the buried pipe is an incompressible constant viscosity fluid, this paper uses the simplified governing equation. In unsaturated soils, the thermal conductivity of the soil is a function of the volumetric water content and changes as the moisture migration progresses [9]. According to the mass conservation equation in the porous medium, the mass transfer governing equation could be obtained.

$$\rho_\omega \frac{\partial \theta}{\partial T} = -\nabla(J_\omega + J_v) = \left[\rho_\omega(D_{\omega\theta}\nabla\theta + D_{\omega T}\nabla T) + \frac{D_e}{R_v T}\left(\nabla P_v - \frac{P_v}{T}\nabla T\right)\right] \quad (1)$$

where θ is volumetric water content, $D_{\omega\theta}$ is the mass diffusivity under moisture gradient, $D_{\omega T}$ is the mass diffusivity under temperature gradient, P_v is the vapor pressure, and D_e is the vapor diffusion coefficient. In saturated soils, a material volume contains both solid and fluid components. It is assumed that the fluid and solid can reach the local thermal equilibrium instantaneously, and the heat capacity and thermal conductivity are considered to be constant. Therefore, if the porosity is φ, then the corresponding energy equation of the two parts could be constructed based on the volume average method.

$$\left[(1-\varphi)(\rho c)_s + \varphi(\rho c_p)_f\right]\frac{\partial T_s}{\partial t} + (\rho c_p)_f V \cdot \nabla T = \nabla\left[(1+\varphi)\lambda_s \cdot \nabla T + \varphi \cdot \lambda_f \cdot \nabla T\right]$$
$$+ (1+\varphi)q_s + \varphi q_f \quad (2)$$

where subscripts f and s are, respectively, the terms of the fluid phase and the solid skeleton, T is the temperature, λ is the thermal conductivity, c is the specific heat of the solid, q is the heat per unit volume generated by the internal heat source, and c_p is the specific heat of the fluid at constant pressure.

2.2 Physical Model Construction

The U-shaped buried pipe used HDPE pipe with an external radius of 16 mm. In order to better approach the actual heat transfer situation, a three-dimensional unsteady layered soil model was established. The grid and coordinates of the model were shown in Fig. 1, and the soil stratification was shown in Fig. 2.

2.3 Initial Conditions

The physical properties of the layered soil were given in Table 1, and the initial conditions of each case were given in Table 2. Soil saturation is the degree to which

Fig. 1 Grid and coordinates of the model

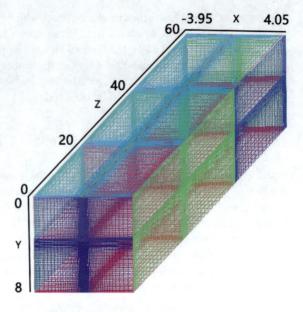

Fig. 2 Schematic diagram of layered soil

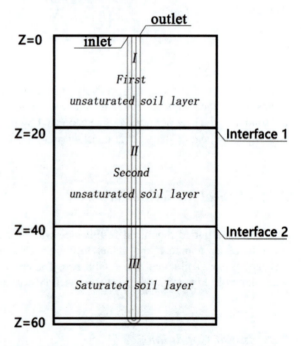

water in the soil is filled with pores and is numerically equal to the ratio of the volume of water to the volume of the pores. In order to explore the effects of moisture migration in layered soils with different saturation on the running

Table 1 Physical property parameter of layered soil

Parameter	Soil I layer	Soil II layer	Soil III layer
Dry density (kg/m^3)	2132	2132	2132
Specific heat capacity (J/kg·K)	1144	1144	1144
Thermal conductivity (W/kg·K)	3.0056	3.0056	3.0056
Saturated hydraulic conductivity (m/s)	8×10^{-6}	8×10^{-6}	8×10^{-6}
Soil porosity (m^3/m^3)	0.4	0.4	0.4

Table 2 Initial conditions

	I Saturated degree of soil (m^3/m^3)	II Saturated degree of soil (m^3/m^3)	III Saturated degree of soil (m^3/m^3)
Case 1	0.4	0.7	1
Case 2	0.4	0.7	1
	U-tube internal fluid inlet temperature (K)	The initial temperature of the soil (K)	Considered the vertical migration
Case 1	313.5	288.5	No
Case 2	313.5	288.5	Yes

performance of ground heat exchangers, different initial saturations were set for each layer of soil. In the actual project, the thermal properties of the soil directly affect the heat transfer of the buried heat exchanger. In this paper, the influence of vertical moisture migration on soil temperature field was studied by the control variable method. Therefore, it was assumed that the soil thermophysical parameters of the layers were the same.

3 Result and Discussions

The changes of soil saturation value in case 1 and 2 after 720 h operation were shown in Fig. 3, and the difference value in soil saturation between case 2 and 1 was shown in Fig. 4. The case 1 only considered horizontal moisture migration. With the continuous inflow of heat from the side of the U-shaped tube, under the influence of temperature gradient, the moisture gradually migrated outward from the side of the U-shaped tube. The closer to the U-tube inlet, the greater the temperature gradient of the soil and the more significant the moisture migration. For soil I with the initial saturation value of 0.4, the saturation value dropped from 0.4 to 0.385 and the maximum decrease value was 3.7%. Vertical moisture migration was considered in case 2, and soil saturation was stratified in soil I. The moisture was migrated downward by gravity so that the saturation value of case 2 was lower than that of case 1 in the shallow soil, the maximum difference value was 0.06, and

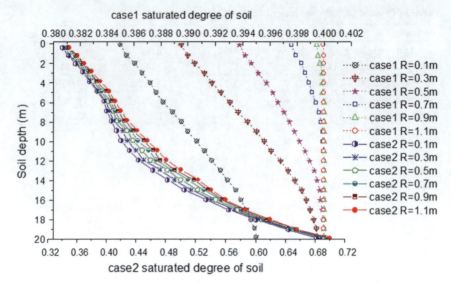

Fig. 3 Soil saturation value in soil I after 720 h operation

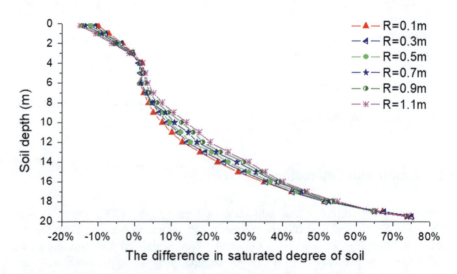

Fig. 4 Soil saturation difference value between case 2 and 1 in soil I after 720 h operation

the percentage difference value was 14.7%. The deep soil was affected by the soil II which has an initial saturation value of 0.7, and the moisture permeated through the interface 1 to the soil I. The farther from the soil interface, the smaller gradient of the moisture migration. The saturation value of case 2 was obviously larger than that of case 1 in the deep. At this time, the soil which closes to the buried pipe still had outward moisture migration due to the temperature gradient and the saturation value was the minimum value of the same depth range.

The changes of soil saturation value in case 1 and case 2 after 720 h operation were shown in Fig. 5, and the difference value in soil saturation between case 2 and case 1 was shown in Fig. 6. For case 1, with the continuous inflow of heat from the side of the U-shaped tube, under the influence of temperature gradient, the closer to

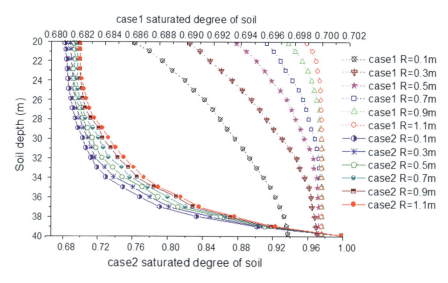

Fig. 5 Soil saturation value in soil II after 720 h operation

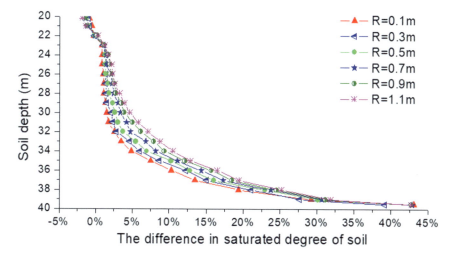

Fig. 6 Soil saturation difference value between case 2 and 1 in soil II after 720 h operation

the U-tube inlet, the greater the temperature gradient of the soil and the more significant the moisture migration. For soil II with initial saturation value of 0.7, the saturation value dropped from 0.7 to 0.686 and the maximum decrease value was 2.0%. Figs. 3 and 5 show the comparison of the saturation value changes, and it could be seen that the greater the initial saturation value, the smaller, the decrease in saturation value near the buried pipe. However, the decreased value of soil II compared with soil I was more obvious at 1.1 m from the buried pipe, so the higher the initial saturation value, the larger the radius of moisture migration. For case 2, the moisture was migrated downward by gravity so that the saturation value of case 2 was lower than that of case 1 in the shallow soil, the maximum difference value was 0.016, and the percentage difference value was 2.2%. The deep soil was affected by the soil III which has an initial saturation value of 1, and the moisture permeated through the interface 2 to the soil II. The farther from the soil interface, the smaller gradient of the moisture migration. The saturation value of case 2 was obviously larger than that of case 1 in the deep soil. At this time, the soil which closes to the buried pipe still had outward moisture migration due to the temperature gradient and the saturation value was the minimum of the same depth range. Figs. 3 and 5 show the comparison of the saturation changes of case 2, and it could be seen that the moisture migration of interface 1 was faster than the interface 2. The maximum increase in saturation value at a distance of 10 m from the interface 1 was 15.2% in soil I, and the maximum increase in saturation value at a distance of 10 m from the interface 2 was 5.7% in soil II. This was because from the soil I to the soil III, the temperature of the fluid in the U-tube was lowered and the moisture migration under the temperature gradient was slowed down. At the same time, the moisture of each soil had a tendency of downward migration which was affected by gravity and it also caused the moisture migration of saturated soil to the unsaturated soil to slow down.

The changes of soil temperature field in case 1 and case 2 after 720 h operation were shown in Fig. 7. The temperature field of case 1 appeared to be significantly stratified due to the different soil saturation of each layer. For unsaturated soils, the soil saturation value increased and the specific heat capacity of the soil increased. In addition, the fluid temperature in the U-tube of the soil II was lower than that of the soil I, so the soil thermal diffusion coefficient of the soil II decreased. The saturation value of saturated soil III was maintained as 1 due to seepage, and the seepage flowed from the radially distal end of the soil to the buried pipe and then flowed out, taking away the heat in the soil, so the soil temperature changed minimally in the soil III.

For soil I in case 2, the soil saturation value of case 2 was lower than case 1 at the shallowest depth. At this time, there were many gas phases, the heat changed rapidly, and the soil temperature was high. The maximum temperature difference value was 0.05 °C higher than case 1. The temperature difference value between the two cases decreased with increasing soil depth. On the contrary, the deep soil

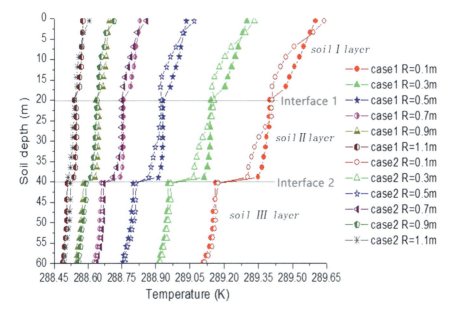

Fig. 7 Soil temperature value after 720 h operation

saturation was affected by the moisture migration at the interface 1, the saturation value gradually increased and was larger than case 1, the heat storage capacity was enhanced, the soil thermal migration coefficient was decreased, and the maximum temperature difference value was 0.09 °C lower than case 1. Similar to soil I, for soil II in case 2, the soil saturation was lowest at the shallowest depth. The maximum temperature difference was 0.0012 °C higher than case 1. On the contrary, the saturation value of the deep soil was affected by the moisture migration at interface 2, the saturation value was larger than case 1, the soil thermal migration coefficient was decreased, and the maximum temperature difference value was 0.082 °C lower than case 1. In saturated soil III, the soil temperature of case 2 was slightly higher than that of case 1, because the thermal diffusivity coefficient of case 2 was lower than that of case 1 in soil I and soil II and the soil heat storage capacity in case 2 was better, resulting in the temperature of the fluid in the buried tube of case 2 was higher than that of case 1 at interface 2. For soil I layer and soil II layer, the soil I layer saturation value changed more at the same distance from the respective lower interface and the temperature value of the buried pipe decreased with the depth of soil. Therefore, the soil heat storage capacity changed more significantly, and the soil temperature difference with case 1 was larger in soil I layer.

4 Conclusions

In the numerical simulation of a layered model spanning unsaturated soil and saturated soil, only horizontal moisture migration was considered, as heat continued to flow from the U-tube side, and the moisture gradually migrated outward from the U-tube side under the influence of the temperature gradient. The soil temperature gradient was greater near the inlet of the U-tube, and the moisture migration was more pronounced. The changes of initial soil saturation value in the soil interface stratified the temperature field. The increased value of saturation increased the soil heat storage capacity and decreased the soil thermal diffusion capacity. The exothermic capacity of the ground heat exchanger was enhanced under the seepage, so the soil temperature in the saturated soil layer changed little.

When considered the vertical moisture migration, the soil temperature field distribution fluctuated greatly due to the change of soil saturation value in each layer. The moisture of soil near the U-tube still migrated outward due to the temperature gradient, and the saturation value was the minimum of the same depth range. The shallow soil moisture in the same soil layer migrated downward, which enhanced the soil thermal diffusion capacity. Due to the different soil saturation value of each layer, the moisture migration at the interface was significant. The increase of saturation value near the interface increased the soil heat storage capacity and decreased the soil thermal diffusion capacity. Because the soil thermal diffusivity when flowing through the upper unsaturated soil layer was generally smaller than that only considered horizontal moisture migration, and the soil heat storage capacity was stronger resulting in a higher temperature of the fluid in the buried pipe at the interface between the saturated and unsaturated soil. The soil temperature was slightly higher in the saturated soil layer when considered the vertical moisture migration.

Acknowledgements The project is supported by National Natural Science Foundation of China (Number 41702242).

References

1. Angelotti, A., Alberti, L., La Licata, I., Antelmi, M.: Energy performance and thermal impact of a borehole heat exchanger in a sandy aquifer: influent of the groundwater velocity. Energy Convers. Manag. **77**, 700–708 (2014)
2. Li, M., Alvin, C.K.L.: New temperature response functions (G functions) for pile and borehole ground heat exchangers based on composite-medium line-source theory. Energy **38**(1), 255–263 (2012)
3. Jame, R.C., Jeffrey, D.S.: A computationally efficient hybrid time step methodology for simulation of ground heat exchangers. Geothermics **40**(2), 144–156 (2011)
4. Seama, K.F., Marc, A.R.: Examination of thermal interaction of multiple vertical ground heat exchangers. Appl. Energy **97**, 962–969 (2012)

5. Georgios, A.F., Paul, C., Panayiotis, P.: Single and double U-tube ground heat exchangers in multiple-layer substrates. Appl. Energy **102**, 364–373 (2013)
6. Hans, J., Jan, C., Hugo, H.: The influence of soil moisture transfer on building heat loss via the ground. Build. Environ. **39**(7), 825–836 (2014)
7. Lee, C.K., Lam, H.N.: A modified multi-ground-layer model for borehole ground heat exchangers with an inhomogeneous groundwater flow. Energy **47**(1), 378–387 (2012)
8. Mostafa, H., Hassan, M.B., Esmail, M.M.: Investigation of buoyancy effects on heat transfer between a vertical borehole heat exchanger and the ground. Geothermics **48**(10), 52–59 (2013)
9. Yang, W.B., Chen, Y.P., Shi, M.H., et al.: Numerical investigation on the underground thermal imbalance of ground-couple heat pump operated in cooling-dominated district. Appl. Therm. Eng. **58**(1–2), 626–663 (2013)

Performance Analysis of a Hybrid Solar Energy, Heat Pump, and Desiccant Wheel Air-Conditioning System in Low Energy Consumption Building

Shaochen Tian and Xing Su

Abstract In hot and humid climate zone of China, a huge amount of building energy is consumed by conventional air-conditioners (CACs) in dehumidification especially in low energy consumption buildings where sensible load can be largely reduced by adopting high-performance building envelopes, but the moisture load is not changed. In this paper, a hybrid solar energy, heat pump, and desiccant wheel (HSHDW) system was proposed and the performance of the system was analyzed based on simulation studies. It was found that the evaporation temperature of the system should be kept at 12.8 and 9.7 °C on a typical day of summer and transition season separately. Compared with CACs, HSHDW system can provide a more comfortable thermal and humid indoor environment. The daily energy consumption of the HSHDW system is lower than that of the CAC, achieving great energy savings, especially for a typical weekday in summer, when a 16.8% daily energy saving is obtained.

Keywords Desiccant wheel · Solar energy · Heat pump

1 Introduction

Nowadays, energy problems have been mentioned constantly. In China, about 50–60% of building energy is consumed by heating, ventilation, and air-conditioning (HVAC) systems [1]. In hot and humid climate zone of China, dehumidification is required in summer and transition season. In transition season,

---

S. Tian · X. Su (✉)
School of Mechanical Engineering, Tongji University, Shanghai, China

Key Laboratory of Performance Evolution and Control for Engineering Structures of Ministry of Education, Tongji University, Shanghai, China
e-mail: suxing@tongji.edu.cn

S. Tian
e-mail: 792272829@qq.com

cooling is not necessarily due to outdoor air temperature which is mostly about 25 °C. In low energy consumption buildings, heat gain through envelopes is reduced by increasing the thermal properties of the materials. However, moisture load which mainly comes from fresh air and people cannot be decreased. For CACs, dehumidification is achieved by cooling the process air below dew point. In this system, low evaporation temperature is required and the amount of energy consumed by compressor is large. Compared with CAC, desiccant wheel (DW) could adsorb water vapor of air without cooling. The main problem is that the energy consumed to regenerate DW is large due to high regeneration temperature requirement of DW. When solar energy is applied to regenerate DW, DW dehumidification system is more suitable for passive houses in hot and humid climate zone.

In the past few years, several investigations have been conducted to evaluate the feasibility of the application of DW cooling systems. According to the research of Goodarzia et al. [2], process air humidity ratio and regeneration air temperature had great influence on the performance of the DW cooling system. Ge et al. [3] proposed that the electricity consumption of solar-driven two-stage rotary desiccant evaporative cooling system can be significantly reduced compared with conventional vapor compression system. Ge et al. [4] evaluated the performance of a combined air source heat pump and DW system in residential buildings of Shanghai. Condenser dissipated heat is applied to regenerate DW, and more than 50% energy requirement could be reduced in transition season. Beccali et al. [5] applied cooling water to precool the fresh air, and the regeneration energy of DW came from the two sensible heat exchangers with process air and PVT panel. Rjibi et al. [6] investigated a desiccant cooling system powered by solar energy in an insulated greenhouse. The coupled system showed best performance when regeneration temperature of DW is 60 °C.

In this paper, a HSHDW system is proposed and adopted in a hypothetical apartment of low energy consumption residential building in Shanghai to evaluate the performance of the system. Firstly, indoor sensible load of a typical day in summer and transition season is calculated by using DeST software. Then, by using TRNSYS software, the operation mode of the HSHDW system in different season is optimized and the hourly energy consumption is simulated. Combined with the indoor sensible load calculation results, the indoor thermal environment is simulated and compared with the CAC.

2 Methods

2.1 System Description

The schematic figure of HSHDW system is shown in Fig. 1. Compared with the existing systems [7], the system has two evaporators and two condensers and the precooling evaporator is in front of the DW. In this system, the evaporation

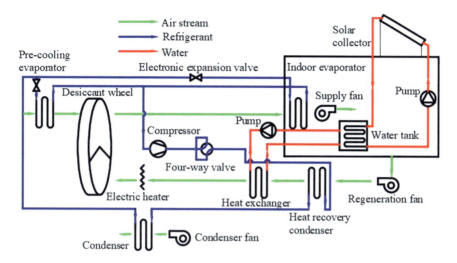

Fig. 1 Schematic figure of HSHDW system

temperature of HP can be increased. Fresh air firstly flows through the precooling evaporator, in which it is cooled and dehumidified. Then, the process air is dehumidified further by the DW. After that, it is mixed with indoor return air. Before supplied into the room, the mixed air is cooled to the supply air temperature by the indoor evaporator. The refrigerant flow rate of the two evaporators is adjusted by the two electronic expansion valves. On the regeneration side, indoor exhaust air is heated by the heat recovery condenser and hot water successively. The two condensers are in series in the refrigerant loop, and the condenser is necessary to dissipate the excessive heat after the condensing heat is recovered by regeneration air through the heat recovery condenser. If the regeneration air temperature is not high enough to regenerate the DW, the auxiliary electric heater is activated to heat regeneration air further. Hot water passes through the heat exchanger to heat regeneration air of DW. Decided by the regeneration temperature requirement of DW, the flow rate of the hot water is variable.

In solar collection system, circulating water passes through the solar collector and heated by solar energy. Then, the high-temperature hot water passes through the coil of the water tank, and the water which is used to heat regeneration air is heated.

2.2 Modeling and Validation

The main components of HSHDW system are DW, HP, and flat-plate solar collector. The DW model in TRNSYS is proposed by Howe [8] and corrected by Schultz [9]. The required regeneration air temperature could be calculated when inlet air conditions and humidity ratio set point are known. The simulation results of DW were validated with the experimental results of Antonellis et al. [10]. The maximum relative error of the model is 7.13% for regeneration inlet air temperature and 7.16% for process outlet air temperature. The model of compressor, evaporator, condenser, and expansion valve is according to Sheng et al. [11]. The model of solar collector is based on the instantaneous efficiency equation.

2.3 Building Description

The plan of the residence is shown in Fig. 2. The area is 83 m^2, and the height is 3 m.

In this work, meteorological data including temperature, relative humidity, and solar radiation is used. The required hourly meteorological data is from TRNSYS weather data file. The main meteorological data of a typical day in summer and transition season is given in Table 1.

A passive residential apartment with three people is adopted as the simulation case. On weekdays, the indoor environment controlling period is from 18:00 to 8:00 of the next day. On weekends, the air-conditioning system is activated all the time. Referring to the technical guideline [12], the main building characteristics are given in Table 2.

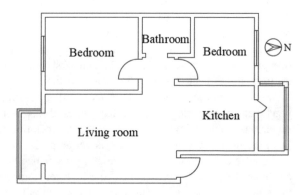

Fig. 2 Plan of the residence

Table 1 Weather condition in summer and transition season

Parameter		Summer	Transition season
Air temperature (°C)	Max	34.8	24.4
	Min	28.5	19.3
Air humidity ratio (g/kg)	Max	24.06	14.69
	Min	22.16	13.66
Solar radiation (W/m^2)	Max	1169	1147
	Min	473	338

Table 2 Building information

Item	Factor
Wall	Heat transfer coefficient: 0.27 W/m^2
Window	Heat transfer coefficient: 1.5 W/m^2
	SHGC: 0.15
People	Sensible load production rate: 134 W/person
	Moisture production rate: 109 g/h/person
	Fresh air rate: 30 m^3/h/person
Design temperature	26 °C
Design relative humidity	57%
Supply air rate	1800–2000 m^3/h
Heating load production of light	3 W/m^2
Heating load production of machine	2 W/m^2

The hourly indoor sensible load changes and the DeST software are adopted to calculate the indoor sensible load of a typical day in summer and transition season separately.

2.4 System Performance Simulation

Based on the weather conditions and the indoor sensible load calculated before, the performance of the HSHDW system was evaluated by using TRNSYS software. The model of HSHDW system in TRNSYS software is shown in Fig. 3.

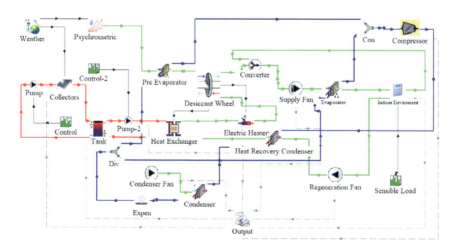

Fig. 3 HSHDW system model in TRNSYS software

On weekdays, the solar collector is in operation during the daytime and the water stored in the tank is heated by solar energy. When the residents are at home, the air-conditioning system is activated and the hot water is used to heat regeneration air of DW. On weekends, the air-conditioning system is in operation the whole day and the solar collection system is only running during the daytime.

3 Results

Hourly indoor sensible load is first calculated. Then, the indoor thermal and humid environment and energy consumption of the HSHDW system and CAC are compared.

3.1 Building Load Calculation

The hourly indoor sensible load calculation results are shown in Fig. 4. Influenced by low outdoor air temperature and radiation intensity, indoor sensible load in transition season is much lower than summer.

Indoor latent load is constant due to only the moisture production of human which is considered. In the HSHDW system, fresh air deals with all of the latent load and the humidity ratio set point of DW can be calculated as 9.1 g/kg.

3.2 Operation Mode Optimization

Regeneration temperature requirement of DW increases with the increasing of dehumidification rate. The dehumidification rate of DW and precooling evaporator is optimized for different seasons. Dehumidification rate of DW can be reduced by

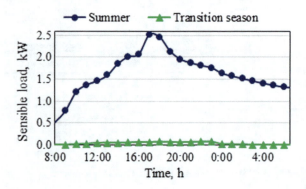

Fig. 4 Hourly indoor sensible load

increasing the moisture removal capacity of precooling evaporator, which can be realized by reducing the evaporation temperature of HP.

In order to make full use of solar energy, the evaporation temperature of HP is adjusted based on the principle that the electric heater is only activated at the last hour of weekend. The optimized evaporation temperature of HP is 12.8 °C on typical summer weekend, and DW deals with 35–40% of the dehumidification load. The supply airflow rate is 2000 m^3/h. In transition season, solar radiation intensity is lower than summer. The evaporation temperature of HP should be adjusted to 9.7 °C, and indoor evaporator is out of operation. DW deals with 47–59% of the dehumidification load. The supply air rate changes to 1800 m^3/h. The operation mode of weekdays is the same as that of weekends in different season. The evaporation temperature of CAC system should be kept at 8.9 °C to meet the dehumidification requirement both in summer and transition season.

3.3 Indoor Thermal and Humid Environment

Indoor thermal and humid environment of each hour was marked on the psychrometric chart with the thermal comfort zone according to ASHRAE [13]. Based on the appendix A of ASHRAE 55, the average air temperature can be used in place of operative temperature. The zone within the dashed line and the solid line is the comfort zone of 0.5 clo and 1.0 clo separately. Indoor thermal and humid environment is compared between the HSHDW system and CAC of different season (see Fig. 5).

In summer, indoor environment of the HSHDW system is mostly within the comfort zone of 0.5 clo. In transition season, indoor thermal environment of the

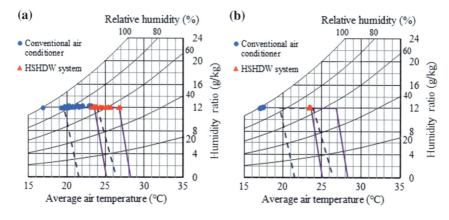

Fig. 5 Indoor thermal and humidity environment in summer (**a**) and transition season (**b**)

HSHDW system concentrates at the boundary of the two zones. For CAC, indoor thermal environment deviates far from the comfort zone especially in transition season.

3.4 Daily Energy Consumption

The daily energy consumption of the HSHDW system and CAC of typical weekday and weekends in summer and transition season is shown in Fig. 6. Compared with CAC, the energy-saving potential of the HSHDW system is 16.8 and 14.5% for weekday and weekend separately in summer. In transition season, HSHDW system can save 13.6% and 12.0% on weekday and weekend, separately.

4 Discussion

The simulation results show that CAC is not suitable for low energy consumption residential buildings especially in transition season because indoor air temperature deviates far from the comfort zone. For HSHDW system, the supply air rate has to be reduced to maintain indoor thermal environment condition in transition season.

The evaporation temperature of HSHDW system is increased due to the deep dehumidification which is achieved by DW. With the increasing of evaporation temperature and the applying of solar energy, energy consumed by air-conditioning system is reduced. On weekday, the hot water heated by solar energy is sufficient to heat regeneration air to the required temperature. On weekend, operation time of the system is extended. Thus, the energy-saving potential of the HSHDW system on weekend is lower than that on weekday.

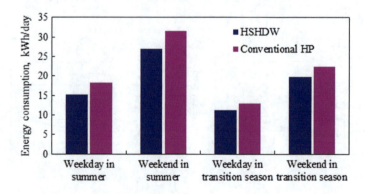

Fig. 6 Daily energy consumption of the HSHDW system and CAC

5 Conclusions

In the present work, the performance of HSHDW system was simulated and analyzed. In summer, the evaporation temperature is 12.8 °C, and the evaporation temperature is adjusted to 9.7 °C in transition season. HSHDW system has better performance in controlling indoor thermal environment than CAC in both seasons. Indoor air temperature and humidity ratio are both within the thermal comfort zone according to ASHRAE standard. The daily energy consumption of the HSHDW system is lower than that of the CAC for the four conditions. The highest and lowest daily energy saving is achieved on weekday in summer and weekend in transition season, respectively, which are 16.8 and 12%. The HSHDW system could be an alternative to the CACs in low energy consumption residential buildings due to the high system performance and more comfortable thermal and humid environment.

Acknowledgements This research is supported by the China's National Key R&D Program during the 13th Five-Year Plan Period (Grant No.2017YFC0702600).

References

1. Jiang, Y.: Current building energy consumption in China and effective energy efficiency measures. HV&AC **35**(5), 30–40 (2005)
2. Goodarzia, G.: Performance evaluation of solid desiccant wheel regenerated by waste heat or renewable energy. Energy Procedia **110**, 434–439 (2017)
3. Ge, T.S., Ziegler, F., Wang, R.Z., Wang, H.: Performance comparison between a solar driven rotary desiccant cooling system and conventional vapor compression system (performance study of desiccant cooling). Appl. Therm. Eng. **30**, 724–731 (2010)
4. Ge, F.H.: Energy savings potential of a desiccant assisted hybrid air source heat pump system for residential building in hot summer and cold winter zone in China. Energy Build. **43**, 3521–3527 (2011)
5. Beccali, M.: Energy and economic assessment of desiccant cooling systems coupled with single glazed air and hybrid PV/thermal solar collectors for applications in hot and humid climate. Sol. Energy **83**, 1828–1846 (2009)
6. Rijibi, A.: The effects of regeneration temperature of the desiccant wheel on the performance of desiccant cooling cycles for green houses thermally insulated. Heat Mass Transf. **54**(11), 3427–3443 (2018)
7. Ge, T.S., Dai, Y.J., Wang, R.Z.: Review on solar powered rotary desiccant wheel cooling system. Renew. Sustain. Energy Rev. **39**, 476–497 (2014)
8. Howe, R.R.: Model and performance characteristics of a commercially-sized hybrid air conditioning system which utilizes a rotary desiccant dehumidifier. University of Wisconsin-Madison, Madison, America (1983)
9. Schultz: The performance of desiccant dehumidifier air-conditioning system using cooled dehumidifiers. University of Wisconsin-Madison, Madison, America (1983)
10. Antonellis, S.D.: Desiccant wheels effectiveness parameters: correlations based on experimental data. Energy Build. **103**, 296–306 (2015)
11. Sheng, Y.: Simulation and energy saving analysis of high temperature heat pump coupling to desiccant wheel air conditioning system. Energy **83**, 583–596 (2015)

12. MHURC: Technical guidelines for passive ultra-low energy consumption green buildings (trial implementation) (residential buildings). Beijing (2015)
13. ASHRAE: ANSI/ASHRAE Standard 55: Thermal environmental conditions for human occupancy. ASHRAE, Atlanta (2013)

Feasibility of Hybrid Ground Source Heat Pump Systems Utilizing Capillary Radiation Roof Terminal in the Yangtze River Basin of China

Lu Xing, Chen Ren, Hanbin Luo, Yin Guan, Dongkai Li, Lei Yan, Yuhang Miao and Pingfang Hu

Abstract This paper proposes applications of a hybrid ground source heat pump system combining cooling tower as auxiliary cooling source and capillary radiation roof as indoor terminal in an office in the Yangtze River Basin of China. By simulations in TRNSYS environment, the system feasibility was investigated and compared to conventional HVAC system (water chiller + gas boiler). The results reveal that with the help of the radiant terminal, the hybrid system provides finer indoor comfort, costs less, and is more energy-saving. However, it is less environmentally friendly. An analytic hierarchy process method is developed to perform a comprehensive system evaluation based on all three factors. The hybrid system performance evaluation result is 0.966 located in the adaptation region; the

---

L. Xing · C. Ren · Y. Guan · D. Li · Y. Miao
School of Energy and Power Engineering, Huazhong University of Science and Technology, Wuhan, China
e-mail: lxing@hust.edu.cn

C. Ren
e-mail: rrenchen@163.com

Y. Guan
e-mail: yinguan@hust.edu.cn

D. Li
e-mail: 253668941@qq.com

Y. Miao
e-mail: myh11@hust.edu.cn

H. Luo (✉)
School of Civil Engineering and Mechanics, Huazhong University of Science and Technology, Wuhan, China
e-mail: luohbcem@hust.edu.cn

L. Yan · P. Hu (✉)
School of Environmental Science and Engineering, Huazhong University of Science and Technology, Wuhan, China
e-mail: pingfanghu21@163.com

L. Yan
e-mail: 309532907@qq.com

conventional HVAC system is 0.746 located in the general adaptation area. Overall, the hybrid GSHP system is more adaptable and it has a promising potential for future applications in this area.

Keywords Hybrid ground source heat pump (HGSHP) · Capillary radiation roof · System performance analyses · Analytic hierarchy process (AHP)

1 Introduction

Ground source heat pump (GSHP) system is a new type of HVAC system which utilizes heat pump technology and renewable geothermal energy to condition the buildings. It is energy-saving and reduces environmental pollutions compared to conventional HVAC systems. For the GSHP systems, it is critical to balance the heat it extracts from and rejected to the ground in heating and cooling seasons to maintain a stable and efficient system performance over the years. Therefore, typically the GSHP system is connected to an auxiliary cooling source (cooling tower) or heating source (solar collector) and becomes the hybrid GSHP system.

Several researchers studied the hybrid GSHP system performance based on experimental work or developed simulation models. Lee [1] investigated the transient performance characteristics of hybrid GSHP system by experiment. It was found out that the averaged system coefficient of performance (COP) was increased by 7.2% compared to the GSHP system at the optimized conditions. Man [2] developed a computational model to study the hybrid GSHP system operational performance when coupled with a cooling tower. In hot-weather areas such as Hong Kong, the hybrid GSHP system helps solving the mitigation of the soil thermal imbalance problem.

Meanwhile, several researches suggested coupling the HVAC systems with different types of terminals such as capillary radiation roof or radiant floor. This helps reaching higher comfort level for the buildings, increasing the system energy efficiency, and reducing the building energy consumption. Villarino [3] developed a HVAC system model for an office building which consisted by a ground-coupled heat pump (GCHP), radiant floor (RF) and mechanical ventilation; the values obtained in simulation model presented a deviation of 2% respected experimental results. Song [4] investigated a dehumidification system coupled with radiant floor cooling terminal in Korea, by cooling and dehumidifying the outdoor air before it flowed into an apartment, they found the solution for floor surface condensation problems. There are several other researchers [5–7] which found out that coupling the radiant floor terminals system efficiency was significantly improved, and system can transfer peak load and reduce system operating costs.

Previous researches mainly focus on modeling or experimenting the hybrid GSHP system, studying the optimal system control strategies, and concluding that the system is more energy efficient than conventional HVAC systems. There have been few studies which evaluate the hybrid GSHP system performance and investigate the system feasibility based on more comprehensive perspective. In our study, we develop a TRNSYS model for studying the hybrid GSHP system implemented in an office building in the Yangtze River Basin of China—Wuhan City. The system utilizes cooling tower as an auxiliary cooling source and radiant capillary roof as indoor terminal. The system parameters such as indoor air temperature, indoor air relative humidity, indoor PPD-PMV, heat pump entering fluid temperature (EFT), and system coefficient of performance (COP) are estimated based on the developed model. The analytic hierarchy process (AHP) method has been developed and is used for comprehensively system evaluation and for studying the system feasibility. The hybrid GSHP system evaluation results are compared to conventional HVAC system evaluation results.

2 Methods

Building loads are calculated based on the weather files and building information. A hybrid GSHP system is designed in TRNSYS environment. The simulated indoor air temperature, indoor air relative humidity, heat pump EFT, and system COP are used for observing the overall system performance. The system is then evaluated in three aspects: economical cost, energy consumptions, and environmental impact. One evaluation method—AHP method—is used for studying the system performance comprehensively. A traditional HVAC system (water chiller + gas boiler) is also modeled and evaluated. The two systems are compared, and the results are discussed.

The typical 11-story office building is built in Wuhan with a construction area of 1400 m^2 and construction volume of 15,646 m^3. The building wall and window materials are designed according to Chinese GB50189-2015—energy efficiency design standards for public buildings. The estimated building loads are used as inputs both for the hybrid GSHP system model and the traditional HVAC (water chiller + gas boiler) model developed in TRNSYS. In Wuhan, the office building is cooling-dominant. In order to balance the heat injection and extraction to/from the soil through the GSHP ground heat exchangers, an auxiliary cooling device—cooling tower—is added to the conventional GSHP system. Since the office building is cooling-dominant, a typical vertical GSHP system is connected to an auxiliary cooling device—cooling tower, and the schematic diagram of the hybrid GSHP system is shown in Fig. 1. The hybrid system utilizes capillary radiation roof as terminal.

The inputs to the hybrid GSHP model include parameters for the roof terminal, vertically ground heat exchangers, soil, circulating fluids, and equipment. The model outputs result such as indoor air temperature and indoor air relative humidity, indoor PPD-PMV, heat pump EFT, and system COP. These results are presented and discussed later. The capillary radiation roof terminal is embedded in the building model by setting roof thermal properties and dimension parameters. Vertical ground heat exchangers are used in the hybrid GSHP system.

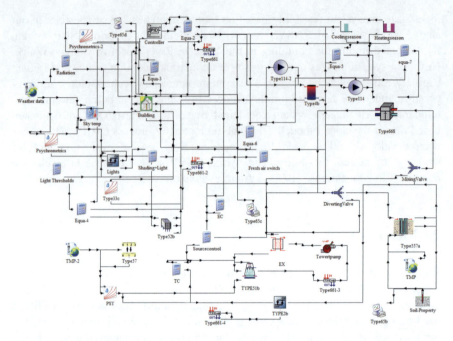

Fig. 1 Hybrid ground source heat pump system coupled with capillary radiation roof model

The U-shaped PE tube inner diameter and outer diameter are 26 mm and 32 mm, respectively. The tube pitch is 100 mm, and the drilling diameter is 200 mm. The borehole depth is 100 m, and the hole pitch is 6 m. The designed borehole number is 161 based on the TRNSYS model results.

3 Results

The hybrid GSHP model results including indoor air condition, system COPs, and heat pump EFTs are presented here. Moreover, the system economical cost, energy consumption, and environmental impact are also calculated. An analytic hierarchy process (AHP) method is used so as to comprehensively evaluate the system. A conventional HVAC system (water chiller + gas boiler) is also modeled. The hybrid GSHP system evaluation results are then compared to the conventional HVAC system evaluation results. Model results shows that in summer, the hybrid GSHP system maintained indoor air temperatures within the range of 25–26 °C; in winter they vary within the range of 19–20 °C as designed.

Condensation is the main concern of applying the capillary radiant roof terminal for building heating and cooling. The roof surface temperature needs to be 2–3 °C above the indoor air dew point temperature to avoid the condensation. In the model, a condensation detector was set at the roof surface, signal 1 indicates condensation occurs, and signal 0 indicates the opposite. In summer, the air relative humidity fluctuates within the range of 40–50% which meets the relevant specifications requirements range from 40 to 70%. In winter, the indoor air relative humidity fluctuated between 55 and 65%, which for most of the time meets the requirement range from 30 to 60%. Meanwhile, the condensation signals stay at 0 which indicate no condensation occurs and radiant terminal is feasible for the investigated office building.

The PMV-PPD value for the hybrid GSHP system is calculated. For most of the time, the PMV results range from −0.5 to 0.5 and the PPD results are less than 10%; during these times, the system would operate and meet the Thermal Comfort Class I requirements. Occasionally, during the cooling and heating season, for about 7.5% of the time within a year, PPD results rise to about 15%, within the Thermal Comfort Class II. This is because that the water in the capillary is temperature-controlled, there is a time delay when conditioning the room through radiation heat transfer process if the outdoor air temperature suddenly rises or drops. A higher indoor comfort level would possibly be achieved if control strategy be improved in the future.

The hybrid GSHP system annual average coefficient of performance (COP) amounts to 4.65 in summer and 3.27 in winter. Meanwhile, the heat pump entering fluid temperatures (EFTs) are lower than 36 °C, which indicated that hybrid GSHP system operates at a relatively stable condition. After ten years' operation, heat pump entering fluid temperatures increase from 34.5 to 35.6 °C. The heat pump EFTs curve reaches a steady-state condition, which demonstrates with the auxiliary cooling tower to balance the ground heat extraction and rejection rate; the hybrid GSHP system performance is stable and sustainable.

4 Discussion

The hybrid GSHP system operates and maintains the building indoor condition at a satisfying comfort level. The economical cost, energy consumptions, and environmental evaluation of the system are assessed. A traditional HVAC system model was developed in TRNSYS. This water chiller unit for cooling and natural gas boiler for heating (WB) system is evaluated and compared to the hybrid GSHP system results.

The system dynamic annual expenses are used as comprehensive economic indicators, which are composed of three parts: system initial investment per year, system operating cost, and maintenance fees. Hybrid GSHP system requires more initial investment than WB system; it costs less in system operation and maintenance. Dynamic annual cost of hybrid GSHP system is 107.2 million Yuan, 14.5% lower than that of the WB system (water chiller + gas boiler) which is 125.4 million Yuan.

Table 1 presents the energy consumption comparisons of the hybrid GSHP system and WB system. The primary energy consumption of two systems is summarized from the model results. The hybrid GSHP system uses only electricity as its energy source, while the WB system consumes both the electricity and natural gas. The natural gas consumptions have been converted to electricity consumption. The WB system utilizes the water chiller for cooling and gas boiler for heating. The energy utilization factor is 0.83 and 0.44, respectively, which are much lower than the hybrid GSHP system. Therefore, the hybrid GSHP system, compared with the WB system, can save 24.7% of the energy in cooling season - summer and 57.2% in heating season - winter. Annually, the hybrid GSHP system energy -saving rate reaches 43.2%. Meanwhile, the hybrid GSHP has a primary energy ratio of 1.01 per year, while the conventional system only has 0.64. For the energy-saving comparison, the hybrid GSHP has obvious advantage; it is more in line with requirements of society for energy-saving purposes.

Energy consumption at the meantime causes environmental pollution, which is harmful to mankind. Environmental pollution is mainly caused by the combustion fuel. For the system which uses electrical energy as the driving power, the pollution is caused by the thermal power plant which generates the electricity. For the direct or indirect (electricity) use of fuel combustion as a driving energy, the pollutant emission values of two systems—hybrid GSHP and WB system—can be calculated as given in Table 2.

Previously, the system evaluation is performed using only one indicator of the several, which are economic impact, energy consumptions, and environmental impact. As discussed above, the hybrid GSHP system costs less and consumes less energy; however, it produces relatively more pollution. In order to evaluate the system from an overall perspective rather than from one, a more comprehensive evaluation method—analytic hierarchy process (AHP) method—is used and introduced here.

Table 1 System energy consumption comparisons

System	Hybrid GSHP		Water chiller + Gas boiler	
	HGSHP (cooling)	HGSHP (heating)	Water chiller (cooling)	Gas boiler (heating)
Electricity energy consumption (MWh)	1479.3	1120.1	1963.7	2619.4
Energy utilization factor	1.10	1.04	0.83	0.44
Primary energy ratio	1.07	0.64		

Table 2 System pollutant emissions (ton)

System	CO_2	SO_2	NO_X	Ash
Hybrid GSHP	960.45	7.48	2.72	52.06
Water chiller + Gas boiler	980.68	6.09	2.33	42.36

AHP is a method to stratify complicated problems and increase the accuracy of the system evaluation, comparing two systems at each level to improve decision-maker's discrimination for each target difference. Elements at the same level act as guidelines to dominate elements in next level; at the same time, they are dominated by elements of previous level, such as target level, criterion level (or indicator level, subindicator level), and so on. The target level is system evaluation. Indicator layer is divided into three factors: energy-saving, economy, and environment; subindicator layer is divided into primary energy ratio, dynamic annual cost, and pollutant emissions.

The numbers of layers need for system evaluation and importance of each subindex are determined accordingly. Both the hybrid GSHP system and the WB system performances are evaluated by three main factors: the energy-saving, economics, and environment. According to the importance of the three factors in indicator layer, the judgment matrix A of indicator layer has been established, and the judgment matrix B is established to evaluate the environmental impact determined in more detail by different pollutants emission, as given in Tables 3 and 4, [8]. The basic quantitative evaluation of the AHP method is as given in Table 5.

Table 6 presents the criteria for system feasibility determined by the evaluation values; Table 7 presents the hybrid GSHP system and WB system evaluation results. For comparison of the hybrid GSHP system and WB system, the evaluation value for system with the better performance is 1 and its value is taken as the reference value. Evaluation value for other system is given according to its ratio to the reference value.

The hybrid GSHP system is more economical, energy-saving, and less environmental than WB system. The hybrid GSHP system is overall more adaptable than the WB system considering all three factors and their corresponding weights. Overall, the hybrid GSHP system performance comprehensive evaluation result is 0.966, located in the adaptation region; the conventional system is 0.746 in the general adaptation area. The hybrid GSHP system coupled with capillary radiation roof terminal has good potential for future wider applications in Yangtze River Basin of China.

Table 3 Judgment matrix of matrix A

A	Energy-saving	Economics	Environment
Energy-saving	1	3	2
Economics	1/3	1	1
Environment	1/2	1	1

Table 4 Judgment matrix of matrix B

B	Ash	SO_2	NO_X	CO_2
Ash	1	2	2	2
SO_2	1/2	1	1/2	1
NO_X	1/2	2	1	2
CO_2	1/2	1	1/2	1

Table 5 Relative importance determination of AHP method

Scale	Mean
1	Two elements are equally important
3	One element is slightly more important than the other
5	One element is obviously more important than the other
7	One element is more important than the other
9	One element is much more important than the other
2, 4, 6, 8	A compromise between two adjacent judgments
Reciprocal of above numbers	Inverse comparison

Table 6 Criteria for determining system feasibility

Partition level	Adaptation	General adaptation	Reluctantly adapted	Not suited
Comprehensive index value	0.9–1.0	0.7–0.9	0.5–0.7	0–0.5

Table 7 System comprehensive evaluation results

System	Primary energy ratio	Weight	Dynamic annual cost	Weight	Pollutant emissions	Weight	Comprehensive value
Hybrid GSHP	1	0.5485	1	0.2106	0.8597	0.2409	0.966
Water chiller + Gas boiler	0.5936	0.5485	0.854	0.2106	1	0.2409	0.746

5 Discussion

In this paper, application of GSHP system combined with capillary radiation roof has been investigated in an office building in Wuhan, which is a cooling-dominated area. In comparison with the conventional HVAC system—water chiller for cooling and natural gas for heating (WB)—the main conclusions are given as follows:

(1) Hybrid GSHP system provides better indoor comfort (PMV and PPD) that meets level I requirements; WB system provides indoor comfort that meets level II partially times.
(2) AHP method is used for evaluating system comprehensive performances. Hybrid GSHP system performance evaluation result is 0.966, and that of WB system is 0.746. Hybrid GSHP system is more adaptable and has good potential for future applications.

To be noted, when designing the hybrid GSHP system utilizing radiation roof terminal, condensation needs to be avoided in order to guarantee the system effectiveness and to meet customer's indoor comfort requirements.

Acknowledgements This research was supported by National Key Research and Development Program of China (Project No.2018YFF0300300) and National Natural Science Foundation of China (Project No. 51678262).

References

1. Lee, J.S.: Transient performance characteristics of a hybrid ground-source heat pump in the cooling mode. Appl. Energy **123**, 121–128 (2014)
2. Man, Y.: Study on hybrid ground-coupled heat pump system for air-conditioning in hot-weather areas like Hong Kong. Appl. Energy **87**, 2826–2833 (2010)
3. Villarino, J.I.: Experimental and modelling analysis of an office building HVAC system based in a ground-coupled heat pump and radiant floor. Appl. Energy **190**, 1020–1028 (2017)
4. Song, D.: Performance evaluation of a radiant floor cooling system integrated with dehumidified ventilation. Appl. Therm. Eng. **28**, 1299–1311 (2008)
5. Zhang, S.Y.: Performance evaluation of existed ground source heat pump systems in buildings using auxiliary energy efficiency index: cases study in Jiangsu. China. Energy Build. **147**, 90–100 (2017)
6. Sebarchievici, C.: Performance assessment of a ground-coupled heat pump for an office room heating using radiator or radiant floor heating systems. Procedia Eng. **118**, 88–100 (2015)
7. Romaní, J.: Experimental evaluation of a heating radiant wall coupled to a ground source heat pump. Renew. Energy **105**, 520–529 (2017)
8. Xu, W.: Handbook of Ground-Source Heat Pump Engineering. Construction Industry Press, China (2011)

Study on Energy Evolution Characteristics of Metro Environmental Control Equipment in Different Periods

Jie Song, Yi Zheng, Lihui Wang, Shan Zhang, Renyi Gao, Chang Liu and Xuecheng Zou

Abstract With the increase in the operating period of metros in large- and medium-sized cities, the single equipment of metro environmental control system, such as water chillers, cooling towers, and pumps, will appear new equipment access, medium-term equipment maintenance, long-term equipment replacement, and other problems with the increase in operation period. For these problems, it is imperative to provide professional guidance on the performance parameters of the operation and maintenance of the subway environmental control system for different years. In this paper, the evolution law of chillers, cooling towers, and pumps in environmental control system of metro station under different operating years is analyzed. Finally, the changing trend of performance test of single equipment with operating years is obtained. This paper can provide a professional and pertinent basis for the establishment of performance evaluation standards for access, maintenance, and replacement of metro environmental control system equipment in the future.

Keywords Metro · Water chiller · The cooling tower · The water pump · Evolution characteristics

1 Introduction

In the big and medium cities in China, more and more subway stations are constructed to meet the requirements of the people's traffic. The thermal comfort and energy-saving environments of the metro stations rely on the environmental control instruments systems including the chillers, cooling towers, pumps, etc. Owing to

---

J. Song · Y. Zheng · X. Zou
The Technological Centre of Shanghai Shentong Metro Group Co., Ltd, Shanghai, China

L. Wang (✉) · S. Zhang · R. Gao
University of Shanghai for Science and Technology, Shanghai, China
e-mail: 66amy99@126.com

C. Liu
Shimao Group, Shanghai, China

the long-term operation demands of the metro station with more than one century, the energy efficiency situation of the above environmental control equipment is the key point, relevant to the subway's sustainable development.

The exiting relevant studies focus on the characteristics of the above instruments with the influencing factors by field testing or computer simulation. As to the chillers, Chan et al. [1] focused on the influencing factors of condenser effluent temperature, improved the performance coefficient evaluation model, and verified the model with the field tested parameters. As to the cooling tower, Rahmati et al. [2] measured cooling water temperature, air flow rate, and water flow rate of mechanical ventilation wet cooling tower and found that the number of fillers had a great influence on the performance of cooling tower. As to the pumps, Li and Wang [3] calculated the pressure loss of the pump and rebuilt the system.

Different from the above existing study works, this paper focuses on the performance evolution characteristics of the chillers, the cooling towers, and the pumps varied with the running periods in the metro stations according to the scientific field measurements. This paper can provide a reference for the access of new technologies and products of metro environmental control, the rapid diagnosis of equipment, and system emergencies.

2 Method and Methodology

The evaluation index of the energy efficiency of the chillers, the cooling tower, and the pumps is introduced in the following, as well as the testing methods in the field measurements.

2.1 Evaluation Index and the Measuring Method of the Chillers

Coefficient of performance (COP) of the chillers and other indexes. COP is commonly used as the most acceptable index to evaluate the energy efficiency of the chillers. The attenuation ratio of the cooling capacity α is defined as the ratio of the minus of the measured cooling capacity and the rated cooling capacity over the rated ones, as shown in formula (1).

$$\alpha = \frac{Q_1 - Q_2}{Q_1} \quad (1)$$

where Q_1 is the rated cooling capacity, kW. Q_2 is the measured cooling capacity after conversion, kW [4].

The attenuation rate of the chillers' COP β is defined as the percentage of the minus of between the rated COP and the measured COP over the rated COP

Table 1 Instruments used in the water chillers' measurements and their accuracies

Instrument type	Mercury thermometer	Ultrasound flow meter	Clamp meter
Model	SYWDJ100	FDT-40E	Ulid UT210A
Parameter range	0–100 °C	0–5 m/s	0–200 A
Accuracy	±0.1 °C	±2%	±2%

under the similar air-conditioning load rate and operating conditions, as shown in formula (2).

$$\beta = \frac{COP_1 - COP_2}{COP_1} \qquad (2)$$

where COP_1 is the rated ones and COP_2 is the measured ones, converted according to the reference [5].

There are the following illustrations in the measuring program of the chillers. Firstly, the measuring parameters of the chillers to criticize the energy efficiency include the temperatures of the supplying and returning chilled water, the flow rate of the chilled water, and the electric current of the chillers. All of these parameters should be measured simultaneously with appropriate testing frequencies. Secondly, as to the measuring conditions, the periods should be selected when both the outside weather conditions and the chiller's working conditions are stable. Thirdly, the accuracy of testing instruments should be guaranteed. Table 1 presents the testing instruments used in the flowing chillers' testing.

Fourthly, as to the testing steps of the screw chillers, on the one hand, the temperature of the supplying and returning chilled water can be read with the mercury thermometers on the chillers with reading frequency 10–15-min interval. Besides, the operation electric current can be obtained from the chillers' panel. On the other hand, to test the flow rate of the chilled water with the ultrasound flow meter, the straight part of the water pipe should be selected, as least 10 times of the pipe's diameter away from the elbow valve and other accessories, and the pipe surface should be smooth and exposed to facilitate to be tested. The testing interval is also 5–10 min within the monitoring periods.

2.2 Evaluation Index and the Measuring Method of the Cooling Towers

The performance index of cooling tower is unit fan power consumption ratio a_1 and thermal performance of cooling tower η.

There are the following illustrations in the measuring program of the cooling towers.

Firstly, the measuring parameters of the cooling towers include the temperatures of the intake and outtake cooling water, the flow rate of the cooling water, and the electric current of the fan in the cooling towers. Secondly, the accuracy of testing instruments should be guaranteed. The testing instruments shown in Table 1 fit to the testing requirements of the cooling towers too. Thirdly, the temperature of the intake and outtake cooling water can be read with the mercury thermometers on the cooling water pipes.

2.3 Evaluation Index and the Measuring Method of the Water Pumps

Pump efficiency is the evaluation index of water pump η.

Relevant measuring method of the water pumps. Firstly, the measuring parameters of the water pumps include the pressures, elevations of the pressure gauges, and diameters both in the pump inlet side and the pump outlet side, as well as the electric current of the pump and the corresponding flow rate. All of these parameters should be measured simultaneously with appropriate frequencies. Secondly, the accuracy of testing instruments should be guaranteed. The testing instruments including the ultrasound flow meter and the clamp meter are shown in Table 1. Thirdly, the flow rate testing method could refer to those of the chillers and the cooling towers. Besides, the operation electric current of the pumps also can be obtained from the electronic control cabinets in the environmental control room of the subway stations.

3 Results and Discussion

3.1 Analysis of the Evolution Characteristics of Water Chillers

The attenuation ratio of the chiller's cooling capacity is defined as the ratio of the measured cooling capacity of the chiller over the rated cooling capacity, where the measured cooling capacity has already been converted into the nameplate condition according to the cooling capacity conversion formula [4]. Based on the results of the field measurements, the average cooling capacity of the chiller's of the initial, mid-term, and the long term are running periods 476.68 kW, 411.06 kW, and 452.28 kW, respectively. The corresponding average attenuation ratios of the cooling capacity in the above three stages are 10.94, 5.36, and 14.17% as shown in Fig. 3. It can be seen from Fig. 3 that the chiller in medium-term operation period has a lower attenuation ratio of the cooling capacity with 5%, owing to the higher load ratio with the higher air-conditioning load in metro stations. The higher cooling rate of the initial unit is mainly due to the lower air-conditioning load rate of the new subway station in summer. The main reason for the lower air-conditioning load rate of the initial units is in the following. On the one hand, lots of natural cool sources, kept in storage in the surrounding rock shield of the newly built subway stations with 20–25 °C soil temperature in the new subway station in Shanghai, lead to the lower air-conditioning load in summer. On the other hand, the equipment selection principle of the subway chiller is based on the most unfavorable situation in the long run (the period when the air-conditioning has the largest cooling load). The above two reasons couple together and lead to the lower load rate and higher attenuation ratio of the cooling capacity in the initial operation period. Furthermore, the higher attenuation ratio of the cooling capacity in the long-term operation period is induced by the increased wears of the mechanical components such as the compressors in the chillers, and the decreased heat transfer effect of the heat exchangers such as condenser and the evaporator in the chillers.

The COP of the chiller is a key indicator of its energy efficiency. Figure 2 shows the evolution trends of COPs, which has been converted from the measured conditions into the nameplate conditions already, with years. It can be seen from Fig. 2 that the COP of the chiller in the mid-term period is highest with more than 4.5, owing to the higher air-conditioning load rate and better energy operation efficiency of the units. On the contrary, the lower COP of the chillers in the initial operation period, below 4, is induced by the lower air-conditioning load rate. Furthermore, although the air-conditioning load rate of the chillers in the long-term running period is high enough, both the mechanical wears of the chillers' compressors and the decrease in the heat-exchange effect of the chillers' relevant components increase with the years, resulting in the lower COP values of the chillers in the long-term operation period (Fig. 1).

The attenuation rate of the chillers' COP is defined as the percentage of the minus between the rated COP and the measured COP over the rated COP under the similar air-conditioning load rate and operating conditions. Figure 1 shows the attenuation rate of the chillers' COP in various operation years, compared with the COP of the new chillers under the same air-conditioning situation. Firstly, it can be demonstrated from Fig. 1 that the attenuation rate of the chillers' COP in the mid-term operation period is lowest with slightly higher than 15%. Secondly, in the initial running period, owing to the short operation period, its attenuation rate of the chillers' COP is lower too with the low air-conditioning load, nearly 18%. Thirdly, in the long-term operation period, both the mechanical wears of the

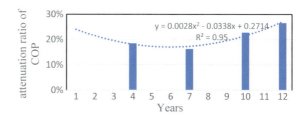

Fig. 1 Trend of attenuation ratio of COP with operating years

Fig. 2 Trend of COP with operating years

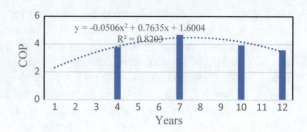

Fig. 3 Trend of attenuation ratio of the chiller's cooling capacity with operation years

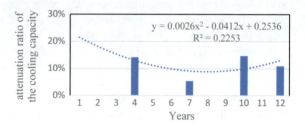

chillers' compressors and the decreases of the heat-exchange effect of the chillers' relevant components increase so much that the attenuation rate of the chillers' COP increases to 20–25% or so.

3.2 Analysis of the Evolution Characteristics of Cooling Tower

The thermal performance value of the cooling tower is a useful index to evaluate its performance (the higher, the better). The measured thermal performance of the cooling tower with different operating years, converted to the rated condition with the designing water temperature drop 5 °C already, is demonstrated in Fig. 4, including 81.46% in the initial running period, 117.73% in the mid-term operation period, and 72.70% in the long-term running period. Thus, it can be concluded that the best cooling performance of the cooling tower belongs to that in the mid-term operation period. Based on the field measurement data, with the similar subway station air-conditioning load, the difference temperature of the cooling water in a mid-term running station and the long-term running station is 3.95 and 2.24 °C, respectively. It can be seen that the heat and moisture transfer effect of the cooling tower in the long-term running subway station is significantly reduced, compared with that in the mid-term running metro station.

The thermal performance attenuation of the cooling tower is defined as the ratio of the minus of the measured thermal performance and the rated thermal performance of the cooling tower over the rated ones. Figure 5 presents the trend of thermal performance attenuation of the cooling tower with various operating years,

Fig. 4 Trend of thermal performance of cooling towers with different operating years

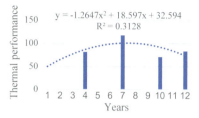

Fig. 5 Trend of thermal performance attenuation of cooling towers with different operating years

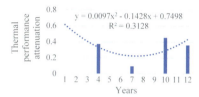

including the average performance attenuation of the cooling tower in the initial running period 37.47%, that in the mid-term period 9.63%, and that in the long-term 44.19%. This is consistent to the characteristics of the thermal performance value of the cooling tower. The cooling tower has lower thermal performance attenuation in the mid-term operation period, and higher thermal performance attenuation in the initial and the long-term operation period.

Figure 6 gives out the tested results of the power consumption per unit air volume of the fan in the cooling tower in the subway stations with various running years. The average values of the power consumption per unit air volume of the fan in the cooling tower in the initial, mid-term, and long-term running period are 0.037, 0.0434, and 0.0389 respectively. It can be seen from that all of the power consumption per unit air volume of the fan in the cooling tower in different running periods fluctuates up and down, with the standard reference value of 0.036, without the obvious corresponding relationship with the running periods.

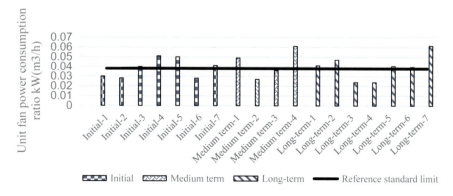

Fig. 6 Power consumption ratio of the cooling tower fan varies with different operating periods

As the non-heat-exchange equipment, the fan in the cooling tower is maintained regularly, such as adjusting its tightness of the belts, resulting in its power consumption per unit air volume having no obvious linear change with the operating periods

3.3 Analysis of the Evolution Characteristics of Pumps

In most of the measured subway stations, the two chilled water pumps for using with standby one are its standard configuration. The two chilled water pumps are connected to the two chillers in series firstly, responsible for the chilled water circulation for the using side. It can be seen from Fig. 7 that the pump efficiencies of the chilled water pumps fluctuate up and down around the standard reference value of 0.85 with the average pump efficiency in the initial period 1.12, that in the mid-term period 0.91, and that in the long-term period 0.82. There is no obvious linear relationship between the pump efficiencies and the operation years, owing to regular maintenance to the chilled water pumps as the non-heat-exchange equipments.

The two cooling water pumps are connected to the two chillers in series firstly, responsible for the cooling water circulation from the chillers to the cooling towers. It can be seen from Fig. 8 that the pump efficiencies of the cooling water pumps fluctuate up and down around the standard reference value of 0.85 with the average pump efficiency in the initial period 0.96, that in the mid-term period 1.21, and that in the long-term period 0.79. There are no obvious linear relationships between the pump efficiencies and the operation years, owing to regular maintenance to the cooling water pumps as the non-heat-exchange equipments.

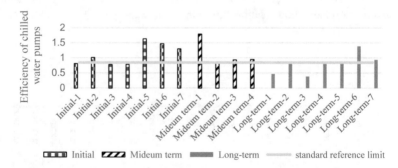

Fig. 7 Trend of chilled water pump efficiency with different operating periods

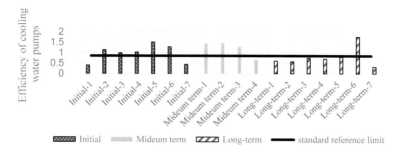

Fig. 8 Trend of cooling water pump efficiency with different operating periods

4 Conclusion

1. By calculating the cooling capacity, COP, cooling capacity attenuation, and COP attenuation of water chillers, the curve fitting analysis is carried out. The results show that the COP value of chillers is low in the initial stage. In the medium and long terms, COP values have obvious attenuation characteristics. The cooling capacity of water chillers should be limited according to the actual average value.
2. By calculating the thermal performance of the cooling tower and the power consumption ratio of the unit fan, the curve fitting analysis is carried out. The results show that the initial thermal performance of the cooling tower is low. Medium- and long-term values have obvious attenuation characteristics. The change trend of unit fan power consumption ratio is not obvious.
3. After data analysis, the change trend of pump with operation life is not obvious. The energy efficiency ratio of the pumps is at a high level, and there is no linear relationship with the operation period.

Acknowledgements This work was supported by National Natural Science Funds (51878408), Shanghai alliance project of the association for the advancement of science (LM201735), and the research project of Shanghai Shentong Metro Group Co. Ltd. (JS-BZ16R011).

Permissions The measuring programs of the chillers, cooling towers, and pumps in environmental control system of metro station under different operating years are permitted by the relevant Shanghai Metro Operating Companies.

References

1. Chan, K.T., Yu, F.W.: Thermodynamic-behavior model for air-cooled screw chillers with a variable set-point condensing temperature. Appl. Energy **83**(3), 265–279 (2006)
2. Rahmati, et al.: Investigation of heat transfer in mechanical draft wet cooling towers using infrared thermal images: an experimental study. Int. J. Refrig. **88**(4), 229–238 (2017)
3. Li, H., Wang, S.: Probabilistic optimal design concerning uncertainties and on-site adaptive commissioning of air-conditioning water pump systems in buildings. Appl. Energy **8**, 53–65 (2017)
4. Energy-Saving Monitoring of Pumping Liquid Delivery System. China Standard Press, Standard No. GB/T 16666-2012 (2012)
5. Cai, C.W.: Conversion equation of chiller performance and its application. In: National Technical Exchange Conference of HVAC Technology Information Network (2005)

Energy-Saving and Economic Analysis of Anaerobic Reactor Heating System Based on Biogas and Sewage Source Heat Pump

Shouwen Sheng and Fang Wang

Abstract Anaerobic treatment technology is a common technology for treating starch wastewater. Relevant research shows that heating the wastewater to the optimum operating temperature can ensure the work efficiency of anaerobic reactor. However, the water from anaerobic reactor has a high temperature and generates lots of waste heat. Therefore, we propose a biogas generator set–sewage source heat pump combined heating system. Compared with the obsolete traditional heating method of single heat source (biogas boiler), this new method can not only save energy consumption, but also heat the anaerobic reactor continuously and steadily. We analyzed the energy-saving and economy of the biogas generator set–sewage source heat pump heating system through theoretical calculation. The results show that compared with the single heat source biogas boiler heating system, the system has remarkable energy-saving effect, the primary energy-saving is about 163.00–358.35 kW, and the annual cost of the combined heat source heating system is about 72% of that of the biogas boiler heating system.

Keywords Anaerobic reactor · Sewage source heat pump · Energy-saving · Economic analysis

1 Introduction

Potato starch is an important industrial raw material. A large amount of wastewater will be produced in the process of starch production [1]. The wastewater belongs to high concentration acidic organic wastewater, and its chemical oxygen demand

---

S. Sheng · F. Wang (✉)
School of Energy and Power Engineering, Nanjing University of Science and Technology, Nanjing 210094, People's Republic of China
e-mail: wfnust@126.com

F. Wang
Nanjing University YanCheng Institute of Environmental Technology and Engineering, YanCheng, People's Republic of China

(COD) content is very high. Therefore, starch wastewater must be effectively treated to meet the discharge standards before it can be discharged [2]. Anaerobic treatment technology is a common technology for treating starch wastewater [3]. In order to maintain high treatment efficiency, its temperature needs to be maintained at about 35 °C, so it is necessary to heat the water of the anaerobic reactor [4].

In practical engineering, the heating method of anaerobic reactor mostly uses a single heat source, just like biogas boiler [5]. However, the above heating methods have some shortcomings, such as low efficiency, environmental damage, and so on. In addition, the wastewater treated by the reactor has three main characteristics: large flow rate, large heat storage, and high temperature [6]. In this context, the sewage source heat pump system is proposed to recover the waste heat from the anaerobic reactor for heating the reactor imported wastewater to achieve the purpose of energy-saving and environmental protection. Besides, biogas generators consume biogas generated by anaerobic reactor to provide electricity. Taking the anaerobic reactor as an example, the energy-saving and economy of the system are analyzed by theoretical calculation.

2 Methods

This paper designed the biogas generator set–sewage source heat pump combined heating system (hereinafter referred to as composite heat source) shown in Fig. 1.

The system consists of a biogas generator set, a compression heat pump, a shell-and-tube heat exchanger, and a biogas generator set exhaust heat exchanger. The system makes up the shortage of a single heat source heating anaerobic reactor and ensures the anaerobic reactor reaches a suitable temperature. The sewage source heat pump hot water system recovers waste heat from wastewater treated of reactor

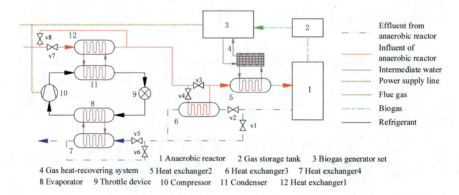

Fig. 1 System diagram

export and uses the collected waste heat to heat the wastewater of reactor inlet. Also, the biogas generator set uses the biogas produced by the anaerobic reaction to generate electricity and provides to the sewage source heat pump. The waste heat generated by the generator set continues to heat the influent of the anaerobic reactor.

According to the different inlet temperatures, the operation mode of the system can be divided into two kinds. Mode 1: using waste heat recovered from biogas generator set combined sewage source heat pump to heat imported wastewater, That is to say, the valves v1, v5, v7, and v3 in Fig. 1 are opened, and the valves v2, v6, v8, and v4 in Fig. 1 are closed, And the heat pump system is operated. Mode 2: using waste heat recovered from export water and waste heat recovered from biogas generator set to heat imported wastewater. Similarly, the valves v1, v5, v7, and v3 in Fig. 1 are closed, the valves v2, v6, v8, and v4 in Fig. 1 are opened, and the sewage source heat pump system is closed.

2.1 Biogas Production

For the biogas production under different COD concentrations, the following formula is used to calculate [7]:

$$V_G = V_{w,i} \times \frac{COD_i}{1000} \times \eta_{rem} \times \eta_{comv} \quad (1)$$

where V_G is the amount of biogas produced by the anaerobic reactor, m³/d; $V_{w,i}$ is the amount of starch wastewater treated by the anaerobic reactor, m³/d; COD_i is COD concentration, mg/L; η_{rem} is COD removal rate of anaerobic reactor, %, the value here is 80%; η_{comv} is the yield of biogas, m³/kg COD, the value here is 0.6.

2.2 Performance of Biogas Generator Set

The biogas power generation system is mainly composed of gas engine, generator, and heat recovery device. The power generation efficiency η_1 is 30%, and the waste heat utilization efficiency η_2 is 50% [8]. The calorific value of biogas combustion H_L is 21544 kJ/m³. The power generation Q_d (kW) and waste heat Q_{yr} (kW) [7]:

$$Q_d = \eta_1 \times V_G \times H_L \quad (2)$$

$$Q_{yr} = \eta_2 \times V_G \times H_L \quad (3)$$

3 Results

This paper takes an anaerobic reactor in Anhui Province as the research object. The flow rate of the anaerobic reactor is 5.79 kg/s. The annual water temperature of the reactor is shown in Table 1. This paper takes the concentration of 10,000 mg/L COD as an example to make an economic analysis of this new heat system.

This paper assumes that the heat dissipation from the anaerobic reactor to the surrounding environment is neglected.

3.1 Determination of System Parameters

According to the performance parameters of anaerobic reactor (The inlet flow is 5.79kg/s, and the required temperature is 35 °C) and local climate parameters (shown in Table 1), the sewage source heat pump system are designed, specific parameters are as follows: Compressor rated power is 43.32kW, refrigerant flow rate is 1.67kg/s, evaporation temperature is 10 °C, condensation temperature is 50 °C.

3.2 Biogas Production and Waste Heat Recovery in the System

The results of biogas production and waste heat recovery in the system are as follows: gas production is 2400 m^3/d, and generation capacity is 179.53 kW.

3.3 Operation Conditions of the System

System Operation Mode I: According to the water temperature in Anhui Province, the system runs according to the mode one in January to April and October to December. In the case of mode 1, through calculation, natural gas must be used as auxiliary fuel to meet the requirements in this mode. The volume of natural gas

Table 1 Annual water temperature of the location of the reactor (°C)

Month	Water temperature	Month	Water temperature	Month	Water temperature
1	4.96	5	14.28	9	14.28
2	3.97	6	15.49	10	12.65
3	9.43	7	16.15	11	8.55
4	11.59	8	15.39	12	4.11

required is calculated according to the following formula, where $t_{w,e}$ is the temperature of sewage after heating:

$$V_t = \frac{c \times m_w \times (35 - t_{w,e})}{\eta_2 \times 33495} \qquad (4)$$

System Operation Mode II: According to the above analysis, the heating system runs from May to September. Through calculation, in this mode, no additional auxiliary fuel is needed. Under this mode, the system saves 179.53 kW of electricity.

3.4 Energy-Saving and Economy

Analysis of Energy Conservation. In order to quantify the energy-saving of the biogas generator set–sewage source heat pump combined heating system, this paper compares it with the energy consumed by the biogas boiler as a single heat source heating system, and all of them are converted into primary energy.

The primary energy consumption of the two heating methods is shown in Fig. 2. It can be seen that the primary energy consumed by the biogas boiler heating system in each month is greater than that of the composite heat source. The primary energy consumption of the biogas boiler heating system is 761.44–1253.45 kW in the whole year, while the composite heat source is 589.44–940.56 kW. Compared with the biogas boiler as a single heat source heating system, the energy-saving effect of the composite source heating system is remarkable, and the primary energy saved is 163.00–358.35 kW. In November, the energy saved was the most, while in July, the energy saved was the least. The average energy saved in the whole year was 273.83 kW, which was remarkable.

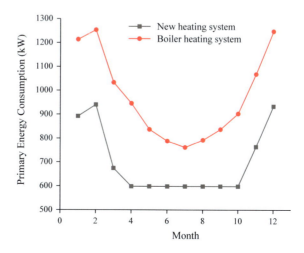

Fig. 2 Primary energy consumption of new system and biogas boiler heating system

Economic analysis. Economic analysis is generally divided into two methods: cost present value and annual cost analysis. In this paper, the annual cost method is used for economic analysis.

The annual cost includes initial investment and annual operating cost. Cost year value calculation formula [9]:

$$F = K \times \frac{(1+i)^N}{(1+i)^N - 1} + C \tag{5}$$

where F, annual operating expenses calculated according to the dynamic method; K, initial investment in the system; i, investment discount rate, %; N, equipment life, year; C, the annual operating cost of the system.

Initial investment includes equipment price, installation cost, and other costs. 30 and 40% of the equipment investment are used for auxiliary equipment and installation cost. The auxiliary equipment and installation cost of the new system are 50 and 45% of the equipment investment.

Annual operating costs include labor costs, maintenance costs, and electricity and fuel costs. Labor costs estimated at 16,000 RMB a person every year. We needs five people to maintain the combined system to run steadily, and traditional heating system needs three people to maintain. Maintenance cost is 2.5% of initial investment, and auxiliary equipment operation cost is 5% of initial investment. The fuel cost is the biogas consumption and natural gas consumption of the two systems (Table 2).

In table, A, equipment price; B, auxiliary equipment cost; D, installation cost; E, total initial investment; G, labor cost; H, maintenance cost; I, auxiliary equipment operation cost; L, fuel cost; M, surplus electricity cost; and O, total annual operating expenses.

The annual cost is the sum of the initial investment and the annual operating expenses. According to Eq. (5), the investment discount rate is $i = 10\%$, and the service life is 15 years. The annual cost of the two heating methods is listed in Table 3.

It can be seen from the table that the annual cost of the biogas generator set–sewage source heat pump combined heating system is 718,400 RMB, and the

Table 2 Calculation results of initial investment and annual operating costs (unit: 10,000 RMB)

	New system	Biogas boiler		New system	Biogas boiler
A	54.5	8.2	H	0.588	0.349
B	27.5	2.46	I	5.314	0.697
D	24.53	3.28	L	110.6	92.05
E	106.28	13.94	M	66.64	–
G	8	4.8	O	57.862	97.896

Table 3 Comparison of annual cost values (unit: 10,000 RMB)

	Compound source heating system	Boiler heating system
Initial investment	106.28	13.94
Annual operating expenses	57.86	97.896
Annual cost	71.84	99.73

annual cost of the biogas boiler heating system is 997,300 RMB. Therefore, the composite heat source has advantages over the biogas boiler heating system in terms of economy.

4 Conclusions

In this paper, anaerobic reactor is taken as the research object. The biogas generator set–sewage source heat pump combined heating system was developed. The anaerobic reactor is heated by the methane generated by the reactor, and the waste heat from the wastewater is recovered from the reactor. Taking an anaerobic reactor in Anhui Province as an example, the energy-saving and economy of the system were studied and analyzed. The conclusions and results are as follows:

(1) In terms of energy-saving, compared with the heating mode of the biogas boiler, the primary energy saved by the system is 163.00–358.35 kW, which saves the most energy in November and the least in July. The average annual primary energy saved by the system is 273.83 kW, and the effect is remarkable.
(2) In economic aspect, compared with single biogas boiler heat source, the initial investment of the biogas generator set–sewage source heat pump combined heating system is higher, but its operation cost is lower. The annual cost of the system is 718400 RMB, and that of biogas boiler heating system is 997300 RMB; that is to say, the annual cost of compound heat source is about 72% of that of single biogas boiler heat source.

Acknowledgements The project is supported by the project "Research on Waste Heat Recovery System of Sweet Potato Starch Wastewater Anaerobic Reactor" (NDYC-KF-2017-08).

References

1. Deng, S.B., Bai, R.B., Hu, X.M., et al.: Characteristics of a bioflocculant produced by Bacillus mucilaginosus and its use in starch wastewater treatment. Appl. Microbiol. Biotechnol. **60**(5), 588–593 (2003)
2. Guo, X.Y., Nian, Y.G., Yan, H.H., et al.: Current status and application prospect of resource utilization of starch wastewater. Acta Environ. Eng. **6**(2), 117–126 (2016)

3. Zhu, G., Li, J., Liu, C., et al.: Simultaneous production of bio-hydrogen and methane from soybean protein processing wastewater treatment using anaerobic baffled reactor (ABR). Desalination Water Treat. **53**(10), 2675–2685 (2015)
4. Ariunbaatar, J., Panico, A., Esposito, G., et al.: Pretreatment methods to enhance anaerobic digestion of organic solid waste. Appl. Energy **123**, 143–156 (2014)
5. Zhou, D.F, Jin, Y.A., Yang, C.H., et al.: Heating biogas slurry with waste heat of boiler to improve gas production. Biogas, China, **27**(6), 28–30 (2009)
6. Frijns, J., Hofman, J., Nederlof, M.: The potential of (waste) water as energy carrier. Energy Convers. Manag. **65**(1), 357–363 (2013)
7. Gao, C.M.: Biogas power generation and waste heat utilization. Urban Manag. Technol. **7**(5), 217–219 (2005)
8. Lübken, M., Wichern, M., Schlattmann, M., et al.: Modelling the energy balance of an anaerobic digester fed with cattle manure and renewable energy crops. Water Res. **41**(18), 4085–4096 (2007)
9. Wu, X.H., Sun, D.X.: Energy efficiency analysis of urban primary sewage heat exchanger. Renew. Energy **25**(02), 73–75 (2007)

Air-Conditioner Usage Patterns in Teaching Buildings of Universities by Data Mining Approach

Xinyue Li, Shuqin Chen, Jiahe Li and Hongliang Li

Abstract Due to the stochastic usage of the air-conditioner and an unpredictably large stream of people in teaching buildings, it is much difficult to accurately simulate the energy consumption of the air-conditioner and formulate energy-saving strategy. This study is based on the real-time usage data of air-conditioning system which were collected by the energy consumption monitoring platform in a university from December 2016 to November 2017. With the data mining approach, typical air-conditioner usage patterns and the proportion of each pattern are proposed in this study. Next, a method used to classify the operation to the specific usage patterns is presented and proved by the cross-validation method.

Keywords Typical usage pattern · Data mining · Teaching building · Usage schedule

---

X. Li · S. Chen
College of Civil Engineering and Architecture, Zhejiang University,
Hangzhou 310058, China
e-mail: lixinyue@zju.edu.cn

S. Chen
State Key Laboratory of Subtropical Building Science, South China University of Technology, Guangzhou 510640, China
e-mail: hn_csq@126.com

J. Li
Zhejiang Bluetron Industry Internet Information Technology Co.,Ltd,
Hangzhou 310053, China

H. Li (✉)
College of Control Science and Engineering, Zhejiang University,
Hangzhou 310027, China

Zhejiang Excenergy Energy-saving Technology Co.,Ltd,
Hangzhou 310052, China
e-mail: lihongliang_zju@zju.edu.cn

1 Introduction

As a result of universities' scale expanded, the energy demand has a soaring increase. There are more than 2000 universities in China, which consume 8% of total energy consumption of society [1]. Zhao et al. [2] analysed energy consumption data and air-conditioner spends most of the energy, with the part of 41.76%. With the large stream of people, the air-conditioner usage in teaching buildings is more random than other buildings like office buildings.

With regard to the usage of teaching buildings, Hu et al. [3] investigated on a teaching building in Beijing which showed an obvious difference of using rate in different seasons. To optimize the using strategy of air-conditioners in self-study classrooms, energy consumption could reduce 18.8%. Liu [4] analysed factors influencing classroom usage in a case study. Results showed the highest using rate of classrooms could be seen in the morning and using rates were various between classrooms. Song et al. [5] developed an energy efficiency-based course timetabling algorithm. This study showed the optimal timetable produces 4% energy-saving, which proved behaviour of occupants directly impact energy consumption from another angle. In this case, managing air-conditioner usage with predefined typical usage patterns could have many benefits on energy-saving. However, there were few attempts to combine the air-conditioner usage and classroom occupancy characteristics.

The progress in digital technology, in particular, building automatic system, opens up the possibility of collecting mega real-time air-conditioner usage data. These data can present long-term air-conditioner usage, while other methods, like measuring sample rooms by apparatus, are hard to extensively demonstrate the air-conditioner usage in the large-scale area. With these data, the inherent rules related to operation behaviour of air-conditioners can be found by data mining approach, which have been widely used in the building science area. Ren et al. [6] investigated the behaviour of thermostat settings and heating system operations in residences. Six room temperature patterns were revealed by data mining method. Simona et al. [7] proposed a data mining framework to discover occupancy patterns and better understand the energy usage in office buildings. Yu et al. [8] emphasized advantages of data mining approaches over statistical analysis techniques in building operation analyse. His study pointed out that statistical techniques mainly depend on the domain knowledge of the analyst and chosen methods, which could miss much useful information.

However, research on air-conditioner usages of teaching building is not as much as that of other building types, like office buildings. Tan et al. [9] analysed energy consumption of different building types on campus. The result showed monthly energy consumption of all teaching buildings on the campus significantly depends on the seasons and behaviour of occupants. Zhang [10] compared the cooling and heating load between various types of buildings on campus. This study indicated the wildest fluctuation could be seen in teaching buildings. However, these studies focus on entire buildings or teaching building area instead of each room or terminal unit, which can directly demonstrate air-conditioner usage of occupants. Commonly, only energy consumption of the integral air-conditioning systems or subsystems was recorded by energy consumption monitoring platforms in universities, but that of each room or terminal units is hard to measure. In this case, research on terminal units' usages is rare, which is closely related to the energy consumption of the air-conditioner system. Thus, analysing the air-conditioner usage of each room in teaching buildings can have many benefits on energy simulation and energy conservation.

In this study, the air-conditioner usage of every single classroom was taken into consideration. With the data mining approach, typical air-conditioner usage patterns and their proportion are proposed. A method used to identify different usage patterns is presented, which also can quantitatively describe typical patterns. The results can provide the guidance for the implementation of energy-saving policy in universities.

2 Data set and Preprocess

2.1 Data set

The data set used in this study was collected by the energy consumption monitoring platform in a university located in Zhejiang Province from December 2016 to November 2017, which include 2.2 million records. The interval of data collecting is half an hour. This energy-saving platform monitors the usage of air-conditioners in 1105 rooms of 35 teaching buildings. VRV systems are used in these buildings. In each classroom, occupants in the classroom can control the indoor unit of the VRV system by control panels. The class schedule of the university was presented in Table 1.

2.2 Data Process

In this study, only the work time, from 8:00 to 22:00, was taken into consideration. In the preprocessing part, real-time air-conditioner usage data of each room were converted into daily ON/OFF sequence. The hourly state will be marked as 1 if the

Table 1 University schedule

Time	Activities
8:05–12:10	Forenoon classes
12:10–13:10	Lunch break
13:10–16:40	Afternoon classes
16:40–18:00	Break
18:00–20:25	Night classes

Table 2 Season division of the whole year

Season	Month	Air-conditioner mode	Schedule
Early/late summer	6, 9	Cooling	School days
Summer	7, 8	Cooling	Vacation
Transition season	4, 5, 10, 11	Cooling, heating	School days
Early/late winter	12, 3	Heating	School days
Winter	1, 2	Heating	Vacation

air-conditioner is used in this hour; otherwise, the state is marked as 0. The sum of daily air-conditioners operation time is also recorded.

As different air-conditioner using behaviour in different month, in this study, the whole year was divided into 5 typical seasons based on the climate and cooling/heating demand. Seasons and cooling/heating modes are given in Table 2.

3 Methods

In this study, cluster analysis was used to identify typical air-conditioner usage patterns. The decision tree method was used to establish a model to recognize the typical usage pattern.

The PAM algorithm was used in cluster analyses in this study to find typical air-conditioner usage patterns. Compared with k-means algorithm, a commonly used clustering algorithm, the PAM algorithm chooses data points as centres and can be used with arbitrary distances, while k-means only minimizes the squared Euclidean distances. This method is more robust than k-means, which can reduce the negative effect of extreme value. The result of cluster analyses will be more accurate.

Decision tree method can provide a tree-like conditional flowchart. Unlike other black-box methods like SVM, the condition provided by decision tree method is understandable and interpretable even for the person who does not have professional knowledge. In this study, this method is used to identify the new air-conditioner operation record into a typical usage pattern.

4 Results

4.1 Usage Profiles

Figure 1 shows the air-conditioner using rate in different seasons. The using rate was calculated by the number of daily records which the air-conditioner was used divided by the number of all daily records. Overall, using rates in summer and

Fig. 1 Using rate during the whole year

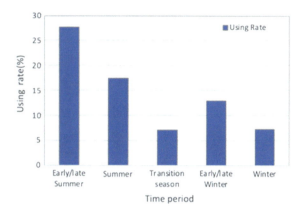

early/late summer were obviously higher than in other seasons. In school days, the highest using rate can be seen in early/late summer (27.6%), followed by early/late winter (13.1%). The air-conditioner was rarely used in transition season (7.2%). In summer and winter vacation, the using rate in summer (17.5%) was higher than winter (7.4%). Comparing with heating/cooling rates, during different seasons as shown in Fig. 2, the cooling demand was greater than heating in teaching buildings. Although heating and cooling usages coexisted in the transition season, the using rate of cooling (77.9%) was significantly higher than the heating counterpart (22.1%).

4.2 Usage Clusters

In the light of various using behaviour of air-conditioners, each season and air-conditioners' running mode have different number of clusters (K). In this study, within sum of squares (WSS) criterion was used to find the best size of clusters. This criterion evaluates cluster by WSS of the tentative cluster result. WSS will decrease with the increase of K, the size of cluster. The best size of clusters is the value that the decline rate of WSS was significantly slow down. The best cluster size (K) of each season is given in Table 3.

After defining the size of clusters, the attributions used to cluster are the daily ON/OFF sequence and the used time. Five clusters were identified in different seasons. The hourly air-conditioner using proportion of each cluster was calculated and showed in Fig. 3. Table 4 presents characteristics of each cluster.

1. Cluster 1, with the largest proportion in the whole year (35.7%) and the shortest average used time (2.2 h) which was matched a lesson-long time. This pattern, with fluctuated using rate between 7.5 and 24.2% in whole-day and no significant peak, reflected on-demand usage and defined as "On-demand usage pattern".

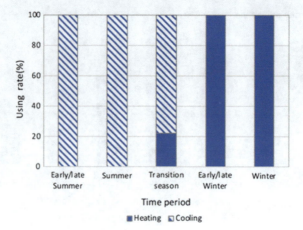

Fig. 2 Heating/cooling rate in each season

Table 3 Best K of each season

Season	Best K
Transition season (heating)	3
Early/late summer	5
Summer	3
Transition season (cooling)	4
Early/late winter	5
Winter	3

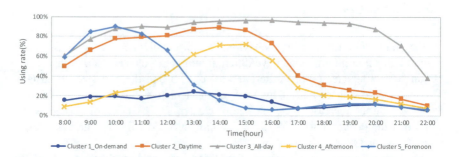

Fig. 3 Hourly using proportion of each cluster

Table 4 Statistics and definitions of each cluster

Clusters	Proportion	Used time (h)	Pattern definition
1	0.357	2.24	On-demand usage pattern
2	0.322	8.39	Daytime usage pattern
3	0.134	12.64	All-day usage pattern
4	0.129	4.81	Afternoon usage pattern
5	0.058	5.45	Morning usage pattern

2. Cluster 2 illustrated the usage in the daytime and was defined as "Daytime usage pattern". The using rate in the daytime was above 50% and had a steep decline after 17:00. The average used time (8.4 h) coincides with the daytime schedule.
3. Cluster 3 had the longest average used time (12.6 h) with the using rate above 60%. This cluster was defined as "All-day usage pattern". The using rate had a sharp decline after 20:00, which is the end time of night classes.
4. Cluster 4 had a gradual increase using rate after 12:00 and peaked at 15:00 before falling to the low level at 17:00, which is the same time as the end of afternoon lessons. This cluster was defined as "Afternoon usage pattern".
5. Cluster 5 was defined as "Forenoon usage pattern". The using rate of this pattern started at over 60% and peaked at 10:00. There is a rapid decrease at 12:00 which is the beginning of the lunch break. The used time of this cluster is 5.4 h.

Table 5 shows the percentage of patterns in different seasons. Five usage patterns were all typical in school days, which revealed that the usage in school days was more complicated than that on vacations. The largest part in school days was "On-demand" usage pattern. The cooling usage in transition season had 4 typical patterns, and "Forenoon" usage pattern was not typical, while "Afternoon" usage pattern had the second large part because of the higher outdoor temperature in a day. With fewer lessons, usage on vacation was simpler (only 3 typical patterns).

4.3 Typical Pattern Recognition

After quantitatively describing typical patterns, the decision tree method was used to classify the unlabelled data to a predefined pattern according to the time sequence

Table 5 Percentage of each pattern in different seasons

Season	Air-conditioner mode	On-demand	Daytime	All-day	Forenoon	Afternoon
Summer	Cooling	0.36	0.45	0.19	–	–
Early/late summer	Cooling	0.36	0.26	0.07	0.11	0.20
Transition season	Cooling	0.28	0.3	0.11	–	0.31
Winter	Heating	0.44	0.38	0.17	–	–
Early/late winter	Heating	0.29	0.24	0.15	0.16	0.16
Transition season	Heating	0.48	0.29	0.23	–	–

"–" Means not a typical pattern

and the used time. In the light of different typical usage pattern under cooling and heating modes, cooling and heating data were split to establish the decision tree model.

To increase reliability and avoid overfitting, this study used the cross-validation method which randomly selects 3/4 data as the training set, and others as the test set. The model of cooling mode was shown in Fig. 4, while that of heating mode was shown in Fig. 5. Leaf nodes demonstrated the result of classification. The value of each bar in leaf nodes indicated the possibility of which pattern the sample belongs. Compared to heating mode, the decision tree of cooling mode had more nodes, which illustrated air-conditioner usage for cooling was more complicated than the heating mode. Confusion matrixes which established based on test set were shown in Table 6.

The row of the confusion matrix represents the actual pattern, and the columns represent the predictions. "All-day usage pattern" in cooling mode as an example, 563 instances were correctly classified on the diagonal of the table, while 133 instances were incorrectly classified as "Daytime usage pattern". According to the result of the test set, the accuracy of the cooling mode was 88.0% and the figure of the heating mode was 85.5%. It showed this method had desirable performance.

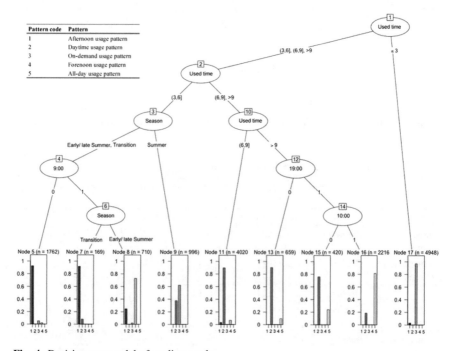

Fig. 4 Decision tree model of cooling mode

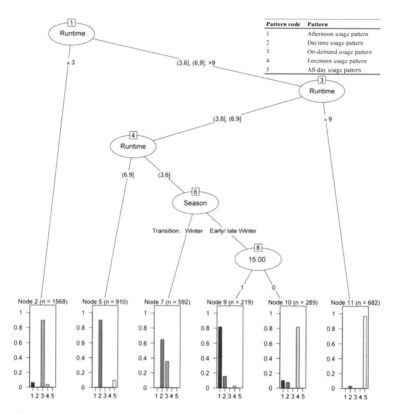

Fig. 5 Decision tree model of heating mode

Table 6 Resulting confusion matrix of cooling/heating mode

	Afternoon	Daytime	On-demand	Forenoon	All-day
Afternoon	626/52	5/16	33/0	12/3	0/0
Daytime	56/0	1498/417	0/83	94/0	67/26
On-demand	56/29	119/0	1790/458	0/21	0/0
Forenoon	54/11	0/8	8/0	187/85	0/0
All-day	0/0	133/9	0/0	0/0	563/203

5 Discussion

The air-conditioner using rate has significant differences between different rooms even though in the same building on the same day. Teaching buildings have more factors affecting the operation of air-conditioner, such as the curriculum table, than other buildings which have regular patterns like office buildings. The energy simulation on teaching buildings will not match the real situation because fixed and homogeneous schedule or simply outdoor temperature ON/OFF control was used. Thus, the wrong energy using strategy and energy-saving management will be made.

In this study, typical usage patterns of air-conditioners were proposed from huge real-time air-conditioner operation data, by which the complicated usage can be represented. The features of typical usage patterns can be a guidance for air-conditioner energy simulation of teaching buildings. Furthermore, the percentage of each pattern can be a reference for the selection of HVAC system. For example, "On-demand" usage pattern has the highest percentage in the whole year. In this case, the device performance under partial load should be taken into consideration. In the light of only one university as a sample of this study, the results may not be completely universal, but the data mining method in this study is considered general.

The typical pattern recognition method which mentioned in this study can identify the proportion of each cluster instantly. With the different used time of each usage pattern, energy consumption is various. Energy consumption intensity can be evaluated according to the identification result of new air-conditioner running data. Thus, appropriate energy-saving rules can be made.

6 Conclusions

This study analyses the air-conditioner usage in a university during different seasons for the whole year. The main conclusions are:

1. Data mining is a powerful tool to analyse big data. In this study, this method was used to find typical pattern and classify undefined air-conditioner usage behaviour into a typical pattern, which can represent complicated air-conditioner operation.
2. Five typical usage patterns of air-conditioners were presented. The percentages of "On-demand" and "Daytime" usage patterns are significantly higher than others. The season and school schedule will influence the distribution of different patterns.
3. The typical pattern recognized model is based on the decision tree method, which can instantly classify real-time air-conditioner using data into the typical pattern. With the dynamic proportion of each pattern, the energy-saving strategy can be made.

The important aspect of this study used typical air-conditioner usage patterns to represent complex usage in teaching buildings, which can benefit energy simulation.

Acknowledgements The study is supported by China National Key R&D Program (2018YFC0704400) and State Key Lab of Subtropical Building Science, South China University of Technology (2018ZB17).

References

1. Tan, H., et al.: Research on building campus energy management. Build. Energy Environ. **29**(1), 36–40 (2010)
2. Zhao, M., et al.: Energy-saving potential analysis of a college building in Shanghai based on the result of energy consumption audit. Build. Energy Efficiency **45**(4), 100–104 (2017)
3. Hu, T., et al.: Influence of usage rate on indoor thermal environment of university's self-learning classrooms in Beijing in heating season. HV&AC **8**, 98–101 (2015)
4. Liu, X.: Study of the classroom resource utilization efficiency of the local university—a case of Henan University, Master diss., Henan University (2012)
5. Song, K., et al.: Energy efficiency-based course timetabling for university buildings. Energy **139**, 394–405 (2017)
6. Ren, X., et al.: Data mining of space heating system performance in affordable housing. Build. Environ., pp 891–13 (2015)
7. Simona, D., et al.: Occupancy schedules learning process through a data mining framework. Energy Build. **88**, 395–408 (2015)
8. Yu, Z., et al.: A novel methodology for knowledge discovery through mining associations between building operational data. Energy Build. **47**, 430–440 (2012)
9. Tan, X., et al.: Characteristics of energy use of typical buildings and its behavioural influencing analysis in one university campus in hot summer and cold winter China. J. Nanjing Univ. Sci. Technol. **43**(1), 101–107 (2019)
10. Zhang, X.: Prediction and analysis method of air-conditioning load in regional buildings, Master diss., Zhejiang University (2018)

Design Optimization of Radiation Cooling Terminal for Ultra-low-Energy Consumption Office Buildings

Zhengrong Li, Xiangyun Chen and Dongkai Zhang

Abstract Ultra-low-energy consumption buildings, combining high-comfort and low-energy advantages due to their high-performance envelope and good air tightness, are the development direction of energy-efficient buildings in China. The advantages of radiant air conditioners such as energy-saving and good thermal comfort are suitable for low-energy buildings. However, different laying methods of radiant terminals have a great impact on building energy consumption and comfort performance. This paper focuses on ultra-low-energy consumption office buildings in hot summer and cold winter zone as the research object and analyzes the effects of radiation area ratios and positions of the radiant panels. Refer to the existing ultra-low-energy consumption buildings in the hot summer and cold winter zone, determine the parameters of the envelope structure, and build a model. Airpak was used to simulate different radiation area ratios and heights of radiant surfaces for indoor thermal environments. The results indicate that when the bottom edge of the radiant panel is laid along the ground, the higher the radiation area ratios is, the more uniform the PMV distribution is. At the same time, the stratification of the temperature distribution is also weakened. Properly increasing the laying height of the radiant panel can make indoor PMV and temperature distribution more uniform and reduce the vertical temperature difference, and the radiation cooling capacity can be utilized better. The research results provide reference for the design of radiant air conditioners for ultra-low-energy consumption buildings.

Keywords Ultra-low-energy consumption building · Radiant cooling · Vertical temperature difference · Asymmetrical radiant

---

Z. Li · X. Chen (✉) · D. Zhang
School of Mechanical Engineering, Tongji University, Shanghai, China
e-mail: 416057216@qq.com

1 Introduction

In recent years, the development of ultra-low-energy buildings has received increasing attention. How to develop ultra-low-energy buildings has also become an urgent problem to be solved. At this stage, China is in need of research and development of high-performance building parts and equipment to provide technical accumulation for the ultimate goal of zero-energy buildings [1]. Radiant heating and cooling (RHC) system has many applications in China [2]. Because of its good thermal comfort, low energy consumption, quiet operation, space saving, etc., it is a good choice for ultra-low-energy building air-conditioning terminal equipment. Yujia Yang et al. studied the effects of different radiant cooling surface laying methods on indoor thermal environment by comparing seven cases [3]. Miaoyu Zhang studied the influence of different radiant panel laying rates on indoor temperature distribution under ceiling radiant cooling conditions and floor radiant cooling conditions [4]. Jiang Ling studied the RHC room with different radiation laying position, radiation area ratios, and different surface temperatures by means of experiments and CFD simulation to investigate the average interior wall temperature and indoor air temperature [5].

Although the predecessors have a lot of research on the different laying positions of the radiant panels, they tend to focus on the laying of the ceiling and the laying of the floor, and only a rough comparison is made to the laying of the side walls. The area of the side wall is often larger than that of the ceiling and the floor. When the wall temperature is much higher than the dew point temperature, the partial laying can still meet the load demand, but different laying heights have different effects on the indoor thermal environment. Therefore, this paper analyzes the impact of the installation height and the laying rate of the radiation side panels on the indoor thermal environment in ultra-low-energy consumption buildings through simulation.

2 Methods

This study used the CFD software Airpak. As shown in Fig. 1, a room with a size of 4.6 m × 4.6 m × 2.8 m is built, and the external wall faces south, and the location is Shanghai.

The room has lamps, computers, monitors, and desks. There are four people in the room, and there is an air inlet and an air outlet. By default, all adjacent rooms do not transfer heat. The indoor air flow simulation in this study uses the indoor zero equation [6], and the radiation heat transfer simulation uses a discrete ordinate (DO) model. In this simulation, the ceiling, the floor, and the inner wall are all insulated walls, and the parameters of the external wall and the window, as shown in Table 1, are summarized through the investigation of ultra-low-energy consumption buildings in the hot summer and cold winter zone. In the simulation, all

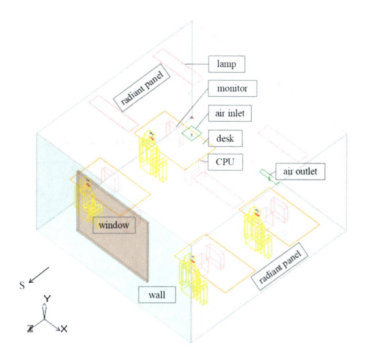

Fig. 1 Physical model of Case 1

the latent heat loads are assumed by the fresh air, the supply air temperature is 26.5 °C, and the wind speed is 0.4 m/s [7]. It is calculated [7, 8] that the indoor heat source in winter conditions can meet the heat load demand due to the enclosure structure, so there is no need to consider the use of radiant heating; in summer conditions, the use of radiant cooling should be considered. The summer cold load of the model is composed as shown in Table 2. After the model completed, increasing the grid density until the calculated results with the increase in the number of grid has no obvious change.

In order to study the effects of radiation area ratios and specific laying methods on indoor thermal comfort, the case design is divided into the following two categories: (1) Influence of radiation area ratios (ratio of the total area of the radiant panel to the area of the wall to be laid) on indoor thermal environment under the cooling condition of two facing vertical walls; (2) Influence of different laying heights on indoor thermal environment under partial laying and cooling conditions of two facing vertical walls (the two facing vertical wall in this paper refers to the east wall and the west wall, the same below).

Table 1 Performance of envelope

Envelope	Heat transfer coefficient
Exterior wall	0.21 W/m^{2}K
Window	1.1 W/m^{2}K

Table 2 Conditions of cooling load

Solar radiation	Envelope	Lighting	CPU	Monitor	Human	Total
159.2 W	35.5 W	192 W	220 W	220 W	1.2 met	1107.5 W

Table 3 Parameters of Case 1–5

	Radiation area ratios %	Radiation surface unit area heat transfer (W)	Mean surface temperature of the radiation surface (°C)	Height of the surfaces (m)
Case 1	100	42.9	21.9	0–2.8
Case 2	80	53.8	20.7	0–2.24
Case 3	70	61.2	19.9	0–1.96
Case 4	60	71.5	18.8	0–1.68
Case 5	50	85.9	17.3	0–1.4

The simulation achieves cold load demand by setting the surface mean temperature of the radiant panel under the radiation area ratios. The heat transfer amount of the radiation surface at different temperatures was calculated according to the following formula [9], thereby determining the mean temperature of the radiation surface, and the results are shown in Table 3.

Radiation surface unit area heat transfer:

$$q = q_f + q_d \tag{1}$$

where q_f is the radiation heat transfer amount per unit area of the radiating surface, and q_d is the convective heat transfer amount per unit area of the radiating surface.

Radiation heat transfer amount per unit area of the radiating surface:

$$q_f = 5 \times 10^{-8} \left[(t_{pj} + 273)^4 - (t_{fj} + 273)^4 \right] \tag{2}$$

Convective heat transfer amount per unit area of the radiating surface under wall cooling:

$$q_d = 1.78 |t_{pj} - t_n|^{0.32} (t_{pj} - t_n) \tag{3}$$

where t_{pj} is the mean surface temperature of the radiating surface, and t_{fj} is the area-weighted mean temperature of the indoor non-heating surface. Here, for convenience calculation, t_{fj} is approximated to the indoor air design temperature, and tn is the indoor air temperature.

Set the first type of case according to the calculation result, as shown in Table 3.

In order to study the influence of the laying height on the indoor thermal environment when the radiant panels are partially laid on the two facing vertical walls, and considering the condensation problem of ultra-low-energy buildings, the radiation area ratio is 70%. The second type of case parameters is shown in Table 4.

Table 4 Parameters of Cases 6–8

	Radiation area ratios %	Height of the surfaces(m)
Case 6	70	0.2–2.16
Case 7		0.4–2.36
Case 8		0.6–2.56

This paper analyzes the indoor thermal environment of the case from two aspects: indoor PMV distribution and indoor temperature distribution. They are evaluated by the following indicators: (1) PMV model. Tian and Love collected experimental data from 82 participants to show that the PMV model is suitable for the actual thermal sensation prediction of the radiant cooling system [10]. The closer the PMV is to 0, the higher the environmental comfort. In this paper, PMV is determined under the thermal resistance of clothing 0.5 clo, relative humidity of 50%, and human body metabolic rate of 1.2 met; (2) Vertical temperature difference. When the temperature difference between the head and the feet of the human body is lower than 3 °C, the human body feels comfortable [11].

3 Results

3.1 Influence of Radiation Area Ratios

By intercepting the temperature distribution of Case 1–5 at $X = 0.75$ m, as shown in Fig. 2, we can see that there are different degrees of temperature stratification. The air temperature below the abdomen of human body is lower than that above the abdomen of human body. This phenomenon is more obvious in Cases 4 and 5.

Figure 3 is drawn by taking PMV with cross sections $Y = 0.1, 0.6, 1.1, 1.7, 2.3$, and 2.7 m for each case. As can be seen from Fig. 3, although Cases 1–5 have different radiation area ratios, it can basically satisfy the $|PMV| < 1$. The distribution of PMV of Case 1 is more evenly distributed. The distribution of PMV in Case 4 and Case 5 is very uneven, and large PMV gradient occurs.

3.2 Influence of Different Laying Heights

Figure 4 shows the temperature distribution of Cases 1, 3, 6–8 at $X = 0.75$ m. It can be seen from Fig. 4 that the temperature distribution of Case 8 is more uniform than that of Case 1. The temperature difference between cross section $Y = 0.1$ m and $Y = 1.1$ m of Case 1, 3, 6–8 is shown in Table 5. It can be seen that the higher the laying height, the smaller the vertical temperature difference. When the laying height is 0.6 m, the vertical temperature difference is even smaller than Case 1.

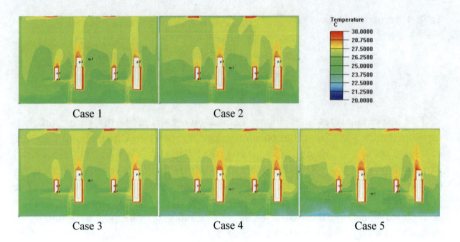

Fig. 2 Temperature distribution of Cases 1–5 at $X = 0.75$ m

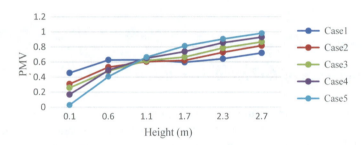

Fig. 3 PMV of different heights of Cases 1–5

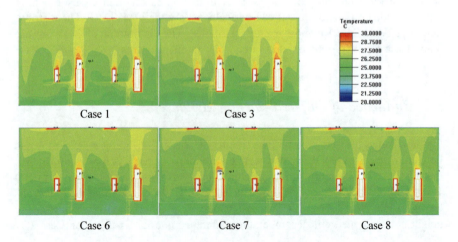

Fig. 4 Temperature of different heights of Cases 1, 3, 6–8 at $X = 0.75$ m

Design Optimization of Radiation Cooling Terminal …

Table 5 Temperature difference between cross section $Y = 0.1$ m, 1.1 m of Cases 1, 3, 6–8

Temperature parameters (°C)		Case 1	Case 3	Case 6	Case 7	Case 8
Height of cross sections	0.1 m	25.6	24.7	24.8	25.1	25.4
	1.1 m	26.7	26.7	26.4	26.4	26.3
Vertical temperature difference		1.1	2	1.6	1.3	0.9

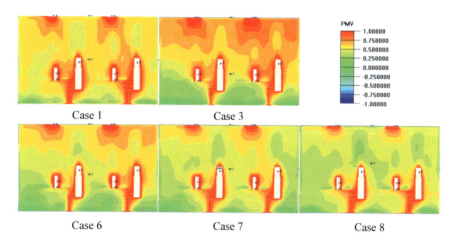

Fig. 5 PMV distribution of Cases 1, 3, 6–8 at $X = 0.75$ m

Comparing the PMV distribution of Case 6–8 at $X = 0.75$ m (Fig. 5), it can be seen that the higher the laying height when the radiation area ratio is 70%, the more uniform the indoor PMV distribution and the numerical value is closer to zero (neutral). It can be shown from Fig. 5 that when the radiation area ratio is 70% and the laying height is higher than 0.2 m, the effect could be better than the radiation area ratio is 100%.

4 Discussion

The analysis of the first kind of case shows that the Cases 1–5 obtained by numerical simulation can meet the load demand. And, it can be found that the distribution of PMV and air temperature is more uniform and the vertical temperature difference is smaller when the proportion of radiation area is larger when the bottom of the radiation panel is laid along the ground under the same air supply condition. Because of the high density of cold air, the air temperature near the ground is lower, and the air temperature near the ceiling is higher. In addition, asymmetric radiation has an effect on the vertical temperature, which leads to lower air temperature in the lower part of the room and higher air temperature in the upper part of the room. The smaller the proportion of radiation area is, the more obvious the effect is.

By analyzing the second kind of case, it can be seen that increasing the laying height of cold radiation panel properly in the case of 70% radiation area ratio can make PMV distribution more uniform and less vertical temperature difference. This is because the radiation temperature of the lower part of the room increases and the air temperature also increases, while the radiation temperature of the upper part of the room decreases and the air temperature also decreases. Therefore, when the proportion of radiation area is 70%, the higher the laying height is, the more uniform the distribution of indoor PMV is, and the value tends to zero. However, the indoor thermal environment is not only related to the laying height of the radiation panel, but also to the height of the radiation panel and the mean temperature of the radiation surface. Therefore, it is not clear whether the height of the radiant panels on both sides of the vertical side wall is the higher the better in all cases.

5 Conclusions

By analyzing and comparing Case 1–8, the following conclusions can be drawn.

(1) When the bottom of radiant panel is laid along the ground, the higher the radiation area ratio, the more uniform the PMV distribution in the room, while the stratification of the indoor temperature distribution is also weakened.
(2) When the radiation area ratio is 70%, increasing the laying height can make PMV distribution and air temperature distribution more uniform, and the cooling quality is no less than that of 100%.
(3) Appropriately increasing the laying height when the radiant panels are partially laid on the two facing vertical walls can reduce the vertical temperature difference and make the PMV distribution in the vertical direction more uniform, and the radiation cooling capacity can be better utilized.

Acknowledgements This research was supported by the National Key R&D Program of China for the 13th Five-Year Plan (No.2017YFC0702600).

References

1. Xu, W., Sun, D.: Research on energy consumption index of China's passive ultra-low-energy buildings. Dynamic (ecological city and green building) (1), 37–41 (2015)
2. Hu, R., Niu, J.L.: A review of the application of radiant cooling and heating systems in mainland china. Energy Build. **52**(none), 11–19 (2012)

3. Yang, Y., Wang, Y., Yuan, X., et al.: Simulation study on the thermal environment in an office with radiant cooling and displacement ventilation system. In: 10th International Symposium on Heating, Ventilation and Air Conditioning, pp. 3146–3153. Jinan (2017)
4. Zhang, M.: Environmental human thermal comfort studies and non-uniform radiant cooling design optimization (2014)
5. Ling, J.: Study on the calculation method of radiant heating and cooling load (2016)
6. Zhao, B., Li, X.T., Yan, Q.S.: Simulation of indoor air flow in ventilated room by zero-equation turbulence model. Tsinghua Sci. Technol. **41**(10), 109–113 (2001)
7. GB 50736-2012.: Technical guide for civil building heating, ventilation and air conditioning design specifications. China building industry press (2012)
8. Yaoqing, Lu: Practical Heating and Air Conditioning Design Manual, 2nd edn. China building industry press, China (1993)
9. JGJ142-2012.: Technical specification for radiant heating and cooling (2012)
10. Tian, Z., Love, J.A.: A field study of occupant thermal comfort and thermal environments with radiant slab cooling. Build. Environ. **43**(10), 1658–1670 (2008)
11. ISO 7730.: Ergonomics of the thermal environment—analytical determination and interpretation of thermal comfort using calculation of the PMV and PPD indices and local thermal comfort criteria (2005)

Study on Operating Performance of Ground-Coupled Heat Pump with Seasonal Soil Cool Storage System

Chao Lyu, Jiachen Zhong, Ping Jiang, Zhiyi Wang, Feng Yu, Yueqin Liu and Maoyu Zheng

Abstract Ground-coupled heat pump (GCHP) is an energy-saving and environmentally friendly air-conditioning technology. In the hot summer and cold winter zone, building's cooling load is much bigger heating load. In order to solve the problem of soil heat unbalance caused by heat injection to the soil exceeding heat extraction from the soil, a ground-coupled heat pump with seasonal soil cool storage (GCHPSSCS) system was put forward. The system uses outdoor fan coil to store air cool energy into the soil in winter or transitional season, which can achieve a shift-seasonal use of natural cold source. The dynamic operating performance of the system is analyzed by computer simulation. The soil temperature increases by 1.02 and 2.72 °C after the GCHPSSCS system operates 1 year and 10 years, respectively. And the annual soil temperature rising gradually decreases. The coefficient of performances (COPs) of heating, cooling, and cool storage in the 10th year of operation are 4.66, 5.71, and 3.78, respectively. Compared with the GCHP system and the GCHP with fan coil-assisted cooling (GCHPFCAC) system, the GCHPSSCS system can reduce the annual rising of soil temperature and improve the cooling COP, which has obvious effect on maintaining soil heat balance and improving operating efficiency of the system.

Keywords Ground-coupled heat pump (GCHP) · Seasonal soil cool storage · Soil heat balance · Soil temperature · Coefficient of performance (COP)

C. Lyu (✉) · J. Zhong · P. Jiang · Z. Wang · F. Yu
School of Civil Engineering and Architecture, Zhejiang Sci-Tech University, Hangzhou, China
e-mail: lvchao-929@163.com

Y. Liu
Hangzhou RUNPAQ Technology Co., Ltd, Hangzhou, China

M. Zheng
School of Architecture, Harbin Institute of Technology, Harbin, China

1 Introduction

Ground-coupled heat pump (GCHP) technology is one of the most promising air-conditioning technologies. In order to make GCHP system operate efficiently in long term, it is very important to keep the annual heat balance of the soil. Research on GCHP and its related composite systems is a hot research topic in the field of building energy conservation and heating, ventilation, and air-conditioning. Domestic and overseas scholars have conducted extensive and in-depth research [1–3]. Although the importance of soil heat balance has gradually been noticed, how to realize the long-term heat balance of the soil is still a problem. Its research depth is not enough, and the practicality also needs to be further strengthened.

In the hot summer and cold winter zone, where the cooling load in summer is greater than the heating load in winter, an auxiliary cold source is necessary for heat elimination or cool replenishment to the soil. Soil cool storage is a combination of cool storage technology and GCHP system, which is an innovation and development of GCHP system. Soil cool storage can be divided into short-term and long-term cool storage. Short-term cool storage, which is a cycle of day and night, stores cool energy into the soil at night when electricity using is small, and releases cool energy during the day when electricity using is large. It can shift peaks and fill valleys on the power system. But cool storage generally requires running heat pump, so the system's energy-saving and economy cannot be improved much. Long-term cool storage is a seasonal cool storage with annual cycle. In the non-cooling season, such as the transition season and heating season, the outdoor air temperature is low. Air cool energy can be stored into the soil directly, which does not need to run the heat pump, and taken out for cooling during the cooling season. At present, there are few research contents on seasonal soil cool storage. Several scholars [4–6] proposed a method of using the cooling tower or fan coil (FC) stores cool energy into the soil in the transition season, and conducted some simulation studies. But the research is still in the initial exploration stage.

From the perspective of soil heat balance, a ground-coupled heat pump with seasonal soil cool storage (GCHPSSCS) system is put forward, which uses outdoor FC to store air cool energy into the soil in winter or transition season. The system is suitable for heating and cooling of buildings in the hot summer and cold winter zone. It can achieve the seasonal utilization of natural cold source and is economical and energy-saving.

2 Methods

2.1 Basic Idea of GCHPSSCS System

In the hot summer and cold winter zone, GCHP system with cooling tower as an auxiliary cold source is often used. However, the cooling tower is large, the control strategy is complex, and the initial investment, operating cost, and energy consumption are high. Compared with the cooling tower, FC saves space and has a simple and flexible structure, low initial investment and operating cost, and significant economic benefits. In addition, when storing cool energy into the soil in winter or transition season, the cooling tower cannot operate at a low outdoor temperature, since it is prone to frosting or even freezing. Therefore, the GCHPSSCS system is proposed, and its schematic diagram is shown in Fig. 1.

The following six operating modes can be realized: GCHP cooling, air source heat pump (ASHP) cooling, GCHP and ASHP combined cooling, GCHP heating, GCHP heating and air cool storage, air cool storage.

The system is flexible and can achieve the seasonal utilization of natural cold source. The cool storage does not need to run the heat pump, so the operating energy consumption is small. In addition, the cooling tower access section (dashed line in Fig. 1) can be reserved in the system. So that when the system is applied to an area with excessive cooling load, the cooling tower can take part of the heat elimination to ensure that the system has sufficient cooling capacity. And the cooling tower can also be used to store cool energy into the soil.

The system in the following research adopts three operating modes: GCHP cooling, GCHP heating, and air cool storage.

2.2 Building Overview

A laboratory of a scientific research office building in Hangzhou in the hot summer and cold winter zone is selected as the research object. The room area is 135 m^2, the

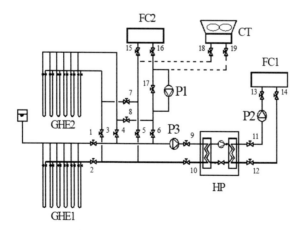

Fig. 1 Schematic diagram of GCHPSSCS system. GHE1, GHE2: ground heat exchangers; HP: heat pump; FC1: indoor fan coil; FC2: outdoor fan coil; CT: cooling tower (reserved); P1–P3: circulating pumps; 1–19: valves

height is 5.85 m, the south wall is the outer wall, the rest are the inner wall, the wall thickness is 240 mm, and the window-wall ratio of the south wall is 0.3. The integrated shading coefficient is 0.46, the heat transfer coefficient of the wall is 0.58 W/(m^2 K), and the heat transfer coefficient of the outer window is 2.4 W/(m^2 K).

2.3 Cooling and Heating Load

The operating time of the system for heating and cooling are as follows: (1) Heat pump cooling: June 1–September 30, 152 d; (2) Heat pump heating: December 1–February 15 (the following year), 77 d; (3) During the period when heat pump operates, it runs from 8:00 to 18:00 every day for 10 h. The annual total cooling and heating load are 18,680.8 kWh and 7338.1 kWh, respectively.

2.4 Cool Storage Quantity and Time

It is calculated that the annual heat injection to the soil is 22,024.3 kWh, and the annual heat extraction from the soil is 5682.2 kWh. Their difference is 16,342.1 kWh, which is the cool storage quantity that should be stored into the soil. Assuming that the cool storage loss of the soil is 20%, the actual cool storage quantity should be 19,610.5 kWh.

The cold storage time should be determined according to the specific meteorological parameters of each place. When the outdoor dry bulb temperature is lower than the soil temperature, it is effective to use FC to cool the soil. If there is a certain temperature difference between the outdoor air and the soil (e.g., 5 °C), the cool storage mode can be turned on. Taking Hangzhou as an example, the data of China Meteorological Data Service Center show that the average annual soil temperature is 17.8 °C. By analyzing the hourly air temperature of typical meteorological year, it can be seen that from the beginning of the year to the end of March, the outdoor dry bulb temperature during the daytime is lower than the soil temperature by 5 °C. And from the beginning of the year to the end of April, the outdoor dry bulb temperature during the night is lower than the soil temperature by 5 °C. So there is a long time for cool storage.

The time of seasonal cool storage is set to be just after the end of heating, that is, from February 16 to April 30, for a total of 74 d. During the period from February 16 to March 31, the outdoor temperature is relatively low, which is conducive to cool storage. Therefore, the system stores cool energy for 24 h throughout the day. From April 1 to April 30, the temperature difference between outdoor air and soil is small. Therefore, the intermittent cool storage method is adopted during this period. Cool storage starts from 6 pm to 6 am (the next day) and runs for 12 h every day. The total running time for cool storage is 1416 h.

2.5 Computer Simulation Model of the System

The computer simulation model of GCHPSSCS system is established by using TRNSYS software, including the model and related control of main equipment such as heat pump, ground heat exchanger, and fan coil. And the dynamic operation performance of the system will be simulated and analyzed. The TRNSYS model of GCHPSSCS system is shown in Fig. 2.

3 Results

The operation results of the GCHPSSCS system were simulated.

The initial soil temperature was set at 17.80 °C, and the system was simulated for operating 1 year. The soil temperature variation is shown in Fig. 3, and the soil temperatures at each period are shown in Table 1. It can be seen that the soil temperature decreased by 2.66 °C during the heating period, decreased by 1.01 °C during the cool storage period, increased by 4.82 °C during the cooling period, and increased by 1.02 °C during the whole year.

The system was further simulated for operating 10 years. The soil temperature increased by 2.72 °C after 10 years, and the annual soil temperature rising

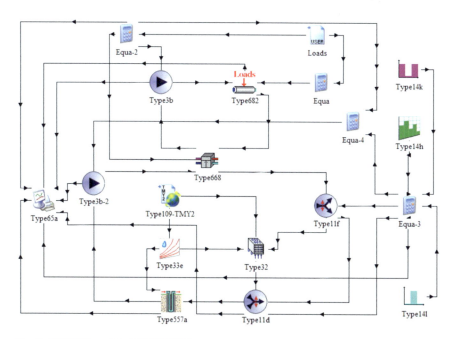

Fig. 2 TRNSYS model of GCHPSSCS system

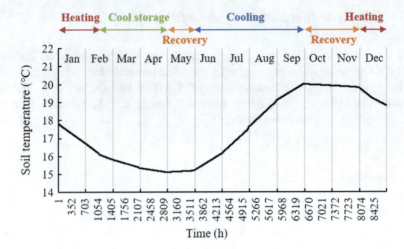

Fig. 3 Soil temperature variation during the system operates in the first year

gradually decreased. The coefficient of performance (COP) of the system did not change much. The COPs of heating, cooling, and cool storage in the 10th year were 4.66, 5.71 and 3.78, respectively. The efficiency is relatively stable.

4 Discussion

The GCHPSSCS system was compared and analyzed with the GCHP system and the GCHP with fan coil-assisted cooling (GCHPFCAC) system.

The systems were simulated for operating 10 years, and the soil temperature variation of the three systems is shown in Fig. 4. The soil temperature of GCHP, GCHPFCAC, and GCHPSSCS system increased by 8.75 °C, 4.28 °C, and 2.72 °C, respectively. And the annual soil temperature rising of all the three systems gradually decreased. The soil temperature of the GCHP system increased most, and that of the other two systems increased less (especially the GCHPSSCS system). It can be seen that the use of FC is effective in reducing the increase of soil temperature.

The COP comparison when the three systems operate in the 10th year is shown in Table 2. The GCHPSSCS system has the lowest heating COP but the highest cooling COP. This is because the cool storage will reduce the soil temperature, which is disadvantageous for heating but advantageous for cooling. For a cooling load-dominated area like Hangzhou, the increase of cooling COP is more effective in improving the overall operating efficiency of the system, which will reduce the total energy consumption.

The analysis shows that using FC to store cool energy into the soil can reduce the annual rising of soil temperature, which has obvious effect on maintaining soil heat balance and improving system operating efficiency.

Table 1 Soil temperatures at each period during the system operates in the first year (°C)

Periods	Dates	Starting temperature	End temperature	Temperature variation
Heating	1.1–2.15	17.80	16.12	−1.68
Cool storage	2.16–4.30	16.12	15.11	−1.01
Recovery	5.1–5.31	15.11	15.20	+0.09
Cooling	6.1–9.30	15.20	20.02	+4.82
Recovery	10.1–11.30	20.02	19.80	−0.22
Heating	12.1–12.31	19.80	18.82	−0.98

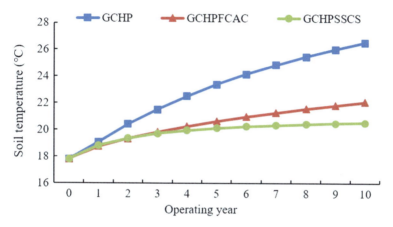

Fig. 4 Soil temperature variation of three systems in 10 years

Table 2 COP comparison when the three systems operate in the tenth year

Systems	Heating COP	Cooling COP
GCHP	4.91	5.47
GCHPFCAC	4.83	5.56
GCHPSSCS	4.66	5.71

5 Conclusions

The operating performance of the GCHPSSCS system has been studied. Some conclusions obtained are as follows:

(1) The GCHPSSCS system is suitable for heating and cooling of buildings in the hot summer and cold winter zone. It can achieve a shift-seasonal use of natural cold source and is economical and energy-saving.
(2) The soil temperature increases by 1.02 and 2.72 °C after the GCHPSSCS system operates 1 year and 10 years, respectively. And the annual soil temperature rising gradually decreases.

(3) The COP of heating, cooling, and cool storage of the GCHPSSCS system in the 10th year are 4.66, 5.71, and 3.78, respectively. The efficiency is relatively high.
(4) Compared with the GCHP system and the GCHPFCAC system, the GCHPSSCS system can reduce the annual rising of soil temperature and improve the cooling COP, which has obvious effect on maintaining soil heat balance and improving operating efficiency of the system.

Acknowledgements The support provided by General Project of Zhejiang Natural Science Foundation (Number LY18E060009 and LY18E060008), ZSTU Innovative Research Team for Green Retrofit Technologies in Urban Underground Space, the Fundamental Research Funds of Zhejiang Sci-Tech University (Number 2019Q057), and Low Carbon Building's Energy and Environment Engineering and Technology Research Center of Zhejiang Province (Number 2011E10033) is greatly appreciated and acknowledged.

References

1. Chiasson, A.D., Spitler, J.D., Rees, S.J., et al.: A model for simulating the performance of a pavement heating system as a supplemental heat rejecter with closed-loop ground-source heat pump systems. ASME J. Solar Energy Eng. **122**, 183–191 (2000)
2. Rad, F.M., Fung, A.S., Leong, W.H.: Feasibility of combined solar thermal and ground source heat pump systems in cold climate, Canada. Energy Build. **61**, 224–232 (2013)
3. Alaica, A.A., Dworkin, S.B.: Characterizing the effect of an off-peak ground pre-cool control strategy on hybrid ground source heat pump systems. Energy Build. **137**(2), 46–59 (2017)
4. Xie, L., Liu, J., Liu, K.: Characteristic study on cool storage in transition period of cooling tower-ground coupled heat pump in hot-summer and cold-winter zone. Build. Sci. **29**(8), 83–89 (2013)
5. Li, K., Chen, C., Shang, P., et al.: A novel hybrid ground source heat pump system for cooling load dominated area. Build. Sci. **31**(8), 58–64 (2015)
6. Mu, K.: Air thermal to compensate the imbalance of heat absorption in winter and heat releasing in summer of ground source heat pump system. Master Thesis of Harbin Institute of Technology, Harbin, China (2010)

Controlling Technique and Policy of Adjacent Rooms Pressure Difference in High-Level Biosafety Laboratory

Peng Gao, Guoqing Cao, Ziguang Chen and Yuming Lu

Abstract In this paper, the main technical indicators and pressure disturbance factors of the high-level biosafety laboratory are introduced firstly. Then the pressure-difference controlling mode of the high-level biosafety laboratory built in China is focused, and the advantages and disadvantages of each mode combined with the field monitoring results are illustrated, respectively. Among them, the innovativeness of the constant air volume controlling mode in laboratory A breaks through the restraint that it cannot be used in the laboratory with high airtightness. Besides, the pressure-difference controlling mode of double regulating valves in laboratory B provides a reference for the laboratory construction with high tightness of laboratory envelope structure and large air supply and exhaust volume.

Keywords High-level biosafety laboratory · Pressure-difference controlling · Constant air volume controlling · Double adjusting valves · Variable air volume controlling

1 Introduction

Biosafety laboratories are microbiology laboratories and animal laboratories that meet biosafety requirements through protective barriers and management measures. The level of biosafety laboratory is divided into four degrees according to the biological hazards and protective measures, namely BSL (Biosafety level) 1–4 and ABSL (Animal biosafety level) 1–4. In this paper, the advanced biosafety laboratory refers to the class b2 laboratory in ABSL-3 and the BSL-4, as well as the ABSL-4, in which all the pathogenic microorganisms are highly infectious.

P. Gao · G. Cao (✉) · Z. Chen
Institute of Building Environment and Energy, China Academy of Building Research, Beijing, China
e-mail: cgq2000@126.com

Y. Lu
University of Science and Technology Beijing, Beijing 100206, China

Adjacent rooms pressure-difference controlling can prevent the hurt by pathogenic microorganisms to operating and experimental environment effectively [1, 2]. However, there are still a few precedents for biosafety accidents caused by biological pollutants produced in laboratories before. Therefore, the key to ensure biosafety in laboratories is to prevent the harmful effects of pathogenic microorganisms on operators and the experimental environment through pressure differential controlling.

2 Main Technical Indicators of the High-Level Biosafety Laboratory

2.1 Absolute and Relative Pressure Difference

According to the provisions of the Chinese standard architectural and technical code for biosafety laboratories (GB 50346-2011), and laboratories: General Requirement of Biosafety (GB 19489-2008) [3, 4], the minimum negative pressure difference of the b2 laboratory in ABSL-3, BSL-4, and ABSL-4 to the atmospheric environment is −80 Pa, −60 Pa and −100 Pa, respectively, and the minimum negative pressure difference of the b2 laboratory in ABSL-3, BSL-4, and ABSL-4 to the adjacent room in the outdoor direction is the same as −25 Pa.

2.2 The Requirements of Absolute and Relative Pressure Difference for Working Condition Verification

In the section 10.1.12 of the standard architectural and technical code for biosafety laboratories (GB 50346-2011), it is stipulated that verification of the working condition should be carried out in the biosafety laboratory. When there are multiple working conditions, engineering inspection for each working condition should be carried out and the safety of the system during the condition conversion should be verified. In addition, items such as system start–stop, standby unit switching, backup power switching, and electrical, self-control, and fault alarms should all be verified for reliability. During the period of working condition, verification is carried out; the absolute pressure in the core workroom is never allowed to reverse; the relative pressure of the core working room to the adjacent buffer room is allowed to reverse of lower value in a short time (≤ 1 min) due to the limitation by the conditions of design, equipment, construction, and commissioning, but it should be noted that this situation should also be avoided as much as possible.

3 Pressure Disturbance Factors of the High-Level Biosafety Laboratory

3.1 Airtightness of the Envelope Structure

The requirement of airtightness of the envelope structure in high-level biosafety laboratory is relatively high. Relative study [4] pointed out that under the action of the same amount of air disturbance, the pressure difference of the airtight laboratory can reach 3.6 times that of the ordinary laboratory, which is the better the airtightness, the faster the pressure difference changes. The rate is increased, that is, the better the airtightness, the greater the fluctuation of the differential pressure. Due to the airtightness of the enclosure structure in the high-level biosafety laboratory is extremely high; the control of pressure difference is much more difficult than that of ordinary laboratories.

3.2 Key Protective Equipment

Key protective equipment in high-level biosafety laboratory can cause pressure disturbance, mainly because the equipment with local exhaust adds a certain volume of fluid to the environment at the start/stop time. Such key protective equipment includes: II-B2 biological safety cabinet, animal isolation equipment, independent ventilation cage (IVC), positive pressure protective clothing, chemical shower disinfection device, and so on.

3.3 Working Condition Conversion

The high-level biosafety laboratory can not only operate under normal conditions, but also maintain a certain period of normal operation or maintain a certain negative pressure isolation environment in an emergency. Usually, there will be spare blowers and spare exhaust fans in the HVAC system, this means that when the spare blower or spare exhaust fan is switched, the absolute negative pressure is maintained in the core laboratory, and the relative pressure difference of the core laboratory to adjacent room in the outdoor direction is as far as possible without reversal.

3.4 Other Factors

Other factors such as door switching, temperature fluctuation, atmospheric pressure fluctuation, pipe working condition, and damper selection can also affect system pressure. It is also a factor that must be considered when selecting the pressure-difference controlling method and system configuration form.

4 Comparison and Analysis of Different Controlling Technique and Policy of Adjacent Rooms Pressure Difference in High-Level Biosafety Laboratory

4.1 Common Pressure-Difference Controlling Mode

The pressure-difference controlling mode of the high-level biosafety laboratory built in China is shown in Table 1.

Table 1 The comparison of controlling mode of pressure difference in different high-level biosafety laboratory built in China [3, 5]

Item		Laboratories in this study	
		Laboratory A (in Wuhan)	Laboratory B (in Harbin)
Controlling of pressure difference		Constant air volume	Double regulating valves
Envelope structure		Stainless steel	Reinforced concrete pouring
Key protective equipment		II-A2 biological safety cabinet, animal isolation equipment, independent ventilation cage (IVC), positive pressure protective clothing, chemical shower disinfection device	II-B2 biological safety cabinet, positive pressure protective clothing, chemical shower disinfection device
Laboratory layout		Chemical shower disinfection device directly adjacent to the core laboratory	Chemical shower disinfection device directly adjacent to the core laboratory
System form	Exhaust	Three units of exhaust fans are installed in each system, and the exhaust fans numbered a and b are large fans of the same specifications, which are operated at the same time and are mutually standby	Each purification air-conditioning system is equipped with one exhaust fan unit, and two exhaust fans are installed inside, which are alternate and alternately operated
	Supply	Two sets of air supply units are installed in each system, which are operated simultaneously and are used alternately	Each purification air-conditioning systems is equipped with an air supply unit and is internally equipped with two fans that operate alternately and alternately

(continued)

Table 1 (continued)

Item	Laboratories in this study	
	Laboratory A (in Wuhan)	Laboratory B (in Harbin)
Controlling policy	The relatively high negative pressure difference of the room is maintained by the permanent small exhaust fan, and the relatively high negative pressure value is used to weaken the influence of the pressure disturbance	Double regulating valves: The air volume is adjusted through the large flow regulating valve and the pressure difference is adjusted through the small flow regulating valve

4.2 Analysis of Controlling Technique and Policy of Adjacent Rooms Pressure Difference

4.2.1 Laboratory A

The constant air volume (CAV) method is an indirect type of pressure-difference controlling. It does not adjust the indoor pressure difference during operation, while it maintains a certain air volume difference by supplying and exhausting air (equivalent to the laboratory air leakage under the set pressure difference) to control the pressure difference. At present, the mechanical constant air volume valve or the automatic fixed air volume valve is basically used.

However, the constant air volume controlling method of laboratory A is also different from the ordinary constant air volume controlling method. Laboratory A is a stainless steel enclosure structure with small air leakage and high airtightness. It is difficult to precisely control a small air volume difference only by a mechanical constant air volume valve or an automatic constant air volume valve, so that the pressure-difference controlling cannot be achieved. The indicators such as pressure difference and air volume are not focused on in laboratory A, while through initial commissioning to determine the air supplying/exhausting volume of the corresponding room, the frequency of the supplying/exhausting fans, and increase the permanent small exhaust fan to achieve a relatively high negative pressure. In actual operation, the relatively high negative pressure value is completely relied on to weaken the influence of the pressure disturbance factor. At the same time, the pressure difference in the core working room is monitored. Only when the system is running for a period of time and the pressure in the core working room deviates too much from the set value, the pressure difference of the core working room is adjusted by adjusting the fan's frequency.

The program draws on the French design philosophy and the system is very simple. The advantage is that it does not depend on the pressure difference in the room, the control of the air volume will not be greatly affected under the action of pressure disturbance factors, and the pressure difference will not be greatly fluctuated. However, there are some shortcomings for this controlling method. For example, the adaptability of this controlling method is poor when equipped with key protective equipment. It should be avoided to equip the II-B2 type biological safety cabinet to avoid fluctuations in pressure difference caused by local exhaust equipment.

According to the actual test results, this controlling mode can achieve a relatively stable state and can also show better stability when dealing with strong pressure disturbance factors. Taking the alternative exhaust fan switching which is the most influential pressure-difference disturbance factor as the example, to monitor the pressure difference in one core room. During the switching process, the permanent-moving small exhaust fan always keeps running. The influence of pressure disturbance factors is completely weakened by relatively high negative pressure values. There is no reversal of the absolute pressure difference in the room. And the relative pressure difference in core working room to the adjacent buffer room is reversed for a short time (≤ 1 min). The details are shown in Fig. 1.

4.2.2 Laboratory B

The pressure-difference controlling mode of double regulating valves means that two air volume regulating valves are installed in parallel on the air supply or exhaust duct (Figs. 2 and 3), and the pressure difference between the two is adjusted by the two air volume adjusting valves. The two valves are generally divided into a large flow regulating valve and a small flow regulating valve. Usually, the large one is a constant air volume (CAV) regulating valve, which is mainly used for adjusting the air volume and the small one is a variable air volume (VAV) regulating valve and mainly used to adjust the pressure difference.

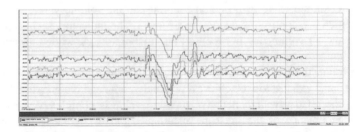

Fig. 1 The reliability verification result of system backup exhaust fan switching process (the light blue line indicates the absolute pressure zero point, the purple line indicates the absolute pressure of the positive pressure room, and the remaining curves indicate the absolute pressure of the controlled room including the core room and the adjacent buffer)

Fig. 2 The controlling principle of double regulating valves

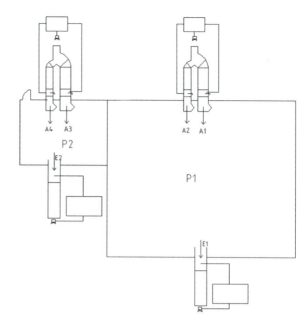

Fig. 3 The photo of double regulating valves

Laboratory B is a reinforced concrete pouring structure. Due to the high tightness of the high-level biosafety laboratory enclosure structure and a large amount of air supplying/exhausting, it might be difficult to control accuracy and sensitivity with a single large-caliber airflow regulator.

The system mode of laboratory B is "adjusting the amount of supplying air and fixed the amount of exhausting air." The double regulating valve is installed in the air duct of core room. The airflow rate is adjusted by a large flow regulating valve to meet the requirements of the number of laboratory air exchanges, and the pressure is coarsely adjusted. The small flow regulating valve is used to finely adjust the air supply to realize the adjustment of the pressure difference, thereby realizing the laboratory differential pressure control efficiently and quickly, and improving the laboratory differential pressure control precision.

This design concept takes into account the high airtightness of the high-level biosafety laboratory, the small amount of air leakage, and a large amount of air supplying/exhausting. However, the selection of the small flow regulating valve is also limited. The pressure difference of the laboratory should be within the range of the small flow regulating valve, and the small flow regulating valve should be kept as far as possible in the linearly better area, and the initial investment is relative to other forms be high.

According to the actual test results, this control mode can achieve fairly stable and sensitive control effects, and can also show better stability when dealing with strong pressure disturbance factors. Taking the alternative exhaust fan switching which is the most influential pressure-difference disturbance factor as the example, to monitor the pressure difference in one core room. During the switching process, the permanent-moving small exhaust fan always keeps running. The influence of pressure disturbance factors is completely weakened by relatively high negative pressure values. There is no reversal of the absolute pressure difference in the room. And the relative pressure difference in core working room to the adjacent buffer room is reversed for a short time (≤ 1 min). The details are shown in Fig. 4.

4.2.3 Summary

The above two pressure-difference controlling methods and system configuration forms have their own advantages and disadvantages, which can be used as a reference for laboratory construction in the future. In general, the controlling method

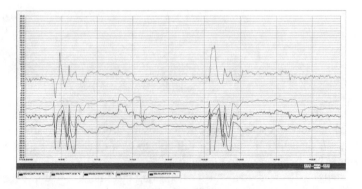

Fig. 4 The reliability verification result of system backup exhaust fan switching process (the light blue line indicates the absolute pressure zero point, the pink line indicates the absolute pressure of the positive pressure room, and the remaining curves indicate the absolute pressure of the controlled room including the core room and the adjacent buffer)

of laboratory A is the simplest; the controlling methods of laboratory B and laboratory C are more complicated, but the controlling precision will be higher. For the construction of high-level biosafety laboratories, pressure control methods, system forms, and so on, it is necessary to consider "controlling effectiveness," "operational safety" and "construction cost" and "operating costs." The appropriate plan can be determined after a comprehensive trade-off [6].

In addition to the above pressure-difference controlling methods, Chinese researchers have also proposed other methods, such as artificial additional air leakage controlling method and variable air volume pressure-difference controlling method (fuzzy PID method). The artificial air leakage controlling policy is similar to the laboratory A controlling mode, that is, artificially increasing the laboratory air leakage. The variable air volume pressure-difference controlling method (fuzzy PID method) is an upgrade of the variable air volume pressure-difference controlling method (conventional PID method). The PID parameters can be adjusted in real time according to different states of the pressure difference, and good stability and fastness can be ensured at the same time. These two methods have yet to be further practised in the construction of high-level biosafety laboratories.

5 Conclusions

On the basis of a long-term field measurement of indoor and outdoor PM2.5 mass concentrations, particle number concentrations and outdoor meteorological parameters from five unoccupied offices located in Beijing, China, a method calculating key design parameters of indoor PM2.5 filtration load, such as the penetration factor (P), deposition rate (k), and air infiltration volume through a unit length of external window (ql), has been proposed. Based on the field measured data, corresponding values of ql, P and k have been proposed for various external window airtightness levels. Furthermore, the corresponding indoor PM2.5 filtration loads were quantified and the results revealed that higher airtightness level windows can help to significantly decrease the indoor PM2.5 filtration load.

Acknowledgements This work was sponsored by the 13th Five-Year Key Project, Ministry of Science and Technology of China (No. 2017YFC0702800).

References

1. Cao, G., Wang, R., Li, Y., Gao, P.: Causes and control strategies of pressure fluctuation in high-level biosafety laboratory. HV&AC **48**(1), 7–12 (2018). (in Chinese with English abstract)
2. Zhang, Z., Zhao, M., Yi, Y., Qi, J.: Pressure difference control for door opening/closing in airtight biosafety laboratories. Mil. Med. Sci. **37**(1), 19–23 (2013). (in Chinese with English abstract)

3. Ministry of Housing and Urban-Rural Development of the People's Republic of China: Biosafety Laboratory Building Technical Specification, GB 50346-2015
4. State administration for market regulation of People's Republic of China: Laboratories: General Requirement of Biosafety GB19489-2008
5. Ministry of Housing and Urban-Rural Development of the People's Republic of China: Technology code for evaluating biosafety performance of laboratory equipment. RB/T 199-2015
6. U.S. Department of Health and Human Services: Biosafety in Microbiological and Biomedical Laboratories (BMBL), 5th edn. Maryland, U.S. (2009)

Investigated on Energy-Saving Measures of HVAC System in High-Level Biosafety Laboratory

Peng Tan, Guoqing Cao and Ziguang Chen

Abstract This paper analyzes the characteristics of typical main rooms of biosafety laboratories and the potential energy-saving measures. We calculated the air-conditioning design load for different temperature and humidity ranges in a real high-level biosafety laboratory. The analysis results showed the air exchange rates were fallen into 15–30 times/h, and these ranges can be seen as the best design values without considering other biosafety equipment. In the case, the envelope structure is tightly rigorous, and the number of air change rates can be further reduced. In the next phase of the laboratory specification revision, it can be considered to reduce the required minimum air supply volume, which can bring objective energy-saving benefits.

Keywords High-level biosafety laboratory · Air-conditioning load · Reasonable air changes per hour (ACH) · Temperature and humidity · Energy analysis

1 Introduction

Biosafety laboratories are microbiology laboratories and animal laboratories that meet biosafety requirements through protective barriers and management measures. Architectural and technical code for biosafety laboratories [1] stipulates that all fresh air systems should be used in the III and IV level biosafety laboratories. The operating energy consumption of the all fresh air system is much greater than the system which can use return air.

P. Tan · G. Cao (✉) · Z. Chen
Institute of Building Environment and Energy Efficiency, Academy of Building Research China, 100013 Beijing, China
e-mail: cgq2000@126.com

This paper analyzes the energy-saving measures of high-level biosafety laboratories by analyzing the air-conditioning load characteristics, indoor heat and humidity parameters, air-conditioning ventilation forms of typical biosafety tertiary laboratories.

2 Typical Main Room Load Calculation

This paragraph takes a typical main room as an example. The laboratory belongs to the b1 category of BSL-3 specified in the specification [1], which is, "the effective use of safety isolation devices to operate conventional airborne pathogenic biological factors." The laboratory is located in Shanghai. The room design temperature is 22 °C (summer) or 20 °C (winter), and relative humidity is 40%.

2.1 Indoor Load of Typical Main Room

In the main room, there is an IIA2 biosafety cabinet, two CO_2 incubators, a low-temperature refrigerator with a working temperature of −80 °C, a desktop computer, and an electron microscope. There are 2 experimenters in the room. The labor intensity is mild, and the clustering factor is 1.0. The room area is 25 m^2, the floor height is 2.8 m, the walls on both sides are inner walls, the temperature difference is 2 °C, the roof is the inner floor, the temperature difference is 3 °C, and the heat transfer coefficient of the inner wall and inner floor is 1.5 W/m^2 °C. The indoor lighting heat dissipation index is estimated at 8 W/m^2.

The main equipment power estimation in the room is shown in Table 1. The heat dissipation amount is estimated at 0.4 times of the rated power, and the simultaneous use factor of 0.7 is taken. The indoor cooling load is calculated by the cold load coefficient method.

The maximum internal heat and cold load in the summer is 2.11 kW, and the wet load is 0.33 kg/h. The indoor heat and humidity ratio of the main room is $\varepsilon = 23{,}018$ kJ/kg. Such a large heat and humidity ratio is almost vertical on the

Table 1 Estimation of heat dissipation of main equipment in a typical main room

No.	Equipment	Rated power (W/set)	Amount
1	IIA2 biosafety cabinet	700	1
2	CO_2 incubator	500	2
3	low temperature refrigerator	700	1
4	desktop computer	500	1
5	electron microscope	200	1
6	Total	3100	6

enthalpy humidity chart. Here we can draw a conclusion that for the typical main room load, waste heat exclusion is the priority.

By convention, the indoor heat load calculation in winter does not consider indoor residual heat. The heat load that may exist is the heat load on the inner wall and the inner floor. Because of the small amount, the heat load can be offset with the indoor residual heat, so the calculation is further simplified.

2.2 Typical Main Room Air-Conditioning Process

In summer condition, in order to remove the residual moisture contained in the fresh air, the fresh air should be handled to a lower temperature td, and the temperature of td is too low to be sent to the room directly; it must be reheated to ensure that the temperature difference of the supply air is not too large (generally no more than 5 °C).

The summer air-conditioning process is cool and dehumidifies first, and then heat by equal humidity to be sent into the room, and the temperature of supply air rises to the indoor design point along the heat moisture ratio line and then be discharged outside.

The winter treatment process is heating and then humidifying (the humidification process has two methods: equal humidification and isothermal humidification). According to the above, the indoor temperature is approximately equal to the supply air temperature.

In order to simplify the calculation, the temperature rise of the duct and the fan is ignored in the calculation below.

2.3 Typical Main Room Air-Conditioning Heat Load

According to the air treatment process, the summer air-conditioning load includes cooling load and reheating load, and the winter air-conditioning load includes heating load and steam humidifying heat load. Therefore, we assume that the ventilation rate of the typical main room is 25 times/h, so the air supply volume is calculated as 1750 m^3/h (Fig. 1).

Air-conditioning system load of summer
It can be seen from the above, the summer indoor heat and humidity load (2.11 kW, 0.33 kg/h), the indoor heat and humidity state point (22 °C, 40%), and the outdoor heat and humidity state point (dry-bulb 34.8 °C, wet ball 28.1 °C [2]) and air supply volume (1750 m^3/h) of the typical core room, the air density is taken as 1.2 kg/m^3, thereby calculating the air treatment process, drawing on the enthalpy humidity chart shown in Fig. 2 [3] (Tables 2 and 3).

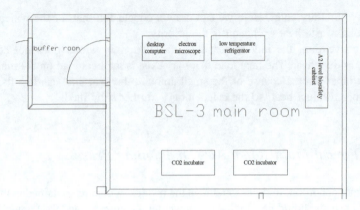

Fig. 1 Typical core workspace interior layout

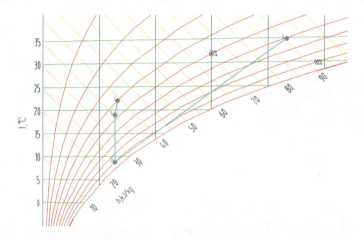

Fig. 2 Air handling line on the enthalpy humidity chart of summer

Table 2 Summer air-conditioning treatment status point (indoor 22 °C, 40%)

Air status	Dry-bulb temperature (°C)	Wet bulb temperature (°C)	Relative humidity (%)	Moisture content (g/kg)	Specific enthalpy (kJ/kg)
Indoor air	22.0	13.8	40.0	6.6	39.0
Outdoor air	34.8	28.1	60.6	21.6	90.5
Supply air	19.0	12.5	46.6	6.4	35.4
Machine dew point	8.9	8.0	90.0	6.4	25.1

Table 3 Winter air-conditioning treatment status point (indoor 20 °C, 40%)

Air status	Dry-bulb temperature (°C)	Wet bulb temperature (°C)	Relative humidity (%)	Moisture content (g/kg)	Specific enthalpy (kJ/kg)
Indoor air	20.0	12.4	40.0	5.7	34.7
Outdoor air	−2.2	−3.3	75.0	2.2	3.3
Preheating point	20.0	8.5	15.5	2.2	25.8

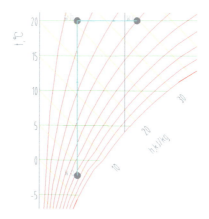

Fig. 3 Air handling line on the enthalpy humidity chart of winter

When the air supply temperature is set to 19.0 °C, the air supply temperature difference is 3 °C, which meets the comfort requirements. Since the selected number of ACH is relatively large, the demand for exhausting heat in the room can also be satisfied. We can calculate the energy required for the cooling process to cool (cooling load) $\dot{Q}_s$ is 38.2 kW, and the energy required for the reheating process (reheat load) $\dot{Q}_h$ is 6.0 kW.

Air-conditioning system load of winter

The indoor heat and humidity state point (20 °C, 40%), and an outdoor heat and humidity state point (2.2 °C, 75% [2]) and air supply volume (1750 m³/h) of the typical core room, the air density is taken as 1.2 kg/m3, and the air supply is steam humidified, drawing on the enthalpy humidity chart shown in Fig. 3 [3].

We can calculate the energy required for the heating process $Q_{w.h}$ is 13.1 kW, and the energy required for the humidification $\dot{Q}_{w.w}$ is 5.2 kW.

3 Analysis of Energy-Saving Methods

3.1 Indoor Heat and Humidity Parameters Selection

It can be seen from the above that the air-conditioning system load is affected by the indoor heat and humidity parameters, especially in summer, and the reheating process doubles this effect. The biosafety laboratory has a wide temperature and humidity range (18–25 °C, 30–70%) [1]; in order to quantitatively analyze this effect, we analyze the heat load changes of the air-conditioning system at different indoor state points.

By analyzing the data in Table 4, the following conclusions can be drawn:

a. The higher the indoor temperature setting in summer, the lower the system cooling load;
b. The higher the indoor relative humidity setting value in summer, the lower the system cooling load;
c. In the summer air-conditioning system cooling load, the reheating load accounts for about 10%;
d. Within the scope of the specification, summer indoor relative humidity has a greater impact on the system's cooling load.

3.1.1 Relationship between indoor air parameters and design heat and humidity load in winter

By analyzing the data in Table 5, the following conclusions can be drawn:

a. The higher the indoor temperature setting in winter, the higher the system heat load;
b. The higher the indoor relative humidity setting value in winter, the higher the system heat load;
c. In the heat load of the air-conditioning system in winter, the humidifying heat load accounts for about 30%;

Table 4 Relationship between indoor air condition point and air-conditioning system design cooling load in summer (atmospheric pressure: 100540 Pa)

No.	Dry-bulb temperature (°C)	Relative humidity (%)	Cooling load (kW)	Reheat load (kW)	System cooling load (kW)
1	20.0	40.0	40.5	6.0	46.5
2	20.0	60.0	32.3	2.2	34.5
3	22.0	40.0	38.2	6.0	44.1
4	22.0	60.0	29.3	2.2	31.5
5	24.0	40.0	35.7	6.1	41.8
6	24.0	60.0	26.3	2.3	28.6

Table 5 Relationship between indoor air condition point and air-conditioning system design heating load in winter (atmospheric pressure: 102,540 Pa)

No.	Dry-bulb temperature (°C)	Relative humidity (%)	Preheating heat load (kW)	Winter humidification heat load (kW)	System heating load (kW)
1	20.0	40.0	13.1	5.2	18.3
2	20.0	60.0	13.1	9.5	22.6
3	22.0	40.0	14.3	6.4	20.7
4	22.0	60.0	14.3	11.3	25.6
5	24.0	40.0	15.5	7.6	23.1
6	24.0	60.0	15.5	13.2	28.6

d. Within the scope of the specification, the indoor relative humidity in winter has a greater impact on the system's cooling load.

3.2 Relationship between indoor heat and humidity parameters and annual hot and cold load

The above calculations show that the choice of indoor air status points affects the design load of the air-conditioning system. Next we estimate the annual energy consumption value by the meteorological parameters of the typical meteorological year.

3.2.1 Relationship between indoor air condition point and annual cooling load of air-conditioning system

Through the outdoor hourly weather parameters of the typical meteorological year, the ratio of the dew point of the treated dew point in Fig. 2 is used as the boundary line for whether the air conditioner needs a cold source, and the system cooling load is added hourly to get the annual cooling load.

3.2.2 Relationship between indoor air condition point and annual heat load of air-conditioning system

By analyzing the outdoor hourly meteorological parameters of the typical meteorological year, the ratio of the indoor air state point in Fig. 3 is taken as the boundary line of the air-conditioning heating condition, and the system heat load is added hourly to get the annual cooling load.

Table 6 Relationship between indoor air status point and annual cooling load of air-conditioning system (atmospheric pressure: 100,540 Pa)

No.	Dry-bulb temperature (°C)	Relative humidity (%)	Indoor air specific enthalpy (kJ/kg)	Annual cooling period (h)	System annual cooling load (GJ)
1	20.0	40.0	35.0	6833	577.9
2	20.0	60.0	42.5	4868	300.0
3	22.0	40.0	39.0	6165	508.8
4	22.0	60.0	47.6	4399	246.7
5	24.0	40.0	43.3	5512	445.5
6	24.0	60.0	52.9	3935	199.1

3.3 Relationship between indoor heat and humidity parameters and air condition load

According to the above calculation, the total annual load of the system is greatly affected by the indoor air state point, and effective energy-saving can be achieved by reasonably setting the indoor state point (Table 6).

3.4 Determining a Reasonable ACH of the Main Room

The influence of indoor parameters on fresh air cooling load and reheating cooling load is analyzed. The code requires that the ACH of the main room is not less than 12 (100,000 class) or 15 (10,000 class). There are many factors influencing the amount of air supply. Compared the P3 laboratories in China and abroad, the design air volume varies greatly. This kind of biosafety project has two differences from the general air-conditioning project. The first is absolute negative pressure. Maintaining absolute negative pressure requires the tightness of the envelope structure. At the same time, it needs more clean air to dilute the air entering from various gaps. To maintain the cleanliness of the room, on the other hand, the biosafety laboratory's self-control of pressure is relatively complicated, and the air supply volume also changes with the adjustment of the pressure gradient (Table 7).

We can take 25 times/h as the representative data from a biosafety level 3 laboratory of China in operation. Indoor particle concentration has a significant downward trend with the increase of the number of ACH; when the number of AHC is less than 15 times/h, the average value of the dust concentration is higher than 352 pc/L, then the cleanliness is not satisfied the requirement of 10,000 class, but still meet the requirements of cleanliness of 100,000 class.

The absolute negative pressure in the main room is very low (generally lower than −60 Pa). The cleanliness of the room is also affected by the tightness of the envelope structure. In view of the fact that biosafety laboratories are not very

Table 7 Relationship between indoor air status point and annual heat load of air-conditioning system (atmospheric pressure: 102,540 Pa)

No.	Dry-bulb temperature (°C)	Relative humidity (%)	Indoor air specific enthalpy (kJ/kg)	Annual heating period (h)	System annual heat load (GJ)
1	20.0	40.0	34.7	3851	114.4
2	20.0	60.0	42.1	4543	179.7
3	22.0	40.0	38.7	4231	148.4
4	22.0	60.0	47.1	4969	229.7
5	24.0	40.0	42.9	4612	187.4
6	24.0	60.0	52.4	5413	287.4

dependent on the cleanliness of the room (only few main rooms are designed as level 10,000 class, the rest are 100,000 class). In the case of a good enclosure structure, the number of air changes in the core workroom of the laboratory can still have a large surplus space. Specification [1] requires that there should be no visible leakage in all gaps between the main laboratory of the BSL-3 and the adjacent buffer. In the next phase of the revision of the laboratory specification, it can be considered to reduce the required minimum air supply volume, which can bring objective energy-saving benefits.

4 Conclusions

a. 15–30 times/h of ACH is relatively common for the biosafety laboratory main rooms (not considering biosafety equipment), and it is not too large to design.
b. In order to achieve better energy-saving purposes, the number of air changes in the core workroom can be reduced to less than 12, but the tightness of the intact envelope structure is required.
c. The choice of indoor temperature and humidity in the biosafety laboratory has a great impact on the total annual load of the system. Relative humidity has a higher impact on the load than temperature. If there is no special process requirement, and the personnel comfort and specification requirements are met, the indoor parameters of energy-saving should be selected as much as possible.
d. This paper only analyzes and calculates the load of a single main room under specific meteorological conditions, and obtains qualitative conclusions. When a specific project is applied, it should be accurately calculated using the annual energy analysis method.

Acknowledgements The project is supported by the 13th Five-Year Key Project, Ministry of Science and Technology of China (Number 2017YFC0702800).

References

1. Ministry of Housing and Urban-Rural Development of the People's Republic of China. Biosafety laboratory building technical specification, GB 50346-2015
2. Ministry of Housing and Urban-Rural Development of the People's Republic of China. Design code for heating, ventilation and air conditioning of civil buildings, GB50736-2012
3. Lu, Y., et al.: HVAC, pp. 110–111. China Building Industry Press, Beijing (2002)

The Optimization Design of Sewage Heat Exchanger in Direct Sewage Source Heat Pump System

Zhaoyi Zhuang, Jun Xu, Jian Song and Wenzeng Shen

Abstract With the continuous application of sewage source heat pump technology, research on key technologies in direct sewage source heat pump systems (DSSHPS) is becoming more and more important. The sewage heat exchanger (evaporation in winter, condensation in summer) is the core component of the system, and its correct thermodynamic calculation and reasonable structural design are more critical. In this paper, the sewage flooded evaporator and condenser are taken as examples. Considering the flow characteristics of sewage, the heat exchanger adopts smooth inner wall tube; combines with practical engineering; designs and calculates the two heat exchangers in the direct sewage source heat pump unit by using MATLAB software. The results show that increasing the flow rate of sewage in heat exchanger can increase the flow rate, and increasing the flow rate of sewage in pipe can improve the heat transfer coefficient of sewage side and reduce the fouling thermal resistance. However, increasing the flow velocity in the tube will increase the flow resistance, so it is necessary to comprehensively consider the flow resistance and heat transfer area of the heat exchanger when determining the flow velocity to improve the heat transfer effect. In the sewage heat exchanger, it is more reasonable to use the sewage flow of 4, and the tube arrangement is more reasonable to concentrate the heat exchange tube in the shell center.

Keywords Sewage flooded refrigerant evaporator · Optimization of pipe layout · Simulation calculation

---

Z. Zhuang (✉) · J. Xu · J. Song · W. Shen
School of Thermal Engineering, Shandong Jianzhu University, Jinan 250101, China
e-mail: hit6421@126.com

Z. Zhuang
Shandong Zhongrui New Energy Technology Co., Ltd, Jinan 250101, China

1 Introduction

Heat pump technology is one of the important technologies to solve the problem of energy conservation and emission reduction in building heating and air conditioning [1], it is called "green air conditioning technology" in the twenty-first century, but "the heat pump is good, the heat source is difficult to find" [2]. Urban primary sewage is a kind of renewable resource with rich low-level heat energy, which has been applied to some extent in some countries in China, Japan, and Northern Europe [3]. In recent years, the domestic sewage source heat pump technology has developed rapidly, and its trend has gradually shifted from indirect to direct systems [4]. The correct and rational design of the sewage evaporator and condenser is a key issue for the safe and efficient operation of the system. Due to the high viscosity of sewage and the serious pollution on the heat exchange surface, the resistance on the sewage side is larger than that of the clean water, and the heat exchange coefficient is smaller than that of the clean water [5]. In order to reduce the risk of clogging and pollution, sewage heat exchangers can not use bellows, inner ribs, inserts, and other measures to enhance heat transfer like the water heat exchanger, but only smooth inner wall tubes. Therefore, the use of previous heat exchanger design experience and parameters in DSSHPS has been unable to meet actual engineering needs. According to the author's experimental research and engineering practice in recent years, this paper gives the design method of sewage heat exchanger (evaporation in winter, condensation in summer) in direct sewage source heat pump system, which can provide a reference for peer designers.

2 Design Calculation Method for Sewage Heat Exchanger

The design of sewage heat exchanger mainly includes two basic parameters: performance and structure. Among them, the performance parameters include: heat exchange quantity Q, logarithmic mean temperature difference Δt_m, sewage flow V, heat exchanger resistance ΔH; structural parameters include: Base tube size of heat exchange tube d_i d_0, single tube length of heat exchange tube l, the total number of heat transfer tubes N, the shell inner diameter of heat exchanger D_i.

2.1 Thermal Design

The design calculations for flooded evaporators and condensers are basically the same as for other types of heat exchanger, including thermal calculations and structural design. To carry out the structural design to complete the arrangement of the tube rows, first calculate the parameters such as the heat exchange area and the flow rate of the refrigerant water according to the known conditions.

Flow of Sewage

$$V = 3.6Q/c_p\rho(t_{in} - t_{out}) \tag{1}$$

In the formula: V is the sewage flow rate, m³/h; Q is the heat transfer capacity of the heat exchanger, W; c_p is the specific pressure of the sewage, kJ/kg °C; ρ is the density of the sewage, kg/m³; t_{in} and t_{out} are for the temperature of the sewage into and out of the heat exchanger, °C.

Logarithmic Mean Temperature Difference

$$\Delta t_m = (t_{in} - t_{out})/\ln(t_{in} - t_e)/(t_{out} - t_e) \text{ or } \Delta t_m = (t_{out} - t_{in})/\ln(t_c - t_{in})/(t_c - t_{out}) \tag{2}$$

In the formula: Δt_m is the logarithmic mean temperature difference of the heat exchanger, °C; t_e and t_c are the temperature of evaporation and the temperature of condensation, °C.

The Heat Transfer Coefficient of the Sewage Side in the Pipe

In the heat exchanger of the unit, the convective heat transfer coefficient of sewage under forced turbulence in the pipe is [6]:

$$h_s = 0.0158 \lambda_s \text{Re}_s^{0.86} \text{Pr}_s^{0.34}/d_i \tag{3}$$

In the formula: λ_s is the thermal conductivity of sewage, W/(m K); d_i is the inner diameter of the heat exchange tube, m; Re_s is the Reynolds number of the sewage side, $\text{Re}_s = u_s^{1.08} d_i^{0.92}/v_s$; u_s is the flow rate of sewage in the pipe, m/s; v_s is the kinematic viscosity of the sewage, m²/s; Pr_s is the Prandtl number of the sewage side, $\text{Pr}_s = c_{p,s} \mu_s d_i^{0.08}/(\lambda_s \cdot u^{0.08})$; $c_{p,s}$ is the constant pressure specific heat capacity of sewage, kJ/kg °C; μ_s is the dynamic viscosity of sewage, N S/m².

Refrigerant Side Heat Transfer Coefficient

The external heat transfer coefficient $h_{r,e}$ of the tube bundle during evaporation and the external heat transfer coefficient of the tube bundle during condensation $h_{r,c}$ can be obtained by the calculation formula in the literature [7]. For a common tube for efficient evaporation and condensation, when evaporating, $h'_{r,e} = \zeta h_{r,e}$; when condensing, $h'_{r,e} = \xi h_{r,e}$, according to the literature [8], the coefficients 1, 2 can be approximated as 0.8.

Fouling Thermal Resistance

When the inside of the pipe is sewage flow, it is known from the experimental conclusion [9] that when the dirt reaches asymptotic stability at a certain flow rate in the coated nanotube copper pipe, the relationship between the thermal resistance value of the fouling and the flow velocity in the pipe is

$$R_{f,s} = 9.67 u_s^{-0.22} \times 10^{-5} \tag{7}$$

In the formula: $R_{f,s}$ is the thermal resistance of the sewage on the sewage side of the pipe, (m² K)/W.

Total Heat Transfer Coefficient of Heat Exchanger

Based on the known heat transfer coefficient of sewage side inside the pipe and refrigerant side outside the pipe, combined with the thermal resistance of the dirt (ignoring the thermal resistance of the pipe wall), the total heat transfer coefficient K of the heat exchanger can be obtained.

$$1/K = A(1/h_i A_i + R_f/A_i + 1/A_o h_r) \qquad (8)$$

In the formula: R_f is the thermal resistance of the pipe inside the dirt, (m² K)/W; A is the heat exchange area outside the pipe based on the outer diameter of the tube envelope, m²; A_i is the heat exchange area inside the pipe based on the standard inner diameter, m²; A_o is the heat exchange tube The total area outside the pipe, m².

Energy Equation

$$Q = KF\Delta t_m \qquad (9)$$

In the formula: Q is the heat transfer capacity of the heat exchanger, W; F is the total heat exchange area of the heat exchanger, m².

Resistance Equation

$$\Delta H = \left(fL/d + \sum \zeta\right) u^2 / 2g \qquad (10)$$

In the formula: ΔH is the on-way resistance of heat exchanger, m; f is the resistance coefficient of the sewage; We know from the experimental results [7] $f = 0.276(Re)^{-0.238}$; L is the total length of single tube series in the heat exchanger, m; $\sum \zeta$ is the local resistance coefficient sum.

2.2 Optimal Arrangement of Tube Rows

Generally, when the flooded evaporators are designed, the center height of the uppermost heat exchange tube on the tube plate is about 2/3 times the inner diameter of the shell. Considering that the heat exchanger meets the requirements of evaporation and condensation at the same time, the inner diameter of the uppermost heat exchange tube on the tube sheet is designed to be 3/4 times the inner diameter of the shell, and the center height of the lowermost heat exchange tube of the tube sheet is 1/5. When the heat exchange tube is evenly covered on the tube sheet, the inner diameter of the housing:

Total number of heat transfer tubes required:

$$N' = VZ/900 u\pi d_i^2 \qquad (11)$$

The inner diameter of the shell when the heat exchange pipe is uniformly covered on the tube plate:

$$D'_i = (1.1\sqrt{N'} - 1)s + d_0 + 2e \tag{12}$$

In the formula: s is the pipe center distance, take $s/d_0 = 1.5$; e is the distance from the inner wall of the casing to the outer surface of the outermost heat exchange tube of the tube bundle, mm.

Determine the inner diameter of the heat exchanger shell: the number of heat transfer tubes on the tube plate should be greater than N', take $D_i = \gamma D'_i$, $\gamma = 1.1$–1.2, and initially given during design.

Calculate the maximum diameter of tube bundle of heat exchanger tubesheet:

$$D_{max} = D_i - 2e \tag{13}$$

Calculate the number of tubes on the centerline of the heat exchanger tubesheet:

$$n_c = (D_{max} - d_o)/s + 1 \tag{14}$$

Calculate the number of tubes in the semicircle above the centerline on the tube plate of the heat exchanger:

$$H_1 = D_i/2\sqrt{3}s \tag{15}$$

Calculate the number of inner semicircle tubes below the centerline on the tube plate of the heat exchanger:

$$H_2 = \sqrt{3}D_i/5s \tag{16}$$

Calculate the total number of heat exchange pipes on the heat exchanger tube plate:

$$N = n_c + n_c(H_1 + H_2) - (H_1 + 1)H_1/2 - (H_2 + 1)H_2/2 \tag{17}$$

The comparison between N and N' must be satisfied $N \geq N'$ (considering the volume of the heat exchanger, should not be too large), otherwise, adjust the value of γ to adjust the size of D_i inner diameter of the shell.

3 Design Procedure and Application of Sewage Heat Exchanger

The performance parameters of heat exchanger Q and ΔH are the parameters that should be given in advance in the design and calculation. The logarithmic average temperature difference Δt_m was, respectively, determined by the inlet and outlet temperature of sewage, the evaporation temperature t_e, and the condensation temperature t_c; The sewage flow rate V is determined by the heat exchange amount and the temperature difference of the sewage. When the temperature difference is uncertain, the sewage flow rate should be given in advance; Therefore, four performance parameters of Q, ΔH, Δt_m, and V of the heat exchanger need to be given in advance. The remaining three parameters, N, L, and d, are the main parameters to be solved in the heat exchanger design. They determine the size of the heat exchanger area. In general, the given pipe diameter d is used to find the remaining two parameters.

4 Examples and Analysis

4.1 Thermal Calculation Results

In this paper, we combine the system with a direct sewage source heat pump system project; the heat exchange in design of the system waste water flooded evaporator is 300 kW. We adopt high-efficiency evaporative condensation pipe, which is made of navy copper pipe. The inside of the pipe is coated with nanocoating, which is smooth. The size of the sewage heat exchange pipe is shown in Table 1. According to the above design calculation by *MATLAB* program, the output of the sewage flooded evaporator is shown in Table 2.

Table 1 Structural parameters of sewage heat exchange tubes

External surface parameters				Inner surface parameters		
Outer diameter/ mm	Fin height/ mm	Fin number/ m^{-1}	Inner diameter/ mm	Fin height/ mm	Internal surface area (m^2/m)	Wall thickness/ mm
19.05	0.62	1575	16.54	–	–	0.635

Table 2 Evaporator performance parameters at different sewage flow rates

u (m/s)	1.0	1.2	1.4	1.6	1.8	2.0	2.2	2.4	2.6
h_s (W/(m^2 K))	3240	3818	4388	4949	5504	6052	6595	7133	7667
$R_{fs} \times 10^{-5}$	9.67	9.29	8.98	8.72	8.5	8.3	8.13	7.98	7.84
K (W/(m^2 K))	1816	2129	2349	2453	2620	2803	2961	3111	3254
F (m^2)	35.53	30.3	27.47	26.3	24.62	23.02	21.79	20.74	19.83
L (m)	2.86	2.95	3.1	3.33	3.55	3.7	3.8	3.94	4.14
Z	4	4	4	4	4	4	4	4	4
N'	208	172	148	132	116	104	96	88	80
D'_i (m)	0.48	0.44	0.41	0.39	0.37	0.35	0.34	0.32	0.31
ΔH (m)	1.73	2.46	3.38	4.52	5.84	7.3	8.82	10.6	12.7

It can be seen from Table 2 that the thermal resistance of the sewage heat exchanger mainly concentrates on the sewage side and the fouling thermal resistance inside the pipe. Increasing the flow rate inside the pipe can increase the heat transfer coefficient of the sewage side and reduce the thermal resistance of the dirt, which also increases the flow resistance in the heat exchanger and increases the energy consumption for the operation of the unit. Taking the flow resistance and heat transfer area into consideration, the designed velocity in the sewage pipe was selected as 1.8 m/s in this paper, and the corresponding design parameters of the heat exchanger are shown in Table 2. The evaporator tube row is arranged in four flows, horizontal arrangement or up and down. Each process has 29 tubes. The effective length of single tube is 3.55 m, the total number of roots is 116, the diameter of the shell is 0.37 m, and the heat exchanger resistance is 5.84 m.

4.2 Optimal Arrangement of Tube Rows

Combined with the above calculation results, the tube row is optimized according to the tube row optimization design procedure and is compared with the traditional flooded evaporator when the existing performance parameters and structural dimensions, heat exchange tubes, tube spacing, and other parameters of the sewage flooded evaporator are unchanged. The structural parameters of the evaporator were compared and analyzed, and the detailed data is shown in Table 3.

We can see that from Table 3, using the tube row optimization design method described herein, the inner diameter of the sewage heat exchanger housing can be reduced from 0.45 to 0.43 m, and the total number of heat exchange tubes is increased from 116 to 124. Thereby increasing the effective heat exchange area while reducing the volume of the evaporator, the number of semicircular tubes above the center line of the tube sheet is increased from the original three rows to four rows, and the number of rows of the lower semicircular tubes is reduced from the original nine rows to five rows. The heat transfer tubes are concentrated to the center of the housing, this kind design facilitates the flow of the refrigerant vapor above the housing and the discharge of the refrigerant liquid below.

5 Conclusion

In this paper, we discuss the thermal design and structural design method of sewage heat exchanger. The design procedure of the two heat exchangers of the unit is compiled. The following conclusions can be drawn through the analysis of actual examples: the flow rate can be increased and the heat exchange can be improved by increasing the number of sewage. The sewage in the sewage heat exchanger is suitable for four processes; improve the overall performance of the sewage heat exchanger, mainly how to strengthen the heat transfer coefficient of the sewage side

Table 3 Optimization design result of sewage full-fluid evaporator tube row

Ordinary evaporator		Evaporation and condensation type evaporator	
Name	Numerical value	Name	Numerical value
Total number of heat exchange tubes	116	Total number of heat exchange tubes required N'	116
Inner diameter of the shell D'_i (m)	0.37	Inner diameter of the shell D'_i (m)	0.37
Inner diameter of the shell D_i (mm)	0.45	Inner diameter of the shell D_i (m)	0.43
The number of tubes on the centerline of the heat exchanger tubesheet Nc	16	The number of tubes on the centerline of the heat exchanger tubesheet n_c	15
The number of tubes in the semicircle above the centerline on the tube plate of the heat exchanger	3	The number of tubes in the semicircle above the centerline on the tube plate of the heat exchanger h_1	4
The number of inner semicircle tubes below the centerline on the tube plate of the heat exchanger	9	The number of inner semicircle tubes below the centerline on the tube plate of the heat exchanger h_2	5
		Total number of heat exchange tubes N	124

and reduce the thermal resistance of the dirt inside the pipe; The heat exchanger tube arrangement is more reasonable and compact while taking into account the evaporation condensation. The optimization method provides a theoretical basis for the design of the direct sewage source heat pump heat exchanger.

Acknowledgements This work was supported by National Natural Science Foundation of China (Grant No. 51708339), the China Postdoctoral Science Foundation Funded Project (Grant No.2017M612303).

References

1. Zhuang, Z.Y., Zhang, C.H., Pan, Y.W., et al.: Research on the related key technique of district heating and cooling using direct sewage heat pump system. Renew. Energy Resour. **29**(3), 141–145 (2011)
2. Liu, Z.B., Ma, L.D., Zhang, J.L.: Application of a heat pump system using untreated urban sewage as a heat source. Appl. Therm. Eng. **62**, 747–757 (2014)
3. Meggers, F., Leibundgut, H.: The potential of wastewater heat and exergy: decentralized high temperature recovery with a heat pump. Energy Build. **43**(4), 879–886 (2011)
4. Zhuang, Z.Z., Qi, J., Zhang, C.H., et al.: Key techniques of direct sewage source heat pump systems. HVAC **41**(10), 96–101 (2011)

5. Tu, A.M., Zhu, D.S., Guo, X.F.: Comparative analysis of centralized heating and cooling system based on sewage-source heat pump system with a traditional system in hot summer and warm winter zone. HVAC **46**(12), 90–95 (2016)
6. Cheng, X.S., Peng, D.G., Li, S.L., et al.: Model and influencing factors analysis of external parameters of sewage source heat pump. J. Chongqing Jianzhu Univ. **39**(4), 26–32 (2017)
7. Wang, Z.W., Li, Y., Sun, H.Y., et al.: Test and research on the fouling in condenser and evaporator of direct sewage source heat pump system. J. XI'an Univ. Arch. Technol. **46**(1), 567–571 (2014)
8. Qian, J.F., Ren, Q.F., Xu, Y., et al.: Experiment on the antiscale and descaling and heat transfer enhancement of acoustic cavitation sewage heat exchanger. J. Harbin Inst. Technol. **50**(2), 166–172 (2018)
9. Shang, Y.M., Bi, H.Y., Wu, Z.: Study on anti clogging performance of large tube heat exchanger equipment of circulating fluidized bed. J. Dalian Univ. **37**(3), 7–9 (2016)

The Optimization of Sensitivity Coefficients for the Virtual in Situ Sensor Calibration in a LiBr–H$_2$O Absorption Refrigeration System

Peng Wang, Kaihong Han, Liangdong Ma, Sungmin Yoon and Yuebin Yu

Abstract The correct data or information from the building sensing networks plays a vital role in the operation algorithms. The sensor errors usually show a negative effect on the performance of control, diagnosis, and optimization of building energy systems. Thus, the physical working sensors periodically need to be removed to be calibrated by the reference sensors, which will disrupt the normal operation of building systems from time to time. The virtual in situ sensor calibration (VIC), based on the Bayesian inference and Markov chain Monte Carlo methods (MCMC), is an effective approach to handle the systematic and random errors of various working sensors simultaneously. This technology uses the distance function and system models to estimate the true measurements and addresses most of the practical problems in a traditional calibration process. However, the sensitivity coefficient in the definition of distance function is one of the determining factors in the calibration accuracy and how to define it still remains uncertain. Therefore, this study employed the genetic algorithm (GA) to optimize this parameter in a LiBr–H$_2$O absorption refrigeration system. The results revealed that the systematic and random errors of temperature and mass flow rate were reduced considerably with the help of optimized sensitivity coefficients and most of the measurements approached to their true values after the calibration.

Keywords Sensor network · Virtual in situ calibration · Sensitivity coefficient optimization · Bayesian MCMC · Genetic algorithm

---

P. Wang · K. Han · L. Ma
School of Civil Engineering, Dalian University of Technology, Dalian City, China

P. Wang · Y. Yu
Durham School of Architectural Engineering and Construction, University of Nebraska-Lincoln, Omaha, USA

S. Yoon (✉)
Division of Architecture and Urban Design, Incheon National University, 119 Academy-Ro, Yeonsu-Gu, Incheon 22012, Republic of Korea

Institute of Urban Science, Incheon National University, 119 Academy-Ro, Yeonsu-Gu, Incheon 22012, Republic of Korea
e-mail: syoon@inu.ac.kr

1 Introduction

The energy consumptions of heating, ventilation and air conditioning, and refrigeration systems in modern buildings account for 30% of the total commercial building energy consumption [1–3]. Many advanced technologies have been developed recently to solve this problem such as the automated fault detection and diagnosis [4–7]. However, if some of the existing physical sensors are malfunctioning, almost all the approaches will become useless because of the inaccurate raw measurements. A large number of previous studies have demonstrated that the erroneous sensors showed a negative effect on the performance of control, diagnosis, and optimization of building energy systems [8–12]. The correct data or information from the building sensing networks plays a vital role in the control and operation algorithms.

Our research group recently proposed a virtual in situ sensor calibration (VIC) method to handle the problems mentioned above [13]. The suggested VIC approach employed the Bayesian parameter estimation and related system models to calibrate both the systematic and random errors of sensing networks inside building energy systems [14, 15]. The previous study indicated that the sensitivity coefficients in the definition of distance function play a crucial role in the calibration results and how to define them properly still remains uncertain. Therefore, the optimization of sensitivity coefficients forms the main purpose of the current study.

2 Reviving Calibration Strategy for Vic

The VIC approach is able to correct the systematic and random errors of multiple physical sensors simultaneously. This method uses a distance function ($D(x)$) to represent the difference between the outputs of reliable system (Y_R) and benchmark (Y_b), as defined in Eqs. (1)–(3). Different working stages (W) are included to cover all the measurement ranges and working environments of various sensors. The benchmark outputs are all established by available system models (f) with calibrated measurements (Y_c) and unknown parameters (x_u).

$$D^W(x) = \sum_{t=1}^{T}\sum_{l=1}^{L} n_l s_l (Y_{R,l} - Y_{b,l})^2 \qquad (1)$$

$$Y_b = f(Y_{c1}, Y_{c2}, \ldots, Y_{cr}, x_{u1}, x_{u2}, \ldots, x_{uq}) \qquad (2)$$

$$Y_c = g(M, x) \qquad (3)$$

The Bayesian inference uses the likelihood function and prior distribution to estimate the posterior probability of each variable, as in Eqs. (4)–(6). All the local distance functions, as formulated in Eq. (1), are plugged into the likelihood function to form a Gaussian distribution individually. This likelihood function will reach its maximum when every single distance function in a given local domain is minimized. As a normal distribution, the posterior distribution of each variable (offsetting constants and unknown parameters) will get a larger probability with the reduction of distance function, which also means a lower error of working sensor.

$$P(x|Y) = \frac{P(Y|x) \times \pi(x)}{P(Y)} \quad (4)$$

$$P(Y|x) = \prod_{t=1}^{T} \frac{1}{\sigma\sqrt{2\pi}} \exp\left[-\frac{1}{2\sigma^2} D^W(x)\right] \quad (5)$$

$$P(Y) = \int P(Y|x)\pi(x)\mathrm{d}x \quad (6)$$

3 Sensitivity Coefficients Optimization

The reviving calibration strategy uses a two-step strategy to calibrate the greedy and poor variables. During the whole process, the distance function in the first step needs to be defined with several sensitivity coefficients in order to "balance" and "target" different greedy variables, as in Eq. (7).

$$\begin{aligned} D(x) = \sum_{t=1}^{T} \Big[& s_1 \times n_1 \times \left(Q_{e,c} - Q_{e,b}\right)^2 + s_2 \times n_2 \\ & \times \left(Q_{c,c} - Q_{c,b}\right)^2 + s_3 \times n_3 \times \left(W_{p,c} - W_{p,b}\right)^2 \Big] \end{aligned} \quad (7)$$

As mentioned in Sect. 2, the smaller the distance functions become, the more accurate the calibration results will be. Therefore, the sensitivity coefficients (s) could be treated as the optimization parameters to minimize the distance function by using genetic algorithm (GA). The distance function could be slightly modified as a fitness/objective function for genetic algorithm with the sensitivity coefficients as the optimization parameters, as defined in Eq. (8). Various combinations of s_1, s_2, and s_3 in Eq. (7) will generate a large number of offsetting constants (x) for each working sensor and further result in different values of objective function in Eq. (8).

$$O(x) = \sum_{t=1}^{T} \left[n_1 \left(Q_{e,c} - Q_{e,b}\right)^2 + n_2 \left(Q_{c,c} - Q_{c,b}\right)^2 + n_3 \left(W_{p,c} - W_{p,b}\right)^2 \right] \quad (8)$$

where $O(x)$ is the objective function of GA.

4 Case Study: The Effects of Sensitivity Coefficients on the Calibration Results

4.1 VIC Application

This study uses a single-effect lithium bromide absorption refrigerator to investigate the effect of sensitivity coefficients on the calibration accuracy. As shown in Fig. 1, the system is mainly composed of a condenser, generator, evaporator, absorber, and solution pump. These components are able to mathematically represent the building energy systems, as they are based on similar conservation equations of mass and energy. The VIC method was applied to three temperature sensors and two mass flow rate sensors in the inner cycle (T_4, T_8, T_{10}, m_r, and m_1). In the given calibration domain, various systematic and random errors were defined for each sensor in two working stages, as listed in Table 1. The offsetting constants (x) involved in the correction functions (Eq. (3)) were used to represent the systematic error of various sensors. The prior distributions of all variables were set as a normal distribution with a zero mean and their standard deviations were defined by the random errors of raw measurements. As depicted in Table 2, Case 1 is the standard two-step reviving calibration with no sensitivity coefficients, while Case 2 used the sensitivity coefficients optimized by the genetic algorithm based on the objective function defined in Eq. (8).

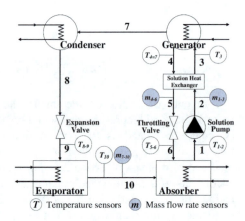

Fig. 1 Target system for the virtual in situ sensor calibration [16]

Table 1 Working stages and sensor errors for the local calibration domain

Sensor	Working stage	Systematic error	Random error
T_4	80–84 °C	3 °C	0.7
T_8	40–44 °C	3 °C	0.7
T_{10}	7–11 °C	−3 °C	0.7
m_r	0.075–0.095 kg/s	0.0075 kg/s	0.0035
m_1	0.6–1.0 kg/s	0.075 kg/s	0.035

Table 2 Sensitivity coefficients optimized by GA for reviving calibration strategy

Sensitivity coefficients	First step of reviving calibration (for greedy variables with high sensitivities)		Second step of reviving calibration (for poor variables with low sensitivities)	
	Case 1	Case 2	Case 1	Case 2
s_1 for Q_e	1	1.00	1.00	
s_2 for Q_c	1	1.30	1.00	
s_3 for W_p	1	2.45	–	

4.2 The Effect of Sensitivity Coefficients on the Calibration Accuracy

With an increase in the number of data sets, the calibration accuracy improved for both cases, as illustrated in Fig. 2. The offsetting constants (systematic error) of both temperatures and mass flow rates became very stable after five data sets, which could be regarded as convergent results. The shapes of posteriori distribution became much narrower than the prior one, which indicated that the all random errors had been significantly reduced after the calibration. As there were no sensitivity coefficients to adjust the relative importance of each working sensor in the distance function in Case 1, the errors of T_4, T_8, T_{10}, m_r, and m_1 reached 19.9, 13.3, −24.6, −1.33, and 40.0%, respectively, as listed in Table 3. This accuracy could not be accepted for the calibration process, which further indicated the essential role of sensitivity coefficients. As a contrast, the calibration results of Case 2 were much better than those of Case 1, as it considered the greedy and poor variables in the distance function simultaneously. As listed in Table 3, the errors of T_4, T_8, T_{10}, m_r, and m_1 decreased to −2.1, 4.0, −0.8, 0.3, and −4.2%, respectively, which is very reasonable for most of actual situations.

It is shown that the relationship between different variables plays a determining role in the calibration process. That's why it is so crucial to optimize the sensitivity coefficients of distance function to balance (increase or decrease) the relative importance of each working sensor. In a two-step reviving calibration process, the main purpose is to minimize the differences between the corrected measurements

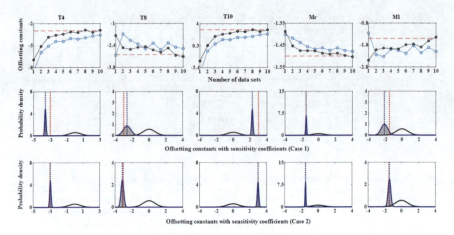

Fig. 2 Calibration results of different cases for the designed working stage (greedy variables: T_8, m_r and m_1; poor variables: T_4 and T_{10})

and benchmark outputs to achieve the smallest distance function. With the help of genetic algorithm, it is possible to obtain the best sensitivity coefficients, but the greedy and poor variables should not be considered separately, as they are strongly coupled with each other during the two-step calibration process. Therefore, it is recommended to utilize the objective function of GA based on both steps to optimize the sensitivity coefficients.

4.3 The Global Sensitivity Analysis with Different Sensitivity Coefficients

Figure 3 describes the global sensitivities of various variables for the designed working stage. In both cases as shown in Fig. 3a and b, the total effect indexes of poor variables (T_4 and T_{10}) are almost zero, which indicates that these parameters show very limited effect on the system models in the given domain during the first step of reviving calibration. In the second step of reviving calibration, all the greedy variables have already been eliminated (killed) and the poor variables are revived and able to balance each other without introducing the sensitivity coefficients any more, as illustrated in Fig. 3c. It is also demonstrated that the sensitivities of poor variables could not be improved no matter how the sensitivity coefficients are optimized. On the contrary, the sensitivity coefficients are crucial to balance (increase or decrease) the relative importance of different greedy variables (T_8, m_1, and m_r). It is strongly suggested to distinguish the greedy and poor variables using the global sensitivity analysis before the calibration starts and then use a two-step reviving calibration strategy with optimized sensitivity coefficients to achieve a more accurate result.

Table 3 Predictions of systematic error in various sensors for the designed working stage

Sensor	Correction function and variable x	True values	All cases Systematic error (X^a)	Priors		Cases 1 Posteriors			Cases 2 Posteriors		
				M^b	SD^c	M	SD	Error (%)	M	SD	Error (%)
T_4	$T_4 = T_4 + x_4$	82	−3	0	0.7	−3.597	0.0815	19.9	−2.936	0.0839	−2.1
T_8	$T_8 = T_8 + x_8$	42	−3	0	0.7	−2.600	0.457	13.3	−3.121	0.166	4.0
T_{10}	$T_{10} = T_{10} + x_{10}$	9	3	0	0.7	2.263	0.0848	−24.6	2.976	0.0895	−0.8
m_r	$m_r = m_r + a^d \cdot x_r$	0.086	−1.5	0	0.7	−1.480	0.0587	−1.33	−1.505	0.0482	0.3
m_1	$m_1 = m_1 + b^d \cdot x_1$	0.754	−1.5	0	0.7	−2.100	0.402	40.0	−1.437	0.162	−4.2

[a] True offsetting constants
[b] M: median
[c] SD: standard deviation
[d] $a = 1/200$ and $b = 1/20$ to balance the prior and post distributions of temperatures and mass flow rates

As described in Fig. 3 and Table 3, the error of m_1 is the largest among all the three greedy variables (T_8, m_r, and m_1) for working stage 1, which has a considerable negative effect on the poor variables in the second step of reviving calibration. m_1 also plays a decisive role in determining T_8, as m_1 could be regarded as the gradient coefficient of T_8. That is to say m_1 dominates in the two-step reviving calibration process and the sensitivity coefficients in Table 2 could be used to balance each sensor. An increase in the sensitivity coefficient (s_3) is conducive to improve the relative importance of m_1 in the system models and may result in smaller errors for all the working sensors. For example, the average prediction accuracies of sensor errors in Case 2 are better than those of Case 1. m_r is the gradient coefficient both poor variables (T_4 and T_{10}) and a small change in m_r may significantly affect the calibration results in the second step. Therefore, a reasonable improvement in the sensitivity coefficient (s_2) is very beneficial to correct T_4 and T_{10}. That is why the systematic errors of poor variables in Case 2 agreed better with their true values. For the designed working stage, the sensitivity coefficient (s_3) for m_1 should be as large as possible to balance the greedy variables, but the sensitivity

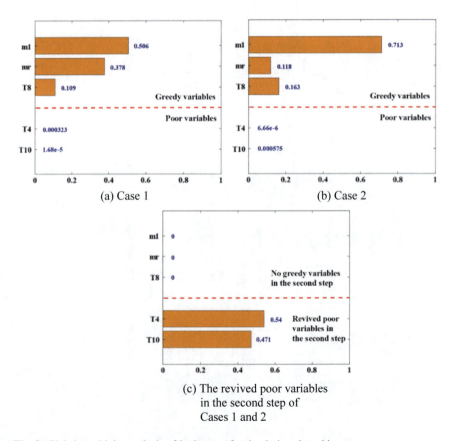

Fig. 3 Global sensitivity analysis of both cases for the designed working stage

coefficient (s_2) for m_r has to be enhanced at a reasonable range to improve the accuracy of poor variables at the same time. Therefore, Case 2 gave the more accurate results for all the sensors, while Case 1 provided worse estimations.

5 Conclusions

The virtual in situ sensor calibration, based on the Bayesian inference and Markov chain Monte Carlo methods, is an effective approach to calibrate all the working sensors in building energy systems. This study used the genetic algorithm (GA) to optimize the sensitivity coefficient. Three fitness functions of GA have been suggested and discussed in order to explore the most effective way to generate the sensitivity coefficients. The results revealed that the systematic and random errors of all sensors (for both temperature and mass flow rate) reduced considerably with the aid of optimized sensitivity coefficients. Some detailed conclusions have been drawn as follows:

(1) The sensitivity coefficients reflect the sensitivity of every single variable in the local distance function and is very crucial to adjust (increase or decrease) the sensitivities of some critical parameters, which may further improve the calculation accuracy, especially for the poor variables with low sensitivities.
(2) The accurate results of greedy variables may not always result in precise poor variables, as the relationship between different variables considerably affect the whole calibration process. Therefore, the sensitivity coefficients are of great importance to improve the calibration accuracy.
(3) GA provides a very good global solution for the optimization of sensitivity coefficients and it is strongly recommended to utilize the distance function of both steps as the objective function to optimize the sensitivity coefficients. But this method has to be used with the reviving calibration strategy of VIC so as to achieve more accurate results.

Acknowledgements This work was supported by the National Natural Science Foundation of China (Grant No. 51806029), National Key R&D Program of China (Grant No. 2017YFC0704200), China Postdoctoral Science Foundation Funded Project (Grant No. 2016M590221), and Fundamental Research Funds for the Central Universities (Grant No. DUT18RC(4)054).

References

1. Amasyali, K., El-Gohary, N.M.: A review of data-driven building energy consumption prediction studies. Renew. Sustain. Energy Rev. **81**, 1192–1205 (2018)
2. Kingma, B., van Marken Lichtenbelt, W.: Energy consumption in buildings and female thermal demand. Nat. Clim. Chang. **5**(12), 1054 (2015)

3. Huo, T., Ren, H., Zhang, X., Cai, W., Feng, W., Zhou, N., Wang, X.: China's energy consumption in the building sector: a statistical yearbook-energy balance sheet based splitting method. J. Clean. Prod. **185**, 665–679 (2018)
4. Bruton, K., Raftery, P., O'Donovan, P., Aughney, N., Keane, M.M., O'Sullivan, D.: Development and alpha testing of a cloud based automated fault detection and diagnosis tool for air handling units. Autom. Constr. **39**, 70–83 (2014)
5. Wang, J., Zhang, Q., Yu, Y., Chen, X., Yoon, S.: Application of model-based control strategy to hybrid free cooling system with latent heat thermal energy storage for TBSs. Energy Build. **167**, 89–105 (2018)
6. Wang, J., Zhang, Q., Yu, Y.: An advanced control of hybrid cooling technology for telecommunication base stations. Energy Build. **133**, 172–184 (2016)
7. Wang, J., Zhang, Q., Yu, Y.: Intelligent control of hybrid cooling for telecommunication base stations. Heat Transf. **4**, 5 (2016)
8. Zhang, R., Hong, T.: Modeling of HVAC operational faults in building performance simulation. Appl. Energy **202**, 178–188 (2017)
9. Roth, K.W., Westphalen, D., Llana, P., Feng, M.: The energy impact of faults in US commercial buildings, (2004)
10. Verhelst, J., Van Ham, G., Saelens, D., Helsen, L.: Economic impact of persistent sensor and actuator faults in concrete core activated office buildings. Energy Build. **142**, 111–127 (2017)
11. Yoon, S., Yu, Y.: Hidden factors and handling strategies on virtual in-situ sensor calibration in building energy systems: prior information and cancellation effect. Appl. Energy **212**, 1069–1082 (2018)
12. Yoon, S., Yu, Y., Wang, J., Wang, P.: Impacts of HVACR temperature sensor offsets on building energy performance and occupant thermal comfort. In: Building Simulation, Springer, 1–13
13. Yu, Y., Li, H.: Virtual in-situ calibration method in building systems. Autom. Constr. **59**, 59–67 (2015)
14. Yoon, S., Yu, Y.: A quantitative comparison of statistical and deterministic methods on virtual in-situ calibration in building systems. Build. Environ. **115**, 54–66 (2017)
15. Yoon, S., Yu, Y.: Extended virtual in-situ calibration method in building systems using Bayesian inference. Autom. Constr. **73**, 20–30 (2017)
16. Yoon, S., Yu, Y.: Strategies for virtual in-situ sensor calibration in building energy systems. Energy Build. **172**, 22–34 (2018)

Research on the Energy-Saving Coefficient and Environmental Effect of the Surface Water Source Heat Pump System

Ying Xu, Yuebin Wu, Liang Chen and Qiang Sun

Abstract Comparing to the atmosphere, the surface water has a reproducible and clean energy source. So, it is important to evaluate the energy-saving and environmental protection value. In this paper, the coefficients of energy-saving and energy-using are defined, according to the operation mode of HVAC. The effect of surface water source heat pump system and the influence on the environment are also estimated. The result shows that the energy-saving coefficient of surface water source heat pump system is between 0.37 and 1, and the refrigeration coefficient is from 0.29 to 0.42. Traditional air-cooling heat pump could lead to urban heat island, so we quantify the energy-saving potential of surface water source heat pump. At last, we quantify its energy-saving and environmental benefits.

Keywords Surface water source heat pump · Energy-using coefficients · Energy-saving coefficients · Heat island effect · Environmental effect introduction

---

Y. Xu
School of Energy and Architecture Engineering,
Harbin University of Commerce, Harbin, China
e-mail: joexying@126.com

Y. Wu (✉) · L. Chen
Laboratory of Cold Region Urban and Rural Human Settlement Environment Science
and Technology, Ministry of Industry and Information Technology, Harbin, China

School of Architecture, Harbin Institute of Technology, Harbin, China
e-mail: ybwu@hit.edu.cn

L. Chen
e-mail: chainline@163.com

Q. Sun
School of Civil Engineering, Northeast Forestry University, Harbin, China
e-mail: sunqiang@nefu.edu.cn

1 Introduction

The surface water such as rivers, lakes, seawater and urban sewage has large area and comfortable temperature. The surface water is an available method to relieve the energy consume on HAVC and environmental pollution [1]. So, it is important to quantitative define surface water heat pump system's energy-saving and environmental value, and there are few researches study these topics. Some authors discussed the energy-saving and environmental value of urban sewage as heat and cold source, which include the amount of primary saved energy sources and reduction in harmful gas, in cooler regions [2]. But the main surface water is in temperate zones, where have different amount of primary energy consumption and exhaust emissions. Therefore, this thesis puts forward two concepts about the energy-saving and energy-using coefficients, which can quantify the energy-saving coefficient of surface water source heat pump system and its effect on environment.

2 Methods

Air pollution partly stems from coal burning [3], so the primary energy efficiency of heating system is not only an index of energy saving, but also an environmental protection indicator [4]. However, different areas have different heat and cold loads and have different temperature of its surface water. In order to quantify its energy-saving coefficient and effect on environment, we had introduced a concept of energy-saving coefficient of water source heat pump,

$$\Delta J_w = \frac{J_t^f - J_w^f}{\Delta t}, \text{ or } \Delta J_w = \frac{J_t^f - J_w^f}{\Delta t'}, \tag{1}$$

$$y = \frac{J_w^f}{J_w}, \text{ or } y = \frac{J_t^f}{J_w}, \tag{2}$$

$$j = \frac{\Delta J_w}{J_w/\Delta t}, \text{ or } j = \frac{\Delta J_w}{J_w/\Delta t'}, \tag{3}$$

$$j = \frac{J_t^f - J_w^f}{J_w}, \tag{4}$$

where Δt and $\Delta t'$ are refer to the water temperature drop and rise, J_w is the used energy, J_w^f is the totally expended primary energy of the surface water source, J_t^f is the totally expended primary energy of other kinds of heating or air-condition system which get same heating or air-condition effect, j and y are refer to energy-saving and energy-using coefficients.

The available energy of surface water source, and indoor heating or cooling loads are different from heating conditions to refrigerate conditions, so we give the following equations.

2.1 Heating Conditions

$$J^h = \frac{\varepsilon_h}{\varepsilon_h - 1} J_w, \tag{5}$$

$$J_w^f = \frac{1}{\varepsilon_h - 1} J_w \cdot \frac{1}{E_w}, \tag{6}$$

$$J_t^f = \frac{J^h}{E_t} = \frac{\varepsilon_h}{E_t(\varepsilon_h - 1)} J_w, \tag{7}$$

in which ε_h is heating coefficient of surface water heat pump, E_w is the energy utilization rate of heat pump units, E_t is the energy utilization rate of other system or fuel. Thus,

$$y = \frac{J_w^f}{J_w} = \frac{1}{\varepsilon_h - 1} \cdot \frac{1}{E_w} \tag{8}$$

Energy-saving coefficient of other systems

$$j = \frac{J_t^f}{J_w} = \frac{\varepsilon_h}{\varepsilon_h - 1} \cdot \frac{1}{E_t} \tag{9}$$

The saved energy of surface water heat pump is

$$\Delta J_w \Delta t = \left[\frac{\varepsilon_h}{E_t(\varepsilon_h - 1)} - \frac{1}{\varepsilon_h - 1} \cdot \frac{1}{E_w} \right] J_w \tag{10}$$

Its energy-saving coefficient is

$$j = \left[\frac{\varepsilon_h}{E_t(\varepsilon_h - 1)} - \frac{1}{\varepsilon_h - 1} \cdot \frac{1}{E_w} \right] \tag{11}$$

2.2 For Refrigerate Conditions

When the available cooling energy of surface water is J_w, the available cooling energy is

$$J^c = \frac{\varepsilon_c}{1+\varepsilon_c} J_w, \tag{12}$$

The primary energy consumption of the system is

$$J_w^f = \frac{1}{1+\varepsilon_c} J_w \cdot \frac{1}{E_w}, \tag{13}$$

$$J_t^f = \frac{J^c}{E_t} = \frac{\varepsilon_c}{1+\varepsilon_c} \frac{1}{E_t} J_w, \tag{14}$$

$$y = \frac{J_w^f}{J_w} = \frac{1}{1+\varepsilon_c} \frac{1}{E_w}, \tag{15}$$

$$y = \frac{\varepsilon_c}{1+\varepsilon_c} \frac{1}{E_t}, \tag{16}$$

$$\Delta J_w \Delta t' = \left(\frac{\varepsilon_c}{1+\varepsilon_c} \cdot \frac{1}{E_t} - \frac{1}{1+\varepsilon_c} \cdot \frac{1}{E_w} \right) J_w, \tag{17}$$

Then, the energy-saving coefficient of refrigeration is

$$j = \frac{\varepsilon_c}{1+\varepsilon_c} \cdot \frac{1}{E_t} - \frac{1}{1+\varepsilon_c} \cdot \frac{1}{E_w}. \tag{18}$$

2.3 Heat Island Effect on Value-Added Energy Consumption

Water chillers, air chillers, direct-fired absorption chiller and other refrigeration and air-conditioning units all discharge waste heat into the atmosphere, often forming the urban heat island effect, especially in summer. The following equations are satisfied when the atmosphere is used as a cold source,

$$C_J = A_J(T_J - T_N) + B_J, \quad H_J = C_J \frac{1+\varepsilon_J}{\varepsilon_J} \tag{19}$$

where $A_J(T_J - T_N)$ is the cooling load of building envelope,

$$N_J = \frac{C_J}{\varepsilon_J} = A_J \frac{1}{\varepsilon_J}(T_J - T_N) + \frac{B_J}{\varepsilon_J}, \tag{20}$$

$$C_S = A_S(T_S - T_N) + B_S, \quad H_S = C_S \frac{1 + \varepsilon_S}{\varepsilon_S}, \tag{21}$$

$$N_S = \frac{C_S}{\varepsilon_S} = A_S \frac{1}{\varepsilon_S}(T_S - T_N) + \frac{B_S}{\varepsilon_S}, \tag{22}$$

where T_J, T_S and T_N are the suburb, city and indoor temperature, respectively. C_J and C_S are the air-conditioning cooling loads corresponding to T_J and T_S. N_J and N_S are the driving power corresponding to T_J and T_S. A_J and A_S are building envelope characteristic values. B_J and B_S are air-conditioning load independent of building structure. ε_J and ε_S are the refrigeration coefficient. In the same building, $A_J = A_S$ = constant and $B_J = B_S$.

By (22)/(23), we can get,

$$\frac{N_S}{N_J} = \frac{\varepsilon_J}{\varepsilon_S} \frac{A_S(T_S - T_N) + B_S}{A_J(T_J - T_N) + B_J}, \tag{23}$$

$$B_J = \xi \cdot C_J \tag{24}$$

$$B_J = B_S = A_J(T_J - T_N) \frac{\xi}{1 - \xi} \tag{25}$$

$$\frac{N_S}{N_J} = \frac{\varepsilon_J}{\varepsilon_S} \frac{A_S(T_S - T_N) + \frac{\xi}{1-\xi} A_J(T_J - T_N)}{A_J(T_J - T_N)\left(1 + \frac{\xi}{1-\xi}\right)} \tag{26}$$

$$\frac{N_S}{N_J} = \frac{\varepsilon_J}{\varepsilon_S} \left[(1 - \xi)\frac{T_S - T_N}{T_J - T_N} + \xi\right] \tag{27}$$

The CPU of cooling system decline with the outdoor temperature increase.

$$\varepsilon_J = [1 + \Delta\varepsilon(T_S - T_J)]\varepsilon_J \tag{28}$$

we substitute (28) into (27)

$$\frac{N_S}{N_J} = \frac{1}{1 - \Delta\varepsilon(T_S - T_J)} \left[(1 - \xi)\frac{T_S - T_N}{T_J - T_N} + \xi\right] \tag{29}$$

To give an example, if $T_N = 26\,°C$, $T_J = 37\,°C$ and $T_S = 40\,°C$, $\frac{N_S}{N_J} = 1.21$ (let $\xi = 0.5$, $\Delta\varepsilon = 2\%$). We could see the energy that traditional heat pump consumed is 20% more than that consumed by water source heat pump.

2.4 Quantify Energy-Saving and Environmental Benefits

We list the energy-saving and energy-using coefficients into Table 2 by using Table 1 which from relevant manufacturer.

These charts show that the surface water source heat pump can save more energy than other systems. According to these charts' data, water source heat pump consumes 1.15 times primary energy, while other systems need 1.5–2.2 times primary energy. The surface water source refrigeration consumed about 0.2 times primary energy, less than other cooling systems.

Using the energy-saving coefficient we can calculate the amount of saved energy, for example, if each cubic meter surface water source increase (decrease) in 4 °C, there will be 16.72 MJ useful energy. For the heating, it can save 6.19–16.72 MJ primary energy, while the figure for cooling is 4.85–7.02 MJ. It also means that surface water source heat pump can save 1.55–4.18 MJ/m^3 °C (heating) and 1.21–1.76 MJ/m^3 °C (refrigeration). These data also offer information about environmental benefits.

In Table 4, we give the data that surface water source heat pump decrease in the emission of contamination, by using Table 3.

Table 1 Energy utilization rate of some heating and air-conditioning system

System name	Heating E_t	Refrigerate E_t	Driving E_w
Coal fuel + water cooler	0.7 × (1–10%) = 0.63	0.31 × 4×(1–10%) = 1.12	33% (1–5%) =0.31
Heating net + water cooler	0.75 × (1–10%) = 0.68	0.31 × 4×(1–10%) = 1.12	33% (1–5%) =0.31
Direct-fired absorption chiller	0.9 × (1–10%) = 0.81	0.9 × 1.2 × (1–10%) = 0.97	0.9
Air source heat pump	0.31 × 3.2 × (1–10%) = 0.89	0.31 × 3.5 × (1–10%) = 0.98	33% (1–5%) =0.31
Water source heat pump	0.31 × 3.8 × (1–10%) = 1.06	0.31 × 4.5 × (1–10%) = 1.26	33% (1–5%) =0.31

Table 2 Energy-saving and energy-using coefficients of surface water source heat pump system

System name	Heating q_y	Refrigerate q_y	Heating q_j	Refrigerate q_j
Coal fuel + water cooler	2.15	0.73	1	0.14
Heating net + water cooler	2.0	0.73	0.85	0.14
Direct-fired absorption chiller	1.68	0.84	0.53	0.25
Air source heat pump	1.52	0.83	0.37	0.24
Water source heat pump	1.15	0.59	–	–

Table 3 Contamination emission standard

Contamination	Rated (g/Gcal)	Contamination	Rated
SO_2	34	CO_2	4.82 kg/Gcal
NO_x	1.8	Dust	2.2 g/Gcal

Table 4 Pollutants reduction quantity of surface water source heat pump system

Contamination	Heating	Refrigeration
SO_2 (g/m^3 °C)	(12.6–34) × 10^{-3}	(9.84–14) × 10^{-3}
NO_x (g/m^3 °C)	(0.67–1.8) × 10^{-3}	(0.52–0.76) × 10^{-3}
CO_2 (kg/m^3 °C)	(1.79–4.82) × 10^{-3}	(1.4–2.03) × 10^{-3}
Dust (g/m^3 °C)	(0.82–2.2) × 10^{-3}	(0.64–0.93) × 10^{-3}

3 Conclusions

Firstly, using the energy-saving and energy-using coefficients predicate surface water source heat pump's energy-saving and environmental benefits, which can be clear and intuitive. Secondly, traditional air-cooling heat pump could lead to urban heat island, so we quantify the energy-saving potential of surface water source heat pump, the relative added energy value may get to 20% because of the heat island effect. Finally, we got that energy-saving quantity of heating is 1.55–4.18 MJ/m^3 °C and refrigeration is 1.21–1.76 MJ/m^3 °C and the amount of reduced carbon dioxide are (1.79–4.82) × 10^{-3} kg/m^3 °C (heating) and (1.4–2.03) × 10^{-3} kg/m^3 °C (refrigeration).

Acknowledgements This work was financial supported by "the Fundamental Research Funds for the Central Universities, (Grant No.2572018BJ14) and National Natural Science Foundation of China (Grant No.51808102).

References

1. Dadzie, J., Runeson, G.: Sustainable technologies as determinants of energy efficient upgrade of existing buildings. In: GP IEEE Conference 2018, LCSGCE, IEEE, Malaysia, 145–149 2018
2. Anmin, T.: Application of urban sewage source heat pump in building energy saving. Heat. Refrig. **5**(11), 22–25 (2017)
3. Lou, S.J., Yang, Y., Wang, H.L., Smith, S.J.: Black carbon amplifies haze over the North China plain by weakening the East Asian winter monsoon. Geophys. Res. Lett. **1**(46), 452–460 (2019)
4. Gu, Y., Deng, H.: The feasibility analysis of wastewater source heat pump using the urban wastewater heat. Res. J. Appl. Sci. Eng. Technol. **4**(18), 3501–3504 (2012)

Application of New Evaporative Cooling Air-Conditioning System in a Data Center in Xinjiang

Xiang Huang, Zhicheng Guo, Zhenwu Tian, Jingwen Xuan and Jincheng Yan

Abstract The principle, characteristics, operation mode and performance test results of the evaporative cooling air–water air-conditioning system are introduced in this paper, in which the evaporative water chiller is adopted as the main cooling source and the evaporative cooling fresh air handling unit as the auxiliary cooling source in a data center room in Xinjiang. The system utilizes local dry air energy and low-temperature air as natural cooling sources. It not only realizes 100% free cooling throughout the year, but also effectively solves the high-temperature crash problem and freezing in the cooling tower of traditional water-cooled air-conditioning system in winter in Xinjiang based on direct evaporative cooling technology, indirect evaporative cooling technology and ethylene glycol-free cooling technology. According to the summer test of the system, the system is stable, and the refrigerating coefficient of performance of the air-conditioning system is 6.65, and the summated refrigerating coefficient of performance (SCOP) is 16.64, which has a significant effect on reducing the energy consumption of data center.

Keywords Data center · Free cooling · Evaporative cooling air–water air-conditioning system

1 Introduction

The energy consumption of data centers in China accounted for about 2% of the total social energy consumption, 10% of the building energy consumption [1] and the energy consumption of the air-conditioning system accounts for about 40% of the total energy consumption of data centers [2, 3]. It can be seen that the refrigeration and air-conditioning industry are under heavy responsibilities. The sensible heat load accounted for most of the cooling load in data center, and it needs to be cooled all year-round. Thus, evaporative cooling air-conditioning technology has great potential in data center cooling [4].

At present, the application form of evaporative cooling air-conditioning technology in the data center air-conditioning system is single at home and abroad, and the running time is short, and the energy-saving advantages of evaporative cooling air-conditioning technology cannot be fully realized [5, 6]. Generally, there are two application modes of air-side evaporative cooling technology form in the system. The first type is closely combined with the civil structure of the building, and the air-conditioning system blends in the building so that the air-side evaporative cooling mode under appropriate working conditions can be operated well [7]. The second type is a functional section of the air-conditioning unit. The air-conditioning unit is generally placed around the data center buildings, on the roof or in the air-conditioning room, and the cool air produced by the air-conditioning unit is sent into the data center room by setting up an air duct [8].

A new evaporative cooling air-conditioning system for a data center room in Xinjiang is the first practical application of evaporative cooling air–water air-conditioning system in the field of data center room at home and abroad. The system fully utilizes local dry air energy and low-temperature air as free cooling source. It not only realizes 100% free cooling throughout the year, but also effectively solves the problem of high-temperature crash in traditional air-cooled air-conditioning system and freezing in the cooling tower of traditional water-cooled air-conditioning system in winter in Xinjiang based on direct evaporative cooling technology, indirect evaporative cooling technology and ethylene glycol-free cooling technology. The principle and operation mode of the system are introduced below, and the summer application effect is tested and analyzed.

2 Principle

2.1 Working Principle of Evaporative Water Chiller

Xinjiang is hot and dry in summer, and the air contains abundant dry air energy. It is a highly applicable area of evaporative cooling technology. In the long cold winter, the air contains abundant cold energy, which is very suitable for the application of ethylene glycol-free cooling technology. Combined with the above natural and technical conditions, the evaporative water chiller can be used in cooling year-round by integrating the cooling coil with ethylene glycol-free cooling instead of the conventional dry cooler to meet the annual cooling requirements of the data center. The chiller adopts double-side air inlet mode, and each functional section is symmetrically arranged. It consists of evaporative cooling section and ethylene glycol-free cooling section. The structure principle of the evaporative water chiller is shown in Fig. 1, and the physical picture is shown in Fig. 2.

The evaporative cooling section is responsible for the transition season and summer cooling demand of the data center room. It consists of cooling coil, single-stage vertical tube-type indirect evaporative cooler and padding tower. Based on the cooling principle of evaporative water chiller, cold water is produced and supplied to the end of air-conditioning system. In summer, the evaporative cooling section can realize two-stage pre-cooling (opening the cooling coil and vertical tube-type indirect evaporative cooler) for the air entering the padding tower. When

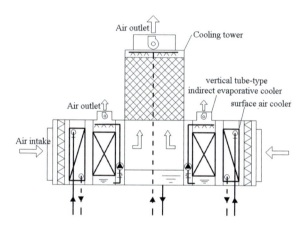

Fig. 1 Structural schematic diagram

Fig. 2 Physical picture

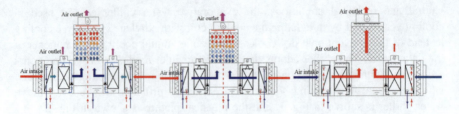

Fig. 3 Working principle of evaporative water chiller

the wet-bulb temperature is low in the transition season, the first stage pre-cooling (opening cooling coil and closing vertical tube-type indirect evaporative cooler) of air entering padding tower can be realized. Under the premise of ensuring that the supply water temperature of chiller meets the requirement, the energy-saving operation is realized. The working principle of evaporative cooling section is shown in Fig. 3.

The ethylene glycol section is mainly responsible for the cooling demand of the data center room in the low-temperature season, with two cooling coils as the main heat exchangers. Outdoor low-temperature fresh air enters the cooling coils to cool the ethylene glycol aqueous solution in the coil. Cooled ethylene glycol aqueous solution is supplied to the end of air-conditioning system, while heated fresh air with higher temperature is discharged through the exhaust fan of the vertical tube-type indirect evaporative cooler and padding tower. The working principle of the ethylene glycol section is shown in Fig. 3.

2.2 Working Principle of the Operation Mode

When the outdoor ambient air wet-bulb temperature is 0–18.2 °C, the operation mode is switched to the water-side evaporative cooling operation mode. Air-conditioning water system flow: the evaporative cooling section of evaporative water chiller has been running to produce cold water for the primary side of a plate heat exchanger, which can cool the circulating water of plate heat exchanger secondary side, and then the return water with higher temperature is sprayed into the padding tower of evaporative cooling section; thus, the water is continued to be cooled, and the primary water system is formed. The cold water produced on the secondary side of plate heat exchanger is supplied to the air-conditioning unit in the data center room. After absorbing heat, part of the return water with higher temperature is flowed into the cooling coil of the evaporative cooling section to pre-cool the outdoor fresh air into the padding tower. After the temperature of return water is increased again and mixed with the other part of return water, the mixed return water is flowed into the secondary side of the plate heat exchanger and then cooled by the cold water on the primary side of the plate heat exchanger, so

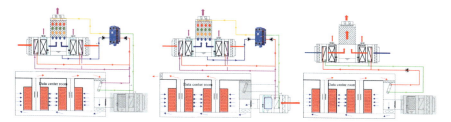

Fig. 4 Working principle of the operation mode

that the secondary water system is formed. The air distribution in the data center room is internal circulation. The working principle of the water- side evaporative cooling operation mode is shown in Fig. 4.

When the outdoor wet-bulb temperature of ambient air is higher than 18.2 °C, the operation mode is switched to the water-side and air-side composite evaporative cooling operation mode. Air-conditioning water system flow: the primary water system of this operation mode is the same as the water-side evaporative cooling operation mode. The cold water produced on the secondary side of the plate heat exchanger is supplied to the evaporative cooling fresh air handling unit. After absorbing heat, part of the return water with higher temperature is flowed into the cooling coil of the evaporative cooling section to pre-cool the outdoor fresh air into the chiller. After the temperature of return water is increased again and mixed with the other part of return water, the mixed return water is flowed into the secondary side of the plate heat exchanger and then cooled by the cold water on the primary side of the plate heat exchanger, so that the secondary water system is formed. The air distribution in the data center room is external circulation. The working principle of the water-side and air-side composite evaporative cooling operation mode is shown in Fig. 4.

When the dry-bulb temperature of outdoor ambient air is less than 3 °C, the operation mode is switched to ethylene glycol-free cooling operation mode. Air-conditioning water system flow: the ethylene glycol-free cooling section of evaporative water chiller has been running to provide cold water mixed with 45% ethylene glycol concentration solution for the air-conditioning unit. The ethylene glycol water solution with increased temperature after absorbing heat flows back to the cooling coil in ethylene glycol-free cooling section and continues to be cooled by outdoor low-temperature fresh air, thus the water system cycle is formed. The air distribution in the data center room is internal circulation. The working principle of the ethylene glycol-free cooling operation mode is shown in Fig. 4.

3 Test and Analysis

The project is located in Urumqi, with a construction area of 10738.2 m², a building height of 23.3 meters, five floors above ground and a maximum of 1500 IT cabinets. The total cooling load of the air-conditioning system is 2767 kW. There are 16 evaporative water chillers, 44 evaporative cooling fresh air handling units and 22 special high-temperature chilled water air-conditioning units for data center room in the system.

3.1 Performance Test of Evaporative Water Chiller

From Fig. 5, it can be seen that the average dry-bulb temperature, relative humidity and wet-bulb temperature of ambient air are 33.5 °C, 22.3% and 17.9 °C, and the average water supply temperature of the unit is 15.4 °C, which meets the requirement that the water supply temperature is not more than 16 °C and is 2 °C lower than the measured ambient air wet-bulb temperature, which can make a technological breakthrough.

3.2 Performance Test of Evaporative Cooling Fresh Air Handling Unit

From Fig. 6, it can be seen that the average ambient dry-bulb temperature is 33.3 °C, relative humidity is 23%, wet-bulb temperature 18 °C, and the average outlet air temperature of the unit is 14.6 °C, which is 3 °C lower than the ambient air wet-bulb temperature.

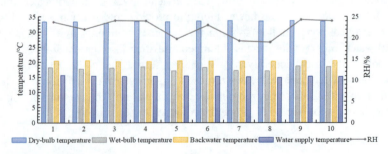

Fig. 5 Measured data of water chillers

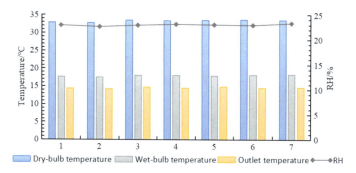

Fig. 6 Measured data of fresh air handling unit

3.3 Performance Test of the Free Cooling Air-Conditioning System

During the summer measurement period, the IT equipment was not fully assembled, only 25% of them are assembled, and the cooling load of the system was about 690 kW. At this time, three evaporative water chillers, one plate heat exchanger, one primary water pump, one secondary water pump and five special high-temperature chilled water air-conditioning units are equipped for the data center room. Under the above test conditions, the water consumption, EER of the unit, COP and SCOP of the system are measured synchronously [9]. The performance parameters are shown in Table 1. The results show that the energy-saving and water-saving effects of the system are remarkable.

3.4 Measurement and Analysis of Environment in Data Center Room

From Fig. 7, it can be seen that the average temperature and relative humidity of the four test points along the floor height in the central area of the cabinet air inlet are 23.5 °C and 42.8%, respectively. The dew point temperature is calculated to be 10.1 °C. In space of different height and same channel, the temperature increases

Table 1 Measured values of performance parameters

	Water consumption m³/h	EER kW/kW	COP kW/kW	SCOP kW/kW
Water chiller	0.34	15.8	–	–
Fresh air handling unit	0.06	18.1	–	–
System	–	–	6.65	16.64

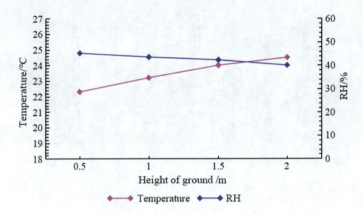

Fig. 7 Temperature and relative humidity field in air inlet area of the cabinet

slightly with the increase of height, and the air flow in the cold and hot channel does not recirculate with each other, which meets the environmental requirements of the data center room [10].

4 Conclusions

The evaporative cooling air–water air-conditioning system based on direct evaporative cooling technology, indirect evaporative cooling technology and ethylene glycol-free cooling technology can achieve 100% free cooling throughout the year when it is applied to the data center room in dry area.

Because of the intensive internal–external cooling composite heat transfer technology, the water supply temperature of evaporative water chiller is 2 °C lower than the ambient wet-bulb temperature, and the outlet temperature of evaporative cooling fresh air handling unit is 3 °C lower than the wet-bulb temperature of the inlet air. The breakthrough has been made in the unit for obtaining sub-wet-bulb temperature cold water and cold air.

The refrigerating coefficient of performance of the air-conditioning system is 6.65; the system coefficient of the summated refrigerating coefficient of performance (SCOP) is 16.64. The water consumption of evaporative water chiller and evaporative cooling fresh air handling unit is 0.34 and 0.06 m^3/h, respectively. Therefore, the energy-saving and water-saving effects of the system are remarkable.

Acknowledgements Supported by National key research and development program(Grant No.2016YFC0700404).

References

1. Yin, P.: Research on data centers (1): current situation and problem analysis. HVAC **46**(8), 42–52 (2016)
2. Li, T., Huang, X., et al.: Test and analysis of data center air condition systems in five provinces of northwest China. HVAC **48**(6), 8–12 (2018)
3. Niu, X., Xia, C., et al.: Air conditioning system design with lake water cooling technology of a data center in Qiandao Lake. HVAC **46**(10), 14–17 (2016)
4. Huang, X., Fan, K., et al.: Discussion on application the data center of evaporative cooling technology. Refrig. Air-Cond. **13**(8), 16–22 (2013)
5. Zhang, H., Shao, S., et al.: Research advances in free cooling technology of data centers. J. Refrig. **37**(4), 47–56 (2016)
6. Yin, P.: Research on data centers (7): Natural cooling. HVAC **47**(11), 49–60 (2017)
7. Mu, Z., Wang, Y.: Air conditioning design for Ningxia Zhongwei cloud computing data center. HVAC **46**(10), 23–26 (2016)
8. Yang, L., Huang, X., et al.: Feasibility of hybrid air conditioning unit applied to data centers. HVAC **46**(10), 9–13 (2016)
9. GB50189-2015: Design standard for energy efficiency of public buildings. China Architecture & Building Press (2015)
10. GB50174-2017: Code for design of data center, China Planning Press (2017)

Trial-and-Error Method for Variable Outdoor Air Volume Setpoint of VAV System Based on Outdoor Air Damper Static Pressure Difference Control

Pengmin Hua, Tianyi Zhao, Wuhe Dai and Jili Zhang

Abstract A variable outdoor air (OA) volume setpoint control method is proposed to solve the problem of mismatch between OA supply and demand in multi-zone VAV system. The method firstly obtains the OA volume setpoint according to the indoor CO_2 concentration change and then combines the OA damper static pressure difference control method to discretize the continuous OA volume setpoint to obtain different grades of OA volume setpoint, and finally optimize the OA volume setpoint according to total volume demand of terminals. The experiment was done to verify the feasibility of this new method and the result shows that the proposed method can be well fitted with other control loops in the air-conditioning system to ensure the demand of indoor OA volume, and this method is also suitable for online applications.

Keywords VAV system · Variable OA volume setpoint · Online control · Trial-and-error method

1 Introduction

Indoor air quality has been a hotspot because of its effect to people's working mood and efficiency. Variable air volume (VAV) air-conditioning system has been widely used in large public buildings because of its advantages of creating a good indoor environment, flexible control and energy saving [1]. The control of OA volume is the key to ensure indoor air quality to meet the requirements, and the distribution of OA volume has always been the difficulty in its control. HVAC system accounts for 40% of total building energy consumption [2], outdoor air energy consumption in VAV system accounts for more than 30% of that of air-conditioning system [3].

P. Hua · T. Zhao (✉) · W. Dai · J. Zhang
Dalian University of Technology, Dalian 116024, China
e-mail: zhaotianyi@dlut.edu.cn

W. Dai
e-mail: 745149719@qq.com

It has been a focus of OA volume control research to reduce energy consumption while guarantee outdoor air supply in various regions. And a lot of research in this area has been done at home and abroad. In 2015, Lin et al. suggested that the demand-controlled ventilation control strategies should been considered for CO_2-based demand-controlled ventilation for multiple-zone single duct VAV systems with terminal reheat [4]. Kim et al. proposed a stratification control method for the minimum airflow of VAV terminal unit and proved that the proposed method was more effective than the existing control method [5]. In 2017, Wang combines indoor positioning system with VAV system to optimize indoor ventilation rate and the experiments show that the proposed method has a certain energy-saving potential [6]. It has been confirmed indirectly that variable OA volume setpoint is beneficial to reduce outdoor air energy consumption, but the researches which study OA volume control method directly from the perspective of OA volume setpoint is few. In this paper, an online control method which is based on outdoor air damper static pressure difference control method is proposed for VAV system with variable setpoint of OA volume.

2 Trial-and-Error Method for Variable OA Volume Setpoint

2.1 Principle of Static Pressure Difference Control Method of Outdoor Air Damper

The static pressure difference of outdoor air damper is obtained via setting static pressure monitoring points before and after outdoor air damper. After determining the setpoint of OA volume, keep the return damper fully open at the maximum supply air volume and adjust the position of outdoor air damper to make OA volume reach setpoint. At this time, the position of outdoor air damper and the static pressure difference before and after outdoor air damper are the setpoint needed to maintain the setpoint of OA volume. Finally, keep the position of outdoor air damper unchanged, change the position of return air damper with the change of supply air volume to maintain the static pressure difference setpoint of outdoor air damper unchanged to realize the control of OA volume. The control principle is shown in Fig. 1.

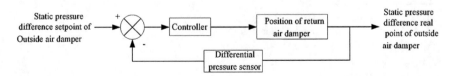

Fig. 1 Principle of static pressure difference control method of outdoor air damper

2.2 Method for Determining Variable Setpoint of OA Volume

Determine the range of the setpoint of OA volume [$OaSet_{min}$, $OaSet_{max}$], $OaSet_{min}$ is the minimum setpoint of OA volume and $OaSet_{max}$ is the maximum. According to the principle of static pressure difference control method of outdoor air damper, if the setpoint of OA volume changes continuously, the actuator needs to operate continuously to realize the control, leading to instability of total system. Therefore, in this paper, the trial-and-error method has been used to discretize the setpoint of OA volume to achieve stable control by determining the initial position of outdoor air damper and the setpoint of pressure difference in each grade. In this method, firstly, a fixed range of CO_2 concentration is defined as the target concentration range to judge whether the setpoint of OA volume changes or not. If the CO_2 concentration of room is greater than the upper limit of concentration, the setpoint of OA volume is increased with fixed step size, and if it is less than the lower limit of concentration, the setpoint of OA volume is reduced with fixed step size. Delay a period of time after the change of OA volume setpoint to determine whether it needs to be changed again. The step size is defined as the difference between two adjacent OA volume grades and the control principle is shown in Fig. 2.

The trial-and-error method determines the setpoint of OA volume based on the feedback of CO_2 concentration of controlled area. In fact, the demand of OA volume varies with the change of total volume demand of terminals. When the total volume demand of terminals decreases to a certain extent, a smaller setpoint of OA volume can be selected to achieve control, which is not only saves energy consumption, but also makes the control of outdoor air more stable. At this time, indoor CO_2 concentration is no longer used as the criterion to determine the setpoint of OA volume.

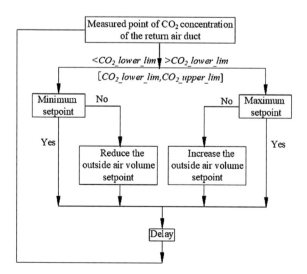

Fig. 2 Logic diagram of variable setpoint of OA volume by trial-and-error method

3 Results

3.1 Introduction of Test System

A comprehensive test-bed for VAV air-conditioning system is established which includes complete VAV air-conditioning system equipment, sensors, actuators and automatic control system. The control system of the test-bed adopts direct digital controller (DDC) and the point-to-point connection between sensors and actuators and controllers constitutes a distributed system, which is arranged as shown in Fig. 3. Physical photos of the test-bed are shown in Fig. 4, in which 1 is outdoor air damper, 2 is return air damper, 3 is exhaust air damper, 4 is filter, 5 is return fan, 6 is orifice flowmeter of return air, 7 is surface cooler, 8 is supply fan, 9 is orifice flowmeter of supply air and 10 is room. C1 is the indoor temperature control loop, C2 is the supply air control loop, C3 is the outdoor air control loop and C4 is the supply air temperature control loop.

3.2 Test Scheme

Firstly, determine the range of air supply volume is [910, 1650] m³/h, then discretize the air supply volume into three grades, that is [910, 1280, 1650] m³/h. Determine the range of outdoor air volume is [450, 650] m³/h, then discretize the OA volume into three grades, that is [450, 550, 650] m³/h. The setpoint of OA volume corresponding to the lower limit of total volume demand of terminals in this experiment is 150 m³/h, which is determined not to change with indoor CO_2 concentration. The setpoint of OA volume is optimized by the change range of total volume demand of terminals and the corresponding relationship is shown in Eq. (1).

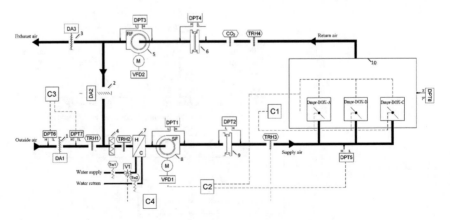

Fig. 3 Principle diagram of VAV air-conditioning system

(a) Air-cooled heat pump units   (b) Air handling unit   (c) VAV terminal unit

Fig. 4 Physical photos of the test-bed

$$OaSet \in \begin{cases} \{450, 550, 650\} & 1280 < QSet_{all} \leq 1650 \\ \{450, 550\} & 910 < QSet_{all} \leq 1280 \\ \{150\} & QSet_{all} \leq 910 \end{cases} \quad (1)$$

The ideal concentration range of CO_2 in the controlled area is [900, 1100] ppm. The setpoint of OA volume is adjusted according to the rule of Eq. (2), *OaSet* is the current setpoint of OA volume and *OaReSet* is the reset setpoint of OA volume. The step size of setpoint of OA volume is 100 m³/h, and the adjustment period is 3 min. The procedure flow chart of the trial-and-error method of setpoint of OA volume is shown in Fig. 5.

$$OaReSet = \begin{cases} OaSet + 100 & CO_2 > 1100 \\ OaSet & 900 \leq CO_2 \leq 1100 \\ OaSet - 100 & CO_2 < 900 \end{cases} \quad (2)$$

In static pressure difference control method of outdoor air damper, the PI value of regulating return air damper is −150, 20, and the control parameters of variable setpoint of OA volume are shown in Table 1. In the experiment, CO_2 release and indoor temperature setpoints are used as disturbances of control. First open doors and windows before the test, and then test indoor and outdoor CO_2 concentration with air quality analyzer, finally close doors and windows to start the test when the two CO_2 concentration approaches. The experiment lasted for 60 min, and CO_2 (180 L) was controlled by a CO_2 flowmeter in the laboratory during the first 20 min of the experiment. At the 30th minute, change the terminal temperature setpoint and the rules are as shown in Table 2.

3.3 Result Analysis

The setting and measured points of OA volume which change with the indoor CO_2 concentration and terminal demand total air volume $QSet_{all}$ are shown in Fig. 6.

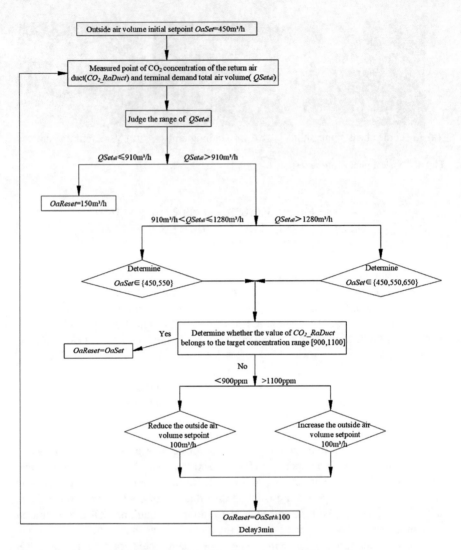

Fig. 5 Flow chart of trial-and-error method for OA volume variable setpoint

Table 1 Control parameters of OA volume variable setpoint

Setpoint of OA volume (m³/h)	Setpoint of static pressure difference of outdoor air damper (Pa)	Position of outdoor air damper (%)
650	53	58
550	87	46
450	115	36
150	45	20

Table 2 Temperature setting rules of variable air volume terminals

Setting time	Temperature setpoint of BOX-A (°C)	Temperature setpoint of BOX-B (°C)	Temperature setpoint of BOX-C (°C)
Initial time	21	21	20
30 min of test	22.5	22.5	22.5

Point A–H is the inflection point of setpoint of OA volume under the constraints of indoor CO_2 concentration and total volume demand of terminals, the number of changes is 8, and the maximum setpoint is 650 m^3/h and the minimum is 150 m^3/h. The mean setpoint and measured setpoint of OA volume are 419 and 429 m^3/h, respectively, and the mean value of relative deviation of OA volume is 11.1%. There is a positive deviation of OA volume when the setpoint is 650 m^3/h. The mean measured setpoint of OA volume is 161 and 175 m^3/h, respectively, in the two periods with OA volume setpoint of 150 m^3/h (Figs. 7 and 8).

It is shown in Table 3 that the mean measured and setting points of static pressure difference of supply air volume are basically the same. The mean measured and setting points of static pressure difference of OA volume are basically the same too. The control effect of supply air volume control loop and OA volume control loop is better. During the experiment, the time that the indoor positive pressure is maintained above 5 Pa accounted for 94% of total time, the average temperature of the air supply is 26.82 °C, and the overall effect of air supply temperature is not bad.

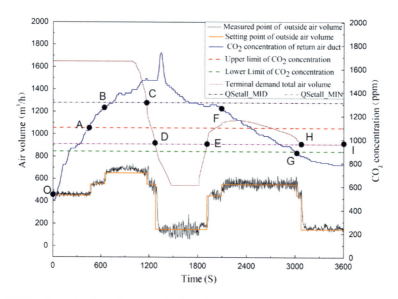

Fig. 6 OA volume in trial-and-error method for OA volume variable setpoint

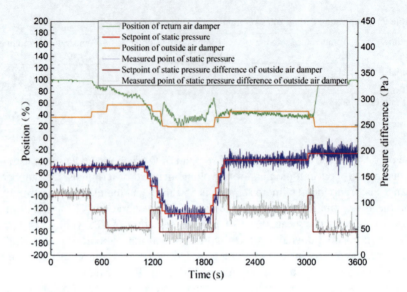

Fig. 7 Control effect of OA volume and static pressure difference

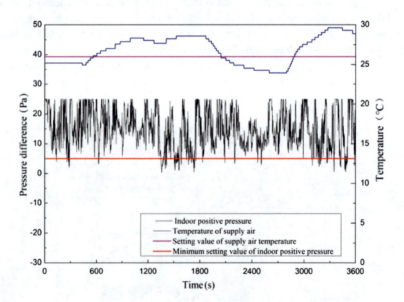

Fig. 8 Indoor positive pressure and supply air temperature

Table 3 Control index of supply air volume and outdoor air volume

Type	Evaluating indicator	Value	Type	Evaluating indicator	Value
Supply air volume	Change number of setpoint of static pressure	14	Outdoor air volume	Change number of setpoint of static pressure	8
	Mean setpoint of static pressure (Pa)	159.69		Mean setpoint of static pressure (Pa)	74.01
	Mean measured value of static pressure (Pa)	159.73		Mean measured value of static pressure (Pa)	75.25
	Mean relative deviation of static pressure setpoint (%)	4.81%		Mean relative deviation of static pressure setpoint (%)	14.39%

4 Conclusions

An online control method for outdoor air volume variable setpoint which is based on static pressure difference control method of outdoor air damper is proposed in this paper. This method cannot only meet the demand of indoor air quality, but also save energy consumption of outdoor air. Experiment has been done to study its online application effect and the control method proposed in this paper can be well coupled with other control loops in the air-conditioning system to meet the control requirements and it is also suitable for online application.

Acknowledgements The project is supported by National Key Research and Development Project of China entitled New generation Intelligent building platform techniques (Number 2017YFC0704100). The project is supported by basic research business fees of central colleges and universities (Number DUT17ZD232). The project is supported by Liaoning Natural Science Foundation Guidance Plan (Number 20180551057). The project is supported by Dalian High-level Talent Innovation Support Program (Youth Technology Star) (Number 2017RQ099).

References

1. Tukur, A., Hallinan, K.P.: Statistically informed static pressure control in multiple-zone VAV systems. Energy Build. **135**, 244–252 (2017)
2. Zhu, Y., Jin, X., Du, Z., Fang, X., Fan, B.: Control and energy simulation of variable refrigerant flow air conditioning system combined with outdoor air processing unit. Appl. Therm. Eng. **64**, 385–395 (2014)
3. Chao, C.Y.H., Hu, J.S.: Development of a dual-mode demand control ventilation strategy for indoor air quality control and energy saving. Build. Environ. **39**, 385–397 (2004)

4. Lin, X., Lau, J.: Demand-controlled ventilation for multiple-zone HVAC systems—part 2: CO_2-based dynamic reset with zone primary airflow minimum set-point reset (RP-1547). Sci. Technol. Built Environ. **21**, 1100–1108 (2015)
5. Kim, H.J., Kang, S.H., Cho, Y.H.: A Study on the control method without stratification of single duct VAV terminal units. J. Asian Arch. Build. Eng. **14**, 467–474 (2015)
6. Wang, W., Chen, J., Huang, G., Lu, Y.: Energy efficient HVAC control for an IPS-enabled large space in commercial buildings through dynamic spatial occupancy distribution. Appl. Energy **207**, 305–323 (2017)

An Extension Theory-Based Fault Diagnosis Method for an Air Source Heat Pump

Yudong Xia, Qiang Ding, Shu Jiangzhou, Yin Liu and Xuejun Zhang

Abstract Fault diagnosis for air source heat pumps (ASHPs) is essential to maintain system's operational efficiency and safety. This paper reports a new extension theory-based fault diagnosis method for an ASHP system. An experimental ASHP was set up in environment chambers. Using the experimental ASHP, abnormal operations under five single faults imposed, including compressor valve leakage, reversing valve leakage, condensing airflow faulting, refrigerant liquid line restriction, and refrigerant charge fault, were implemented, and the related fault data obtained. The extension diagnosis method based on the extended correlation function and the matter-element model was then proposed to identify the different fault types. The diagnosis results showed that the proposed fault diagnosis method was able to detect the malfunction types occurring in the experimental ASHP system correctly and promptly.

Keywords Air source heat pump · Matter-element model · Fault diagnosis · Extension theory

1 Introduction

Heat pump unit is an environmentally friendly and reliable device to maintain indoor thermal comfort for both space heating and cooling. During a cooling season, it transfers heat from the indoor space to a heat sink, in the same way as an air conditioner does. During a heating season, it extracts heat from a heat source such as ambient air and wastewater and delivers the extracted heat energy to a heated indoor space. Air source heat pumps (ASHPs) are more widely employed in

---

Y. Xia · Q. Ding (✉) · S. Jiangzhou · Y. Liu
School of Automation, Institute of Energy Utilization and Automation, Hangzhou Dianzi University, Hangzhou 310018, China
e-mail: dingqiang@hdu.edu.cn

Y. Xia · X. Zhang
Institute of Refrigeration and Cryogenics, Zhejiang University, Hangzhou 310027, China

residential buildings than geothermal heat pumps for cooling or heating due to its advantages of simpler configuration and lower maintenance cost. However, under long-term operation, faults may be occurred and thus cause the abnormal operation of ASHPs, consequently, leading to performance degradation. For example, the degradation of 10–13% in COP for a residential heat pump would be resulted in when a condenser was blocked by 30% of its flow area, and a 30% refrigerant undercharge reduced cooling capacity by almost 15% on average [1]. Therefore, diagnosing faults precisely and timely are essential for the applications of ASHPs, contributing to both energy saving and environment protection. A number of fault diagnosis methods have been applied in HVAC&R system [2–4]. However, in terms of heat pumps or air conditioners, comparatively few studies on fault diagnosis based on data-driven methods may be identified [5, 6].

On the other hand, the extension theory has been successfully applied to fault diagnosis in power systems [7]. Since this theory allows classification problems without learning process, it should be also suitable for ASHP fault diagnosis application. Therefore, the design of an extension theory-based fault diagnosis method for an ASHP system is reported in this paper. Firstly, the descriptions of the experimental ASHP and tests setups are presented. Then the development of the new extension theory-based fault diagnosis method is detailed. This is followed by reporting the diagnosis results. Finally, discussions and conclusions are given.

2 Experimental System Descriptions and Tests Arrangement

2.1 Experimental System Descriptions

As a data-driven fault diagnosis method, process history data including normal and fault data were required to matter-element development. Therefore, a real experimental ASHP system where different operating faults could be manually imposed was established. Two environment chambers were used to simulate indoor and outdoor operation conditions. The experimental ASHP is schematically shown in Fig. 1. As seen, the experimental ASHP mainly consisted of a compressor, a tube-louver-fin evaporator, an air-cooled tube-plate-fin condenser, a capillary tube, and a reversing valve. The evaporator was placed in the indoor environment chamber, and the others were installed in the outdoor environment chamber. The nominal cooling capacity of the ASHP was 7.5 kW. In addition, three extra electronic valves were installed in the experimental system to achieve the operating faults implementation.

The experimental ASHP system was fully instrumented for measuring all of its operating parameters, including refrigerant temperature, air temperature, and refrigerant pressure. The refrigerant temperature sensors in various locations in the

An Extension Theory-Based Fault Diagnosis Method ...

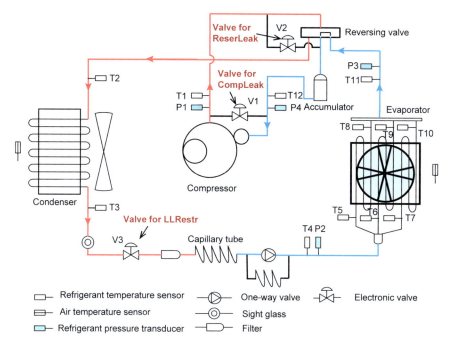

Fig. 1 Schematic diagram of the experimental ASHP

system as shown in Fig. 1 were type T thermocouples with reported uncertainties of ±0.5 °C, and the temperature sensors for air were of platinum resistance type with a pre-calibrated accuracy of ±0.1 °C. Refrigerant pressure was measured using pressure transmitters with an accuracy of ±0.3% of full-scale reading. The arrangement of the instruments in the experimental ASHP system is illustrated in Fig. 1. All measurements were computerized and transferred to a data acquisition system for logging and recording.

2.2 Tests Arrangement

Six sets of test, including one set of normal operation and five sets of abnormal operation, were carried out using the experimental ASHP system to generate enough data for matter-element development. The definition and implementation of the five types of operating fault are listed in Table 1. Compressor valve leakage will cause a reduction in refrigerant mass flow rate circuited in the system. Normally, a leak occurs at the suction or discharge valves for the reciprocating type or between the high-pressure and low-pressure portions of the scroll type. Therefore,

Table 1 Faults descriptions and implementations

No.	Fault name	Symbol	Implementation
F1	Compressor valve leakage	CompLeak	Controlling the opening of V1
F2	Reserving valve leakage	ReserLeak	Controlling the opening of V2
F3	Air-side fouling of the condenser	CondFoul	Blocking the condenser face area
F4	Refrigerant undercharge/overcharge	ChargFault	Adding/removing the refrigerant based on a correct charge
F5	Liquid line restriction	LLRestr	Controlling the opening of V3

CompLeak was implemented using a hot gas bypass form the discharge line to the suction line of the compressor, and the fault level was regulated by controlling the opening of V1. ReserLeak happens from the hot gas (high pressure side) to suction gas passages (low pressure side), and thus, this type of fault was realized by V2 installed between the hot gas and suction gas passages. CondFoul was simply implemented by blocking portions of the condenser face area with paper sheets. ChargFault was implemented by adding or removing the refrigerant from a correctly charged system. LLRestr can be resulted from a dirty refrigerant dryer or filter and thus was implemented by closing V3 installed by the liquid refrigerant line. During the experiments, the ASHP was operated in cooling mode and the indoor air dry-bulb and wet-bulb temperatures were maintained at 27 and 19 °C, respectively, and the outdoor at 35 and 24 °C through controlling the indoor and outdoor environment chambers.

3 Extension Theory-Based Fault Diagnosis Method Development

3.1 Basics of Extension Theory

The matter-element model and extended mathematics are the main principles of extension theory. It can indicate the alterable relations between quality and quantity by matter-element transformation. The matter-element is one of the main theories in extension theory. A matter-element contains three essential factors, i.e., name of the matter, N, its characteristic, c, and the value related to the characteristic, v. The matter-element can be expressed as follows (13):

$$R = (N\ c\ v) \qquad (1)$$

The multi-dimensional matter-element can be written as:

$$R = (N\ c\ v) = \begin{bmatrix} R_1 \\ R_1 \\ R_2 \\ \ldots \\ R_n \end{bmatrix} = \begin{bmatrix} N & c_1 & v_1 \\ & c_2 & v_2 \\ & \ldots & \ldots \\ & c_n & v_n \end{bmatrix} \quad (2)$$

In Eq. (2), $R_j = (N\ c_j\ v_j)$ ($j = 1, 2, 3, \ldots, n$) is the sub-matter-element of R.

Another important principle of extension theory is the correlation function. Assuming $X_0 = \langle a, b \rangle$ and $X = \langle f, g \rangle$ are two intervals in the real number field, and $X_0 \subset \leftarrow X$, where X_0 and X are the classical (concerned) and neighborhood domains, respectively. Then the correlation function can be defined as:

$$K(x) = \begin{cases} -\rho(x, X_0) & x \in X_0 \\ \dfrac{\rho(x, X_0)}{\rho(x, X) - \rho(x, X_0)} & x \notin X_0 \end{cases} \quad (3)$$

where

$$\rho(x, X_0) = \left| x - \frac{a+b}{2} \right| - \frac{b-a}{2} \quad (4)$$

$$\rho(x, X) = \left| x - \frac{f+g}{2} \right| - \frac{g-f}{2} \quad (5)$$

Using the correlation function, the membership grade between x and X_0 could be evaluated. When $K(x) \geq 0$, it indicates the degrees to which x belongs to X_0. When $K(x) < 0$, it describes the degree to which x does not belong to X_0.

3.2 Proposed Fault Diagnosis Method Development

3.2.1 Matter-Element Model of Different Types of Fault for the ASHP

As mentioned above, using the extension theory for fault diagnosis, matter-element model should be developed initially. The matter-element model for the ASHP included six operational conditions, i.e., a normal operation and five abnormal operations. Eight operating parameters were selected as the feature parameters for characterizing the different types of operating fault. The eight parameters were the refrigerant temperature at the evaporator inlet, T_{re_in}, the refrigerant temperature at the evaporator outlet, T_{re_out}, the refrigerant temperature at the middle of the evaporator, T_{re_mid}, the refrigerant temperature at the condenser inlet, T_{rc_in}, the compressor discharge temperature, T_{dis}, the compressor suction temperature, T_{suc}, the compressor suction pressure, P_{suc}, and the compressor discharge pressure, P_{dis}. The matter-element model of every operating fault can be expressed as follows:

$$R_F = (F \quad \mathbf{c} \quad V_F) = \begin{bmatrix} R_{F0} \\ R_{F1} \\ R_{F2} \\ R_{F3} \\ R_{F4} \\ R_{F5} \end{bmatrix} = \begin{bmatrix} F_j & T_{re\_in} & V_{j1} \\ & T_{re\_out} & V_{j2} \\ & T_{re\_mid} & V_{j3} \\ & T_{rc\_in} & V_{j4} \\ & T_{dis} & V_{j5} \\ & T_{suc} & V_{j6} \\ & P_{suc} & V_{j7} \\ & T_{dis} & V_{j8} \end{bmatrix} \quad j = 0, 1, 2, \ldots, 5 \quad (6)$$

where F_0 indicates the normal operation and F_1–F_5 the five different types of fault as described in Table 1, and thus, there were totally six categories. V_F is the classic regions of every feature, which are assigned by the lower and upper boundary of each feature data. Consequently, using the experimental data the matter-element models for the six categories can be obtained as follows:

$$R_{F0} = \begin{bmatrix} F_0 & T_{re\_in} & \langle 5.681, 5.910 \rangle \\ & T_{re\_out} & \langle 7.235, 7.662 \rangle \\ & T_{re\_mid} & \langle 6.154, 6.458 \rangle \\ & T_{rc\_in} & \langle 78.770, 80.102 \rangle \\ & T_{dis} & \langle 83.995, 80.102 \rangle \\ & T_{suc} & \langle 0.402, 0.415 \rangle \\ & P_{suc} & \langle 0.401, 0.412 \rangle \\ & P_{dis} & \langle 1.532, 1.552 \rangle \end{bmatrix} \quad R_{F1} \begin{bmatrix} F_1 & T_{re\_in} & \langle 6.142, 6.418 \rangle \\ & T_{re\_out} & \langle 6.514, 6.925 \rangle \\ & T_{re\_mid} & \langle 6.408, 6.688 \rangle \\ & T_{rc\_in} & \langle 79.662, 79.812 \rangle \\ & T_{dis} & \langle 82.558, 84.687 \rangle \\ & T_{suc} & \langle 7.206, 7.682 \rangle \\ & P_{suc} & \langle 0.416, 0.425 \rangle \\ & T_{dis} & \langle 1.541, 1.567 \rangle \end{bmatrix}$$

$$R_{F2} = \begin{bmatrix} F_2 & T_{re\_in} & \langle 6.328, 6.602 \rangle \\ & T_{re\_out} & \langle 7.360, 7.856 \rangle \\ & T_{re\_mid} & \langle 6.651, 6.938 \rangle \\ & T_{rc\_in} & \langle 78.097, 78.302 \rangle \\ & T_{dis} & \langle 83.013, 83.275 \rangle \\ & T_{suc} & \langle 8.010, 7.312 \rangle \\ & P_{suc} & \langle 0.427, 0.432 \rangle \\ & P_{dis} & \langle 1.539, 1.547 \rangle \end{bmatrix} \quad R_{F3} = \begin{bmatrix} F_3 & T_{re\_in} & \langle 7.584, 7.986 \rangle \\ & T_{re\_out} & \langle 5.805, 6.332 \rangle \\ & T_{re\_mid} & \langle 7.875, 8.279 \rangle \\ & T_{rc\_in} & \langle 68.571, 70.296 \rangle \\ & T_{dis} & \langle 75.604, 75.693 \rangle \\ & T_{suc} & \langle 6.784, 6.946 \rangle \\ & P_{suc} & \langle 0.432, 0.439 \rangle \\ & P_{dis} & \langle 1.610, 1.718 \rangle \end{bmatrix}$$

$$R_{F4} = \begin{bmatrix} F_4 & T_{re\_in} & \langle 5.497, 5.667 \rangle \\ & T_{re\_out} & \langle 7.184, 7.506 \rangle \\ & T_{re\_mid} & \langle 5.873, 6.349 \rangle \\ & T_{rc\_in} & \langle 80.621, 80.796 \rangle \\ & T_{dis} & \langle 85.784, 85.603 \rangle \\ & T_{suc} & \langle 8.940, 9.356 \rangle \\ & P_{suc} & \langle 0.405, 0.411 \rangle \\ & P_{dis} & \langle 1.547, 1.557 \rangle \end{bmatrix} \quad R_{F5} = \begin{bmatrix} F_5 & T_{re\_in} & \langle 4.156, 4.358 \rangle \\ & T_{re\_out} & \langle 7.486, 7.616 \rangle \\ & T_{re\_mid} & \langle 4.898, 5.434 \rangle \\ & T_{rc\_in} & \langle 82.701, 83.256 \rangle \\ & T_{dis} & \langle 87.042, 88.453 \rangle \\ & T_{suc} & \langle 10.201, 10.346 \rangle \\ & P_{suc} & \langle 0.355, 0.367 \rangle \\ & P_{dis} & \langle 1.521, 1.530 \rangle \end{bmatrix}$$

In addition, one can set a matter-element model to express the neighborhood domain of every feature for describing the possible range of all fault set. The value range of neighborhood domain $V'_F = <f, g>$ could be determined from the maximum and minimum values of every feature, which can be expressed as follows:

$$R_F = (F \quad \mathbf{c} \quad V_F) = \begin{bmatrix} F & T_{re\_in} & \langle 4.156, 7.986 \rangle \\ & T_{re\_out} & \langle 5.905, 8.156 \rangle \\ & T_{re\_mid} & \langle 4.898, 83.256 \rangle \\ & T_{rc\_in} & \langle 69.571, 83.256 \rangle \\ & T_{dis} & \langle 75.604, 88.453 \rangle \\ & T_{suc} & \langle 0.355, 0.439 \rangle \\ & P_{dis} & \langle 1.521, 1.728 \rangle \end{bmatrix}$$

3.2.2 Fault Diagnosis Procedure for the Proposed Extension Theory-Based Method

After establishing the matter-element model for the ASHP at different operational conditions, the operating fault could be classified by evaluating the correlation function between the matter-element tested and the matter-element model developed. The diagnosis procedure is presented as follows:

Step 1 Establishing the matter-element model of every faulting category for the ASHP as indicated by Eq. (6).
Step 2 Setting the matter-element of the ASHP to be tested using the testing data.

$$R_F = (F_x \quad \mathbf{c} \quad V_F) = \begin{bmatrix} F_x & T_{\text{re\_in}} & v_{f1} \\ & T_{\text{re\_out}} & v_{f2} \\ & T_{\text{re\_mid}} & v_{f3} \\ & T_{\text{rc\_in}} & v_{f4} \\ & T_{\text{dis}} & v_{f5} \\ & T_{\text{suc}} & v_{f6} \\ & P_{\text{suc}} & v_{f7} \\ & P_{\text{dis}} & v_{f8} \end{bmatrix} \quad (7)$$

Step 3 Evaluating the correlation functions between the matter-element of the tested ASHP with the developed matter-element models of each fault type using the following equation.

$$K_{jk}(v_{fk}) = \begin{cases} \dfrac{-\rho(v_{fk}, V_{jk})}{|V_{jk}|} & v_{fk} \in V_{jk} \\ \dfrac{\rho(v_{fk}, V_{jk})}{\rho(v_{fk}, V'_{jk}) - \rho(v_{fk}, V_{jk})} & v_{fk} \notin V_{jk} \end{cases} \quad j = 0, 1, 2, \ldots, 5; k = 1, 2, \ldots, 8$$

(8)

where

$$|V_{jk}| = \left| \frac{b_{jk} - a_{jk}}{2} \right| \quad (9)$$

$$\rho(v_{fk}, V_{jk}) = \left| v_{fk} - \frac{a_{jk} + b_{jk}}{2} \right| - \frac{b_{jk} - a_{jk}}{2} \quad (10)$$

$$\rho(v_{fk}, V'_{jk}) = \left| v_{fk} - \frac{f_{jk} + g_{jk}}{2} \right| - \frac{g_{jk} - f_{jk}}{2} \quad (11)$$

Step 4 Calculating the correlation degrees for each category:

$$\lambda_j = \sum_{k=1}^{8} w_{jk} K_{jk} \quad (j = 0, 1, 2, \ldots, 5) \quad (12)$$

where w_{jk} is the weight for each feature parameter. In this paper, the weights were evaluated using entropy weight method, and the weight values were w_{jk} = (0.135, 0.086, 0.128, 0.165, 0.145, 0.095, 0.126, 0.118).

Step 5 Normalizing the relation degrees for every fault category to be between 1 and −1.

$$\lambda'_j = \begin{cases} \lambda_j / |\lambda_{\max}| & (\lambda_j > 0) \\ \lambda_j / |-\lambda_{\max}| & (\lambda_j < 0) \end{cases} \quad (13)$$

Step 6 Selecting the maximum value from the normal relation degrees (or 1) to recognize the fault type.

$$\text{if}\left(\lambda'_j = 1\right), \quad \text{then } (F_x = F_j) \tag{14}$$

4 Fault Diagnosis Results

Based on the procedure mentioned above, the operating fault for the ASHP could be diagnosed. Sixty sets of testing data were selected for validating the feasibility of the proposed fault diagnosis method. Totally, 57 out of 60 sets of testing data were diagnosed successfully with 95% accuracy rate. Table 2 lists the results of three testing data selected arbitrarily for demonstrating the feasibility of the proposed diagnosis method. For instance, in Test 1, the normal relation degree was 1 for the fault type F3, indicating that the fault of CondFoul occurred. For Test 2 and Test 3, the recognized fault types were CompLeak and ReserLeak, respectively.

In order to further demonstrate the advantages for the proposed fault diagnosis method, more data were tested, and the test results were compared with that using artificial neural network (ANN) based fault diagnosis method. The accuracy rates for each fault type are shown in Table 3. As seen, fault type of CompLeak had the highest accuracy rate of 91.25%, and ChargFault had the lowest at 88.43%. Furthermore, in comparison with the ANN-based diagnosis method, as shown in Fig. 2, the proposed method had a better diagnosis performance as expressed in terms of a higher accuracy rate and a shorter testing time.

Table 2 Three selected cases of diagnosis results

Test no.	Normal relation degree of each fault type, λ'_j						Diagnosis result
	F0	F1	F2	F3	F4	F5	
1	−0.717	−1	0.688	1	−0.097	−0.424	F3
2	−1	1	−0.291	−0.555	0.095	−0.440	F1
3	−0.621	−0.532	1	−1	−0.722	−0.939	F2

Table 3 Diagnosis accuracy rate for each fault type using the proposed method

Fault type	Accuracy rate (%)
CompLeak	91.25
ReserLeak	90.00
CondFoul	89.45
ChargFault	88.43
LLRestr	89.74

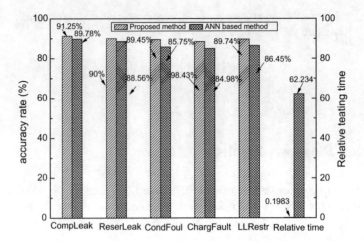

Fig. 2 Diagnosis results in comparison with ANN-based diagnosis method

5 Conclusions

An extension theory-based fault diagnosis method for an ASHP in cooling mode has been introduced, and the development results are reported in this paper. Five types of fault were implemented using an experimental ASHP system to generate enough data for matter-element model development. The matter-element models for six categories including one normal operation and five abnormal operations were developed. Through calculating the correlation degrees between the matter-element generated by the testing data with the matter-element models for the six categories, the operation condition or the operating fault could be recognized. The diagnosis results showed that the diagnosis accuracy rate for the proposed method was at approximately 90% for every fault type. In comparison with the conventional ANN-based fault diagnosis method, a better diagnosis performance in terms of a higher accuracy rate and a shorter testing time could be achieved using the extension theory-based fault diagnosis method.

Acknowledgements The financial supports for the Natural Science Foundation of Zhejiang Province (Project No. LQ19E060007) are gratefully acknowledged.

References

1. Cho, J.M., et al.: Normalized performance parameters for a residential heat pump in the cooling mode with single faults imposed. Appl. Therm. Eng. **67**(1–2), 1–15 (2014)
2. Katipamula, S., Brambley, M.R.: Methods for fault detection, diagnostics, and prognostics for building systems—a review, part II. Hvac&R Res. **11**(2), 169–187 (2005)

3. Katipamula, S., Brambley, M.R.: Methods for fault detection, diagnostics, and prognostics for building systems—a review, part I. Hvac&R Res. **11**(1), 3–25 (2005)
4. Reddy, A.T., et al.: Evaluation of the suitability of different chiller performance models for on-line training applied to automated fault detection and diagnosis (RP-1139). HVAC&R Res. **9**(4), 385–414 (2003)
5. Kim, M., et al.: Design of a steady-state detector for fault detection and diagnosis of a residential air conditioner. Int. J. Refrig. **31**(5), 790–799 (2008)
6. Chao, K.H., Ho, S.H., Wang, M.H.: Modeling and fault diagnosis of a photovoltaic system. Electr. Power Syst. Res. **78**(1), 97–105 (2008)
7. Cai, W.: The extension set and incompatibility problem. J. Sci. Explor. **1**, 81–93 (1983)

Effect of Lewis Factor on Performance of Closed Heat Source Tower under Spraying Conditions

Fenglin Zhang, Nianping Li, Haijiao Cui, Shengbing Li, Meng Wang and Meiyao Lu

Abstract The purpose of this paper is to analyze the effect of the Lewis factor (Le) on the performance of closed heat source tower (HST) under spraying conditions. Generally, Le is assumed to be one between air and water. However, in the closed HST, antifreeze solution is used as a working fluid. The water vapor pressure of the antifreeze solution is lower than that of the pure water under the same temperature, which has an impact on Le. Therefore, it is vital to analyze the effect of Le on the performance of closed HST under spraying conditions and determine the range of Le in actual working conditions. In this paper, the mathematical model of the closed HST under spraying conditions was established and validated by experiments. Then the performance of the closed HST under different conditions (Le = 1 and Le ≠ 1) was analyzed. And the range of the Le within the scope of the experiment was determined.

Keywords Closed heat source tower · Heat absorption efficiency · Numerical simulation · Heat transfer performance · Lewis factor

1 Introduction

The closed heat source tower (HST) is the unit of the heat pump system to extract heat from the surroundings. It can absorb both sensible and latent heat and use antifreeze solution to prevent the unit from freezing. So the HST integrated heat pump system has great energy-saving potential in subtropical region.

The Lewis number (Le = a/D) is defined as the ratio of thermal diffusivity to mass diffusivity. The Le is often assumed to be one for the water–air system. When Le = 1, $h/h_{md} = c_p$ can be obtained, where h is the heat transfer coefficient, h_{md} is the mass transfer coefficient, c_p is the specific heat at constant pressure [1]. However, Lewis pointed out that the relation, Le = 1, applied only for water–air

---

F. Zhang · N. Li (✉) · H. Cui · S. Li · M. Wang · M. Lu
College of Civil Engineering, Hunan University, Changsha, Hunan 410082, China
e-mail: linianping@126.com

systems, but not for all liquid–air systems [2]. The HST under spraying conditions works under low temperature and high humidity. The water vapor pressure of the antifreeze solution is lower than that of the pure water under the same temperature, and therefore, Le $\neq$ 1.

A series of theoretical researches were carried out. The effect of Le on the performance prediction of wet-cooling towers was investigated in 2005 [3]. The results showed that the influence of Le on tower performance diminished under the higher inlet air temperature and relative humidity. Wei established a mathematical model for the counterflow mechanical ventilation cooling tower assuming Le $\neq$ 1 [4]. The heat and mass transfer (HAMT) of cross-flow HST (glycol solution) was experimentally studied by Huang et al. The results showed that the coefficients mainly depended on flow rates of air and solution, and the Le in the research was 0.91–1.12 [5].

In this paper, a mathematical model of the closed HST under spraying conditions with calcium solution ($CaCl_2$) assuming Le $\neq$ 1 was established. The range of the Le within the scope of the experiment was determined. The results provided a theoretical basis for the optimization design of the closed HST under spraying conditions.

2 Methods

The mathematical model of the closed HST is simplified by the following assumptions:

(1) The spraying liquid film uniformly covers the surface of the finned tube heat exchanger.
(2) Concentration of the solution is constant.
(3) The liquid and gas parameters change only in the flow direction.
(4) No thermal losses from walls.
(5) When the tube number exceeds 3, it can be considered as a counterflow fin heat exchanger.

2.1 The Mathematical Model of the Closed HST

The heat absorption efficiency (η) indicates the proximity of the actual heat transfer capacity of the HST to the ideal capacity, as expressed in Eq. (1) [6].

$$\eta = \frac{t_{wo} - t_{wi}}{t_{aw} - t_{wi}} \tag{1}$$

where t_{aw} is the wet-bulb temperature of the inlet air, t_{wi} is the inlet solution temperature, and t_{wo} is the outlet solution temperature. Then the HAMT analysis was carried out by taking the heat transfer microelement dz in the direction of airflow, as shown in Fig. 1.

Fig. 1 Infinitesimal schematic diagram of closed HST

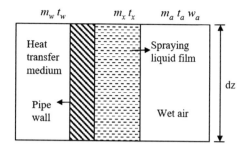

The control equations based on the energy and mass conservation are as follows:

(1) Heat transfer rate of the solution

$$m_w c_{pca} dt_w = k_{fw}(t_x - t_w) A_i A dz \quad (2)$$

where k_{fw} is the total heat transfer coefficient of the CaCl$_2$ to the liquid film, W/(m^2 K), A_i is the heat transfer area of the inner surface of the coil heat exchanger unit volume, m^2/m^3, z is the distance from the infinitesimal of coil heat exchanger to the top of the coil heat exchanger.

(2) Mass conservation in the air side

$$m_a dw_a = h_{md}(w_a - w_x) A_0 A dz \quad (3)$$

where m_a is wet air mass flow rate, kg/s; w_a is the humidity, kg/kg; w_x is the humidity on the surface of the solution, kg/kg; A_0 is the air-side heat transfer area of the outer surface of per unit volume, m^2/m^3.

(3) Sensible heat loss of wet air:

$$m_a c_{pa} dt_a = h(t_a - t_x) A_0 A dz \quad (4)$$

(4) Ignore the evaporation loss of the liquid film and apply the energy conservation equation to the liquid film:

$$c_{px} m_x t_x = [r_0 h_{md}(w_a - w_x) + h(t_a - t_x)] A_0 A dz + k_{fw}(t_w - t_x) A_i A dz \quad (5)$$

where m_x is the mass flow rate of solution, kg/s; r_0 the latent heat of vaporization of water at 0 °C, J/k. The value was set as Eq. (6), and the control equations can be calculated using Eq. (7).

$$a_1 = \frac{K_{fw} A_i A}{m_w c_{pca}}, \quad a_2 = \frac{h_{md} A_0 A}{m_a}, \quad a_3 = \frac{h A_0 A}{m_a c_{pa}},$$
$$a_4 = \frac{r_0 h_{md} A_0 A}{m_x c_{pca}}, \quad a_5 = \frac{h A_0 A}{m_x c_{pca}}, \quad a_6 = \frac{K_{fw} A_i A}{m_x c_{pca}} \quad (6)$$

$$\frac{dt_w}{dz} = a_1(t_x - t_w),$$
$$\frac{dw_a}{dz} = a_2(w_a - w_x)$$
$$\frac{dt_a}{dz} = a_3(t_a - t_x),$$
$$\frac{dt_x}{dz} = a_4(w_a - w_x) + a_5(t_a - t_x) + a_6(t_w - t_x)$$

(7)

The boundary conditions are

$$t_a|_{z=0} = t_{ai}; \quad w_a|_{z=0} = w_{ai}; \quad t_{xi} = t_{xo}; \quad t_w|_{z=h} = t_{wi}$$

In the simulation, T_{w0} was assumed to be T_{w01}. Then the control equation was calculated and the corresponding solution outlet temperature T_{w02} was obtained. The convergence criterion can be determined using Eq. (8).

$$\varphi = 2\left|\frac{T_{w01} - T_{w02}}{T_{w01} + T_{w02}}\right| < 10^{-6} \tag{8}$$

2.2 Calculation Procedure

As the control equations are typical linear homogeneous differential equations, and therefore, the classical fourth-order Runge–Kutta method can be used for calculation. The solving process of the governing equation is given as follows:

(1) Input the air and the solution inlet parameters.
(2) Calculate the internal surface heat transfer coefficient and fin efficiency according to the temperature of the solution in the tube, the dry-bulb and wet-bulb temperature of the air, and the concentration of the spraying solution;
(3) Calculate the air-side convective coefficient h_{ow}, h_{md}, and other parameters;
(4) Calculate a_1–a_6 in the control equation, set the step size and accuracy.
(5) Set the solution outlet temperature T_{wo1} and calculate the air outlet dry-bulb and wet-bulb temperature and solution outlet temperature using the fourth-order Runge–Kutta method;
(6) Determine whether $\varphi < 10^{-6}$, and if yes, proceed to the next step; if not, return to step (5);
(7) Output the calculation result.

2.3 Model Validation Results

Compare the results calculated by the model with those of the experiments of the reference [6]. The model was then validated by the experiments. The experimental conditions were dry-bulb temperature 2–10 °C, relative humidity 60–100%, airflow rate 1400–4400 m³/h, and t_{wi} −10–2 °C.

The experimental data were dealt with energy balance, which meant the heat loss of air Q_a was equal to the heat absorption Q_w of the CaCl$_2$ solution. The unbalance rate was expressed as Eq. (9). The data satisfying R = ±5% were adopted in the experiment.

$$R = \left| \frac{2(Q_a - Q_w)}{Q_a + Q_w} \right| \times 100\% \tag{9}$$

The heat absorption efficiency was used to validate the simulation system. The validation results are presented in Figs. 2, 3, 4, and 5. It can be observed that the simulated value and the experimental value agree well. The maximum deviation is 4.6%. Within the scope of the experiment, the value of the Le was about 0.77–0.86.

3 Results and Discussion

As shown in Figs. 6 and 7, the heat absorption efficiency decreases with the air (t_{ai}) and the solution inlet temperature (t_{wi}). The decreasing trends of two curves are close, and the relative error $\left(\varepsilon = \left| \frac{\eta_{Le \neq 1} - \eta_{Le=1}}{\eta_{Le \neq 1}} \right| \right)$ between the two curves is within 7%, which indicates that under different Le numbers (Le = 1 or Le ≠ 1), t_{ai} and t_{wi} have little impact on the heat absorption efficiency, especially at high temperatures. It can

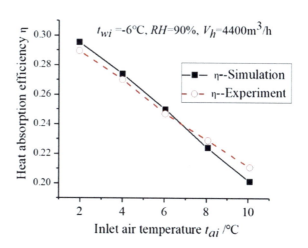

Fig. 2 Relationship between t_{ai} and η

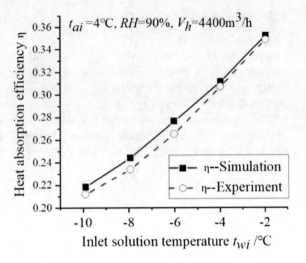

Fig. 3 Relationship between t_{wi} and η

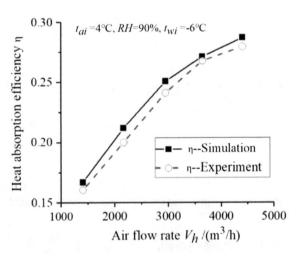

Fig. 4 Correlation between V_h and η

be traced back to the fact that sensible heat transfer capacity of the closed HST increases significantly with the increasing t_{ai}, while the latent heat transfer changes very little. So the Le will approach to 1 with less relative error. Under the dew point temperature, the latent heat transfer mainly depends on the air humidity and the mass transfer coefficient will have little change. Therefore, the assumption of Le = 1 will have little impact on the performance calculation of HST under high t_{ai} and t_{wi}.

As shown in Fig. 8, when t_{ai}= 4 °C, RH= 90%, t_{wi}= −6 °C, the relative error (ε) between the two curves is gradually reduced with the increasing airflow rate. When V_h is 1400 m³/h, ε is 10.4%, while it decreases to 6.9% when V_h is 4400 m³/h. It is indicated that Le has a greater effect on the closed HST performance under lower airflow rate. It is mainly because that the wet air turbulence will reduce under the

Fig. 5 Correlation between RH and η

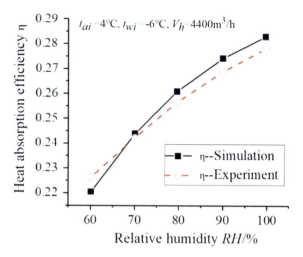

Fig. 6 t_{ai} correlated with η

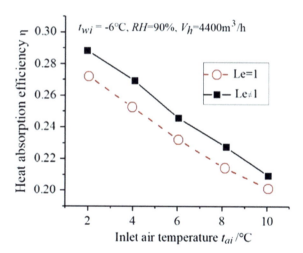

lower airflow rate, which leads to an insufficient sensible heat transfer. However, when reaching the dew point temperature, the latent heat transfer will not decrease significantly, which makes the Le significantly reduced. So under low airflow rate, it is necessary to adopt the actual value of Le for the performance evaluation.

As shown in Fig. 9, when t_{ai}= 4 °C, V_h= 4400 m³/h, t_{wi}= −6 °C, ε between two curves gradually increases with the relative humidity. When the relative humidity varies from 60 to 100%, ε increases from 7.1 to 10.6%. It is indicated that Le has a greater effect on the closed HST performance under higher relative humidity. Because at the dew point temperature, the sensible heat transfer remains constant while, at the same time, the latent heat transfer will decrease with increasing relative humidity. So the Le will decrease and the latent heat transfer will be underestimated. Therefore, under high relative humidity, the assumption of Le = 1 is not appropriate.

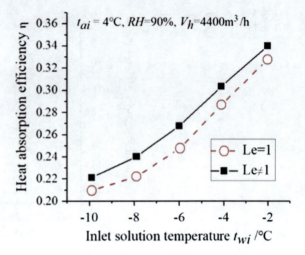

Fig. 7 t_{wi} correlated with η

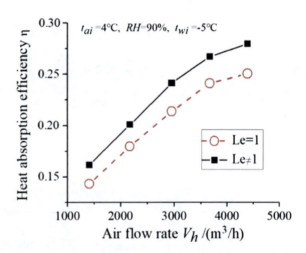

Fig. 8 V_h correlated with η

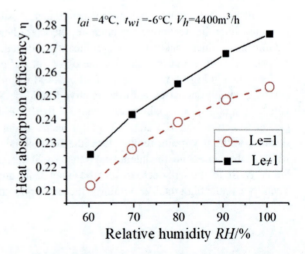

Fig. 9 RH correlated with η

4 Conclusions

(1) Within the scope of the experiment, the value of the Le was about 0.77–0.86.
(2) If Le = 1, the latent heat transfer under the spraying conditions will be underestimated, and the deviation of the heat absorption efficiency could reach 6.7–10.6%.
(3) The developed model considering Le $\neq$ 1 is more accurate, and the accuracy is within 4.6%.
(4) Le has a greater effect on the closed HST performance under lower airflow rate and higher relative humidity.

Acknowledgements This work was financially supported by the China National Key R&D Program "Solutions to heating and cooling of buildings in the Yangtze River region" (Grant No.2016YFC0700305) and the Natural Science Foundation of China (No. 51878255).

References

1. Merkel, F.: Verdunstungskühlung. VDI-Zeitchrift **70**, 123–128 (1925)
2. Lewis, W.K.: The evaporation of a liquid into a gas. Int. J. Heat Mass Transf. **5**, 109–112 (1962)
3. Kloppers, Johannes C., Kröger, Detlev G.: The Lewis factor and its influence on the performance prediction of wet-cooling towers. Int. J. Therm. Sci. **44**, 879–884 (2005)
4. Wei, Z., Qi, X.: Thermodynamic numerical model of counter flow wet cooling tower. Manuf. Autom. **35**, 105–108 (2013)
5. Huang, S., Lv, Z., Liang, C., Zhang, X.: Experimental study on heat and mass transfer coefficient of cross-flow heat source tower. J. Eng. Therophysics **V38**, 914–919 (2017)
6. Li, S., Li, N., Cui, H., Cheng, J.: Experimental study on heat transfer characteristics of heat source tower under low temperature and high humidity conditions. Sci. Technol. Eng. **17**, 271–275 (2017)

An Experimental Study on the Thermosiphon Loop with a Microchannel Heat Sink Operating with the Phase Change Emulsion

Xiaoxu Cai, Shugang Wang, Jihong Wang, Tengfei Zhang and Xiaozhou Wu

Abstract The chip cooling has become more appealing due to the increasing integration of integrated circuits. The physical parameters of a stable phase change emulsion are obtained, and the thermosiphon loop system with a microchannel heat sink is built. The results revealed the system using a 0.5% mass concentration emulsion (c_m) runs stably at a low heating power ($Q \leq 8.5$ W). An increase of c_m (= 1.0 and 1.5%) enables the system to present a periodic start-up characteristic at 5.0 and 7.5 W, while the system with $c_m = 2.0\%$ failed to start, and the heating power of 8.5 W makes the system with $c_m = 2.0\%$ achieve a steady state. Moreover, the hysteresis of peak temperature of adjacent measuring points is weakened by increasing c_m or Q. Compared with the deionized water, heat transfer coefficients of the microchannel heat sink with $c_m = 0.5$ and 2.0% were increased by 16.0% and 17.1%, respectively.

Keywords Microchannel heat sink · Thermosiphon loop · Phase change emulsion

1 Introduction

The trend of data centers to high power density and super large-scale is creating a demand for more efficient refrigeration systems, and the heat flux of many electronic devices has reached more than 100 W/cm^2 [1]. When electronic components are maintained at a higher temperature due to the insufficient heat dissipation, the failure rate will increase, and even bring great losses to the data center. Therefore, the heat dissipation is a key problem affecting the reliability of electronic devices. However, traditional heat dissipation methods of electronic devices, such as pure

---

X. Cai · S. Wang (✉) · J. Wang · T. Zhang · X. Wu
Faculty of Infrastructure Engineering, Dalian University of Technology, Dalian, China
e-mail: sgwang@dlut.edu.cn

thermal conduction, natural convection and forced convection, have failed to meet the cooling demand of high-heat-flux electronic devices.

The thermal conductivity of heat pipes was much higher than that of any known metal within certain limitations [2]. As heat pipes have the advantage of achieving efficient heat dissipation with low heat transfer temperature difference [3], they are employed in heat dissipation of electronic devices in the form of thermosiphon [4], loop heat pipe [1], miniature heat pipe [5, 6] and pulsating heat pipe [7]. As a solution to the requirement of large-scale and long-distance heat transfer of heat pipes, the design of separated heat pipes has attracted an attention of scholars. In addition, the microchannel cooling method is studied to solve the heat dissipation problem of chips with high power density [8–11]. The concept of phase change cooling of microchannel first appeared in a kind of silicon-based rectangular microchannel water cooling device for VLSI circuit [12], and a lot of experimental and theoretical researches have been made in recent years. Microchannel heat dissipation structure, of compact structure and high heat transfer coefficient, has been widely used. Khodabandeh [5] conducted an experiment of the two-phase loop thermosiphon, in which the heat flux of 31.15 W/cm^2 could be dissipated by adding a millimeter-scale channel in the flat-plate heating block. Chang et al. [13] focused experimentally on the factors influencing the thermal resistance of evaporator and condenser of two-phase closed-loop thermosiphon system. In addition to optimizing the heat pipe structure, improving the thermal properties of the working medium is another effective method to improve the heat pipe performance. Chang et al. [14] designed a two-phase loop thermosiphon with the evaporation section as the visible section, and observed the flow characteristics and working instability of the working medium with water as the medium. In order to avoid the problem of unstable flow, the solid–liquid phase change slurry is adopted due to a small volume change. The numerical results of Charunyakorn et al. [15] showed that the forced convection heat transfer performance of microencapsulated slurry in a circular tube was about 2–4 times higher than that of single-phase fluid. Goel et al. [16] studied experimentally that Stefan number is the main influencing factor on the heat transfer performance of laminar flow in the tube of microcapsule suspension under the condition of constant wall heat flow boundary. The experimental results of Inaba et al. [17] indicated that the average heat transfer coefficient of microcapsule solid–liquid phase slurry was 2–2.8 times that of single-phase water. Ho et al. [18] experimentally investigated the phase change nanocapsules suspension can significantly enhance the heat transfer performance of natural circulation. However, the phase change emulsion becomes an alternative to the microcapsule slurry which is difficult to prepare and easy to break. Paraffin phase change emulsion is mostly used in energy storage systems, but has not been used in a thermosiphon loop.

Therefore, a loop thermosiphon system with a rectangular microchannel heat sink is designed by using phase change emulsion as working medium. A stable phase change emulsion is prepared and its physical parameters are tested, and the heat transfer characteristics of the system are obtained. The loop thermosiphon system has certain potential applications in chip cooling.

2 Preparation and Thermal Properties of Phase Change Emulsion

2.1 Preparation of Phase Change Emulsion

Phase change emulsion is prepared by high-speed shearing of RT24, deionized water and emulsifier including Span80 and Tween80. By changing the parameters such as HLB value, emulsifying time, emulsifying agent content, temperature of the constant-temperature bath, the phase change emulsions were prepared and observed by standing methods. HLB value of 9, emulsifying time of 2 h (10,000 rpm), emulsifier content of 10%, and the bath temperature of 60 °C were used to prepare the stable phase emulsion. The phase change emulsion, with even particle size distribution, is milky white, and of good fluidity.

2.2 Thermal Properties of Phase Change Emulsion

By measuring the physical parameters of phase change emulsion in Figs. 1, 2 and 3, it is found that the density of the phase change emulsion decreases slightly with the increase of temperature and mass concentration. The thermal conductivity increases with an increase of temperature and slightly changes with the concentration. When the temperature is higher than 10 °C, the viscosity decreases with an increase of temperature and slightly changes with an increase of the mass concentration. While the temperature is 10 °C, its value increases sharply and increases obviously with the increase of concentration.

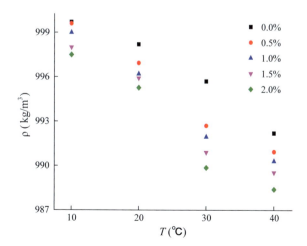

Fig. 1 Density

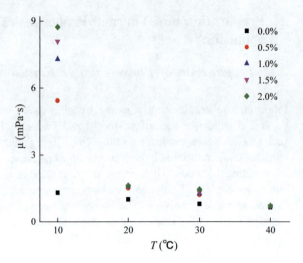

Fig. 2 Viscosity

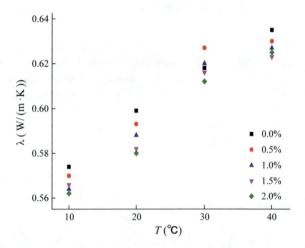

Fig. 3 Thermal conductivity coefficient

3 Experimental Apparatus

The experimental system used in this paper is mainly composed of a thermosiphon loop with a microchannel heat sink, a heating system, a cooling system and a data acquisition system. The system schematic diagram is shown in Fig. 4, and the thermosiphon loop was very well insulated with a layer of Armaflex insulation material ($\lambda = 0.034$ W/(m K), 24 mm thick), to minimize heat losses (or gains).

The thermosiphon loop is mainly composed of the heating section, the cooling section and the expansion chamber accommodating the volume change of the phase change emulsion during operating, and is connected by pipes with the same inner diameter, as shown in Fig. 4.

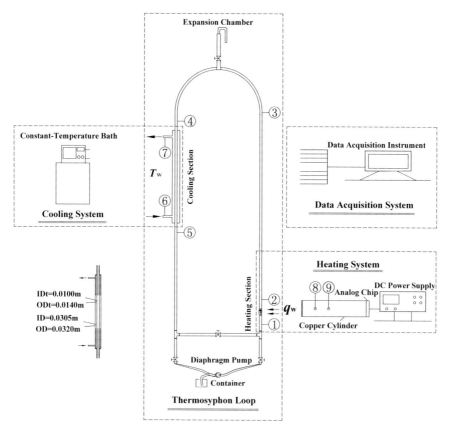

Fig. 4 Schematic diagram of the experiment system and measuring points

The basic design parameters of the thermosiphon loop are shown in Table 1. The details are listed in Fig. 5. The loop pipeline is connected by 316 stainless steel pipes with smooth inner wall, and the diaphragm pump is used to fill the system with fluid.

The heating system includes the ceramic heating plate instead of a chip and a DC power supply to control the heating power, and a pure copper column with the same cross-sectional area as the microchannel heat sink is used to conduct heat. The cooling system maintaining a boundary condition of constant wall temperature is used to cool the phase change emulsion with cold water provided by a constant-temperature bath Set to 5 °C. The data acquisition system is employed to continuously record the temperature changes of each point measured by thermocouples (PT100) and output data. See Table 2 for the layout of the measuring points.

Table 1 Basic design parameters of thermosiphon loop

Parameters		Values
Heating section	Microchannel heat sink size (mm)	10.00 × 10.00 × 1.00
	Microchannel size (mm)	0.65 × 0.50
	Fin thickness (mm)	0.15
Cooling section	Double-pipe heat exchanger length (m)	0.50
	Outer diameter of double-pipe heat exchanger (mm)	32.00
	Wall thickness of double-pipe heat exchanger (mm)	1.50
Expansion chamber	Length (mm)	120.00
	Diameter (mm)	19.05, 9.50
	Volume (ml)	40
Pipes	Total length (m)	4.20
	Inner diameter (mm)	10.00
	Wall thickness (mm)	2.00
	Volume (ml)	370
	Length of vertical pipe (m)	1.50
	Length of horizontal pipe (m)	0.46
	Curvature of bends (m)	0.23
Height difference of heating and cooling sections (m)		1.00

4 Results and Discussions

The thermal resistance of the system is calculated by formulas Eqs. (1)–(3).

$$R_{\text{PHP}} = (T_h - T_c)/Q \quad (1)$$

$$T_h = (T_1 + T_2)/2 \quad (2)$$

$$T_c = (T_4 + T_5)/2 \quad (3)$$

where R_{PHP} is the thermal resistance of the loop thermosiphon system, (°C/W), T_h is the average temperature of inlet and outlet of the microchannel heat sink, (°C), T_c is the average temperature of the inlet and outlet of the cooling section, (°C), T_1 and T_2 are the inlet and outlet temperatures of the heating section, (°C), and T_4 and T_5 are the inlet and outlet temperatures of the cooling section, (°C), respectively.

When no fluid is filled in the system, the temperature changes of the empty pipe run were observed at the heating power of 5.0 W in Fig. 6. Temperatures of the inlet and outlet of the heating section rise to 30 °C after running about 1.5 h, and the temperatures rise no more than 5 °C during the following three hours. The

(a) Microchannel heat exchanger

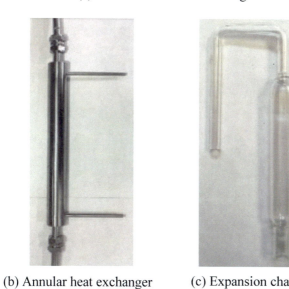

(b) Annular heat exchanger   (c) Expansion chamber

Fig. 5 Major parts of thermosiphon loop

thermal resistance of the empty pipe run is about 3.77 °C/W. Then, the experiments with the deionized water and emulsions are carried out following the empty pipe operation.

Table 2 Operation summary

	0.5%	1.0%	1.5%	2.0%
5.0 W	Successful	Periodic	Periodic	Failed
7.5 W	Successful	Periodic	Periodic	Failed
8.5 W	Successful	Successful	Successful	Successful

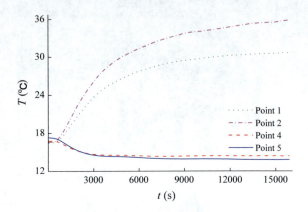

Fig. 6 Temperature change during empty pipe operation

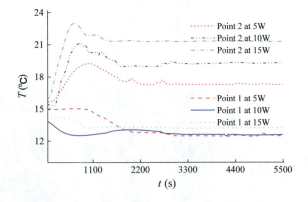

Fig. 7 Temperature change of deionized water at different heating power

Figure 7 depicts the start-ups of the system with deionized water at different heating powers. In according with the above formulas, the thermal resistances are 0.06 °C/W, 0.034 °C/W and 0.089 °C/W, respectively, which are lower than that of the empty pipe run (3.77 °C/W), when the operations reach steady states at the heating powers of 5, 10 and 15 W. It is obvious that the system with deionized water can start successfully and runs stably. The operations of the 0.5% mass concentration emulsion are the same as that of deionized water at 5.0, 7.5 and 8.5 W. Figure 8 describes the temperature changes of Points 2 and 3 of the system with a 1.0% mass concentration phase change emulsion at different heating powers.

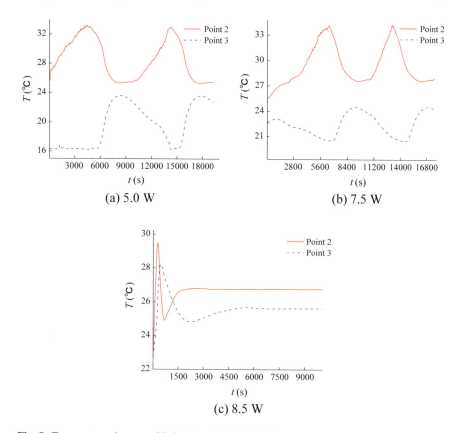

Fig. 8 Temperature changes of Points 2 and 3 with 1.0% mass concentration emulsion

It is observed that all runs have been started successfully, and the temperature of Point 2 changes earlier than that of Point 3. In Fig. 8a, b, the two runs have the similar start-up, and the temperature change trends at the two points are opposite. The system start-up presents periodicity, and the flow rate is not uniform. In Fig. 8c, the lag time of peak temperature at two adjacent points is shortened when the heating power is 8.5 W, and the steady state can be achieved, i.e., an increase in the density difference of the heating and cooling sections generated by increasing the heating power provides a greater driving force.

The operation results of all the experimental runs are summarized in Table 2. It concludes that, the system is started successfully by only increasing the mass concentration of phase change emulsion but also considering whether the heating power can drive the system to operate.

Figure 9 presents the influences of the heating power and mass concentration on the lag time of temperature peak of two adjacent measuring points in the loop. Figure 9a shows that when the mass concentration of emulsion is 0.5%, the lag time decreases with an increase of the heating power. According to Fig. 9b,

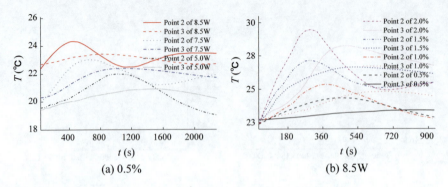

Fig. 9 Temperature changes of points 2 and 3

Table 3 Input parameters and experimental results for runs

c_m (%)	Q (W)	P1 (°C)	P2 (°C)	P4 (°C)	P5 (°C)	v (m/s)	P6 (°C)	h (W/(m²·K))
1.0	5.0	15.19	25.43	19.02	7.91	0.0026	48.81	1754.39
1.0	7.5	14.73	28.08	18.98	7.98	0.0031	61.87	1853.45
0.0	8.5	15.24	20.10	19.56	13.51	0.0143	68.94	1657.89
0.5	8.5	14.28	25.09	20.47	7.26	0.0057	62.76	1973.30
1.0	8.5	14.53	28.90	20.27	7.47	0.0044	70.79	1693.41
1.5	8.5	14.60	25.96	20.90	7.10	0.0041	65.71	1871.01
2.0	8.5	15.18	25.78	20.07	7.20	0.0039	62.58	1999.77

with a decrease of mass concentration, the lag time decreases and the peak value becomes larger at 8.5 W. Compared with deionized water, the lower flow velocity of phase change emulsion due to the higher viscosity, results in the extension of the time interval to reach the peak temperature of the loop measuring points.

The steady-state temperatures of main measuring points are listed in Table 3. The results indicate that the velocity of the 1.0% mass concentration phase change emulsion increases with an increase of the heating power, and that the velocity decreases with an increase of the mass concentration when the heating power maintains at 8.5 W. Compared with deionized water, the heat transfer coefficients with 0.5 and 2.0% phase change emulsions are increased by 16.0% and 17.1%, respectively, at $Q = 8.5$ W. Both the viscosity and latent heat of emulsions have effects on heat transfer performance of the system, and operation is more stable when $c_m = 0.5\%$.

5 Conclusions

The stable phase change emulsion is firstly prepared, and its physical parameters are measured. The thermosiphon loop system with a microchannel heat sink operating with phase change emulsions is constructed. The experimental results indicate that the system can reach a steady state with a 0.5% mass concentration emulsion or at a heating power of 8.5 W, otherwise the system presents a periodic start-up. The hysteresis of peak temperature of adjacent measuring points decreases with an increase of the heating power and the emulsion mass concentration. Compared with the deionized water, heat transfer coefficients of the microchannel heat sink were increased by 16.0% and 17.1%, respectively, when the emulsion mass concentrations are 0.5 and 2.0%. Therefore, improving the fluidity of the emulsion will further improve the heat transfer performance of the thermosiphon loop with a microchannel heat sink.

Acknowledgements The project is supported by National Natural Science Foundation (Number 51678102, 51508067).

References

1. Maydanik, Y.F., et al.: Review: loop heat pipes with flat evaporators. Appl. Therm. Eng. **67**(1–2), 294–307 (2014)
2. Grover, G.M., et al.: Structures of very high thermal conductance. J. Appl. Phys. **35**(6), 1990–1991 (1964)
3. Chan, C.W., et al.: Heat utilisation technologies: a critical review of heat pipes. Renew. Sustain. Energy Rev. **50**, 615–627 (2015)
4. Lin, L., et al.: High performance miniature heat pipe. Int. J. Heat Mass Transf. **45**(15), 3131–3142 (2002)
5. Khodabandeh, R.: Heat transfer in the evaporator of an advanced two-phase thermosyphon loop. Int. J. Refrig. **28**(2), 190–202 (2005)
6. Samba, A., et al.: Two-phase thermosyphon loop for cooling outdoor telecommunication equipments. Appl. Therm. Eng. **50**(1), 1351–1360 (2013)
7. Thompson, S.M., et al.: An experimental investigation of a three-dimensional flat-plate oscillating heat pipe with staggered microchannels. Int. J. Heat Mass Transf. **54**(17–18), 3951–3959 (2011)
8. Brunschwiler, T., et al.: Forced convective interlayer cooling in vertically integrated packages. In: 11th Intersociety Conference on Thermal and Thermomechanical Phenomena in Electronic Systems, pp. 1114–1125, Orlando, FL, USA (2008)
9. Hirshfeld, H., et al.: High heat flux cooling of accelerator targets with micro-channels. Nucl. Instrum. Methods Phys. Res. Sect. A **562**(2), 903–905 (2006)
10. Colgan, E.G., et al.: A practical implementation of silicon microchannel coolers for high power chips. IEEE Trans. Compon. Packag. Technol. **30**(2), 218–225 (2007)
11. Koşar, A., Peles, Y.: Thermal-hydraulic performance of MEMS-based pin fin heat sink. J. Heat Transf. **128**(2), 121–131 (2006)
12. Tuckerman, D.B., Pease, R.F.W.: High-performance heat sinking for VLSI. IEEE Electron Device Lett. **2**(5), 126–129 (1981)

13. Chang, C.C., et al.: Two-phase closed-loop thermosyphon for electronic cooling. Exp. Heat Transf. **23**(2), 144–156 (2010)
14. Chang, S.W., et al.: Sub-atmospheric boiling heat transfer and thermal performance of two-phase loop thermosiphon. Exp. Therm. Fluid Sci. **39**, 134–147 (2012)
15. Charunyakorn, P., et al.: Forced convection heat transfer in microencapsulated phase change material slurries: flow in circular ducts. Int. J. Heat Mass Transf. **34**(3), 819–833 (1991)
16. Goel, M., et al.: Laminar forced convection heat transfer in micro capsulated phase change material suspensions. Int. J. Heat Mass Transf. **37**(4), 593–604 (1994)
17. Inaba, H., et al.: Heat transfer characteristics of latent microcapsule-water mixed slurry flowing in a pipe with constant wall heat flux (numerical analysis). Nippon Kikai Gakkai Ronbunshu, B Hen/Trans. Jpn. Soc. Mech. Eng., Part B **68**(665), 161–168 (2002)
18. Ho, C.J., et al.: Thermal performance of water-based suspensions of phase change nanocapsules in a natural circulation loop with a mini-channel heat sink and heat source. Appl. Therm. Eng. **64**(1–2), 376–384 (2014)

Experimental Study on Effect of Water Flow Rate on Heating Performance of a Series Bathing Wastewater Source Heat Pump Hot Water Unit

Liangdong Ma, Tixiu Ren, Tianyi Zhao and Jili Zhang

Abstract To maximize the recovery of heat energy from bathing wastewater, a two-series-connected bathing wastewater source heat pump hot water unit (SCBWSHP) with large temperature difference was proposed. Its rated design parameters are: the inlet and outlet temperatures of the bathing wastewater are 30 °C and 6 °C, respectively. The tap water inlet temperature is 10 °C, and the hot water outlet temperature is 45 °C. The effects of the mass flow rate of the tap water and the bathing wastewater on heating performance of the SCBWSHP were investigated experimentally. The results showed that the COP of the whole unit is 4.5 at the ratings. When the tap water mass flow rate increases from 600 to 900 kg/h, the whole unit COP increased by 18.7%, from 3.20 to 4.88. And the COP of the whole unit increases from 4.03 to 4.83 as the bathing wastewater mass flow rate rises from 600 to 1000 kg/h, increased by 16%.

Keywords Bathing wastewater · Heat pump unit · Large temperature difference · Coefficient of heating performance · Experimental research

Nomenclature

c_p Specific heat at constant pressure, J/(kg K)
COP   Coefficient of heating performance
m_c Mass flow rate of tap water, kg/s
m_e Mass flow rate of bathing wastewater, kg/s
Q_c Heating capacity, W
Q_e Heat quantity removed from wastewater, W
t_c The temperature of tap water, K
t_e The temperature of bathing wastewater, K
T_c Condensing temperature, K
T_e Evaporating temperature, K
W_e Electric motor power of the compressor, W

L. Ma · T. Ren · T. Zhao (✉) · J. Zhang
Faculty of Infrastructure Engineering, Dalian University of Technology, Dalian, China
e-mail: zhaotianyi@dlut.edu.cn

Subscripts

H   High-temperature-zone-cycle unit
L   Low-temperature-zone-cycle unit
Z   The whole unit

1 Introduction

The wastewater discharged from large-scale public bath is an available heat source for heat pump because of its large discharge, high temperature up to 30 °C, and stable discharge time throughout the year [1, 2].

In the past few decades, many researchers have carried out investigation on recovering waste heat from bathing wastewater using heat pumps [3]. These researches demonstrated that recovering heat from bathing wastewater through heat pump has advantages of energy conservation, emission reduction, and low operating cost, as compared to conventional water heating methods. In these studies, the heat pump system is conventional system and the hot water temperature increased usually by 7–10 °C. In order to achieve the required temperature, the tap water is heated repeatedly, resulting in the hot water temperature on the condenser side gradually rises and the wastewater temperature on the evaporator side gradually decreases, eventually lead to unstable operation of the system. To maximize the recovery of heat energy from bathing wastewater, large temperature difference heat pump is proposed. However, it will lead to lower operation efficiency and higher energy consumption. To solve these problems, researchers proposed some series schemes for heat pump units [4, 5]. Song [4] proposed that under the same water temperature change, the series scheme can help save more heat exchanger area and compressor power consumption compared with parallel connection. Sun et al. [5] concluded that the energy-saving rate of the water source heat pump units in series can be increased by 20.4% compared with the single unit. Even compared with the single unit with large temperature difference, 17.3% of energy for the two-series-connected heat pump can also be saved.

To improve the utilization efficiency of the waste heat, for the bathing wastewater temperature, the annual temperature ranges of tap water, and the temperature requirement of bathing water, a two-series-connected bathing wastewater source heat pump hot water unit (SCBWSHP) with large temperature difference was proposed by the authors [6]. The optimal design parameter of the SCBWSHP was investigated by numerical simulation. And based on the simulation results, the research team developed an experimental prototype of the SCBWSHP with heating capacity of 35 kW [7]. In this paper, the effect of the tap water flow rate and the bathing wastewater flow rate on the heat performance of the SCBWSHP was investigated by the experimental method and the results can be used as a key reference for design and operation of bathing wastewater source heat pump systems in the future.

2 Methods

The schematic of performance experimental system for the SCBWSHP is shown in Fig. 1. The SCBWSHP is composed of two heat pump units, a low-temperature-zone-cycle unit (LTZCU), and a high-temperature-zone-cycle unit (HTZCU). Each heat pump unit contains a compressor, condenser, expansion valve, and evaporator, and all other necessary accessories, R134a was chosen as refrigerant. The thermodynamic cycle process of the two units is independent of each other and the coupling operation of the two units is realized by the bathing wastewater on the evaporator side and the hot water on the condensing side.

In this system, the tap water through the condensers of the LTZCU and the HTZCU in turn, and the bathing wastewater through the evaporators of the HTZCU and the LTZCU in sequence, where the tap water is heated by the condensers and the bathing wastewater is cooled by the evaporators. In the rated design conditions, the tap water is heated from 10 to 45 °C and the bathing wastewater is cooled from 30 to 6 °C. The design heating capacity of the system is 35 kW, and it is obtained by simulation that the COP of the LTZCU, the HTZCU, and the whole unit is 6.3, 5.2, and 5.66, respectively, under the design conditions [6].

The specifications of the main components of the SCBWSHP are given in Table 1, and the test instrumentations are provided in Table 2. The performance of the SCBWSHP was evaluated using COP. Equation (1) is the definition of the COP of the LTZCU, the HTZCU, and the whole unit. The heating capacity is the heat

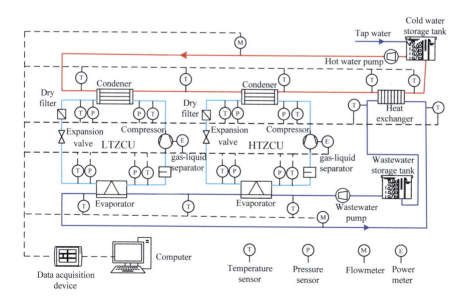

Fig. 1 Schematic diagram of the SCBWSHP system

absorbed by hot water as shown in Eq. (2). The heat removed from wastewater is calculated using Eq. (3).

$$\text{COP}_L = \frac{Q_{c,L}}{W_{e,L}}, \text{COP}_H = \frac{Q_{c,H}}{W_{e,H}}, \text{COP}_z = \frac{Q_{c,z}}{W_{e,z}} = \frac{Q_{c,z}}{W_{e,L} + W_{e,H}} \quad (1)$$

$$Q_{c,L} = m_c c_p(t_{c,m} - t_{c,i}), Q_{c,H} = m_c c_p(t_{c,o} - t_{c,m}), Q_{c,z} = Q_{c,L} + Q_{c,H} \quad (2)$$

$$Q_{e,L} = m_e c_p(t_{c,m} - t_{c,o}), Q_{e,H} = m_e c_p(t_{e,i} - t_{e,m}), Q_{e,z} = Q_{e,L} + Q_{e,H} \quad (3)$$

Based on the data in Table 2, uncertainties in error of COP of the LTZCU, the HTZCU, and the whole unit are calculated by error analysis method [8, 9]. The maximum uncertainty of COP_L, COP_H, and COP_z is 2.2%, 2.3%, and 1.2%, respectively.

3 Results and Discussion

3.1 Effect of Tap Water Flow Rate

Figure 2 gives the effect of tap water flow rate on operating performance parameters of the SCBWSHP when the tap water inlet temperature is 10 °C, and the inlet and outlet temperature of wastewater is 30 °C and 6 °C, respectively. As shown in Fig. 2a, the heating capacity of the LTZCU is always lower than that of the HTZCU, which is consistent with the design parameters (see Table 1). Under the rated condition, the total heating capacity of the SCBWSHP is 17% lower than the design value which is 35 kW. The heat removed from wastewater increases and the power

Table 1 Specifications of the SCBWSHP developed in the experimental system

Parameters	The LTZCU	The HTZCU	The SCBWSHP
Heat transfer rate of the condenser, kW	16.47	19.55	36.02
Heat transfer rate of the evaporator, kW	12.79	15.21	28.00
Inlet and outlet temperature of hot water in the condenser side, °C	10/26	26/45	10/45
Inlet and outlet temperature of wastewater in the evaporator side, °C	17/6	30/17	30/6
Condensing temperature, °C	31	50	/
Evaporating temperature, °C	1	12	/
Power consumption of the compressor, kW	3.75	4.50	8.25
The coefficient of performance	6.3	5.2	5.66
Mass flow rate of hot water, kg/s	0.240	0.240	0.240
Mass flow rate of wastewater, kg/s	0.293	0.293	0.293

Table 2 Parameters of the measurement instruments

Parameters	Instruments	Range	Measurement errors
Temperature	RTD(PT-100)	−50–300 °C	±0.1 °C
Pressure	Pressure transmitter	0–2.5 Mpa	±0.5% of full scale
Pressure	Pressure transmitter	0–1.0 Mpa	±0.5% of full scale
Flow rate	Electromagnetic flow meter	0–3 m³/h	±0.5% of full scale
Electrical power	Multifunctional power meter	0–10 kW	±0.5% of full scale

consumption decreases and the descending gradient of $W_{e,H}$ is greater than that of $W_{e,L}$. The power consumption of the HTZCU is higher compared with that of the LTZCU when the tap water flow rate is less than 700 kg/h, but it gradually reduced to lower than that of the LTZCU with the increase of the tap water flow rate. These results show that the tap water flow rate has a greater influence on the heating performance of the HTZCU. Just as shown in Fig. 2b, the COP_H, COP_L and COP_Z increase by 28.2%, 9.8%, and 18.7%, respectively, with the tap water flow rate increases from 600 to 900 kg/h. During the experiment, the bathing wastewater mass flow rate is 786 kg/h when the tap water mass flow rate is 745 kg/h, and the COP of the SCBWSHP is 4.46.

The increase of the tap water flow rate decreases the condensing temperature of the LTZCU and the HTZCU, thus COP_L and COP_H increase. It is found from Fig. 3 that as the tap water flow rate changes from 600 to 900 kg/h, the condensing temperature $T_{c,L}$ of the LTZCU changes from 49.2 to 43.6 °C, and the condensing temperature $T_{c,H}$ of the HTZCU changes from 57.1 to 44.9 °C but the evaporating temperatures of the two units are almost constant. In this condition, the decrease of condensing temperature increases the heat extraction of per unit refrigerant from wastewater and decreases the power consumption of per unit refrigerant, so the heat removed from wastewater increases and the power consumption decreases.

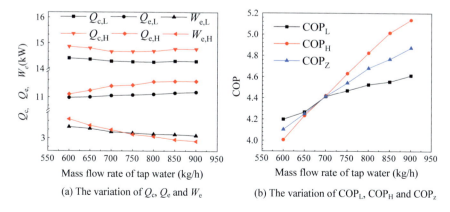

Fig. 2 Effect of tap water mass flow rate on performance of the SCBWSHP ($t_{c,i} = 10$ °C)

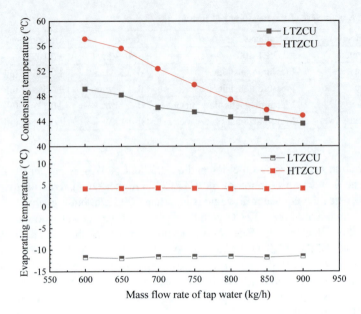

Fig. 3 Effect of tap water mass flow rate on condensing temperature and evaporating temperatures ($t_{c,i} = 10\ °C$)

Compared with the design parameters at the rated conditions (see Table 1), the condensing temperature of the LTZCU and the HTZCU in the experiment increased by about 14 °C and 2 °C, respectively, and the evaporating temperature of the LTZCU and the HTZCU decreased by about 13 °C and 9 °C, respectively. This is because the total heat transfer coefficient of the condenser and evaporator under actual operating conditions decreases possibly compared to the data provided by the heat exchanger manufacturer, leading to the increase in the heat transfer temperature difference of the heat exchanger.

3.2 Effect of Bathing Wastewater Flow Rate

Figure 4 shows the effect of the bathing wastewater flow rate on operating performance parameters of the SCBWSHP when the bathing wastewater inlet temperature is 30 °C, and the inlet and outlet temperatures of the tap water are 10 °C and 45 °C, respectively. It is found from Fig. 4a that the heating capacity for the LTZCU and HTZCU increases by 24.4% and 12.2%, respectively, as the bathing wastewater flow rate increases from 600 kg/h to 1000 m³/h, and the heat removed from the wastewater for the LTZCU and HTZCU increases by 31.7% and 14.6%, respectively; however, the power consumption of the compressors is almost constant. Figure 4a shows that the effect of the mass flow rate of wastewater on the

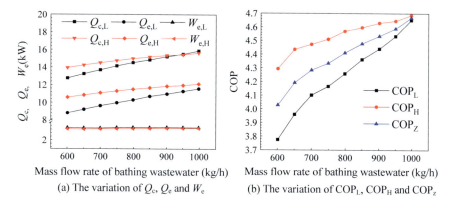

Fig. 4 Effect of bathing wastewater mass flow rate on heating performance of the SCBWSHP ($t_{e,i}$ = 30 °C)

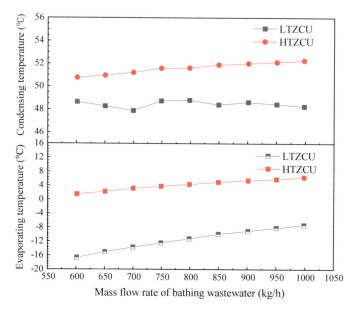

Fig. 5 Effect of bathing wastewater mass flow rate on condensing temperature and evaporating temperatures ($t_{e,i}$ = 30 °C)

heating performance of the LTZCU is larger than that of the HTZCU. It can be seen from Fig. 4b that the COP_L increases by 23%, from 3.77 to 4.65, the COP_H increases by 9%, from 4.29 to 4.68, and the COP_Z increases by 16%, from 4.03 to 4.67, when the mass flow rate of bathing wastewater changes from 600 to 1000 kg/h. This is because the increase of the bathing wastewater mass flow rate causes the increase of the refrigerant mass flow rate and evaporating temperature of the two units, which leads to the growth of recovery waste heat and the heating capacity. Therefore, the heating performance of the SCBWSHP goes up.

Figure 5 shows the variation of evaporating temperature of the SCBWSHP, the evaporating temperature $T_{e,L}$ of the LTZCU changes from −16.6 to −7.3 °C, and the evaporating temperature $T_{e,H}$ of the HTZCU increases from 1.5 to 6.4 °C. It can also be seen from Fig. 5 that the condensing temperature of the LTZCU and the HTZCU is almost invariable with increasing of mass flow rate of bathing wastewater. So the temperature difference between condensing temperature and evaporating temperature reduces for the LTZCU and the HTZCU, resulting in the increase of COP_L and COP_H.

During the experiment, the mass flow rate of tap water is 745 kg/h when the bathing wastewater mass flow rate is 786 kg/h, and COP of the SCBWSHP is 4.39, which is consistent with the experimental results of the effect of tap water mass flow rate on performance of the SCBWSHP.

4 Conclusions

In this paper, an experimental setup is built to investigate the heating performance of a two-series-connected bathing wastewater source heat pump hot water unit (SCBWSHP). The effects of the mass flow rate of tap water and bathing wastewater on the heating performance of the SCBWSHP are studied based on the experimental results. It can be concluded from this investigation that:

(1) With increasing of the tap water mass flow rate, the heat removed from wastewater increases, the power consumption decreases, and the heating capacity is almost constant, resulting in the increase of the coefficient of heating performance of the SCBWSHP. As the tap water mass flow rate increases from 600 to 900 kg/h, the COP_H, COP_L, and COP_Z increase by 28.2%, 9.8%, and 18.7%, respectively.
(2) With increasing of the bathing wastewater flow rate, the heating capacity and the heat removed from the wastewater increase; however, the power consumption of the compressors is almost invariable, which leads to the increase of the coefficient of heating performance of the SCBWSHP. When the bathing wastewater mass flow rate changes from 600 to 1000 kg/h, the COP_L, COP_H, and COP_Z increase by 23%, 9%, and 16%, respectively.

Acknowledgements The authors are grateful to the financial support by the National Natural Science Foundation of China (No. 51676026).

References

1. Shen, C., et al.: A field study of a wastewater source heat pump for domestic hot water heating. J. Build. Serv. Eng. Res. Technol. **34**(4), 433–447 (2012)
2. Liu, L., et al.: Application of an exhaust heat recovery system for domestic hot water. Energy **35**(3), 1476–1481 (2010)
3. Hepbasli, A., et al.: A key review of wastewater source heat pump (WWSHP) systems. Energy Convers. Manag. **88**, 700–722 (2014)
4. Song, G.J.: Study on characteristic of water source heat pumps in series. Tianjin University, Tianjin (2005). (in Chinese)
5. Sun, T.Y., et al.: Study on heating performance of water source heat pump unit with double series. Energy Conserv. **382**(7), 64–67 (2014). (in Chinese)
6. Ma, L.D., et al.: The performance simulation analysis of the sewage-source heat pump heater unit dealing with large temperature difference. Procedia Eng. **205**, 1769–1776 (2017)
7. Zheng, X.Z.: Experimental study on performance of large temperature difference bathing wastewater heat pump hot water unit. Dalian University of Technology, Dalian (2018). (in Chinese)
8. Kline, S.J., McClintock, F.A.: Describing uncertainties in single sample experiments. Mech. Eng. **75**, 3–8 (1953)
9. Kline, S.J.: The purpose of uncertainty analysis. ASME J. Fluids Engineering **117**, 153–160 (1985)

Field Test Analysis of a Novel Continuous Running Dual-Channel Condensation Gasoline Vapor Recovery System

Mengmeng Wu and Lin Cao

Abstract In China, the emission reduction and control of volatile organic compounds (VOCs) have become the focus of air pollution prevention and control. Condensation method was applied for gasoline vapor recovery due to its simple recycling principle, mature key technologies, and high safety. However, frost on the surface of the heat exchanger caused unstable operation of the recovery system. A novel continuous running dual-channel gasoline vapor recovery system was presented and tested in this paper. With dual channels of A and B switching cooling and defrosting conditions, the system could achieve recovering gasoline vapor while defrosting. The typical gasoline vapor was selected for the analysis of the condensation separation process of the system, and the condensation temperatures of the three stages were designed at 0, −25, and −75 °C, respectively. The temperature and pressure difference of the gasoline vapor were investigated during the condensation process. By the field measurement, the gasoline vapor could be cooled to −75 °C, and the ability of the system to operate stably and continuously was proven.

Keywords Gasoline vapor · Dual channels · Condensation method

1 Introduction

Volatile organic compounds (VOCs) were key precursors of PM2.5 and O_3, so controlling the emission of VOCs was beneficial to reducing the concentration of pollutants in the atmosphere. However, in the operations (e.g., storage,

---

M. Wu · L. Cao (✉)
School of Energy and Power Engineering,
Nanjing University of Science and Technology, Nanjing, China
e-mail: caolin1212@126.com

L. Cao
Postdoctoral Work Station, Guangdong Jirong Air Conditioning Equipment Co., Ltd, Jieyang, China

transportation, loading, and unloading) of gasoline, VOCs were easily evaporated and discharged from refineries, petrochemical plants, and gasoline service stations [1], which not only caused environmental pollution and risk of fire accidents, but also resulted in energy resource wastage and economic loss [2]. Hence, it was necessary to choose the proper technology according to the different concentrations and components of VOCs to control the emission of VOCs and recover the valuable resources. Various methods had been developed for recovering the gasoline vapor, which included adsorption [3–5], absorption [6], condensation [7, 8], and membrane [9]. And the trend of VOCs recovery process would be an integrated technology with various recovery technologies [10]. The condensation separation method remained more energy efficient for the recovery of high concentration and boiling point VOCs (e.g. toluene, octane, acetone) [11]. Thus, the condensation separation method was especially suitable to be chosen as the front process of the integrated technology [12].

The condensation process generally consisted of three stages: the pre-cooling, inter-cooling, and supercooling stages [13]. Many scholars [14–16] investigated the relationship between the recovery efficiency of the gasoline vapor and the energy consumption of the three-stage cooling process and the influence of various factors (e.g., pressure, temperature, flow rate) on the recovery efficiency. They concluded that the recovery efficiency of the gasoline vapor mainly depended on the condensation temperature and the heat transfer efficiency of the condenser. It meant that higher vapor recovery efficiency required lower temperature and higher energy consumption. Zhao and Du found that setting the double evaporator could simplify the refrigeration system, and setting the gas–gas heat exchanger could recover the residual cooling of the exhaust vapor [17]. Lee and Li et al. improved the temperature by pressurizing, which reduced the energy consumption of the system [18, 19]. As can be seen, there were many studies on improving efficiency and reducing energy consumption of the system. However, there still was another problem. The gasoline vapor was actually mixture of gasoline vapor and air, and thus, the water vapor would frost on the surface of the heat exchanger during the condensation process, which affects the energy consumption and stability of the system. Currently, little mention was made on defrosting issue.

In this paper, a novel continuous running dual-channel gasoline vapor recovery system was tested as the front process of the integrated technology. Both the pre-cooling and inter-cooling stages were single-stage refrigeration cycles, while the supercooling stage was cascade refrigeration cycle. The typical sample of the gasoline vapor was selected for the analysis of the condensation separation process of the system. And the condensation temperatures of the pre-cooling, inter-cooling, and supercooling stages were designed at 5, −25, and −75 °C, respectively. Proof of running results, the condensation recovery system could work stably and economically.

2 Process Description

As shown in Fig. 1, the whole system can be divided into the following processes:

Dual-channel unit: Both channels A and B had three groups of shell-tube evaporators (PE-A, IE-A, SE-A; PE-B, IE-B, SE-B) in the system, and they always ran under the opposite conditions. For example, when the gasoline vapor entered to channel A and exchanged heat with refrigerant in evaporators, the high-temperature refrigerant vapor flowed through the evaporators of channel B to defrost.

Gasoline vapor recovery: The gasoline vapor temperature would decrease after passing through the evaporator, and the part that condensed into liquid was separated, while the vapor continued to the next stage. The gasoline vapor would eventually be cooled to −75 °C. In addition, the temperature sensor was placed at the gasoline vapor outlet of each evaporator to record the gasoline vapor outlet temperature.

Refrigeration cycle: The criteria of refrigerant selection were first the safety, and nonflammable refrigerant was preferred. For the pre-cooling stage, R22 was selected as refrigerant. Combined with environmental protection, R404A was selected as refrigerant of the inter-cooling stage and the high-temperature cycle of supercooling stage. For the low-temperature cycle of cascade refrigeration cycle, ultra-low-temperature refrigerant was required. R23 was selected as refrigerant due to its excellent performance at low temperature. In addition, there was a heat exchanger designed as the economizer for economizing function in every stage of the refrigeration system (not shown in Fig. 1).

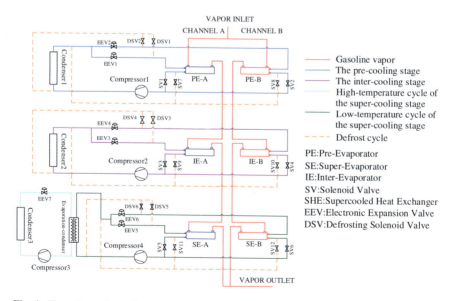

Fig. 1 Flow sheet of gasoline vapor recovery process

Defrost cycle: The principle of defrosting was to use high-temperature refrigerant vapor to melt the frost layer. For instance, when the evaporators of channel B were in defrosting, the high-temperature and pressure refrigerant gas that discharged from the compressor flowed directly into the evaporators of channel B to achieve defrosting, while the channel A operated cooling and vice versa.

3 Field Test Analysis

The system was used to recover gasoline vapor produced by the refinery which was located in Panjin City, Liaoning Province, China. The typical sample of the gasoline vapor was taken and analyzed. The inlet volumetric flow and temperature of the gasoline vapor were set at 1000 m^3/h and 25 °C, and the components were shown in Table 1. The condensation characteristics of the hydrocarbon components and water vapor were shown in Fig. 2 (N_2 and O_2 were not considered). According to the condensation characteristics of the components, the condensation temperature of each stage could be determined. The temperature of the pre-cooling stage was set at 0 °C to condense the water vapor and heavy hydrocarbon components. The temperature of the supercooling stage should be set at −75 °C to ensure the outlet vapor concentration met the requirement. The temperature of the inter-cooling stage should be set at −25 °C to keep the total cooling duty of the three-stage condensation process at a low level. According to the production arrangement of the refinery, the gasoline vapor condensation recovery system continuously operated for 8 h every day, and the real-time data was recorded every minute.

Table 1 Components of the gasoline vapor

Components	Molecular formula	Volumetric fraction (%)	Molar mass (g/mol)
Methane	CH_4	0.68	16
Ethane	C_2H_6	1.24	30
Propane	C_3H_8	1.7	44
Isobutane	C_4H_{10}-2	8.7	58
N-butane	C_4H_{10}-1	11.2	58
Isopentane	C_5H_{12}-2	6	72
N-pentane	C_5H_{12}-1	0.68	72
C5+	C5+	3.3	86
Nitrogen	N_2	51.9	28
Oxygen	O_2	13.8	32
Water vapor	H_2O	0.8	18

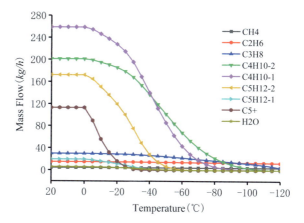

Fig. 2 Condensation characteristics of gasoline vapor

3.1 Analysis of the Gasoline Vapor Outlet Temperature

Figure 3 illustrates the variation of the gasoline vapor outlet temperature of the pre-cooling stage with the running time. It could be easily seen that both the channel A and channel B have the operation cycle of 4 h, with two hours for cooling and two hours for defrosting. The initial temperature of the gasoline vapor in Fig. 2 was about 38 °C instead of the inlet temperature, which was due to the increase of the gasoline vapor pressure after passing through the fan. When the evaporator was in cooling, the gasoline vapor could be cooled to about 0 °C. Both the curves of A and B had temperature fluctuations when the outlet temperature was lower than −5 °C. The reason was that the evaporation temperature was lower than the set value at this time, and the compressor stopped working to avoid excessive compression ratio. The components with higher molar mass like C5+ would be separated by condensation at this stage. In addition, a small amount of water vapor will condense into liquid. When the evaporator was in defrosting, the high-temperature refrigerant vapor passed through the evaporator, so the temperature at the gasoline vapor outlet would rise, finally to about 20 °C.

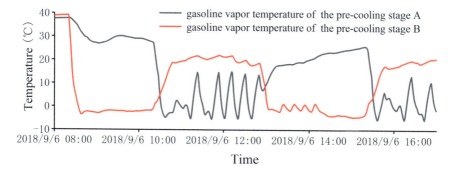

Fig. 3 Variation of the gasoline vapor outlet temperature of the pre-cooling stage

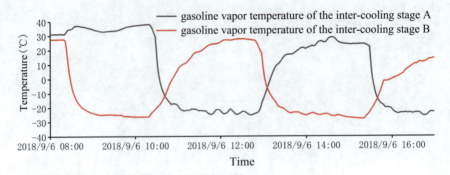

Fig. 4 Variation of the gasoline vapor outlet temperature of the inter-cooling stage

The trends of the inter-cooling and supercooling stages were almost same as the pre-cooling stage. As shown in Fig. 4, when the inter-evaporator was in cooling, the gasoline vapor outlet temperature rapidly reduced to −25 °C. The components like isopentane would be separated by condensation at this stage, and water vapor in the mixture would easily frost on the surface of evaporator. When the inter-evaporator was switched from the cooling condition to the defrosting condition, the temperature at the gasoline vapor outlet would increase finally to about 20 °C. As indicated by Fig. 5, when the super-evaporator was in cooling, the gasoline vapor outlet temperature dropped rapidly to −75 °C. The components like n-butane and isobutane would be separated at this stage. When the super-evaporator was in defrosting, the temperature at the gasoline vapor outlet would increase finally to about −35 °C.

After the gasoline vapor passing through the three-stage condensation, most components were separated from vapor, and the remaining vapor needed to be separated by the next process of the integrated technology. In summary, the system as the front process of the integrated technology of gasoline vapor recovery was able to operate continuously and stably and could finally make the gasoline vapor be cooled to −75 °C.

3.2 Analysis of the Gasoline Vapor Pressure Difference

The change in pressure difference between the gasoline vapor inlet and outlet was presented in Fig. 6. The pressure difference was mainly affected by the gasoline vapor flow and the thickness of the frost layer. At the beginning, the evaporators of channel B were in cooling, while the evaporators of channel A did not work. So the resistance of the channel B increased with the increase of the frost layer thickness and finally reached 9 kPa. When the gasoline vapor switched to the channel A for condensation, the pressure difference would reduce to about 3 kPa immediately. In the early stage after switching channels, the vapor flow would remain at a low level

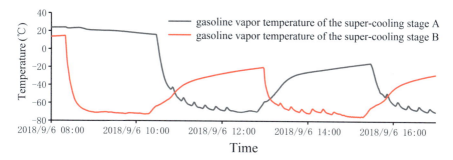

Fig. 5 Variation of the gasoline vapor outlet temperature of the supercooling stage

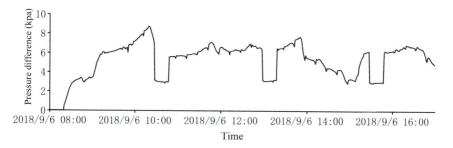

Fig. 6 Pressure difference of the gasoline vapor inlet and outlet

to ensure the efficient and stable operation of the system. Then the pressure difference increased rapidly to about 5 kPa due to the increase of the gasoline vapor flow and frost layer thickness. The frost layer thickness would continue to increase; however, the actual gasoline vapor flow may increase or decrease, so there were some fluctuations in the curve.

4 Conclusion

A novel continuous running dual-channel gasoline vapor recovery system was tested and analyzed in this paper. By setting two sets of evaporators, the system could perform both defrosting and cooling conditions, which solved the problem that unstable operation of the condensation method gasoline vapor recovery system due to the frost on the surface of the heat exchanger. The field test results indicated that the system could operate stably and efficiently. And after passing through the three cooling stages, the outlet temperatures of gasoline vapor could reach 0, −25, and −75 °C, respectively. When the system operated stably, the pressure difference between the gasoline vapor inlet and outlet increased with the increase of the gasoline vapor flow and the thickness of the frost layer. After switching channels, the pressure difference would reduce to 3 kPa rapidly. The test results could provide technical support for the promotion and application of condensation method in gasoline vapor recovery and had a positive effect on VOCs pollution reduction.

Acknowledgements This work is supported financially by the National Natural Science Foundation of China (grant numbers 51606096)

Permissions Appropriate permissions from the refinery responsible authorities were obtained for study in continuous running dual-channel condensation gasoline vapor recovery system.

References

1. Liang, J., Sun, L., Li, T.: A novel defrosting method in gasoline vapor recovery application. Energy **163**, 751–765 (2018)
2. Huang, W., et al.: Investigation of oil vapor emission and its evaluation methods. J. Loss Prev. Process Ind. **24**, 178–186 (2011)
3. Wang, S., et al.: Enhanced adsorption and desorption of VOCs vapor on novel micro-mesoporous polymeric adsorbents. J. Colloid Interface Sci. **428**, 185–190 (2014)
4. Scholten, E., et al.: Electrospun polyurethane fibers for absorption of volatile organic compounds from air. ACS Appl. Mater. Interfaces. **3**(10), 3902–3909 (2011)
5. Yu, L., et al.: Adsorption of VOCs on reduced graphene oxide. J. Environ. Sci. **67**, 171–178 (2018)
6. Chiang, C., et al.: Absorption of hydrophobic volatile organic compounds by a rotating packed bed. Ind. Eng. Chem. Res. **51**(27), 9441–9445 (2012)
7. Huang, W., et al.: Simulation of condensation process for gasoline vapor recovery. Adv. Mater. Res. **396–398**, 582–585 (2012)
8. Zhao, Z., Du, K.: Numerical investigation of condensation of gasoline vapor with turbulent flow in vertical tubes. J. SE Univ. (Engl. Ed.) **26**(2), 302–306 (2010)
9. Yang, W., et al.: Study on membrane performance in vapor permeation of VOC/N_2 mixtures via modified constant volume/variable pressure method. Sep. Purif. Technol. **200**, 273–283 (2018)
10. Liu, Y., et al.: Development status and trend of VOCs recovery technology. Mod. Chem. Ind. **31**(3), 21–25 (2011)
11. Belaissaoui, B., Moullec, Y., Favre, E.: Energy efficiency of a hybrid membrane/condensation process for VOC (volatile organic compounds) recovery from air: a generic approach. Energy **95**, 291–302 (2016)
12. Huang, W., et al.: Integrated technology of condensation and adsorption for volatile organic compounds recovery. Chem. Eng. **40**(6), 13–17 (2012)
13. Shi, L., Huang, W.: Sensitivity analysis and optimization for gasoline vapor condensation recovery. Process Saf. Environ. Prot. **92**, 807–814 (2014)
14. Huang, W., Peng, Q., Li, B.: Optimization of oil and gas condensation recovery technology based on Aspen software. Chem. Eng. Oil Gas **38**(4), 313–316 (2009)
15. Wang, M., Wang, T., Yang, Y.: The optimization of light gas recovery process with condensation method. Oil Gas Storage Transp. **32**(3), 329–333 (2013)
16. Kong, X., et al.: Optimal operation of gasoline vapor recovery equipment by condensation. Energy Eng. **59**(5), 59–64 (2008)

17. Zhao, Z., Du, K.: Study on optimization for condensation gasoline vapor recovery technique. In: The 6th National Refrigeration and Air Conditioning New Technology Seminar (2010)
18. Li, C., Li, J., Liu, X.: A discussion of the recovery technique of volatile oil-gas mixtures. Pet. Process. Petrochem. **41**(9), 85–90 (2010)
19. Lee, S., Choi, I., Chang, D.: Multi-objective optimization of VOC recovery and reuse in crude oil loading. Appl. Energy **108**, 439–447 (2013)

Matching Characteristics of Two Heat Exchangers for the Direct Sewage Source Heat Pump System

Zhaoyi Zhuang, Jian Song, Jun Xu and Wenzeng Shen

Abstract According to the switching mode of the cooling and heating function, the urban sewage source heat pump unit can be divided into water-side (external) switching unit and machine-side (internal) switching unit. For the design of sewage source heat pump unit, this paper first discusses the standard working conditions of sewage source heat pump unit, analyses the relationship between the performance of external and internal switching units and the matching of two heat exchangers and compares the characteristics and selection methods of internal and external switching units.

Keywords Sewage source heat pump · Internal/external switching system · Output ratio · Area ratio · Matching characteristics

1 Introduction of Direct Sewage Source Heat Pump System

According to the switching modes of heating and cooling function in winter and summer, direct sewage source heat pump system (DSSHPS) can be divided into two modes [1], that is external switching system and internal switching system. The internal switching system uses four-way reversing valve to switch inside units and realizes the purpose of air conditioning in winter and summer by changing the flow direction of refrigerant. The external switching system realizes the winter and summer switching by opening and closing different valves on the pipeline outside the unit [2]. Therefore, the external switching system is also called a pipeline switching system. Considering that the unit is only switched once a year, the

Z. Zhuang (✉) · J. Song · J. Xu · W. Shen
School of Thermal Engineering, Shandong Jianzhu University, 250101 Jinan, China
e-mail: hit6421@126.com

Z. Zhuang
Shandong Zhongrui New Energy Technology Co., Ltd, 250101 Jinan, China

ordinary manual valve group can be used to switch to avoid problems such as stuck four-way valve and leakage. Meanwhile, it is low in cost, safe and reliable and easy to operate [3].

2 Standard Design Conditions of the Direct Sewage Source Heat Pump System

In comparison with the conventional clean water source heat pump unit, the heat transfer performance of the two heat exchangers of the DSSHPS is quite different, which results in a large difference between the heating capacity and the cooling capacity of the DSSHPS. At the same time, the cold and heat loads in different regions are also inevitably different. The combination of the two characteristics will guide the design of the DSSHPS [4].

Because of the different methods of switching the cooling and heating functions between the external switching unit and the internal switching unit, the evaporation and condensation heat exchangers of the unit are quite different, which leads to the difference of the cooling and heating capacity of the unit [5]. In addition, when designing and calculating the two heat exchangers of direct sewage source heat pump units, we can design them according to heating conditions (heating design) or refrigeration conditions (cooling design), which will also lead to differences in cooling and heating capacity of sewage source heat pump units.

The sewage heat exchanger of the internal switching unit flows sewage in winter and summer, but the effect of heat exchange is quite different in winter and summer. The sewage heat exchanger acts as an evaporator in winter and as a condenser in summer. The water heat exchanger flows clear water in winter and summer, but it acts as a condenser in winter and as an evaporator in summer. Because the sewage just flows through one heat exchanger, two heat exchangers of the internal switching unit work efficiently on the clear water side; on the contrary, the effect on the refrigerant side is not ideal.

Therefore, the characteristics of the internal and external switching heat exchangers of the direct sewage source unit are given in Table 1.

The standard design condition should be determined first in the design of SSHPS units. Based on the sewage temperature of Chinese urban between 9 and 15 °C in winter and between 20 and 26 °C in summer, the following conditions are determined as the standard design conditions for sewage source heat pump units [6]:

Table 1 Two heat exchanger characteristics of internal and external switching heat pumps

The unit	Heat exchanger	Cooling in summer	Heating in winter
External switching	Evaporator (E)	Water (q)	Sewage (w)
	Condenser (C)	Sewage (w)	Water (q)
Internal switching	Sewage heat exchanger (W)	Condensation (c)	Evaporation (e)
	Water heat exchanger (Q)	Evaporation (e)	Condensation (c)

Table 2 Heat exchanger parameters of internal and external switching units in standard condition

The unit	HE	Cooling in summer	Heating in winter
External switching	E	K_{qe} = 2880 W/(m²·°C) U_{qe} = 14.4 kW/m²	K_{we} = 2740 W/(m²·°C) U_{we} = 13.7 kW/m²
	C	K_{wc} = 2720 W/(m²·°C) U_{wc} = 13.6 kW/m²	K_{qc} = 3020 W/(m²·°C) U_{qc} = 16.55 kW/m²
Internal switching	W	K_{wc} = 2470 W/(m²·°C) U_{wc} = 12.35 kW/m²	K_{we} = 2620 W/(m²·°C) U_{we} = 13.1 kW/m²
	Q	K_{qe} = 4320 W/(m²·°C) U_{qe} = 21.6 kW/m²	K_{qc} = 4170 W/(m²·°C) U_{qc} = 20.85 kW/m²
φ		φ_c = 1.2222	φ_h = 1.2857

Heating standard condition: the inlet temperature of the sewage is 12 °C; the outlet temperature is 5 °C; the average heat transfer temperature difference of evaporation is 5 °C; the inlet temperature of circulating water is 40 °C; the outlet temperature is 45 °C; the average heat transfer temperature difference of condensation is 5 °C; COP_h = 4.5.

Cooling standard condition: the inlet temperature of the sewage is 20 °C; the outlet temperature is 27 °C; the average heat transfer temperature difference of evaporation is 5 °C; the inlet temperature of circulating water is 12 °C; the outlet temperature is 7 °C; the average heat transfer temperature difference of condensation 5 °C; COP_C = 5.5.

In addition, this paper defines the COP of the heat pump unit according to the following relationship: $Q_c = \frac{COP}{COP-1} Q_e$, and this paper assumes $\varphi = \frac{COP}{COP-1}$.

Assuming that the product of the heat transfer coefficient and the average heat transfer temperature difference in the heat exchanger is a parameter, that is $U = K\Delta t_m$ (heat flux density). Various measures for strengthening heat transfer can be taken in the water-side heat exchanger tube of the internal switching unit, but it is not suitable to take measures for strengthening heat transfer in the water-side heat exchanger tube of the external switching unit. Therefore, both K_{qe} and K_{qc} of the water heat exchanger of the internal switching unit are larger than those of the external switching unit [7]. The parameters of the two heat exchangers of the unit under standard design conditions are given in Table 2.

3 Matching Design of Two Heat Exchangers of Direct Sewage Source Heat Pump System

3.1 Matching Design of Two Heat Exchangers of External Switching System

The area size of the two heat exchangers of the SSHPS units directly affects the cooling capacity or heat production of the unit, and the ratio of the area of the two heat exchangers determines the ratio of the standard cooling capacity to the heating capacity of the unit. The heat pump units with better performance require the ratio of cooling capacity to heating capacity to match the ratio of the cold loads to the heat loads of the building, and it is adjustable within a certain range, which puts requirements on the matching of the two heat exchangers of the unit [8].

In this paper, the ratio of cooling capacity to heating capacity of heat pump unit is defined as the output ratio, that is $\eta_w = \Phi_c/\Phi_h$; the ratio of the heat exchange area between the condenser and the evaporator is the area ratio of the two heat exchangers, that is $\xi w = A_c/A_e$.

For traditional water source heat pump units, both water heat exchangers flow clear water (the medium in the two heat exchangers is the same in winter and summer), whether it is internal switching or external switching unit. Further more, the heat transfer coefficient of the two heat exchangers is basically the same in winter and summer. Since the medium in the two heat exchangers of the SSHPS is different, and the heat transfer coefficient of the sewage side heat exchanger is small, which leads to the SSHPS units will have a different output-area ratio relationship than the conventional heat pump units. The following results are obtained in Table 3: the area of two heat exchangers, the cooling capacity, the heating capacity, the relationship between output ratio and area ratio under the cooling design and heating design of the external switching sewage source heat pump unit. The meaning of $\Phi_c(c)$ in the conclusion of Table 3 is that the cooling capacity needs to be calculated according to the area of the condenser, and the condenser limits the amount of cooling capacity; the meaning of $\Phi_h(E)$ is that the heating capacity needs to be calculated according to the area of the evaporator, and the evaporator limits the amount of heating capacity.

3.2 Matching Design of Two Heat Exchangers of the Internal Switching System

The two heat exchangers of the external switching unit and the internal switching unit have different functional requirements, working medium and heat exchange surface structure. Therefore, the internal switching sewage source heat pump units will also have a different output-to-area ratio relationship than the external switching units.

For the internal switching unit, the output ratio is defined as the ratio of cooling capacity to heating capacity, that is $\eta_r = \Phi_c/\Phi_h$; the area ratio of two heat exchangers is the ratio of area of the sewage heat exchanger to the freshwater heat exchanger, that is $\xi w = A_w/A_q$; the following results are obtained in Table 4: the area of two heat exchangers, the cooling capacity, the heating capacity, the relationship between output ratio and area ratio under the cooling design and heating design of the internal switching sewage source heat pump unit. The meaning of $\Phi_c(W)$ in the conclusion of Table 4 is that the cooling capacity needs to be calculated according

Table 3 Design of external switching sewage source heat pump unit and its power rate η-area rate ξ

	Function	Evaporator (E)		Condenser (C)	
Cooling design	Cooling	Water	$A_{e1} = \frac{\Phi_c}{U_{qe}}$	Sewage	$A_c = \frac{\varphi_c \Phi_c}{U_{wc}}$
	Heating	Sewage	$Q_{h1} = \frac{U_{we}\Phi_c}{U_{qe}}$ (✕) $\Rightarrow$ $Q_{h2} = \frac{U_{qc}\varphi_c\Phi_c}{U_{wc}\varphi_h}$ $\Leftarrow$	Water	$\Phi_{h1} = \varphi_h \frac{U_{we}\Phi_c}{U_{qe}}$ (✕) $\Phi_{h2} = U_{qc} \frac{\varphi_c\Phi_c}{U_{wc}}$
			$\eta_{w1} = \frac{\Phi_c}{\Phi_{h1}} = \frac{U_{qe}}{\varphi_h U_{we}}$ (✕)		$\eta_{w2} = \frac{\Phi_c}{\Phi_{h2}} = \frac{U_{wc}}{\varphi_c U_{qc}} < 1$
			$\xi_{w1} = \frac{A_c}{A_{e1}} = \frac{\varphi_c U_{qe}}{U_{wc}}$ (✕)		$\xi_{w2} = \frac{A_c}{A_{e2}} = \frac{\varphi_h U_{we}}{U_{qc}}$
		\multicolumn{4}{l|}{If the area of E is A_{e1}, E and C will be saturated. As result, the heat isn't enough, and the number of Φ_h will decline. If we increase A_e from A_{e1} to $A_{e2} = \frac{Q_{h2}}{U_{we}} = \frac{U_{qc}\varphi_c\Phi_c}{U_{we}U_{wc}\varphi_h}$, the number of Φ_h will increase;, otherwise, η_w will decline from η_{w1} to η_{w2}. In other words, if $\xi_{w2} < \xi_w < \xi_{w1}$ 时, $\eta_{w2} < \eta_w < \eta_{w1}$}			
Heating design	Heating	Sewage	$A_e = \frac{\Phi_h}{\varphi_h U_{we}}$	Water	$A_{c1} = \frac{\Phi_h}{U_{qc}}$
	Cooling	Water	$\Phi_{c1} = U_{qe}\frac{\Phi_h}{\varphi_h U_{we}}$ $\Rightarrow$ $\Phi_{c2} = \frac{U_{wc}\Phi_h}{\varphi_c U_{qc}}$ (✕) $\Leftarrow$	Sewage	$Q_{c1} = \varphi_c \frac{U_{qe}\Phi_h}{\varphi_h U_{we}}$ $Q_{c2} = U_{wc}\frac{\Phi_h}{U_{qc}}$ (✕)
			$\eta_{w1} = \frac{\Phi_{c1}}{\Phi_h} = \frac{U_{qe}}{\varphi_h U_{we}}$		$\eta_{w2} = \frac{\Phi_{c2}}{\Phi_h} = \frac{U_{wc}}{\varphi_c U_{qc}}$ (✕)
			$\xi_{w1} = \frac{A_{c2}}{A_e} = \frac{\varphi_c U_{qe}}{U_{wc}}$		$\xi_{w2} = \frac{A_{c1}}{A_e} = \frac{\varphi_h U_{we}}{U_{qc}}$ (✕)
		\multicolumn{4}{l|}{If the area of C is A_{c1}, E and C will be saturated. As result, heat dissipation isn't enough, and the number of Φ_c will decline. If we increase A_c from A_{c1} to $A_{c2} = \frac{Q_{c2}}{U_{wc}} = \frac{U_{qe}\varphi_c\Phi_h}{U_{we}U_{wc}\varphi_h}$, the number of Φ_c will increase; otherwise η_w will increase from η_{w2} to η_{w1}. In other words, if $\xi_{w2} < \xi_w < \xi_{w1}$, $\eta_{w2} < \eta_w < \eta_{w1}$}			
Conclusion	(1) If $\xi_w \leq \xi_{w2} = \frac{\varphi_h U_{we}}{U_{qc}}$		$\eta_w \equiv \eta_{w2} = \frac{U_{wc}}{\varphi_c U_{qc}}$		$\Phi_c = \Phi_c(C)$, $\Phi_h = \Phi_h(C)$
	(2) If $\xi_{w2} \leq \xi_w \leq \xi_{w1}$		$\eta_{w2} \leq \eta_w = \frac{\xi_w U_{wc}}{\varphi_h \varphi_c U_{we}} \leq \eta_{w1}$		$\Phi_c = \Phi_c(C)$, $\Phi_h = \Phi_h(E)$
	(3) If $\xi_w \geq \xi_{w1} = \frac{\varphi_c U_{qe}}{U_{wc}}$		$\eta_w \equiv \eta_{w1} = \frac{U_{qe}}{\varphi_h U_{we}}$		$\Phi_c = \Phi_c(E)$, $\Phi_h = \Phi_h(E)$

to the area of the sewage heat exchanger, and the sewage heat exchanger limits the amount of cooling capacity; the meaning of $\Phi_h(Q)$ is that the heating capacity needs to be calculated according to the area of the water heat exchanger, and the water heat exchanger limits the amount of heating capacity.

Table 4 Internal switching sewage source heat pump unit and its power rate η-area rate ξ

	Function	Sewage heat exchanger (W)		Water heat exchanger (Q)	
Cooling design	Cooling	Condensation	$A_w = \frac{\varphi_c \Phi_c}{U_{wc}}$	Evaporation	$A_{q1} = \frac{\Phi_c}{U_{qe}}$
	Heating	Evaporation	$Q_{h1} = \frac{U_{qc}\Phi_c}{\varphi_h U_{qe}}$ (※) $\Leftarrow$	Condensation	$\Phi_{h1} = U_{qc}\frac{\Phi_c}{U_{qe}}$ (※)
			$Q_{h2} = U_{we}\frac{\varphi_c \Phi_c}{U_{wc}}$ $\Rightarrow$		$\Phi_{h2} = \varphi_h U_{we}\frac{\varphi_c \Phi_c}{U_{wc}}$
		$\eta_{r1} = \frac{\Phi_c}{\Phi_{h1}} = \frac{U_{qe}}{U_{qc}}$ (※)		$\eta_{r2} = \frac{\Phi_c}{\Phi_{h2}} = \frac{U_{wc}}{\varphi_h \varphi_c U_{we}} < 1$	
		$\xi_{r1} = \frac{A_w}{A_{q1}} = \frac{\varphi_c U_{qe}}{U_{wc}}$ (※)		$\xi_{r2} = \frac{A_w}{A_{q2}} = \frac{U_{qc}}{\varphi_h U_{we}}$	
		If the area of Q is A_{q1}, Q and W will be saturated. As a result, heat dissipation isn't enough, and the number of Φ_h will decline. If we increase A_q from A_{q1} to $A_{q2} = \frac{Q_{h2}}{U_{qc}} = \frac{U_{we}\varphi_c \Phi_c}{U_{wc}U_{qc}}$, the number of Φ_h will increase; otherwise, η_r will decline from η_{r1} to η_{r2}. In other words, if $\xi_{r2} < \xi_r < \xi_{r1}$, $\eta_{r2} < \eta_r < \eta_{r1}$			
Heating design	Heating	Evaporation	$A_{w1} = \frac{\Phi_h}{\varphi_h U_{we}}$	Condensation	$A_q = \frac{\Phi_h}{U_{qc}}$
	Cooling	Condensation	$Q_{c1} = \varphi_c U_{qe}\frac{\Phi_h}{U_{qc}}$ $\Leftarrow$	Evaporation	$\Phi_{c1} = U_{qe}\frac{\Phi_h}{U_{qc}}$
			$Q_{c2} = U_{wc}\frac{\Phi_h}{\varphi_h U_{we}}$ (※)		$\Phi_{c2} = \frac{U_{wc}\Phi_h}{\varphi_c \varphi_h U_{we}}$
		$\eta_{r1} = \frac{\Phi_{c1}}{\Phi_h} = \frac{U_{qe}}{U_{qc}}$		$\eta_{r2} = \frac{\Phi_{c2}}{\Phi_h} = \frac{U_{wc}}{\varphi_h \varphi_c U_{we}}$	
		$\xi_{r1} = \frac{A_{w2}}{A_q} = \frac{\varphi_c U_{qe}}{U_{wc}}$		$\xi_{r2} = \frac{A_{w1}}{A_q} = \frac{U_{qc}}{\varphi_h U_{we}}$	
		If the area of W is A_{w1}, W and Q will be saturated. As a result, heat dissipation isn't enough, and the number of Φ_C will decline. If we increase A_w from A_{w1} to $A_{w2} = \frac{Q_{c2}}{U_{wc}} = \frac{U_{qe}\varphi_c \Phi_h}{U_{wc}U_{qc}}$, the number of Φ_c will increase; otherwise, η_r will increase from η_{r2} to η_{r1}. In other words, if $\xi_{r2} < \xi_r < \xi_{r1}$, $\eta_{r2} < \eta_r < \eta_{r1}$			
Conclusion	(1) If $\xi_r \leq \xi_{r2} = \frac{U_{qc}}{\varphi_h U_{we}}$ $\eta_r \equiv \eta_{r2} = \frac{U_{wc}}{\varphi_h \varphi_c U_{we}}$			$\Phi_c = \Phi_c(W)$, $\Phi_h = \Phi_h(W)$	
	(2) If $\xi_{r2} \leq \xi_r \leq \xi_{r1}$ $\eta_{r2} \leq \eta_r = \frac{\xi_r U_{wc}}{\varphi_c U_{qc}} \leq \eta_{r1}$			$\Phi_c = \Phi_c(W)$, $\Phi_h = \Phi_h(Q)$	
	(3) If $\xi_r \geq \xi_{r1} = \frac{\varphi_c U_{qe}}{U_{wc}}$ $\eta_r \equiv \eta_{r1} = \frac{U_{qe}}{U_{qc}}$			$\Phi_c = \Phi_c(Q)$, $\Phi_h = \Phi_h(Q)$	

4 The Effect of Matching of Two Heat Exchangers on the Output Ratio of Heat Pump System

Under the standard design conditions of the SSHPS units, the output ratio–area ratio of the external switching unit and the internal switching unit is given in Table 5 and Fig. 1.

Table 5 $\eta - \xi$ of internal and external switching sewage source heat pump unit

		ξ_2	ξ_1	η_{min}	η_{max}	$\eta\ (\xi_2 \leq \xi \leq \xi_1)$
External switching	Formula	$\dfrac{\varphi_h U_{we}}{U_{qc}}$	$\dfrac{\varphi_c U_{qe}}{U_{wc}}$	$\dfrac{U_{wc}}{\varphi_c U_{qc}}$	$\dfrac{U_{qe}}{\varphi_h U_{we}}$	$\dfrac{\xi_w U_{wc}}{\varphi_h \varphi_c U_{we}}$
	Numerical value	1.06	1.29	0.67	0.82	$0.63\ \xi_w$
Internal switching	Formula	$\dfrac{U_{qc}}{\varphi_h U_{we}}$	$\dfrac{\varphi_c U_{qe}}{U_{wc}}$	$\dfrac{U_{wc}}{\varphi_h \varphi_c U_{we}}$	$\dfrac{U_{qe}}{U_{qc}}$	$\dfrac{\xi_r U_{wc}}{\varphi_c U_{qc}}$
	Numerical value	1.24	2.14	0.60	1.04	$0.48\ \xi_r$

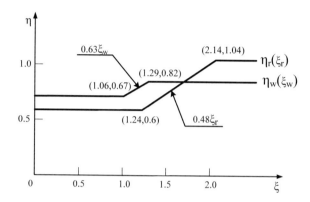

Fig. 1 Relational comparison of $\eta - \xi$ for internal and external switching sewage source heat pump units

5 Conclusion

It can be seen from Fig. 1 that the commonalities of internal and external switching sewage source heat pump units are as follows:

There is a range of the area ratio of the two heat exchangers. In this range, the heat exchangers are operated saturated with no waste of heat exchange area when the heat pump units operating under the standard design condition; for units operating without this range, there is always a certain heat exchanger that has more than enough area for operation under standard conditions and that is only a waste of investment.

The output ratio is independent of the area ratio without the range of suitable area ratio, and the maximum value and the minimum value are, respectively, adopted.

A relatively small area ratio should be adopted in northern areas and big area ratio in the south areas for the SSHPS units.

The differences between internal and external switching sewage source heat pump units are as follows:

The internal switching unit has a larger area ratio than the external switching unit, that is, the area of the two heat exchangers of the internal switching type sewage source heat pump unit is larger.

The internal switching unit has a larger output ratio than the external switching unit, and a smaller output ratio, that is to say, the cooling and heating capacity of the internal switching sewage source heat pump unit is much different.

Acknowledgements This work was supported by National Natural Science Foundation of China (Grant No. 51708339), the China Postdoctoral Science Foundation Funded Project (Grant No. 2017M612303).

References

1. Zhuang, Z.Z., Qi, J., Zhang, C.H., et al.: Key techniques of direct sewage source heat pump systems. HVAC **41**(10), 96–101 (2011)
2. Hepbasli, A., Biyik, E., Ekpen, O.: A key review of wastewater source heat pump (WWSHP) systems. Energy Convers. Manag. **88**, 700–722 (2014)
3. Cheng, X.S., Peng, D.G., Li, S.L., et al.: Model and influencing factors analysis of external parameters of sewage source heat pump. J. Chongqing Jianzhu Univ. **39**(4), 26–32 (2017)
4. Tu, A.M., Zhu, D.S., Guo, X.F.: Comparative analysis of centralized heating and cooling system based on sewage-source heat pump system with a traditional system in hot summer and warm winter zone. HVAC **46**(12), 90–95 (2016)
5. Meggers, F., Leibundgut, H.: The potential of wastewater heat and exergy: decentralized high temperature recovery with a heat pump. Energy Build. **43**(4), 879–886 (2011)
6. Qian, J.F., Ren, Q.F., Xu, Y., et al.: Experiment on the antiscale and descaling and heat transfer enhancement of acoustic cavitation sewage heat exchanger. J. Harbin Inst. Tech. **50**(2), 166–172 (2018)
7. Chao, S., Yang, L., Wang, X., et al.: An experimental and numerical study of a de-fouling evaporator used in a wastewater source heat pump. Appl. Therm. Eng. **70**, 501–509 (2014)
8. Liu, Z.B., Ma, L.D., Zhang, J.L.: Application of a heat pump system using untreated urban sewage as a heat source. Appl. Therm. Eng. **62**, 747–757 (2014)

Single-Phase Heat Transfer and Pressure Drop of Developing Flow at a Constant Heating Flux Inside Horizontal Helical Finned Tubes

Zhixian Ma, Nan Zhao, Anping Zhou and Jili Zhang

Abstract An experimental investigation was carried out to determine the friction factor and heat transfer coefficients under the uniform heat flux boundary condition. Heat transfer and pressure drop data were obtained from two different types of internal helical finned tubes with diameters of 22.48 and 16.662 mm, a fin height to diameter ratio of 0.0222 and 0.0534, number of starts of 60 and 38 and helix angles of 45 and 60°. Reynolds numbers ranged between 300 and 30,000, while Prandtl number was in the order of 12.1 to 47.5. Results show that friction factors of single-phase flow in the internal helical finned tube were higher than the counterpart of the plain tube. The heating boundary condition has a significant effect on friction factors of internal helical finned tubes in the laminar and transition flow regions. Transitions of single-phase flow in the internal helical finned tube under uniform heat flux were delayed when compared with the adiabatic result. In turbulent region, heat transfer results showed an overall increase when compared with the smooth tubes. With the same Re or q, the heat transfer enhancement of tested tubes was in the range of 1.0–6.8 in the turbulent region and the heat transfer enhancement increase with Reynolds number.

Keywords Internal helical-finned tube · Adiabatic · Secondary transition · Developing flow · Friction factor · Experiment · Single-phase

1 Introduction

Since Osborne Reynolds demonstrated the laminar-turbulent transition occurs at Re = 2300 in a circular pipe in which he examined the behavior of water flow [1], it was found that the transition occurs between Re = 2000 and 13,000, depending on the initial conditions such as initial disturbance amplitude and surface roughness [2].

Z. Ma (✉) · N. Zhao · A. Zhou · J. Zhang
Institute of Building Energy, Dalian University of Technology, 116024 Dalian, China
e-mail: mazhixian@dlut.edu.cn

© Springer Nature Singapore Pte Ltd. 2020
Z. Wang et al. (eds.), *Proceedings of the 11th International Symposium on Heating, Ventilation and Air Conditioning (ISHVAC 2019)*, Environmental Science and Engineering, https://doi.org/10.1007/978-981-13-9524-6_84

Thus, it is common sense to design water chiller units and heat exchangers in such a way that they do not operate within the transition region. This is mainly due to the unstable flow as well as the insufficient information in this region. Due to design constraints or change in operating conditions, however, exchangers are often forced to operate in this region, such as the ice storage system of water–ethylene glycol mixture and variable flow air-conditioning system. This is even worse for internal helical finned tubes as much less information within this region is available.

Since the internal helical finned tube was invented [3], extensive work has been performed to investigate the heat transfer and pressure drop characteristics inside the internal helical finned tube over a range of geometric parameters (0.1 mm < e < 2.06 mm, 4 mm < d_i < 25.78 mm, 0° < β < 79.4°, 1 < N_s < 82, 0.00644 < e/d_i < 0.0850) and between Reynolds number of 250 and 100,000 [4–8]. However, how the thermal boundary condition affects the heat transfer and friction factor should be explored, especially in the transition region.

Therefore, the purpose of this study experimentally determines the heat transfer and pressure drop characteristics of the enhanced tubes in the transitional flow region for the different heating flux conditions. Heated cases were considered across the entire flow regions from laminar to turbulent, with the special attention was given to the transition flow region. The fluid in the tube is believed to be the nature and forced convection. The emphasis in this paper was therefore determining the flow and heat transfer characteristics under the uniform heat flux boundary.

2 Methods

2.1 Experiment Setup

Figure 1 shows the schematic of the experimental setup. As is shown, it consisted of a fluid flow loop, the test section, and a chiller unit. The working fluid circulated from a 1200-liter storage tank using a centrifugal pump through the flow meters to the test section and then went to the heat exchanger to discharge the heat to the refrigerant. The end of the stream was the storage tank which contained 50% water–ethylene glycol mixture. Six industrial immersion electric heaters (with capacity of 2.0 kW each) were lowered into the tank to heat the fluid to a preselected temperature and thermostatically controlled by a thermostat. The centrifugal pump with a maximum flow rate of 8.0 m³/h and the chiller unit with a capacity of 9.0 kW were attached to the test bench. The experimental test section consisted of a smooth tube of inner diameter 16.34 mm and three enhanced tubes with diameters from 16.662 to 22.48 mm. A cross section of one of the tubes is given in Fig. 2, and the geometric properties of the tubes are given in Table 1.

Thirty-eight thermocouples stations were designated at closer intervals near the entrance and exit, and at wider intervals downstream of the tube. As for enhanced tube, for each station, one thermocouple was used placed at the bottom of fins and was squeezed by peripheral micro fins with thermal silicone grease in the remainder

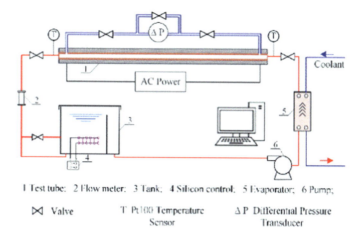

Fig. 1 Schematic of the test bed

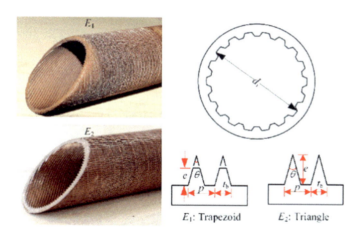

Fig. 2 Test tubes

Table 1 Geometric properties of the tubes tested

Tube	l (mm)	d_o (mm)	d_i (mm)	N_s (–)	p (mm)	e (mm)	β (°)	t_b (mm)	θ (°)
R^a	2945	18.34	16.34	–	–	–	–	–	–
E_1^a	2643	25.15	22.48	60	1.18	0.5	45	0.61	43.1
E_2^a	2945	16.669	16.662	38	1.38	0.89	60	0.72	43.8

[a]R refers to the plain tube, E_1 and E_2 refer to the enhanced tubes

of the base. Twenty-eight thermopiles were spaced at the outer wall of enhanced tubes. The remaining thermocouples were placed at outer wall of the smooth tube. Average inlet and outlet fluid bulk temperatures were measured by six four-wired

Pt100 temperature sensors, and the temperature difference measurement error was calibrated to less than 0.04 °C. The test section was insulated from the environment by covering it with 80-mm-thick aluminum silicate and 50-mm-thick elastomeric foam sheets for fire insulation and heat preservation. All thermocouples were calibrated against the average temperature of two Pt100 detectors placed at the inlet and outlet of test tubes. Calibrations were done with the water–ethylene glycol mixture circulated at temperatures of 10–80 °C, with heat loss in good satisfactory.

Tests were conducted at inlet temperature of 10 ± 0.04 and 50 ± 0.04 °C. During the runtime, heat transfer tests were experimented under a constant heat flux. All data points were recorded upon the computer reaching a steady-state condition, which was when the change in the average temperature of the fluid at the outlet was within ± 0.04 °C over two minutes. And twenty groups of data were recorded for each stable experimental condition. Moreover, heat transfer between the test section and the environment has been stabilized and the wall temperatures have fluctuated at a small range.

Due to energy balance would be in no good satisfactory in low Reynolds number. Hence, data with heat balance within 10% were allowed. In the turbulent, the data with energy balance within 5% were allowed for analysis.

2.2 Data Reduction and Error Analysis

All fluid properties were evaluated at the fluid bulk temperature. The bulk temperature was determined as Eq. (1).

$$T_b = (T_i + T_o)/2 \tag{1}$$

The heat transfer rate in the test tube was determined as Eq. (2).

$$\dot{Q} = \dot{m} c_p (T_o - T_i) = UI \tag{2}$$

where U was the heating voltage and I was the heating current.

The average Reynolds number was determined from the measured average flow rate u_m, the inner tube diameter d_i, and the fluid viscosity μ. They were defined as follows:

$$\text{Re} = \rho u_m d_i / \mu \tag{3}$$

$$u_m = 4V/\pi d_i^2 \tag{4}$$

where V was the fluid volumetric flow rate.

The heat flux applied to the test tube was determined by the heat transfer rate and the tube inner heat transfer area.

$$\dot{q} = \dot{Q}/(\pi d_i l) \tag{5}$$

where l was the heat transfer length of the test tube.

The resistance through the wall R_w was determined from:

$$R_w = \ln(d_o/d_i)/(2\pi k_{cu} l) \tag{6}$$

where k_{cu} is the thermal conductivity of the copper tube. The average outer wall surface temperature was determined by the average temperature of thermocouples per station. The inner wall surface temperature T_{wi} was determined by

$$T_{wi} = T_{wo} - \dot{Q} R_w \tag{7}$$

The average heat transfer coefficient and Nusselt number, based on the inner diameter, were estimated using Eq. (8).

$$h = \dot{q}/(T_{wi} - T_b) \tag{8}$$

The average and local Nusselt numbers were determined based on the heat transfer coefficient and thermal conductivity of the fluid as:

$$\text{Nu} = hd/k \text{ or } \text{Nu}_x = h_x d/k_x \tag{9}$$

The friction factor was obtained using Darcy–Weisbach equation, shown as follows:

$$f = 2\Delta p d_i/(\rho l u_m^2) \tag{10}$$

The uncertainty analysis of the experiment was performed by the method suggested by Kline and Mclintock [9] over the whole experimental procedures. The maximum uncertainties of friction factors were 10.32% at low Reynolds numbers due to higher uncertainties of the flow meter at lower flow rates. As the Reynolds number increased, the friction factor uncertainties reduced to uncertainties lower than 1.0%.

3 Results

3.1 Friction Factor

The diabatic friction factors for the enhanced tubes are given in Figs. 3 and 4. For reference purposes, the laminar Poiseuille relation in terms of the Darcy friction factor as well as the Filonenko equation for turbulent flow is plotted as black solid

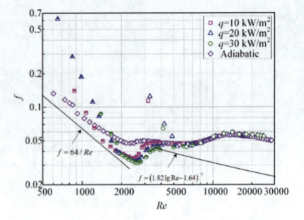

Fig. 3 Diabatic friction factor results of E_1

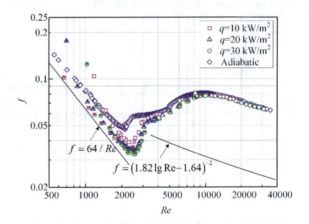

Fig. 4 Diabatic friction factor results of E_2

lines. These results show that there is an overall increase in friction factor for the two tested enhanced tubes compared with the equivalent smooth tubes, as did the adiabatic results [10, 11]. In the laminar region, diabatic friction factors outdid 1.4–6.4 times compared to the smooth tube counterparts. When the flow approached to the transition, the diabatic friction factors were lower than those of the adiabatic flow; but when the flow is staying away from the transition, the diabatic friction factors were higher than those of the adiabatic flow. The turbulent results are very similar to those of the adiabatic friction factors, also having the same secondary transition region with Reynolds numbers range from 3000 to 10,000. Transition under uniform heat flux conditions was delayed when compared with the adiabatic result and the secondary flow advanced the flow stability, and the greater the heat flux is, the later the transition appeared.

3.2 Heat Transfer

Figures 5 and 6 illustrate the developing heat transfer results as *j*-factors for the two enhanced tubes.

In laminar region, it was noted that *j*-factors decreased with the increase of Re and increased with the heat flux. The *j*-factor of E_1 outdid 4.2–8.2 times compared to the theoretical value (Nu = 4.36/Re/Pr1/3), while the *j*-factor of E_2 outdid 4.0–7.2 times. After the transition, the *j*-factor of the test enhanced tubes had a rapid growth with the increase of Re. And the turbulent results show that there is a definite increase in heat transfer with the use of the enhanced tubes.

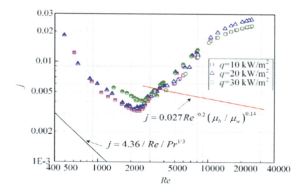

Fig. 5 Results of heat transfer of E_1 (opening symbols: Pr = 12; filled symbols: Pr = 47)

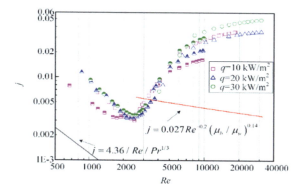

Fig. 6 Results of heat transfer of E_2 (opening symbols: Pr = 12; filled symbols: Pr = 47)

4 Conclusions

Effect of different thermal boundary conditions and heat flux on pressure drop and heat transfer of the two enhanced tubes with different structure were investigated. The investigation covered the laminar, transitional and turbulent flow regions. Friction factors of single-phase flow in the internal helical finned tube were higher than the counterpart of the plain tube. The increase of friction factors was in the range of 1.0–4.6 times higher than the plain tube. Heating boundary affected more in the laminar and transition flow regions. Transition of uniform heat flux was delayed when compared with the adiabatic result. In turbulent region, the difference of friction factors under heating and adiabatic boundary conditions was negligible. Heat transfer results showed an overall increase when compared with the smooth tubes. With the same Re or q, the heat transfer enhancement of tested tubes was in the range of 1.0–6.8 in the turbulent region and the heat transfer enhancement increases with Reynolds number.

Acknowledgements The project is supported by the National Natural Science Foundation of China (Grant No. 51606029) which is greatly acknowledged.

References

1. Reynolds, O.: An experimental investigation of the circumstances which determine whether the motion of water shall be direct or sinuous, and of the law of resistance in parallel channels. Philos. Trans. R. Soc. A: Math., Phys. Eng. Sci. **174**, 935–982 (1883)
2. Tam, L.M., Ghajar, A.J.: Transitional heat transfer in plain horizontal tubes. Heat Transf. Eng. **27**(5), 23–38 (2006)
3. Fujie, K., Itoh, M., Innami, T.: NO. 4044797, 30 Aug 1977
4. Tam, H.K., et al.: Experimental investigation of heat transfer, friction factor, and optimal fin geometries for the internally microfin tubes in the transition and turbulent regions. J. Enhanc. Heat Transf. **19**(5), 457–476 (2012)
5. Meyer, J.P., Olivier, J.A.: Transitional flow inside enhanced tubes for fully developed and developing flow with different types of inlet disturbances: part II–heat transfer. Int. J. Heat Mass Tran. **54**(7–8), 1598–1607 (2011)
6. Meyer, J.P.: Heat transfer in tubes in the transitional flow regime (2014)
7. Siddique, M., Alhazmy, M.: Experimental study of turbulent single-phase flow and heat transfer inside a micro-finned tube. Int. J. Refrig. **31**(2), 234–241 (2008)
8. Li, X., et al.: Experimental study of single-phase pressure drop and heat transfer in a micro-fin tube. Exp. Therm. Fluid Sci. **32**(2), 641–648 (2007)
9. Kline, S.J., Mclintock, F.A.: Describing uncertainties in single-sample experiments. Mech. Eng. **75**, 3–8 (1953)
10. Wang, Y., et al.: Experimental determination of single-phase pressure drop and heat transfer in a horizontal internal helically-finned tube. Int. J. Heat Mass Tran. **104**, 240–246 (2017)
11. Ma, Z., et al.: Experimental investigation on the friction characteristics of water-ethylene glycol mixture flow in internal helical finned horizontal tubes. Exp. Therm. Fluid Sci. **89**, 1–8 (2017)

Experimental Study on Influence of Outdoor Ambient Temperature on Heating Performance of Two-Stage Scroll Compression Air Source Heat Pump System

Yiling Wu and Lin Cao

Abstract The air source heat pump (ASHP) system was a substitute for fossil fuels for space heating. But in cold climate, some problems that high discharge temperature, low heating capacity and the coefficient of performance (COP) of heat pump water heaters limited its wider application. In this paper, a two-stage scroll compression air source heat pump (TSCASHP) system coupled with dual-cylinder scroll compressor was developed, and the heating performance under low ambient temperature was quantitatively evaluated. The experiment was conducted under a variety of ambient temperatures. The internal parameters of temperature and pressure were analyzed to evaluate the heating performance of the system. Test results showed that the system could provide reliable and stable heating at low ambient temperature. The system could generate hot water of 41 °C in the extreme low temperature of −18 °C with the COP of 2.10. By analyzing internal parameters of the system, the mass flow rate was considered the main influencing factor of system operation at low ambient temperature. This study extended the operating range of ASHP system and also provided experimental basis and technical support for clean heating projects in cold regions.

Keywords Air source heat pump system · Two-stage scroll compression · Low ambient temperature

---

Y. Wu · L. Cao (✉)
School of Energy and Power Engineering, Nanjing University of Science and Technology, 210094 Nanjing, China
e-mail: caolin1212@126.Com

L. Cao
Postdoctoral Work Station, Guangdong Jirong Air Conditioning Equipment Co., Ltd., 522000 Jieyang, China

1 Introduction

1.1 A Subsection Sample

Shown by some relevant studies, urban heating energy consumption in northern China is 191 million tons of standard coal in 2016, accounting for 21% of building energy consumption [1]. And more than 80% of urban heating energy consumption uses the coal as the main source of energy. This leads to a series of environmental and energy problems. Therefore, the major project of clean heating is carried out. The air source heat pump system has been approved to be one of the most economical and environment-protecting ways to provide heat. Air source heat pump has been widely used in east China and south China and other regions in China. However, due to the influence of outdoor temperature in cold regions in north China, there are some problems such as performance attenuation and unsteady operation. The main reasons are that: (1) the compressor is out of the working range; (2) the heating performance reduces; (3) the heat production does not match the heat load; and (4) the physical properties of lubricating oil changes at low ambient temperature [2, 3]. To solve these problems, considerable research has been proposed to enhance the heating performance, such as two-stage compression, refrigerant injection, additional heat source, applying electronic expansion valve, inverter technology, cascade system, mixed-refrigerants, variable refrigerant flow and so on.

In the past decades, the domestic and foreign scholars have made a lot of theoretical and experimental researches on the (quasi-) two-stage compression using injection technique [4–6]. However, due to the limitations of the compressor structure, the problem that heating capacity cannot satisfy the heating demand may arise during quasi-two-stage compression under extreme conditions. Two-stage compression system is more suitable for use in cold regions, relative to quasi-two-stage compression system [7]. Torrella et al. [8] analyzed the effects of refrigerant injection on the intermediate pressure and system performance using the thermodynamic cycle theory. It was indicated that the intermediate pressure, the compressor power and the system performance increase with the increase in the injection rate. Jiang et al. [9] presented a numerical simulation model of two-stage compression heat pump system with intercooler. And the simulation results showed that the heating capacity increased linearly with an increase in frequency of low-stage compressor, and the optimum heating capacity and COP were determined when the quality of the injection point was very close to unity. Jin et al. [3] analyzed the compression process of the two-stage compression system, the change of the intermediate pressure with time, the performance of the intermediate pressure changing condition and the influence of the intermediate pressure change on the system. But this did not take into account the influence of the intermediate injection process. In later studies, they [10] analyzed the coupling relation among the injection parameters and found that the specific enthalpy of the injection refrigerant determines the direction of the change in the state point, and the mass flow rate of the injection refrigerant determines the degree of the change in the state point of the system. Sun et al. [11] studied heating performance of refrigerant injection heat pump with a single-cylinder inverter-driven rotary compressor. Further they proposed that the injection pressure ratio could be adopted as the control parameter to adjust the injection mass flow ratio in order to achieve the high heating performance and safe operation. There are few researches on the variations of heating performance with the lower ambient temperature for dual-cylinder scroll compressor systems at present.

To evaluate the effects of lower ambient temperature on the system heating performance, an R22 two-stage compression air source heat pump system with single-cylinder scroll compressor was built in this study. The heating performance of the R22 TSCASHP system was measured and analyzed by comparison. The experimental results would be evaluated for use in the further optimizing of the injection structure and the control strategy and also provided experimental basis and technical support for clean heating projects in cold regions.

2 Experimental Setup and Test Procedure

2.1 Dual-Cylinder Scroll Compressor

As shown in Fig. 1, two scroll compressor chambers were installed on the both ends of the cylinder, driven by a variable frequency motor in the intermediate chamber. Thus, the frequencies of the two scrolls were same. The refrigerant was injected into the low-stage working chamber. The low-stage discharge refrigerant vapor was cooling by the injection gas in the mix chamber. Then, the refrigerant vapor would be further compressed, and the working chamber pressure will reach the discharge pressure. It should be noted that the injection gas would also increase the mass flow.

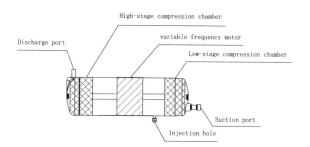

Fig. 1 Structure of scroll compressor

2.2 Test Model and Facility

A test bed of the two-stage scroll compression air source heat pump with economizer was built. The refrigerant injection technique coupled with the inverter rotary compressor was applied in the system under the testing condition. Figure 2 illustrates the schematic diagram of TSCASHP test bed. Figure 3 represents the thermodynamic P–h diagram of TSCASHP system.

The experimental setup was initially equipped with a dual-cylinder scroll compressor, a four-way reversing valve, a double-pipe heat exchanger, a finned tube heat exchanger, an economizer, a main electronic expansion valve (EEV1), an economize electronic expansion valve (EEV2) and other common refrigeration components. The dual-cylinder scroll compressor was driven by a DC inverter. The rated heating capacity and rated power of the compressor are 16.0 and 5.5 kW, respectively, while the outdoor dry bulb temperature 7 °C, relative humidity 86.82%, 41 °C, and the water temperature in and out of the water temperature, 3 °C. The four-way reversing valve was used to switch the system between refrigeration mode, heating mode and defrosting mode. In the heating condition, the finned tube heat exchanger is the evaporator to absorb heat from the environment, and the double-pipe heat exchanger is the condenser to prepare heating water. The main refrigerant is undercooled after heat transfer with the auxiliary refrigerant in economizer. The function of EEV1 was to control the superheated degree of the suction vapor in the range between 3 and 5 °C. The EEV2 was equipped at the inlet of the economizer to control the injection pressure and the injection mass flow rate. And the TSCASHP could turn into one-stage compressor air source. The EEV1 was automatically adjusted depending on the superheated degree of the suction vapor.

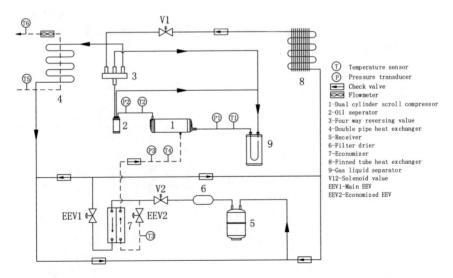

Fig. 2 Schematic diagram of the experimental setup

Fig. 3 Pressure-enthalpy diagram

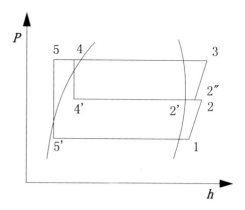

2.3 Test Condition

In the present study, only the heating model without frosting condition was considered. The system was set up in a psychrometer testing room in refrigeration equipment manufacturers of Nanjing.

The experimental condition is given in Table 1. This setup used R22 as the working fluid. The heating performance of the SCRCVI system was comprehensively investigated by varying the ambient temperature T_a. The inlet and outlet water temperatures were regulated by the thermostatic mixing valve on the return pipe. The indoor temperature is precisely controlled at the operating temperature by automatically adjusts the heater and the piston chiller.

Table 1 Experiment conditions

Operating model	Indoor		Outdoor	
	$T_{in,w}$	$T_{out,w}$	TDB/°C	TWB/°C
Heating	36	41	−12	−14
			−18	–
			−24	–
			−30	
Refrigerant type				R22
Refrigerant charge				7.1 kg
Compressor frequency f				50–90 Hz
Superheat of the compressor injection gas				3–5 K

2.4 Data Analysis Method

The system heating capacity and coefficient of performance are calculated from Eqs. (1)–(2). The power consumption W was acquired by a power meter.

Heating capacity

$$Q_h = \frac{mc(T_{out,w} - T_{in,w})}{3600} \quad (1)$$

where Q_h is heating capacity, m water flow, c specific heat capacity of water, $T_{out,w}$ outlet water temperature and $T_{in,w}$ inlet water temperature.

Coefficient of performance

$$COP = \frac{Q_h}{W} \quad (2)$$

where COP is coefficient of performance, and W is total power consumption.

The calibration of all the measuring instruments was conducted before testing. The measurement uncertainties for the key parameters are shown in Table 2. The theory of propagation of errors including systematic and random errors was applied to analyze the accuracy of the test data. With the purpose of reducing random error, five measurement values under setting condition would be collected in the interval of 3 min, as the system state parameters of air side and refrigerant side kept steady for about 10 min.

Table 2 Instrumentation and propagated uncertainties

No.	Instrument	Type	Measure point	Range	Uncertainty
1	Temperature sensor	Platinum thermistor	T1, T2, T3, T4	−50–300 °C	±0.15 °C
2	Thermocouple	T-type	T5, T6	−200–350 °C	±0.5 °C
3	Pressure transducer	Ceramic membrane	P1, P2, P3	0–6000 kPa	±0.2% of full scale
4	Mass flow meter	–	–	0–600 kg/h	±0.5% of full scale
5	Power meter	–	–	0–20 kW	±0.5% of full scale

3 Results and Discussion

The researches on the heating performances of the TSCASHP system were carried out by the ambient temperature T_a, as well as the compressor frequency f was fixed at 80 Hz. In preliminary tests, the opening of EEV1 and EEV2 was controlled automatically to control the superheated degree of the suction vapor in the range between 3 and 5 K and the discharge temperature at about 80 °C, respectively. All the tests were measured in the same psychrometer testing room.

The variations on the heating capacity, power consumption and COP of the TSCASHP system are shown in Fig. 4. The experimental results demonstrated that the Q_h, W and COP reduced as the ambient temperature decreased. For the ambient temperature from −12 to −30 °C, the heating capacity Q_h decreased about 52.3% (675 W). And yet, the power consumption W just reduced by 14.2% (73 W). Corresponding to the variation tendency of the Q_h and W, the system COP declined gradually from 2.51 to 1.40. As a result, this system could reach an acceptable COP (>2.1) at the ambient temperature above −18 °C. With the further decrease in the ambient temperature, it is necessary to develop a more optimal control program of EEV2 to improve the COP of the system. The trend could be explained by the variations of the system parameters. The temperature and pressure in the system could be obtained by experimental sensors.

Figure 5 shows the variation of the discharge pressure P_{dis}, the suction pressure P_{suc} and the injection pressure P_{inj} at the ambient temperature from −12 to −30 °C. As the outlet temperature of heating water constant, the P_{dis} of the compressor did not change with the ambient temperature. However, as the ambient temperature decreased, the evaporation pressure decreased, the P_{suc} decreased, the P_{inj} decreased, and the pressure ratio increased, the mass flow decreased.

Figure 6 shows the variation of discharge temperature T_{dis} and suction temperature T_{suc} at the ambient temperature from −12 to −30 °C. The T_{suc} decreased as the ambient temperature decreased controlled by EE1 to keep the superheated degree of the suction vapor in the range between 3 and 5 K. Since the P_{sub} and T_{sub}

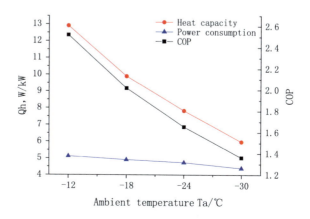

Fig. 4 Variation of Q_h, W and COP with ambient temperature

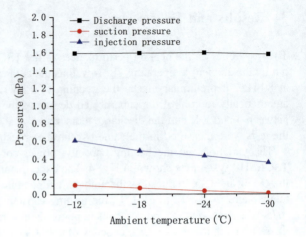

Fig. 5 Variation of P_{dis}, P_{suc} and P_{inj} with ambient temperature

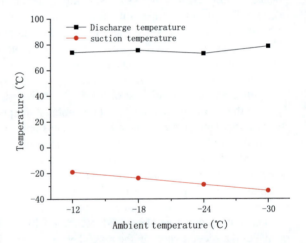

Fig. 6 Variation of T_{dis} and T_{suc} with ambient temperature

decreased, the suction vapor-specific volume increases, which led to a further decrease in mass flow. In the meanwhile, the T_{dis} was stabilized at about 80 °C controlled by EE2, which means the mass flow of injection gas decreased.

As a result, the total mass flow decreased, which led to a decline in heating capacity Q_h and compressor consumption W eventually. Therefore, improving refrigerant mass flow is one direction of optimizing air source heat pump in low ambient temperature.

4 Conclusion

The heating performances of the R22 economizer two-stage scroll compression heat pump system with a dual-cylinder scroll compressor were studied under various ambient temperatures. The test results indicated that the TSCASHP system could provide reliable and stable heating at low ambient temperature from −12 to −30 °C and enhance the heating performance effectively at lower ambient temperature. The system could reach an acceptable COP (>2.1) at the ambient temperature above −18 °C. This means that the TSCASHP system has a good application prospect in the cold region. And the COP could improve by exploring an optimal control strategy of vapor injection. The internal parameters, pressure and temperature of the system were future studied. The result shows that refrigerant mass flow rate decrease was the main influencing factor of system operation. The TSCASHP system could be much improved by exploring an optimal control strategy of gas injection.

Acknowledgements This work is supported financially by the National Natural Science Foundation of China (Numbers 51606096).

References

1. Zhang, X., Qi, Y.: Annual Review of Low-Carbon Development in China. Social Sciences Academic Press, China (2017)
2. Wang, F., et al.: Research progress and prospect of air source heat pump in low temperature environment. J. Refrig. **34**(5), 47–54 (2013)
3. Jin, X.: Interstage matching and coupling characteristics in a two-stage compression heating pump system with variable capacity. Doctoral dissertation, Dalian University of Technology (2013)
4. Wang, B., et al.: Numerical research on the scroll compressor with refrigeration injection. Appl. Therm. Eng. **28**(5), 440–449 (2008)
5. Mathison, M., et al.: Performance limit for economized cycles with continuous refrigerant injection. Int. J. Refrig. **34**(1), 234–242 (2011)
6. Kim, J., et al.: Experimental study of R134a/R410A cascade cycle for variable refrigerant flow heat pump systems. J. Mech. Sci. Technol. **29**(12), 5447–5458 (2015)
7. Oquendo, F., et al.: Performance of a scroll compressor with vapor-injection and two-stage reciprocating compressor operating under extreme conditions. Int. J. Refrig. **63**, 144–156 (2016)
8. Torrella, E., et al.: Experimental evaluation of the inter-stage conditions of a two-stage refrigeration cycle using a compound compressor. Int. J. Refrig. **32**(2), 307–315 (2009)
9. Jiang, S., et al.: Simulation on a two-stage compression heat pump with focus on optimum control. Lect. Notes Electr. Eng. **262**, 381–397 (2014)
10. Jin, X., et al.: Numerical research on coupling performance of inter-stage parameters for two-stage compression system with injection. Appl. Therm. Eng. (2018)
11. Sun, J., et al.: Experimental investigation of the heating performance of refrigerant injection heat pump with a single-cylinder inverter-driven rotary compressor. J. Therm. Anal. Calorim. (2018)

The Field Survey on Local Heat Island Effect of Precision Air-Conditioning

Mo Chen, Zhixian Ma and Mingsheng Liu

Abstract The local heat island effect (LHIE) plays an important role in the energy consumption of precision air-conditioning system, the data center cooling system. This paper conducted a field survey on the LHIE caused by precision air-conditioning system and its effect on the energy efficiency of the cooling system. The inlet and outlet air temperature and air velocity of the air cooled condenser (outdoor component of precision air conditioner), and also the environment temperature, are measured and applied in the analysis. The results show that the average inlet air temperature of outdoor units of the precision air-conditioning is 8 °C warmer than the environment air temperature, and the inlet airspeed is slow. The existence of the LHIE will be resulting 24% more energy consumption of the tested precision air-conditioning system. This paper verified the existence of LHIE and offers a reference for further study of it.

Keywords Precision air-conditioning · Local heat island effect · Energy consumption

1 Introduction

Data center has become a big consumer of the country electricity consumption, and energy optimization of it becomes a key approach of low-carbon cycle green development in data center [1]. Data center cooling equipment energy consumption accounts for about 50% of the total energy consumption, which is the key area of data center energy saving [2–4].

Usually, the precision air-conditioning in the date center not only operates all year round, but also has a large cooling capacity. With the increase of the cooling capacity, the outdoor units of precision air-conditioning release a large amount of heat to the external environment, making the ambient air temperature around the

---

M. Chen · Z. Ma · M. Liu (✉)
Institute of Building Energy, Dalian University of Technology, Dalian, China
e-mail: liumingsheng@dlut.edu.cn

building rise, leading to the flow of hot air, which is called the local heat island effect. LHIE will affect the cooling load in the data room and *COP* of the precision air-conditioning system. Therefore, there is a need to investigate the LHIE and its effect on the performance of the data center cooling system.

Many scholars had done research on the airflow field around the outdoor unit of air-conditioning. Zhou et al. [5] conducted a study on the thermal environment around the outdoor units in high-rise buildings and analyzed the effects of outdoor airspeed, outdoor air temperature, and other factors on the thermal environment around the outdoor units of the air-conditioning. You et al. [6] analyzed and discussed the outdoor units based on the theory of hot airflow analysis, and combined the specific engineering practice to optimize the heat dissipation of the outdoor units. Chow et al. of City University of Hong Kong simulated and studied the velocity field and temperature field around the outdoor units of the split air-conditioning installed at the concave corner of the building, and analyzed the influence of the shape of concave corner on the refrigeration performance of the air-conditioning. Hang et al. used CFD technology to simulate and optimize the airflow organization of the outdoor air-conditioning in a high-rise building. According to the simulation results of its velocity field, temperature field and pressure field, it is found that different air outlet modes and the shape of tuyere have a great influence on the airflow. Many foreign scholars [7] had also done research on the direction of the thermal environment around the outdoor units of the air-conditioning, mainly for the layout of the outdoor units, considering the airspeed of the outdoor units, wall height, outdoor natural air direction (or no outdoor natural air), and other factors, and numerical simulation and experimental methods were used to analyze the influence of various factors on the thermal environment around the outdoor units.

However, most of the studies only focus on theoretical analysis of LHIE rather than actual investigation, and few of the measured data can confirm the existence of LHIE. So, we conducted the field survey in a Data Center of Beijing to confirm the existence of LHIE and analyzed the influence of LHIE on the energy consumption of the precision air-conditioning, and also confirmed the necessity of eliminating LHIE. Eliminating LHIE can improve the performance of precision air-conditioning and reduce the energy consumption of the air-conditioning, which plays a vital role in energy conservation and environmental protection in the whole society.

2 Methods

We conducted the field survey in the Data Center of Beijing on September 14, 2018. The weather in Beijing was sunny. It was 26 °C. The test instruments we used were fluke thermocouple thermometer and hot wire anemometer. We tested the inlet and outlet air temperatures and inlet and outlet air speeds of the two groups of outdoor units, respectively.

As shown in Fig. 1, the first outdoor unit group was placed on the second and third floors of the building with four outdoor units placed in two rows and two columns. The outdoor unit group on the third floor was directly above the outdoor unit group on the second floor. The side of the inlet air of all units was on the back and the side of the outlet air was on the front. This kind of arrangement leads to the existence of local heat island effect in all the outdoor units of precision air-conditioning. Occasionally, when the local heat island effect is severe, the compressors would stop.

As shown in Figs. 2 and 3, the second outdoor unit group was placed in a row. The side of the inlet air besides the environment and the side of the outlet air toward a wall of the machine room. When all the outdoor units were on operation, the inlet air of the outdoor units was the air of environment, which of the temperature was 26 °C. The outdoor air entered the outdoor unit after passing through the filter and was discharged from the outlet side after the heat exchange was completed. Due to the low density of the hot air, the exhaust air direction flowed out to the upper air. Because the inlet side and the outlet side were separated, the outlet air would not flow back into the inlet side, which solved the problem of the local heat island effect of the outdoor unit group.

Fig. 1 First outdoor unit group

Fig. 2 Second outdoor unit group

3 Test Results and Discussion

The following conclusions can be drawn by analyzing the test data of the first outdoor unit group:

According to the data, the inlet air temperature on the second floor was higher than that on the third floor and the inlet air velocity on the second floor was slower than that on the third floor. And the air circulation effect on the second floor was not as good as that on the third floor. The reason for the existence of this phenomenon was that the parapet wall of the second floor was too high and the inlet air of the outdoor unit on the second floor was not enough, resulting in a very high inlet air temperature and slow inlet air velocity (Tables 1 and 2).

Although the inlet air temperature of the outdoor unit on the third floor was lower than that of the second floor, the inlet air temperature was still very high. The reason for this phenomenon was that the exhaust of the outdoor unit on the second floor raised and was inhaled by the outdoor unit on the third floor, resulting in excessively high inlet air temperature of the outdoor unit on the third floor.

The following conclusions can be drawn by analyzing the test data of the second outdoor unit group:

The arrangement of the outdoor unit group like picture 2 had an advantage. Since the inlet side and the outlet side were separated, the outlet air would not flow

Fig. 3 Second outdoor unit group

Table 1 Test data of the first outdoor unit group

	Inlet air temperature (°C)	Inlet air velocity (m/s)	Outlet air temperature (°C)	Outlet air velocity (m/s)
One of the units on the second floor	33.22	0.67	39.39	0.24
One of the units on the third floor	29.44	1.77	/	/

Since the lower unit on the third floor was not started during the testing, and the other units was too high, the data of the side of outlet air were not measured

Table 2 Test data of the second outdoor unit group

	Inlet air temperature (°C)	Outlet air temperature (°C)
One of the units	26	41.9

back into the inlet side, which solved the problem of the local heat island effect of the outdoor unit group.

However, this arrangement of the outdoor unit group like picture 2 had a disadvantage that all the exhaust sides of the outdoor units were facing a wall of the machine room where the precision air-conditioning were placed, which caused all the exhaust air to blow to the wall. Due to exhaust air temperature was high and the temperature in the machine room was low, and the heat transfer from the exhaust air to the machine room, it increased the cooling load of the machine room, which would cause more energy consumption of the precision air-conditioning.

The increase in air temperature around the building will directly lead to an increase in cooling load in the computer room, which will increase the consumption of electricity of the precision air-conditioning. At the same time, the increase of air temperature near the condensers, outdoor units of precision air-conditioning, will increase the condensation temperature, thus reducing the operating efficiency of precision air-conditioning system. As reported, 1 °C rise in the condensation temperature will result 3% decline in COP of the precision air conditioner [8] or even worse, the operation of precision air-conditioning is interrupted due to too high condensation temperature.

The results show that the average inlet air temperature of outdoor units of the precision air-conditioning is 8 °C warmer than the environment air temperature, and the inlet air speed is slow. The existence of the LHIE will result 24% more energy consumption of the tested precision air-conditioning system.

4 Conclusions and Recommendations

Through the field survey on the Data Center of Beijing, it is confirmed that LHIE does exist. It is a common phenomenon that LHIE is caused by the unreasonable arrangement of the outdoor units of the air-conditioning, and LHIE plays an important role in the energy consumption of the air-conditioning; so, it is necessary to eliminate LHIE.

Acknowledgements The authors acknowledge the support from the Bes. Tech, Inc for the field survey of this study.

Permissions Appropriate permissions from responsible authorities were obtained for this study in the field survey on the Data Center.

References

1. Zhang, L.F., et al.: Research on influence of cold channel closure on airflow distribution and energy efficiency in the data center computer room. Electr. Power Inf. Commun. Technol. **16**(5), 63–67 (2018)

2. Liu, C., et al.: Analysis and protection strategy of power data center security based on cloud computing and SDN technology. Electron. Des. Eng. **24**(9), 136–138, 143 (2016)
3. Wang, Q., et al.: Discussion on the feasibility for distributed energy in internet data center. Power Syst. Clean Energy **29**(9), 87–91 (2013)
4. Wang, X.Y.: Discussion on cloud computing data center construction. Comput. Eng. Softw. **35**(2), 129–130 (2014)
5. Zhou, D., et al.: Numerical simulation of thermal environment for outdoor units at building re-entrant. J. Guangzhou Univ. (Nat. Sci. Ed.) **9**(6), 17–22 (2010)
6. You, B., et al.: The analysis and optimization on heat dissipation of outdoor unit of air conditioner. J. Shunde Polytech. **7**(4), 35–38 (2009)
7. Xue, H., et al.: Prediction of temperature rise near condensing units in the confined space of a high-rise building. Build. Environ. **42**(7), 2480–2487 (2007)
8. Gong, G.C., et al.: The influence of condensation heat emission on environment of outdoor air conditioner. In: 2006 National Academic Annual Conference on HVAC Refrigeration 809 (2007)

Study on Operation Strategy of Cross-Season Solar Thermal Storage Heating System in Alpine Region

Haoran Li, Hanyu Yang, Enshen Long, Xin Liu and Yin Zhang

Abstract Based on the cross-season solar thermal storage heating system (CSTSHS) in a typical Alpine town in the west of China, this paper analyzes and compares the electric auxiliary capacity, power consumption indicators in the heating season, and the solar guarantee rate under three operation strategies (e.g., thermal storage priority, electro-thermally assisted priority, and hybrid control mode). The results show that although the electric auxiliary capacity of the studied system under thermal storage priority control mode is larger than demand, it can make full use of solar energy. Under the electro-thermally assisted priority control mode, the required electric auxiliary capacity can be reduced, but the power consumption indicators increase in the heating season, and the solar guarantee rate is low. The hybrid control mode can effectively reduce the electric auxiliary capacity and the power consumption indicators in the heating season at the same time, which has obvious advantages. This paper can offer reference and guidance for cross-season solar thermal storage system operation.

Keywords Alpine region · Cross-season solar thermal storage · Operation mode · Electric auxiliary capacity · Power consumption indicators · Solar guarantee rate

---

H. Li · H. Yang · E. Long (✉) · X. Liu · Y. Zhang
School of Architecture and Environment, Sichuan University, Chengdu, China
e-mail: longes2@163.com

H. Li
e-mail: hrhrli@163.com

E. Long
MOE Key Laboratory of Deep Earth Science and Engineering, Sichuan University, Chengdu, China

1 Introduction

Western Sichuan, Tibet, and other Alpine regions are all located in severe cold climatic zone in China, where winter is long and cold, making heating a basic demand in these areas. Because of the high altitude in this area, solar energy resources are abundant, so solar thermal storage heating has high application potentials. According to the time span of thermal storage and use, the solar thermal storage heating system can be divided into short-term thermal storage and cross-season thermal storage [1]. In the Alpine region, the CSTSHS using the sensible heat storage of water has more application prospects [2]. CSTSHS stores the solar energy in the water tank in the non-heating season to store heat for the heating season. After entering the heating season, the heat stored in the water tank is released to heating users through the urban pipe network.

On the other hand, because of the instability of solar energy [3], combined with the initial investment and the area of land needed to be considered in actual projects, it is generally difficult to meet the heating demand of the whole heating season for CSTSHS while the demand for heat use in winter is large and the solar radiation is weak. During cloudy and rainy time, the heating system cannot work effectively as that in sunny days. At present, the main way to solve this problem is to increase the auxiliary heat source capacity and optimize the system configuration [4–6]. When the solar energy is insufficient, it is replenished by auxiliary heat source. The other is to meet the heating demand through efficient system operation strategy. Wang [7] adopts the operation strategy of temperature difference control in the system design. When the temperature of the regenerator reaches the requirement of direct supply temperature, it is directly supplied to the user. When the temperature doesn't reach the requirement, the temperature can be supplied to the user after the temperature reaches the requirement by auxiliary heat source. Li [8] put forward three kinds of operation strategies for solar-ground source heat pump cross-season thermal storage system: temperature control, temperature difference control, and time control. Most of the existing projects are controlled by temperature difference. Although this method is effective in practical engineering, it is found that this method is relatively simple if the heat-collecting system needs to start and stop temporarily [9].

Even though many scholars have carried on some research about the system operation strategy, the coupling mechanism of solar energy, auxiliary heat source, and heat storage is still a lack of deep investigation. Taking the CSTSHS project in typical towns at high altitude in the western China as an example, three different operation strategies (e.g., thermal storage priority, electro-thermally assisted priority, and hybrid control mode) are put forward and compared. Through the analysis and comparison of different control strategies, this paper aims to explore the matching control strategy of heating system that is more suitable for the Alpine region, so as to reduce the electric auxiliary capacity, improve the system operation efficiency, and increase the solar guarantee rate. This paper can offer reference and guidance for cross-season solar thermal storage system operation.

2 Methods

2.1 Research Objective

In this paper, a proposed central heating project in Alpine region is selected. The heating area of the town is about 550,000 m², including residential buildings, hotel buildings, commercial buildings, and office buildings. The CSTSHS is composed of solar collector, water tank, auxiliary heat source, thermal pipe network, and heating end. The collector selects the trough solar collector with high heat-collection efficiency and less heat loss, with an opening of 2550 mm. The total area of the collector is determined to be 38,300 m² by synthesizing all kinds of factors, such as occupation area and initial investment. In order to achieve good thermal storage and heating effect, the system uses buried water tank storage heat. The volume of the storage water tank is 60,000 m³. In addition, because the local hydropower is abundant, the electricity price is low. The auxiliary heat source is electric heating boiler. The heating season is set from October 15 to April 15 of the following year with reference to climate conditions and national policies in the same region.

2.2 Methods

The characteristic temperature method (CTM method for short) [10] is used to simulate and predict the hourly load of dynamic heating system. The method has been proved to be reliable by experiment and software comparison in a large number of literatures [11]. Based on the local annual hourly meteorological database, the shape coefficient of various representative buildings, the ratio of window to wall area, and the thermal characteristics, the hourly heat load curve of the heating season is obtained according to the superposition of the law of work and rest. The design heating load of the system is 60,633 kW, and the heating load index is up to 82.75 W/m², due to the poor envelope thermal insulation and air tightness of the existing buildings.

The radiation captured by the trough collector comes from direct radiation, so the solar radiation absorbed by the slot reflector per unit opening area can be calculated by direct solar radiation and angle of incidence.

$$H_T = H_{DN} \cdot \cos\theta \cdot \rho \cdot \tau \cdot \alpha \tag{1}$$

where H_T is direct radiation, W/m²; H_{DN} is solar direct radiation intensity, W/m²; θ is solar incident angle, °; ρ is condensing mirror efficiency; τ is glass tube transmittance; and α is metal absorption tube coating absorptivity.

In the horizontal position of the collector and the east–west tracking mode, the cosine value of the incident angle can be calculated by the height and azimuth of the sun.

Fig. 1 Comparison between hourly solar radiation and heating load in heating season

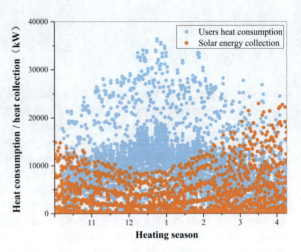

$$\cos\theta = \sqrt{1 - \cos^2\alpha_s \cdot \cos^2\gamma_s} \tag{2}$$

where α is solar altitude angle, °; and γ is sun azimuth angle, °.

Finally, the total radiation absorption of the system can be calculated:

$$Q_x = A \cdot H_T \tag{3}$$

where A is collector area, m².

Figure 1 shows the comparison between hourly solar radiation and hourly heat load in the heating season predicted by local meteorological conditions. It can be seen that the collected solar energy is small when the heating load demand is the largest, resulting in a large difference between supply and demand. However, the size of the heat storage tank directly affects the area of the system and the initial investment. Therefore, the auxiliary heat of electric heating boiler is a better matching scheme. However, different operation strategies, such as the electric auxiliary capacity, the consumption of electricity in heating season, and the solar guarantee rate, vary greatly. Therefore, optimizing the system operation control strategy will become the key to reduce the initial investment of the system, reduce the operation cost, and maximize the utilization of renewable energy.

3 Results

Combined with the actual situation, this paper puts forward three operation strategies for the CSTSHS, such as heat storage priority control mode, electro-thermally assisted priority control mode, and hybrid control mode.

3.1 Heat Storage Priority Control Mode

The heat storage priority control is that the heat stored in the water tank is first used in the early heating season for heating. The water tank is in the same state of storage and release. After the heat of the water tank is used up, solar energy directly provides users with heat in real time. When the solar energy is insufficient, the heat will be supplied by electric boiler. Figure 2 shows the comparison of the hourly available solar energy in the heating season with the hourly heat load. Compared with Fig. 1, the heat supplied by the "water tank + solar energy" mode at the early stage of heating can fully meet the hourly heating demand until the middle of November due to the utilization of the solar energy stored in non-heating season. Since then, the water tank has no extra stored heat anymore. The solar collector provides heat for the users in real time. However, due to the large heating demand in the middle and late periods of heating season, it cannot be met by relying on the collector alone. So, the electric boiler is needed to supplement the heat. Figure 3 shows the real-time electrical load supplement during the heating season. It can be seen from the figure that in this mode, the power demand in heating season is large. Since the maximum heat load of the building often occurs at night, the maximum heat load of 35,000 kW should be matched with the installed electric auxiliary heat capacity.

3.2 Electro-Thermally Assisted Priority Control Mode

The electro-thermally assisted priority control mode is to set a small critical load value and use the water tank and electric boiler for heating in the initial heating season. When the heating demand is less than or equal to the critical value, all the heat is supplied by the electric boiler. When the heating demand is higher than the

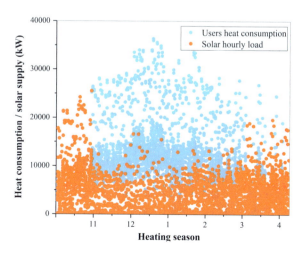

Fig. 2 Solar hourly load satisfaction during heating season

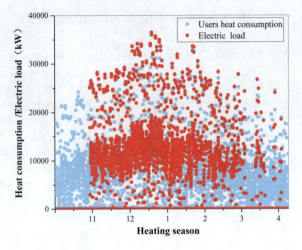

Fig. 3 Hourly power load replenishment during heating season

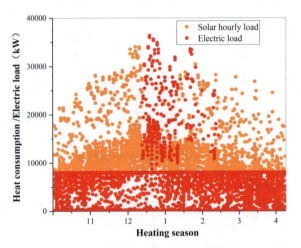

Fig. 4 Hourly electric load and solar load under critical load of 8000kW

critical value, the electric boiler only guarantees the critical value. The extra heat is replenished by the water tank (solar energy). The operation effect of the system is different when different critical values are set under the determined collector area and the volume of the regenerator. The following section starts with the planned 8000 kW installed capacity and gives the detailed analysis.

Critical Value = 8000 kW. Figure 4 shows the hourly electric load and solar load under critical load of 8000 kW. It can be seen that from the beginning of heating season to the middle of December, the electric boiler can meet the heat demand at the end. In the middle period of heating season, the heat demand increases and exceeds the critical value of 8000 kW. At this point, the heat stored in the water tank previously and real-time solar energy are used to supplement heating for users. Since the use of heat in the water tank in the initial period of heating season is reduced, the temperature of the water tank decreases slowly. Therefore, in the middle period of heating, when the heat demand is large, the water tank still has heat output. The service time of the water tank has been extended to December 29, which also makes the heat stored in the water tank plays a load shifting effect at the peak time, alleviating the contradiction between the large heating demand and the low solar radiation at the end of heating season. However, at this critical value, the water tank cannot be used continuously until the end of the heating season. When the heat demand reaches the maximum, the water tank has no heat storage. In Fig. 4, there are still many moments when the power load exceeds the critical value. The system still needs 35000 kW of electric auxiliary heat to meet the heating demand which cannot effectively reduce the configuration of electric boilers by setting the critical value at 8000 kW.

Critical Value = 11,000 kW. In order to effectively reduce the installed capacity of electric-assisted heating, the critical value can be set higher. Figure 5 shows the hourly electric load and solar load under critical load of 11,000 kW. It can be seen from the figure that the hourly heat load in the heating season is all below 11,000 kW, which eliminates the peak power load in the heating season. The total heat supplied from water tank, real-time solar energy, and electric boiler can meet the demand when the heat load reaches the maximum in the middle period of heating season. In this mode, the electric boiler equipped with 11,000 kW can meet the requirements. However, in the late period of heating, the solar radiation increases. If the electric boiler is still preferred to provide heat for users, the water tank will start to store heat and the temperature will rise, which increases the power consumption of electric auxiliary heat.

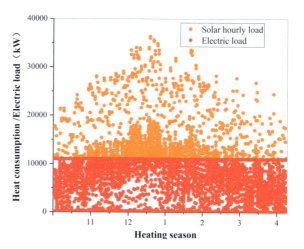

Fig. 5 Hourly electric load and solar load under critical load of 11000 kW

Fig. 6 Daily variation of water temperature during heating season

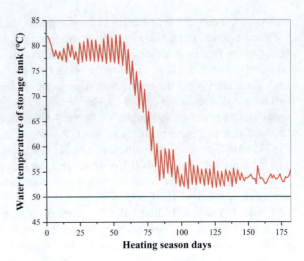

3.3 Hybrid Control Mode

To make full use of the load shifting function of the water tank and fulfill the potential of electric auxiliary heating device, this paper puts forward the model of hybrid control. This model can heat the water tank in advance by solar and electric boiler to ensure the heating requirement of the most unfavorable day for three days. Water temperature is maintained above a critical temperature until the coldest moment.

At the beginning of the heating season, the priority is given to the use of water tank for heating. When the water temperature reduces to the critical temperature, the electric boiler heats the water tank immediately to maintain the higher water temperature so as to ensure sufficient heat supply in the coming of the coldest period.

The designed water tank of this system is 60,000 m^3. Hot water absorbs (releases) 2.52×10^5 MJ of heat per rise (decrease) of 1 °C. If the 8000 kW electric boiler is used for auxiliary heating, it can provide 192,000 kWh (6.912105 MJ) heat when running 24 h a day. If you do not consider solar energy supplement and user consumption, the water temperature of water tank can increase by 8.22 °C when electric boiler runs for three days continuously in advance. Through the corresponding relationship between the temperature of the water tank and the heat, the maximum temperature difference of the water tank can be obtained. The results show that the guaranteed temperature is 68 °C under the condition of three days in advance guarantee. Figure 6 shows the daily temperature variation of the water tank

Table 1 Comparison between economy and solar energy guarantee rate of different operation strategies

Operation mode	Heat storage priority control mode	Electro-thermally assisted priority control mode	Hybrid control mode
Electric auxiliary capacity (kW)	35,000	11,000	8000
Power consumption indicators in the heating season (kWh/m²)	24.56	32.33	24.90
Solar guarantee rate (%)	40.37	27.89	38.29

in hybrid control mode. It can be seen that the water temperature remains over 75 °C in the initial heating season. During the coldest period, even if the electric boiler is guaranteed to run at full load for 72 h in three days, the water temperature cannot be guaranteed over 68 °C. The water temperature begins to drop to about 52 °C. But, it still meets the temperature requirement of heating. After the coldest period, due to the increase of solar radiation and the decrease of heat demand, the electric boiler can keep the water temperature over 50 °C without running during the whole period of time. Therefore, in this mode, the electric auxiliary capacity can be reduced to 8000 kW effectively.

Based on the previous simulation and analysis, the electric auxiliary capacity, power consumption indicators in the heating season, and the solar guarantee rate under different operation strategies can be obtained, as given in Table 1.

Through the comparison of three operation strategies, it can be seen that in the heat storage priority control mode, although the power consumption indicators in the heating season is the lowest and the solar guarantee rate is the highest, the demand of the electric auxiliary capacity under this mode is the largest, even over tripling than the others, which is undesirable from the economic perspective. Although the installed capacity can be reduced effectively in the electro-thermally assisted priority control mode, the excessive operation time of the auxiliary heating system leads to high power consumption indicators in the heating season and low solar energy guarantee rate, which is a lack of flexibility and controllability. Compared with the former two modes, the hybrid control mode has the lowest installed capacity, and the power consumption indicators and solar energy guarantee rate keep the same as that under the heat storage priority control mode, which reaches a compromising operation strategy that combines the advantages of the former two modes for both system economic and flexible reasons.

4 Conclusions

In this paper, a proposed central heating project in Alpine region is selected for cross-season solar thermal storage heating system. After establishing mathematical model, three different operation strategies are put forward. By comparing and analyzing the electric auxiliary capacity, power consumption indicators in the heating season, and the guarantee rate of solar energy, it can be found that different control strategies have great influence on the economic performance of the system and the power consumption of the system during heating season. The solar energy guarantee rate may also exist the big difference.

Compared with the electro-thermally assisted priority control mode, heat storage priority control mode, and the hybrid control mode have lower (about 24 kWh/m^2) and a higher solar energy guarantee rate (about 39%).

The hybrid control mode makes full use of the load shifting effect of water tank and fulfills the potential of electric auxiliary heating. Compared with the heat storage priority control mode, the hybrid control mode can reduce the electric auxiliary capacity from 35,000 to 8000 kW. It has an excellent comprehensive performance and engineering application prospect.

Through the research in this paper, the cross-season solar thermal storage heating system in the Alpine area is optimized. The initial investment of equipment and the operating cost of the system is reduced, which has certain guiding significance for the application of solar energy heating.

References

1. Wang, H.J., et al.: Cross-season Regenerative Solar centralized heating Technology. Solar Energy, (2005)
2. Han, X., et al.: Technical progress of cross-season regenerative solar heating system. Refrig. Air Cond. (Sichuan) **3**, 228–233 (2012)
3. Xue, Y.B., et al.: Application Technology of Renewable Energy Building. China Construction Industry Press (2012)
4. Wang, L.: Calculation and Research on Trans-seasonal Solar Water Tank Regenerative Heating System. North China Electric Power University (2012)
5. Duan, C.W., Gao, Y.: Simulation analysis of solar energy seasonal regenerative heating system. Appl. Energ. Technol. (2016)
6. Wang, X.Y.: Simulation Study of Solar Energy Trans-seasonal Energy Storage Heating System in Central Plains. Zhongyuan University of Technology (2017)
7. Wang, X.: Study on optimal Design of Solar Hot Water Regenerative Heating System. Chinese Academy of Architectural Sciences (2010)
8. Li, Z. M.: Study on operation mode of solar-ground source heat pump system for cross-season heat storage. Sci. Technol. Horiz. **04**, 174–175 (2017)
9. Liu, M.S., et al.: A summary of cross-season water regeneration solar centralized heating engineering and its optimization. Build. Therm. Energy Vent. Air-Cond. 34(06), 26–30 22 (2015)
10. Long, E.S.: Building energy consumption gene theory system. Electromechanical Inf. 19, 35–37 (2007)
11. Li, Y.R., et al.: Study on indoor temperature variation characteristics of batch air-conditioner with different terminal forms during start-up process. Refrig. Air Cond. (Sichuan). 29(05), 533–537 (2015)

Experimental Research on Performance of VRF-Based Household Radiant Air-Conditioning System

Danyang Wang, Jianbo Chen, Chenyue Yan and Meng Zhao

Abstract Taking the variable refrigerant flow (VRF)-based household radiant air-conditioning (A/C) system as the research object, this paper established the experimental test platform of the VRF-based household radiant A/C system in an existed psychometric room, simulated winter and summer operating conditions, and conducted a performance test study of the system, to verify the stability of the VRF-based household radiant A/C system and obtain the relationship between the parameters of the system and the outdoor air temperature. The experimental results showed that the VRF-based household radiant A/C system could operate stably in both winter and summer experimental conditions. In summer mode, the total cooling capacity (TCC) of the system increased by 31.32%, as the outdoor air temperature varied from 29 to 43 °C. Also, the electric power consumption of the VRF outdoor unit increased by 45.4%, and the system EER decreased to a minimum of 3.55. In winter mode, when the outdoor air temperature changed from 7 to −5 °C, the total heating capacity (THC) of the system increased by 27.70%, while the electric power consumption of the VRF outdoor unit increased by 42.41%, and the system COP decreased to a minimum of 3.48.

Keywords VRF · Household radiant air-conditioning system · Temperature and humidity · Power consumption · System stability

---

D. Wang (✉) · J. Chen · C. Yan · M. Zhao
School of Environment and Architecture, University of Shanghai for Science and Technology, 200093 Shanghai, China
e-mail: 18217136527@163.com

J. Chen
e-mail: cjbzh@vip.sina.com

C. Yan
e-mail: yanchenyue666@163.com

M. Zhao
e-mail: 1045935947@qq.com

1 Introduction

There is a great demand for high-comfort residential buildings in the more developed areas of the Yangtze River Basin in China. However, the area in the Yangtze River Basin has a large annual humidity, especially in the plum rains season, which has high requirements for dehumidification.

In the current situation of increasing social energy, considering the problems above, L. Zhao proposed a VRF-based household radiant A/C system [1]. The radiant A/C system could independently control the temperature and humidity, thereby avoiding the energy waste problem caused by the temperature and humidity coupling processing of the conventional air conditioner. Based on this, the performance test of VRF-based household radiant A/C system was carried out, and the energy consumption of the VRF outdoor unit under different conditions was analyzed to verify its system stability.

2 The Experimental Setup

2.1 The Experimental Prototype of the VRF-Based Household Radiant A/C System

The radiant A/C system studied in this paper mainly combines the radiant cooling/heating technology with the variable refrigerant flow (VRF) air-conditioning system. Its principle is to provide the cold and heat source for the radiant terminals by connecting the VRF outdoor unit to the refrigerant–water plate heat exchanger (RWPHE). And the VRF outdoor unit is connected with the outdoor air dehumidifier (OAD) to provide fresh air for the radiant A/C system to dehumidify. The schematic diagram is shown in Fig. 1.

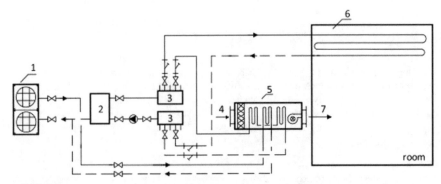

1. VRF outdoor unit   2. Refrigerant-water plate heat exchanger   3. Water headers
4. Outdoor air   5. OAD   6. Radiant terminals   7. Supply air

Fig. 1 Schematic diagram of VRF-based household radiant A/C system

As shown in Fig. 1, the VRF-based household radiant A/C system mainly includes the refrigerant system, water system, and wind system. The refrigerant system delivers the refrigerant to the refrigerant–water plate heat exchanger and the OAD to perform heat exchange with the water and handle the fresh air, respectively, including cooling and dehumidifying fresh air in summer and heating fresh air in winter. And the water system is to supply water to the radiant terminals and to pre-cool/heat and reheat the fresh air.

In summer, the radiation system handled the indoor sensible cooling load, and the OAD handled the indoor latent cooling load and the cooling load from outdoor air while meeting the fresh air requirements of the indoor personnel and the indoor air quality. In winter, the radiation system mainly handled the indoor heating load. The OAD needs to handle the heating load from outdoor air and provide fresh air to the room, but no dehumidification was required.

2.2 Experiment Conditions

The load object of this experiment is an apartment located in Shanghai. The total area is 140 m^2 with an indoor space height of 2.7 m. The heat transfer coefficient of the exterior wall is 1.25 W/(m^2 K), and the heat transfer coefficient of the window is 2.70 W/(m^2 K). When the indoor dry-bulb temperature (T_{db}) was set to 27 °C and the wet-bulb temperature (T_{wb}) was set to 19.5 °C in summer, the indoor cooling load of the building was calculated to be 13.9 kW. Similarly, when the indoor dry-bulb temperature was set to 22 °C and the wet-bulb temperature was set to 15 °C in winter, the indoor heating load of the building was 7.4 kW [2, 3]. The volume of fresh air in the building was determined to be 190 m^3/h according to the requirements of health and dehumidification [4, 5]. According to the above results, the components of the VRF-based household radiant A/C system are selected and shown in Table 1.

The experiment was carried out in an existed psychometric room. The size of the chamber was 6.2 m × 6.8 m × 3.2 m (W × L × H). The chamber plan is shown in Fig. 2. As seen, the psychometric room was divided into an indoor space and an outdoor space. The two parts were equipped with air handling unit (AHU), and they were mainly used to simulate the conditions of indoor and outdoor environments.

The VRF outdoor unit was installed on the outdoor side of the psychometric room and provided radiant water supply. The OAD was installed on the indoor side. And then the VRF outdoor unit and the OAD were connected according to the schematic diagram of Fig. 1. During the test, the nominal airflow rate of the OAD was 190 m^3/h. The rated water flow of the refrigerant–water plate heat exchanger was 2.88 m^3/h. The summer outlet water temperature and the winter outlet water temperature of the refrigerant–water heat exchanger were 16–18 °C and 32–35 °C, respectively.

Each experimental condition was determined by relevant national standard [6], as shown in Table 2.

The instrumentation used in the experiment can be found in the reference [7].

Table 1 Parameters of components

VRF outdoor unit	Operating mode	Type	Rated EER/COP	Rated capacity/kW	Rated input power/kW	Number of unit
	Summer mode	ADX060 MDGHA	3.60	15.5	4.31	1
	Winter mode		4.10	18.0	4.39	
Refrigerant–water plate heat exchanger	Operating mode	Type	Rated capacity/kW	Rated input power/W	Rated operating current/A	Rated water pressure drop/kPa
	Summer mode	AM160 FNBDEH	14	10	0.05	26
	Winter mode		16			
OAD	Nominal airflow rate m³/h	Fan power/W	Refrigerant line cooling capacity/W	Waterway cooling capacity/W	Pipe diameter	Water flow m³/h
	190	22	1900	1600	DN15	0.35

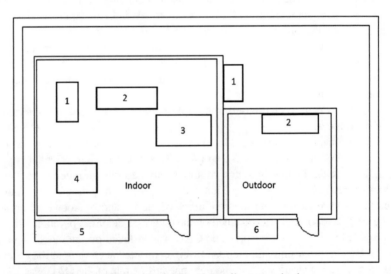

1. Humidifier 2. AHU 3. OAD 4. Radiant terminals
5. Instrument control cabinet 6. Computer

Fig. 2 Layout plan of the psychometric room

Table 2 Experimental conditions of the radiant A/C system

Items	Indoor air states/°C		Outdoor air states/°C		Inlet water temperature of radiant terminal/°C
	T_{db}	T_{wb}	T_{db}	T_{wb}	
The summer mode	27.0	19.5	29	20	18.0
			32	22	
			35	24	
			38	25	
			41	25	
			43	26	
The winter mode	22.0	15.0	7	6	32.0
			4	3	
			2	2	
			1	0	
			−2	−3	
			−5	−6	

2.3 Data Reductions

The experiment mainly tested and recorded the parameters such as the inlet and outlet air temperature and humidity of OAD, the inlet and outlet water temperature and the water flow rate of the radiant terminals, and the power consumption of the VRF outdoor unit under different outdoor air temperature. The processing formulas used in the experiment were as follows:

$$Q_F = \frac{L\rho c \Delta t}{3600} \qquad (1)$$

Where Q_F is the cooling (heat) capacity of the radiant terminals, kW; L is the water flow of the radiant terminals, m³/h; ρ is the density of water, 1000 kg/m³; C is the specific heat capacity of water, 4.18 kJ/(kg °C); and Δt is the temperature difference between the supply and return water of the radiant terminals, °C.

$$Q_X = \frac{G_s \rho (h_x - h_s)}{3600} \qquad (2)$$

Where Q_X is the cooling (heat) capacity of OAD, kW; G_S is the indoor fresh air supply volume, 190 m³/h; h_x is the enthalpy value of the outdoor fresh air, kJ/kg; and h_s is the enthalpy value of the supply air, kJ/kg.

The total cooling (heat) capacity of the system can be given by:

$$Q_Z = Q_F + Q_X \qquad (3)$$

3 Results

3.1 The Summer Mode

Figures 3, 4, and 5 present the test results in summer mode. As seen in Fig. 3, when the outdoor air temperature increased from 29 to 43 °C, the supply air dry-bulb temperature of the OAD was basically maintained at about 19 °C, and the wet-bulb temperature was maintained at about 14.5 °C, with small fluctuation range. The humidity ratio of the supply air was also basically between 8.8 and 9.4 g/kg, and the fluctuation range was also small, which can basically meet the room radiation dehumidification requirements. The system can run stably and reliably within the experimental temperature range of the summer mode.

As seen in Fig. 4, the radiant terminal cooling capacity increased from 8.0 to 10.7 kW, as the outdoor air temperature varied from 29 to 43 °C, while the cooling capacity of the OAD increased from 1.07 to 2.52 kW.

According to Fig. 5, when the outdoor air temperature changed from 29 to 43 °C, the TCC of the system increased from 9.07 to 13.22 kW, with amplitude of 31.32%. While the VRF outdoor unit electric power consumption increased by 45.4% from 2.01 to 3.67 kW, and the EER of the system decreased from 4.53 to 3.55. When the outdoor temperature was lower than 38 °C, the EER decreased slowly. When the outdoor temperature was higher than 38 °C, the descent rate of EER was accelerated.

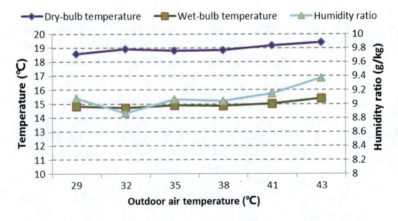

Fig. 3 Supply air temperature and humidity ratio of the OAD in summer

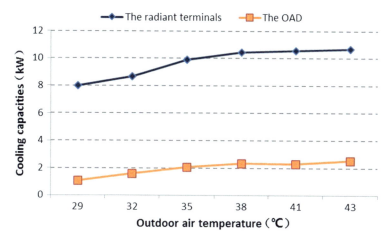

Fig. 4 Cooling capacities of the radiant terminals and the OAD in summer

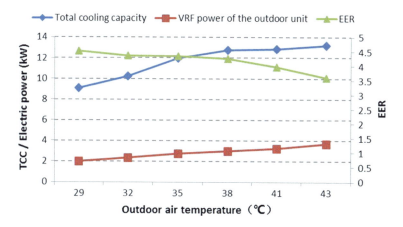

Fig. 5 Performances of the VRF-based radiant A/C system in summer

3.2 The Winter Mode

Figures 6, 7, and 8 present the test results in winter mode. According to Fig. 6, the supply air dry-bulb temperature of the OAD was fluctuated in the range of 31.8–32.9 °C with an average value of 32.5 °C with the outdoor air temperature changed from 7 to −5 °C. The fluctuation range was small. Therefore, the system can run stably and reliably in winter mode.

As seen in Fig. 7, the heating capacity of radiant terminals increased from 4.07 to 6.33 kW, as the outdoor air temperature varied from 7 to −5 °C, so did the OAD from 2.11 to 2.54 kW.

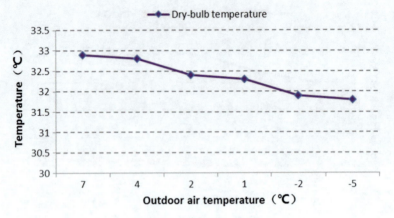

Fig. 6 Supply air dry-bulb temperature of the OAD in winter

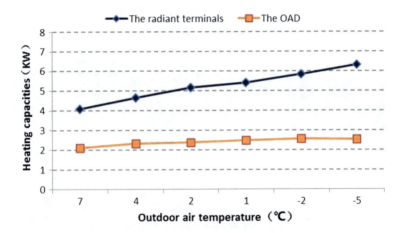

Fig. 7 Heating capacities of the radiant terminals and the OAD in winter

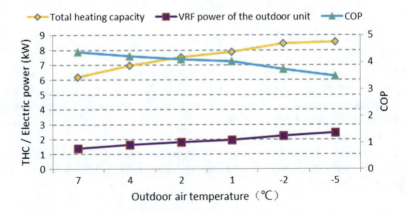

Fig. 8 Performances of the VRF-based radiant A/C system in winter

According to Fig. 8, the system THC increased from 6.17 to 8.5 kW, with an amplitude of 27.70%, as the outdoor air temperature decreased from 7 to −5 °C. While the VRF outdoor unit electric power consumption increased by 42.41% from 1.41 to 2.45 kW, leading to a decreases system COP from 4.53 to 3.55.

4 Conclusions

In this paper, the performance test of VRF-based household radiant A/C system under different conditions was carried out, and conclusions can be drawn as follows:

(1) The VRF-based household radiant A/C system can operate stably in both winter and summer experimental conditions. In summer mode, the supply air dry-bulb temperature of the OAD was basically maintained at about 19 °C, and the wet-bulb temperature was maintained at about 14.5 °C. In winter mode, the supply air dry-bulb temperature of the OAD was basically maintained at about 32.5 °C.
(2) In summer mode, when the outdoor air temperature changed from 29 to 43 °C, the TCC of the system increased by 31.32%, while the electric power consumption of the VRF outdoor unit increased by 45.4%, and the EER of the system decreased to its minimum value, i.e., 3.55.
(3) In winter mode, when the outdoor air temperature varied from 7 to −5 °C, the THC of the system increased by 27.70%, while the electric power consumption of the VRF outdoor unit increased by 42.41%, and the COP decreased to its minimum value, i.e., 3.48.

References

1. Zhao, L., et al.: A simulation study for evaluating the performances of different types of house-hold radiant air conditioning systems. Appl. Therm. Eng. **131**, 553–564 (2018)
2. Lu, Y.: Practical Heating and Air Conditioning Design Manual. 2nd edn, Architecture & Building Press, Beijing, China (2007)
3. GB 50736-2012, Design code for heating ventilation and air conditioning of civil buildings
4. Sui, X., Zhang, X.: Minimum fresh air requirement of air conditioning system for residential buildings in hot summer and cold winter zone. HVAC (10), 99–104 + 81 (2008)
5. Spengler, J.D., Chen, Q.: Indoor air quality factors in designing a healthy building. Annu. Rev. Energy Environ. **25**(25), 567–600 (2000)
6. GB/T7725-2004, Room air conditioners
7. Zhao, L., et al.: The development and experimental performance evaluation on a novel household variable refrigerant flow based temperature humidity independently controlled radiant air conditioning system. Appl. Therm. Eng. **122**, 245–252 (2017)

Precise Control for Heating Supply to Households Based on Heating Load Prediction

Ruiting Wang, Fulin Wang, Zhaohan Nan, Minjie Xiao and Aijun Ding

Abstract Commonly, no control measures are mounted at the heating terminals in a household in district heating systems in China, so the problem of unbalanced heating pipe network causes problems of overheating, under-heating, energy loss, etc. At the same time, the mandatory installation of calorimeters in residential buildings, which aims to solve the inefficient energy-using behavior by charging heating cost according to heat amount used. However, charging heating cost according to heat amount was not successfully adopted by heating company. As a result, the aim of energy saving was not achieved, and giant investment was wasted as well. Aiming to solve these problems, this paper proposed a precise heating terminal control system utilizing the heating amounts measured by the calorimeter, which can solve the unbalance problem of heating supply and can precisely control the heat supplied to a household to meet the individual heating requirement of the household as well. The proposed control system takes advantage of the data predicting ability of artificial neural network (ANN) to predict heating requirement using the information of weather, desired indoor temperature, building envelope, etc. By comparing the predicted heating load with the heating amount measured by the calorimeter, the on/off of heating water valve is controlled to make the supplied heating amount match the predicted heating load. In this paper, the heating load prediction by ANN is described and the results show that the load prediction is accurate enough for the purpose of achieving precise control of heating amount supplied to the terminal users.

Keywords Precise heating control · Heating system · Heating load prediction · Neural network

---

R. Wang · F. Wang (✉) · Z. Nan · M. Xiao
Beijing Key Laboratory of Indoor Air Quality Evaluation and Control,
School of Architecture, Tsinghua University, 100084 Beijing, China
e-mail: flwang@tsinghua.edu.cn

A. Ding
Kechuang Jieneng Mechanical and Electrical Engineering Co. Ltd, 264003 Yantai, China

1 Introduction

Commonly, no control measures are mounted at the terminal heating system in a household in China. Therefore, problems caused by unbalanced heating pipe network often occur. For example, some households are overheated and have to open windows in winter to cool down room and as a result, a lot of heating energy was wasted. On the contrary, some households cannot obtain enough heat so the indoor temperature is very low and the room occupants need wear very thick clothes, which caused a lot of occupant complaints and even health problems. On the other hand, for the purpose of promoting energy-efficient behavior of room occupants, the calorimeters are mandatorily installed at all newly built residential buildings, which aim to prevent the problems of heating energy waste, such as open window during heating period, by letting the households pay more heating fee if they use more heating energy.

However, the actual situation is that almost all heating companies do not charge heating fee according to heating amount but according to floor area. This situate cannot achieve the original purpose of promoting energy-efficient behavior of room occupants, and the giant investment of calorimeter, which is over 3 billion RMB [1], will be totally wasted as well if the calorimeter cannot be utilized.

Aiming to utilize the heating amounts measured by the calorimeter to solve the existing unbalance of heating energy supplying and to precisely control the heat supplied to a household to meet the individual heating requirements of different households, the paper proposes a new control system, which takes advantage of the heating loads prediction by artificial neural network (ANN) and heating amounts measured by calorimeters.

Artificial neural network is a well-developed model of machine learning, widely used in a variety of fields of science and engineering. It has also shown great potential in energy application of the building. Researches have been done in assessing the effectiveness of some artificial neural network models in predicting building energy consumptions [2].

The proposed heating load prediction method uses information of weather, desired indoor temperature, building envelope, etc., and neural network to predict the heating load of a household. The predicted heating load is compared with the heating amount measured by the calorimeter and to control the on/off of water valve to make the supplied heating amount match the predicted heating load. This method is expected to offer a precise heating control of residential building, which could not only solve the presently existing problem of unbalanced heating, but also precisely control the indoor temperature to meet the individual desired indoor temperatures. The proposed control system has another advantage that it makes possible to achieve individualized indoor temperature settings for different households without using indoor temperature sensors.

The following parts of this paper firstly introduce the method of building ANN model and then discuss the heating load prediction results. In the end, conclusions are summarized.

2 Method

2.1 Multilayer Perceptron Artificial Neural Network Model

Artificial neural networks widely used in a variety of fields of science and engineering, including prediction of energy consumption and building load [3]. Multilayer perceptron (MLP) is one of the popular learning algorithms in ANN. Backpropagation is used in the model to calculate weights [4]. By training on a dataset involving a set of features $X(x_1, x_2, \ldots, x_n)$ and a target output y, the algorithm can learn a function approximator which allows an internal connection to be built for regression. Between the input layer and output layer, one or more hidden layers extract useful data features and rearrange the connection. Figure 1 gives an example of MLP model with one hidden layer.

The input layer connects the hidden layer with a weight matrix A, a bias array $b_1, b_2, \ldots, b_m$, and an activation function $\sigma(g)$, so that $t_i = \sigma\left(\sum_{j=1}^{n} x_j a_{ji} + b_i\right)$. Between hidden layers, the features of data convey likewise till the last hidden layer, from which the output layer receives values and transforms them to the desired output y. The learning process lies in the renew and adjustment of weights and accordant bias between each layer, which enable the ANN to minimize the difference between the predicted output and target output [5, 6].

It could be seen that, without an activation function, the correlation between each layer is linear. To introduce nonlinear elements into regression, in this study, the selected activation is sigmoid function, featured by:

$$f(x) = \frac{1}{1 + e^{-x}} \quad (1)$$

The shape of the function is described by Fig. 2:

Fig. 1 Structure of MLP ANN model

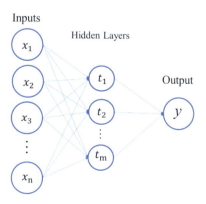

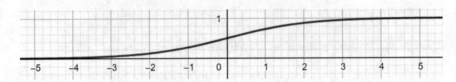

Fig. 2 Sigmoid function used as the ANN activation function

2.2 Prediction Using Back Propagation Neural Network

(1) Input and output parameters

Heating and cooling loads of a household have a variety of contributory factors, including building envelope, the outdoor climate parameters, the indoor occupancy as well as heat source. However, for ordinary residential building where occupant density is relatively small and thermal equipment uncommon, the most influential parameters fall in the first two categories.

The raw dataset of contributory fact of the experimental building required for training process was attained from actual building structure and the climate database of Shandong in a meteorology year. Thirteen parameters are available for input features, which are: sequential hour of the year, the room types, dry bulb temperature (°C), humidity (g/kgDA), solar radiation (W/m^2), diffuse solar radiation (W/m^2), direct solar radiation (W/m^2), air exchange rate, two parameters representing orientation of the building, floor position (ground floor, middle floor or top floor), thickness of insulating layer (mm), and room area (m^2).

Hourly heating loads are calculated by DeST, a software simulator for building environment, HVAC system, and energy consumption, developed by Tsinghua University [7]. The reliability of DeST had been verified by earlier researches [8].

Considering the practical condition of field experiment, the correlation between the parameters and the effectiveness of simulation, ten of which are selected as input features, i.e.: sequential hour of the year, the room type, dry bulb temperature (°C), humidity (g/kg), solar radiation (W/m^2), air exchange rate, two parameters representing orientation of the building, floor position (ground floor, middle floor or top floor), and thickness of insulating layer (mm).

The only output of the model is heating load, which will be used to be compared to the heating amount measured by the calorimeter to control the on/off state of water valve automatically, so that the goal of energy saving would be archived.

(2) Determine parameters and preprocessing

The Python programming language and the scikit-learn package [5] are used for training. Multilayer perceptron regressor optimizes the squared-loss using limited-memory Broyden–Fletcher–Goldfarb–Shanno (BFGS) method (LBFGS) or stochastic gradient descent. The model consists of two hidden layers with, respectively, 5 neurons and 2 neurons. The activation function is sigmoid function. Ten percent of the data is selected randomly to test the accuracy of trained model.

For preprocessing, all data were normalized into the range of zero to one, within which the activation function has an appropriate gradient. Such normalization is also applied to new inputs to predict heating loads.

(3) Evaluation of Predicting Performance

The performance of the prediction of heating loads is measured by the function 'score(X, y [, sample_weight]),' which returns the coefficient of determination R^2 to show how accurate the model is. The coefficient of R^2 is determined by formula below:

$$R^2 = 1 - \frac{SS_{res}}{SS_{tot}} \qquad (2)$$

Where,

SS_{res} is the sum of squares of residuals; $SS_{res} = \sum_i (y_i - f_i)^2$;
SS_{tot} is the total sum of squares: $SS_{tot} = \sum_i (y_i - \bar{y})^2$;
y_i is the target output;
f_i is predicted output.

In an effective model where y is always correctly predicted, the total score of R^2 tends to be close to the best possible result, i.e. 1. On the contrary, an arbitrarily predicted y will receive a low score, which is close to 0.

3 Results

3.1 Predicting Accuracy

Evaluated by the coefficient of determination, R^2, the scores of 5-time training range from 0.9349 to 0.9368, with an average score of 0.9358, indicates a satisfying result. For field application, the trained model with the highest score was used. With the weight matrix and bias extracted, a prediction of the whole dataset is available. Figure 3 gives an example to show the variance of predicted output and the target output in three different time periods, scattering in November, December, and January. It can be seen that, the predicted y matches pretty well with the target y generally. However, for some extreme heating load, the matching is not so good.

3.2 Sensitivity Analysis

Considering the easiness of parameter obtaining in field application, the training of the model utilized ten parameters. The inputs of direct solar radiation and diffuse

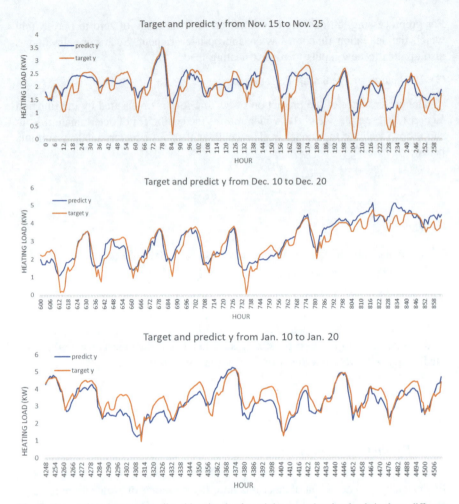

Fig. 3 Comparison between predicted heating loads and the target heating loads in three different time periods

solar radiation were eliminated due to the difficulties in obtaining these parameters. However, the actual impact of such elimination remains unclear. To distinguish and justify this simplification, cases with and without direct solar radiation and diffuse solar radiation as input parameters were studied. Meanwhile, the performance of different activation function was also taken into consideration. Table 1 shows the different settings and the different scores of each setting.

As shown in tables, for Case 1 and Case 2, the variance of R^2 is within the range of acceptance, so the simplification of input parameters is reasonable. For activation functions, the scores differ little with sigmoid and tanh function. The R^2 drops drastically when no activation is used. It could be concluded that nonlinear model

Table 1 Sensitivity analysis of ANN model with different inputs and activation functions

No.	Input parameters	R^2
Case 1	10 parameters without direct solar radiation and diffuse solar radiation	0.93681
Case 2	12 parameters including direct solar radiation and diffuse solar radiation	0.93942
Activation Function		
Case 3	Sigmoid	0.93681
Case 4	Tanh	0.93873
Case 5	No activation	0.76181

performs much better in prediction than linear model. Therefore, the setting of ANN prediction parameters as Case 1 and Case 3 is feasible and operative.

4 Future Work

The future work following the former described heating load prediction is to develop a control system that uses the well-trained model to predict heating loads and compare them with the heating amounts measured by the calorimeter to decide the on/off of heating water valve according to the logic shown in Fig. 4.

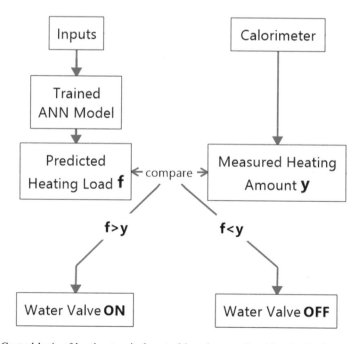

Fig. 4 Control logic of heating terminal control based on predicted heating load

Parameters about building envelope and environment constitute the most part of input parameters. However, for field application of different households with different window opening habits, airtightness of envelopes might vary significantly. To solve this problem, on-learning adaptive ANN might be a promising solution. A research by Fan et al. [9] shows the potential of using deep learning algorithms for cooling load prediction. Several existing models work well in predicting 24-h ahead building cooling load profiles. Therefore, the heating load prediction can be online-tuned using history data of heating loads to achieve a reasonable prediction.

Further, field experiment will be conducted to verify the performance of the proposed control system. More improvement will be made in the future research in optimizing the parameters and the control logic.

5 Conclusions

In order to solve the problem of unbalanced heating pipe network and utilized the idle calorimeters, this study proposes a control system, which aims to precisely control the indoor temperature to meet the individual desire of each household. The essential part of the control system is the prediction of heating loads. Multilayer perceptron artificial neural network algorithms are used in the study to predict the required heating amount. The result turns out to be satisfying. The coefficient of determination R^2 is used to evaluate the prediction accuracy. The average R^2 of prediction is 0.9358.

Sensitivity analysis is conducted to simplify the input parameters and ignore the ones that do not impact the accuracy of prediction significantly. It is reasonable to ignore the direct solar radiation and diffuse solar radiation in inputs. In addition, nonlinear model performs much better in prediction than linear model.

This study proposes a feasible and operative solution for the widely existing problem in China. Further study will be carried out in field experiment to validate the model and test the performance of this control system.

Acknowledgements This research is supported by Innovative Research Groups of the National Natural Science Foundation of China (grant number 51521005).

References

1. Lu, Y.: Heating charge of switching from floor area accordance to heating amount unsuccessful, People's Daily, December 25 (2013)
2. Shi, L.: Application of artificial network to predict the hourly cooling load of an office building. In: IEEE 2009 International Joint Conference on Computational Sciences and Optimization, Sanya, Hainan, China 2009

3. Zhao, H., Magoulès, F.: A review on the prediction of building energy consumption. Renew. Sustain. Energy Rev. **16**, 3586–3592 (2012)
4. Ian, G., Yoshua, B., Aaaron, C.: Deep Learning, p. 196. MIT Press, Boston, USA 2016
5. Rafiq, M.Y., Bugmann, G., Easterbrook, D.J.: Neural network design for engineering applications. Comput. Struct. **79**, 1541–1552 (2001)
6. Pedregosa, et al.: Scikit-learn: machine Learning in Python. J. Mach. Learn. Res. **12**, 2825–2830 (2011)
7. DeST Developing Group: Building Environmental System Simulation and Analysis-DeST. China Architecture and Building Press, Beijing, China (2006)
8. Zhou, X., Hong, T.Z., Yan, D.: Comparison of HVAC system modeling in EnergyPlus, DeST and DOE-2.1E. In: Building Simulation, vol. 7, pp. 21–33 (2014)
9. Fan, C., Xiao, F., Zhao, Y.: A short-term building cooling load prediction method using deep learning algorithms. Appl. Energy 195, 222–233 (2017)

Experimental Study on the Influence of Fouling Growth on the Flow and Heat Transfer of Sewage in the Heat Exchange Tube

Shunzhi Chen, Liangdong Ma, Zhiyuan Zhang and Jili Zhang

Abstract Fouling growth in pipes is a basic problem in the design and application of sewage source heat pump systems. After investigation, in the design of the operating parameters of the sewage source heat pump actual project, the flow rate of the sewage in the industrial sewage source heat pump is 0.2–1.8 m/s, the sewage inlet temperature is 18–23.5 °C, and the condensation temperature is 30–35 °C. Aiming at the basic problem of fouling growth in sewage pipe, a fouling growth test bench in heat exchanger tube under condensation condition of sewage source heat pump was built. Based on the actual collected urban sewage, the effects of fouling growth on the flow and heat transfer characteristics of the heat exchanger were studied under the condition of equal wall temperature. The experimental results show that the larger the initial Reynolds number, the smaller the thermal resistance of the pipe fouling, and the longer the time for the stable growth of the dirt. In terms of heat transfer, pipe fouling growth affects the overall heat transfer coefficient of the pipe, but the initial Reynolds number also affects the overall heat transfer of the pipe. At the initial Reynolds number between 7150 and 16,091, there is a point that minimizes the effect of pipe fouling growth on the overall heat transfer of the pipe. The research results in this paper can provide guidance for the optimal design of sewage source heat pump system.

Keywords Untreated urban sewage · Fouling heat resistance · Experimental research

1 Introduction

Sewage source heat pump technology is one of the emerging cold and heat source technologies for winter heating and summer air conditioning in northern China. Its cold and heat source, urban domestic sewage, has great potential for energy

---

S. Chen · L. Ma (✉) · Z. Zhang · J. Zhang
Institute of Building Energy, Dalian University of Technology, Dalian, China
e-mail: liangdma@dlut.edu.cn

utilization. However, after the system has been running for a period of time, the pipeline fouling is seriously deposited in the pipeline and the heat exchanger, the heat exchange capacity of the heat exchanger is weakened [1], and the energy consumption of the heat pump system is increased (the comprehensive performance coefficient of some systems is even lower than 3.0). There is a problem that energy saving technology does not save energy.

Research on fouling thermal resistance, especially hard fouling, in the 1980s, Kern–Seaton equation describing the growth characteristics of fouling was proposed. Subsequently, Zubair et al. proposed four forms of fouling thermal resistance changing with time, namely linear growth type, falling rate type, power-law type, and asymptotic type. Due to the complexity of fouling deposition process, the prediction of fouling is still uncertain at present. In recent years, fouling monitoring experimental research has also developed rapidly. Albert et al. studied the effects of surface roughness and flow section shrinkage on heat transfer in crystallization fouling [2]. Xu used automatic monitoring technology to study the influence of water quality characteristics and heat exchange pipe geometric characteristics on the fouling formation process in the pipe [3]. In view of the crystallization of $CaSO_4$, Lee studied its effect on the thermal resistance of corrugated plate heat exchangers [4]. Please note that the first paragraph of a section or subsection is not indented. The first paragraph that follows a table, figure, equation, etc. does not have an indent, either.

In the field of sewage pipe flow, many scholars at home and abroad have carried out systematic theoretical and experimental studies on the flow characteristics of sewage in recent 10 years. In 2008, Sun gave an expression for calculating the drag coefficient of sewage along the pipeline under laminar and turbulent conditions through experiments [5]. In engineering design, Sun recommended that the resistance coefficient of raw sewage should be 2.252 times as much as that of clean water under the same working condition through experimental study [6].

In terms of heat transfer in sewage pipes, the convective heat transfer coefficient in sewage pipes measured by Zubair is slightly lower than that in clean water, which is about 0.85–0.9 times of that in clean water pipes under the same conditions [7]. Xu theoretically deduced and experimentally verified the heat transfer characteristics of sewage under laminar flow conditions and obtained the heat transfer criterion correlation under laminar flow conditions [8]. The research team (IBE) conducted a field test of a sewage source heat pump air conditioning system demonstration project, comprehensively analyzed the sewage heat exchanger sewage flow, heat transfer coefficient, fouling thermal resistance, and heat pump unit heating coefficient with time characteristics [9].

Based on the above-mentioned multi-faceted research, the sewage growth and flow resistance test system in the heat exchange tube under the condensing condition of the sewage source heat pump were built. Aiming at the summer refrigeration conditions of the direct sewage source heat pump unit, a dirt growth test and flow resistance test platform for the sewage heat exchange tube was established. Thereby, continuous monitoring of the inlet and outlet temperature, pressure difference, sewage flow rate, system condensation pressure and condensation temperature, evaporation pressure and evaporation temperature, and heating power of the heat exchange tube sewage are realized. According to the qualitative sewage (the concentration of suspended solids in the sewage is maintained constant), the growth characteristics of the bottom layer of the fouling inside the pipe, the thermal conductivity of the flow resistance in the pipe, and the change characteristics of the heat transfer in the pipe are studied experimentally under different sewage flow rates. The research results in this paper can provide guidance for the optimal design of sewage source heat pump system in the future. On this basis, based on non-Newtonian fluid mechanics, the turbulent heat transfer heat of the sewage under the action of the bottom layer of the sewage was analyzed.

1.1 Test Platform for Monitoring Fouling Growth

The test system is composed of four parts: refrigerant vapor circulation system, sewage circulation system, cooling water system, and data acquisition system.

Refrigerant vapor circulation system: The system consists of an evaporator, a condenser, and a conduit connecting the evaporator and the condenser, as shown in Fig. 1a by the cycle of the red line segment connection. Figure 1b is a partial enlarged view of this part. When the system works, the liquid refrigerant in the evaporator absorbs the heat generated by the electric heater and changes into saturated vapor. Under the action of evaporation pressure, refrigerant vapor enters the shell side of the condenser, namely the outside of the sewage heat exchange pipe. As the sewage cools, the refrigerant vapor is cooled to a liquid. Under the action of gravity, the refrigerant condensate flows back to the evaporator from the condenser, thus forming a cycle of refrigerant vapor.

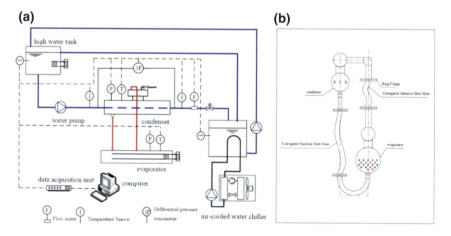

Fig. 1 Schematic diagram of test system and local refrigerant steam circulation system

Sewage circulation system: The system consists of a high water tank, a low water tank, a condenser heat exchange tube, and a water pump, as shown by the cycle of the blue line segment connection in Fig. 1a. The constant temperature sewage (20 °C) sends the sewage to the heat exchange tube of the condenser through the water pump. After the sewage absorbs the heat of the refrigerant vapor, the temperature of the sewage rises and flows into the low water tank. The heated sewage is cooled in the lower tank to a temperature slightly lower than the constant temperature of the sewage. The cooled sewage in the low water tank is then sent to the high water tank by the action of the water pump, and the sewage is heated to the required constant temperature by the electric heater. The flow rate from the low water tank to the high water tank is greater than the flow rate of the sewage into the heat exchange tube. The excess part of the sewage flows back into the low water tank through the overflow port of the high water tank to form a sewage circulation system.

Cooling water system: This system is as shown by the cycle of the black line segment connection in Fig. 1a. The system provides cooling water to remove the added heat from the sewer system to maintain thermal balance on the test bench.

Data acquisition system: This system is as shown by the cycle of the dotted line segment connection in Fig. 1a. It adopts Keithley 2700 data acquisition system, Keithley 7708 data acquisition. The measurement range and accuracy of each sensor in the experimental system are shown in Table 1.

The test conditions are designed according to the operating parameters of several sewage source heat pump actual projects. The flow rate of sewage in the industrial sewage source heat pump is 0.2–1.8 m/s, the sewage inlet temperature is 18–23.5 °C, and the condensation temperature is 30–35 °C. The experimental conditions of this experiment are shown in Table 2. The experimental working fluid adopts the self-contained primary domestic sewage. Based on a series of physical property parameter testing methods in the literature [10], the specific data are shown in Table 3.

The fouling growth in sewage tube was analyzed from two aspects of flow and heat transfer. In terms of flow, the resistance coefficient along the path is used to evaluate it, as shown in Formula (1). In terms of heat transfer, fouling heat resistance and comprehensive heat transfer coefficient of pipelines were selected to evaluate it. The specific derivation process is as in Formulae (2) and (3). The physical parameters involved in the experiment, such as density, viscosity, thermal conductivity, and specific heat capacity of sewage, are all samples of the experimental working medium and measured by the instrument.

Table 1 Parameters of the measurement instruments

Parameters	Instruments	Range	Measurement errors
Temperature	RTD(PT-100)	−50–300 °C	±0.05 °C
Pressure	Pressure difference transmitter	0–2.5 Mpa	±0.1% of full scale
Pressure	Capacitance pressure transducer	0.3–1.0 Mpa	±0.25% of full scale
Flow rate	Electromagnetic flow meter	0–1 m^3/h	±0.5% of full scale

Table 2 Test condition

Test number	t_{in} (°C)	t_0 (°C)	d (mm)	u_m (m/s)	Initial Re
1	20.0	30.0	23.06	0.3	7150
2	20.0	30.0	23.06	0.5	11,747
3	20.0	30.0	23.06	0.7	16,091

Table 3 Basic parameters of test sewage

Contaminant concentration g/kg	ρ kg/m³	$\mu \times 10^6$ Pa s	$\lambda \times 10^2$ W/(m K)	Cp J/(kg K)
1.21	999.2	1077	61.18	4011.5

$$\lambda = \frac{2 \cdot d \cdot (p_1 - p_2)}{\rho \cdot L \cdot u_m^2} \tag{1}$$

$$Q_w = m_w c_{p,w}(t_{in} - t_{out}), \quad K_f = \frac{Q_w}{A \Delta t_m},$$

$$A = \pi dL, \quad \Delta t_m = \frac{(t_0 - t_{in}) - (t_0 - t_{out})}{\ln \frac{(t_0 - t_{in})}{(t_0 - t_{out})}} \tag{2}$$

$$K_f = \frac{1}{R_w + R_f + \frac{1}{h_{in,f}}}, \quad R_w = \frac{\delta_w}{\lambda_w},$$

$$\frac{1}{K_c} = R_w + \frac{1}{h_{in,c}}, \quad R_f = \frac{1}{K_f} - \frac{1}{K_c} \tag{3}$$

The thermal balance and reliability of the experimental tubes of the platform were verified, respectively. The average heat loss coefficient of two pipes with an inner diameter of 23.06 mm was 2.6 and 2.5%, respectively. The experimental results and the Dittus–Boelter heating fluid formula calculated the convective heat transfer coefficient errors of the tubes were 2.97 and 3.84%, respectively. The error between the experimental results and the Darcy formula calculated in the tube is 6.1 and 5.8%, respectively.

2 Results and Discussion

2.1 Analysis of Fouling Feat Resistance Change in Sewage Tubes

The curve of fouling thermal resistance at the same pipe diameter and different flow rates is shown in Fig. 2. In the initial stage of contact between the test sewage and

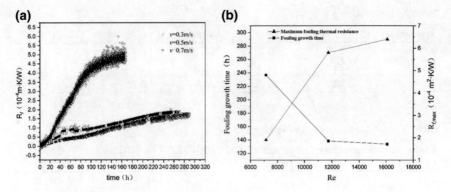

Fig. 2 **a** Variation of fouling thermal resistance with time under different sewage flow rates; **b** variation of thermal resistance of dirt at different initial Reynolds number

the heat exchange tube (about 10 h ago), the thermal resistance of the dirt is near 0, indicating that the fouling layer is not formed during this time, so this period is called the induction period of the dirt growth, and its length is positively correlated with the flow rate. After the induction period, the slope of the fouling thermal resistance curve reaches the maximum, which indicates that the fouling growth rate reaches the peak at this time. With the passage of time, the fouling thermal resistance gradually tends to be flat. The stability value of fouling thermal resistance is negatively correlated with the initial Reynolds number, and the stabilization time of fouling thermal resistacne is positively correlated with the initial Reynolds number, that is, the larger the Reynolds number at the beginning of the test, the longer the stability value of fouling thermal resistance takes. The above variation curve is shown in Fig. 2. The final thermal resistance values of the three experiments are 4.8×10^{-4} m^2 W/K, 1.85×10^{-4} m^2 W/K and 1.71×10^{-4} m^2 W/K, respectively.

2.2 Analysis of Flow and Heat Transfer Characteristics in Sewage Heat Exchange Tubes

As shown in Fig. 3a, as the initial flow rate increases, the trend of resistance increases along the pipe becomes slower. In the initial stage of contact between the test sewage and the heat exchange tube (about 10 h ago), the resistance coefficient along the heat transfer tube did not change. It can be considered that there is no dirt generation during this time, and this time is called the induction period of the dirt growth. When the initial flow rate is 0.3 m/s, the resistance in the tube increases rapidly after the induction period; when the initial flow rate is 0.7 m/s, the variation in the resistance along the tube is not obvious. As shown in Fig. 3b, the stability value of the heat transfer tube along the path resistance coefficient is negatively correlated with the initial Reynolds number of the test. That is, the larger the initial

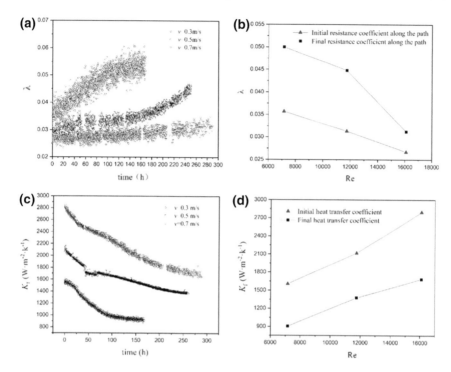

Fig. 3 a The curve of resistance coefficient along with the pipe inside the pipe with time under different flow rates; b variation of resistance coefficient along the pipeline of different initial Re; c total heat transfer coefficient with time under different flow rates; d variation of total heat transfer coefficient of different initial Re

Reynolds number of the test, the smaller the stable value of the final resistance coefficient along the path, and the smaller the variation of the resistance coefficient along the path. Therefore, if a high flow rate water flow is used to flush the pipe, the effect of cleaning the pipe wall scale layer can be achieved.

As shown in Fig. 3c, all three curves are in a downward trend, and as the initial flow velocity decreases, the curve changes more violently with time. As the initial flow rate increases, it can be seen that the change in the overall heat transfer coefficient of the pipeline presents two stages. At a small flow rate, the overall heat transfer coefficient tends to stabilize as it goes through a rapid decline phase; while the flow rate increases to a certain extent, the overall heat transfer coefficient has two rapid decline stages and finally stabilizes. According to the expression of total heat transfer coefficient in Formula (3), the heat transfer characteristics of sewage heat exchanger tubes are affected by two factors: the growth of fouling resistance on the wall and the change of convective heat transfer coefficient on the surface of fouling layer after the formation of fouling layer. Therefore, as shown in the results shown in Fig. 3d, the Reynolds number is between 7150 and 16,091, and there is a point that the heat transfer of the pipeline is minimized in the case of dirt growth.

3 Conclusion

Based on the original domestic sewage, the paper studied the fouling growth in the heat exchange tube and the changes in flow and heat transfer characteristics under the influence of the pipeline. The specific results are as follows:

(1) The thermal stability of the fouling is negatively correlated with the initial Reynolds number of the test, that is, the larger the Reynolds number at the beginning of the test, the smaller the stable value of the final fouling thermal resistance. Similarly, the greater the Reynolds number at the beginning of the test, the longer it takes for the final stable value of the fouling thermal resistance.
(2) The resistance coefficient of the pipeline along the path is greatly affected by the initial Reynolds number. The larger the initial Reynolds number, the smaller the initial value of the resistance coefficient along the path, and correspondingly, the variation of the resistance coefficient along the path will be smaller until the scale layer is stabilized. Therefore, a high flow rate of water can serve to clean the wall of the pipe and inhibit the growth of dirt.
(3) Pipe fouling growth affects the overall heat transfer coefficient of the pipe, but the initial Reynolds number also affects the overall heat transfer of the pipe. Through experiments, it can be found that there is a point between the initial Reynolds number between 7150 and 16,091, which can minimize the influence of pipe fouling growth on the overall heat transfer of the pipe.

References

1. Zhang, J.L., et al.: Study progress in intaking water, defouling and heat transfer in sewage water source heat pump air conditioning systems. Heat. Vent. Air Cond. **39**(7), 41–47 (2009)
2. Albert, F., et al.: Roughness and constriction effects on heat transfer in crystallization fouling. Chem. Eng. Sci. **66**(3), 499–509 (2011)
3. Xu, Z.M., et al.: Experimental investigation on fouling characteristics of the discrete double inclined ribs tube. Proc. Chin. Soc. Electr. Eng. **32**(32), 64–68 (2012)
4. Lee, E., et al.: Thermal resistance in corrugated plate heat exchangers under crystallization fouling of calcium sulfate ($CaSO_4$). Int. J. Heat Mass Transf. **78**, 908–916 (2014)
5. Wu, X.H.: Research on characteristic of flow and heat transfer and development of fouling of city sewage. Harbin Institute of Technology, Harbin (2008). (in Chinese)
6. Zubair, S.M., et al.: A maintenance strategy for heat transfer equipment subject to fouling: A probabilistic approach. J. Heat Transfer **119**, 575–580 (1997)
7. Xu, Y.: Heat transfer characteristic of urban sewage. J. Harbin Inst. Technol. **41**(10), 70–73 (2009)
8. Liu, Z.B., et al.: Application of a heat pump system using untreated urban sewage as a heat source. Appl. Therm. Eng. **62**, 47–757 (2014)
9. Liu, Z.B.: Characteristics of sewage flow and heat transfer in intaking water and heat transfer process of sewage heat pump. Dalian University of Technology, Dalian (2014). (in Chinese)
10. Liangdong, M., Zhiyuan, Z., Jialin, S., et al.: Test of basic physical parameters of urban untreated sewage. J. Southeast Univ. (Nat. Sci. Ed.) (2018)

Study on Exhaust Uniformity of a Multi-terminal System

Yirui Wang and Jun Gao

Abstract Accompanied by the growth of high-rise residential, the main problems of the residential flue (uneven exhaust; deficiency exhaust; fumes and odor intrusion; fire hazard; and automatic control system design difficulty) are more prominent, which will lead to the fumes and cause great menace to human health. In order to improve the exhaust performance of the flue, this paper designed and installed a variable pressure plate in the centralized flue, and analyzed various factors (presence of variable pressure plate, opening rate, position distribution of open terminals) affecting the exhaust uniformity of the multi-terminal system through experiment. The results show that the installation of the plate in the reference condition can effectively improve the uniformity. The variance analysis and range analysis of the obtained data show that for the multi-terminal system without the variable pressure plate, the influence of opening rate on its uniformity is far greater than the position distribution; and the influence of terminal position distribution is much more than the opening rate in the multi-terminal system with the transformer plate.

Keywords Multi-terminal system · Exhaust uniformity · Variable pressure plate · Opening rate · Position distribution of open terminals

1 Introduction

With the accelerating process of urbanization in China, the urbanization rate of permanent resident population in China has increased from 17.9% in 1978 to 58.52% in 2017 [1]. Faced with the contradiction between the housing needs of the newly increased urban population and urban land resource constraints, high-rise residential construction has become a trend. China's total housing stock of urban housing has reached 18.84 billion square meters by 2010, and 17.43% of households were medium- and high-rise residential buildings and above [2].

Y. Wang · J. Gao (✉)
School of Mechanical Engineering, Tongji University, Shanghai, China
e-mail: gaojun-hvac@tongji.edu.cn

The growing number of high-rise buildings has created an urgent problem—kitchen ventilation. The kitchen ventilation mode of high-rise residence experienced the evolution of natural ventilation, axial flow exhaust fan, range hood, and so on. It was not until the 1990s that the use of communal shafts for centralized fumes emission emerged [3]. That is, to select a reasonable source of power to capture indoor pollutants, discharge fumes to the centralized flue and then let it to the atmosphere. After 30 years of research and improvement, the system still has the following major problems: (1) uneven exhaust [4]; (2) deficiency exhaust [5]; (3) fumes and odor intrusion [6]; (4) fire hazard [7]; and (5) automatic control system design difficulty. Aforementioned problem causes the fumes that Chinese cooking produces to be diffused indoors, which cause great menace to human health. Therefore, it is very important to design a reasonable centralized flue and improve the fumes' emission performance of the system to optimize the living environment.

At present, many scholars at home and abroad are committed to optimizing exhaust uniformity by optimizing local components [8–10]. On the basis of previous studies and in consideration of the problem of fumes' recirculation, the study adopts the centralized flue system, installs the variable pressure plate [11] in the main flue duct at the user exit to improve the evenness of terminal exhaust, and forms a set of multi-terminal centralized exhaust system, as shown in Fig. 1. Standards stipulate that the seventh to ninth floors of residential buildings are medium- and high-rise residential buildings, and the tenth and above floors are high-rise residential buildings. The research object of this study is high-rise residential buildings, with 30 electric opening and closing valves at the terminal [12].

The uniformity of exhaust air is affected by many factors (presence of variable pressure plate, open rate, distribution of open terminals, air leakage, etc.). In order to further clarify the improvement effect of ventilation uniformity of the system and the main factors affecting the uniformity, the study conducted experiments on

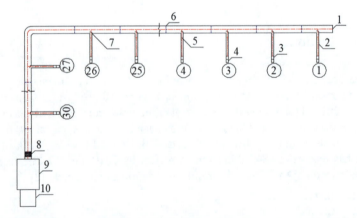

Fig. 1 Schematic diagram of multi-terminal exhaust system (1—the main duct; 2—the branch; 3—duct connector; 4—air valve controller; 5—velocity sampling hole; 6—static pressure sampling hole; 7—variable pressure plate; 8—flexible connection; 9—exhaust fan; 10—exhaust outlet)

various factors through numerical simulation and experiments, so as to better improve the ventilation performance of the system and put forward a complete set of design methods.

2 Methods

In this study, CFD simulation and experiment were used to complete the analysis of factors affecting the uniformity of exhaust air in the system, and the resistance characteristic formula of the variable pressure plate in the system and the local resistance coefficient of the branch were obtained through the preliminary experiment for the calculation of the shape parameters of the plate and the setting of boundary parameters in the simulation. Finally, the simulation results are consistent with the experimental analysis results to ensure the reliability of the research. Therefore, the result and conclusion mainly select the experimental data.

2.1 Determination of Basic Parameters

The test bench is built in 1:1 scale according to the Chinese national building standard design atlas 16J916-1 *residential exhaust*. The total length of the main flue is 98 m, about 30 floors, and the section size of the main is 450 × 550 mm. Each floor of the branch is 1.5 m long, the spacing of which is 3 m, and the section size of the sub is 150 × 150 mm. Due to site constraints, a square elbow is placed between the 26th and 27th floors of the flue. Flue wall is galvanized steel sheet whose equivalent roughness height K is 0.15 mm. λ is on-way resistance coefficient which can be calculated according to Альтшуль formula. Airflow is approximate dry air in 20 °C, 1.01×10^5 Pa. Local resistance coefficient of the branch is 3.75 when the terminal is opened.

2.2 Calculation of Shape Parameters of Variable Pressure Plates

According to the basic parameters of the system and the conversion principle of dynamic and static pressure in aerodynamics, the size of the variable pressure plate used in the system can be calculated. After the test and simulation, the resistance characteristics of the variable pressure plate in the system can be calculated by formula as follows:

$$\tag{1}$$

where ζ_{cf} is coefficient of confluence resistance which is equal to the resistance loss of the flow from i to $i + 1$ divided by the dynamic pressure of the flow downstream; ζ_{str} is coefficient of straightway resistance which is equal to the resistance loss of the flow from $i - 1$ to $i + 1$ divided by the dynamic pressure of the flow downstream; n_q is flow ratio which is equal to branch flow divided by the downstream main flow.

$P_1 = 110$ Pa (total pressure) needs to be preset, which is obtained according to CFD simulation. Then, select the cell shown in Fig. 2, and set out equations ($i = 2$, ... 0.30):

$$P_i = P_{i-1} - P_{f,(i-1)-i} \tag{2}$$

$$P_i - \zeta_{str,i} P_{d,(i+1)} = P_a - P_{f \cdot i} - P_{m \cdot i} - \zeta_{cf,i} P_{d,(i+1)} \tag{3}$$

$$P_{i+1} = P_i - \zeta_{str,i} P_{d,(i+1)} - P_{f,i-(i+1)} \tag{4}$$

where P_i is the total pressure at point i, P_{i+1} and P_{i-1} are in the same way; $P_{d,(i+1)}$ is dynamic pressure at point $i + 1$; $P_{f,(i-1)-i}$ is the on-way resistance, the subscript $(i - 1)-i$ is the distance from $i - 1$ to i, $i-(i + 1)$ is in the same way, and $P_{f,i}$ is the on-way resistance along the branch; $P_{m,i}$ is the local resistance at the branch; P_a is the atmospheric gauge pressure.

When $i = 30$, P_{31} is located downstream 3 m of the connection of branch No. 30 and the main flue. And the calculation formula of fan pressure head is as follows:

$$P_{fan} = P_{31} - \sum \zeta P_{d,sum} \tag{5}$$

where P_{fan} is the pressure head of the fan; $\sum \zeta$ includes the local resistance of flexible connection and exhaust outlet.

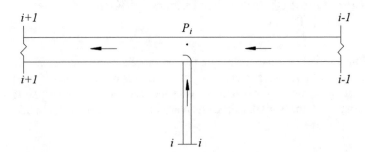

Fig. 2 Schematic diagram of unit segment in variable pressure plate calculation

Table 1 Summary of variable pressure plate shape parameters

No.	h (mm)	No.	h (mm)	No.	h (mm)	No.	h (mm)	No.	h (mm)
1	110	7	125	13	120	19	95	25	75
2	110	8	125	14	115	20	95	26	75
3	115	9	125	15	110	21	90	27	75
4	120	10	125	16	110	22	85	28	65
5	120	11	120	17	105	23	85	29	65
6	120	12	120	18	105	24	80	30	65

Particularly, there is a square bend at the confluence point of the 26th floor 4 m along the flow direction, and the calculation formula corresponding to P_{27} is as follows:

$$P_{27} = P_{26} - \zeta_{str,26} P_{d,27} - P_{f,26-27} - \zeta_{sq} P_{d,27} \tag{6}$$

where ζ_{sq} is local coefficient of the square bend.

By combining the above equations, the coefficient of confluence resistance and straightway of the 30 transformer plates can be obtained. Then, according to the Eq. (7) between the straightway resistance coefficient of the transformer plate ζ_{str} and the shape parameter h of the transformer plate, the h_i ($i = 1,\ldots 0.30$) can be obtained, as shown in Table 1 (The selected reference design condition is: open rate 0.6; even open).

$$h \cdot n_q = 1.65 \left(1 - \frac{\zeta_{str} n_q}{3.68}\right)^{-1.25} \tag{7}$$

2.3 Working Condition Settings

Exhaust air uniformity of the system is affected by multiple factors (whether there is a variable pressure plate, terminal opening rate, distribution of open terminals, air leakage, etc.), so the design conditions of experiments are shown in Fig. 3. And set the system without plate as the group without plate. During setting working conditions, select four open rates which are 0.6, 0.4, 0.2, and 0.13 (30 terminals are set in the system, and the number of open terminals corresponding to 0.13 is 2), and select four open terminal distribution positions of open centrally at the top, open centrally on the bottom, both sides open and even open. Check the air tightness of the system before the experiment to avoid air leakage.

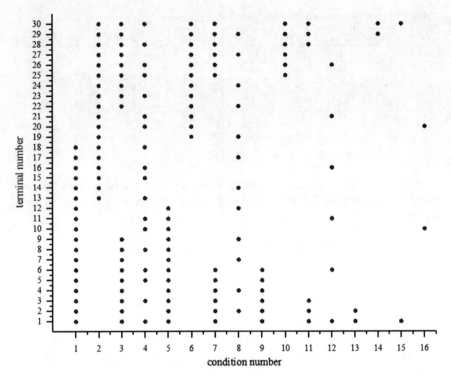

Fig. 3 Schematic diagram of open terminal in each working condition (gray dots indicate open terminals)

2.4 Other Details

During the experiment, the ambient wind speed was controlled to be less than 2 m/s, and the total exhaust volume was controlled to be the number of open terminals multiplied by the terminal reference exhaust air volume, which is set at 500 m^3/h. Handheld hot-wire anemometer (KANOMAX 6006) was used to measure the flow velocity at each open branch, and the portable differential manometer (KIMO MP110) was used to measure the static pressure on each floor. Boundary conditions and material settings in CFD numerical simulation are the same as those in experiments. According to the working conditions listed in Fig. 3, the experiments and simulations of the group without plate and the group with plate (transformer plates were installed) were successively completed, and the range analysis and variance analysis [13, 14] of the obtained data were carried out to obtain the significance level of each factor.

3 Results and Discussion

Based on total air volume control, the distribution of exhaust air volume at each terminal of the system under different working conditions was obtained through simulation and experiment. Compared with the group without plate, the results show that the variable pressure plate has a better effect on the uniformity of exhaust air under the benchmark condition, as shown in Fig. 4. Due to the different distances from different floors to concentrated power in high-rise residential buildings, the exhaust volume distribution presents the situation that users on lower floors have less exhaust air volume, while users on higher floors have more exhaust air volume, and the air volume distribution is extremely uneven, which can be seen from the data of the group without plate in Fig. 4 (working condition: open rate 0.6, even open; group 1: simulation group without variable pressure plate; group 2: simulation group with variable pressure plate; group 3: experiment group with variable pressure plate). The experimental and simulation results are consistent, which proves that the use of the transformer plate can effectively improve the evenness of exhaust air at the terminals in the reference design condition. Due to the inevitable influence of wind field and air leakage in the experiment, compared with the simulation, the experimental data show larger fluctuations.

The root mean square error of terminal air volume S^* was used as an index to further study the influence of two factors (opening rate and distribution of open terminals) on the uniformity of exhaust air in the system before and after the installation of the transformer, as shown in Fig. 5. In the group without plate, the uniformity was better when the opening rate is small, and the terminals are concentrated at the bottom. In the group with plate, the terminals are concentrated at the bottom or uniformly distributed with good uniformity. In addition, the effect of variable pressure plate on improving system uniformity under high opening rate is

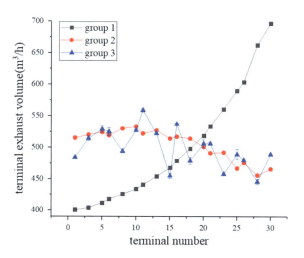

Fig. 4 Distribution of exhaust air volume

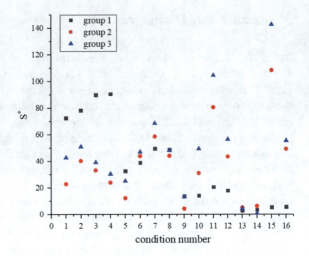

Fig. 5 Root mean square error of exhaust air volume

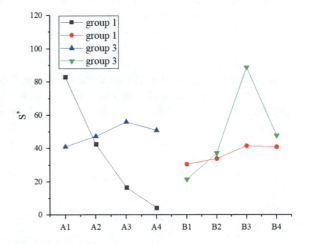

Fig. 6 Range analysis of influencing factors

obvious, which is mainly related to the selection of reference conditions when designing the shape parameters of variable pressure plate.

In order to further clarify the main influencing factors, the range analysis and variance analysis results of experimental and simulated data are shown in Figs. 6 and 7 (A: opening rate; B: terminals distribution). In Fig. 6, A1: 0.6; A2: 0.4; A3: 0.2; A4: 0.13 (30 terminals are set in the system, and the number of open terminals corresponding to 0.13 is 2); B1: open centrally at the top; B2: open centrally on the bottom; B3: two sides open; B4: even open. This indicates that, in general, high-rise residential buildings, the influence of terminal opening rate on the evenness is much greater than the position distribution (black broken line and red broken line). The influence of the position distribution of the open end on the uniformity of the system is much greater than that of the opening rate (blue broken line and green broken line). These can also be seen in the contribution rate of variance analysis.

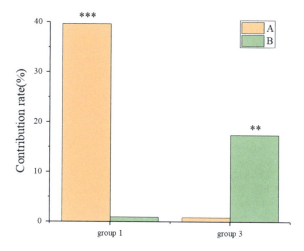

Fig. 7 Variance analysis of influencing factors

4 Conclusions

In this study, the main factors affecting the exhaust air uniformity of the multi-terminal system (whether there is a component to improve the uniformity—variable pressure plate; terminal opening rate; position distribution of open terminals) were simulated and tested. The results show that:

1. In the reference working condition, variable pressure plate improves the uniformity by a significant amount, but when there is a large difference between the actual condition and the reference condition, variable pressure plate has a negative effect on the uniformity. Therefore, when designing the shape parameters of the transformer plate, it is suggested to carry out sufficient investigation and choose the typical working condition (the maximum terminal opening rate under the non-guarantee rate of 5% is met) as the reference working condition.
2. For the multi-terminal system without the variable pressure plate, the influence of opening rate on its uniformity is far greater than the position distribution; and the influence of terminal position distribution is much more than the opening rate in the multi-terminal system with the transformer plate. This indicates that for the multi-terminal system with centralized power (which is not convenient to install components), the opening rate can be appropriately controlled to improve the ventilation uniformity of each household. For the newly built multi-terminal system with centralized power, if its location distribution is relatively fixed, the ventilation uniformity of the system can be improved by installing transformer.
3. Due to the interference of external wind field environment and air leakage, compared with the simulation, the experimental data show larger fluctuations.

Acknowledgements This research has been supported by the National Natural Science Foundation of China (NSFC) for its financial support for the research project 'research on the uniform and adaptive smoke exhaust dynamics and design methods of high-rise residential kitchens' (No. 51578387).

References

1. China Statistics Bureau.: Statistical bulletin of the national economic and social development of the People's Republic of China in 2017. http://www.stats.gov.cn/tjsj/zxfb/201802/t20180228_1585631.html. Last accessed 07 Apr 2019 (2017)
2. Liu, H.Y., Yang, F., Xu, Y.J.: Analysis of urban housing situation in China based on 2010 census data. Journal of Tsinghua University(Philosophy and Social Sciences), **28**(06), 138–147 + 158 (2013)
3. You, M.T.: Aerodynamic Performance of Residential Exhaust System. Shenyang University of Architecture, Shenyang (2014)
4. Xiong, J.: The Research about the Characteristics and Optimization of High-Rise Residential Kitchens Centralized Exhaust System. Chongqing University, Chongqing (2016)
5. Liu, N.: Research on Centralized Smoke Exhaust (Wind) System in High-rise Residential Buildings. Xi'an University of Architecture and Technology, Xi'an (2009)
6. Wang, H.C.: Experimental and Simulation Study on Diversion Components of a Centralized Flue Gas Airway System in High-Rise Residential Kitchens. Xi'an University of Architecture and Technology, Xi'an (2013)
7. Yang, T.: Consideration on fire hazard and fire extinguishing device in kitchen. J. Tongling Univ. **10**(06), 102–103 (2011)
8. Wang, X.X.: Research on Design Method of Centralized Flue Gas System in High-Rise Residential Kitchens. Tongji University, Shanghai (2006)
9. Shao, Z.M.: Simulation Research on Centralized Exhaust System in High-Rise Residential Kitchens. Xi'an University of Architecture and Technology, Xi'an (2010)
10. ASHRAE Handboook.: HVAC Applications, Chapter 33 Kitchen Ventilation. American Society of Heating, Refrigerating and Air-conditioning Engineers, Atlanta, USA (2015)
11. Wu, L.: Research on Design of Central Fume Exhaust for Residential Kitchens in High-rise Buildings based on Controlled Flue Area. Tongji University, Shanghai (2017)
12. GB50352-2005, Code for design of civil buildings
13. Liu, S.L., Liu, W.T.: Experimental development process of a new fluid-solid coupling similar-material based on the orthogonal test. Processes **6**(11), 17 (2018)
14. Xie, Y.L.: Study on Spray Cooling Effect in Long and Narrow Space Strong Disturbance Process. Tongji University, Shanghai (2018)

Topology Description of HVAC Systems for the Automatic Integration of a Control System Based on a Collective Intelligence System

Zhen Yu, Huai Li and Wei Liu

Abstract This paper proposes a collective intelligence system (CIS) that uses a decentralized and self-organizing approach to build a smart control system for HVAC equipment. Using standard control units, the control system can automatically identify the building space, HVAC equipment, sensors and actuators in the identified spaces without human intervention. To support the automatic system integration and relation identification, a novel topology description based on graph theory of the building space and HVAC system is proposed. The description method was examined by its application to a typical building layout with a water distribution system and a ventilation system. The potential of using the CIS for HVAC system control is further explored, and the benefits are discussed.

Keywords Control system · Topology description · HVAC · Collective intelligence system (CIS) · Automatic integration

1 Instruction

1.1 A Subsection Sample

The HVAC system is one of the major energy consumers in buildings. For many types of buildings, over 40% of the energy is spent on heating, cooling and ventilation. Proper control of the HVAC system is crucial for energy saving in buildings. The building automation system has become a standard configuration in modern public buildings in recent years [1]. Many research reports and control system providers claim that the building automation system can save up to 30% of the energy consumption. Despite the common understanding that the energy-saving

---

Z. Yu (✉) · H. Li
China Academy of Building Research, Beijing, China
e-mail: yuzhen@chinaibee.com

W. Liu
College of Urban Construction, Nanjing Tech University, Nanjing, Jiangsu, China

potential of the control system is essential, surveys reveal that the engineering practice quality of the HVAC control system is far from ideal. The reasons behind the performance gap between the promising potential and inferior practice have drawn much attention from academia. The current concept and technology of building automation come from the factory automation industry. However, unlike the factory automation system, the building automation system (BAS) is not considered a critical system in buildings because the manual control can often provide acceptable service quality for occupants. The BAS, particularly the HVAC control system, generally has much lower budget and less expertise than the factory automation. During the construction process of buildings, the installation and integration of the control system are often one of the last phases to begin and finish, and the time limit pressure hinders the high-quality implementation of the BAS. The less reliable field devices, rigid time and budget limit, and lack of sufficient expertise limit the BAS from better performance.

Despite the resource limits, the HVAC control requires experienced engineers to implement complex configuration and time-consuming project-oriented programming. Although many studies have been performed to improve the performance of the HVAC control system, and many new methods have been introduced to the industry [2–5], the conventional centralized control system architecture has not significantly changed. The architecture of BAS inherited from factory automation may not be suitable for today's development. Based on the recent development of the ICT industry, the HVAC control industry must develop a suitable technology and a new system architecture that can reduce the system integration cost and enhance system reliability.

2 Methods

2.1 Requirements for the New Architecture of the HVAC Control System

Addressing the essential issues that have troubled for many years in the conventional centralized building automation system, the next generation of HVAC control system should have the following crucial properties: self-identification, self-integration, self-adaptation and self-programming.

2.2 Introduction of CIS Concept

A decentralized, flat-structured building automation system controlled by smart nodes was proposed [6]. The new collective intelligence system (CIS) uses standard distributed smart nodes, called CPN, to manage the automation system. Each CPN controls a subspace in the building, and all CPNs have identical structure, hardware and software. A CPN has a limited number of communication ports that connect to the neighbouring CPNs. Unlike the traditional centralized system, all CPNs have equal relations to one another. Each CPN represents a smart zone or a smart device. A smart zone includes a building subspace, HVAC equipment sensors and actuators in the space. A smart device represents a chiller, a boiler or other building devices. The CPNs can only exchange information with their neighbouring nodes. There is no central computer that works as a higher-level coordinator or controller. Using the proposed control architecture, the CIS can decentralize the building automation system into a flat-structured system, which enables the new system to self-identify, self-integrate, self-adapt and self-program.

2.3 Identification and Integration Process

The process of control system identification and integration based on the CIS is shown in Fig. 1. First, the architecture design document with space information is used as the input, and a design tool is used to identify the space topology. Second, the HVAC design document and drawings are used to automatically find the HVAC equipment in the building spaces and its system information. The locations of CPNs are decided according to the acquired building and system information. The CPNs should be installed following the decision. When the CPNs have been properly installed, the identified information is downloaded into the CPNs. The CPNs search the sensors, actuators and equipment in the building subspace and integrate the control system without much human intervention. Then, the suitable control strategy for the available sensor, actuator and equipment is suggested for approval. As soon as the permission is granted, the control program immediately generates

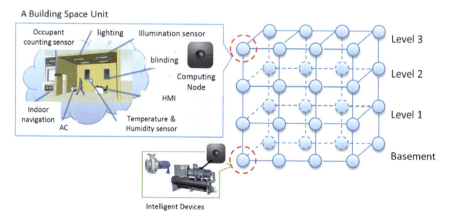

Fig. 1 Structure of the CIS [6]

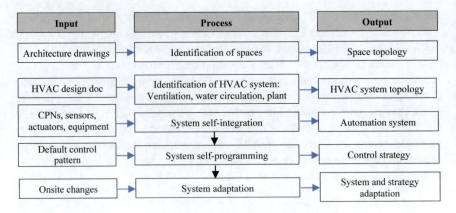

Fig. 2 Identification and integration process based on the CIS concept

the code without project-oriented programming. During the operation phase, the CPN actively monitors the possible control element change and adapts following the above processes. The control system continues updating suggestions about the system adaptation to the administrator (Fig. 2).

3 Results

3.1 Topology Description of Buildings

The CIS uses standard CPNs to control the subspaces of buildings. The distribution of CPNs reflects the space distribution of buildings, and the links among the CPNs reflect the adjacent and linkage relationships among building spaces. The topology of the buildings is stored in the CPN networks in a distributive format. The topology information is the basic for further system integration because the core concept of collective intelligence is to integrate the system based on the building space. Unlike the centralized system, there is no central computer or CPN that has the general information of the entire building. Table 1 provides an analysis of the building topology description based on the CPN network.

Figure 3 shows a typical building layout to explain how the proposed system works. The building spaces are divided into subspaces according to their functions and spaces [7].

Figure 4 describes the topology of the building based on the CPN network installed in building spaces. The solid lines indicate that the building spaces are adjacent to each other, and the dotted lines indicate that they are connected. This information is useful for building automation applications such as indoor navigation and emergency evacuation.

Table 1 Building topology description based on the CIS

Engineering application	Building automation, e.g. HVAC control, lighting control, fire and safety management, business intelligence
Standard unit	Building subspace: Each CPN located in a building subspace represents the building subspace
Relations between standard unit	Adjacent: Two building subspaces share a wall or virtual wall Connected: Two building subspaces have a pass in between
Mapping between CPN and building topology	CPN—Building subspace (room, corridor, atrium, room not connected to corridor) Wire—Neighbourhood between building subspaces Virtual wire—Connectivity between building subspaces
Information stored in CPN	Name of the subspace, position (X-Y-Z), floor, space type (room, corridor, atrium, room not connected to corridor), linkage information (neighbour CPN)
Topology information	Distributed topology information, one CPN only known its neighbour directly. General topology information of the building stored in the CPN network

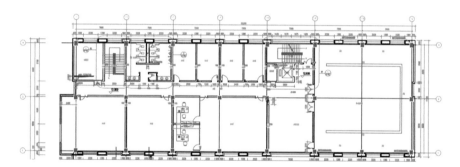

Fig. 3 Typical building layout

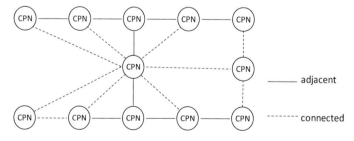

Fig. 4 Topology description of a building based on the CIS

3.2 Topology Description of the HVAC Ventilation System

Table 2 provides an analysis of the topology description of the ventilation system based on the CPN network. Because the ventilation system is located or attached to the building space, a virtual layer of information about the ventilation system is attached to the building topology description stored in the CPN network.

Figure 5 shows a typical ventilation system, where a PAU provides fresh air to the building spaces in Fig. 3. Figure 6 shows the topology description of the ventilation system.

Table 2 Topology description of the ventilation system based on the CIS

Engineering application	Ventilation system control
Standard unit	Ventilation space: Ventilation system is mapped to building subspace
Relations between standard unit	Connected: Two ventilation spaces belong to same ventilation system and connected by ductwork
Mapping between CPN and ventilation topology	CPN—Ventilation subspace (room, corridor, atrium, room not connected to corridor) Wire—Neighbourhood between building subspaces Virtual wire—duct of ventilation system
Information stored in CPN	Name of the ventilation space, ventilation element type (PAU, AHU, VAV-box, diffuser), linkage information (neighbour CPN)
Topology information	Distributed topology information, one CPN only known its neighbour directly. General topology information of the ventilation system stored in the CPN network. Each CPN saves its air pressure, flow rate and resistance

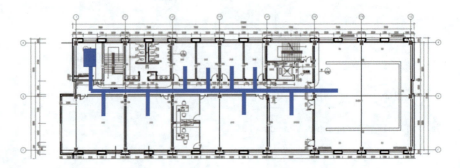

Fig. 5 Typical ventilation system

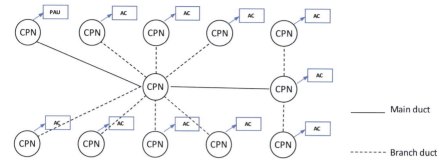

Fig. 6 Topology description of the ventilation system based on the CIS

3.3 Topology Description of the HVAC Water Circulation System

Because the water circulation system is located or attached to the building space, a virtual layer of information about the water circulation system is attached to the building topology description in the CPN network as the ventilation system. Figure 7 shows a typical topology of the water circulation system associated with the ventilation system and building space in Figs. 3 and 5. The information about the water flow resistance (s), flow rate (G) and pressure (P) is distributively stored in the CPN network.

4 Discussion

Acquirement of necessary information: There are two methods to acquire necessary information about the topology information of a building space or HVAC system based on the CIS: (1) the building design information can be downloaded to

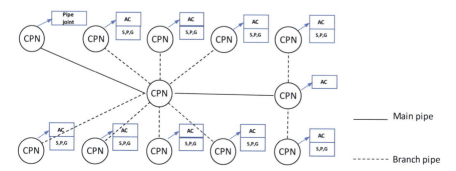

Fig. 7 Topology description of the water circulation system based on the CIS

the CIS. Each CPN knows in which building subspace it is located and the attached HVAC system information to the building subspace. This information can be downloaded to the CIS as soon as the CPNs are installed. (2) The CIS identifies the building space topology using the CPN location and connection information with minimal human configuration help. The HVAC system information comes from the location information, pre-decided rules and human configuration help. The first approach is more common and robust in engineering practice.

Potential optimal control applications: The CIS only requires minimal human interference for the HVAC control system configuration, integration and programming. This feature strongly enhances the reliability of the HVAC control system and reduces the associated cost. Furthermore, because the space topology information and HVAC system information are distributively stored in the CIS, the implementation of system-level optimization becomes more convenient than the traditional centralized system. Many useful optimal control applications using the CIS concept are reported in recent publications [8–12]. Many difficult problems such as the fault diagnosis, supervisory optimal control of a chiller plant and parallel equipment optimal control can be elegantly handled with much less project-specific effort than the traditional centralized control system.

5 Conclusions

This paper has introduced the collective intelligence system (CIS), which can support the self-identification, self-integration, self-programming and self-adaptation of the HVAC control system. The working process of the identification and integration of the CIS system has been described.

To support control applications using the CIS concept, this paper presents a topology description of the building space and HVAC system based on graph theory. The description method was examined by its application to a typical building layout, a water distribution system and an air distribution system. A proper topology description of the space and system will help the system-level control applications of the HVAC control system. The potential of using the collective intelligent control for HVAC system control has been briefly further discussed.

Acknowledgements This work was supported by National Key Research and Development Project of China (No. 2017YFC0704100 entitled New Generation Intelligent Building Platform Techniques). We appreciate Dr. Ziyan Jiang for the helpful discussion.

References

1. Wang, S., Ma, Z.: Supervisory and optimal control of building HVAC systems: a review. HVAC&R Res. **14**(1), 3–32 (2008)

2. Braun, J.E.: Reducing energy costs and peak electrical demand through optimal control of building thermal mass. ASHRAE Trans. **96**(2), 876–888 (1990)
3. Dalamagkidis, K., Kolokotsa, D., Kalaitzakis, K., Stavrakakis, G.S.: Reinforcement learning for energy conservation and comfort in buildings. Build. Environ. **42**, 2686–2698 (2007)
4. Henze, G.P., Dodier, R.H.: Adaptive optimal control of a grid-independent photovoltaic system. ASME J. Solar Energy Eng. **125**, 34–42 (2003)
5. Yu, Z., Dexter, A.: Hierarchical fuzzy control of low-energy building systems. Sol. Energy **84**(4), 538–548 (2010)
6. Zhao, Q., Xia, L., Jiang, Z.: Project report: new generation intelligent building platform techniques. Energy Inf, pp. 1–2 (2018)
7. Ślusarczyk, G., Łachwa, A., Palacz, W., Strug, B., Paszyńska, A., Grabska, E.: An extended hierarchical graph-based building model for design and engineering problems. Autom. Constr. **74**(Supplement C), 95–102 (2017)
8. Dai, Y., Jiang, Z., Shen, Q., Chen, P., Wang, S., Jiang, Y.: A decentralized algorithm for optimal distribution in HVAC systems. Build. Environ. **95**(2016), 21–31 (2016)
9. Wang, S., Xing, J., Jiang, Z., Li, J.: A decentralized sensor fault detection and self-repair method for HVAC systems. Build. Serv. Eng. Res. Technol. **39**(6), 667–678 (2018)
10. Wang, Y., Zhao, Q.: A distributed algorithm for building space topology matching. In: 2018 International Conference on Smart City and Intelligent Building. Springer, Hefei, China (2018)
11. Zhang, Z., Zhao, Q., Yang, W.: A distributed algorithm for sensor fault detection. In: 14th IEEE International Conference on Automation Science and Engineering, pp. 756–761. Munich, Germany (2018)
12. Zhao, T., Zhang, J.: Development of a distributed artificial fish swarm algorithm to optimize pumps working in parallel mode AU-Yu, Hao. Sci. Technol. Built Environ. **24**(3), 248–258 (2018)

Theoretical Analysis of a Novel Two-Stage Compression System Using Refrigerant Mixtures

Zuo Cheng, Baolong Wang, Wenxing Shi and Xianting Li

Abstract Influenced by the international environmental situation, refrigerant mixtures are regarded as potential alternatives in the field of air conditioning and heat pumps. Meanwhile, as an effective means, two-stage system is widely applied to improve the system performance of heat pump under large compression ratio. While according to the previous study, the two-stage compression system using refrigerant mixtures has great potentials for making further efficiency improvement. In this paper, a novel two-stage compression system using refrigerant mixtures is proposed and analyzed and compared with traditional two-stage system. The results show that the heating capacity of the novel system is 4.6–8.9% higher than flash tank system and 1.4–3.6% higher than intermediate heat exchanger system; The COP of the novel system is 6.1–6.4% higher than flash tank system and 3–3.9% higher than intermediate heat exchanger system.

Keywords Refrigerant mixture · Two-stage · Auto-cascade

1 Introduction

Influenced by ozone depletion and global warming crisis, most refrigerants are being phased down owing to their high GWP. While according to the latest authoritative research [1], none refrigerant that can accomplish all requirements at the same time, and the mixed refrigerants provide more flexibility in searching for new alternatives. So refrigerant mixtures are required.

At the same time, the key issue in the heat pump system that has been focused but still needs to be addressed is the poor performance under large pressure ratio. Two-stage or quasi two-stage system is widely considered as good solutions to this problem [2] because it can increase the heating capacity and improve the system

Z. Cheng · B. Wang (✉) · W. Shi · X. Li
Department of Building Science, School of Architecture, Tsinghua University, 100084 Beijing, China
e-mail: wangbl@tsinghua.edu.cn

energy efficiency at low ambient temperature. Ma et al. [3] analyzed the performance of improved quasi two-stage system under different condensing and evaporating temperature, it was founded that this technology can help system run smoothly for a long time when the ambient temperature was near −15 °C and exhibited larger COP than conventional heat pump system. There are also some researchers concentrated in the two-stage system with refrigerant mixtures. Jung et al. [4] simulated the multi-stage heat pumps with a heat exchanger economizer. R22/R142b/R134a, R32/R134a, R125/R134a mixtures were studied. The results indicated that the three-stage heat pump is up to 27.3% more energy efficient than the conventional single-stage with pure refrigerant. Zheng and Wei [5] proposed a novel two-stage system using R290/R600a (50/50) and found that the COP of this novel system is 2.6% higher than flash tank (FT) system. Our previous study [6] analyzed the quasi two-stage system with R32/R1234ze(E) mixtures and found that the system with intermediate heat exchanger (IHX) is better than FT system. Because more low-pressure composition is sucked into compressor, which caused lower evaporating pressure in the FT system owing to the gas-liquid phase separation of refrigerant mixtures in the flash tank. Meanwhile, lower evaporating pressure can brings lower capacity and efficiency. During the gas-liquid phase separation, more high-pressure composition is present in the gas phase and more low-pressure composition exists in liquid phase. In the FT system, liquid phase composition enters into evaporator, while the gas phase composition is injected and mixed with the discharged stream from the first-stage compressor.

The main purpose of this paper is to present a novel two-stage system which can exchange the destination of the liquid and gas fluids in the FT system and reach higher efficiency in the two-stage system using refrigerant mixtures. Then the novel system with the thoughts of auto-cascade is analyzed and compared with the traditional two types of two-stage compression systems.

2 System Description

Except for the traditional evaporator, two-stage compressor and condenser, flash tank and two intermediate heat exchangers (IHXs) contain in the system. When leaving from the condenser, the refrigerant mixture is throttled in expansion (EX) valve-1, the gas-liquid separation happens in the separation pressure of flash tank. Then the liquid fluids are separated to three fluids, the first liquid fluid (6') is mixed with gas fluid (5) (This process is to guarantee the quantity of refrigerant enters into evaporator), the mixed two-phase fluid (7) enters into IHX-1, release heat to the throttled second liquid fluid (14). Through this process, original more low-pressure composition (15) is injected to the intermediate chamber, the mixed fluid (8) goes into next intermediate heat exchanger IHX-2, proceeding the next heat release to the throttled third liquid fluid (12), then the third liquid fluid absorb heat to saturated gas in evaporating pressure (13), and the original mixed fluid goes to subcooled liquid (9), through the EX valve-4, it enters into evaporator, absorb

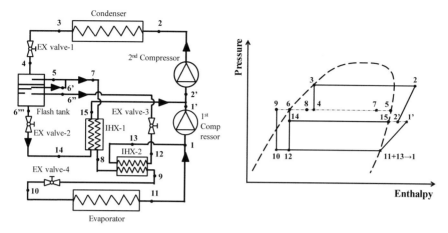

Fig. 1 Schematic and p-h diagram of novel system

heat from ambient to another saturated gas point (11), which will be mixed with previous saturated gas (13) to point 1, and then sucked by the compressor.

It should be noted that through the whole novel cycle, more high-pressure composition exists in evaporator, which increases the evaporating pressure and will improve the whole efficiency (Fig. 1).

3 Method

To simplify the model, the following assumptions are made:

- The superheated degree at the evaporator outlet is fixed at 5 K and the subcooling degree at the condenser outlet is set as 3 K.
- The pinch temperature difference in IHX is set as 2 K.
- The pressure drop in the injection pipe is neglected.
- The injected refrigerant is saturated vapor.
- The throttling process is isenthalpic.

Because the main purpose is to analyze the performance of the novel system, the thermodynamic method is adopted in this research. The working conditions are listed in Table 1. In this research, R32/R1234ze(E) mixtures are chosen as typical refrigerant pairs for its great potential in the future. The composition of 50/50% (Mol) is selected as the charged composition for its great separation characteristic in flash tank.

Table 1 Working conditions of the thermodynamic analysis

Condensing temperature (°C)	Evaporating temperature (°C)	Superheated degree (K)	Subcooling degree (K)
50	−10, −5, 0, 5, 10	5	3

The basic evaluation parameters are listed in the following equations.

$$Q_c = m_{tot} \cdot (h_2 - h_3) \tag{1}$$

$$W = W_{hp} + W_{lp} = m_{hp} \cdot (h_{1'} - h_1) + m_{lp} \cdot (h_2 - h_{2'}) \tag{2}$$

$$\text{COP} = \frac{Q_c}{W} \tag{3}$$

where the heating capacity Q_c is the product of the refrigerant mass flow rate and enthalpy difference at the condenser. The power consumption W is the sum of the high-pressure stage power consumption W_{hp}, and low-pressure stage power consumption W_{lp}. The COP is defined as the ratio of Q_c and W.

The simulation flowcharts of the novel cycle are presented in Fig. 2. In the solving procedure, the evaporating temperature T_e, condensing temperature T_c, superheated degree ΔT_{sup}, subcooling degree ΔT_{sub}, and the refrigerant's charged concentration is input at the beginning. After that, the injection ratio r_{inj} and separation pressure p_{sep} are assumed consecutively because these two parameters can be considered as a design parameter, which exist optimal value to reach the maximum COP. And then the ratio of part of liquid fluid (6") to the total liquid fluid r_{rec} and injection pressure p_{inj} are adjusted, respectively to reach 2 K pinch temperature difference in IHX.

Based on the aforementioned parameters, the mass flow rate of refrigerant flowing into the evaporator to the intermediate loop and the concentrations of refrigerant flowing into the evaporator and the intermediate loop can be obtained. Then, by the low-stage cylinder model, the low-stage mass flow rate m_{lp}, low-stage discharge enthalpy $h_{1'}$, power consumption W_{lp} are all calculated. After mixed with intermediate refrigerant, the suction enthalpy of high-stage cylinder $h_{2'}$ can be gotten. On the basis of high-stage cylinder model, the high-stage mass flow rate m_{hp}, high-stage discharge enthalpy h_2, power consumption W_{hp} and corresponding COP are also obtained. Finally, after scanning all the r_{inj} and p_{sep}, the maximum COP and corresponding heating capacity can be calculated. The REFPROP 9.1 [7] is linked to calculate the thermodynamic properties of the refrigerants. The mixing model [8] applicable for R32 and R1234ze(E) mixtures is adopted.

This section should describe research methodologies, such as theory, numerical and experimental techniques, research procedure, and data analytical approach. Due to limited space available for a conference paper, please use references for general knowledge used. This would allow you to highlight the method particularly used for the investigation. In other words, the method should be sufficient in detail so that other interested could repeat your investigation without problem.

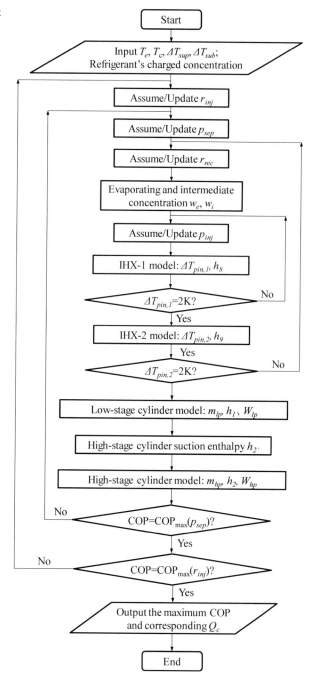

Fig. 2 Simulation flowchart

Because the main purpose is to analyze the performance of the novel system, the thermodynamic method is adopted in this research. The working conditions are listed in Table 1. In this research, R32/R1234ze(E) mixtures are chosen as typical refrigerant pairs for its great potential in the future. The composition of 50/50% (Mol) is selected as the charged composition for its great separation characteristic in flash tank.

4 Results

According to the results, it can be found that the separation pressure p_{sep} is the most effective parameter to the system performance. So in this section, the effect of p_{sep} to the system performance is mainly discussed. After that, the performance of the novel system is compared with traditional FT and IHX two-stage compression system.

4.1 Effect of Separation Pressure

COP and heating capacity of this system among different evaporating temperature (−10–10 °C) is presented in Fig. 3. It can be found that the COP and heating capacity increases with the increase of evaporating temperature. Moreover, with the increase of separation pressure, the heating capacity decreases continuously, which can be well explained by the concentration of refrigerant mixtures presented in Fig. 4. This figure shows the concentration of R32 in evaporator and intermediate loop. It can be obtained the concentration of R32 in evaporator are both higher than 50% (the originally charged concentration) and decreases with the increases of separation pressure because in lower separation pressure, the amount of gas is larger and the high-pressure component R32 is more volatile. On the contrary, the concentration of R32 increases with the increases in separation pressure. With lower concentration of R32 in evaporator, the heating capacity will decrease.

As for the COP, it is closely linked with injection pressure, the injection firstly increases with the increases of separation pressure, then decreases, which causes the corresponding optimal COP.

4.2 Comparison with Traditional Two-Stage System

After analyzing the common performance of this novel system, an additional comparison of this system (DCI) and traditional FT and IHX system is presented in Fig. 5a and b, it can be found that the lower the evaporating temperature, the larger the increasing degree of this novel system compared with another two traditional

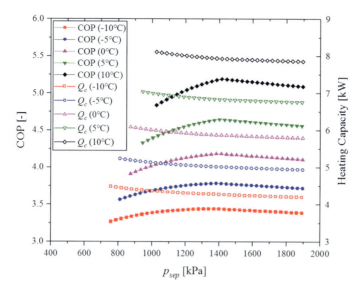

Fig. 3 COP and heating capacity with different separation pressure

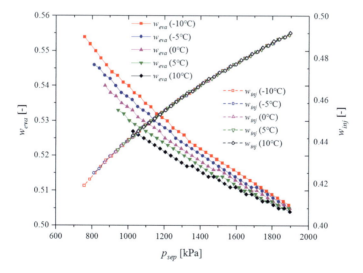

Fig. 4 R32 concentration in evaporator and intermediate loop

systems. For heating capacity, the novel system is 4.6–8.9% higher than FT system and 1.4–3.6% higher than IHX system; For COP, the novel system is 6.1–6.4% higher than FT system and 3–3.9% higher than IHX system.

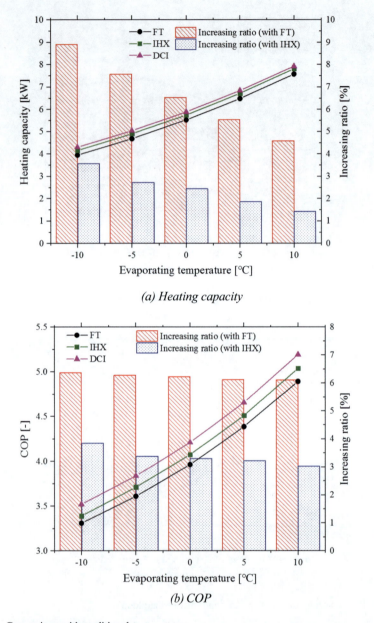

Fig. 5 Comparison with traditional two systems

5 Conclusions

In this paper, a novel two-stage compression system using refrigerant mixtures is proposed and analyzed, the basic operating parameters are presented. Meanwhile, comparison of this system with traditional two-stage system is also carried out. According to the above results, these conclusions are drawn:

- Molar fraction of R32 in evaporator is higher than 0.5 (charged fraction) in this novel system; with the increase of separation pressure, heating capacity decreases, the optimal COP exists;
- The heating capacity of the novel system is 4.6–8.9% higher than FT system and 1.4–3.6% higher than IHX system;
- The COP of the novel system is 6.1–6.4% higher than FT system and 3–3.9% higher than IHX system.

Acknowledgements This research is funded by the China National Key R&D Program "Solutions to Heating and Cooling of Buildings in the Yangtze River Region" (Grant No. 2016YFC0700304) and the National Natural Science Foundation of China (Grant No. 51676104).

References

1. McLinden, M.O., Steven Brown, J., Brignoli, R., Kazakov, A.F., Domanski, P.A.: Limited options for low-global-warming-potential refrigerants. Nat. Commun. **8**, 14476 (2017)
2. Xu, X., Hwang, Y., Radermacher, R.: Refrigerant injection for heat pumping/air conditioning systems: literature review and challenges discussions. Int. J. Refrig. **34**, 402–415 (2011)
3. Ma, G., Chai, Q., Jiang, Y.: Experimental investigation of air-source heat pump for cold regions. Int. J. Refrig. **26**, 12–18 (2003)
4. Jung, D., Kim, H.J., Kim, O.: Study on the performance of multi-stage heat pumps using mixtures. Int. J. Refrig. **22**, 402–413 (1999)
5. Zheng, N., Wei, J.: Performance analysis of a novel vapor injection cycle enhanced by cascade condenser for zeotropic mixtures. Appl. Therm. Eng. **139**, 166–176 (2018)
6. Wang, B., Cheng, Z., Shi, W., Li, X.: Numerical study of gas injected heat pump using zeotropic R32/R1234ze(E) mixture: comparison of two type economizers. Appl. Therm. Eng. **142**, 410–420 (2017)
7. Lemmon, E.W., Huber, M.L., McLinden, M.O.: NIST reference database 23: reference fluid thermodynamic and transport properties-REFPROP, version 9.1. Standard Reference Data Program (2013)
8. Kunz, O., Wagner, W.: The GERG-2008 wide-range equation of state for natural gases and other mixtures: an expansion of GERG-2004. J. Chem. Eng. Data **57**, 3032–3091 (2012)

Optimal Control of Water Valve in AHU Based on Actual Characteristics of Water Valve

Xia Wu, Yan Gao and Bin Wang

Abstract Reasonable operational control of AHU (Air Handler Units) requires the actual working characteristics of the water valve. However, in the current engineering, it is usually based on the simulation of the hypothetical working characteristics of the water valve. The hypothetical characteristics of the equipment are often different from their actual characteristics. As a result, the expected control scheme is not effective in the actual operation. In this paper, applying the operational control scheme for AHU, which was based on the hypothetical characteristics of the water valve, to control the actual water valve, the effects obtained might be different from the expected effects when there were different room load characteristics. Therefore, an optimized method for AHU control based on the actual characteristics of the water valve was proposed according to different room load characteristics.

Keywords Water valve in AHU · Optimal control · Load characteristics

1 Introduction

AHU is an important piece of air handling equipment. Whether it works well or not will affect the thermal comfort of the room. In appropriate control, strategies will lead to energy wasting. So it is meaningful to study how to develop appropriate control strategies for AHU, improving control stability and reducing energy consumption.

D. Xu established a parameter self-tuning fuzzy controller system based on the incremental adjustment principle and designed the rule base [1]. Z. Cao combined neural network and PID control and set parameters online using neural network

X. Wu (✉) · Y. Gao · B. Wang
Beijing University of Civil Engineering and Architecture, Beijing, China
e-mail: wangyiwuxia@163.com

Y. Gao
Beijing Advanced Innovation Center for Future Urban Design, Beijing, China

algorithm [2]. X. Li et al. proposed a room temperature prediction control method, which could improve the stability of the room temperature [3]. Ehsan proposed an online adaptive scheme for adjusting PID parameters [4]. Hamed et al. designed a decoupled PID-Fuzzy controller to ensure stability against AHU regulation based on model parameter uncertainty [5]. Dae et al. developed an evaporation pressure control method to improve comfort and energy efficiency [6]. Sohair used the fuzzy logic system to distribute the flow so that each room could be better controlled [7].

It can be seen that current researches mainly focused on PID controllers. Besides, optimal control strategies were proposed. However, these studies ignored the actual working characteristics of the equipment in AHU, especially water valves', which might have an important impact on the control effects. Therefore, in this paper, the opening-relative flow characteristic of the water valve was measured and established based on the AHU in an air-conditioning system. Then, a simulation platform of AHU was built in MATLAB/Simulink. The optimal control method for the water valve in AHU based on its actual characteristic was studied by using PI control method of return air temperature in two cases with the same design loads but different loads changing.

2 Model Establishment

2.1 Model of the Water Valve in AHU

The flow characteristics of water valves can be divided into linear flow characteristics, equal percentage flow characteristics and quick open flow characteristics. The water valve studied in this paper is a continuous regulating one used in an existing AHU system, which theoretical working characteristic in the user manual is linear flow (see Fig. 1), and the mathematical model in Eq. (1).

$$Q = \alpha \tag{1}$$

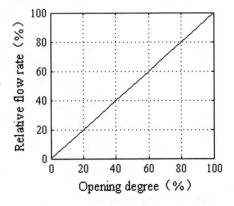

Fig. 1 Hypothetical working characteristic curve of water valve

Fig. 2 Actual working characteristic curve of water valve

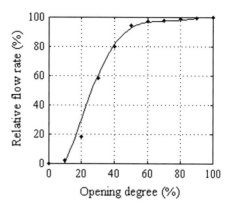

Where Q is the relative flow, which is, the ratio of the real-time flow through the water valve to the maximum flow, %; α is the opening degree of water valve, %.

According to the actual measurement, the relative flow characteristic of the valve can be plotted as the actual working characteristic curve (see Fig. 2). It can be seen that there is almost no flow when the valve opening is below 10%. The range of flow change is mainly concentrated in 10–50%. It is similar to the type of quick opening water valve. The actual mathematical model of water valve obtained by fitting is Eq. (2).

$$Q = 14.40\alpha^6 - 64.40\alpha^5 + 110.39\alpha^4 - 88.95\alpha^3 + 31.25\alpha^2 - 1.62\alpha - 0.07 \quad (2)$$

2.2 Model of the AHU System

A simulation model of sub-AHU system has been built in MATLAB/Simulink, including models of blower and return fan, air valve, mixed air model of outdoor fresh air and indoor return air, cooling coil, room, water valve controller and water valve. These models have been integrated into the AHU system (see Fig. 3).

3 Operation Control and Optimization

The AHU was controlled by the return air temperature control method here. The fan ran at constant frequency. The controller received the room temperature by the return air temperature sensor and compared it with the setting value, 26 °C. According to the deviation, opening of the water valve was adjusted to change the chilled water flow, and then air supply temperature was changed to ensure the room temperature at 26 °C.

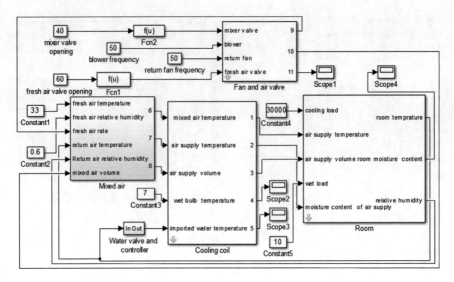

Fig. 3 AHU system model

At present, PI controllers are widely used in air-conditioning systems. The most commonly PI parameters tuning method is experience trial and error method: Increasing the proportional coefficient K_p can reduce the static error. But if the value is selected too much, the stability of the system is reduced; Increasing K_i can reduce the oscillation of the system operation, but the possible negative effect is that the static error elimination time will be longer [8]. In this paper, empirical trial and error method has been used for PI parameters setting.

When the design loads were the same, two cases with different loads changing were studied, corresponding to CASE 1, CASE 2 (see Fig. 4). The valve's opening, air supply, and room temperature, energy consumption have been studied. According to the load value, the design air supply volume was 8800 m3/h, and chilled water was 6000 kg/h.

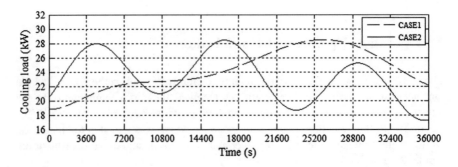

Fig. 4 Cooling load characteristics of CASE 1 and CASE 2

3.1 Case 1

First, the load characteristic of CASE 1 was gentle within 1 h. Under the hypothetical curve of the water valve, after setting, K_p and K_i parameters were determined to be 50 and 0.1, and ideal control effects were obtained. The water valve's opening, supply air, and room temperature changed over time were shown in Figs. 5 and 6. The room temperature was stabilized at 26 °C after about 2000 s.

When the system was running actually, the actual model of the water valve was different from the hypothetical curve. If the actual water valve was controlled according to the hypothetical curve and simulated control parameters, then $K_p = 50$, $K_i = 0.1$, effects could be observed by the following simulations and calculations as shown in Figs. 7 and 8. The room temperature stabilized at 26 °C after about 2000 s.

Comparing control effects under the hypothetical working characteristic curve of the water valve with it under the actual model, there was no significant difference, and the room temperature was well controlled. Therefore, it was considered that the actual model of the water valve did not have a significant effect on control effects

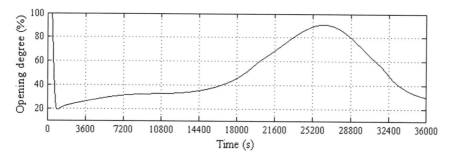

Fig. 5 Opening degree for hypothetical water valve in CASE 1

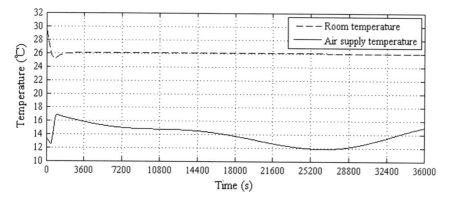

Fig. 6 Air supply and room temperature for hypothetical water valve in CASE 1

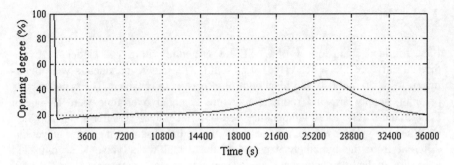

Fig. 7 Opening degree for actual water valve in CASE 1

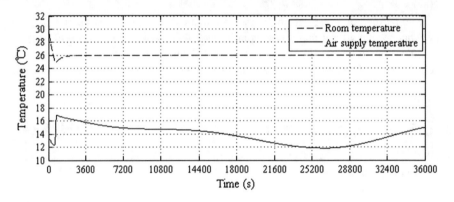

Fig. 8 Air supply and room temperature for actual water valve in CASE 1

under the load characteristic of CASE 1. When the AHU was in actual operation control, PI parameters set under the hypothetical curve of the water valve could also be used.

3.2 Case 2

Then, the load characteristic of CASE 2 varied greatly within 1 h. Under the hypothetical working curve of the water valve, after setting, K_p and K_i parameters were determined to be 10 and 0.3. Ideal control effects were obtained (see Figs. 9 and 10). The room temperature was stabilized at 26 °C after about 5000 s.

The actual water valve was still controlled according to the hypothetical curve and the simulated control parameters, then $K_p = 10$, $K_i = 0.3$. The results were shown in Figs. 11 and 12. At this point, opening degree of the water valve oscillated greatly in the range of 0–100%, especially in the vicinity of 10,000 and 22,000 s. The room temperature fluctuated up and down at 26 °C for a long time,

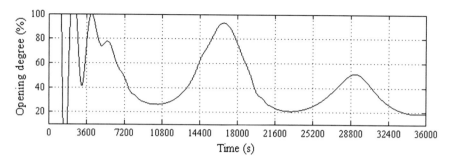

Fig. 9 Opening degree for hypothetical water valve in CASE 2

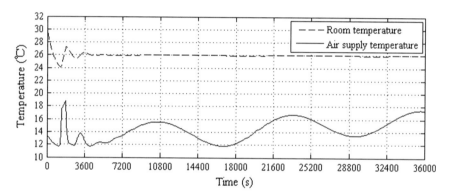

Fig. 10 Air supply and room temperature for hypothetical water valve in CASE 2

and the degree of deviation was ± 3 °C. The actual control effects had a significant deviation from the ideal effects.

According to the analysis, due to the obvious fluctuation of the room load, original control parameters were not effective in controlling the actual water valve. Excessive K_i value led to overshoot. In addition, excessive K_p value also caused oscillation. After reducing parameters, K_p and K_i values were set to 8, 0.01, and the control effects were shown in Figs. 13 and 14. Although the room temperature fluctuated within a small range up and down at 26 °C, the deviation degree was within ± 0.5 °C. The overall control effects were significantly optimized.

Due to the different opening degrees of the water valve, the resistance may be different. So, the energy consumption of transmission may be different as well, which was caused by the change of the end resistance. In order to compare the energy consumption of the water valve before optimization and after, it was calculated according to Eq. (3) [9].

$$N = \frac{\zeta \rho}{2A^2} Q^3 \qquad (3)$$

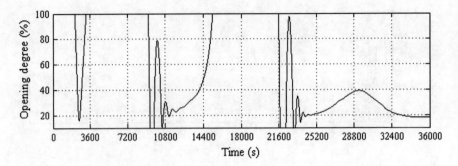

Fig. 11 Opening degree for actual water valve in CASE 2

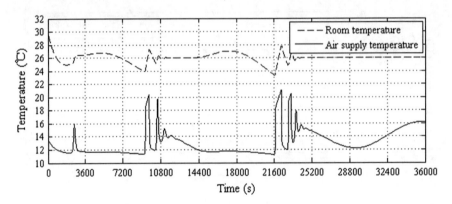

Fig. 12 Air supply and room temperature for actual water valve in CASE 2

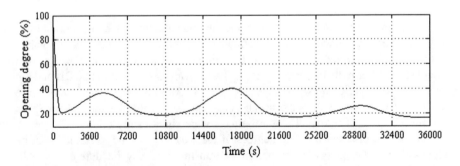

Fig. 13 Opening degree for actual water valve in CASE 2 after optimization

where N is energy consumption of the water valve, W; Q is water flow rate, m³/s; ζ is the water valve resistance coefficient, value here by 15; ρ is fluid density, kg/m³; A is the sectional area of the pipeline, which is 0.00126 m².

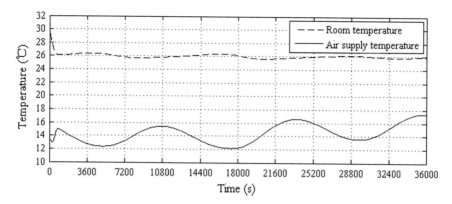

Fig. 14 Air supply and room temperature for actual water valve in CASE 2 after optimized

The calculation results showed that before optimized, the average room temperature in 36,000 s was 26 °C, and the energy consumption of the water valve was 12.40 W. A after optimized, the average room temperature in 36,000 s was still 26 °C, and the energy consumption of the water valve was 8.90 W. It was reduced by 28.23% while the average room temperature was constant.

4 Conclusions

If the control parameters were simulated according to the hypothetical linear working curve of the water valve and implanted into the field controller to control the water valve with a distinct field working curve, it might cause the oscillation of the water valve and large fluctuation of the room temperature. After reducing PI control parameters properly, the control effects were optimized. Therefore, due to the different actual working characteristics of the water valve installed, reasonable control parameters will change when there are different load requirements, which need to be determined after detailed simulation before implantation.

Acknowledgements The authors wish to thank support from National Key Research and Development Plan (2016YFC0700703) and High Level Innovation Team of Beijing Municipal Education Commission (Grant No. IDHT20180512).

References

1. Xu, D.: Application and research on parameter auto-tuning fuzzy-pid control technique based on self-learning technique. Master Degree Dissertation, Chongqing University, Chongqing (2006)

2. Cao, Z.: Research on air-conditioning system control strategy based on artificial neural network. Master Degree Dissertation, XiHua University, Chengdu (2007)
3. Li, X., et al.: Predication control for indoor temperature time-delay using Elman neural network in variable air volume system. Energy Build. **154**, 545–552 (2017)
4. Ehsan, G.N., et al.: Online adaptive robust tuning of PID parameters. Proc. IFAC **45**(3), 625–630 (2012)
5. Harmed, M., et al.: PID-Fuzzy control of air handling units in the presence of uncertainty. Int. J. Therm. Sci. **109**, 123–135 (2016)
6. Dea, K.L., et al.: Method to control an air conditioner by directly measuring the relative humidity of indoor air to improve the comfort and energy efficiency. Appl. Energy **215**, 290–299 (2018)
7. Sohair, F.R., et al.: Management of air-conditioning systems in residential buildings by using fuzzy logic. Alex. Eng. J. **54**, 91–98 (2015)
8. Yang, B., et al.: Composite tuning method for partition PID control parameters. Autom. Instrum. **7**, 44–49 (2016)
9. Fu, Y., et al.: Energy consumption of valves in conventional water systems and construction and application of separate power systems. Hv Ac **35**(9), 6–10 (2005)

Study on Heat Recovery Air Conditioning System with Adsorption Dehumidification by Solar Powered

Yi Liu and Liu Chen

Abstract Air conditioning system with solid wheel adsorption dehumidification has advantages of environmental protection and energy-saving. It mainly processed fresh air. Aiming at the problem of high consumption for regenerative heating in the conventional system, this paper proposes a heat recovery air conditioning system with adsorption dehumidification driven by solar energy, which mainly processes fresh and returns air. Three types of heat recovery schemes are studied. According to the weather conditions of 7 humid regions in China in summer, the operation plans and energy consumption of the system are discussed. The results show that the typical air conditioning heat recovery system with two-stage heat recovery of exhausted air is the most energy-saving, and the payback period of the system is only 4 years; the total energy consumption of the front-heat recovery system is lower than the back-heat recovery system.

Keywords Solar · Desiccant wheel · Heat recovery · Energy consumption

1 Introduction

Adsorption dehumidification air conditioning system with a desiccant wheel uses a non-polluting combination of working fluids. It also can be driven by low-grade heat resources, such as solar energy, industrial waste heat. It has more development prospects in dehumidification air conditioning systems because it addresses the problems caused by the combined treatment of heat and humidity in traditional condensing dehumidification [1, 2] Goodarzia et al. [3] analyzed the feasibility of systems involving desiccant wheels use solar thermal energy or waste heat to regenerate desiccant material. Sumathy et al. [4] studied the performance and

---

Y. Liu · L. Chen (✉)
School of Energy, Xi'an University of Science and Technology, 710054 Xi'an, China
e-mail: chenliu@xust.edu.cn

Y. Liu
e-mail: liuyipeace@163.com

energy consumption for the combined system of the rotating dehumidification equipment and the traditional compression refrigeration air conditioning. Fong [5] pointed that the dehumidification cooling air conditioning system driven by solar was more energy efficient than the traditional air conditioning system, and experimentally tested the proposed three fresh air systems in Hong Kong.

Adsorption dehumidification air conditions usually process fresh air. The heat recovery air conditioning system with adsorption dehumidification by solar-powered was proposed, which simultaneously processes fresh air and return air. In order to optimize the system, the influence of the heat recovery on the energy consumption of the system is also studied.

2 Heat Recovery Air Conditioning System with Adsorption Dehumidification by Solar Powered

The typical heat recovery air conditioning system with adsorption dehumidification by solar-powered compared with the conventional adsorption desiccant air conditioning system, the following energy-saving optimization: (1) the plate-type heat exchanger is added in front of the desiccant wheel to precool air. The process air is first cooled and then dehumidified, which can reduce the desiccant temperature to increase the water vapor partial pressure difference of the desiccant surface air. Regeneration temperatures are reduced [6]. (2) The plate-type heat exchanger is added behind the desiccant wheel to reduce the temperature of the processed air after dehumidification. In the process of dehumidification, the air temperature increases due to the heat dissipated during the adsorption process. Heat exchanger is added behind the desiccant wheel to recover the heat of adsorption. (3) Exhausted air is a regeneration air, which reduces the moisture content of the regenerated air and increases the partial pressure difference with the water vapor on the surface of the desiccant. It makes desorption more effective [7].

According to the different heat recovery positions, the heat recovery air conditioning system with adsorption dehumidification by solar-powered is divided into three types: the typical heat recovery system; the front-heat recovery system; the back-heat recovery system.

2.1 The Typical Heat Recovery System

Schematic diagram of the typical heat recovery systems is shown in Fig. 1. First, heat transfer between low-temperature exhausted air and process air in the front-heat exchanger. Next, in back-heat exchanger, the air heat exchanges with air which has been dehumidified. This reached the two-stage heat recovery of the exhausted air. The regenerative heat comes from solar energy and the cold water of the air cooler comes from natural cold sources.

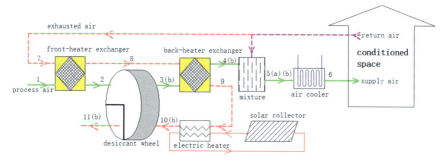

Fig. 1 Schematic diagram of the typical air conditioning system with adsorption dehumidification by solar-powered

The enthalpy-humidity chart of the typical heat recovery system is shown in Fig. 2a. (1) The process air (point 1), which is first sensible cooled to the point 2 by the front-heat exchanger. Then, the processed air passes through the process side of the desiccant wheel, where air is dehumidified to the point 3 by adsorption of water vapor from the desiccant wheel. Next, the air heat exchange with an exhausted air in the back-heat exchanger, where air is sensibly cooled to the point 4. The processed air is mixed with return air (point 7) to point 5 in the mixture. Finally, air is cooled by the air cooler to an appropriate temperature; then, the air at point 6 is supplied to the conditioned space. (2) The regeneration air is heated to point 9 by two-stage heat recovery, and then is heated to require regeneration temperature of point 10 by the solar collector. This high-temperature air passes through the regeneration side of the desiccant wheel, where the wet desiccant is regenerated by desorption of water, and the generation air is humidified. The regeneration heat can be heated by electric heaters when solar collectors provide insufficient heat.

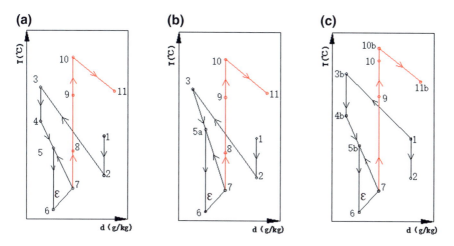

Fig. 2 Enthalpy-humidity chart of the heat recovery air conditioning system

2.2 The Front-Eat Recovery System

Schematic diagram of the front-heat recovery system is shown in Fig. 1 which removes the back-heat exchanger. The air treatment on side of process air is shown in Fig. 2b, 1 → 2→3 (mixed with 7) → 5a → 6. After dehumidification, the high temperature and low humidity air are directly mixed with the indoor return air. The process is as shown in Fig. 2b, 7 → 8→10 → 11.

2.3 The Back-Heat Recovery System

Schematic diagram of the back-heat recovery system is shown in Fig. 1 which removes the front-heat exchanger. The air treatment on side of process air is shown in Fig. 2c, 1 → 2→4b (mixed with 7) → 5b → 6. The air treatment on side of regeneration air is shown in Fig. 2c, 7 → 9→10 → 10b → 11b. The regeneration temperature of point 10 cannot meet the regeneration requirements because the process air is not pre-cooled before dehumidification. So the regeneration air is heated by the electric heater and then sent to the desiccant wheel.

3 Mathematical Model of Energy Consumption

3.1 The Energy Consumption of Regeneration Heat(E_r)

$$Q_r = E_r \tag{1}$$

$$Q_r = m_r c_r (t_c - t_d) \tag{2}$$

where, Q_r is the heat regeneration heat, kW; t_d, t_c are the inlet and outlet air temperatures of the regenerative electric heater, °C; c_r is the specific heat of regeneration air, kJ/(kg k);

3.2 The Energy Consumption of Cooling Device (E_c)

$$E_c = Q_c / \text{COP} \tag{3}$$

$$Q_c = m(h_b - h_a) \tag{4}$$

where, Q_c is the heat load of the air cooler, kW; COP is the coefficient of performance of the air cooler; h_b, h_a is the air inlet and outlet enthalpy of the air cooler, respectively, kJ/kg.

3.3 Total Energy Consumption (E)

If the electric energy of the pump and fan and the runner drive is not counted, the total energy consumption is defined as:

$$E = E_c + E_r \qquad (5)$$

4 Energy Analysis

Three types of heat recovery air conditioning systems with adsorption dehumidification are applied to seven typical cities in China, which based on a typical building. Take the first floor of a shopping mall as an example, the building area is 400 m² and the indoor cooling load index is 200 w/m². The indoor design temperature is 26 °C, and the air supply temperature difference is 6 °C. Supply air volume is 6.78 kg/s and the fresh air volume is 2.76 kg/s.

The operation conditions were as follows: the high-temperature chiller for dehumidification air conditioning system COP_c is 6.34, the low-temperature chiller for the conventional air conditioning system with electric compression COP is 4.6 [8]. Heat transfer efficiency η is 0.6 [9]. The desiccant wheel inlet air of the typical and the front-heat recovery systems is pre-cooled. So the regeneration temperature is only required to be 80 °C. The regeneration temperature of the back-heat recovery system without pre-cooling is required to be 100 °C [10]. The solar collector provides 80 °C of regenerative hot air for the desiccant wheel [11] which is supplemented by regenerative electric heater when the heat is insufficient.

4.1 Outdoor Meteorological Parameters

Seven cities are selected in humid areas. The outdoor meteorological parameters of each city are shown in Table 1 [12].

4.2 State Point Solution

The air handling process of each system is known (as shown in Fig. 2). Haikou is taken as an example. The point solution results are shown in Table 2.

Table 1 Outdoor meteorological parameters of each city

Parameter	Tg (°C)	RH (%)	H (kJ/kg)	dg (g kg^{-1})
Harbin	30.7	62	74.9	17.2
Beijing	33.5	61	84.7	19.9
Chengdu	31.8	73	87.5	21.7
Changsha	35.8	61	94.3	22.7
Guangzhou	34.2	68	93.9	23.1
Shanghai	34.4	69	90.5	23.8
Haikou	35.1	68	98.6	24.6

4.3 The Results of Energy Analysis Calculation

The values of each state point are obtained and substituted into the energy consumption model. The energy consumption results of each city are shown in Table 3.

4.4 Energy Saving Evaluation

The total energy-saving rate was compared to the energy savings of each heat recovery system and the conventional air conditioning system with electric compression. The relationship between the total energy saving rate (γ) and the moisture content of each city is shown in Fig. 3.

Table 2 shows the total energy consumption of the three types of heat recovery air conditioning systems with adsorption dehumidification by solar-powered is

Table 2 State point parameter value of Haikou

Parameter	Tg (°C)	RH (%)	dg (g kg^{-1})	h (kJ/kg)
1	35.1	68	24.4	98.0
2	29.6	92.8	24.4	92.2
3	67.4	5.3	9.2	92.2
4	45.9	14.8	9.2	70.1
5	32.4	37.1	11.2	62.2
6	20	71.3	11.2	46.4
7	26	60	12.5	58.2
8	31.5	59.8	12.5	63.8
9	53	14	12.5	86
10	80	4.2	12.5	113.9
5a	41.8	22.2	11.2	71.9
3b	73	4.2	9.2	98
4b	44.8	15.7	9.2	69
5b	33.3	34.6	11.2	63.1
10b	100	0.8	12.5	134.6

Table 3 Energy consumption of each city (Unit: KW)

Parameter	Conventional		Typical		Front		Back	
	E_c	E	E_c	E	E_c	E	E_c	E
Harbin	28.7	34.8	13.8	13.8	18.7	18.7	13.9	32.5
Beijing	34.6	40.7	15.7	15.7	22.3	22.3	15.0	33.6
Changsha	40.4	46.5	17.2	17.2	25.7	25.7	16.6	35.2
Chengdu	38.4	42.5	15.8	15.8	23.0	23.0	16.0	34.6
Shanghai	41.3	47.4	16.8	16.8	26.4	26.4	18.0	36.6
Guangzhou	40.1	46.2	16.5	16.5	24.3	24.3	16.0	34.6
Haikou	42.6	48.7	16.7	16.7	27.2	27.2	17.1	35.7

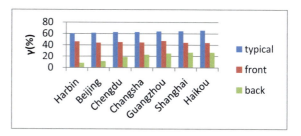

Fig. 3 Energy saving rate of each heat recovery system and the conventional system

lower than the conventional system in the same city. The typical heat recovery system is most energy-saving. The total energy consumption of the front-heat recovery system is lower than the back-heat recovery system. A major reason is that the regeneration air needs to be heated by regenerative electric heater from 80 to 100 °C, increasing the energy consumption of regeneration heating of 18.6 kW.

Figure 3 shows that in cities with low outdoor air humidity, such as Beijing and Harbin, the energy-saving rate of typical system is around 60%, and the energy-saving rate of back-heat recovery system is about 10%. In areas with high outdoor humidity in Shanghai, Changsha, Guangzhou, Haikou, and Chengdu, the energy-saving rate of typical systems is about 65%, and the energy-saving rate of post-heat recovery systems is around 20%. The energy-saving rate of the front-heat recovery system is about 45%, which is not much different in cities with different humidity.

4.5 Economic Analysis

Taking Haikou City as the representative, the economics of three types of heat recovery systems compared with traditional electric compression systems were

Table 4 Energy consumption of each city

Parameter	Conventional	Typical	Front	Back
Solar collector area (m²)	–	60	110	60
Power consumption (kW)	48.7	16.7	27.2	35.7
Initial cost (RMB)	45,690	70,500	85,500	66,500
Operation cost (RMB)	13,149	4509	7344	9639
payback period (year)	–	4	9	8

analyzed. According to the hourly solar radiation intensity relation of the corresponding maximum load, the solar energy area required by the system is defined as:

$$S = Q_r(\eta_{he} \times \eta_{sc} \times I) \quad (6)$$

where, Q_r is the heat of regeneration, kW; η_{he}, the efficiency of steam-water heat exchanger, 0.65; η_{sc} is the efficiency of a solar collector, 0.65; I is the average solar radiation intensity of the maximum load, 1350 (w/m²) [13].

The evaluation index of system economy includes initial investment, operation cost and payback period. In Haikou, it is assumed that the system runs at full load for 300 h per year. The average industrial electricity price is 0.9 RMB/kWh. The dynamic payback period which considered the time value of capital is evaluated. The dynamic payback period n is defined as:

$$n = \log_{(i+1)} \frac{\Delta P}{\Delta P - \Delta R \cdot i} \quad (7)$$

where, ΔP is the recovery of the capital for the system to save electricity, RMB; ΔR is the initial cost added to the system, RMB; i is the discount rate, which should not be lower than the bank's interest rate, 5.9% [14] (Table 4).

The data show that the initial cost of the typical system is high, but the operation cost is low, and the payback period of the system is only 4 years. The regeneration air of the front-heat recovery system only passes through the front heat recovery. The payback period of the system is increased to 9 years because the solar collector plate area of the system is increased compared to the typical system. The initial cost of the post-heat recovery system is only a heat recovery device, but it has a high operation cost and its recovery period is 8 years.

5 Conclusion

In the study, a heat recovery air conditioning system with adsorption dehumidification by solar-powered was proposed, which is divided into three types. By comparison with the conventional air conditioning system with electric compression, the typical air conditioning heat recovery system with two-stage heat recovery of exhausted air is the most energy-saving, and the payback period of the system is only 4 years; the total energy consumption of the front-heat recovery system is lower than the back-heat recovery system.

Acknowledgements This work is supported by the Natural Science Foundation of China (NSFC) (Number 51176104).

References

1. Yutang, F., et al.: Review on adsorbent materials of rotary-type dehumidifier. Chem. Ind. Eng. Prog. (2005)
2. JEONG, J., et al.: Performance analysis of four-partition desiccant wheel and hybrid dehumidification air-conditioning system. Int. J. Refrig. [s. l.] **33**, 496–509 (2010)
3. Goodazria, G., et al.: Performance evaluation of solid desiccant wheel regenerated by waste heat or renewable energy. Energy Procedia. [s. l.] **110**, 434–439 (2017)
4. Sumathy, K., et al.: Study on a novel hybrid desiccant dehumidification and air conditioning system. In: Sustainability in Energy and Buildings (2009)
5. Fong, K.F., et al.: Advancement of solar desiccant cooling system for building use in subtropical Hong Kong. Energy Build. **42**(12), 2386–2399 (2010)
6. Chen, S.H., et al.: Energy saving measures and energy consumption analysis of hybrid air condition system using desiccant wheel, Cryo. Supercond. **46**(4), China (2018)
7. Ge, T.S.: Study on rotary two-stage dehumidification air conditioning, China (2008)
8. T. Shi, et al. Humidity control and cooling ceiling (1998)
9. Yang, C.Z., et al.: Analysis of the system COP of different THIC air conditioning systems in humid regions in summer season. J. Hunan Univ. (Nat. Sci.). **43**(5), 144–150, China (2010)
10. Chen, L.: Temperature and humidity independent control system applied to deep well cooling, Met. Mine. **32**(10), 133–137, China (2014)
11. Ding, Y.F., et al.: Dedicated outdoor air systems by rotary wheel removing moisture load based on solar energy regeneration. Fluid Mach. **34**(8), 63–66, China (2006)
12. GB 50736-2012 Code for design of heating, ventilation and air conditioning for civil buildings, pp. 102–177. China Building Industry Press, Beijing (2012)
13. Zhang, X.B., et al.: Assessment and application analysis of solar energy resources in Haikou, vol. 4, pp. 125–126. Science and Technology Innovation Herald, China (2012)
14. Ge, T.S., et al.: Feasibility study of solar driven two-stage rotary desiccant cooling system, China (2009)

Study on the Optimal Cooling Power for the Internally Cooled Ultrasonic Atomization Dehumidifier with Liquid Desiccant

Ruiyang Tao, Zili Yang, Yanming Kang and Zhiwei Lian

Abstract In this work, performance simulations on the internally cooled ultrasonic atomization system (IC-UADS) were carried out to clear the effects of the input cooling power on promoting the dehumidification performance. The performance model was established based on the conservation laws of mass and energy, together with the sensible heat balance equation, and then well-validated within the constructed IC-UADS. Three indices, namely the dehumidification effectiveness, the moisture removal rate, and the specific moisture removal rate per cooling power were employed. It was found that though the dehumidification performance was growing continuously with the rise of cooling power, the increase rate was descending significantly. Considering the balance between the cooling power consumption and the dehumidification performance improvement, the optimal power input, around 0.53 kW for the current IC-UADS, was figured out for energy-efficient performance improvement. The results may help clarify the necessary cooling power input and optimize the operation of the internally cooled liquid desiccant systems.

Keywords Internal cooling · Ultrasonic atomization · Liquid desiccant · Optimal cooling · Dehumidification

---

R. Tao · Z. Yang · Y. Kang
Department of Civil and Energy Engineering, College of Environmental Science and Engineering, Donghua University, Shanghai, China
e-mail: ruiyangtry@163.com

Y. Kang
e-mail: ymkang@dhu.edu.cn

Z. Yang (✉) · Z. Lian
Department of Architecture, Shanghai Jiao Tong University, Shanghai, China
e-mail: ziliy@dhu.edu.cn

Z. Lian
e-mail: zwlian@sjtu.edu.cn

Nomenclature

C_p	Specific thermal capacity, (kJ/(kg K))
d	Humidity ratio, (g/kg dry air)
G	Mass flow rate, (kg/s)
h	Enthalpy, (kJ/kg)
Mol	Molar mass, (g/mol)
t	Temperature, (°C)
n	Desiccant mass fraction, (%)
p	Pressure, (Pa)

Subscripts

a	Air
c	Cooling
d	Dry air
equ	Equilibrium
i	Inlet
l	Liquid desiccant
o	Outlet

1 Introduction

Liquid desiccant dehumidification system (LDAC), which has presented great potentials for energy saving of buildings [1–3], has been drawing plenty of attention. Within the system, the dehumidifier, where the air moisture is absorbed directly by the desiccant solution, is playing a vital role in the system's overall performance. Also, the dehumidification performance of the LDAC can be enhanced substantially via the internal cooling [4, 5]. Numerous studies were conducted to clarify the dehumidification performance of the system under the effects of various operating conditions, such as the inlet parameters of the airstream and the desiccant solution [6, 7]. However, it is still unclear if there exist any optimal ranges of the input internal-cooling power. In view of this, this work conducted simulations via a well-validated model to investigate the dehumidification performance of the IC-UADS under various cooling powers. The results may help figure out the optimal cooling input and optimize the operation of the internally cooled LDAC.

2 Methods

2.1 The IC-UADS System

In this paper, an internally cooled ultrasonic atomization liquid desiccant dehumidification system (i.e., the IC-UADS) was constructed. The system schematic is shown in Fig. 1. As can be seen in Fig. 1, the IC-UADS consists of four parts, namely, the ultrasonic atomization system, the air handling system, the desiccant solution system, and the internal cooling system. With the ultrasonic atomization system, the desiccant solution can be atomized into numerous fine droplets (with diameter around 50 μm, according to the manufacturer). Thus, the specific contact area between the liquid desiccant solution was expanded drastically, and good dehumidification performance was presented [8]. More detailed information and the specifications of the UADS system can be found in the authors' previous work [8].

2.2 Performance Indices

Three indices, namely the moisture removal rate, the dehumidification effectiveness, and the specific moisture removal rate per cooling power, were employed here to evaluate the dehumidification performance of the IC-UADS under various cooling inputs.

The dehumidification effectiveness (DE) is assessed via the practical air humidity ratio change to its possible decrease to the equilibrium value of the initial liquid desiccant solution, as shown in Eq. (1).

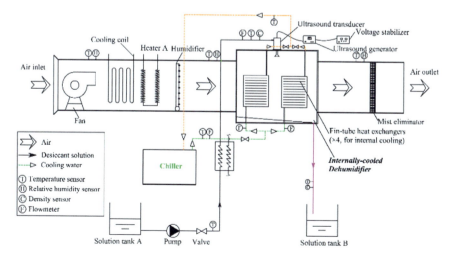

Fig. 1 System setup of the IC-UADS

$$DE = \frac{d_{a,i} - d_{a,o}}{d_{a,i} - d_{equ}} \times 100\% \quad (1)$$

The moisture removal rate (MRR) is defined as Eq. (2) and is to evaluate how much moisture was removed from the humid air per second.

$$MRR = G_{a,i} \times (d_{a,i} - d_{a,o}) \quad (2)$$

The specific MRR per cooling power input (SpcMRR) can be defined as the ratio of MRR to the consumed cooling power and is to value the efficiency of the power input on promoting the dehumidification performance, as shown in Eq. (3):

$$SpcMRR = \frac{MRR}{CP_{in}} \quad (3)$$

2.3 The Performance Prediction Model

To predict the dehumidification performance of the IC-UADS, a model based on the conservation law of mass and energy, as well as the sensible heat transfer between the internal cooler to the airstream and desiccant solution, was established. The model's governing equations can be separately described as Eqs. (4–6):

$$G_{a,d} \cdot (d_{a,i} - d_{a,o}) = [G_l + G_{a,d} \cdot (d_{a,i} - d_{a,o})] \cdot (1 - n_{l,o}) - G_{a,d} \cdot (1 - n_{l,i}) \quad (4)$$

$$G_{a,d} \cdot h_{a,i} + G_l \cdot h_{l,i} + G_w \cdot h_{w,i} = G_{a,d} \cdot h_{a,o} + [G_l + G_{a,d} \cdot (d_{a,i} - d_{a,o})] \cdot h_{l,o} + G_w \cdot h_{w,o} \quad (5)$$

$$C_{p,l} \cdot (G_{l,o} \cdot t_{l,o} - G_{l,i} \cdot t_{l,i}) - C_{p,a} \cdot (G_{a,o} \cdot t_{a,o} - G_{a,i} \cdot t_{a,i}) - Q_{c,w} = 0 \quad (6)$$

where $Q_{c,w}$ was the input cooling power and other symbols can be found in the Nomenclature. Meanwhile, owing to the well-atomized desiccant droplets, which were flowing together with the airstream as parallel flow, the vapor pressure of the desiccant droplets may reach a balance with the airstream in the same position, as depicted by Eq. (7);

$$p_{l,o}(n_{l,o}, t_{l,o}) = \frac{Mol_{a,d} \times d_{a,o} \times p_{AT}}{Mol_{a,d} \times d_{a,o} + Mol_{a,q}} \quad (7)$$

In addition, the characteristics equations of the airstream and the desiccant solution can be expressed in Eqs. (8–11). Then, by combining the governing

Table 1 Operation conditions for the experimental study

Parameters	$G_{l,i}$	$t_{l,i}$	n_i	$G_{a,i}$	$t_{a,i}$	d_i	$V_{w,i}$	$t_{w,i}$	$G_{l,i}/G_{a,i}$
Unit	kg/h	°C	%	kg/h	°C	g/kg	L/min	°C	–
Nominal condition	48	25	26	92	33	18	4.3	18	0.5

equations and the characteristics equations, the outlet parameters of the dehumidifiers can be figured out. Thus, the dehumidification performance can be obtained by Eqs. (1–3) shows.

$$h_{a,i} = 1.01 t_{a,i} + d_{a,i}(2501 + 1.85 t_{a,i}) \tag{8}$$

$$h_{a,o} = 1.01 t_{a,o} + d_{a,o,ideal}(2501 + 1.85 t_{a,o}) \tag{9}$$

$$h_{l,i} = f(t_{l,i}, n_{l,i}) \tag{10}$$

$$h_{l,o} = f(t_{l,o}, n_{l,o}) \tag{11}$$

To validate the feasibility and accuracy of the established model, experimental runs with the nominal conditions shown in Table 1 were carried out within the IC-UADS. The validation results can be introduced in the following section.

3 Results and Discussion

3.1 Model Validation

In this study, 58 experimental runs were carried out to validate the prediction model. Then performance simulations were conducted for various cooling power to clarify the possible energy-efficient cooling input. The validation results are shown in Fig. 2.

As can be seen in Fig. 2, good consistency was achieved between the experimental results and the predicted performance. In the present study, the average deviation between the predicted MRR and the experimental results was merely 1.6% with the maximum difference of most cases less than 15%. In the meantime, the predicted DE was in line with the experimental ones with an average deviation of less than 3.9%. This manifests that the established prediction model is of good reliability and accuracy in predicting the dehumidification performance of the IC-UADS.

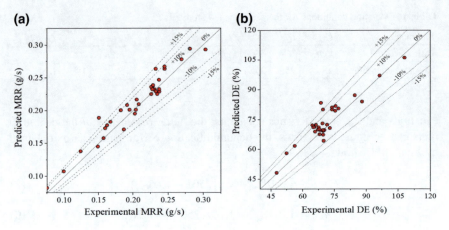

Fig. 2 Model validation with experimental data **a** MRR and **b** DE

3.2 Effects of Cooling Power on the Dehumidification Performance

The dehumidification performance of the IC-UADS was investigated with the proposed prediction model under the effects of the cooling power input. The results can be illustrated and discussed as follows.

Under nominal conditions of airstream and desiccant solution
Figure 3 shows the dehumidification performance under various cooling power. As evident in the figure, both the MRR and DE were increasing continuously, well-fitted as exponential trends, with the growth of the cooling input into the

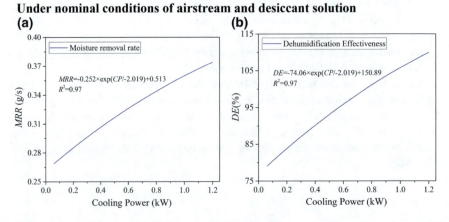

Fig. 3 Dehumidification performance of the IC-UADS **a** MRR and **b** DE under various cooling power

IC-UADS. This is owing to the effective removal of the heat released by the desiccant solution when absorbing the moisture from the humid air. As the consequence, the temperature rises of the desiccant solution, as well as the airstream, is curved. With the continuous rise of the cooling power, the airstream and the desiccant solution may even be cooled inside the dehumidifier. This benefits the stronger dehumidification capability of the desiccant solution and better dehumidification performance of the IC-UADS.

However, with the further rise of the input cooling power, the growth rate is descending significantly. Besides, considering the extra energy cost of the cooling source, it is necessary to figure out a potentially optimal value of the cooling power to realize the balance between the cooling power cost and the performance improvement of the IC-UADS.

To clarify the efficiency of the cooling power on promoting the dehumidification performance, the SpcMRR, defined in Eq. (3) as the ratio of the MRR and the exhausted cooling power, was figured out for the IC-UADS and illustrated in Fig. 4a. As can be seen in Fig. 4a, with the growth of the input cooling, the SpcMRR was dropping fast. This indicates that a shrinking contribution from the internal cooling power to improve the dehumidification performance. As to the present IC-UADS, the SpcMRR decreased to a relatively low status when cooling power (CP) exceeded 0.8 kW and seems leveled off even though the CP reached 1.2 kW. This manifests that there may exist an optimal power input to realize energy-efficient performance improvement of the IC-UADS.

To clarify this, the slope of the SpcMRR trends (i.e., derivative of SpcMRR concerning cooling power) was plotted to depict the significance of the internal cooling on performance promoting effects, as shown in Fig. 4b. To assess the cooling power's significance on the SpcMRR, slope ≤ -1 was adopted in this work as the judging criteria. As displayed in Fig. 4b, the criteria cooling power for the present IC-UADS was around 0.53 kW, under which the most cost-efficient

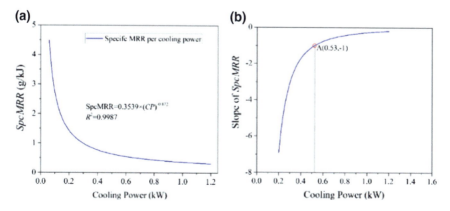

Fig. 4 Trends of specific MRR under per cooling power (**a**) and its slope (**b**)

Table 2 SpcMRR and performance improvement under various cooling power

Internal-cooling power (kW)	SpcMRR (g/kJ)	MRR (g/s)	DE (%)	Improvement (MRR or DE, compared to adiabatic)
0 (Adiabatic)	–	0.20	58.90	–
0.53 (Cost-effective)	0.60	0.32	93.91	59%
1.2 (Maximum studied)	0.31	0.37	109.96	86%

Under varying desiccant mass fraction

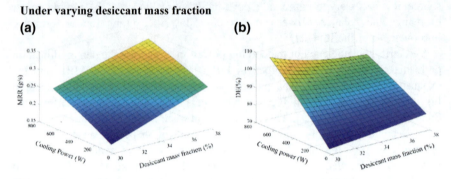

Fig. 5 Effects of cooling power on the dehumidification performance of the IC-UADS with varying desiccant mass fraction

performance improvement of the IC-UADS may be obtained via internal cooling. Its improvement can be summarized and shown in Table 2.

Under varying desiccant mass fraction

Figure 5 illustrated the dehumidification performance of the IC-UADS under the effects of cooling power with different desiccant mass fractions. As presented in Fig. 5, both MRR and DE were increasing considerably with the rise of the input cooling power under various desiccant mass fractions. To analyze the cost-effectiveness of the cooling power with varying desiccant mass fractions, the SpcMRR and its slope were figured out and shown in Fig. 6.

As can be seen in Fig. 6a, The SpcMRR was decreasing significantly with the rise of the cooling power input under various desiccant mass fractions. Meanwhile, the impact has been exerted by the desiccant mass fraction seems to be limited. With the slope trends shown in Fig. 6b, it can be referred that the varying rate of the SpcMRR appears decreasing significantly with the cooling power but remain stable under various desiccant mass fractions. This manifests that great significance on the cost-effectiveness of the IC-UADS was exerted by the cooling power under various desiccant mass fraction. Considering the space limit of the present paper, the effects of cooling power under other working conditions will be further presented in the future work.

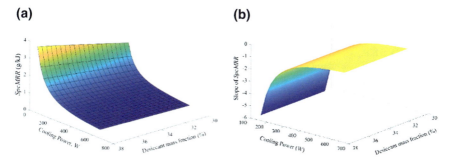

Fig. 6 Trends of SpcMRR under per cooling power (**a**) and its slope (**b**)

4 Conclusions

In this paper, simulation runs were carried out to clarify the effects of the internal-cooling power on the dehumidification performance of the IC-UADS with the established prediction model. The model was built based on the heat and mass conservation laws for the internally cooled ultrasonic atomization system, and thorough experiments were carried out to validate the model. The results show that though the dehumidification performance would continue to increase with the rise of the cooling power, the increase rate was descending. Besides, given the ratio of the cooling power and the consequent dehumidification performance, there exist optimal ranges of the cooling input where the dehumidification performance under unit cooling power may become the most cost-efficient. The results may help inspire the optimal running of the internally cooled liquid desiccant dehumidification systems.

Acknowledgements This work is financially supported by the Shanghai Sailing Program (No. 19YF1401800), the Fundamental Research Funds for the Central Universities of China (No. 2232018D3-36), and the China Postdoctoral Science Foundation (No. 2018M630385).

References

1. Luo, Y., Wang, M., Yang, H., Lu, L., Peng, J.: Experimental study of internally cooled liquid desiccant dehumidification: application in Hong Kong and intensive analysis of influencing factors. Build. Environ. **93**, 210–220 (2015). https://doi.org/10.1016/j.buildenv.2015.05.022
2. Qi, R., Lu, L.: Energy consumption and optimization of internally cooled/heated liquid desiccant air-conditioning system: a case study in Hong Kong. Energy **73**, 801–808 (2014)
3. Zhang, T., Liu, X., Zhang, L., Jiang, J., Zhou, M., Jiang, Y.: Performance analysis of the air-conditioning system in Xi'an Xianyang International Airport. Energy Build. **59**, 11–20 (2013)
4. Lun, W., Li, K., Liu, B., Zhang, H., Yang, Y., Yang, C.: Experimental analysis of a novel internally-cooled dehumidifier with self-cooled liquid desiccant. Build. Environ. **141**, 117–126 (2018). https://doi.org/10.1016/j.buildenv.2018.05.055

5. Zhang, T., Liu, X., Jiang, J., Chang, X., Jiang, Y.: Experimental analysis of an internally-cooled liquid desiccant dehumidifier. Build. Environ. **63**, 1–10 (2013). https://doi.org/10.1016/j.buildenv.2013.01.007
6. Gao, W., Shi, Y., Cheng, Y., Sun, W.: Experimental study on partially internally cooled dehumidification in liquid desiccant air conditioning system. Energy Build. **61**, 202–209 (2013)
7. Gommed, K., Grossman, G., Prieto, J., Ortiga, J., Coronas, A.: Experimental comparison between internally and externally cooled air-solution contactors. Sci. Technol. Built Environ. **21**(3), 267–274 (2015)
8. Yang, Z., Lin, B., Zhang, K., Lian, Z.: Experimental study on mass transfer performances of the ultrasonic atomization liquid desiccant dehumidification system. Energy Build. **93**, 126–136 (2015). https://doi.org/10.1016/j.enbuild.2015.02.035

Experimental Study and Energy-Saving Analysis on Cooling Effect with Large Temperature Difference and High Temperature of Chilled Water System in Data Center

Zhibo Kang, Zhenhua Shao, Lin Su, Kaijun Dong and Hongxian Liu

Abstract From the perspective of the precision air-conditioning operating condition, the effects of different load powers, different water temperatures of supply-and-return and temperature difference on energy consumption of precision air-conditioning and COP of water chiller were debated in this paper. The experimental results show that when the cooling mode of large temperature difference and high temperature is adopted, a higher water supply temperature and a higher efficiency of the water chiller are obtained. Moreover, the return water temperature is also increased, and the utilization rate of natural cold source can be improved. In order to reduce the energy consumption of the water chiller and increase the COP of the water chiller, natural cold source with gradient utilization can be used, which increased the utilization time of natural cold source. In a word, not only the energy consumption of precision air-conditioning water system but the chiller can be reduced by means of large temperature difference and high temperature.

Keywords Data center · Large temperature difference · Cooling effect

---

Z. Kang · Z. Shao (✉) · L. Su · K. Dong (✉)
Guangdong Provincial Key Laboratory of New and Renewable Energy Research and Development, Guangzhou 510640, China

Guangzhou Institute of Energy Conversion, Chinese Academy of Sciences, Guangzhou 510640, China

CAS Key Laboratory of Renewable Energy, Guangzhou 510640, China
e-mail: shaozh@ms.giec.ac.cn

K. Dong
e-mail: dongkj@ms.giec.ac.cn

Z. Kang · H. Liu
Northeast Electric Power University, Jilin 132012, China

© Springer Nature Singapore Pte Ltd. 2020
Z. Wang et al. (eds.), *Proceedings of the 11th International Symposium on Heating, Ventilation and Air Conditioning (ISHVAC 2019)*, Environmental Science and Engineering, https://doi.org/10.1007/978-981-13-9524-6_97

1 Introduction

Air-conditioning system accounts for nearly 40% of energy consumption in China's data centers, among which the energy consumption of water chiller occupies for more than half, and the energy consumption of air-conditioning water system occupies about 25% of all [1]. Therefore, it is urgent for data centers to reduce the energy consumption and as much as possible. In air-conditioning systems, the supply-and-return water temperature of chilled water for conventional chillers is generally designed at 7/12 °C. However, the current temperature situation is not conducive to reducing the energy consumption of the chiller and pump [2–4].

Water supply-and-return cycle was carried out with the action of pumps, and the pump power consumption is proportional to the water flow. So in the process of increasing different temperature of supply-and-return water, the energy consumption of water pump was saved while the water flow went into reduced. Consequently, the energy saving of pumps can be achieved by means of "the large temperature difference, the small flow rate" [5–7].

Chen et al. [8] analyzed the influence of large temperature difference system on the energy consumption of water pumps and chillers qualitatively. It was realized that the water transportation volume, energy consumption of pumps and initial investment of water system can be reduced while the large temperature difference of water is adopted; whereas, the water flow and heat transfer coefficient of evaporator water side reduced, and the COP of the water chiller was also declined. Fan [9] took an engineering design project as an example to analyze the energy-saving of air-conditioning water system with large temperature difference by 6/13 °C. The results shown that the energy consumption of chillers increases, the energy consumption of pumps decreases and the system that compared with the traditional 5 °C temperature difference has a certain energy saving. Mao [10] studied the large temperature difference water system of new air-conditioning. The results show that the large temperature difference method can greatly reduce the energy consumption of pumps; for water chillers, the increase of temperature difference had little effect on the energy consumption by the water supply temperature unchanged. That is to say, the chilled water system of air-conditioning has a certain energy-saving effect by using large temperature difference operation mode [11].

Therefore, the large temperature differences and the high-temperature cooling ways can be used to explore the impact of chiller and pump energy saving and natural cold resource utilization in this paper.

2 Experimental Research of the Data Center

Figure 1 shows the principle diagram of data center test system. The water chiller transfers cold energy to the precision air-conditioning water system through the PHE, then the heat exchanger by internal gas–water and precision air-conditioning fans transmission cold to the heating load to achieve cooling effect. In this experiment, a precision air-conditioner with room-level refrigerated water is selected as the research object; a conventional tube-fin heat exchanger was designed, and it was placed in the middle of the cabinet and connected with fans to form a precision air-conditioning. The length of heat exchanger is 540 mm, the width is 280 mm and the height is 1780 mm. Galvanized sheet is selected for the outer frame, and it used $\varnothing$ 16 × 0.5 copper tube with aluminum fins in the inner part. The heat exchange area is 80 m^2, the design pressure is 0.6 MPa and the design temperature is 90 °C. Six 48 V/5.16 A fans from AVC manufacturer are selected, and 10 sets of 4 kW/AC220 V rack loads are selected with temperature measurement points. The internal loads of the test cabinet are 5, 10, 13.5 and 15 kW. The ten loads placed on both sides of the precision air-conditioning and each side of the load arrangement is balancing. If the 5 kW load is used for the test, 500 W is selected for each load to ensure the accuracy of the test. Figure 2 shows the internal structure of precision air-conditioning and the arrangement of temperature transducers.

In this experiment, the enclosed air supply inside the cabinet is adopted, which reduces the distance of air supply, decreases the resistance and wind consumption along the way as well, in order to achieve faster and better cooling effect. The main indicators of detection and analysis of cooling effect inside the cabinet are as follows: (1) the inlet and outlet air temperature of the cabinet; (2) supply-and-return water temperature and temperature difference; (3) the water flow of precision air-conditioning.

The dry-bulb return air temperature was designed as 37.8 °C in this experiment. According to GB50174-2017, the range of the air temperature in the cabinet area is determined as shown in Table 1: T_{inlet} = 18 °C/27 °C. This experiment mainly selects 27 °C as the maximum inlet temperature. According to a large number of literature research and market research, the range of outlet air temperature in cabinet is determined as follows: $T_{output} \leq 36$ °C, and the large temperature difference of precision air-conditioning side is selected: $T\Delta$ = 8 °C/14 °C.

From the initial design, when considering the inlet air temperature is higher than the dew point temperature of the environment to make full use of the cold source,

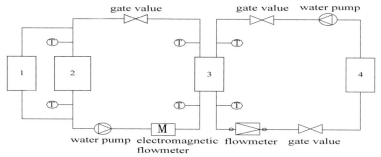

Fig. 1 Schematic diagram of data center test system

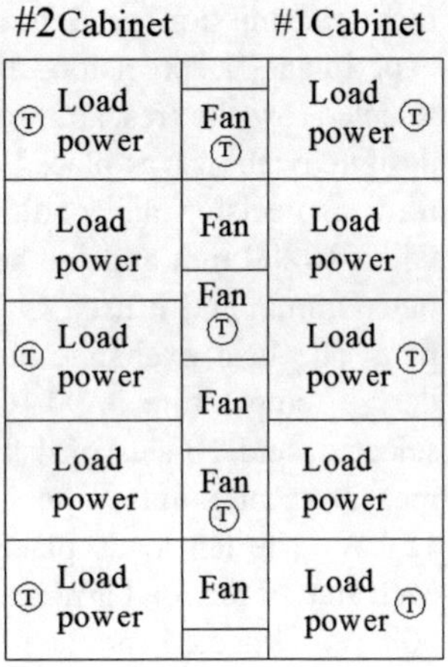

Fig. 2 Fan, load and sensor distribution in precision air-conditioning of data center

#2Cabinet		#1Cabinet
(T) Load power	Fan (T)	Load power (T)
Load power	Fan	Load power
(T) Load power	Fan (T)	Load power (T)
Load power	Fan	Load power
Load power	Fan (T)	Load power
(T) Load power	Fan	Load power (T)

Table 1 Data center temperature conditions and technical requirements

	Temperature conditions (°C)	Technical requirements
Supply water temperature of chilled	7–17	
Return water temperature of chilled	12–23	
Air inlet temperature in the area of the cold passage or cabinet (recommended value)	18–27	Shall not dew
Air outlet temperature in the area of the cold passage or cabinet (allowable value)	15–32	When the environmental temperature and relative humidity of electronic information equipment can be relaxed, this parameter should not dew

now the data center for the supply-and-return water temperature of chilled water by conventional has promoted from 7 °C/12 °C to 10 °C/15 °C. 10 °C/15 °C is chosen as the chilled water supply-and-return water system temperature to make further efforts for the analysis of cooling effect with high-temperature cooling and large temperature difference in water system.

For the evaluation and analysis of energy-saving, the following formula is selected:

$$\text{COP} = \frac{Q}{N} \quad (1)$$

where Q is the refrigerating capacity, kW; N is the energy consumption, kW; COP is the coefficient of performance.

3 Results and Discussion

3.1 Relationship Between Return Water Temperature and Inlet Air Temperature of Cabinet Under Different Temperature Difference of Supply-and-Return Water

Figure 3 shows that the inlet air temperature of cabinets under different loads level off to be 27 °C, which shows that the test coincidences the criterion of the highest inlet air temperature with cabinets is below 27 °C. The return water temperature and the inlet air temperature are similar under different loads and different temperature difference of supply-and-return water and the temperature difference between the two are relatively stable, which is about 2 °C.

Because the conventional fin heat exchanger is chosen in this paper, the coil structure of the heat exchanger is fixed, resulting in the return water temperature and the inlet air temperature of the cabinet are relatively fasten, and it leads that the return water temperature can not be further raised. Thus, the inlet air temperature of the cabinet tends to be stable under the standard of 27 °C, and the return water temperature is also stable in a certain temperature range.

3.2 Relationship Between Inlet Air and Outlet Air Temperature of the Cabinet Under Different Temperature Difference of Supply-and-Return Water

Figure 4 shows that the trend of inlet air temperature is similar to the outlet under different load powers. As the inlet air temperature is stable, the outlet air temperature increasing in path with the loads, and it still conforms to the range of outlet air temperature determined in this paper. The different temperature difference of supply-and-return water has little effect on the outlet air temperature of cabinet, so the large temperature difference method is used for cooling and the cooling effect is remarkable.

3.3 Relationship Between Supply Water, Return Water Temperature and Temperature Difference of Supply-and-Return Water

In Fig. 5a, the water supply temperature of precision air-conditioning decreases gradually while the temperature differences of supply-and-return water increasing under the same load. The temperature of water supply under different loads is similar under the same temperature difference of supply-and-return water. Because

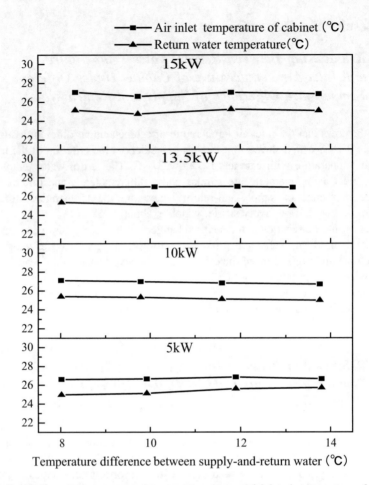

Fig. 3 Relation curves between return water temperature and inlet air temperature of cabinet under different temperature difference of supply-and-return water

the temperature of return water tends to be constant, the temperature of water supply decreases, with the increasing temperature difference of supply-and-return water. The water supply temperature shown in the figure is higher than 10 °C while the standard of water supply temperature sets to 10 °C. It means the cooling effect of high-temperature cooling mode with large temperature differences can be tested under different load power of cabinet under the air outlet temperature standard.

Figure 5b shows that the return water temperature is relatively stable under the same load and different temperature differences of supply-and-return water, and the return water temperature with the same temperature difference is similar under different loads. The setting value of return water temperature is 15 °C, and the actual return water temperature is much higher than 15 °C. It shows that the energy consumption of the chiller improves and the COP decreases while the return water

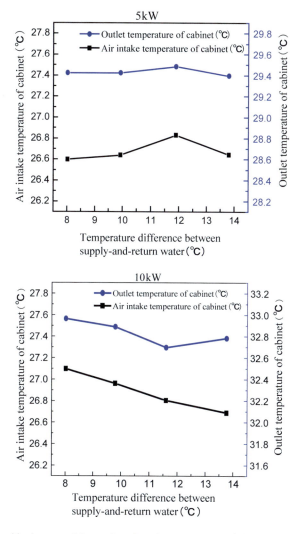

Fig. 4 Relationship between inlet and outlet air temperature of the cabinet under different temperature difference of supply-and-return water

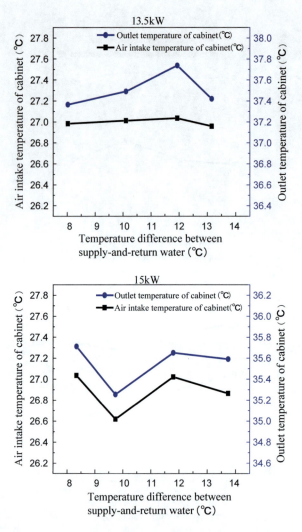

Fig. 4 (continued)

temperature increases. Therefore, the cascade utilization mode of natural cold source is chosen in this experiment. When the return water temperature is higher than the set value, it is preferred to apply natural cool source as first cooling. The return water temperature of the chiller is cooled to the set temperature or even lower, which reduces the energy consumption and improves the COP; meanwhile, the utilization time of natural cold source is increased, and the utilization rate of natural cold source is enhanced. If the natural cold source can not reach the cooling standard, the mechanical refrigeration of water chiller is used to ensure that the refrigeration requirements are met.

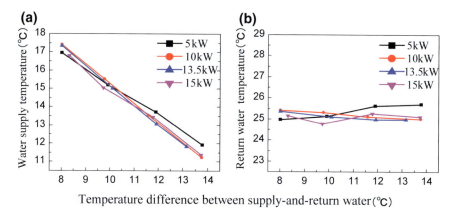

Fig. 5 Relation curve between water temperature and temperature difference of supply-and-return water

4 Conclusions

In this paper, a data center system test system is built. The energy saving and cooling effects of water system are analyzed by changing the loads and temperature differences of supply-and-return water. The results are as follows:

Through the analysis of the energy consumption in data center, the main energy consumption problems existing in the system are clarified, and the targeted improvement schemes are put forward. By means of the relationship between the water flow and the temperature differences of supply-and-return water in precision air-conditioning, the way of "the large temperature difference, the small water flow" is determined to save energy consumption of the water chillers and pumps.

Through the test results and analysis, it is concluded that the heat exchanger designed in this paper still conforms to the inlet air temperature standard under the condition of large supply-and-return water temperature differences, which not only can reduce the energy consumption of the water chillers and pumps but also keep the remarkable cooling effect of the cabinet.

Through the relationship between the water supply temperature and temperature differences of supply-and-return water, it is further determined that the precision air-conditioning in this test meets the demand of large temperature differences and high-temperature cooling.

Through the relationship between the return water temperature and the temperature differences of supply-and-return water, it is determined that when the water supply temperature of chiller rises, natural cooling cascade utilization mode can be used to conduct first cooling through cooling tower, which increases the utilization time of natural cool source and promotes the COP of the water chillers. If it is not enough to meet the cooling requirements with natural cold source, the mechanical refrigeration of water chillers is carried out to ensure that the cooling requirements of cabinets are met.

Acknowledgements This work is supported by Science and Technology Planning Project of Guangdong Province, China (No.2017B090907027), Science and Technology Planning Project of Guangzhou, China201705YH091.

References

1. TAO, Y., ZHANG, J., WANG, H.: Cold water large temperature difference combined air conditioning unit research. National HVAC refrigeration Year Academic Corpus, pp. 252–257. China Construction Industry Press, Beijing (2002)
2. Jiang, X., Xiao, J., Li, X.: Energy consumption comparison of large-scale air-conditioning system with different temperature of chilled-water. Procedia Eng. **205**, 2100–2104 (2017)
3. Zhang, S.: Energy saving analysis of free cooling in chilled water system for data centers. HAVC 2016. 46(5), 80–83 (2016)
4. Lv, J., Wang, T., Zhao, L.: Energy saving analysis of data center air conditioning system based on application of natural cooling technology. J. Refrig. **37**(3), 113–118 (2016)
5. Zhang, H., Zhang, J., Chen, D.: Energy saving optimization control of cold water unit based on COP curve. Build. Energy Environ. **37**(1), 6–10 (2018)
6. Dong, X.: Research of reducing energy consumption of air conditioning chilled water system. Donghua University (2012)
7. Zhu, K., Ren, J., Zhang, J.: Energy consumption analysis of refrigeration system with large temperature difference in cooling water. J. Refrig. **39**(5), 129–134 (2018)
8. Chen, S., Qu, G., Huang, W.: Application of large temperature difference technique of air conditioning chilled water to engineering. HAVC **42**(6), 20–23 (2012)
9. Fan, J.: Discussion on energy-saving design of large temperature difference air conditioning water system. In: Proceedings of the 7th national technical exchange conference on building environment and energy application. Building environment and energy application branch of China survey and design association pp. 3. HAVC magazine (2017)
10. Mao, L.: Design and application of a new type of air conditioning system with large temperature difference. Dalian Ocean University (2017)
11. Duan, Z.: Technical and economic analysis of air-conditioning water system using large temperature difference in temperate regions. Refrig. Air Cond. **28**(6), 672–676 (2014)

Sensitivity and Uncertainty Analysis for Chiller Sequencing Control of the Variable Primary Flow System

Zhenbing Cai and Yundan Liao

Abstract Chiller sequencing control is a key function for the automatic control of chiller plants in air-conditioning system. It switches chillers on/off according to the building cooling load for cooling supply and energy saving. However, when facing uncertainties, sequencing controller may switch-on/off chillers improperly or frequently, leading to unstable operation or energy waste. In order to enhance the robustness of chiller sequencing control under uncertainties, an uncertainty analysis was performed for the primary variable flow system, taking account of possibly confronted uncertain factors in real operations. A chiller plant was constructed using the simulation platform TRNSYS. A typical sequencing control strategy was coded using MATLAB and inserted into TRNSYS by the open linking type. Associated uncertainties such as measurement uncertainty, control uncertainty or threshold uncertainty were extracted from site operational data of a real chiller plant and modeled using statistic methods. The impacts of each uncertainty were evaluated with Monte Carlo simulations. Their sensitivities were also ranked. Further, this study provides an guidance for the robustness enhancement of chiller sequencing control.

Keywords Chiller sequencing control · Uncertainty · Sensitivity

1 Introduction

The central air-conditioning system operated jointly by multiple chillers has been widely used in commercial buildings to improve operation flexibility and energy efficiency in the case of partial load [1, 2]. Chiller plant accounts for more than 60%

Z. Cai · Y. Liao (✉)
Guangzhou University, Guangzhou 510006, China
e-mail: ydliao@gzhu.edu.cn

Y. Liao
Academy of Building Energy Efficiency, School of Civil Engineering, Guangzhou University, Guangzhou 510006, China

of the total energy consumption of the centralized air-conditioning system [3]. Chiller sequencing control properly switches chillers on or off to fulfill the demanded building instantaneous cooling load and maximize coefficient of performance (COP) of the online chillers, which is of great significance to the overall performance of the air-conditioning system.

Various chiller sequencing control strategies have been developed and applied to practical projects. However, uncertainty is widely existent in engineering process [4]. As a result, the developed control strategies may not work well on site. In recent years, Liao et al. systematically studied the uncertainty, divided the uncertainty into four categories, and studied its mechanism and influence in the typical control strategy [5]. Evaluating the impact of different uncertainties on chiller sequencing control will make the optimal control strategy more scientific and practical.

Considering the importance of uncertainty impacts on practical performance and optimization of chiller sequencing control, this paper presents an uncertainty analysis and sensitivity analysis for a typical control strategy that are widely applied for chiller sequencing control of the variable primary flow (VPF) systems. The uncertainties will be firstly identified and modeled with the site data from a real chiller plant. The impacts of these uncertainties will be evaluated by simulation studies using TRNSYS and MATLAB simulation software. Three performance indices will be used to evaluate the impact on system stability, energy efficiency, and cooling supply reliability, respectively. Sensitivity analysis will also be conducted to identify the relative importance of each uncertainty on such chiller sequencing control, which will help to guide the performance enhancement for chiller sequencing control of VPF systems.

2 Methodology for Uncertainty and Sensitivity

2.1 Chiller Sequencing Control Strategy

Chiller sequencing control strategy applied to VPF systems generally uses the calculation of cooling load, the percentage of full-load amperage (PFLA) of the compressor motor of the online chillers, and the CHWS temperature to indicate the change of the building load. However, the large flow fluctuation of the VPF system leads to the large fluctuation of cooling load obtained by direct indicators, which is easy to cause frequent switching actions [6].

In this section, a typical control strategy for the VPF system is introduced. The control strategy combined the PFLA of the online chillers and the CHWS temperature two indirect indices to improve the accuracy of control. Avoid in some cases, the cooling capacity of the chiller can not reach the rated cooling capacity and cause incorrect switch-on/off behaviors. Hence, the switch-on/off criteria are:

- Switch-on criterion: If the PFLA of the online chillers is greater than the switch-on threshold, and this state duration is longer than the time limit. At the condition, if the CHWS temperature is greater than the set temperature, then a chiller will be switched on.
- Switch-off criterion: If the PFLA of the online chillers is lower than the switch-off threshold, and the duration is longer than the time limit, then a chiller will be switched off.

2.2 Uncertainty Sources, Shifting, and Propagation

Chiller sequencing control is affected by many types of uncertainties. Liao et al. classified uncertainty into four categories, including measurement error, control error, operational error, and threshold error [5]. The control strategy described in this paper needs to monitor the CHWS temperature and the PFLA of the online chillers. The measurement uncertainty in this study is assumed to be a normal distribution.

In addition, the COP rated curve is used to correlate the cooling load with the PFLA of the online chillers, thereby correlating the switching thresholds at different indicators. The rated COP curve is obtained by fitting. Therefore, threshold uncertainty is unavoidable. With a given PLR, the errors between the measured COP and the rated COP can be described by a normal distribution. The control strategy suffers form measurement uncertainty of the CHWS temperature (UN_{M_Tem}), measurement uncertainty of the PFLA of the online chillers (UN_{M_I}), and threshold uncertainty (UN_{Th}).

Uncertainties need to be transferred to the input variables related to the control strategy for simulation. Then, the measurement error and threshold error are substituted into the load indicator, viz. the compressor operating current I_{th}. Since it is only affected by the measurement error, the measured CHWS temperature becomes:

$$T_{\text{sup,meas}} = T_{\text{sup}} + \Delta T_{\text{sup,meas}} \qquad (1)$$

where $T_{\text{sup,meas}}$ is the measured CHWS temperature, T_{sup} is the true value, and $\Delta T_{\text{sup,meas}}$ is the measurement uncertainty.

Then, the measurement error and the threshold error are substituted into the compressor operating current.

$$I_{th} = I_{th,\text{meas}}^{id} - \Delta I_{th,\text{meas}} + \frac{\Delta \text{COP}}{\text{COP}_r} \left(I_{th,\text{meas}} - \Delta I_{th,\text{meas}} \right) \qquad (2)$$

where $\Delta I_{th,\text{meas}}$ denote the measurement uncertainty of current, $I_{th,\text{meas}}$ the measured value of current, $I_{th,\text{meas}}^{id}$ is the actual indicator, Q_{th} is the ideal threshold of cooling load, COP_r is the rated COP, and I_{th} is the actual threshold. The measurement and threshold uncertainty are all transformed into the uncertainty of current I, as shown in the following equation:

$$\Delta I = \Delta I_{th,meas} - \frac{\Delta COP}{COP_r}(I_{th,meas} - \Delta I_{th,meas}) \tag{3}$$

At this point, all the uncertainties related to the control strategy have been transferred to the input variables of the control strategy.

2.3 Uncertainty and Sensitivity Analysis Methods

In this study, the Monte Carlo method is used for uncertainty analysis, which is a sample-based method and is widely used in uncertainty analysis. The concept of the Monte Carlo method is simple. First, a large number of output samples are generated based on the distribution pattern of the input sample. Secondly, the output corresponding to each input sample is estimated by simulation, and finally, the output distribution is obtained by analyzing a large number of outputs. The method can effectively obtain the uncertainty of the output propagated by multiple inputs with different uncertainty distributions [7].

The essence of sensitivity analysis is to explain the law that the key indexes are affected by the change of these factors by changing the values of the relevant variables one by one. Sensitivity coefficients can be used to characterize the sensitivity of various factors, which are calculated by Eq. 4.

$$SEN = \frac{EI - EI_{bs}}{EI_{bs}} \div \frac{IV - IV_{bs}}{IV_{bs}} \tag{4}$$

where EI is the evaluation index; IV is the input variable; subscript "bs" is represented as the baseline value.

3 Simulation Setup

3.1 Simulation Platform

A simulation platform for a typical VPF system was constructed by simulation software TRNSYS16 and the control strategy was coded using MATLAB and communicated with TRNSYS 16 through the existing interface. The system consists of four centrifugal chillers and three variable-speed pumps, which were connected in parallel before being connected in series. The simulation sampling time is 1 min. In order to ensure the smooth operation of the chiller plant, 30 min was selected as a safety time. One-week (168 h) operation of the chiller plant was simulated, and each scene of this study is repeated 100 times. Three levels of uncertainty were set, listed in Table 1. The value of μ is 0. The data of uncertainties were added to the measurement of temperature and compressor amperage. Moreover, the measurement uncertainties of compressor amperage and flow rate should be recorded by the meter in full measurement scale and were in percentage.

3.2 Performance Indices

To evaluate the effects of uncertainties, three performance indices were defined, including the total chiller switch number, the energy use, and the accumulated tracking error of the SAT [8]. The total switch number is used to quantify the switch frequency, representing the stability of the system. The energy use is the sum of energy consumption of all the chillers and pumps. Moreover, supply air temperature is closely related to indoor thermal comfort. The accumulated tracking error of the SAT describes the degree of deviation from the set temperature, which reflects the deterioration of indoor thermal comfort. These performance indices can well evaluate the whole performance of the air-conditioning system.

4 Results and Analysis

The benchmark case without uncertainty based on the static condition described in Table 1 was first simulated, and the one-week results were then compared with simulations with uncertainties. In this study, the uncertainty analysis focuses on the simulation results of the worst case that is the uncertainty level 3.

4.1 Impact of the Measurement Uncertainty of CHWS Temperature

When only considering the measurement bias of the CHWS temperature at the level 3, the overall switch number was reduced by 4.3% compared with the benchmark, the accumulated tracking error of the SAT increased by 4.4%, and the energy use

Table 1 Three levels of the uncertainty

Uncertainty	Example	Level 1	Level 2	Level 3
Water temperature sensor	δ	0.0388	0.1163	0.1938
Current meter	δ	0.0039	0.0078	0.0116
COP/COP$_r$	δ	0.0039	0.0116	0.0194

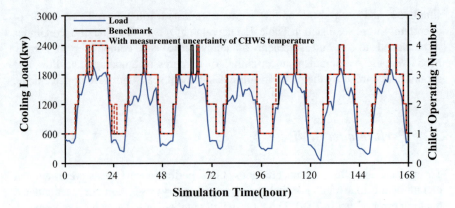

Fig. 1 Switch-on/off actions with the measurement bias of the CHWS temperature

decreased by 0.1%. The impact of UN_{M_Tem} on the three performance indices was limited. The control strategy had good robustness to the UN_{M_Tem}.

It can be seen from Fig. 1 that the UN_{M_Tem} did not affect switch-off actions. The main reason was that the CHWS temperature did not participate in the determination of switch-off, and was only used as an auxiliary judgment of switch-on actions. Occasionally, it may lead to wrong behaviors about switch-on actions, including delayed or advanced switch-on, missed switch-on, and unnecessary switch-on. These behaviors only occurred when the building cooling load was close to the integer times of the rated cooling load of chiller. Once the cooling load exceeds the overall cooling capacity of the online chillers to a certain range, the CHWS temperature will fluctuate greatly and exceed the accuracy of the sensor. In this condition, the threshold condition had already satisfied, and one more chiller will be switched on.

4.2 Impact of the Measurement Uncertainty of Compressor Current

Under the impact of measurement uncertainty of compressor current, the overall switch number was reduced by 8.7% compared with the benchmark, the accumulated tracking error of the SAT increased by 13.4%, and the energy use decreased by 1.28%. The three performance indices changed more greatly than the simulation results of the UN_{M_Tem}.

As shown in Fig. 2, at some time section, delayed switch-on or even lack of switch-on was observed, which is caused by that the UN_{M_I} broke the continuous timer counting and reset the timer to zero. Hence, it led to an increase in the SAT. Although the energy was saved, the indoor thermal comfort had become worse due to the increase of the SAT. The reduction of energy use is meaningless in this

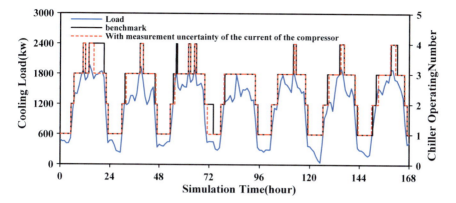

Fig. 2 Switch-on/off actions with the measurement uncertainty of compressor current

condition. Although both were measurement uncertainties and the accuracy of the current sensor was relatively higher at the same level of uncertainty, the accuracy of current had more influence than that of temperature sensor in this control strategy. The mainly reason was that the measurement uncertainty of compressor current can impact switch-off actions, while the measurement uncertainty of CHWS temperature did not.

4.3 Impact of Threshold Uncertainty

The control strategy also suffers from the threshold uncertainty, which has complex effects. It can be seen from Fig. 3 that not only delayed switch-on/off, but also missed switch-on/off and unnecessary switch-off were observed. However, there were no unnecessary switch-on actions. The reason is that, at this time, the online chillers had sufficient capacity to eliminate the building cooling load and the CHWS temperature was maintained at the set value, which means that the switch-on condition will not be met. The figure also shows that when the cooling load frequently fluctuates up and down within the load corresponding to switch-on threshold in a short time, the counting is easily interrupted by the threshold uncertainty, resulting in severe delayed switch-on actions so that the overall switch number was reduced. This delay led to a large rise in the SAT.

4.4 Results of Sensitivity Analysis

In this study, the sensitivity coefficients were calculated using the method described in the "Uncertainty and Sensitivity Analysis Methods" section, which can be used

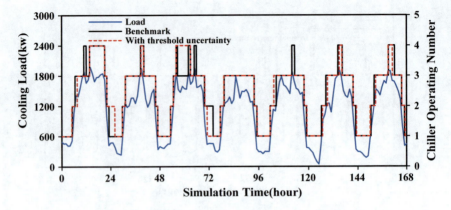

Fig. 3 Switch-on/off actions with threshold uncertainty

to compare the impacts of the three uncertainties in the control strategy. Table 2 shows the calculated sensitivity coefficients, which represented the deviation of the three performance indices from uncertainties that correspond to the three levels. Uncertainty level 1 is representing the design condition when the system has acceptable accuracy of measured data and setup data. However, whatever performance indices were used as the evaluation index at the level 1, the sensitivity coefficients of the UN_{M_Tem} was smaller than that of the UN_{M_I} and the UN_{Th}. These results demonstrated that even under the design condition, the control strategy still has large sensitive to UN_{M_I} and UN_{Th}. In general, under normal operation of the air condition system, the worst cases are uncertainty level 3. The results threshold uncertainty has the greatest impact at the level 3, which means that how to get accurate COP rated curve is significant.

The control strategy had lowest sensitivity to temperature measurement uncertainty. In contrast, the control strategy had much higher sensitivity to the threshold uncertainty and power measurement uncertainty. Hence, when this strategy is

Table 2 Sensitivity coefficients of three levels of the uncertainty

Evaluation indices	Uncertainty	Level 1	Level 2	Level 3
The total chiller switch number	UN_{M_Tem}	0	−1.02	−0.61
	UN_{M_I}	−4.35	−3.26	−2.90
	UN_{Th}	−4.35	−2.90	−3.48
The accumulated tracking error of the SAT	UN_{M_Tem}	0.42	0.97	0.60
	UN_{M_I}	7.03	5.74	4.49
	UN_{Th}	5.80	3.63	4.51
The energy use	UN_{M_Tem}	−0.05	−0.02	−0.01
	UN_{M_I}	−0.46	−0.52	−0.43
	UN_{Th}	0.74	−0.61	−1.06

applied, the temperature sensor with appropriate accuracy should be selected to monitor the CHWS temperature, the current sensor should be as high-precision as possible, and the proper set up of thresholds is required.

5 Conclusion

This paper presented a systematic analysis on uncertainties of a chiller sequencing control strategy for the primary variable flow (VPF) system. In order to investigate the impacts and significance of uncertainties on chiller sequencing control of VPF system, four steps of works, including uncertainty identification, uncertainty modeling, uncertainty shifting, and sensitivity analysis were processed. The uncertainties have been shifted to the input variables of the control model by mathematic derivation. Results showed that the typical control strategy suffers seriously from measurement uncertainty of compressor current and threshold uncertainty, while the chiller water supply temperature measurement uncertainty has lower impact on chiller sequencing control. Hence, improving the accuracy of the current senor and setup thresholds with carefull evaluations on related COP curves is of great significance for this control strategy. Since uncertainties have important impacts on chiller sequencing controls, the future work will focus on enhancing the robustness and control performance of VPF system chiller sequencing control so as to improve the practicability and reliability.

References

1. Chang, Y., et al.: Optimal chiller sequencing by branch and bound method for saving energy. Energy Convers. Manag. **46**(13–14), 2158–2172 (2005)
2. Yu, F., Chan, K.: Optimum load sharing strategy for multiple-chiller systems serving air-conditioned buildings. Build. Environ. 42(4), 1581–159 (2007)
3. Jiang, Y.,: Current building energy consumption in China and effective energy efficiency measures. J. HVAC 35(5), 30–39 (2005)
4. Wright, J.: Uncertainty in model-based condition monitoring. Build. Serv. Eng. Res. Technol. 25, 65–75 (2004)
5. Liao, Y., et al.: Uncertainty analysis for chiller sequencing control. Energy Build. 85, 187–198 (2014)
6. Huang, G., et al.: A data fusion scheme for building automation systems of building central chilling plants. Autom. Constr. **18**(3), 302–309 (2009)
7. Shan, K., et al.: Sensitivity and uncertainty analysis of measurements in outdoor airflow control strategies. HVAC R Res. 19(4), 423–434 (2013)
8. Liao, Y., et al.: Robustness analysis of chiller sequencing control. Energy Convers. Manag. 103, 180–90 (2015)

Theoretical Analysis of Smoke Exhaust System with Ringed Arrangements in the Field of HVAC

Minmin Zhang, Yixue Wu and Meiling He

Abstract As an indispensable part of the building, the design of air conditioning system has attracted many people's attention. Smoke manage system is one of the most important parts of air conditioning and ventilation system, which directly relates to people's safety. In the previous studies, the loop network was seldom used in the research and application of air conditioning design, especially in the mechanical smoke evacuation system. In this paper, the concept of circular arrangement of pipe is put forward, which is used to discharge the smoke. The advantages of circular arrangement of smoke exhaust system are shown by the case analysis. This study can provide the references to the research and design of ringed arrangement in the field of air conditioning and ventilation system in the future.

Keywords Air conditioning system · Smoke exhaust system · Ringed arrangement · Case study

1 Introduction

With people's higher pursuit of comfort and safety, the design of air conditioning system has become a key part of architectural design. In recent years, people have studied a lot of methods and technique in the air conditioning system. Cao [1] introduced how control system works in the monitoring of air conditioner ventilation system and designed the whole system in intelligent buildings. Wang and Liu [2] studied the problems in the operation of air conditioning system in Petroleum Engineering Research and Development Center, and the scheme is given. Ai and Mak [3] developed a design framework which can determine appropriate design parameters, including ventilation period, ventilation frequency and start concentration of ventilation and then provided the detailed design guidelines.

M. Zhang · Y. Wu · M. He (✉)
The Architectural Design & Research Institute of Zhejiang University Co., Ltd, Hangzhou, China
e-mail: 280788517@qq.com

© Springer Nature Singapore Pte Ltd. 2020
Z. Wang et al. (eds.), *Proceedings of the 11th International Symposium on Heating, Ventilation and Air Conditioning (ISHVAC 2019)*, Environmental Science and Engineering, https://doi.org/10.1007/978-981-13-9524-6_99

Zhuang et al. [4] proposed a novel 'adaptive full-range decoupled ventilation strategy' and showed that the proposed ventilation strategy can offer superior energy performance over the full range of internal load and weather conditions. Fong et al. [5] investigated the effect of indoor air distribution strategy on solar air conditioning.

Smoke exhaust system plays an important role in ensuring people's safety in case of fire. Hence, smoke exhaust system is very important in the field of HVAC (Heating, Ventilation and Air Conditioning), and many previous studies have mentioned it. Wang and Fan [6] investigated the common issues of smoke control and management system in engineering practice, and the solutions were suggested. Shen [7] gave the specific technical countermeasures to the problems existing in the installed smoke control and exhaust system. Ji [8] put forward the influence of selection of exhaust fan on the stability of hydraulic condition of air duct system and showed that the relevant parameters should be provided to equipment manufacturers. He et al. [9]. utilized the computational fluid dynamics method employing the software Fire Dynamics Simulator to analyze the effects of heat release rate and exhaust velocity on the entrainment near the vent. Tang et al. [10] revealed the variation characteristics of ceiling maximum temperature with different ceiling smoke extraction rates.

The pipe dimensions of smoke exhaust system are often larger than that of general air conditioning system. Reasonable smoke exhaust system design is not only conducive to ensuring safety but also to saving resources and space. Current smoke exhaust system design is mostly branch pipe network, and almost no ring pipe network is involved. Ringed arrangements in previous researches are mainly applied in water and gas system. Zhao [11] put forward a simple mathematical model of hydraulic calculation for the ring network design of self-spraying system in combination with practical engineering. Yao [12] optimized the loop pipe network layout and diameter and combined theoretical study and optimization algorithm with computer analysis and case analysis. Li [13] expounded the approach of flushing the annular pipes during the commissioning, which based on a data center project in Shanghai. Laajalehto et al. [14] examined the energy efficiency of a new district system which included a ring network. The results showed that the district heating system is easier to control with ring network. Creaco and Franchini [15] presented an automatic algorithm for the identification of the minimum loops in a multi-source looped water distribution network. Hafsi et al. [16] studied a numerical simulation of transient hydrogen-natural gas mixture flow in a looped network on account of a mathematical model that considers the variation of the compressibility factor of the gas mixture with pressure under isothermal gas flow. Based on the design status of smoke exhaust system, this paper proposes the concept of circular arrangement of smoke exhaust system.

2 Methods

2.1 Layout of Pipeline Network

Reference [17] points that there are three types of the pipe network: branch layout, ring layout and radial layout. The trunk and branch of dendritic pipeline network are distinct, which form a dendritic shape. This arrangement has many advantages such as simplicity, low investment, convenient operation and management, etc. When the ring pipe network is applying, its trunk lines are connected in a ring. The investment of this method is high, but it is reliable and safe. As for the radial network, it is similar to branched network. In the field of HVAC, no matter what kind of medium (gas, smoke, water, etc.) is in the pipeline, there is only one source of the branched pipe network. It means that the gas or liquid at each point can only come from one direction. However, the source of the ring network can be multiple.

2.2 Design Requirement of Smoke Exhaust

The latest 'technical standard for smoke management systems in buildings' was implemented last year. Compared with the old standard, the new standard has a great change in the design of the smoke exhaust amount. When the mechanical smoke venting is used, the new standard [18] requires that the smoke exhaust amount (the height of the space not exceed 6 m) should be calculated at least 60 m^3/(h m^2), and the value should be no less than 15,000 m^3/h at the same time. The standard [18] also stipulates that the smoke exhaust amount of the system should choose the maximal value in the same fire partition, which is the sum of the smoke extraction volume from any two adjacent smoke bays. According to the requirement of airspeed and pressure, the size of the trunk pipe should be about 0.65 m^2, while the branch only needs half of it. Thus, compared with air conditioning ventilation system, the air volume and duct size of smoke exhaust system are usually larger, which usually need to occupy more space.

2.3 Distinction Between Ring and Branch Arrangements

The smoke will be sucked away by running the fan when the fire breaks out. In order to operate the smoke exhaust system effectively, it is necessary that the wind pressure of the fan is greater than the resistance (frictional drag and local resistance) of the whole journey. In the design of smoke exhaust system, only two smoke bays are considered to be on fire at the same time. Combining with the design principle of smoke exhaust system, when the number of smoke vents is less than or equal to four, it is found that only a few pipes are thick, and there is no need for ring

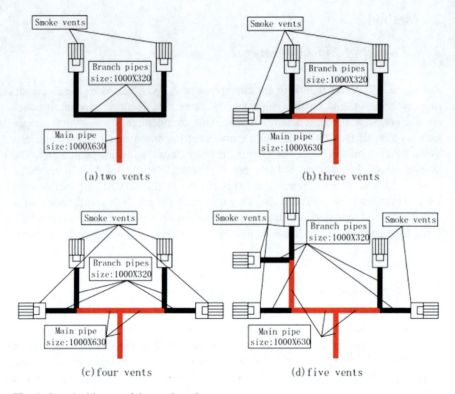

Fig. 1 Branched layout of the smoke exhaust system

arrangement. However, there would be a lot of pipes are thick when the number of smoke vents is more than four. It takes up a lot of space and height. The branched layout of the smoke exhaust system is shown in Fig. 1.

3 Results

3.1 Theoretical Analysis

When the number of smoke vents is more than five, ringed arrangements of the smoke exhaust system have some advantages. In this paper, the smoke exhaust system with six vents is taken as an example, and the area of a single smoke bay should not exceed 250 m². The size of the ring network is one size larger than that of the branch pipe, which means that the size of the ring network is 1000 × 400 and the size of the branch pipe is 1000 × 320. Only two smoke bays are considered to be on fire at the same time, so only two smoke vents need to be open simultaneously, and the amount of the smoke in single smoke vents is 15,000 m³/h. The

Fig. 2 Smoke trend in the ring network (open vent 5 and vent 6)

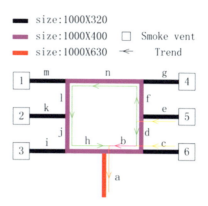

ring network conforms to the characteristics of parallel pipeline. Obviously, when vent 5 and vent 6 are opened at the same time, the smoke speed of tubulation b will reach the maximum value. The smoke trend in the ring network is shown in Fig. 2. The smoke of vent 6 flows from c to b, while the smoke of vent 5 flows in two directions. When the duct size is 1000 × 400 and the smoke speed is controlled within 20 m/s, the maximum amount of smoke is about 28,500 m^3/h. If the total loop length is fixed, the smoke speed in the tubulation b will be lower as the tubulation b and tubulation d get longer. Thus, the size of the pipe can meet the requirement of smoke speed which need less 20 m/s in most of the projects.

3.2 Case Analysis

Figure 3 shows the plan of smoke exhaust system for a standard floor of an office building (branched layout). The area of this standard layer is about 1500 m^2, and the area of office space is about 1280 m^2 (excluding the core tube area). The whole area is divided into six smoke bays, each of which has an area of less than 250 m^2. Therefore, the smoke emission of each smoke bay is 15,000 m^3/h. Considering the layout of other equipment, there are two wells with symmetrical arrangement of left and right. The main pipe size is 1000 × 630, and the branch pipe size is 1000 × 320. Considering the height of story and beam, the size of local pipe is 1500 × 250. In this smoke exhaust system, the smoke speed is calculated by Hongye software. The branch smoke speed is about 12 m/s, and the main smoke speed is about 13.22 m/s, which meets the requirements of the standard. Because there is no clear specification for the calculation of smoke pressure in smoke exhaust system and the calculation of pressure in engineering does not need to be very accurate, the formula in reference [17] is used to estimate the smoke pressure in this system.

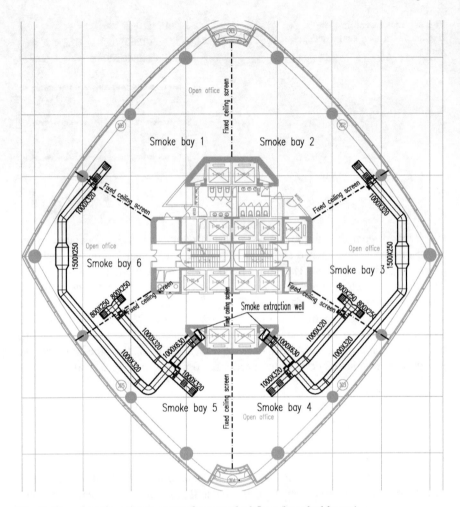

Fig. 3 Plan of smoke exhaust system for a standard floor (branched layout)

$$\Delta P = \Delta p_m \times l(1+k) \qquad (1)$$

where ΔP is pressure loss along unit length ventilation pipe, l is total length of the pipes, k is the ratio of local pressure loss to pressure loss along pipeline network. When the system has a lot of tees and elbows, the value of k is 3–5, otherwise 1–2.

The smoke pressure required for the worst loop in this system with branching arrangement is about 330 Pa.

Figure 4 shows the plan of the smoke exhaust system after optimum design. The area of each exhaust well is about 1 m². When the ringed arrangement is applied in the smoke system, only one well is needed, and the size of the loop network is

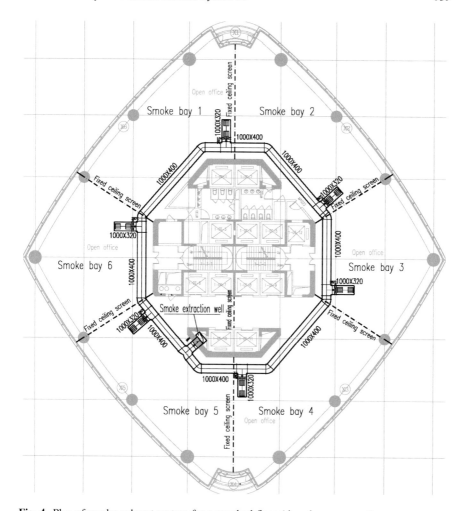

Fig. 4 Plan of smoke exhaust system for a standard floor (ringed arrangement)

1000 × 400, which means a lot of area and height can be saved. The division of each smoke bay in the system is the same as that of the branch layout. Accordingly, the exhaust volume of each smoke bay is also 15,000 m³/h.

The total length of the loop is about 75.4 m, and the branch length is 2.75 m. The calculation of smoke pressure also adopts the estimation method. We define k as 1 when the number of elbows is no more than 2 and as 2 when the number of elbows is more than 2. The design of smoke exhaust system only considers the fire situation of two adjacent smoke control bays. Therefore, in all cases, the maximum smoke speed and the most disadvantageous loop smoke pressure are shown in Table 1. It can be seen from the table that the smoke pressure of the most disadvantageous loop reaches the maximum when smoke bay 1 and smoke bay 2 are on

Table 1 Highest smoke speed and smoke pressure of different situations

Smoke bays	1 and 2	2 and 3	3 and 4	4 and 5	5 and 6	6 and 1
The highest smoke speed	10.5	12.3	16	10.5	15.5	12.7
Smoke pressure	181	177	86	32	96	145

fire at the same time, exceeding 180 Pa. But the pressure is still lower than that of the branch layout. When smoke bay 5 and smoke bay 6 are on fire, the highest smoke speed in the pipeline network is about 15.5 m/s, which is slightly higher than that of the branch arrangement. But it still meets the requirements of the standard. Moreover, the tubulation with high speed is short, so the smoke pressure of the whole loop is relatively lower. The smoke exhaust system with ring arrangement has some advantages over that with branch arrangement.

4 Discussion

Ringed arrangement has been studied and applied in many systems, especially in water system. However, the ventilation system in the field of HVAC has little to do with ring network. The opening rules of vents in smoke exhaust system are relatively simple, which reduces the difficulty of calculating the resistance of ring network. Besides, the smoke extraction volume could be controlled at several fixed values. It is easy to balance the whole system. By the case, we can find that the height of the pipe is reduced, and it is valuable in practical engineering. In other words, it saves a lot of space. So, it can be said that smoke exhaust system with ringed arrangement is very simple and practical. This study certainly has limitations. For example, the size of the pipe is not the minimal and the situation with the lowest smoke pressure that can be used in any systems is also not provided. These problems still need to be solved in the future.

5 Conclusions

Through theoretical and case studies, we can draw the following conclusions:

(1) The ringed arrangement of smoke exhaust system needs certain conditions, which is suitable for the situation of relatively large number of smoke vents.
(2) The ringed arrangement of smoke exhaust system has the advantage of saving ceiling height. According to the case analysis, it can be seen that the ringed arrangement has the potential of reducing smoke pressure and saving the area of well.

(3) The analysis and research in this paper provide a reference for the future application of ringed arrangement in air conditioning system.

Acknowledgements This work was financially supported by the Development and Research on Output Software of HVAC Standard Equipment Table (Project No. UAD201618). We would like to thank our members in our research group and other participants for support to this study.

References

1. Cao, S.: Monitored control and design of air conditioner ventilation system based on BAS. Control. Eng. **19**, 49–51 (2012)
2. Wang, Q., Liu, J.: HVAC system design for petroleum engineering technology research and development center in CNPC Technology Innovation Base. J. HV&AC **47**(7), 56–62 (2017)
3. Ai, Z., Mak, C.: Short-term mechanical ventilation of air-conditioned residential buildings: a general design framework and guidelines. Build. Environ. **108**, 12–22 (2016)
4. Zhuang, C., et al.: Adaptive full-range decoupled ventilation strategy and air-conditioning systems for clean rooms and buildings requiring strict humidity control and their performance evaluation. Energy **168**, 883–896 (2019)
5. Fong, K., et al.: Investigation on effect of indoor air distribution strategy on solar air-conditioning systems. Renew. Energy **131**, 413–421 (2019)
6. Wang, Q., Fang, Y.: Analysis and discussion on common issues of smoke control system in fire acceptance process. J. HV&AC **46**(9), 75–79 (2016)
7. Shen, W.: Research on the countermeasures of enhancing the efficiency of smoke control and exhaust system in buildings. Fire Sci. Technol. **35**(9), 1229–1231 (2016)
8. Ji, X.: Problem analysis on fire smoke exhaust system design and fan selection. J. HV&AC **47**(7), 27–31 (2017)
9. He, L., et al.: Analysis of entrainment phenomenon near mechanical exhaust vent and a prediction model for smoke temperature in tunnel fire. Tunn. Undergr. Space Technol. **80**, 143–150 (2018)
10. Tang, F., et al.: Experimental study on maximum smoke temperature beneath the ceiling induced by carriage fire in a tunnel with ceiling smoke extraction. Sustain. Cities Soc. **44**, 40–45 (2019)
11. Zhao, J.: Studies on the ringed arrangement of sprinkler system. Xi'an University of Architecture and Technology (2008)
12. Yao, W.: Research on simultaneous optimal model and algorithm loop-type pipe network. Northwest A and F University (2011)
13. Li, L.: Design and evaluation of smoke exhaust system of an exhibition center. Fire Sci. Technol. **35**(4), 488–491 (2016)
14. Laajalehto, T., et al.: Energy efficiency improvements utilising mass flow control and a ring topology in a district heating network. Appl. Therm. Eng. **69**, 86–95 (2014)
15. Creaco, E., Franchini, M.: The identification of loops in water distribution networks. Procedia Eng. 119, 506–515 (2015)
16. Hafsi, Z., et al.: A computational modelling of natural gas flow in looped network: effect of upstream hydrogen injection on the structural integrity of gas pipelines. J. Nat. Gas Sci. Eng. **64**, 107–117 (2019)
17. Lu, Y.: Design manual of practical heating and air conditioning. China Architecture and Building Press (2007)
18. GB 51251-2017: Technical standard for smoke management systems in buildings. China Planning Press (2017)